D0982488

Springer Advanced Texts in Chemistry

Charles R. Cantor, Editor

Springer Advanced Texts in Chemistry

Series Editor: Charles R. Cantor

Principles of Protein Structure
G.E. Schulz and *R.H. Schirmer*

Bioorganic Chemistry: A Chemical Approach to Enzyme Action (Second Edition)
H. Dugas

Protein Purification: Principles and Practice (Second Edition)
R.K. Scopes

Principles of Nucleic Acid Structure
W. Saenger

Biomembranes: Molecular Structure and Function
R.B. Gennis

Ralph E. Christoffersen

Basic Principles and Techniques of Molecular Quantum Mechanics

With 85 Figures

Springer-Verlag
New York Berlin Heidelberg
London Paris Tokyo Hong Kong

Ralph E. Christoffersen
The Upjohn Company
Kalamazoo, Michigan 49001, U.S.A.

Series Editor:
Charles R. Cantor
Director of Human Genome Center
Lawrence Berkeley Laboratory
1 Cyclotron Drive
Berkeley, California 94720, U.S.A.

Library of Congress Cataloging-in-Publication Data

Christoffersen, Ralph E., 1937-
Basic principles and techniques of molecular quantum mechanics.

(Springer advanced texts in chemistry)
Includes bibliographical references.
1. Quantum chemistry. I. Title. II. Series.
QD462.C495 1989 541.2'8 89-21815
ISBN 0-387-96759-1

Printed on acid-free paper.

Typeset by Technical Typesetting Incorporated, Baltimore, Maryland.
Printed and bound by R.R. Donnelley, Harrisonburg, Virginia
Printed in the United States of America.

9 8 7 6 5 4 3 2 1

ISBN 0-387-96759-1 Springer-Verlag New York Berlin Heidelberg
ISBN 3-540-96759-1 Springer-Verlag Berlin Heidelberg New York

Series Preface

New textbooks at all levels of chemistry appear with great regularity. Some fields like basic biochemistry, organic reaction mechanisms, and chemical thermodynamics are well represented by many excellent texts, and new or revised editions are published sufficiently often to keep up with progress in research. However, some areas of chemistry, especially many of those taught at the graduate level, suffer from a real lack of up-to-date textbooks. The most serious needs occur in fields that are rapidly changing. Textbooks in these subjects usually have to be written by scientists actually involved in the research which is advancing the field. It is not often easy to persuade such individuals to set time aside to help spread the knowledge they have accumulated. Our goal, in this series, is to pinpoint areas of chemistry where recent progress has outpaced what is covered in any available textbooks, and then seek out and persuade experts in these fields to produce relatively concise but instructive introductions to their fields. These should serve the needs of one semester or one quarter graduate courses in chemistry and biochemistry. In some cases, the availability of texts in active research areas should help stimulate the creation of new courses.

New York, New York CHARLES R. CANTOR

Preface

This book is not a traditional quantum chemistry textbook. Instead, it represents a concept that has evolved from teaching graduate courses in quantum chemistry over a number of years, and encountering students with diverse backgrounds. The conclusion reached from those experiences was that a book was needed in which the principles and contemporary techniques of molecular quantum mechanics were presented in detail, along with mathematical and classical mechanical material sufficient to ensure understanding of the quantum mechanical principles and techniques.

It is therefore the primary purpose of this book to provide a classroom and reference text for graduate and advanced undergraduate students that presents a detailed exposition of the principles and techniques of molecular quantum mechanics. It is intended throughout that rigor and depth of understanding should not be sacrificed simply for the purpose of brevity. Frequent examples are used to illustrate the methods and abstract ideas, and recent developments that have led to significant changes in philosophy or techniques have been included. Through this approach it is hoped that a wide diversity of student background can be accommodated, that the necessary mathematical tools can be learned, and that a detailed understanding of the principles and techniques of molecular quantum mechanics will be obtained.

Implementation of these ideas led to several conclusions regarding the presentation of topics. First, the mathematics that is expected to be new to many chemists has been separated from the new concepts to be encountered. Since vector analysis and matrix techniques are used heavily in contemporary quantum mechanics, and since many students have not encountered them in prior experiences, two early chapters are devoted to a detailed discussion of them. For those students who are already familiar with vectors and matrices, these chapters can be easily skipped. However, separation of the mathematical topics into early chapters allows easy reference when the techniques are used in later chapters. This approach also allows the student to concentrate on one topic at a time, and introduction of new quantum mechanical concepts will not be confused by the simultaneous introduction of new mathematics.

Quantum mechanics itself is introduced using an axiomatic approach, without a detailed discussion of experimental results that led to the need for quantum theory. This leads naturally to the use of operator techniques and matrix representations, and matrix representations are emphasized throughout the text instead of using differential equations and their solutions. This also corresponds well with current approaches, since a large fraction of contemporary research relies heavily on the use of matrix techniques.

One of the distinguishing features of this text is the specific consideration given to the molecular Hamiltonian, where an exposition is given of the terms arising from external electromagnetic fields, both static and time-dependent. In addition, relativistic effects are considered so that, when combined with the terms arising from external electromagnetic fields, the origin of the various terms can be clearly traced.

The level of mathematics and physics assumed is that of advanced calculus, and an introduction to classical mechanics and electromagnetic theory. Since the text is intended to serve as a reference text as well as classroom text, a rather extensive bibliography to review articles, original papers, and other texts that are relevant has been included.

Due to the detailed discussions of the topics included, many worthwhile topics have had to be omitted. For example, topics such as field theory, detailed discussions of atomic problems, semiempirical techniques, creation–annihilation operator techniques, scattering theory, and spectroscopy have not been included. However, the concepts and techniques introduced here are particularly appropriate for subsequent study of the various topics not explicitly included, and that would typically be covered in courses other than the course dealing with principles and techniques of molecular quantum mechanics.

Kalamazoo, Michigan RALPH E. CHRISTOFFERSEN

Acknowledgments

As is to be expected in a major writing project, a number of persons have had a significant influence on the project's progress and completion. In this case, a major debt of gratitude is owed to Gerry Maggiora, who conceptualized the initial ideas for this text jointly with the author, and who participated actively and very substantially in the writing of the first four chapters before other demands and interests took him away from the project.

In addition, William Deskin heavily influenced the author's early chemistry motivation and activities, and Harrison Shull has shaped much of the author's interest and knowledge in quantum chemistry. Along with George Hall and Per-Olov Löwdin, they have indirectly and directly affected both the content and approach used. In addition, many graduate students have participated, whether or not realizing it, in testing the concepts and approaches presented here, including (listed alphabetically) Cary Chabalowski, Tim Davis, Don Genson, K. M. Karunakaran, Joe Loter, Larry Nitsche, Les Shipman, Dale Spangler, and Art Williamson. To each of them, many, many thanks are due.

Checking of the concepts, text and equations was a major project undertaken by Jim Petke. In addition, his careful reading of the manuscript and generous donation of problems at the end of several chapters, plus a number of important conceptual suggestions for later chapters, comprised a major contribution. In addition, the tireless and extraordinarily accurate typing of Jan Peterson has made completion of the project a pleasure. Also, appreciation is expressed to Tony Rappe for carrying out SCF calculations on H_2, as well as to Gary Maciel for helpful comments, and Rod Skogerboe for providing a supportive environment for a portion of the writing.

Finally, and certainly not least, I would like to thank my wife, Barbara, for supporting this project over many years and providing on many occasions the encouragement necessary to continue the project when it would have been abandoned otherwise. It has truly been a team effort.

Contents

Series Preface v
Preface vii

Chapter 1

Experimental Basis of Quantum Theory 1

1-1. Introductory Remarks 1
1-2. Classical Concepts of Linear Momentum, Angular Momentum, and Energy 2
1-3. Energy Levels and Photons 11
1-4. Electron Impact Experiments 13
1-5. Atomic Spectra 16
1-6. Quantization of Angular Momentum 17
1-7. Momentum of a Photon 21
1-8. Wave-Particle Duality 23
Problems 28

Chapter 2

Vector Spaces and Linear Transformations 31

2-1. Vector Spaces 31
2-2. Linear Independence, Bases, and Dimensionality 36
2-3. Inner Product Spaces 42
2-4. Orthonormality and Complete Sets 49
2-5. Hilbert Space 61
2-6. Function Space and Generalized Fourier Series 66
2-7. Isomorphism between Hilbert Space and Function Space 69
2-8. Examples of Complete Sets of Functions 71
2-9. Extension to Continuum Functions 79
2-10. Function Minimization with Constraints 82
2-11. Linear Operators 84
2-12. Algebra of Linear Operators 87
2-13. Special Kinds of Linear Operators 91
2-14. Eigenvalues and Eigenvectors 97
Problems 98

Chapter 3

Matrix Theory 102

3-1. Elements of Matrix Algebra 102

3-2. Determinants 109
3-3. Characterization of Square Matrices 117
3-4. Matrix Inversion 121
3-5. Matrices Having Special Properties 127
3-6. Matrix Representations of Linear Operators and Matrix Transformations 129
3-7. Changes of Basis and Similarity Transformations 138
3-8. Matrix Eigenvalue Problems 147
3-9. Infinite Matrices and Linear Transformations on Hilbert Space 167
3-10. Dirac Notation 169
Problems 170

Chapter 4

Postulates of Quantum Mechanics and Initial Considerations 175

4-1. Quantum Mechanical States and Observables 176
4-2. Time Evolution of a Quantum State 181
4-3. Quantum Theory of Measurement and Expectation Values 185
4-4. Compatible Observables and Commuting Operators 192
4-5. Constants of Motion and Transition Probabilities 198
4-6. Different Pictures of Quantum Phenomena 204
4-7. Hamiltonian Operator Construction: Initial Considerations 211
Problems 216

Chapter 5

One-Dimensional Model Problems 219

5-1. General Comments 219
5-2. Wavefunction Criteria and Boundary Conditions 220
5-3. The Nondegeneracy Theorem 221
5-4. Particle on a Ring 223
5-5. Particle Trapped in a Box 228
5-6. Parity of Eigenfunctions 233
5-7. Square Well Potential 235
5-8. Double Wells and Tunneling 241
5-9. The Harmonic Oscillator 250
5-10. Zero Point Energy and the Uncertainty Principle 265
Problems 267

Chapter 6

Angular Momentum 271

6-1. Introduction 271
6-2. General Angular Momentum Considerations 273
6-3. Orbital Angular Momentum 282
6-4. Spin Angular Momentum 288
Problems 291

Chapter 7

The Hydrogen Atom, Rigid Rotor, and the H_2^+ Molecule 293

7-1. Separation of Motion of Center of Mass 293
7-2. Solution of Equation for Relative Electron Motion of the Hydrogen Atom and Hydrogen-Like Atoms 297
7-3. Wavefunction Shapes 303
7-4. Rigid Rotor 314
7-5. The H_2^+ Molecule 317
Problems 325

Chapter 8

The Molecular Hamiltonian 327

8-1. General Principles and Discussion 327
8-2. Introduction of External Fields 328
8-3. Introduction of Relativistic Effects 333
8-4. The Born-Oppenheimer Approximation 343
Problems 347

Chapter 9

Approximation Methods for Stationary States 352

9-1. The Variation Principle 352
9-2. Accuracy Considerations 355
9-3. Example: The Hydrogen Atom 357
9-4. Example: Variational Treatment of the Helium Atom 359
9-5. The Linear Variation Method 362
9-6. Example: The Hydrogen Atom Revisited 369
9-7. Lower Bounds 372
9-8. Rayleigh-Schrödinger Perturbation Theory 374
9-9. Brillouin–Wigner Perturbation Theory 386
Problems 392

Chapter 10

General Considerations for Many Electron Systems 397

10-1. Early Computational Concepts and Procedures 397
10-2. Symmetry Considerations and Group Theory 402
10-3. Antisymmetry and the Pauli Exclusion Principle 429
10-4. Multielectron Systems and Slater Determinants 431
10-5. Expansion Theorem and Slater Determinant Expansions 435
10-6. Matrix Elements between Slater Determinants 437
10-7. Virial Theorem, Hypervirial Theorem, and Hellmann–Feynman Theorem 447
10-8. Scaling 453
10-9. Coupling of Angular Momenta 460
10-10. Orbital Transformations 471
Problems 477

Chapter 11

Computational Techniques for Many-Electron Systems Using Single Configuration Wavefunctions 481

11-1. Hartree–Fock Theory for Closed Shell Systems 481
11-2. Hall–Roothaan LCAO-MO-SCF Theory for Closed Shell Systems 498
11-3. Hartree–Fock Theory for Open Shell Systems 556
Problems 573

Chapter 12

Beyond Hartree–Fock Theory 576

12-1. Electron Correlation: General Comments 576
12-2. Configuration Interaction 577
12-3. Specialized CI Approaches 624
12-4. Many-Body Perturbation Theory and Coupled Cluster Theory 648
Problems 653

Appendix 1 657
References 659
Index 677

Chapter 1

Experimental Basis of Quantum Theory

1-1. Introductory Remarks

In the years immediately following 1925, a dramatic series of theoretical developments occurred that explained for the first time the apparently anomolous experimental data then existing. Also, these developments provided a conceptual and mathematical framework that has motivated and allowed interpretation of an immense number and diversity of experiments since then. These theoretical developments, called *quantum mechanics*, will be the subject of our attention throughout this entire text.

The notions of quantum mechanics are intended, insofar as possible, to provide a satisfactory theoretical and mathematical basis for the discussion of particles and their interactions at a *microscopic level.* As far as most chemistry is concerned, the microscopic particles of interest are nuclei and electrons, and the discussions to follow will treat these particles as the "basic building blocks" of chemistry, i.e., any substructure of them will be ignored. Given the masses of these particles, along with their charge[1] [and several fundamental constants such as the speed of light (c) and Planck's constant (h)], the ideas of quantum mechanics will be seen to provide, at least in principle, all of the information that can be known about a system of these particles. In other words, with a sufficient quantum mechanical description of a system of electrons and nuclei, one may calculate *any* property of that system, or predict the results of experiments involving that system and external forces (including other systems of particles).

In some cases, by the nature of the experiment, the macroscopic result can be

[1] The charge on an electron will be taken $-e$, and the charge on a nucleus will be taken as $+Ze$, where Z is the atomic number of the nucleus.

correlated directly to an individual molecular "event," and in that case the experiment and quantum-mechanical description are closely related. In other cases, the quantum-mechanical description must be combined with the methods of statistical mechanics to yield macroscopic properties or experimental observations. Thus, insofar as the present theory is complete, knowledge of the masses and charges of the various particles (along with a few fundamental constants) allows calculation of anything of chemical interest.

Unfortunately, as we shall see in later discussions, this ability to calculate "anything of chemical interest" *in principle* has not generally been achieved in practice, except for a few systems. However, the advent of high-speed digital computers, both large scale and in "minicomputer" form, along with the development of suitable techniques for obtaining highly accurate (though approximate) values of various properties, have provided an important set of "tools" for the prediction of physical and chemical properties of molecules. In addition and beyond the computational aspect, quantum mechanics has provided chemists with a powerful conceptual framework for the discussion of chemical problems, which is at least of equal importance as the computational capabilities.

In the sections and chapters to follow, both the computational and conceptual aspects of quantum mechanics, and how they can be used to solve chemical problems of interest, will be discussed. To do this, both mathematical techniques and various conceptual approaches to quantum mechanics will be developed.

In the following discussions of this chapter, two kinds of topics will be presented. First, several concepts from classical mechanics that will be useful in the development of quantum theory will be reviewed briefly. In part, this will allow introduction of notation that will be useful later, and will also identify concepts with which the student should be familiar. Second, experiments of historical interest will be described when they clarify the concepts. However, a historical development has been eschewed in favor of a rapid entrance into the study of the subject.

1-2. Classical Concepts of Linear Momentum, Angular Momentum, and Energy

In preparation for analogies to be made in later discussions, we begin by presenting a number of theorems and concepts from classical mechanics. They include theorems concerning conservation of linear momentum, angular momentum, and energy, and the theorem concerning the separation of energy into external and internal parts. These concepts and theorems will be discussed only briefly, since classical mechanics does not form the major emphasis of this text.[2] However, although striking differences exist, quantum mechanics makes

---

[2] For additional discussion of classical mechanics see, for example, E. A. Hylleraas, "Mathematical and Theoretical Physics," Vol. I, Wiley-Interscience, New York, 1970.

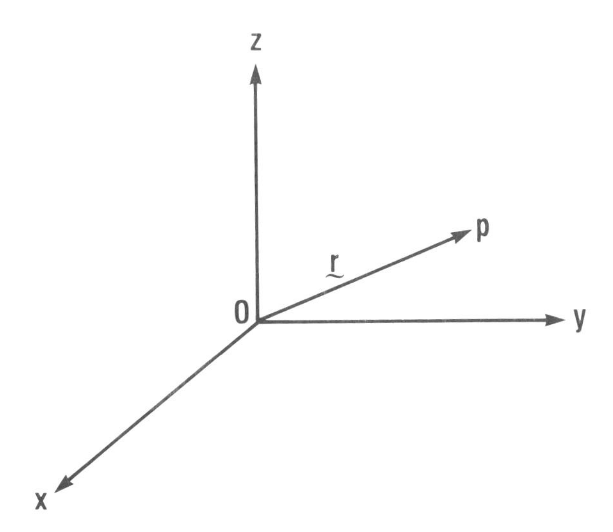

Figure 1.1. Position vector of a particle, P, relative to an origin at point 0.

frequent use of classical concepts, and an introduction to the notation and concepts of classical mechanics is appropriate for our initial discussions.

First consider the motion of a single particle. The position of this particle relative to an arbitrary origin is defined by a vector $\mathbf{r}$, the *position vector* of the particle (see Fig. 1.1).

The *velocity* $\mathbf{v}$ of the particle is the time derivative of $\mathbf{r}$, which we indicate as $\dot{\mathbf{r}}$, and the acceleration of $\mathbf{a}$ is $\dot{\mathbf{v}}$. *Newton's equations of motion* for the particle are then

$$\mathbf{F} = m\mathbf{a}, \tag{1-1}$$

where $\mathbf{F}$ is the total *force* exerted on the particle, and m is its mass. Thus, in general, $\mathbf{F}$ is a function of both position and time. As a specific example, the x component of Eq. (1-1) is

$$F_x = m\,\frac{d^2x}{dt^2}\,. \tag{1-2}$$

The main problem in Newtonian mechanics is to integrate this differential equation, to obtain x as a function of t.

Newton's equations can also be written in somewhat more compact form. First, we define the *linear momentum* ($\mathbf{p}$) to be $m\mathbf{v}$. Then, Newton's equations become

$$\mathbf{F} = \dot{\mathbf{p}}. \tag{1-3}$$

If $\mathbf{F} = 0$, the time derivative of momentum is zero, and the momentum is constant in time. In other words, if no net force is exerted on the particle, its momentum is said to be *conserved* in time, i.e., is a *constant of the motion*.

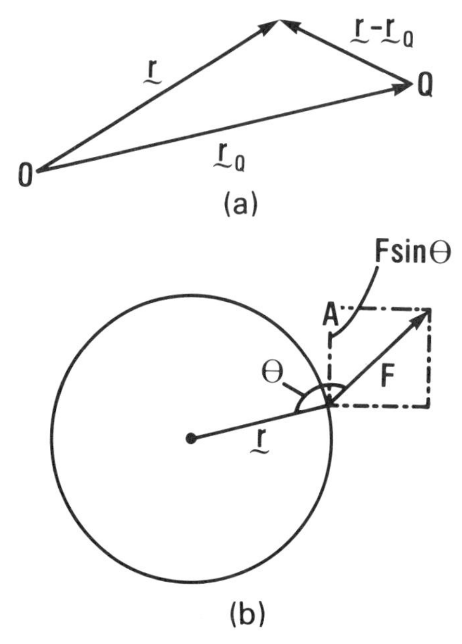

Figure 1.2. (a) Vectors, $\mathbf{r}$, $\mathbf{r}_Q$, used in defining torque and angular momentum. (b) Illustration of a force perpendicular to the radius vector $\mathbf{r}$.

A particular kind of momentum that is frequently of interest has to do with motion about a central origin. In that case, the moment of force ($\mathbf{N}$) about point Q for a particle is called the *torque*, and is defined to be

$$\mathbf{N}_Q = (\mathbf{r} - \mathbf{r}_Q) \times \mathbf{F}, \tag{1-4}$$

as illustrated in Fig. 1.2a. Similarly, the moment of momentum or *angular momentum* ($\mathbf{L}$) about a point Q is defined as

$$\mathbf{L}_Q = (\mathbf{r} - \mathbf{r}_Q) \times \mathbf{p}. \tag{1-5}$$

To illustrate the definitions, consider a particle constrained to move on a circle of radius r. If a force in the plane of the circle is applied to the particle, only the component of force perpendicular to the radius (which is $F \sin \theta$) will be effective in "twisting" the particle about the circle (see Fig. 1.2b). In addition, this "twist" or torque is expected to increase if the particle is farther from the origin (i.e., as r increases.) Combining these observations, we have $N = rF \sin \theta$, in agreement with Eq. (1-4). To illustrate the momentum definition, let ϕ be the angle about the circle. Then the magnitude of the momentum is $p = mv = m(r\dot{\phi})$, which can be seen from an examination of Fig. 1.3. Since this momentum is perpendicular to the radius, the momentum is clearly an *angular* momentum, and is seen from Eq. (1-5) to be $L = mr^2\dot{\phi}$.

It is of interest to show the relationship between the time derivative of the angular momentum and the torque applied to the particle moving on a circle. In

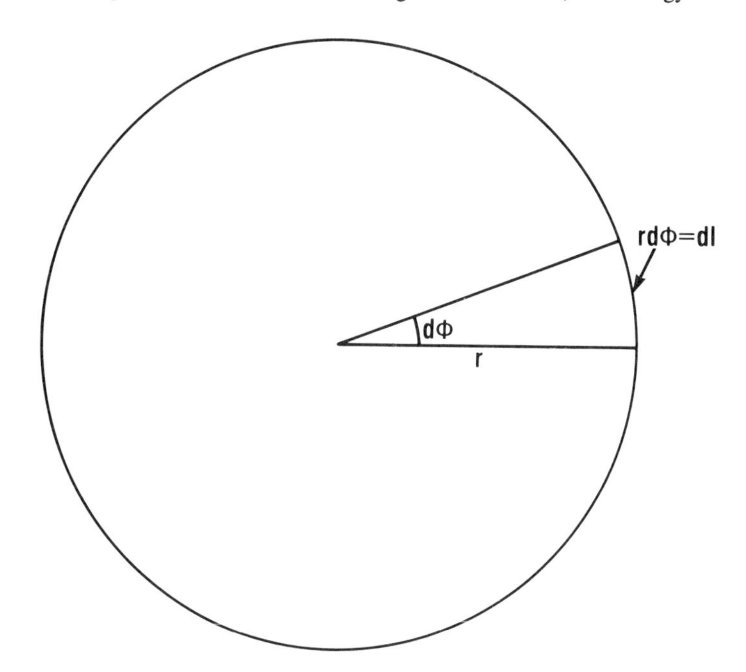

Figure 1.3. Diagram of a particle restricted to moving in a circle of radius r.

particular (assuming Q to be fixed),

$$\dot{\mathbf{L}}_Q = \frac{d}{dt}[(\mathbf{r}-\mathbf{r}_Q)\times\mathbf{p}] = \dot{\mathbf{r}}\times\mathbf{p} + (\mathbf{r}-\mathbf{r}_Q)\times\dot{\mathbf{p}}. \tag{1-6}$$

The first term on the right-hand side vanishes because $\dot{\mathbf{r}}$ is parallel to $\mathbf{p} = m\dot{\mathbf{r}}$. But, by taking the cross-product of $(\mathbf{r} - \mathbf{r}_Q)$ on both sides of Newton's equations [Eqn. (1-3)], one obtains

$$(\mathbf{r}-\mathbf{r}_Q)\times\mathbf{F} = (\mathbf{r}-\mathbf{r}_Q)\times\dot{\mathbf{p}}, \tag{1-7}$$

or, using Eqs. (1-4), (1-6), and (1-7), we have

$$\mathbf{N} = \dot{\mathbf{L}}. \tag{1-8}$$

Expressed in words, we see that the angular momentum will be changed in time if a torque is applied. If the torque is zero, $\dot{\mathbf{L}} = 0$, and $\mathbf{L}$ is constant. Thus, as long as the torque about a point is zero, the angular momentum is conserved, i.e., is constant in time, and hence, is a constant of the motion.

The next concept of interest is that of *work*. The work (w) done in moving a particle from point A to point B by some path is given by the line (path) integral

$$w = \int_A^B \mathbf{F}\cdot d\mathbf{r}. \tag{1-9}$$

In general, the value of w depends on the path taken between points A and B. If, however, $\mathbf{F}\cdot d\mathbf{r}$ is an *exact* differential, say $-dV$, with V a function of position only, the value of w is then independent of path, since

$$w = -\int_A^B dV = V_A - V_B. \tag{1-10}$$

But,

$$dV = \frac{\partial V}{\partial x}\, dx + \frac{\partial V}{\partial y}\, dy + \frac{\partial V}{\partial z}\, dz, \tag{1-11}$$

and the *gradient* of V is the vector

$$\nabla V = \mathbf{i}\,\frac{\partial V}{\partial x} + \mathbf{j}\,\frac{\partial V}{\partial y} + \mathbf{k}\,\frac{\partial V}{\partial z}, \tag{1-12}$$

where $\mathbf{i}$, $\mathbf{j}$ and $\mathbf{k}$ are unit vectors in the x, y, and z directions, respectively. Hence,

$$dV = \nabla V \cdot d\mathbf{r} = \frac{\partial V}{\partial x}\, dx + \frac{\partial V}{\partial y}\, dy + \frac{\partial V}{\partial z}\, dz, \tag{1-13}$$

and we have the result [from Eqs. (1-9) and (1-13)] that

$$\mathbf{F} = -\nabla V. \tag{1-14}$$

Thus, for this particular case, the force field is specified completely by a scalar field V, which is a function of position only. V is called the *potential energy*, and forces derivable from a scalar function as in Eq. (1-14) are said to be *conservative*.

Newton's equations [Eq. (1-1)] for conservative forces can thus be written

$$\nabla V + m\dot{\mathbf{v}} = 0. \tag{1-15}$$

Forming the scalar product with $\mathbf{v}$ gives

$$(\nabla V)\cdot \mathbf{v} + m\dot{\mathbf{v}}\cdot \mathbf{v} = 0,$$

which can be written as

$$\nabla V \cdot \frac{d\mathbf{r}}{dt} + m\mathbf{v}\cdot \frac{d\mathbf{v}}{dt} = 0. \tag{1-16}$$

Using Eq. (1-13), we obtain

$$\frac{dV}{dt} + \frac{1}{2}\, m\, \frac{d(v^2)}{dt} = 0, \tag{1-17}$$

which can be rearranged to give one of the important forms of the equation of motion of a particle:

$$\frac{d}{dt}\left(\frac{1}{2}\, mv^2 + V\right) = 0. \tag{1-18}$$

The quantity $\frac{1}{2}mv^2$ is known as the *kinetic energy* (T) of the particle. Hence, for conservative forces, we see that the sum of kinetic energy and potential energy [which is called the *total energy* (E)] is conserved, i.e., is a constant of the motion:

$$E = T + V = \text{constant in time.} \tag{1-19}$$

All of these results can be generalized to systems of N particles. Let us consider a system with interparticle forces F_k^i (the force on particle k due to particle i) and weaker external forces F_k^e (external force on particle k). It is useful first to define the *center of mass* of the system, given by the vector,

$$\mathbf{R} = \sum_{k=1}^{N} m_k \mathbf{r}_k / M, \tag{1-20}$$

with

$$M = \sum_{k=1}^{N} m_k. \tag{1-21}$$

Then, the *total linear momentum* $(\mathbf{P})$ of the system of particles is given by the sum of the contributions from each particle, i.e.,

$$\mathbf{P} = \sum_{k=1}^{N} m_k \dot{\mathbf{r}}_k = \sum_{k=1}^{N} \mathbf{P}_k = M\dot{\mathbf{R}}. \tag{1-22}$$

In other words, *the total momentum of a system of particles can be thought of as the momentum of a single particle of mass M, moving with the center of mass.* One of the implications of this observation is that the *internal momentum* $(\mathbf{P}_\mathrm{I})$, which is the momentum of the various particles relative to the center of mass, must be zero. This can be seen as follows:

$$\mathbf{P}_\mathrm{I} = \sum_{k=1}^{N} m_k(\dot{\mathbf{r}}_k - \dot{\mathbf{R}}) = \left(\sum_{k=1}^{N} m_k \dot{\mathbf{r}}_k\right) - M\dot{\mathbf{R}} = 0, \tag{1-23}$$

where Eq. (1-22) has been used.

Next, Newton's equations of motion [see Eq. (1-3)] for the kth particle are given by

$$\mathbf{F}_k^e + \sum_{i(\neq k)}^{N} \mathbf{F}_k^i = \dot{\mathbf{P}}_k. \tag{1-24}$$

Summing over all particles, we obtain

$$\mathbf{F}^e + \sum_{k=1}^{N} \sum_{i(\neq k)}^{N} \mathbf{F}_k^i = \dot{\mathbf{P}}, \tag{1-25}$$

where $\mathbf{F}^e = \Sigma_{k=1}^N \mathbf{F}_k^e$ (the *total external force*), and $\dot{\mathbf{P}} = \Sigma_{k=1}^N \dot{\mathbf{P}}_k$. The second term on the left vanishes if the interparticle forces are such that $\mathbf{F}_k^i = -\mathbf{F}_i^k$ (Newton's Third Law of Forces). Then

$$\mathbf{F}^e = \dot{\mathbf{P}}. \tag{1-26}$$

In other words, *the center of mass moves as if it were a particle of mass M acted on by the total external force.* If the total external face is zero, then the total momentum is conserved, i.e., is constant in time.

Before proceeding further, it is useful to point out two alternative ways of expressing Newton's equations of motion that will be useful to us in later discussions. To do this, we note first that it is frequently more convenient to use coordinate systems other than Cartesian coordinates when discussing equations of motion. For example, when dealing with systems having spherical symmetry, it is clearly more advantageous to utilize spherical polar coordinates rather than Cartesian coordinates. In general, these coordinate systems are all related to each other (i.e., we can transform the variables from one coordinate system to another), which can be expressed as

$$\begin{aligned} x_1 &= f(q_1 q_2, \dots, q_{3N}) \\ &\ \vdots \qquad\qquad \vdots \\ z_N &= f(q_1 q_2, \dots, q_{3N}) \end{aligned}$$

where there are N particles having $3N$ coordinates. The q_i are referred to as *generalized coordinates*, since they are used to represent the position coordinates of the particles in any coordinate system.

Using such coordinates, Lagrange derived an alternative formulation of Newton's equations of motion for the case of conservative forces by defining the function

$$L = T - V, \tag{1-27}$$

where T and V are the total kinetic and potential energies of the system, respectively, and L is called the *Lagrangian* of the system. Using such a definition, Newton's equations of motion can be written as

$$\frac{d}{dt}\left(\frac{\partial L}{\partial \dot{q}_i}\right) - \frac{\partial L}{\partial q_i} = 0, \qquad i = 1, 2, \dots, 3N \tag{1-28}$$

where

$$\dot{q}_i = \frac{\partial q_i}{\partial t}. \tag{1-29}$$

If p_i is the component of momentum associated with the coordinate q_i (in which case p_i and q_i are referred to as *conjugate variables*), then they are related by

$$p_i = \frac{\partial L}{\partial \dot{q}_i}, \qquad i = 1, 2, \dots, 3N, \tag{1-30}$$

when the Lagrangian of Eq. (1-27) is used.

A third formulation of equations of motion was formulated by Hamilton, who defined

$$H = \left(\sum_{i}^{3N} p_i \dot{q}_i - L \right) \tag{1-31}$$

$$= T + V, \tag{1-32}$$

where H is known as the Hamiltonian of the system. From Eq. (1-19), we see that H also represents the total energy of the system for the case of conservative forces. When Eq. (1-31) is used, Newton's equations of motion can be written as

$$\frac{\partial H}{\partial p_i} = \dot{q}_i, \qquad i = 1, 2, \ldots, 3N \tag{1-33}$$

and

$$\frac{\partial H}{\partial q_i} = -\dot{p}_i, \qquad i = 1, 2, \ldots, 3N. \tag{1-34}$$

These equations are known as *Hamilton's equations of motion.*[3]

Thus, a number of alternative formulations of the equations of motion exist for a system of N particles, and the particular one to be used will depend on the convenience of the formulation to the problem at hand.

Next, let us consider the total *angular* momentum about point Q, which is given by

$$\mathbf{L}_Q = \sum_{k=1}^{N} (\mathbf{r}_k - \mathbf{r}_Q) \times (\mathbf{p}_k - \mathbf{p}_Q),$$

while the total torque about a point Q due to external forces is given by

$$\mathbf{N}_Q = \sum_{k=1}^{N} (\mathbf{r}_k - \mathbf{r}_Q) \times \mathbf{F}_k^e.$$

Using the information given previously, it is possible now to prove the following theorem: *If the total external torque about the center of mass,* $\mathbf{N}_{\text{CM}}$, *of a system of particles is zero, then* $\dot{\mathbf{L}}_{\text{CM}} = 0$, *and the total angular momentum (about the center of mass) is conserved.*

One can also generalize the conservation of energy theorem, Eq. (1-19), for conservative forces: *If the external and interparticle forces are conservative, the total energy E (the sum of kinetic and potential energy) is conserved.*

To see this, let us first separate the energy into an internal and an external part. Looking at the total kinetic energy, which can be written as (see Fig. 1.4),

---

[3] For additional discussion see, for example, H. Goldstein, *Classical Mechanics,* Addison Wesley, Reading, MA, 1950.

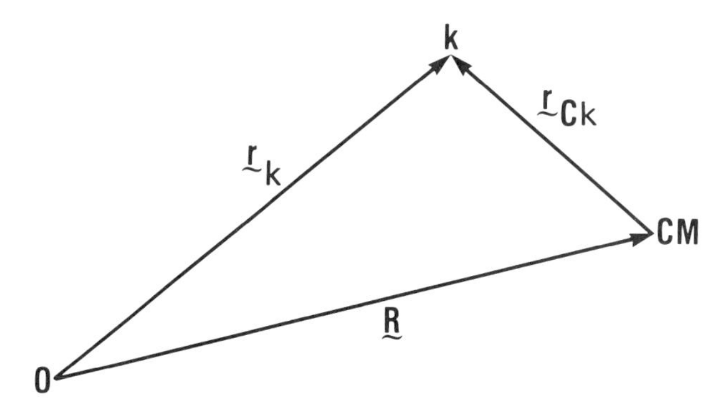

Figure 1.4. Coordinates of particle k, relative to the origin and to the center of mass.

we have

$$T=\frac{1}{2}\sum_{k=1}^{N} m_k\dot{r}_k^2=\frac{1}{2}\sum_{k=1}^{N} m_k(\dot{\mathbf{r}}_{ck}+\dot{\mathbf{R}})^2$$

$$=\frac{1}{2}M\dot{R}^2+\frac{1}{2}\sum_{k=1}^{N} m_k\dot{r}_{ck}^2+\left(\sum_{k=1}^{N} m_k\dot{\mathbf{r}}_{ck}\right)\cdot\mathbf{R}. \qquad (1\text{-}35)$$

The last term of this equation vanishes because the *internal* momentum ($\Sigma_{k=1}^{N} m_k\dot{\mathbf{r}}_{ck}$) is zero [see Eq. (1-23)]. We obtain

$$T=\frac{1}{2}M\dot{R}^2+\frac{1}{2}\sum_{k=1}^{N} m_k\dot{r}_{ck}^2. \qquad (1\text{-}36)$$

Thus, *the total kinetic energy is equal to the kinetic energy of a particle of mass, M, moving with the center of mass, plus the kinetic energy of the particles relative to the center of mass (the internal kinetic energy).*

Since the potential energy is an additive function of, at most, two-body interactions, a convenient separation into external and interparticle forces is automatically achieved. In particular, the total potential energy (V) of a system of N particles is given by

$$V=\sum_{k=1}^{N} V_k^e(\mathbf{r}_k)+\sum_{k<l}^{N} V_{kl}(r_{kl}), \qquad (1\text{-}37)$$

where $V_k^e(\mathbf{r}_k)$ represents the potential energy contribution from the kth particle in the external force field, at point $\mathbf{r}_k$, and $V_{kl}(r_{kl})$ is the interparticle contribution to the potential energy from particles k and l at points $\mathbf{r}_k$ and $\mathbf{r}_l$. Also, it has been assumed that the interparticle forces depend only on the distance between the particles, where $r_{kl} = |\mathbf{r}_k - \mathbf{r}_l|$, and are directed along the line of center of the particles.

If the external forces are slowly varying, then further simplification of Eq. (1-37) is possible. In particular, consideration of a Taylor series expansion of $V_k^e(\mathbf{r}_k)$ about the center of mass $(\mathbf{R})$ is useful in that case, i.e.,

$$V_k^e(\mathbf{r}_k) = V_k^e(\mathbf{R}) + \left(\frac{\partial V_k(\mathbf{r}_k)}{\partial x_k}\right)_{x_k = X} (x_k - X) + \left[\frac{\partial V_k(\mathbf{r}_k)}{\partial y_k}\right]_{y_k = Y} (y_k - Y)$$
$$+ \left[\frac{\partial V_k(\mathbf{r}_k)}{\partial z_k}\right]_{z_k = Z} (z_k - Z) + \cdots. \tag{1-38}$$

If $V_k^e(\mathbf{r}_k)$ is slowly varying, then only the first term on the right-hand side of Eq. (1-38) need be retained, and Eq. (1-37) can be written as

$$V \cong \sum_{k=1}^{N} V_k^e(\mathbf{R}) + \sum_{k<l}^{N} V_{kl}(r_{kl}). \tag{1-39}$$

In other words, for this case, *the potential energy is equal to an external potential energy term (depending only on the center of mass), plus an internal potential energy term.*

In spite of the generality and power of classical physics for the description of macroscopic systems, these concepts were not adequate for the description of microscopic systems. For example, there were at least three areas where major conceptual revisions were needed, including the concepts of energy level quantization, angular momentum quantization, and wave–particle duality. In the next several sections these concepts are described, along with examples of experiments that led to the need for conceptual revision and that verify the appropriatencss of the new concept. However, it should be noted that many experiments led to the need for revisions in classical physics, and only those that illustrate the concepts of interest have been included here.[4]

1-3. Energy Levels and Photons

One of the most important concepts in quantum mechanics, which differs drastically from its classical counterpart, is that of *energy levels*. In particular, it is found experimentally that, in general, it is possible to assign *definite, discrete values of the energy* (called *energy levels*) to microscopic systems, and that such systems may absorb precisely the correct amount of energy to allow a transition from one energy level to another. However, it cannot absorb energy that takes it to an "unallowed" energy. Thus, a system (e.g., a molecule) in some energy level (i.e., that has some allowed energy) may absorb energy, possibly by colliding with another molecule, and undergo a transition to a higher energy level (see Fig. 1.5). It may also be possible for this molecule to lose

[4] For an extensive discussion of experiments and empirical observations that led to the development of quantum theory, see M. Jammer, *The Conceptual Development of Quantum Mechanics,* McGraw-Hill, New York, 1966.

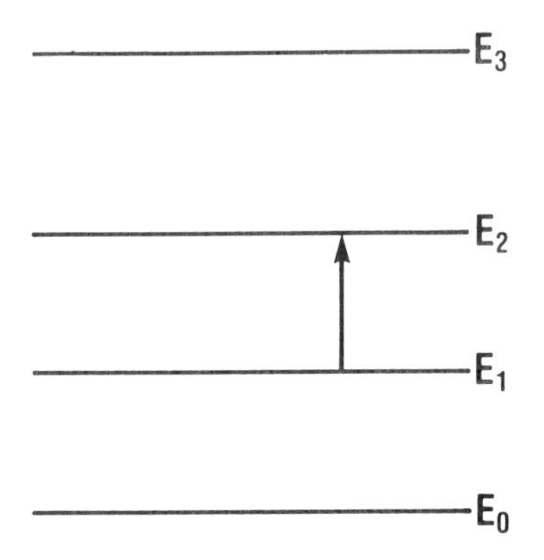

Figure 1.5. Schematic energy level diagram for a microscopic system, depicting a transition (absorption) from level E_1 to E_2.

energy and go to a lower energy level, if there is one. This possible limitation of the energy to discrete values is referred to as energy *quantization*, and we shall see how this effect is observed experimentally in a moment.

Another concept that we shall find very useful is that light can be considered to have *particulate* properties so that we may consider it to be composed of particles called *photons*. Each photon has associated with it an energy (E_p) that is related to the frequency ν of the light wave. Specifically, it is found that

$$E_p = h\nu \tag{1-40}$$

where h is Planck's constant (which has the numerical value 6.6256×10^{-27} erg·sec).

Let us assume that a molecule has allowed energy levels E_0, E_1, E_2, ... If the molecule is initially in energy level E_1, for example, then it can undergo a transition to other levels only by absorbing or emitting light whose frequency corresponds exactly to the energy difference between E_1 and the energy of the state to which the transition is to occur. Thus, if photons whose energy is exactly

$$E_p = E_2 - E_1 \tag{1-41}$$

are used to irradiate the molecule, the absorption of a photon can occur, and the molecule will undergo a transition to energy level E_2. The photon is then considered to be ''annihilated,'' i.e., its energy is absorbed entirely by the molecule. Alternately, a molecule in energy level E_1 may emit a photon of frequency

$$\nu = (E_1 - E_0)/h. \tag{1-42}$$

A photon is thus considered to be ''created,'' and energy is conserved by placing the molecule in energy level E_0. It should be emphasized that only energy *differences* are directly observed, and only those differences corresponding to transitions between ''allowed'' levels are possible.

All of these energy changes can be neatly ''pictured'' by means of an energy level diagram, in which horizontal lines are drawn to depict allowed energies

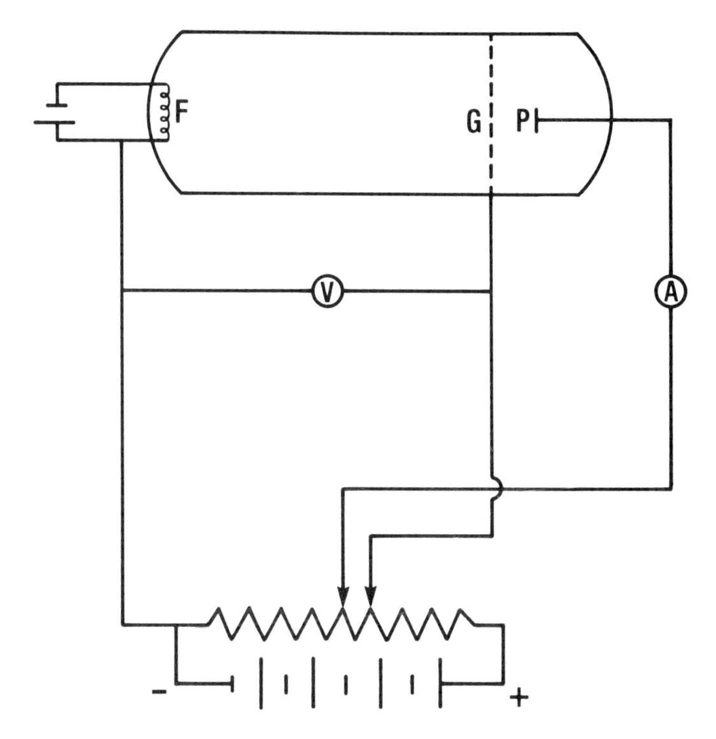

Figure 1.6. Schematic drawing of the Franck-Hertz experiment. (F = filament; G = grid; P = plate; V = voltmeter; A = ammeter.)

(energy levels), using the vertical axis as an energy scale (see Fig. 1.5). A possible energy jump or transition is depicted on the diagram by means of an arrow. For the particular case depicted in Fig. 1.5, the molecule initially had energy E_1 and absorbed energy, either by collision with another molecule or by absorption of a photon, and underwent a transition to level E_2.

1-4. Electron Impact Experiments

The experiments of J. Franck and G. Hertz[5] give one of the most direct verifications of the energy level concept. In brief, the experiment consists of bombarding molecules by electrons, and measuring the loss of kinetic energy of the electrons. In a direct and striking manner, it is found that the electrons lose definite *discrete* amounts of energy. The experimental arrangement is shown in Fig. 1.6.

The main portion of the apparatus consists of a glass vessel to which a

[5] J. Franck and G. Hertz, *Verhand. Deutsch. Phys. Gesell.*, **16**, 457 (1914); *Phys. Z.*, **17**, 409 (1916); *Phys. Z.*, **20**, 132 (1920).

filament, grid, and plate are attached, and which is filled with mercury vapor at low pressure. A small battery provides a current that heats the filament. When a positive potential is applied to the grid, electrons are accelerated from the hot filament to the grid. The amount of energy that is imparted to an electron by an accelerating potential of V volts is

$$E = eV. \tag{1-43}$$

Here E will be in joules if e is in coulombs ($e = 1.603 \times 10^{-19}$ C) and V is in volts. Equation (1-43) provides a very useful unit of energy; namely, the amount of energy imparted to an electron by an accelerating voltage of 1 V is called the *electron volt* (1 eV = 23.05 kcal/mol).

If an electron does not undergo a collision with a mercury atom, it will pass through the grid and be collected at the plate. The plate current is registered by the ammeter A. If, however, an electron collides with a mercury atom near the grid, and if the electron transmits most of its kinetic energy to the mercury atom, the electron will be prevented from reaching the plate, because the plate has a potential that is slightly negative with respect to the grid. With this in mind we look at the result of an experiment (Fig. 1.7).

As the voltage increases from zero to nearly 5 V, we see that the plate current increases, i.e., as V increases, the number of electrons arriving at the plate in unit time increases. But, at 4.9 V, the plate current begins to drop. This in interpreted to mean that electrons having energies of 4.9 eV are colliding with mercury atoms in the vicinity of the grid, are transferring energy to the mercury atom, and consequently do not have sufficient remaining energy to reach the

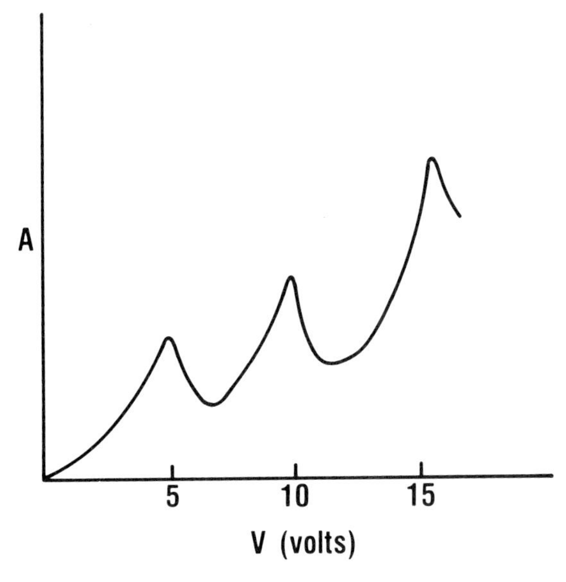

Figure 1.7. Schematic representation of plate current (A) associated with increased voltage, measured with mercury vapor by Franck and Hertz.

plate. (These kinds of collisions, in which energy transfer takes place, are usually referred to as *inelastic collisions.*) In other words, mercury atoms are able to *absorb* 4.9 eV of energy from a collision with an electron. It should be pointed out that, while the mercury atom and the electron may have any value of kinetic energy, the electron cannot transmit an appreciable amount of kinetic energy to the mercury atom as kinetic energy of the atom, because of their disparity in masses. Hence, the energy of the electron is absorbed primarily as *internal* energy of the atom. The plate current maximum at 4.9 eV exhibited in Fig. 1.7 therefore establishes the presence of *internal energy levels* of the mercury atom.

When the voltage is increased further, the region in which electrons undergo inelastic collisions with mercury atoms recedes from the region of the grid, because the distance the electron must travel to gain a kinetic energy of 4.9 eV is shorter. This electron will then be reaccelerated by the grid potential after the collision, and will be able to pass through the grid to the plate. Thus, the plate current will again increase. However, if the voltage is increased to 9.8 V, the electron will have sufficient energy to undergo another inelastic collision in the grid region, with a concomitant drop in plate current. If the experiment is carried out carefully, it is possible to deduce a number of energy levels of the mercury atom.

A direct verification of the relation $E_p = h\nu$ can also be provided in this experiment by observing the *spontaneous emission* of light (called fluorescence[6]) from the tube. No emission is noted below 4.9 V, but above this voltage, the emission of the 2536.6 Å line of mercury is observed. This result can be interpreted in the following way. Electrons with sufficient kinetic energy collide inelastically with mercury atoms, and these atoms are then placed in an energy level that is 4.9 eV above the lowest level. However, an excited mercury atom is *unstable*, and returns to the lowest level after a length of time, with the *emission* of a photon.

Let us compute the wavelength λ of the photon of emitted light from the energy level that is 4.9 eV above the ground state, using Eq.(1-40). The frequency is

$$\nu = \frac{E}{h} = \frac{(4.9\ \text{eV})(1.60 \times 10^{-12}\ \text{erg/eV})}{(6.63 \times 10^{-27}\ \text{erg} \cdot \text{sec})} = 1.18 \times 10^{15}\ \text{sec}^{-1}.$$

Then, the wavelength is seen to be

$$\lambda = \frac{c}{\nu} = \frac{3.00 \times 10^{10}\ \text{cm/sec}}{1.18 \times 10^{15}\ \text{cycles/sec}} = 2.54 \times 10^{3}\ \text{Å}, \tag{1-44}$$

which is precisely what is observed experimentally.

[6] Another kind of emission, called phospherescence, occurs when the lifetime of the upper state is long.

1-5. Atomic Spectra

Although electron impact experiments provide the most direct verification of the energy level concept, the most accurate way to establish these energy levels experimentally is by methods of spectroscopy. For example, by passing an electric discharge through a tube containing hydrogen at low pressure, excited hydrogen atoms (atoms in states above the lowest energy) can be prepared. These atoms omit light of various frequencies that can be analyzed, and the energy levels deduced. One finds that the frequencies of light emitted in this case can be rationalized by the general formula developed by Balmer[7]

$$\nu = cR\left(\frac{1}{n_2^2} - \frac{1}{n_1^2}\right) \tag{1-45}$$

where R is *Rydberg's constant,* c is the velocity of light, and n_1 and n_2 are positive integers. Comparing Eq. (1-45) with Eq. (1-41), we can see that the internal energy levels of the hydrogen atom are given by

$$E_n = -hcR\left(\frac{1}{n^2}\right), \qquad n = 1, 2, \cdots. \tag{1-46}$$

It should be noted that any constant can be added to the energy without changing the physical result, since only energy *differences* are observed. As we shall see later, the value of R can be calculated from theory, and is found to be

$$R = \frac{2\pi^2 \mu e^4}{h^3 c}, \tag{1-47}$$

where μ is the reduced mass (nearly equal to the mass of the proton).

Thus, the various energy levels E_n are characterized by the integer n, which is called the *principal quantum number*. As n goes from its lowest value of 1 (the *ground state*) to infinity, the energy increases from $-hcR$ to zero. Those states with $n > 1$ are referred to as *excited states.* All of the states of the hydrogen atom with negative energies are called *bound states.* The electron in a bound state remains attached to the proton, although as n increases, the region where the electron is most likely to be found will be seen in later considerations to recede from the proton. At $E = 0$ the electron has just been removed from the proton, i.e., the hydrogen atom has been *ionized*. It turns out that further energy may be added in *any* amount whatever. This energy simply goes into kinetic energy of the ionized electron. In other words, above $E = 0$, there is a *continuum* of energies that correspond to *free states* of the electron. In general, such free states (sometimes called unbound states) do not exhibit discrete energies, i.e., there is no energy quantization.

Thus, the hydrogen atom spectrum shows a series of lines (see Fig. 1.8)

[7] For a more extensive discussion see, for example, G. Herzberg, *Atomic Spectra and Atomic Structure,* Dover Publications, New York, 1944, pp. 11–12.

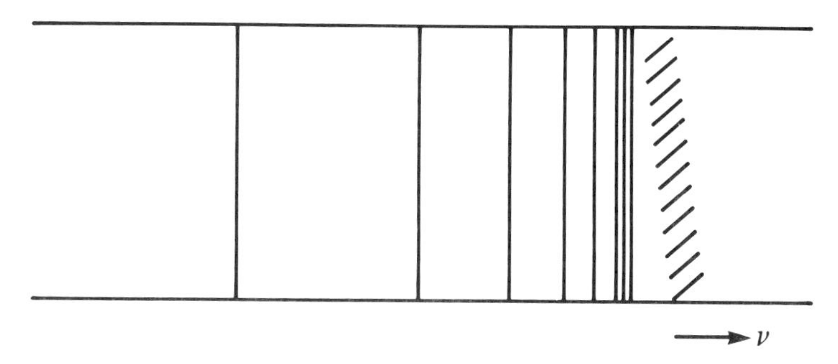

Figure 1.8. Schematic representation of the Balmer series ($n_2 = 2$) of the hydrogen atom spectrum.

corresponding to transitions between energy levels, followed by a continuous darkening of the plate, corresponding to transitions to the continuum.

1-6. Quantization of Angular Momentum

Like the energy, angular momentum also displays discrete values in microscopic systems. In fact, the observed values are always discrete, i.e., only bound states are found, and continuum effects are not observed. Specifically, the square of the angular momentum is found experimentally to have only the following values,

$$l(l+1)\hbar^2, \qquad l=0,\ 1,\ 2,\ \ldots .$$

Here the symbol $\hbar = h/2\pi$ has been introduced, since this combination of constants occurs frequently in discussions of momentum and angular momentum. The integer l is called the (orbital) *angular momentum quantum number.*

A second kind of angular momentum is also observed, and is referred to as *intrinsic angular momentum,* which is called *spin.*[8] It gives rise to a concept that has no classical counterpart. In the case of spin, the quantum number, called the *spin quantum number*, may be either an integer or half-integer (in units of $\hbar$), and has a specific value that is characteristic of the particle. For example, the electron has a spin quantum number of $\pm 1/2$, while the deuteron has a spin quantum number of 1.

The far-infrared absorption spectrum of a diatomic molecule such as HCl provides an example in which the discrete values of angular momentum can be observed experimentally. To interpret the experimental data, we take as a model of HCl two particles of masses m_{H} and m_{Cl}, *fixed* at a distance R apart. This is called the *rigid rotor model*, and the internal energy of such a system is entirely kinetic energy of the two masses. Thus, separating the motion of the center of mass (which can be done in a manner similar to that used in Section 1-2), we

[8] G. E. Uhlenbeck and S. Goudsmit, *Naturwissenschaften,* **13**, 953 (1925).

shall be interested in the internal energy associated with the rotation of a particle of reduced mass (μ), where

$$\mu = \frac{m_{\mathrm{H}} m_{Cl}}{m_{\mathrm{H}} + m_{\mathrm{cl}}} . \tag{1-48}$$

From the discussion following Eq. (1-5), we know that the magnitude of the angular momentum (L)[9] associated with this rotation is given by

$$L = (\mu R^2)\dot{\phi}, \tag{1-49}$$

where ϕ is the angle of rotation of the rigid body about the center of mass. Since there is no potential energy involved, the total energy is given simply by the total kinetic energy, i.e.,

$$E = \frac{L^2}{2(\mu R^2)} , \tag{1-50}$$

which is the analog for angular momentum of the expression $E = p^2/2m$ for linear momentum. By defining the *moment of inertia* (I) as[10]

$$I = \mu R^2, \tag{1-51}$$

and combining Eqs. (1-50) and (1-51) with the experimental observation referred to above that only discrete values of angular momentum are observed, we obtain

$$E_L = \frac{l(l+1)\hbar^2}{2I} , \tag{1-52}$$

for the energy levels of a rigid rotor, where l is an integer.

To understand the observed spectrum, one must also realize that all conceivable transitions between energy levels are not observed. Some transitions occur with very low or near-zero probability, and therefore are not observed. The observed transitions are summarized by a *selection rule*. In the rigid rotor model, the only transitions that turn out to occur with nonzero probability are those in which l changes by one. Thus, for absorption, we have the selection rule

$$\Delta l = +1, \tag{1-53}$$

---

[9] This use of "L" should not be confused with the angular momentum quantum number (l) discussed earlier in this section.

[10] For a system containing N particles, the definition of the moment of inertia is generalized to

$$I = \sum_{i=1}^{N} m_i r_i^2,$$

where r_i is the distance of the ith particle from the center of mass.

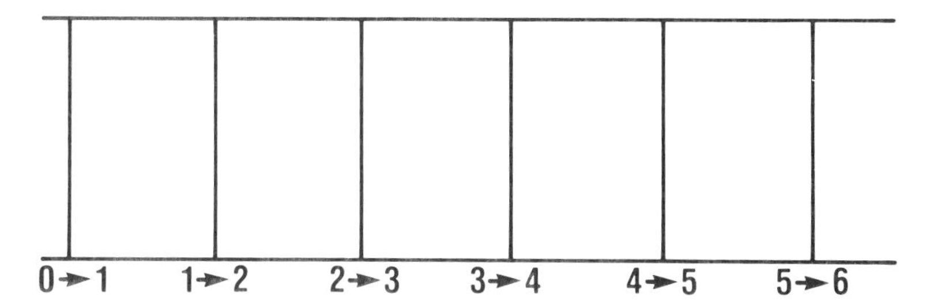

Figure 1.9. Schematic representation of the far infrared spectrum of HCl, showing a series of equally spaced lines.

and

$$\Delta E_{l+1\to l} = E_{l+1} - E_l = (\hbar^2/I)(l+1). \tag{1-54}$$

The observed HCl spectrum is shown in Fig. 1.9, which obviously correlates well with Eq. (1-54).

So far, we have discussed only the magnitude of angular momentum. What about the *direction* of **L**? We now discuss an experiment, first devised by Stern and Gerlach,[11] for measuring the component of **L** along one axis, say L_z. Of particular interest is the experimental result that, even for components of **L**, only *discrete* values of L_z are observed, i.e.,

$$l_z = M\hbar \tag{1-55}$$

where M is called the *magnetic quantum number*, and may take on values

$$M = -l,\ -l+1,\ \ldots\ ,\ +l. \tag{1-56}$$

In other words, M may take on any value between $-l$ and l, differing by integer amounts. For example, if $l = 1$, the possible values of M are -1, 0, 1, or, if $l = 3/2$, the possible values of M are -3, 2, $-1/2$, 1/2, 3/2.

The Stern–Gerlach experiment consists of sending a beam of atoms through an inhomogeneous magnetic field, and noting the deflection of the atoms due to the interaction of the atomic magnetic moments with the external magnetic field (see Fig. 1.10a and b). It should be recalled that a rotating charged particle generates a magnetic moment. Thus, we would expect[12] the orbital angular momentum **L** to be related to the magnetic moment $\boldsymbol{\mu}$. In fact, it will be shown

---

[11] O. Stern and W. Gerlach, *Z. Phys.*, **8**, 110 (1922), *Z. Phys.*, **9**, 349 (1922).

[12] For the case of spin, it is best to regard the spin magnetic moment as an *intrinsic property*, rather than resulting from a particle spinning about its axis, since the classical model yields incorrect numerical results. Thus, although the concepts may be more easily grasped by using analogies from classical mechanics, the shortcomings of such analogies in this case should be remembered.

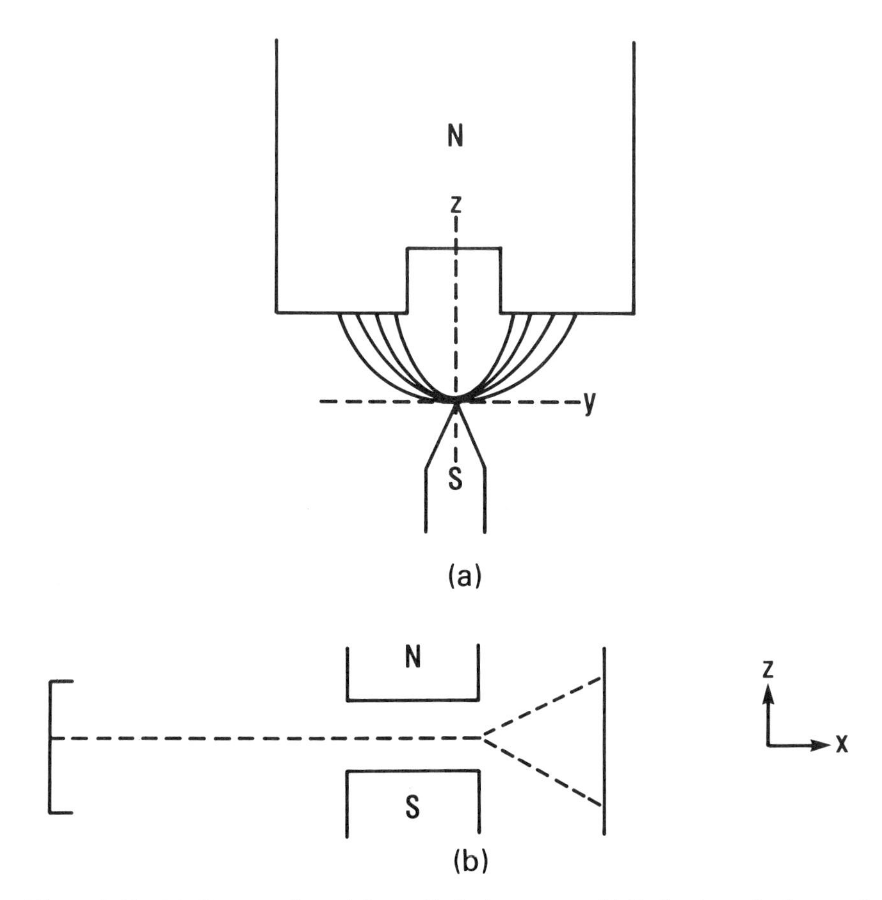

Figure 1.10. (a) Cross-section of Stern–Gerlach magnet. (b) Deflection of a beam of hydrogen atoms.

later that an atomic moment $\boldsymbol{\mu}$ is related to its angular momentum $\mathbf{L}$ by

$$\boldsymbol{\mu} = \gamma \mathbf{L}, \tag{1-57}$$

where γ is a constant called the *magnetogyric ratio* of the nucleus.

There is a theorem from classical mechanics (*Larmor's theorem*)[13] relating to the motion of a system of charged particles placed in a weak magnetic field. It says that *the motion of this system is the same as in the absence of the field, except that the angular momentum vector will precess about the direction of the magnetic field, instead of pointing in a fixed direction in space.* In other words, the angular momentum vector moves uniformly so as to describe a cone whose axis is in the direction of the magnetic field $\mathbf{B}$ (see Fig. 1.11). The angular

[13] See, for example, A. Carrington and A. D. McLachlan, *Introduction to Magnetic Resonance,* Harper & Row, New York, 1967, Chapter 11, for a discussion of this theorem.

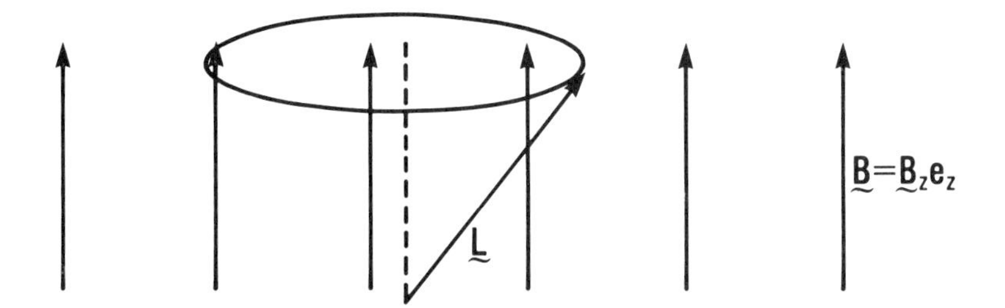

Figure 1.11. Precession of angular momentum **L** in a magnetic field **B**.

velocity for the precessional motion is

$$\boldsymbol{\omega} = -\gamma \mathbf{B}, \tag{1-58}$$

where $\boldsymbol{\omega}$ is called the *Larmor frequency*. Thus, if a magnetic moment $\boldsymbol{\mu}$ is placed in a magnetic field $B_z\mathbf{e}_z$, the moment will begin to precess about the z axis, i.e., the x and y components of $\boldsymbol{\mu}$ will change in time, though the component along the field axis μ_z will not change.

The force exerted on the magnetic moment $\boldsymbol{\mu}$ due to **B** depends upon its inhomogeniety, $\partial B_z/\partial z$, i.e.,

$$\mathbf{F} = \mathbf{e}_z \mu_z \frac{\partial B_z}{\partial z} = \mathbf{e}_z L_z \gamma \frac{\partial B_z}{\partial z}. \tag{1-59}$$

Thus, the moments will be deflected in the z direction, depending on the magnitude of L_z.

This experimental set-up is useful in at least two important ways. First, for atoms or molecules in which the total angular momentum is not zero, it provides an opportunity to measure the magnitude of the components of the (orbital) angular momentum. For example, if a system was prepared in an $L = 1$ state, it would be possible through a Stern–Gerlach experiment to measure each of the values of M that are possible, i.e., $M = -1, 0, +1$.

Of perhaps greater importance is the result that is obtained when systems with $L = 0$ are used in this experiment. In that case (e.g., with silver atoms), where only one value of M is possible ($M = 0$), two states are still observed. This result provides direct experimental confirmation of the existence of another kind of angular momentum (spin), in which only two states are possible. Hence, any theory that is to provide a satisfactory description of microscopic systems must include the possibility of both angular and spin angular momentum.

1-7. Momentum of a Photon

We have already mentioned the particulate nature of light, and have ascribed a value of the energy $E_{\mathrm{p}} = h\nu$ to each particle of light or photon. Let us now consider the momentum associated with the photon. The photon has no rest mass, but does have a mass by virtue of its motion; and we can calculate this mass from the energy–mass equivalence relation

$$E_p = mc^2. \tag{1-60}$$

Since momentum is mass times velocity, we have

$$p = mc = E_p/c. \tag{1-61}$$

Using the fact that $E_p = h\nu$, and the relationship between frequency ν and wavelength of light λ,

$$\lambda = c/\nu, \tag{1-62}$$

we find, at least for the case of photons, that

$$p = h/\lambda. \tag{1-63}$$

Hence, the inextricable connection between the wave and particle characteristics of photons is displayed in Eq. (1-63).

A beautiful demonstration of the momentum of the photon is offered by experiments on the scattering of X-rays by matter. Compton[14] showed that the change of frequency of the scattered X-rays, as a function of scattering angle, can be calculated exactly if one views the microscopic process as due to scattering of a photon by an electron in an atom. When the photon scatters from an electron, its direction is changed, and the electron recoils in a direction to preserve momentum. This recoiling electron has kinetic energy that must come from the photon. Hence, the wavelength of the scattered X-rays is longer than the wavelength of incident X-rays, and depends on the scattering angle in a way that is exactly predicted from conservation of energy and momentum relationships.

As an example of the relation of photon momentum to chemical and spectroscopic properties, we can calculate the amount of kinetic energy that can be transmitted to a photon from an atom emitting it. Let an excited atom have velocity $\mathbf{v}$ and mass M. Suppose the atom now emits a photon with momentum $\mathbf{p}$, and the atom recoils with velocity $\mathbf{v}'$ (see Fig. 1.12). From conservation of momentum, we write

$$M\mathbf{v} = M\mathbf{v}' + \mathbf{p}. \tag{1-64}$$

Squaring this equation on both sides and rearranging slightly, we get

$$\frac{1}{2}Mv^2 = \frac{1}{2}M(v')^2 + \frac{1}{2}p^2/M + v'p\cos\theta. \tag{1-65}$$

If the atom was originally in energy level E_2 and emits a photon of energy $h\nu$ to

---

[14] A. H. Compton, *Phys. Rev.,* **18**, 96 (1921); *Phys. Rev.,* **19**, 267 (1922); *Phys. Rev.,* **21**, 483 (1923).

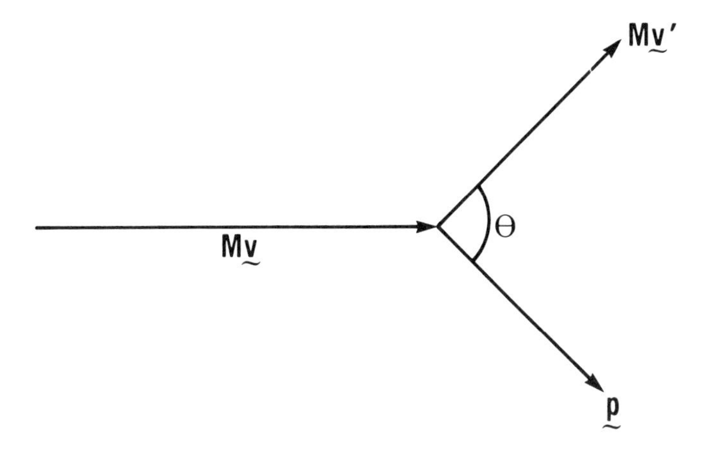

Figure 1.12. Emission of a photon by an atom.

go to the state of energy E_1, we write

$$\left(E_2+\frac{1}{2}Mv^2\right)=\left[E_1+\frac{1}{2}M(v')^2\right]+h\nu, \tag{1-66}$$

to guarantee conservation of energy. Setting $E_2 - E_1 = \nu_0$, Eq. (1-66) can be rearranged to

$$\frac{1}{2}Mv^2=\frac{1}{2}M(v')^2+h(\nu-\nu_0). \tag{1-67}$$

Comparing Eqs. (1-65) and (1-67), and using Eqs. (1-62) and (1-63), we obtain

$$\frac{\nu-\nu_0}{\nu}=\frac{h\nu}{2Mc^2}+\frac{v'}{c}\cos\theta. \tag{1-68}$$

This ratio gives the fractional change of the actual frequency ν that is emitted compared to the value expected (ν_0) if the emission were due entirely to a change in internal energy. In other words, this ratio will be zero if the emission is due entirely to a change in internal energy. In fact, for electronic transitions in atoms, this ratio is found to be of the order 10^{-7}. Hence, the frequencies emitted are due essentially to changes in *internal* energies, and *not* to changes in the kinetic energy of the atom.

1-8. Wave-Particle Duality

While the wave–particle duality for photons that is described in Eq. (1-63) is important both conceptually and computationally, one of the most surprising and far-reaching conceptual revisions needed in quantum theory is the extension of this concept beyond radiation to *all* matter. DeBroglie[15] was the first to propose such a generalization, arguing that *all matter* should obey the relation

$$\lambda = \frac{h}{p}, \tag{1-69}$$

since the same relativistic equations of motion apply to all matter (including photons).

To understand the experimental implications of Eq. (1-69) we note that, for an electron, use of Eq. (1-43), $E = mv^2/2$ and $p = mv$ in Eq. (1-69) gives rise to a value of λ of approximately 1 Å (10^{-8} cm) at a potential of 300 V. This means that, for matter–wave duality to be observed for electrons, an experimental arrangement in which dimensions of the "apparatus" are on the order of λ (i.e., 1 Å) will be needed. Although such an "apparatus" cannot be manufactured using macroscopic materials, an appropriate "microscopic apparatus" is provided in nature by atomic crystals, where the distance between atoms is of the order of Angstroms.

Davisson and Germer[16] were the first to provide such an experimental verification, by diffracting an electron beam from a nickel crystal. In particular, they prepared an electron beam of given momentum as in the impact experiments described in Section 1-4, and the wavelength was determined from the diffraction pattern according to the Bragg relation.[17]

Given the wave–particle duality just noted, it is also important in preparation for our discussion of quantum mechanics to review several properties of waves. In classical physics, an important type of *wave* is simply *some periodic disturbance of a material medium* (e.g., sound waves) or, more generally, *a periodic variation of some field* (e.g., electromagnetic waves). As an example, consider the sinusoidal wave

$$\Psi(x, t) = A \sin\left[2\pi\left(\frac{x}{\lambda} - \nu t\right)\right], \tag{1-70}$$

at the particular time $t = 0$ (see Fig. 1.13). The *wavelength* (λ) is simply the distance between two adjacent crests (or troughs) of the wave, and the *amplitude* (A) is the maximum deviation of the wave from zero. The frequency (ν) is just the number of wavelengths (λ) passing a given point in 1 sec.

Now consider what happens at a later time, e.g., $t = t_1 > 0$. This particular "snapshot" of the wave is also depicted in Fig. 1.13, and we see that while the wavelength and amplitude remain constant in time, the wave itself appears to move from left to right with increasing time. The quantity that is usually used to

---

[15] L. DeBroglie, *Compt. Rend.*, **177**, 507–510 (1923). See also L. deBroglie, *Phil. Mag.*, **47**, 446–458 (1924) for an English summary of earlier papers.

[16] C. Davisson and L. H. Germer, *Phys. Rev.*, **30**, 705–740 (1927).

[17] See, for example, Walter J. Moore, *Physical Chemistry*, Prentice Hall, New York, 1962, p. 655.

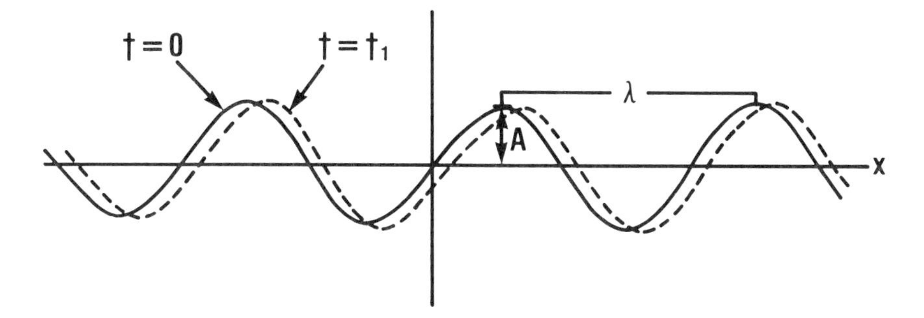

Figure 1.13. A sinusoidal wave at $t = 0$ and $t = t_1$ showing the wavelength λ and amplitude A.

describe how the wave moves as a function of time is called the *phase* (α), and is defined as

$$\alpha = 2\pi \left(\frac{x}{\lambda} - \nu t \right) . \tag{1-71}$$

It is sometimes convenient to describe the motion of a wave using different (but equivalent) terms. In particular, let us define

$$k = \frac{2\pi}{\lambda} , \tag{1-72}$$

and

$$\omega = 2\pi\nu, \tag{1-73}$$

where k is called the *propagation number* (or *wave number*), and ω is known as the *circular frequency* of the wave. In terms of these quantities the wave expressed by Eq. (1-70) can be written as

$$\Psi(x, t) = A \sin (kx - \omega t). \tag{1-74}$$

As observed above, this expression represents a wave moving from *left to right* along the x-axis as a function of time. It is also of interest to note that a wave which is moving from *right to left* along the x-axis as a function of time would have a phase of

$$\alpha = kx + \omega t. \tag{1-75}$$

Another concept that is useful in describing the motion of waves is that of *phase velocity* (c), which measures how fast a point that represents some particular value of $\Psi(x, t)$ moves along the x axis. To illustrate this, let us choose the value $\Psi = 0$, and examine the phase velocity associated with movement of the wave along the x-axis. From Eq. (1-71), we see that one point at which $\Psi = 0$ is when $x = 0$ and $t = 0$ (which corresponds to $\alpha = 0$). Let x_0 be the next point at which Ψ is again zero, and suppose that the time required for the point at $x = 0$ to reach the point $x = x_0$ is $t = t_0$ (sec). For this particular example, we must have

$$n\pi = 2\pi \left(\frac{x_0}{\lambda} - \nu t_0 \right), \qquad n = 0, 1, 2, \dots, \tag{1-76}$$

or

$$0 = \frac{x_0}{\lambda} - \nu t_0, \tag{1-77}$$

where we have chosen $n = 0$ for convenience. The phase velocity (c) is simply the distance travelled by the point in a given time, which is seen from Eq. (1-77) to be

$$c = \frac{x_0 \text{ (cm)}}{t_0 \text{ (sec)}} = \nu\lambda. \tag{1-78}$$

Another important property of waves is that they may be added together (*superposed*) to get a new wave. To illustrate this notion, consider the waves

$$\begin{aligned} \Psi_1 &= A \sin (kx - \omega t) \\ \Psi_2 &= A \sin (kx - \omega t + \delta). \end{aligned} \tag{1-79}$$

These waves differ in phase by the amount δ. If the waves are *in phase* ($\delta = 0$ or a multiple of 2π), it is easily seen that they reinforce each other to give a new wave $\Psi_1 + \Psi_2 = 2\Psi_1$, with amplitude $2A$. If, however, the waves are *out of phase* by e.g., 180° ($\delta = n\pi$, with $n = 1, 3, 5, \dots$), we have $\Psi_1 + \Psi_2 = 0$ (*interference*), and the waves destroy each other.

Thusfar we have interpreted the wave $\Psi(x, t) = A \sin (kx - \omega t)$ as a periodic variation in *one* dimension. Thus, for example, it could represent the vibration of a string, so that Ψ represents the displacement of points along the string from equilibrium, at some time t, along the x axis. However, it is possible to generalize this idea to a wave that propagates in an arbitrary direction, and not restrict its propagation to be only along the x axis. In the general case (see Fig. 1.14), a wave propagating along the direction of a vector $\mathbf{k}$, having components (k_x, k_y, k_z), is given by

$$\Psi(\mathbf{r}, t) = A \sin (k_x x + k_y y + k_z z - \omega t) \tag{1-80}$$

$$= A \sin (\mathbf{k} \cdot \mathbf{r} - \omega t), \tag{1-81}$$

where

$$|\mathbf{k}| = [k_x^2 + k_y^2 + k_z^2]^{1/2}. \tag{1-82}$$

The vector $\mathbf{k}$ is usually referred to as the *propagation vector*. These waves also possess another property of interest. In particular, for a given time t, let us examine at which points the phase of the wave represented by Eq. (1-81) is

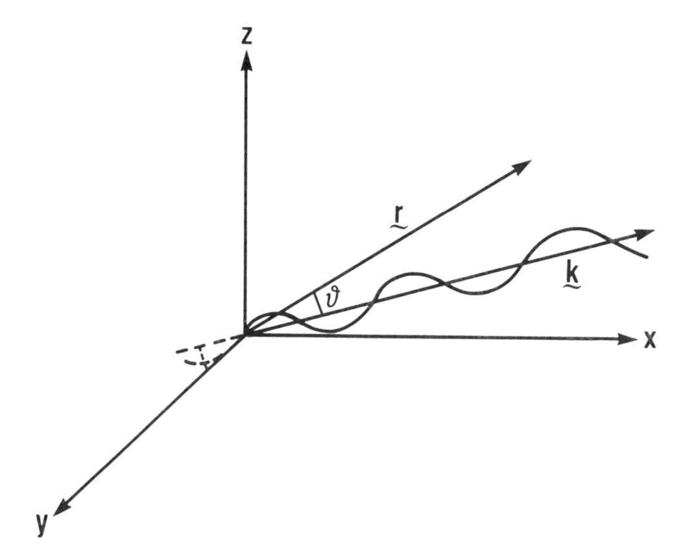

Figure 1.14. Depiction of a wave traveling in the direction of a vector **k**.

constant. Since the equation

$$\mathbf{k} \cdot \mathbf{r} = kd \tag{1-83}$$

represents the equation of a plane perpendicular to **k** and located a distance, d, from the origin, it follows that the phase

$$\alpha = \mathbf{k} \cdot \mathbf{r} - \omega t \tag{1-84}$$

will be constant for all points on that plane for a given time t. Thus, we see that α is a constant whenever $\vartheta = 90°$. In other words, any point in the plane perpendicular to the vector **k** will have the same (constant) phase. For this reason such waves are generally called *plane waves*. For the earlier example of Eq. (1-74), we see that it represents a plane wave travelling in the X-direction, i.e., $k_y = k_z = 0$, and the phase ($\alpha = k_x X - \omega t$) is obviously the same for any y or z.

Another point of interest regarding the description of waves that will be useful later involves the mathematical form of the description. Since

$$e^{i\alpha} = \cos \alpha + i \sin \alpha, \tag{1-85}$$

we see that an alternate description of waves can be obtained by the use of complex exponentials. For example, the plane waves of the preceding discussion can also be represented as the real part of $\Psi(\mathbf{r}, t)$, where

$$\Psi(\mathbf{r}, t) = Ae^{i(\mathbf{k} \cdot \mathbf{r} - \omega t)}. \tag{1-86}$$

Finally, we note that other kinds of waves are also useful in discussions. For example, a *spherical wave* is one in which the phase is constant (for a given t) along *spheres* (instead of planes), eminating from the origin of the wave.

The presence of wave–particle duality referred to earlier also leads to additional complications in the interpretation of events at the microscopic level. These difficulties arise because of the inadequacies of classical concepts to describe these events. For example, the "position" of a *particle* in classical mechanics at a given time is uniquely defined as being some single *point* in space. On the other hand, the concept of "position" when discussing wave phenomena is totally inappropriate in classical mechanics, since a wave extends throughout all space. Thus, the way to describe the "position" of (e.g., an electron that exhibits both wave and particle characteristics is not obvious. As we shall see in later chapters, these two concepts can be reconciled by means of the *uncertainty principle,* which can be thought of as a principle that "blurs" the notion of simultaneously specifying the position and momentum of a particle, while "sharpening" the wave picture somewhat by treating "wave packets," formed by suitable superpositions of waves to give reasonably localized wave packets that can be thought of as resembling particles.

Thus, we see that the conceptual and mathematical structure of quantum mechanics, while retaining many notions from classical mechanics, must contain the capability for significant modification of classical ideas in several areas. In particular, it must be able to accommodate at least the concepts of discrete energy levels, with quantization of both total energies and angular momentum, as well as to allow discussion of the wave–particle duality mentioned above. In order to facilitate the development of concepts such as these and other "nonclassical" notions (e.g., electron spin), a rather powerful set of mathematical techniques is typically employed, much of which is frequently not familiar. Consequently, before introducing the formal definitions of quantum theory in Chapter 4, the next two chapters are designed to introduce the necessary mathematical techniques, so that the introduction of new concepts can be studied without the complication of introducing new mathematics simultaneously.

Problems

1. Using Eqs. (1-29) and (1-30), along with the definition of the Hamiltonian function in Eq. (1-31), show that Hamilton's equations of motion [Eq. (1-33) and (1-34)] can be derived from Lagrange's equations of motion [Eq. (1-28)].
2. One of the important examples to be studied later is known as the "harmonic oscillator," which is a particle whose force attracting it back to its origin is proportional to the distance from the origin. In one dimension, the potential energy represented by this force is given by

$$V=\frac{1}{2}kx^2$$

where k is a constant. Show that substitution of the above potential energy in

Lagrange's equation can be reduced to Newton's equation of motion, i.e.,

$$F=m\,\frac{dx^2}{dt^2}\,.$$

3. Consider a single particle of mass in moving with velocity $\mathbf{v}$. Show that the kinetic energy

$$T=\frac{1}{2}\,m|\mathbf{v}|^2$$

can be written in terms of spherical coordinates as

$$T=\frac{1}{2}\,m(\dot{r}^2+r^2\dot{\vartheta}^2+r^2\,\sin^2\,\vartheta\dot{\varphi}^2).$$

4. Assuming the potential (V) to be a function of coordinates only, i.e., $V = V(r,\ \vartheta,\ \varphi)$, show that an alternate expression for the kinetic energy in the previous problem can be written in terms of conjugate moments (p_r, p_ϑ, p_φ) as:

$$T=\frac{1}{2m}\left(p_r^2+\frac{1}{r^2}\,p_\vartheta^2+\frac{1}{r^2\,\sin^2\,\vartheta}\right).$$

5. Consider a single particle of mass m moving in a circular orbit in the xy plane under the influence of a gravitational potential:

$$V(r)=-\frac{GMm}{r}\,,$$

where M is a mass at the origin and G is the gravitational constant. Derive the equations of motion in terms of spherical coordinates r, ϑ, φ and momenta p_r, p_ϑ, p_φ and show that p_φ is a constant of the motion.
6. Consider an electron of mass m and charge ($-e$) moving in a circular orbit in the xy plane under the influence of a central electrostatic potential:

$$V(r)=-\frac{Ze^2}{r}\,.$$

Derive expressions for p_r, p_ϑ, and p_φ and their time derivatives. Also, using Bohr's hypothesis that angular momentum is quantized, i.e.,

$$p_\vartheta=n\hbar \qquad n=1,\ 2,\ 3,\ \dots\,,$$

derive expressions for the Bohr radius and the total energy of the system.
7. The kinetic energy in cartesian coordinates (x_i) for a system of N particles of mass m is given by:

$$T=\left(\frac{m}{2}\right)\sum_{i=1}^{3N}\dot{x}_i^2.$$

If each of the coordinates is expressed in terms of generalized coordinates [as in Eq. (1-27)], i.e.,

$$\begin{aligned} x_1 &= x_1(q_1, \ldots, q_{3N}) \\ &\vdots \\ x_{3N} &= x_{3N}(q_1, \ldots, q_{3N}), \end{aligned}$$

show that the kinetic energy can be expressed in generalized coordinates as

$$T = \left(\frac{m}{2}\right) \sum_{l}^{3N} \sum_{k}^{3N} b_{lk} \dot{q}_l \dot{q}_k,$$

where the b_{lk} are functions of the coordinates (q_k).

8. Given

$$T = \left(\frac{m}{2}\right) \sum_{l}^{3N} \sum_{k}^{3N} b_{lk} \dot{q}_l \dot{q}_k$$

from the previous problem, show that

$$H = T + V$$

for a conservative system.

Chapter 2

Vector Spaces and Linear Transformations

Having seen that it was necessary to construct a new theory to describe the behavior of sub-microscopic particles, it should not be surprising that mathematical techniques were introduced concurrently to aid in the development of the new theory.

One of the basic mathematical tools of quantum mechanics is vectors, and how they may be manipulated. As we shall see, these vectors are generalizations of the ordinary three-dimensional vectors that are encountered in other areas of chemistry and physics.

In this chapter and the next, we shall develop the idea of vectors, discuss their characteristics in some detail, and show how different representations of vectors lead, in one case, to the solution of differential equations for the desired results and, on the other hand, to the solution of matrix equations. For most systems of interest to chemists, the solution of the appropriate differential equations is not possible in closed form, and approximations are necessary. In these cases, the matrix representation of the problem will typically be most convenient, since the manipulation of matrices is readily adaptable for computer use.

2-1. Vector Spaces

The idea of a vector in a two- or three-dimensional space is a familiar concept in physics and chemistry. It has been used in many ways to describe such things as forces, angular momentum, and dipole moments. These vectors all possess two important properties: magnitude and direction. In addition, these vectors can be manipulated in two important ways. They can be combined (''*added*'') to form a new vector, i.e.,

$$\mathbf{u} + \mathbf{v} = \mathbf{w}, \tag{2-1}$$

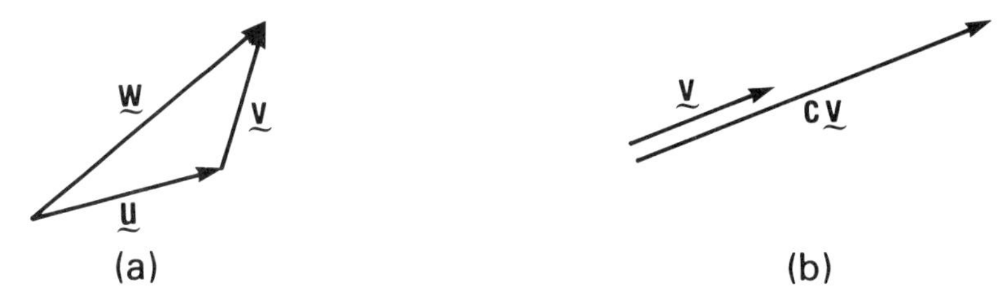

Figure 2.1. (a) Addition of vectors. (b) Scalar multiplication of vectors.

and they can be *multiplied by a scalar*, c, i.e.,

$$c\mathbf{v} = \mathbf{u}, \tag{2-2}$$

where c can be thought of as a number (real or complex). Figure 2.1 illustrates the effect of these operations.

As is also familiar, it is sometimes useful to consider a vector in terms of its *components*. These components represent the vector in some particular coordinate system, and Fig. 2.2 illustrates this point for a three-dimensional vector in a Cartesian coordinate system. The components in this case are called the *coordinates* of the vector. In this particular representation of a vector, addition and scalar multiplication are given by

$$(\mathbf{v} + \mathbf{u}) = (v_x, v_y, v_z) + (u_x, u_y, u_z) = (v_x + u_x, v_y + u_y, v_z + u_z), \tag{2-3}$$

and

$$c\mathbf{v} = c(v_x, v_y, v_z) = (cv_x, cv_y, cv_z). \tag{2-4}$$

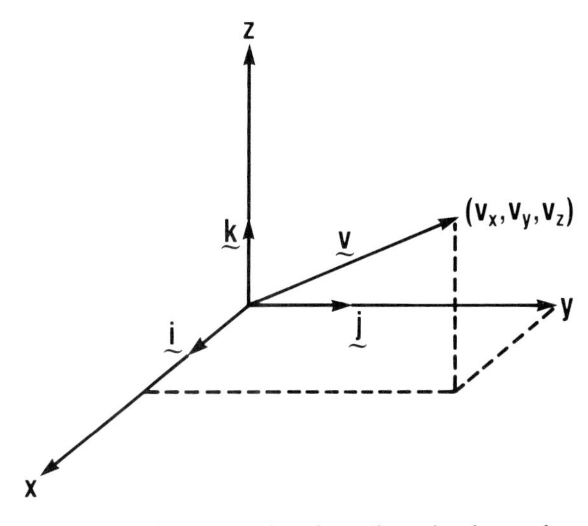

Figure 2.2. Representation of a vector $\mathbf{v}$ in a three-dimensional space by means of its x, y, and z components and unit vectors in the x, y, and z directions.

Several other properites of vectors will also be important in our considerations. The *scalar* or *inner product* between two vectors in three-dimensional space is usually defined as

$$(\mathbf{u}, \mathbf{v}) = u_x v_x + u_y v_y + u_z v_z. \tag{2-5}$$

From this definition, the *length of a vector* **u**, which shall also be referred to as the *norm* of **u**, is taken to be

$$|\mathbf{u}| = [(\mathbf{u}, \mathbf{u})]^{1/2} = [u_x u_x + u_y u_y + u_z u_z]^{1/2}. \tag{2-6}$$

If $|\mathbf{u}| = 1$, the vector **u** is said to be normalized to *unity*.

Sometimes the scalar product is written in the alternative form,

$$(\mathbf{u}, \mathbf{v}) = |\mathbf{u}| \; |\mathbf{v}| \cos \vartheta_{uv}, \tag{2-7}$$

where ϑ_{uv} is the angle between the vectors **u** and **v**.

In addition to the scalar product, vectors can also be multiplied together in another way. The second kind of multiplication is generally referred to as a *vector* or *outer product*, and is defined for two vectors **u** and **v** as[1]

$$(\mathbf{u} \times \mathbf{v}) = |\mathbf{u}| \; |\mathbf{v}| \sin \vartheta_{uv}. \tag{2-8}$$

Thus far, vectors have been discussed from primarily a geometric viewpoint. The concept of a vector, however, is much more general, and the generalization is based on the concept of a *vector space*. This is an abstract space that we shall now describe in a formal manner using the set of postulates given below. Such a generalization is important to carry out, for it will allow us to remove the restrictive notion that vector spaces can contain only the geometric vectors with which we are familiar, and allow consideration of other possibilities (such as functions) as belonging to a vector space.

In particular, a *vector space* (or linear space) V over the real or complex number system is composed of a set of elements, called vectors, together with two operations, addition and scalar multiplication (with real or complex scalars), that satisfy the following postulates:

1. Closure under addition. For every pair of vectors **v**, **u** in V, there is a unique sum $(\mathbf{v} + \mathbf{u})$ that is also in V.
2. Associativity under addition.

$$(\mathbf{v} + \mathbf{u}) + \mathbf{w} = \mathbf{v} + (\mathbf{u} + \mathbf{w}).$$

3. Commutivity under addition.

$$\mathbf{v} + \mathbf{u} = \mathbf{u} + \mathbf{v}.$$

4. Additive identity. A "zero" vector, **0**, exists in V such that $\mathbf{v} + \mathbf{0} = \mathbf{v}$ for all **v** in V.

---

[1] While this definition is appropriate for vectors having three dimensions, generalization of the vector product to N-dimensional vectors requires the use of tensors. In this text, such a generalization will not be needed in general, and is therefore omitted from the discussions here.

5. Additive inverses. A "negative" vector, $-\mathbf{v}$, exists for each $\mathbf{v}$ in V such that $\mathbf{v} + (-\mathbf{v}) = \mathbf{0}$.
6. Closure under scalar multiplication. For every scalar c (real or complex) and for every $\mathbf{v}$ in V there is a unique vector $c\mathbf{v}$ that is also in V.
7. Scalar Multiplication.

   a. $c(\mathbf{v}+\mathbf{u})=c\mathbf{v}+c\mathbf{u}$,

   b. $(c+c')\mathbf{v}=c\mathbf{v}+c'\mathbf{v}$,

   c) $(cc')\mathbf{v}=c(c'\mathbf{v})$.

Several comments are appropriate concerning these postulates. First, no attempt has been made to ensure that only a minimum number of postulates has been used. Instead, those postulates in common usage in mathematics texts have been adopted for convenience.[2] Also, it should be noted that the above postulates that describe a general vector space do not include the scalar product operation. Only special kinds of vector spaces, called inner product spaces, contain this operation as well as the characteristics of a general vector space listed above. Such spaces will be discussed in Section 2-3.

When considering whether certain mathematical objects constitute a vector space, one must be careful to ensure that each of the above postulates is satisfied. It is easily verified that the geometric vectors in two- and three-dimensional space discussed previously satisfy these postulates. Let us now consider examples of other kinds of mathematical objects that also behave as vectors in a vector space, that perhaps are not so obvious.

As a first example, let us consider a set of vectors that is more general than those encountered previously, in which there are more than three components for each vector. In particular, let us assume that there are n components needed to describe each vector, e.g.,

$$\mathbf{v}=(v_1, v_2, \dots, v_n). \tag{2-9}$$

Each of the v_i is restricted only to be a real number. Vectors such as these can be thought of as generalizations of the three-component vectors discussed earlier, except that the geometric visualization in this case is considerably more difficult, since n-dimensions are required. However, the lack of a geometric visualization does not restrict our consideration of these vectors. If another vector is given by

$$\mathbf{u}=(u_1, u_2, \dots, u_n), \tag{2-10}$$

and if addition and scalar multiplication are defined as

$$\mathbf{w}=\mathbf{v}+\mathbf{u}=(v_1+u_1, v_2+u_2, \dots, v_n+u_n), \tag{2-11}$$

and

$$\mathbf{z}=c\mathbf{v}=(cv_1, cv_2, \dots, cv_n), \tag{2-12}$$

[2] See, for example, Paul R. Halmos, *Finite-Dimensional Vector Spaces*, Van Nostrand, New York, 1958, pp. 3–4.

where c is also a real number, then it can easily be shown that such vectors satisfy the basic postulates of a vector space. For example, we are assured of closure under addition since, for each component of $\mathbf{w}$, the sum of two real numbers is involved, which must result in another real number. Hence, each component of $\mathbf{w}$ must be a real number, which guarantees that $\mathbf{w}$ is another vector of the type previously defined. Next, associativity and commutivity under addition are also easily verified, i.e.,

$$\begin{aligned}(\mathbf{v}+\mathbf{u})+\mathbf{w} &= (v_1+u_1,\ v_2+u_2,\ \ldots,\ v_n+u_n)+(w_1,\ w_2,\ \ldots,\ w_n) \\ &= (v_1+u_1+w_1,\ v_2+u_2+w_2,\ \ldots,\ v_n+u_n+w_n) \\ &= (v_1,\ v_2,\ \ldots,\ v_n)+(u_1+w_1,\ u_2+w_2,\ \ldots,\ u_n+w_n) \\ &= \mathbf{v}+(\mathbf{u}+\mathbf{w}). \end{aligned} \tag{2-13}$$

Also,

$$\begin{aligned}(\mathbf{v}+\mathbf{w}) &= (v_1+w_1,\ v_2+w_2,\ \ldots,\ v_n+w_n) \\ &= (w_1+v_1,\ w_2+v_2,\ \ldots,\ w_n+v_n) \\ &= \mathbf{w}+\mathbf{v}, \end{aligned} \tag{2-14}$$

since the order of addition of any two real numbers does not affect the result. If the zero vector is defined as

$$\mathbf{0}=(z_1,\ z_2,\ \ldots,\ z_n), \tag{2-15}$$

where $z_1 = z_2 = \cdots = z_n = 0$, then the additive identity is clearly assured. Concerning the additive inverse of a vector ($\mathbf{v}$), we define

$$\mathbf{w}=(-v_1,\ -v_2,\ \ldots,\ -v_n), \tag{2-16}$$

which gives

$$\mathbf{v}+\mathbf{w}=\mathbf{0}. \tag{2-17}$$

Thus, $\mathbf{w} = -\mathbf{v}$, and an additive inverse exists. In a similar manner, closure under scalar multiplication as given in postulates 6 and 7 can be easily verified.

Considering next an example of a vector space that might not be as obvious, let us examine the space of all polynomial functions of degree $n-1$. One particular polynomial function in this space is given by

$$f(x)=c_0+c_1x+c_2x^2+\cdots+c_{n-1}x^{n-1}, \tag{2-18}$$

where $c_0,\ c_1,\ \ldots,\ c_{n-1}$ are fixed scalars (real or complex) that are independent of x. If addition and scalar multiplication for two such functions, $f(x)$ and $g(x)$ are defined as

$$\begin{aligned} f(x)+g(x) &= (c_0+c_1x+c_2x^2+\cdots+c_{n-1}x^{n-1}) \\ &\quad +(c_0'+c_1'x+c_2'x^2+\cdots+c_{n-1}'x^{n-1}) \\ &= (c_0+c_0')+(c_1+c_1')x+(c_2+c_2')x^2+\cdots+(c_{n-1}+c_{n-1}')x^{n-1}, \end{aligned} \tag{2-19}$$

and

$$cf(x) = c(c_0 + c_1 x + c_2 x^2 + \cdots + c_{n-1} x^{n-1})$$
$$= (cc_0) + (cc_1)x + (cc_2)x^2 + \cdots + (cc_{n-1})x^{n-1}, \quad (2\text{-}20)$$

then it can be shown (see Problem 2-1) that such polynomials also constitute a vector space. It is important to note that the behavior of the coefficients resembles that of the components of vectors. We shall discuss this relationship in more detail shortly.

2-2. Linear Independence, Bases, and Dimensionality

The vector spaces that have just been introduced possess several properties that will be of interest to quantum chemistry. Basic to these properties is the notion of a *linear combination of vectors*, which is nothing more than a sum of vectors

$$c_1 \mathbf{v}_1 + c_2 \mathbf{v}_2 + \cdots + c_n \mathbf{v}_n = \sum_{i=1}^{n} c_i \mathbf{v}_i, \quad (2\text{-}21)$$

in which the c_i are real or complex scalars, and all $\mathbf{v}_i$ are in V. A vector $\mathbf{u}$ is said to be a linear combination of vectors if it can be written as

$$\mathbf{u} = \sum_{i=1}^{n} c_i \mathbf{v}_i. \quad (2\text{-}22)$$

One property that is associated with the notion of linear combination of vectors is connected with the idea of *linear independence*. A set of n vectors $\{\mathbf{v}_i\}$ in a vector space V is said to be linearly independent if and only if the linear combination

$$c_1 \mathbf{v}_1 + c_2 \mathbf{v}_2 + \cdots + c_n \mathbf{v}_n = \mathbf{0} \quad (2\text{-}23)$$

implies that $c_1 = c_2 = c_3 = \cdots = c_n = 0$. If there exists a set of scalars that are not all zero such that Eq. (2-22) is satisfied, then the set $\{\mathbf{v}_i\}$ is said to be *linearly dependent*.

This definition says that, if the $\{\mathbf{v}_i\}$ are linearly independent, the zero vector $\mathbf{0}$, can only be obtained as a linear combination of the $\{\mathbf{v}_i\}$ if all coefficients $c_i = 0$. If, however, the zero vector is obtained such that some $c_i \neq 0$, then at least one of the vectors, say $\mathbf{v}_j$, can be written as

$$\mathbf{v}_j = -\left(\frac{1}{c_j}\right) \sum_{i \neq j}^{n} c_i \mathbf{v}_i, \quad (2\text{-}24)$$

which shows that the vector $\mathbf{v}_j$ depends linearly on the other vectors. Let us consider some examples to illustrate this concept. To begin with, consider the set of vectors

$$\begin{aligned} \mathbf{v}_1 &= (1,\ 0,\ 0) \\ \mathbf{v}_2 &= (0,\ 1,\ 0) \\ \mathbf{v}_3 &= (0,\ 0,\ 2). \end{aligned} \tag{2-25}$$

These vectors are linearly independent, since

$$c_1\mathbf{v}_1 + c_2\mathbf{v}_2 + c_3\mathbf{v}_3 = \mathbf{0} \tag{2-26}$$

implies that $c_1 = c_2 = c_3 = 0$. If, however, the vector

$$\mathbf{v}_4 = (1,\ 1,\ 1) \tag{2-27}$$

is added to the set, it now is a linearly dependent set, since

$$\mathbf{v}_4 = \mathbf{v}_1 + \mathbf{v}_2 + \mathbf{v}_3. \tag{2-28}$$

Next, consider the set of polynomials

$$\begin{aligned} f_1(x) &= 1 + x^3 \\ f_2(x) &= 2 + x \\ f_3(x) &= x - 2x^3, \end{aligned} \tag{2-29}$$

over the range $-\infty \leq x \leq +\infty$. This set is not linearly independent, i.e., it is linearly dependent. The student can verify this by choosing $c_1 = 2$, $c_2 = -1$, and $c_3 = 1$, and examining the linear combination of functions

$$c_1 f_1(x) + c_2 f_2(x) + c_3 f_3(x).$$

A polynomial set that is more often used, and slightly generalized from the previous example, is the set

$$\begin{aligned} f_0(x) &= 1 \\ f_2(x) &= x \\ &\vdots \\ f_{n-1}(x) &= x^{n-1}. \end{aligned} \tag{2-30}$$

This set is linearly independent, since the sum

$$c_0 f_0(x) + c_1 f_1(x) + \cdots + c_{n-1} f_{n-1}(x) = 0 \tag{2-31}$$

implies that $c_i = 0$ for *all* i (see Problem 2-2).

The previous definition of linear independence leads to the following important theorem.

Theorem 2-1. The set of nonzero vectors $\{\mathbf{v}_1, \mathbf{v}_2, \ldots, \mathbf{v}_n\}$ in a vector space V is linearly dependent if and only if one (or more) of the vectors, say $\mathbf{v}_k$, is a linear combination of the proceeding ones.

Proof: (Sufficient condition). Suppose $\mathbf{v}_k$ is a linear combination of the preceding vectors, i.e.,

$$\mathbf{v}_k = \sum_{i=1}^{k-1} c_i \mathbf{v}_i. \tag{2-32}$$

If we choose $c_k = -1$, then we can write

$$c_1\mathbf{v}_1 + c_2\mathbf{v}_2 + \cdots + c_{k-1}\mathbf{v}_{k-1} + (-1)\mathbf{v}_k = \mathbf{0}. \tag{2-33}$$

Since the sum of vectors $\Sigma_{i=1}^{k} c_i\mathbf{v}_i = 0$, where *not* all $c_i = 0$, the set of vectors must be linearly dependent, and the sufficiency condition has been proven.

(Necessary condition). Suppose $\{\mathbf{v}_1, \mathbf{v}_2, \ldots, \mathbf{v}_n\}$ are linearly dependent. Then, there exist constants, d_i, not all zero, such that

$$\sum_{i=1}^{n} d_i\mathbf{v}_i = \mathbf{0}, \tag{2-34}$$

by the definition of linear dependence. Let d_k be the last d_i that is not equal to zero. (Note we can always rearrange the ordering of our vector set $\{\mathbf{v}_i\}$ such that the summation contains the first k terms with nonzero coefficients and the rest of the coefficients zero.) Dividing through by d_k and rearranging Eq. (2-34) we obtain

$$\mathbf{v}_k = \left(\frac{-1}{d_k}\right) \sum_{i=1}^{k-1} d_i\mathbf{v}_i, \tag{2-35}$$

which completes the proof.

There is some terminology that is associated with linearly independent vectors that is conveniently introduced at this point. A set of vectors $\{\mathbf{v}_i\}$ in a vector space V is said to *span* the vector space V if every vector in V can be written as a linear combination of the set $\{\mathbf{v}_i\}$. That is, if $\mathbf{u}$ is in V and the set of n vectors $\{\mathbf{v}_i\}$ span V, there exists a set of scalars $\{c_1, c_2, \ldots, c_n\}$ such that

$$\mathbf{u} = \sum_{i=1}^{n} c_i\mathbf{v}_i. \tag{2-36}$$

Note that a set of vectors need *not* be linearly independent to span V. For example, the four vectors $\{\mathbf{v}_1, \mathbf{v}_2, \mathbf{v}_3, \mathbf{v}_4\}$ given in Eqs. (2-24) and (2-26) are linearly dependent, yet they span the space of all three-component vectors.

The next theorem shows that if we are given a set of vectors that spans (i.e., generate) a vector space, V, there exists within this set of vectors a set of vectors that spans V and is also linearly independent.

Theorem 2-2. Let n vectors span V. Then there exists a subset of these vectors which *both* span V and are linearly independent.

Proof: 1. If the vectors are linearly independent, the theorem is trivially true.

2. Suppose the n vectors spanning V are linearly dependent. By Theorem 2-1 we can write

$$\mathbf{v}_n = \sum_{i=1}^{n-1} c_i\mathbf{v}_i, \tag{2-37}$$

for at least one of the n vectors. If $\mathbf{u}$ is any vector in vector in V, it can be represented as

$$\mathbf{u}=\sum_{i=1}^{n} b_i\mathbf{v}_i, \tag{2-38}$$

since the n vectors $\mathbf{v}_i$ span V. Substituting Eq. (2-37) into Eq. (2-38) and rearranging terms gives

$$\mathbf{u}=\sum_{i=1}^{n-1} (b_i+b_n c_i)\mathbf{v}_i. \tag{2-39}$$

Thus, the $n - 1$ vectors $\mathbf{v}_1, \mathbf{v}_2, \ldots, \mathbf{v}_{n-1}$ span V also. This argument can be repeated until we arrive at the subset $\{\mathbf{v}_1, \mathbf{v}_2, \ldots, \mathbf{v}_k\}$, $k < n$, which spans V and is linearly independent.

A linearly independent set of vectors that spans the entire space is said to be a *basis for a vector space* V. If the set of basis vectors is finite, V is said to be a *finite-dimensional vector space*. If the set of basis vectors is infinite (but denumerable), V is said to be an *infinite-dimensional vector space*.

As an example of this concept, consider once again the three-dimensional Cartesian coordinate system used in standard vector analysis, which is shown in Fig. 2.2. A set of vectors that forms a basis for this space (usually called *unit vectors*) are given by the following component representation:

$$\mathbf{i}=(1, 0, 0); \qquad \mathbf{j}=(0, 1, 0); \qquad \mathbf{k}=(0, 0, 1). \tag{2-40}$$

It is easily seen that these basis vectors are linearly independent. Furthermore, the following equation shows that any vector in this three-dimensional space can be represented using the basis vectors, i.e.,

$$\mathbf{v}=v_x\mathbf{i}+v_y\mathbf{j}+v_z\mathbf{k}. \tag{2-41}$$

If we had chosen a fourth vector $\mathbf{l} = (1, 0, 1)$, the set $\{\mathbf{i}, \mathbf{j}, \mathbf{k}, \mathbf{l}\}$ would also span the three-dimension Cartesian coordinate space. It is clear however, that this set of vectors is linearly dependent, since the subset given in Eq. (2-40) is linearly independent.

The following two theorems exhibit important properties of basis vectors. These theorems demonstrate the uniqueness of the representation of a vector in terms of a specific basis, and the uniqueness of the number of basis vectors in a space of finite-dimension.

Theorem 2-3. If the set of vectors $\{\mathbf{v}_1, \mathbf{v}_2, \ldots, \mathbf{v}_n\}$ in V is a basis, then every vector $\mathbf{u}$ in V has a unique representation as a linear combination of the basis vectors:

$$\mathbf{u}=c_1\mathbf{v}_1+c_2\mathbf{v}_2+\cdots+c_n\mathbf{v}_n. \tag{2-42}$$

Proof: Suppose there exist two representations

$$\mathbf{u}=\sum_{i=1}^{n} c_i\mathbf{v}_i, \tag{2-43}$$

$$\mathbf{u}=\sum_{i=1}^{n} c_i'\mathbf{v}_i, \tag{2-44}$$

for the same vector with respect to the basis. Subtracting Eq. (2-44) from Eq. (2-43), we obtain

$$\mathbf{u}-\mathbf{u}=\mathbf{0}=\sum_{i=1}^{n} (c_i-c_i')\mathbf{v}_i. \tag{2-45}$$

Since the $\{\mathbf{v}_i\}$ form a basis, they are linearly independent. This implies, from the definition of linear independence, that

$$c_i-c_i'=0 \qquad \text{for } i=1, 2, \dots, n. \tag{2-46}$$

Hence, $c_i = c_i'$ for $i = 1, 2, \dots, n$, which proves that the representation is unique.

Theorem 2-4. Any two bases of a finite-dimensional vector space V have the same number of basis vectors.

Proof: Let $\{\mathbf{v}_1, \mathbf{v}_2, \dots, \mathbf{v}_n\}$ and $\{\mathbf{u}_1, \mathbf{u}_2, \dots, \mathbf{u}_m\}$ be two bases in V. To prove the theorem, suppose that $m > n$. Since the $\{\mathbf{v}_i\}$ span V, the $\mathbf{u}_i$ can be written in terms of $\{\mathbf{v}_i\}$. Beginning with $\mathbf{u}_1$, we have

$$\mathbf{u}_1=\sum_{i=1}^{n} a_i\mathbf{v}_i. \tag{2-47}$$

Since $\mathbf{u}_1$ belongs to a linearly independent set $\{\mathbf{u}_i\}$, it is not the zero vector (see Problem 2-3). Hence, at least one of the $a_i \neq 0$. The $\mathbf{v}_i$ can be renumbered such that $a_1 \neq 0$, and we can solve for $\mathbf{v}_1$, obtaining

$$\mathbf{v}_1=a_1^{-1}\left[\mathbf{u}_1-\sum_{i=2}^{n} a_i\mathbf{v}_i\right]. \tag{2-48}$$

Thus, a new set of vectors $\{\mathbf{u}_1, \mathbf{v}_2, \dots, \mathbf{v}_n\}$ has been constructed that spans V. Proceeding in a similar manner, we now write $\mathbf{u}_2$ as a linear combination of this set, i.e.,

$$\mathbf{u}_2=b_1\mathbf{u}_1+\sum_{i=2}^{n} b_i\mathbf{v}_i. \tag{2-49}$$

Not all the b_i, $i = 2, 3, \dots, n$, are equal to zero since, if they were, $\mathbf{u}_2$ and $\mathbf{u}_1$ would be linearly dependent, which violates one of our initial assumptions.

Reordering the $\mathbf{v}_i$, if necessary, such that $b_2 \neq 0$, we solve for $\mathbf{v}_2$,

$$\mathbf{v}_2 = b_2^{-1}\left[\mathbf{u}_2 - b_1\mathbf{u}_1 - \sum_{i=3}^{n} b_i\mathbf{v}_i\right]. \tag{2-49a}$$

This gives a new set of vectors $\{\mathbf{u}_1, \mathbf{u}_2, \mathbf{v}_3, \ldots, \mathbf{v}_n\}$ that also spans V. The process is then repeated until $\mathbf{u}_n$ is reached, which can be written as

$$\mathbf{u}_n = c_1\mathbf{u}_1 + c_2\mathbf{u}_2 + \cdots + c_{n-1}\mathbf{u}_{n-1} + c_n\mathbf{v}_n. \tag{2-50}$$

We solve for $\mathbf{v}_n$, obtaining

$$\mathbf{v}_n = c_n^{-1}\left[\mathbf{u}_n - \sum_{i=1}^{n-1} c_i\mathbf{u}_i\right], \tag{2-51}$$

so that the new set of vectors spanning V is the set $\{\mathbf{u}_1, \mathbf{u}_2, \ldots, \mathbf{u}_n\}$. If we now attempt to write $\mathbf{u}_{n+1}$ as

$$\mathbf{u}_{n+1} = \sum_{i=1}^{n} d_i\mathbf{u}_i, \tag{2-52}$$

we obtain a contradiction, since Eq. (2-52) implies that $\mathbf{u}_{n+1}$ is linearly dependent on the set $\{\mathbf{u}_1, \ldots, \mathbf{u}_n\}$, by Theorem 2-1. But the set of $\mathbf{u}_i$ was assumed to be linearly independent. Therefore, the assumption that $m > n$ leads to a contradiction. Hence, we must have $m \leq n$. If the roles of $\{\mathbf{v}_i\}$ and $\{\mathbf{u}_i\}$ are reversed, it is easily seen that $n > m$ also leads to a contradiction, by reasoning analogous to the above. From this it follows that $m = n$, which completes the proof.

From the previous considerations, we define the *dimension* of a finite-dimensional vector space V as the number of unique elements in the basis.

Now suppose that we have an n-dimensional vector space V over the complex (or real) number system, with basis vectors $\{\mathbf{v}_1, \mathbf{v}_2, \ldots, \mathbf{v}_n\}$. From Theorem 2-3 we know that, *relative to this basis*, every vector in V has a unique representation in terms of n scalars, i.e.

$$\mathbf{u} = c_1\mathbf{v}_2 + c_2\mathbf{v}_2 + \cdots + c_n\mathbf{v}_n. \tag{2-53}$$

However, once the basis has been chosen, we note that all of the information needed to describe the vector $\mathbf{u}$ is contained in the coefficients $c_1, c_2, \ldots, c_n$. Let us form an n-component vector,

$$\mathbf{u}' = (c_1, c_2, \ldots, c_n), \tag{2-54}$$

where the components of $\mathbf{u}'$ are the coefficients of the basis vectors in Eq. (2-53). It is also clear that the formation of the "primed" vector can be carried out for every "unprimed" vector, e.g.,

$$\mathbf{w} = d_1\mathbf{v}_1 + d_2\mathbf{v}_2 + \cdots + d_n\mathbf{v}_n, \tag{2-55}$$

where the corresponding "primed" vector is

$$\mathbf{w}' = (d_1, d_2, \dots, d_n). \tag{2-56}$$

Note, however, that the form of $\mathbf{w}'$ (i.e., the coefficients of the $\mathbf{v}$'s) is dependent on the basis chosen. The manner in which the components change under a basis change will be discussed in Chapter 3. It also follows from Eqs. (2-53) through (2-56) that

$$\mathbf{u} + \mathbf{w} = \sum_{i=1}^{n} (c_i + d_i)\mathbf{v}_i, \tag{2-57}$$

and

$$\mathbf{u}' + \mathbf{w}' = (c_1 + d_1, c_2 + d_2, \dots, c_n + d_n). \tag{2-58}$$

Finally, it follows that

$$b\mathbf{u} = \sum_{i=1}^{n} (bc_i)\mathbf{v}_i, \tag{2-59}$$

and

$$b\mathbf{u}' = (bc_1, bc_2, \dots, bc_n). \tag{2-60}$$

We have now established several very important relationships. First, we have shown that there is a one-to-one correspondence between the unprimed vectors in the vector space V and the primed vectors in vector space V'. In other words, for every vector $\mathbf{u}$ in V there is one and only one corresponding vector $\mathbf{u}'$ in V'. Next, we note that the laws of combination are the same in both spaces, i.e., addition and scalar multiplication are defined in the same manner in both spaces. Such a set of relationships between two vector spaces is called an *isomorphism*, and the two vector spaces are said to be *isomorphic* to each other.

Isomorphisms will be extremely useful to us in our study of quantum mechanics, since the concept assures that whatever properties are deduced for a given vector space must also be true for any other vector space that is isomorphic to it. This means that we can choose whichever vector space is most convenient for the problem at hand.

2-3. Inner Product Spaces

In Section 2-1 we discussed the concept of a scalar product between two three-dimensional geometric vectors. The generalization of this concept to spaces of n-dimensions and even to spaces of infinite dimensions is quite important in quantum mechanics.

The desired generalization to n-dimensional spaces is straightforward, and easily understood if the component form of vectors is employed. In particular, for an n-dimensional space, each vector is specified by n-components,

$$\mathbf{u}=(u_1, u_2, \dots, u_n), \tag{2-61}$$

and

$$\mathbf{v}=(v_1, v_2, \dots, v_n), \tag{2-62}$$

where it is assumed that all of the components are *real*. In this case, the appropriate definition to be used for a *scalar product* (or *inner product*) is

$$(\mathbf{v}, \mathbf{u})=\sum_{i=1}^{n} v_i u_i. \tag{2-63}$$

A vector space on which such a scalar product is defined is called an *n-dimensional Euclidean* vector space.

If we are dealing with an n-dimensional complex vector space, however, the generalization is not so straightforward. Consider, for example, the complex vector[3]

$$\bar{\mathbf{v}}=i\mathbf{v}=(iv_1, iv_2, \dots, iv_n). \tag{2-64}$$

If the inner product of $\bar{\mathbf{v}}$ with itself is formed as in Eq. (2-63), we obtain

$$|\bar{\mathbf{v}}|=|i\mathbf{v}|=(i\mathbf{v}, i\mathbf{v})^{1/2}=[-1(\mathbf{v}, \mathbf{v})]^{1/2}. \tag{2-65}$$

This is imaginary if $\mathbf{v}$ is real. Distance, however, must be a real quantity, so the definition used for an n-dimensional Euclidean space cannot be used for this case. This difficulty can be circumvented by redefining the inner product as

$$(\mathbf{v}, \mathbf{u})=\sum_{i=1}^{n} v_i^* u_i, \tag{2-66}$$

where the asterisk denotes complex conjugation.[4] This redefinition provides a new formulation of the *length* (or norm) *of a vector*, viz.,

$$|\mathbf{v}|=(\mathbf{v}, \mathbf{v})^{1/2}=\left(\sum_{i=1}^{n} v_i^* v_i\right)^{1/2}=\left(\sum_{i=1}^{b} |v_i|^2\right)^{1/2}, \tag{2-67}$$

where $|v_i|^2$ is the absolute value of the v_ith component.[5] Note that the inner product of a complex vector space is not symmetrical, as it is in a real vector space, i.e.,

$$(\mathbf{v}, \mathbf{u})\neq(\mathbf{u}, \mathbf{v}). \tag{2-68}$$

---

[3] Here i represents $\sqrt{-1}$, and should not be confused with the unit vector in the x-direction.

[4] If a complex number is denoted by $z = x + iy$, then its complex conjugate is given by $z^* = x - iy$.

[5] If $z = x + iy$, then $zz^* = |z|^2 = x^2 + y^2$, which can be interpreted as the magnitude of z in the complex plane.

Instead, we now have

$$(\mathbf{v}, \mathbf{u}) = (\mathbf{u}, \mathbf{v})^*, \tag{2-69}$$

which can be proven easily from the definition of an inner product given in Eq. (2-66). Complex n-dimensional vector spaces in which the inner product is defined as in Eq. (2-66) are usually called *n-dimensional unitary vector spaces.*[6]

Before examining some of the properties of such inner-product spaces in more detail, we shall summarize the previous discussion by giving the following formal definition of an inner product space. In addition to satisfying the properties of a unitary or Euclidean vector space, an *inner-product space* also contains an inner product that is defined for all pairs of vectors in the space. Furthermore, this inner-product is a scalar-valued function of a pair of vectors, **v** and **u**, such that

1. $(\mathbf{v}, \mathbf{u}) = (\mathbf{u}, \mathbf{v})^*,$ (2-70)

and

2. $(a\mathbf{v} + b\mathbf{u}, \mathbf{w}) = a^*(\mathbf{v}, \mathbf{w}) + b^*(\mathbf{u}, \mathbf{w}),$ (2-71)

where a and b are scalars over the real or complex number systems,[7]

3. $(\mathbf{v}, \mathbf{v}) \geq 0$ for any $\mathbf{v}$, (2-71a)

and $(\mathbf{v}, \mathbf{v}) = 0$ if and only if $\mathbf{v} = 0$.[8]

Let us consider an example of inner products of vectors which will help crystallize the concept. Consider a four-dimensional Euclidean Vector space with the basis $\{\mathbf{v}_1, \mathbf{v}_2, \mathbf{v}_3, \mathbf{v}_4\}$. Two unique vectors in the space can be given, with respect to the above basis, as

$$\mathbf{u} = a_1\mathbf{v}_1 + a_2\mathbf{v}_2 + a_3\mathbf{v}_3 + a_4\mathbf{v}_4 \tag{2-72}$$

and

$$\mathbf{w} = b_1\mathbf{v}_1 + b_2\mathbf{v}_2 + b_3\mathbf{v}_3 + b_4\mathbf{v}_4. \tag{2-73}$$

Since it has been shown that this space is isomorphic to the space of vectors having four components per vector, **u** and **w** can also be represented as

$$\mathbf{u} = (a_1, a_2, a_3, a_4)$$
$$\mathbf{w} = (b_1, b_2, b_3, b_4).$$

---

[6] Sometimes such spaces are also called *Hermitian* or *complex Euclidean* vector spaces.

[7] In some mathematics textbooks this property would be given as $(a\mathbf{v} + b\mathbf{u}, \mathbf{w}) = a(\mathbf{v}, \mathbf{w}) + b(\mathbf{u}, \mathbf{w})$. We have used the above form to maintain consistency with general usage in physics and chemistry.

[8] The assumption that $(\mathbf{v}, \mathbf{v}) = 0$ if and only if $\mathbf{v} = 0$ will have to be modified when we are considering functions as vectors. The details of this will be discussed shortly.

Applying Eq. (2-63), which defines the inner product in an n-dimensional Euclidean vector space, we obtain simply

$$(\mathbf{u}, \mathbf{w}) = \sum_{i=1}^{4} a_i b_i. \tag{2-74}$$

Several other properties of inner-product spaces of special importance when considering infinite-dimension vector spaces are collected in the following theorem.

Theorem 2-5. In any Euclidean or unitary vector space V, any two vectors $\mathbf{u}$ and $\mathbf{v}$ have the following properties:

1. $|c\mathbf{v}| = |c|\ |\mathbf{v}|,$ (2-75)

   where c is a real or complex scalar,

2. $|\mathbf{v}| > 0,$ (2-76)

   unless $\mathbf{v} = 0$. In this case $|\mathbf{0}| = 0$.

3. $|(\mathbf{u}, \mathbf{v})| \leq |\mathbf{u}|\ |\mathbf{v}|.$ (2-77)

   (Schwartz' Inequality)

4. $|\mathbf{u}+\mathbf{v}| \leq |\mathbf{u}| + |\mathbf{v}|.$ (2-78)

   (Triangle Inequality)

Proof: The proof of (1) and (2) is left as an exercise.[9]

3. Schwartz inequality: Consider the vector

$$\mathbf{w} = a\mathbf{u} - b\mathbf{v}, \tag{2-79}$$

where a and b are arbitrary complex numbers. The square of the norm of $\mathbf{w}$ is

$$0 \leq |\mathbf{w}|^2 = (\{a\mathbf{u} - b\mathbf{v}\}, \{a\mathbf{u} - b\mathbf{v}\})$$
$$= |a|^2 |\mathbf{u}|^2 - a^* b(\mathbf{u}, \mathbf{v}) - b^* a(\mathbf{v}, \mathbf{u}) + |b|^2 |\mathbf{v}|^2. \tag{2-80}$$

Since a and b are arbitrary, let us choose

$$a = |\mathbf{v}|(\mathbf{u}, \mathbf{v})^{1/2} \tag{2-81}$$

and

$$b = |\mathbf{u}|(\mathbf{v}, \mathbf{u})^{1/2}. \tag{2-82}$$

Substituting Eqs. (2-81) and (2-82) into Eq. (2-80) gives

$$0 \leq |\mathbf{v}|^2 |\mathbf{u}|^2 |(\mathbf{u}, \mathbf{v})| - |\mathbf{v}|\ |\mathbf{u}|\ |(\mathbf{u}, \mathbf{v})|^2 - |\mathbf{v}|\ |\mathbf{u}|\ |(\mathbf{v}, \mathbf{u})|^2 + |\mathbf{v}|^2 |\mathbf{u}|^2 |(\mathbf{v}, \mathbf{u})|. \tag{2-83}$$

[9] See Problem 2-5.

Since $|(\mathbf{u}, \mathbf{v})| = |(\mathbf{v}, \mathbf{u})|$ we obtain

$$0 \leq 2|\mathbf{v}|^2|\mathbf{u}|^2|(\mathbf{u}, \mathbf{v})| - 2|\mathbf{v}|\ |\mathbf{u}|\ |(\mathbf{u}, \mathbf{v})|^2. \tag{2-84}$$

Dividing both sides of Eq. (2-84) by $2|\mathbf{v}|\ |\mathbf{u}|\ |(\mathbf{u}, \mathbf{v})|$ and rearranging terms gives

$$|(\mathbf{u}, \mathbf{v})| \leq |\mathbf{u}|\ |\mathbf{v}|, \tag{2-85}$$

which is the desired result.

4. Triangle inequality: Expand the square of the norm of $\mathbf{u} + \mathbf{v}$, i.e.,

$$|\mathbf{u}+\mathbf{v}|^2 = |\mathbf{u}|^2 + |\mathbf{v}|^2 + (\mathbf{u}, \mathbf{v}) + (\mathbf{v}, \mathbf{u}) \tag{2-86}$$

$$= |\mathbf{u}|^2 + |\mathbf{v}|^2 + 2\mathrm{Re}(\mathbf{u}, \mathbf{v}) \qquad (2\text{-}87)^{10}$$

$$\leq |\mathbf{u}|^2 + |\mathbf{v}|^2 + 2|(\mathbf{u}, \mathbf{v})|. \qquad (2\text{-}88)^{11}$$

Using the Schwartz inequality just proved, we obtain

$$|\mathbf{u}+\mathbf{v}|^2 \leq |\mathbf{u}|^2 + 2|\mathbf{u}|\ |\mathbf{v}| + |\mathbf{v}|^2 \tag{2-89}$$

$$\leq (|\mathbf{u}| + |\mathbf{v}|)^2. \tag{2-90}$$

Taking the sqare root of both sides of Eq. (2-90), the required result is obtained, i.e.,

$$|\mathbf{u}+\mathbf{v}| \leq |\mathbf{u}| + |\mathbf{v}|. \tag{2-91}$$

A few examples of the Schwartz and triangle inequalities may be helpful in understanding the essentials of these theorems.

Consider a real, two-dimensional vector space. The inner-product between two vectors $\mathbf{u}$ and $\mathbf{v}$ is

$$(\mathbf{u}, \mathbf{v}) = |\mathbf{u}|\ |\mathbf{v}| \cos \vartheta_{uv}. \tag{2-92}$$

Rearranging terms and taking the absolute value of both sides gives

$$\frac{|(\mathbf{u}, \mathbf{v})|}{|\mathbf{u}|\ |\mathbf{v}|} = |\cos \vartheta_{uv}|. \tag{2-93}$$

If the Schwartz inequality is used to replace the numerator of the left-hand side of Eq. (2-93), we see that the $|\cos \vartheta_{uv}| \leq 1$, which is a well-known result from trigonometry.

Now let us consider a different aspect of the same real two-dimensional vector space used in the previous example. Figure 2-1a shows the addition of

---

[10] Since $(\mathbf{u}, \mathbf{v})$ and $(\mathbf{v}, \mathbf{u})$ are complex conjugates of each other [see Eq. (2-69)], say $(\mathbf{u}, \mathbf{v}) = x + iy$, therefore, adding to two quantities gives $(\mathbf{u}, \mathbf{v}) + (\mathbf{v}, \mathbf{u}) = (\mathbf{u}, \mathbf{v}) + (\mathbf{u}, \mathbf{v})^* = (x + iy) + (x - iy) = 2x = 2\mathrm{Re}(\mathbf{u}, \mathbf{v})$, where "Re" indicates that we take the real part of the complex number.

[11] That $\mathrm{Re}(\mathbf{u}, \mathbf{v}) \leq |(\mathbf{u}, \mathbf{v})|$ can be seen as follows: Letting $(\mathbf{u}, \mathbf{v}) = x + iy$ we see that $|(\mathbf{u}, \mathbf{v})| = \sqrt{x^2+y^2}$, while $\mathrm{Re}(\mathbf{u}, \mathbf{v}) = x$. Since $y^2 > 0$, it follows that $2|(\mathbf{u}, \mathbf{v})| \geq 2\mathrm{Re}(\mathbf{u}, \mathbf{v})$.

two vectors **u** and **v**. The vector sum $\mathbf{w} = \mathbf{u} + \mathbf{v}$ forms the third side of a triangle. The triangle inequality states for this case that the sum of the lengths is always greater than or equal to the length of the third side. In the case of the equality we no longer truely have a triangle but, rather, a straight line.

As a final example, consider the set of complex-valued functions of the real variable x, defined on the interval $a \leq x \leq b$, which are both integrable and square integrable. This means that these functions belong to the class of functions for which $\int_a^b f(x)\,dx < \infty$ and $\int_a^b |f(x)|^2\,dx < \infty$, viz. their integrals exist.[12]

First let us examine some of the postulates of a vector space to see whether functions of this type, in fact, form a vector space. We define addition and scalar multiplication as $f(x) + g(x)$ and $cf(x)$, respectively. Note that these operations are only defined on $a \leq x \leq b$.

To demonstrate closure under addition, we must show that the function obtained by adding $f(x)$ and $g(x)$ on $a \leq x \leq b$ is also integrable and square integrable. This is obvious for the case of integrability, but is not so obvious for square integrability. To demonstrate square integrability, we proceed as follows. Consider $|f + g|^2$ for all x on the interval $a \leq x \leq b$:

$$|f+g|^2 = (f^* + g^*)(f+g) \tag{2-94}$$

$$= |f|^2 + |g|^2 + f^*g + g^*f \tag{2-95}$$

$$= |f|^2 + |g|^2 + 2\mathrm{Re}(f^*g) \tag{2-96}$$

or

$$|f+g|^2 \leq |f|^2 + |g|^2 + 2|f^*g| \qquad (2\text{-}97)^{13}$$

$$\leq |f|^2 + |g|^2 + 2|f|\,|g|. \qquad (2\text{-}98)^{14}$$

Since

$$(|f| - |g|)^2 = |f|^2 + |g|^2 - 2|f|\,|g| \geq 0, \tag{2-99}$$

we obtain, on rearranging Eq. (2-99),

$$2|f|\,|g| \leq |f|^2 + |g|^2. \tag{2-100}$$

Substituting Eq. (2-100) into Eq. (2-98) gives

$$|f+g|^2 \leq 2\{|f|^2 + |g|^2\}. \tag{2-101}$$

---

[12] For a discussion of theories of integration for very general classes of functions, see e.g., F. W. Byron and R. W. Fuller, *Mathematics of Classical and Quantum Physics,* Vol. 1, Addison-Wesley, Reading, MA, 1969, pp. 214–215. A detailed discussion is also given by G. Fano, *Mathematical Methods of Quantum Mechanics*, McGraw-Hill, New York, Chapter 4.

[13] That $\mathrm{Re}(f^*g) \leq |f^*g|$ is shown in Footnote 11.

[14] To show $|f^*g| = |f|\,|g|$, we first note that $|f| = \sqrt{f^*f}$, $|g| = \sqrt{g^*g}$. We then write $|f^*g| = \sqrt{(f^*g)(fg^*)}$, which can be rearranged as $|f^*g| = \sqrt{(f^*f)(g^*g)} = \sqrt{|f|^2|g|^2}$, to give finally $|f^*g| = |f|\,|g| = |fg^*|$.

Note that this holds for every x on the interval $[a, b]$. Therefore, we see that

$$\int_a^b |f+g|^2\,dx \leq 2\left\{\int_a^b |f|^2\,dx + \int_a^b |g|^2\,dx\right\}. \tag{2-102}$$

Since both f and g are square integrable, it then follows that

$$\int_a^b |f+g|^2 < \infty. \tag{2-103}$$

Closure under scalar multiplication is even easier to demonstrate, since the function $cf(x)$ is easily shown to be integrable. Also,

$$\int_a^b |cf|^2\,dx = \int_a^b |c|^2|f|^2\,dx = |c|^2\int_a^b |f|^2\,dx < \infty. \tag{2-104}$$

It is left as an exercise for the student[15] to demonstrate that the other postulates of a vector space are satisfied.

The *inner product* is defined in this case as

$$(f, g) = \int_a^b f^*(x)g(x)\,dx. \tag{2-105}$$

When we are dealing with functions or vectors with infinite numbers of components, we must take care to examine whether the inner product even exists. To show the existence of the inner product defined in Eq. (2-105) for the above class of functions, we first examine the expression[16]

$$|f^*g| = |f|\,|g| \leq \frac{1}{2}(|f|^2 + |g|^2), \tag{2-106}$$

which holds for all x on $[a, b]$. Taking the integral of both sides of Eq. (2-106) over $[a, b]$ gives

$$\int_a^b |f^*g|\,dx \leq \frac{1}{2}\left\{\int_a^b |f|^2\,dx + \int_a^b |g|^2\,dx\right\}. \tag{2-107}$$

Since f and g are square integrable, the term in braces converges, and hence $\int_a^b |f^*g|\,dx < \infty$. In elementary analysis it is shown that, if the integral of the absolute value of a function converges, then the integral of the function itself converges.[17] Therefore, since $\int_a^b |f^*g|\,dx < \infty$, it follows that $\int_a^b f^*g\,dx < \infty$, and the inner product exists. This definition also satisfies the postulates of an

---

[15] See Problem 2-6.

[16] This expression is derived by combining Eq. (2-100) with the fact that $|f^*g| = |f|\,|g|$, which was shown in Footnote 14.

[17] See, for example, *Advanced Calculus*, by A. E. Taylor, Ginn and Co., NY, 1955, Chapter 16.

inner product space, i.e.,

$$1.\ (f, g)=\int_a^b f^*g\,dx=\int_a^b (g^*f)^*\,dx=(g, f)^*. \tag{2-108}$$

$$2.\ (cf+dg, h)=\int_a^b (c^*f^*+d^*g^*)h\,dx$$

$$=c^*\int_a^b f^*h\,dx+d^*\int_a^b g^*h\,dx \tag{2-109}$$

$$=c^*(f, h)+d^*(g, h). \tag{2-110}$$

$$3.\ (f, f)=\int_a^b f^*f\,dx=\int_a^b |f|^2\,dx\geq 0. \tag{2-111}$$

There is, however, a modification of Eq. (2-111) that should be noted for the special case when $(f, f) = 0$. It is not necessary that $f(x) = 0$ for all x on $[a, b]$ for the equality in Eq. (2-111) to hold. In fact, $f(x)$ may be non-zero for a finite number of points on $[a, b]$ and the integral will still equal to zero.[18]

Finally, Schwartz' inequality and the triangle inequality take the following forms

$$\left|\int_a^b f^*g\,dx\right|\leq\left[\int_a^b |f|^2\,dx\right]^{1/2}\left[\int_a^b |g|^2\,dx\right]^{1/2}, \tag{2-112}$$

and

$$\left[\int_a^b |f+g|^2\,dx\right]\leq\left[\int_a^b |f|^2\,dx\right]^{1/2}+\left[\int_a^b |g|^2\,dx\right]^{1/2}, \tag{2-113}$$

respectively.

The previous discussions have given some hint as to what powerful mathematical tools the concepts of vector and inner product spaces can be. We shall now examine another mathematical concept that will simplify our work in a practical as well as a theoretical sense.

2-4. Orthonormality and Complete Sets

The concepts of length and angle between vectors in an inner product space lead naturally to the concepts of orthonormality.[19] To begin with a familiar example, consider the three perpendicular unit basis vectors in three-dimensional

---

[18] The basis for this discussion is the theory of Lebesque integration, since such integrals do not exist in the Riemann sense. Suffice it to say that, if $(f, f) = 0$, this implies that $f(x) = 0$ for *almost all* x in $[a, b]$. See Footnote 12 for references to more detailed treatments of this point.

[19] Note that between the concept of a vector space and that of an inner product space lies the notion of a "normed-vector space." While it follows that an inner product space always is normed, a normed space does not necessarily possess an inner-product. In an inner-product space distances *and* angles are defined. In a normed-vector space only distances need be defined.

Cartesian space, viz., $\{\mathbf{i}, \mathbf{j}, \mathbf{k}\}$. Figure 2-2 shows the relative orientation of these vectors. Since these vectors are oriented 90° with respect to each other, it is easy to see that their inner products are zero, e.g.,

$$(\mathbf{i}, \mathbf{j}) = |\mathbf{i}|\ |\mathbf{j}| \cos \vartheta_{ij} = (1)(1) \cos(90°) = 0. \tag{2-114}$$

Let us also consider the *projections* of the vector $\mathbf{v}$, of Fig. 2-2, onto the three coordinate axes. First we write $\mathbf{v}$ as

$$\mathbf{v} = v_x \mathbf{i} + v_y \mathbf{j} + v_z \mathbf{k}. \tag{2-115}$$

If we now form the inner product of $\mathbf{i}$ with both sides of Eq. (2-115), we obtain

$$(\mathbf{i}, \mathbf{v}) = v_x(\mathbf{i}, \mathbf{i}) + v_y(\mathbf{i}, \mathbf{j}) + v_z(\mathbf{i}, \mathbf{k}).$$

Applying the orthonormality conditions on $\{\mathbf{i}, \mathbf{j}, \mathbf{k}\}$ gives

$$(\mathbf{i}, \mathbf{v}) = v_x. \tag{2-116}$$

Thus, we can think of v_x as the projection of $\mathbf{v}$ onto the basis vector $\mathbf{i}$. We can repeat the process for both $\mathbf{j}$ and $\mathbf{k}$, obtaining

$$(\mathbf{j}, \mathbf{v}) = v_y, \tag{2-117}$$

and

$$(\mathbf{k}, \mathbf{v}) = v_z. \tag{2-118}$$

From the above it is clear that the components (or coordinates) of a vector are nothing more than the projections of this vector onto the various basis vectors. This concept is not restricted to basis vectors that are geometrical vectors, but also includes functions or other mathematical objects that satisfy the postulates of an inner product space.

In general, two vectors $\mathbf{u}$ and $\mathbf{v}$ are said to be *orthogonal* if and only if $(\mathbf{u}, \mathbf{v}) = 0$. If an entire set of vectors is each mutually orthogonal, then we refer to that set as an orthogonal set of vectors. Furthermore, we say that a set of vectors $\{\mathbf{v}_1, \mathbf{v}_2, \ldots, \mathbf{v}_n\}$ is *orthonormal* if

$$(\mathbf{v}_i, \mathbf{v}_j) = \delta_{ij} \tag{2-119}$$

for all i and j, where δ_{ij} is known as the *Kronecker delta*, and is defined as

$$\delta_{ij} = \begin{cases} 0 & i \neq j \\ 1 & i = j. \end{cases} \tag{2-120}$$

Thus, an orthonormal set of vectors is simply a set in which the vectors are all orthogonal and normalized to unity.[20] It is clear that any vector that has a finite norm, $|\mathbf{v}| < \infty$, can be normalized by dividing the vector by its length, i.e.,

$$\mathbf{v}' = \frac{\mathbf{v}}{|\mathbf{v}|}. \tag{2-121}$$

---

[20] Other normalizations are possible, but we shall mean normalization to unity unless otherwise specified.

That this vector is normalized is obvious if we form the scalar product

$$(\mathbf{v}', \mathbf{v}') = \frac{(\mathbf{v}, \mathbf{v})}{|\mathbf{v}|^2} = \frac{|\mathbf{v}|^2}{|\mathbf{v}|^2} = 1, \tag{2-122}$$

and thus

$$|\mathbf{v}'| = [(\mathbf{v}', \mathbf{v}')]^{1/2} = 1. \tag{2-123}$$

Just as in the case of three-dimensional Cartesian basis vectors, the components (or coefficients) of any vector $\mathbf{u}$ and an d-dimensional vector space with an orthonormal basis $\{\mathbf{v}_1, \mathbf{v}_2, \dots, \mathbf{v}_n\}$ can be found as the projections of $\mathbf{u}$ onto the various basis vectors. To accomplish this, we write $\mathbf{u}$ as a linear combination of the orthonormal basis

$$\mathbf{u} = \sum_{j=1}^{n} c_j \mathbf{v}_j. \tag{2-124}$$

Taking the inner product with $\mathbf{v}_i$, we obtain

$$(\mathbf{v}_i, \mathbf{u}) = \sum_{j=1}^{n} c_j(\mathbf{v}_i, \mathbf{v}_j) = \sum_{j=1}^{n} c_j \delta_{ij}, \tag{2-125}$$

or

$$(\mathbf{v}_i, \mathbf{u}) = c_i. \tag{2-126}$$

Now let us establish several properties of interest concerning a set of orthonormal vectors.

Theorem 2-6. A finite orthonormal set of vectors $\{\mathbf{v}_1, \dots, \mathbf{v}_k\}$ is linearly independent.

Proof: Consider the linear combination

$$\sum_{i=1}^{k} c_i \mathbf{v}_i = 0. \tag{2-127}$$

Now form the inner product of one of the other basis vectors, say $\mathbf{v}_j$, with both sides of the equation:

$$\left(\mathbf{v}_j, \sum_{i=1}^{k} c_i \mathbf{v}_i\right) = \sum_{i=1}^{k} c_i(\mathbf{v}_j, \mathbf{v}_i), \tag{2-128}$$

or

$$\sum_{i=1}^{k} c_i(\mathbf{v}_j, \mathbf{v}_i) = \sum_{i=1}^{k} c_i \delta_{ij} = 0. \tag{2-129}$$

This implies that $c_j = 0$. If we repeat this for all the k basis vectors, we obtain $c_j = 0; j = 1, 2, \dots, k$. Therefore, we have shown that if Eq. (2-127) is true, then $c_i = 0$, $i = 1, 2, \dots, k$, which is just the definition of linear independence.

From the above theorem it follows that for an n-dimensional vector space, any set of n orthonormal vectors is a basis for that space. In a finite dimensional vector space, an orthonormal set is said to be *complete* if it is not contained in any larger orthonormal set in that space, or equivalently, if it is a basis for the space. Of course, it also implies that any vector in that space can be represented as a linear combination of the orthonormal basis vectors [see Eq. (12-124)].

Two other important theorems which have their most fruitful applications in infinite-dimensional vector spaces will now be proved first for finite-dimensional vector spaces.

Theorem 2-7. (*Bessel's inequality*) Let $\{\mathbf{v}_1, \mathbf{v}_2, \dots, \mathbf{v}_n\}$ be any finite orthonormal set in a complex inner product space. Then, if $\mathbf{u}$ is any vector in the space, and if $c_i = (\mathbf{v}_i, \mathbf{u})$,

$$\sum_{i=1}^{n} |c_i|^2 \leq |\mathbf{u}|^2. \tag{2-130}$$

Furthermore, the vector

$$\mathbf{u}' = \mathbf{u} - \sum_{i=1}^{n} c_i \mathbf{v}_i \tag{2-131}$$

is orthogonal to each $\mathbf{v}_i$, and hence to the entire space spanned by $\{\mathbf{v}_1, \mathbf{v}_2, \dots, \mathbf{v}_n\}$.

Proof: First we shall prove the assertion given in Eq. (2-130). Form the square of the norm of $\mathbf{u}'$, i.e.,

$$0 \leq |\mathbf{u}'|^2 = (\mathbf{u}', \mathbf{u}') = \left(\mathbf{u} - \sum_{i=1}^{n} c_i \mathbf{v}_i,\ \mathbf{u} - \sum_{i=1}^{n} c_i \mathbf{v}_i\right), \tag{2-132}$$

or

$$0 \leq |\mathbf{u}'|^2 = (\mathbf{u}, \mathbf{u}) - \sum_{i=1}^{n} c_i^* (\mathbf{v}_i, \mathbf{u}) - \sum_{i=1}^{n} c_i (\mathbf{u}, \mathbf{v}_i)$$

$$+ \sum_{i=1}^{n} \sum_{j=1}^{n} c_i^* c_j (\mathbf{v}_i, \mathbf{v}_j). \tag{2-133}$$

Substituting the expression $(\mathbf{v}_i, \mathbf{u}) = c_i$ into the above equation gives

$$0 \leq |\mathbf{u}|^2 - \sum_{i=1}^{n} |c_i|^2 - \sum_{i=1}^{n} |c_i|^2 + \sum_{i=1}^{n} \sum_{j=1}^{n} c_i^* c_j \delta_{ij}, \tag{2-134}$$

$$0 \leq |\mathbf{u}|^2 - 2 \sum_{i=1}^{n} |c_i|^2 + \sum_{i=1}^{n} |c_i|^2 \tag{2-135}$$

$$0 \leq |\mathbf{u}|^2 - \sum_{i=1}^{n} |c_i|^2. \tag{2-136}$$

Hence,

$$|\mathbf{u}|^2 \geq \sum_{i=1}^{n} |c_i|^2. \tag{2-137}$$

To prove the assertion given in Eq. (2-131), we form the inner product $(\mathbf{u}', \mathbf{v}_j)$ for all j, i.e.,

$$(\mathbf{u}', \mathbf{v}_j) = \left(u - \sum_{i=1}^{n} c_i \mathbf{v}_i, \mathbf{v}_j \right) \tag{2-138}$$

$$= (\mathbf{u}, \mathbf{v}_j) - \sum_{i=1}^{n} c_i^* (\mathbf{v}_i, \mathbf{v}_j) \tag{2-139}$$

$$= (\mathbf{u}, \mathbf{v}_j) - \sum_{i=1}^{n} c_i^* \delta_{ij} \tag{2-140}$$

or

$$(\mathbf{u}', \mathbf{v}_j) = c_j^* - c_j^* = 0, \tag{2-141}$$

for $j = 1, 2, \ldots, n$, which completes the proof. Note also that by combining Eqs. (2-132), (2-134), and (2-136), we have

$$|\mathbf{u}'|^2 = |\mathbf{u}|^2 - \sum_{i=1}^{n} |c_i|^2, \tag{2-142}$$

which can be rearranged to give

$$|\mathbf{u}|^2 = \sum_{i=1}^{n} |c_i|^2 + |\mathbf{u}'|^2. \tag{2-143}$$

We have shown above that $\mathbf{u}'$ is orthogonal to the space spanned by $\{\mathbf{v}_1, \mathbf{v}_2, \ldots, \mathbf{v}_n\}$. If our space is n-dimensional, then the orthonormal set of vectors $\{\mathbf{v}_1, \mathbf{v}_2, \ldots, \mathbf{v}_n\}$ is a basis for the space, and $\{\mathbf{v}_1, \mathbf{v}_2, \ldots, \mathbf{v}_n\}$ is complete. This implies that $\{\mathbf{v}_1, \mathbf{v}_2, \ldots, \mathbf{v}_n\}$ is not contained in any larger orthonormal set. Therefore, it is clear that

$$\mathbf{u}' = \mathbf{0}. \tag{2-144}$$

Thus, while it is true that $\mathbf{u}'$ is orthogonal to the set of basis vectors, that fact cannot be used to extend the set, since its inclusion would make the set linearly dependent, and hence would not be a basis. Equation (2-143) now becomes

$$|\mathbf{u}|^2 = \sum_{i=1}^{n} |c_i|^2. \tag{2-145}$$

This result, for finite-dimensional spaces, is an important result, and will be even more important when we discuss infinite-dimensional spaces, since completeness of the basis vectors for such a space is implied by Eq. (2-145). In other words, Bessel's inequality becomes an equality if the basis is complete. We will discuss this in more detail shortly.

Theorem 2-8. (*Parseval's equation*) If $\{\mathbf{v}_1, \mathbf{v}_2, \dots, \mathbf{v}_n\}$ is a finite, orthonormal basis for the inner-product space V', and if the two vectors $\mathbf{u}$ and $\mathbf{w}$ are in V', then

$$(\mathbf{w}, \mathbf{u}) = \sum_{i=1}^{n} (\mathbf{w}, \mathbf{v}_i)(\mathbf{v}_i, \mathbf{u}). \tag{2-146}$$

Proof: Since the vectors $\{\mathbf{v}_1, \mathbf{v}_2, \dots, \mathbf{v}_n\}$ form a basis for V', we can write $\mathbf{u}$ and $\mathbf{w}$ as

$$\mathbf{u} = \sum_{i=1}^{n} a_i \mathbf{v}_i, \tag{2-147}$$

and

$$\mathbf{w} = \sum_{i=1}^{n} b_i \mathbf{v}_i. \tag{2-148}$$

From Eq. (2-126) we see that $a_i = (\mathbf{v}_i, \mathbf{u})$ and $b_i = (\mathbf{v}_i, \mathbf{w})$. Forming the inner product of $\mathbf{u}$ and $\mathbf{w}$ gives

$$\begin{aligned}(\mathbf{w}, \mathbf{u}) &= \left(\sum_{i=1}^{n} b_i \mathbf{v}_i, \sum_{j=1}^{n} a_j \mathbf{v}_j \right) \\ &= \sum_{i=1}^{n} \sum_{j=1}^{n} (b_i^* a_j)(\mathbf{v}_i, \mathbf{v}_j).\end{aligned} \tag{2-149}$$

Since the basis is orthonormal, Eq. (2-149) becomes

$$(\mathbf{w}, \mathbf{u}) = \sum_{i=1}^{n} \sum_{j=1}^{n} (b_i^* a_j) \delta_{ij}, \tag{2-150}$$

or

$$(\mathbf{w}, \mathbf{u}) = \sum_{i=1}^{n} b_i^* a_i. \tag{2-151}$$

Substituting the expressions for the coefficients into Eq. (2-151) gives

$$(\mathbf{w}, \mathbf{u}) = \sum_{i=1}^{n} (\mathbf{w}, \mathbf{v}_i)(\mathbf{v}_i, \mathbf{u}), \tag{2-152}$$

which completes the proof.

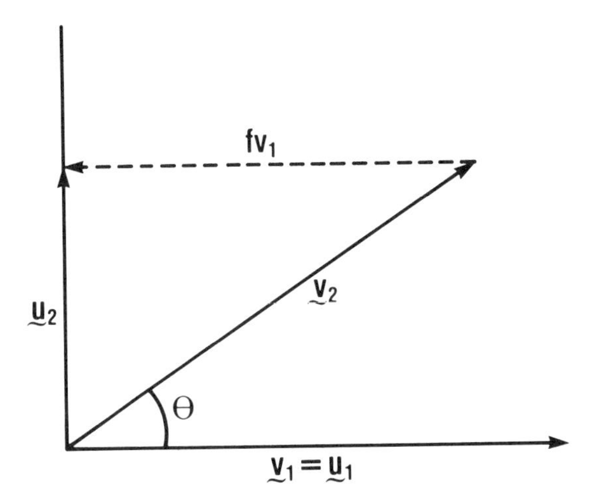

Figure 2.3. Formation of two orthogonal vectors $\mathbf{u}_1$ and $\mathbf{u}_2$ from two linearly independent, but nonorthogonal vectors $\mathbf{v}_1$ and $\mathbf{v}_2$.

Thus far in this section we have spoken at some length about the properties of orthonormal functions. We have not shown, however, that it is always possible to obtain orthonormal sets of vectors in finite-dimensional vector spaces, or how one goes about their construction in practice. In the following discussion we shall show that, given a finite set of linearly independent vectors, there is at least one procedure that always allows us to construct an *orthonormal* set of vectors from this set.[21]

The particular procedure that we shall use to demonstrate this is called the *Gram-Schmidt orthogonalization procedure*. To illustrate this procedure first with a simple example, let us consider the two-dimensional problem of forming two orthogonal vectors $\mathbf{u}_1$ and $\mathbf{u}_2$ from two linearly independent, but nonorthogonal vectors $\mathbf{v}_1$ and $\mathbf{v}_2$. These vectors are depicted in Fig. 2.3. As can be seen from the figure, we can take $\mathbf{u}_1 = \mathbf{v}_1$, and form $\mathbf{u}_2$ by the use of elementary vector analysis, i.e.,

$$\mathbf{u}_2 = \mathbf{v}_2 - f\mathbf{u}_1 . \tag{2-153}$$

Also, since

$$\begin{aligned} f &= \frac{|\mathbf{v}_2| \cos \vartheta}{|\mathbf{v}_1|} = \frac{|\mathbf{v}_1|\,|\mathbf{v}_2| \cos \vartheta}{|\mathbf{v}_1|\,|\mathbf{v}_1|} \\ &= \frac{(\mathbf{v}_1, \mathbf{v}_2)}{(\mathbf{v}_1, \mathbf{v}_1)} , \end{aligned} \tag{2-154}$$

[21] It is assumed that the set of linearly independent vectors possesses a well defined inner product. If this is not true, it is no longer in general possible to construct an orthonormal set.

we have that

$$\mathbf{u}_2 = \mathbf{v}_2 - \left[\frac{(\mathbf{v}_1, \mathbf{v}_1)}{(\mathbf{v}_1, \mathbf{v}_2)} \right] \mathbf{u}_1 . \tag{2-155}$$

We can also see the importance of the linear independence of the original vectors $\mathbf{v}_1$ and $\mathbf{v}_2$ from this simple example. If $\mathbf{v}_1$ and $\mathbf{v}_2$ were linearly dependent, the two vectors would both lie along the same axis (along $\mathbf{v}_1$, for example), and it would be impossible to form two orthogonal vectors, since the length of the vector pointing in the $\mathbf{u}_2$ direction would be zero.

Now we shall generalize this procedure so that it is adaptable to any finite set of linearly independent vectors that belongs to an inner product space.

Theorem 2-9. (*Gram–Schmidt orthogonalization*) If $\{\mathbf{v}_1, \mathbf{v}_2, \dots, \mathbf{v}_n\}$ is a set of linearly independent vectors in an inner product vector space V, then it is always possible to construct a set of orthogonal vectors $\{\mathbf{u}_1, \mathbf{u}_2, \dots, \mathbf{u}_n\}$ that are also in V.

Proof: We shall prove the theorem "by construction." First, choose

$$\mathbf{u}_1 = \mathbf{v}_1 . \tag{2-156}$$

Next, let

$$\mathbf{u}_2 = \mathbf{v}_2 + k\mathbf{u}_1 , \tag{2-157}$$

where we choose k so that $(\mathbf{u}_2, \mathbf{u}_1) = 0$. This gives

$$0 = (\mathbf{u}_2, \mathbf{u}_1) \tag{2-158}$$

$$= (\mathbf{v}_2 + k\mathbf{u}_1, \mathbf{u}_1)$$

$$= (\mathbf{v}_2, \mathbf{u}_1) + k^*(\mathbf{u}_1, \mathbf{u}_1). \tag{2-159}$$

This requires that

$$k^* = -\frac{(\mathbf{v}_2, \mathbf{u}_1)}{|\mathbf{u}_1|^2} , \tag{2-160}$$

or

$$k = -\frac{(\mathbf{u}_1, \mathbf{v}_2)}{|\mathbf{u}_1|^2} . \tag{2-160a}$$

Thus, the second member of the desired orthogonal set is

$$\mathbf{u}_2 = \mathbf{v}_2 - \left[\frac{(\mathbf{u}_1, \mathbf{v}_2)}{|\mathbf{u}_1|} \right] \mathbf{u}_1 . \tag{2-161}$$

To obtain the next member, we let

$$\mathbf{u}_3 = \mathbf{v}_3 + k_1\mathbf{u}_1 + k_2\mathbf{u}_2 , \tag{2-162}$$

and choose k_1 and k_2 to assure that

$$(\mathbf{u}_3, \mathbf{u}_1) = (\mathbf{u}_3, \mathbf{u}_2) = 0. \tag{2-163}$$

The desired orthogonality of $\mathbf{u}_3$ to $\mathbf{u}_1$ implies that

$$0=(\mathbf{u}_3,\ \mathbf{u}_1)=(\mathbf{v}_3,\ \mathbf{u}_1)+k_1^*(\mathbf{u}_1,\ \mathbf{u}_1)+k_2^*(\mathbf{u}_2,\ \mathbf{u}_1)$$
$$=(\mathbf{v}_3,\ \mathbf{u}_1)+k_1^*|\mathbf{u}_1|^2, \tag{2-164}$$

or

$$k_1=-\frac{(\mathbf{u}_1,\ \mathbf{v}_3)}{|\mathbf{u}_1|^2}. \tag{2-165}$$

In a similar manner, it is easily seen that the desired orthogonality of $\mathbf{u}_3$ to $\mathbf{u}_2$ implies that

$$k_2=-\frac{(\mathbf{u}_2,\ \mathbf{v}_3)}{|\mathbf{u}_2|^2}. \tag{2-166}$$

Therefore, the third member of the desired set is

$$\mathbf{u}_3=\mathbf{v}_3-\left[\frac{(\mathbf{u}_1,\ \mathbf{v}_3)}{|\mathbf{u}_1|^2}\right]\mathbf{u}_1-\left[\frac{(\mathbf{u}_2,\ \mathbf{v}_3)}{|\mathbf{u}_2|^2}\right]\mathbf{u}_2. \tag{2-167}$$

Now suppose that we have continued this procedure m times ($m < n$), and, thus, have constructed m orthogonal vectors, $\{\mathbf{u}_1,\ \mathbf{u}_2,\ \ldots,\ \mathbf{u}_m\}$ from $\{\mathbf{v}_1,\ \mathbf{v}_2,\ \ldots,\ \mathbf{v}_m\}$. The general procedure to find $\mathbf{u}_{m+1}$ is as follows. Let

$$\mathbf{u}_{m+1}=\mathbf{v}_{m+1}+\sum_{i=1}^{m} k_i\mathbf{u}_i, \tag{2-168}$$

and determine the k_i from the conditions:

$$(\mathbf{u}_{m+1},\ \mathbf{u}_1)=(\mathbf{u}_{m+1},\ \mathbf{u}_2)=\cdots=(\mathbf{u}_{m+1},\ \mathbf{u}_m)=0. \tag{2-169}$$

Application of these orthogonality conditions gives the following set of equations:

$$\begin{aligned}(\mathbf{u}_{m+1},\ \mathbf{u}_1)&=(\mathbf{v}_{m+1},\ \mathbf{u}_1)+k_1^*|\mathbf{u}_1|^2\\ (\mathbf{u}_{m+1},\ \mathbf{u}_2)&=(\mathbf{v}_{m+1},\ \mathbf{u}_2)+k_2^*|\mathbf{u}_2|^2\\ &\ \ \vdots\\ (\mathbf{u}_{m+1},\ \mathbf{u}_m)&=(\mathbf{v}_{m+1},\ \mathbf{u}_m)+k_m^*|\mathbf{u}_m|^2\end{aligned} \tag{2-170}$$

so that the k_i's are given by

$$\begin{aligned}k_1&=-\frac{(\mathbf{u}_1,\ \mathbf{v}_{m+1})}{|\mathbf{u}_1|^2}\\ k_2&=-\frac{(\mathbf{u}_2,\ \mathbf{v}_{m+1})}{|\mathbf{u}_2|^2}\\ &\ \ \vdots\\ k_m&=-\frac{(\mathbf{u}_m,\ \mathbf{v}_{m+1})}{|\mathbf{u}_m|^2}.\end{aligned} \tag{2-171}$$

This gives $\mathbf{u}_{m+1}$ as,

$$\mathbf{u}_{m+1}=\mathbf{v}_{m+1}-\sum_{i=1}^{m}\left[\frac{(\mathbf{u}_i,\ \mathbf{v}_{m+1})}{|\mathbf{u}_i|^2}\right]\mathbf{u}_i, \tag{2-172}$$

and the procedure is continued until all n orthogonal vectors have been constructed.

If the set of linearly independent vectors $\{\mathbf{v}_1, \ldots, \mathbf{v}_n\}$ spans V, i.e., is a basis of V, then the set $\{\mathbf{u}_1, \ldots, \mathbf{u}_n\}$ is an orthogonal basis of V. Furthermore, suppose that after each new orthogonal vector is constructed, we normalize it. This means that Eq. (2-172) becomes,

$$\mathbf{u}_{m+1}=\mathbf{v}_{m+1}-\sum_{i=1}^{m}[(\mathbf{u}_i,\ \mathbf{v}_{m+1})]\mathbf{u}_i \tag{2-173}$$

and, the normalized $\mathbf{u}'_{m+1}$ is given by

$$\mathbf{u}'_{m+1}=\frac{\mathbf{u}_{m+1}}{|\mathbf{u}_{m+1}|}=\frac{\mathbf{v}_{m+1}-\sum_{j=1}^{m}[(\mathbf{u}_j,\ \mathbf{v}_{m+1})]\mathbf{u}_j}{\left|\mathbf{v}_{m+1}-\sum_{j=1}^{m}[(\mathbf{u}_j,\ \mathbf{v}_{m+1})]\mathbf{u}_j\right|}. \tag{2-174}$$

Let us consider an example to illustrate the procedure. We shall construct an orthogonal basis for the Euclidean 4-space from the following set of four non-orthogonal vectors.

$$\begin{aligned}\mathbf{v}_1&=(1,\ 1,\ 0,\ 0)\\ \mathbf{v}_2&=(0,\ 1,\ 2,\ 0)\\ \mathbf{v}_3&=(0,\ 0,\ 3,\ 4)\end{aligned} \tag{2-175}$$

$$\mathbf{v}_4=(1,\ 1,\ 1,\ 1). \tag{2-175}$$

First, take

$$\mathbf{u}_1=\mathbf{v}_1=(1,\ 1,\ 0,\ 0). \tag{2.176}$$

Then, form

$$\mathbf{u}_2=\mathbf{v}_2-\left[\frac{(\mathbf{u}_1,\ \mathbf{v}_2)}{|\mathbf{u}_1|^2}\right]\mathbf{u}_1. \tag{2-177}$$

The inner product in the above equation is easily seen to be

$$(\mathbf{u}_1,\ \mathbf{v}_2)=(1)(0)+(1)(1)+(0)(2)+(0)(0)=1. \tag{2-178}$$

Also,

$$|\mathbf{u}_1|^2=1^2+1^2+0^2+0^2=2. \tag{2-179}$$

Thus,

$$\mathbf{u}_2 = (0,\ 1,\ 2,\ 0) - \frac{1}{2}(1,\ 1,\ 0,\ 0), \tag{2-180}$$

or

$$\mathbf{u}_2 = \left(-\frac{1}{2},\ \frac{1}{2},\ 2,\ 0\right). \tag{2-181}$$

Next, take

$$\mathbf{u}_3 = \mathbf{v}_3 - \left[\frac{(\mathbf{u}_1,\ \mathbf{v}_3)}{|\mathbf{u}_1|^2}\right]\mathbf{u}_1 - \left[\frac{(\mathbf{u}_2,\ \mathbf{v}_3)}{|\mathbf{u}_2|^2}\right]\mathbf{u}_2. \tag{2-182}$$

The desired inner products are given by

$$\begin{aligned}(\mathbf{u}_1,\ \mathbf{v}_3) &= 0\\ (\mathbf{u}_2,\ \mathbf{v}_3) &= 6\\ |\mathbf{u}_2|^2 &= \frac{9}{2},\end{aligned} \tag{2-183}$$

which yields

$$\mathbf{u}_3 = (0,\ 0,\ 3,\ 4) - \frac{6}{(9/2)}\left(-\frac{1}{2},\ \frac{1}{2},\ 2,\ 0\right),$$

or

$$\mathbf{u}_3 = \left(\frac{2}{3},\ -\frac{2}{3},\ \frac{1}{3},\ 4\right). \tag{2-184}$$

Finally, we choose

$$\mathbf{u}_4 = \mathbf{v}_4 - \left[\frac{(\mathbf{u}_1,\ \mathbf{v}_4)}{|\mathbf{u}_1|^2}\right]\mathbf{u}_1 - \left[\frac{(\mathbf{u}_2,\ \mathbf{v}_4)}{|\mathbf{u}_2|^2}\right]\mathbf{u}_2 - \left[\frac{(\mathbf{u}_3,\ \mathbf{v}_4)}{|\mathbf{u}_3|^2}\right]\mathbf{u}_3. \tag{2-185}$$

The evaluation of the various inner products and the verification that the set $\{\mathbf{u}_1, \mathbf{u}_2, \mathbf{u}_3, \mathbf{u}_4\}$ forms a mutually orthogonal set is left to the student.

For an example in which functions are employed, consider the set

$$\begin{aligned}f_0(x) &= 1\\ f_1(x) &= x\\ f_2(x) &= x^2\\ f_3(x) &= x^3,\end{aligned} \tag{2-186}$$

which form a basis for the space of all real polynomials of degree ≤ 3 on the interval $-1 \leq x \leq +1$. This means that any polynomial $p(x)$ in this space can be written as

$$p(x)=a_3x^3+a_2x^2+a_1x+a_0. \tag{2-187}$$

If the inner product is defined as

$$(f_i, f_j)=\int_{-1}^{+1} f_i(x)f_j(x)\,dx, \tag{2-188}$$

then the Gram–Schmidt procedure can be used to form a corresponding set of orthogonal polynomials on the interval $-1 \leq x \leq 1$. First, take

$$p_0(x)=f_0(x)=1. \tag{2-189}$$

Then, take

$$p_1(x)=f_1(x)-\left[\frac{(f_1, p_0)}{|p_0|^2}\right] p_0(x), \tag{2-190}$$

where

$$(f_1, p_0)=\int_{-1}^{+1} (1)(x)\,dx=0. \tag{2-191}$$

Therefore,

$$p_1(x)=f_1(x)=x. \tag{2-192}$$

Now take

$$p_2(x)=f_2(x)-\left[\frac{(f_2, p_0)}{|p_0|^2}\right] p_0(x)-\left[\frac{(f_2, p_1)}{|p_1|^2}\right] p_1(x), \tag{2-193}$$

where it is easily seen that

$$\begin{aligned} (f_2, p_0)&=\frac{2}{3}, \\ |p_0|^2&=2, \\ (f_x, p_1)&=0, \end{aligned} \tag{2-194}$$

so that

$$p_2(x)=x^2-\frac{1}{3}. \tag{2-195}$$

Finally, form

$$p_3(x)=f_3(x)-\left[\frac{(f_3, p_0)}{|p_0|^2}\right] p_0(x)-\left[\frac{(f_3, p_1)}{|p_1|^2}\right] p_1(x)-\left[\frac{(f_3, p_2)}{|p_2|^2}\right] p_2(x), \tag{2-196}$$

where it is seen that

$$\begin{aligned}(f_3, p_0) &= 0\\ (f_3, p_1) &= 2/5\\ (f_3, p_2) &= 0\\ |p_1|^2 &= 2/3,\end{aligned} \tag{2-197}$$

so that

$$p_3(x) = x^3 - \frac{3}{5}x. \tag{2-198}$$

The orthogonal set is summarized below:

$$\begin{aligned}p_0(x) &= 1\\ p_1(x) &= x\\ p_2(x) &= \frac{1}{3}(3x^2 - 1)\\ p_3(x) &= \frac{1}{5}(5x^3 - 3x).\end{aligned} \tag{2-199}$$

As we shall see later, the polynomials that we have just generated are the first four (unnormalized) Legendrc polynomials.[22]

2-5. Hilbert Space

Thus far we have been concerned primarily with vector spaces of finite dimension. We have also shown that there is an isomorphism between an n-dimensional vector space V and the vector space formed by the collection of complex numbers that are the coefficients of the various basis vectors. Now we shall extend the concept to vectors with an infinite number of components, i.e., where the components represent infinite-dimensional sequences of complex numbers $(v_1, v_2, v_3, \ldots, v_n, \ldots)$. These sequences will, under certain conditions, also form a vector space. In particular, consider the vector

$$\mathbf{v} = (v_1, v_2, \ldots, v_n, \ldots). \tag{2-200}$$

We define the norm of a vector in this space as

$$|\mathbf{v}| = \left[\sum_{i=1}^{\infty} |v_i|^2\right]^{1/2}, \tag{2-201}$$

[22] See Chapter Six, Eq. (6-87).

which is analogous to that given by Eq. (2-67) for an n-dimensional space. Since the above sum contains an infinite number of terms, not all sequences will possess finite norms. However, we shall restrict our considerations to only those sequences possessing finite norms, with the norm defined in Eq. (2-201). Note that this condition is somewhat analogous to that of square integrability of functions discussed earlier.

If we define addition and scalar multiplication as

$$\mathbf{u}+\mathbf{v}=(u_1, u_2, \dots, u_n, \dots)+(v_1, v_2, \dots, v_n, \dots) \tag{2-202}$$

$$=(u_1+v_1, u_2+v_2, \dots, u_n+v_n, \dots), \tag{2-203}$$

and

$$c\mathbf{v}=c(v_1, v_2, \dots, v_n, \dots) \tag{2-204}$$

$$=(cv_1, cv_2, \cdots, cv_n, \dots), \tag{2-205}$$

respectively, where c is some complex scalar, then it is possible to show that such infinite-dimensional vectors do form a vector space V. In fact, such vectors, also form an inner product space V'. The proof of these assertions follows. First, let us consider whether the vectors form a vector space.

To prove *closure under addition*, we must show that $\Sigma_{i=1}^{\infty} |u_i + v_i|^2 < \infty$. To do this, consider $|u_i + v_i|^2$ for all $i = 1, 2, \dots, n, \dots$, i.e.,

$$|u_i+v_i|^2=(u_i^*+v_i^*)(u_i+v_i) \tag{2-206}$$

$$=|u_i|^2+|v_i|^2+u_i^*v_i+v_i^*u_i \tag{2-207}$$

$$=|u_i|^2+|v_i|^2+2\mathrm{Re}(u_i^*v_i) \tag{2-208}$$

$$\le |u_i|^2+|v_i|^2+2|u_i^*v_i| \tag{2-209}$$

$$\le |u_i|^2+|v_i|^2+2|u_i|\ |v_i|. \tag{2-210}$$

Now consider

$$(|u_i|-|v_i|)^2=|u_i|^2+|v_i|^2-2|u_i|\ |v_i|\ge 0, \tag{2-211}$$

which gives

$$|u_i|^2+|v_i|^2\ge 2|u_i|\ |v_i|. \tag{2-212}$$

Substituting Eq. (2-212) into Eq. (2-210), we obtain

$$|u_i+v_i|^2\le 2(|u_i|^2+|v_i|^2). \tag{2-213}$$

Summing over all the elements, $i = 1, 2, \dots, n, \dots$, gives

$$\sum_{i=1}^{\infty} |u_i+v_i|^2\le 2\left[\sum_{i=1}^{\infty} |u_i|^2+\sum_{i=1}^{\infty} |v_i|^2\right]. \tag{2-214}$$

Since $\Sigma_{i=1}^{\infty} |u_i|^2 < \infty$ and $\Sigma_{i=1}^{\infty} |v_i|^2 < \infty$, it follows that

$$\sum_{i=1}^{\infty} |u_i + v_i|^2 < \infty, \tag{2-215}$$

and closure under addition has been proven.

Closure under scalar multiplication is shown by demonstrating that $\Sigma_{i=1}^{\infty} |cv_i|^2 < \infty$. Consider the ith term of $c\mathbf{v}$, i.e., cv_i, and form the product

$$|cv_i|^2 = (c_i^* v_i^*)(cv_i) = |c|^2 |v_i|^2. \tag{2-216}$$

Now form the sum of all these terms,

$$\sum_{i=1}^{\infty} |cv_i|^2 = |c|^2 \sum_{i=1}^{\infty} |v_i|^2 < \infty, \tag{2-217}$$

which establishes closure under scalar multiplication.

The zero vector is taken to be

$$\mathbf{0} = (0, 0, \ldots, 0, \ldots), \tag{2-218}$$

and thc negative (or inverse) of the vector $\mathbf{v}$ is given by

$$-\mathbf{v} = (-v_1, -v_2, \ldots, -v_n, \ldots). \tag{2-219}$$

Using these results, it is easy to verify that the remaining properties of a vector space are satisfied.[23]

To show that we have an inner product space, we must first show that the inner product is defined. Consider

$$(\mathbf{u}, \mathbf{v}) = \sum_{i=1}^{\infty} u_i^* v_i, \tag{2-220}$$

which is defined by analogy to that given for an n-dimensional inner product space [see Eq. (2-66)]. From Eq. (2-212) we can write

$$\frac{1}{2}\{|u_i|^2 + |v_i|^2\} \geq |u_i^* v_i| \qquad \text{for all } i. \tag{2-221}$$

It follows from this property that the sum $\Sigma_{i=1}^{\infty} |u_i^* v_i|$ converges, since

$$\sum_{i=1}^{\infty} |u_i^* v_i| \leq \frac{1}{2}\left\{\sum_{i=1}^{\infty} |u_i|^2 + \sum_{i=1}^{\infty} |v_i|^2\right\}, \tag{2-222}$$

and each sum on the right-hand side of Eq. (2-222) converges. Since the sum of

---

[23] See Problem 2-10.

the absolute value of the terms $|u_i^* v_i|$ converges, the terms themselves also converge, and we have proven the existence of the inner product.

In a fashion analogous to that used earlier, it can be shown that the other postulates of an inner product space are also satisfied.[24] Furthermore, the Schwartz' and triangle inequalities are obeyed, taking the form

$$\left|\sum_{i=1}^{\infty} u_i^* v_i\right| \leq \left[\sum_{i=1}^{\infty} |u_i|^2\right]^{1/2} \left[\sum_{i=1}^{\infty} |v_i|^2\right]^{1/2}, \tag{2-223}$$

and

$$\left[\sum_{i=1}^{\infty} |\mu_i + v_i|^2\right]^{1/2} \leq \left[\sum_{i=1}^{\infty} |u_i|^2\right]^{1/2} + \left[\sum_{i=1}^{\infty} |v_i|^2\right]^{1/2}, \tag{2-224}$$

respectively.

The infinite-dimensional space we have just described, which possesses a finite norm and an inner product, is usually called a *Hilbert space*, $\mathcal{H}$. The main difference (and importance!) of Hilbert spaces from other spaces is that, in these spaces, the concept of a complete set of orthonormal vectors is generalized from that used in finite-dimensional vector spaces. In particular, there will be, in general, an infinite number of vectors that forms the basis, and each vector will have an infinite number of components.

If $\{\mathbf{v}_1, \mathbf{v}_2, \ldots, \mathbf{v}_n, \ldots\}$ is some set of orthonormal vectors, we can write the projection of any vector $\mathbf{u}$ in $\mathcal{H}$ onto the vectors of the orthnormal set as shown by Eq. (2-126), i.e.,

$$\sum_{i=1}^{\infty} (\mathbf{v}_i, \mathbf{u})\mathbf{v}_i = \sum_{i=1}^{\infty} c_i \mathbf{v}_i. \tag{2-225}$$

If the set $\{\mathbf{v}_1, \mathbf{v}_2, \ldots, \mathbf{v}_n, \ldots\}$ is complete, then the expression in Eq. (2-225) will be equal to $\mathbf{u}$. For that to be the case, we need to establish several points. First we must show that the series converges. Instead of proving this point directly, we note that the series in Eq. (2-225) converges if and only if the series

$$\sum_{i=1}^{\infty} |(\mathbf{v}_i, \mathbf{u})|^2 = \sum_{i=1}^{\infty} |c_i|^2 \tag{2-226}$$

converges.[25] To see that the latter series converges, consider Bessel's inequality, Eq. (2-130), in which the limit of the summation goes to infinity for infinite-dimensional spaces,[26] i.e.,

$$|\mathbf{u}|^2 \geq \sum_{i=1}^{\infty} |(\mathbf{v}_i, \mathbf{u})|^2 = \sum_{i=1}^{\infty} |c_i|^2. \tag{2-227}$$

---

[24] See Problem 2-11.

[25] See for example, G. F. Roach, *Green's Functions*, van Nostrand Reinhold, New York, 1970.

[26] The relationship expressed in Eq. (2-227) is frequently referred to as a closure relation.

Since $\mathbf{u}$ is in $\mathcal{H}$, its norm must be finite, i.e.,

$$|\mathbf{u}| < \infty, \tag{2-228}$$

and applying Eq. (2-227) yields

$$\sum_{i=1}^{\infty} |(\mathbf{v}_i, \mathbf{u})|^2 = \sum_{i=1}^{\infty} |c_i|^2 < \infty. \tag{2-229}$$

Therefore, the series of Eq. (2-225) also converges.

This does not, however, complete the proof that the set of orthonormal vectors $\{\mathbf{v}_1, \mathbf{v}_2, \dots, \mathbf{v}_n, \dots\}$ is complete, i.e., forms a basis in $\mathcal{H}$. In particular, we must also show that if $\{\mathbf{v}_1, \mathbf{v}_2, \dots, \mathbf{v}_n, \dots\}$ is complete, we can write $\mathbf{u}$ as

$$\mathbf{u} = \sum_{i=1}^{\infty} (\mathbf{v}_i, \mathbf{u})\mathbf{v}_i = \sum_{i=1}^{\infty} c_i \mathbf{v}_i. \tag{2-230}$$

In other words, the arbitrary vector $\mathbf{u}$ in $\mathcal{H}$ must be expressible as a linear combination of the vectors of the orthonormal set $\{\mathbf{v}_1, \mathbf{v}_2, \dots, \mathbf{v}_n, \dots\}$. To see this, consider the vector $\mathbf{u}'$,

$$\mathbf{u}' = \mathbf{u} - \sum_{i=1}^{\infty} (\mathbf{v}_i, \mathbf{u})\mathbf{v}_i. \tag{2-231}$$

In a fashion exactly analogous to that used in the proof of Theorem 2-7, we can show that

$$(\mathbf{u}', \mathbf{v}_i) = 0, \qquad i = 1, 2, \dots, n, \dots \tag{2-232}$$

which shows that $\mathbf{u}'$ is orthogonal to each member of the orthonormal set of vectors $\{\mathbf{v}_1, \mathbf{v}_2, \dots, \mathbf{v}_n, \dots\}$. Thus, $\mathbf{u}'$ has no component in common with any member of the orthonormal set. It must therefore be the null vector, and Eq. (2-230) is proved.

We shall now prove a *generalized closure relation* that is the equivalent of Parseval's equation, generalized to infinite dimensional vector spaces. If we have a complete set of orthonormal vectors, $\{\mathbf{v}_1, \mathbf{v}_2, \dots, \mathbf{v}_n, \dots\}$, i.e., a basis, in an infinite-dimensional vector space, then we can represent two arbitrary vectors in that space as

$$\mathbf{u} = \sum_{i=1}^{\infty} (\mathbf{v}_i, \mathbf{u})\mathbf{v}_i = \sum_{i=1}^{\infty} a_i \mathbf{v}_i, \tag{2-233}$$

and

$$\mathbf{w} = \sum_{j=1}^{\infty} (\mathbf{v}_j, \mathbf{w})\mathbf{v}_j = \sum_{j=1}^{\infty} b_j \mathbf{v}_j. \tag{2-234}$$

Now form the inner product $(\mathbf{u}, \mathbf{w})$,

$$(\mathbf{u}, \mathbf{w}) = \left(\sum_{i=1}^{\infty} a_i \mathbf{v}_i, \sum_{j=1}^{\infty} b_j \mathbf{v}_j \right) \tag{2-235}$$

$$= \sum_{i=1}^{\infty} \sum_{j=1}^{\infty} a_i^* b_j (\mathbf{v}_i, \mathbf{v}_j). \tag{2-236}$$

Since we have an orthonormal basis

$$(\mathbf{v}_i, \mathbf{v}_j) = \delta_{ij}, \tag{2-237}$$

then

$$(\mathbf{u}, \mathbf{w}) = \sum_{i=1}^{\infty} \sum_{j=1}^{\infty} a_i^* b_j \delta_{ij}. \tag{2-238}$$

Hence,

$$(\mathbf{u}, \mathbf{w}) = \sum_{i=1}^{\infty} a_i^* b_i. \tag{2-239}$$

Substituting in the expressions for the coefficients, we obtain the desired result:

$$(\mathbf{u}, \mathbf{w}) = \sum_{i=1}^{\infty} (\mathbf{u}, \mathbf{v}_i)(\mathbf{v}_i, \mathbf{w}). \tag{2-240}$$

2-6. Function Space and Generalized Fourier Series

It was previously shown that the class of functions that is integrable and square integrable on the interval $[a, b]$ constitutes an inner product space V'.[27] Now, however, it will be useful to take a slightly different view of such functions, which will show the infinite-dimensionality of the space.[28] We will now refer to such a space as a *function space* $\mathfrak{F}$. Since, as shown previously, functions of

[27] We could take a more restrictive view and consider only those functions that are completely continuous on $[a, b]$. We will, however, continue to allow functions that have only the characteristics of being integrable and square integrable on $[a, b]$, and constitute an inner product space V'.

[28] One way to accomplish this is to think of the function $f(x)$ as a vector whose components are given by the different values of $f(x)$ for each x on $[a, b]$, i.e., $f(x_0), f(x_1), f(x_2), \ldots, f(x_n), \ldots$. The number of such components is clearly infinite. However, a careful distinction should be made between the infinite but countable number of complex components of a vector in Hilbert space, and the infinite, but noncountable, number of elements in the "functional" sequence $f(x_0), f(x_1), f(x_2), \ldots$. This particular view of a function space should be used cautiously, for it can lead to misleading or erroneous results.

interest to us possess an inner product and norm, we can speak of orthonormal sets of these functions $\{f_1, f_2, \ldots, f_n, \ldots\}$. Such functions satisfy the condition

$$(f_i, f_j) = \int_a^b f_i^*(x) f_j(x)\, dx = \delta_{ij}. \tag{2-241}$$

Suppose we want to approximate some arbitrary functions $F(x)$, in a least-square sense, using a linear combination of orthonormal functions $\{f_1, f_2, \ldots, f_n, \ldots\}$. We then want to choose the coefficients (c_k) so as to minimize

$$I = \left| F - \sum_{k=1}^{n} c_k f_k \right|^2$$

$$= \int_a^b \left| F - \sum_{k=1}^{n} c_k f_k \right|^2 dx. \tag{2-242}$$

Expanding this quantity gives,

$$I = \int_a^b |F|^2\, dx + \sum_{k=1}^{n} |c_k|^2 - \sum_{k=1}^{n} c_k \int_a^b (F^* f_k)\, dx$$

$$- \sum_{k=1}^{n} c_k^* \int (f_k^* F)\, dx. \tag{2-243}$$

Let

$$\alpha_k = (f_k, F) = \int_a^b f_k^* F\, dx, \tag{2-244}$$

and Eq. (2-243) becomes

$$I = (F, F) + \sum_{k=1}^{n} |c_k|^2 - \sum_{k=1}^{n} c_k \alpha_k^* - \sum_{k=1}^{n} c_k^* \alpha_k. \tag{2-245}$$

Since the α_k are fixed, and since $\Sigma_{k=1}^{n} |\alpha_k|^2 \geq 0$, then a procedure that minimizes $I + \Sigma_{k=1}^{n} |\alpha_k|^2$ also guarantees that I is minimized. We can write $I + \Sigma_{k=1}^{n} |\alpha_k|^2$ as

$$I' = I + \sum_{k=1}^{n} |\alpha_k|^2 = (F, F) + \sum_{k=1}^{n} |c_k|^2 - \sum_{k=1}^{n} c_k^* \alpha_k$$

$$- \sum_{k=1}^{n} \alpha_k^* c_k + \sum_{k=1}^{n} |\alpha_k|^2, \tag{2-246}$$

or

$$I' = (F, F) + \sum_{k=1}^{n} |c_k - \alpha_k|^2. \tag{2-247}$$

Clearly I' is a minimum if we choose

$$\alpha_k = c_k = (f_k, F) = \int_a^b f_k^* F\, dx. \tag{2-248}$$

Making the proper substitutions for α_k into Eq. (2-245), we obtain, upon combining terms,

$$0 \le I = |F|^2 - \sum_{k=1}^{n} |c_k|^2. \tag{2-249}$$

Since $|F|^2$ is not a function of n, we obtain as $n \to \infty$,

$$|F|^2 \ge \sum_{k=1}^{\infty} |c_k|^2 \tag{2-250}$$

which is simply Bessel's inequality again. Since F is a function in $\mathfrak{F}$, we know that its norm is finite and hence $|F|^2 < \infty$, which implies that $\Sigma_{i=1}^{\infty} |c_i|^2 < \infty$. This suggests that for every function of interest in the region $a \le x \le b$, there exists a corresponding convergent sequence $c_1, c_2, \ldots, c_n, \ldots$, in Hilbert space.

If the equality sign of Eq. (2-250) holds, i.e.,

$$|F|^2 = \sum_{k=1}^{\infty} |c_k|^2, \tag{2-251}$$

we say we have a *complete* set of functions, and the *closure relation* is satisfied. This equality sign is obtained if and only if

$$\lim_{n\to\infty} \int_a^b \left| F - \sum_{k=1}^{n} c_k f_k \right|^2 dx = 0. \tag{2-252}$$

If this is the case, we say that the series $\Sigma_{k=1}^{\infty} c_k f_k(x)$ *converges in the mean* to $F(x)$. This *does not mean* that we can write

$$F(x) = \sum_{k=1}^{\infty} c_k f_k(x), \tag{2-253}$$

but rather, we should write

$$F(x) \doteq \sum_{k=1}^{\infty} c_k f_k(x), \tag{2-254}$$

which indicates that $\Sigma_{k=1}^{\infty} c_k f_k(x)$ is a *series representation* of $F(x)$, but it is not necessarily equal to $F(x)$ at *all* x. Eq. (2-254) is usually called a *generalized*

Fourier Series. Strong convergence criteria are needed to obtain the equality for all x.[29]

The proof of the generalized closure relation in $\mathfrak{F}$ needs to be slightly modified from the one used in $\mathcal{H}$. The reason for this is that, as shown in Eq. (2-254), we cannot exactly write $F(x)$ as a series in $f(x)$. However, since the orthonormal set $\{f_1, f_2, \dots, f_n, \dots\}$ forms a basis for $\mathfrak{F}$, we can write

$$\int_a^b F(x)\,dx = \sum_{k=1}^{\infty} c_k \int_a^b f_k(x)\,dx. \tag{2-255}$$

If we consider a function $G(x)$, write the products $G(x)^*F(x)$ and $G(x)^*f_k(x)$ for all k, and then integrate, we obtain

$$\int_a^b G(x)^*F(x)\,dx = \sum_{k=1}^{\infty} c_k \int_a^b G(x)^*f_k(x)\,dx, \tag{2-256}$$

which can be rewritten as

$$(G, F) = \sum_{k=1}^{\infty} c_k(G, f_k). \tag{2-257}$$

Substituting in the expression for the Fourier coefficient (c_k), we obtain finally

$$(G, F) = \sum_{k=1}^{\infty} (G, f_k)(f_k, F), \tag{2-258}$$

which is the relationship sought.

2-7. Isomorphism between Hilbert Space and Function Space

Consider the complete, orthonormal set of functions $\{f_1, f_2, \dots, f_n, \dots\}$ which satisfies the closure relation for any two arbitrary functions $F(x)$ and $G(x)$ in $\mathfrak{F}$. Then, the two functions will have the following Fourier coefficients:

$$\begin{aligned} F(x) &: \{a_1, a_2, \dots, a_n, \dots\} \\ G(x) &: \{b_1, b_2, \dots, b_n, \dots\}, \end{aligned} \tag{2-259}$$

[29] For a good discussion of the different convergence criteria, see F. W. Byron and R. W. Fuller, *Mathematics of Classical and Quantum Physics*, Vol. I, Addison-Wesley Publishers, Reading, MA, 1969.

defined by

$$\begin{aligned} a_k &= (f_k, F) \\ b_k &= (f_k, G), \end{aligned} \tag{2-260}$$

respectively. The closure relation for the difference of the two functions is given by

$$|F-G|^2 = \int_a^b |F-G|^2 \, dx = \sum_{k=1}^{\infty} |a_k - b_k|^2. \tag{2-261}$$

If $F(x)$ and $G(x)$ are different, $|F - G|^2 > 0$, and therefore

$$\sum_{k=1}^{\infty} |a_k - b_k|^2 > 0, \tag{2-262}$$

and hence different functions in $\mathfrak{F}$ have different Fourier coefficients. In other words, every $F(x)$ in $\mathfrak{F}$ is completely characterized by its Fourier coefficients, and from the closure relation, the sum of the squares of the absolute values of these coefficients forms a convergent series. Therefore, the functions belonging to $\mathfrak{F}$ and the infinite sequences of their corresponding Fourier coefficients both form vector spaces. The functions belong to $\mathfrak{F}$ and the infinite sequences belong to $\mathcal{H}$, and the above discussion show the one-to-one correspondence between the two spaces. In other words, an isomorphism exists between the two spaces. This is exactly analogous to the isomorphism between an n-dimensional vector space and the vector space of vector components previously discussed [see pages 41–42].

Note that while every function in $\mathfrak{F}$ corresponds to a definite vector in $\mathcal{H}$, the converse is not true, since vectors exist in $\mathcal{H}$ that do not have all of the characteristics needed to have corresponding functions. For the converse to be true, a much wider class of functions would have to be admitted to our function space.[30]

The above isomorphism has implications in quantum mechanics. Schrödinger wave mechanics deals essentially with a function space, while Heisenberg's matrix mechanics deals essentially with a Hilbert space. The isomorphism between the two spaces establishes the equivalence of the two theories. Consequently, the two descriptions differ only in the manner in which they represent the same physical reality.

---

[30] A more general mathematical treatment of this problem shows that both $\mathcal{H}$ and $\mathfrak{F}$ belong to an *abstract Hilbert* space. This shows their basic similarity since both the functions and sequences belong to the same abstract space. For a detailed discussion of this point, see W. Schmeidler, *Linear Operators in Hilbert Space*, Academic Press, New York, 1965.

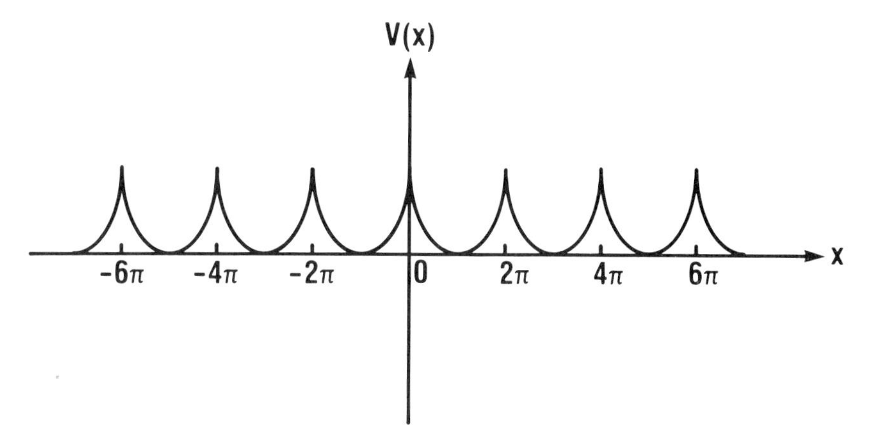

Figure 2.4. Example of a periodic function $V(x) = V(x + 2\pi)$ with period 2π.

2-8. Examples of Complete Sets of Functions

To illustrate the utility of expansions in complete sets of functions, suppose we had a one-dimensional function, $V(x)$, that is periodic as shown in Fig. 2.4. This might be a one-dimensional model of the potential energy in an atomic crystal.

As can be seen from the figure, the function $V(x)$ has a period of 2π, i.e., $V(x + 2\pi) = V(x)$. In order to represent this function, we shall use the following set of functions, which are complete and orthonormal in the region $-\pi \leq x \leq +\pi$:

$$\frac{1}{\sqrt{2\pi}}, \frac{\cos x}{\sqrt{\pi}}, \frac{\sin x}{\sqrt{\pi}}, \frac{\cos 2x}{\sqrt{\pi}}, \frac{\sin 2x}{\sqrt{\pi}}, \ldots . \tag{2-263}$$

When used in a series expansion of a function, this series is known as a *Fourier series*, and it can be written as

$$V(x) = \frac{a_0}{\sqrt{2\pi}} + \frac{1}{\sqrt{\pi}} \sum_{n=1}^{\infty} (a_n \cos nx + b_n \sin nx). \tag{2-264}$$

Although there are several expansions that could be used for this problem, the Fourier series is a particularly convenient one, since the trigonometric functions also have a period of 2π. This means that we need to consider only one region, e.g., the region $-\pi \leq x \leq +\pi$, since the periodicity guarantees that the solution will be valid for the other regions.

Considering henceforth only the region $-\pi \leq x \leq +\pi$, let us assume that we know that $V(x)$ has the following form in that region

$$V(x) = V_0 + kx^2. \tag{2-265}$$

In order to approximate $V(x)$ to various degrees of accuracy using various numbers of terms in the Fourier series, we need to know how to find the coefficients a_n and b_n. To obtain the coefficient a_p, we multiply both sides of Eq. (2-264) by

$$\frac{\cos px}{\sqrt{\pi}},$$

and integrate over $-\pi \leq x \leq +\pi$. The orthonormality of the functions in Eq. (2-263) over that range will guarantee that only one term on the right-hand side of Eq. (2-264) will survive, namely, the term with $n = p$. This gives

$$\frac{1}{\sqrt{\pi}} \int_{-\pi}^{+\pi} V(x) \cos(px)\, dx = a_p, \tag{2-266}$$

which is the form of a generalized Fourier coefficient given by Eq. (2-248).

For the example used in Eq. (2-265), we obtain

$$a_p = \frac{1}{\sqrt{\pi}} \left[V_0 \int_{-\pi}^{+\pi} \cos(px)\, dx + k \int_{-\pi}^{+\pi} x^2 \cos(px)\, dx \right], \qquad p = 1, 2, \ldots . \tag{2-267}$$

The student can easily verify that the first integral in Eq. (2-267) is zero, and that integration by parts in the second integral results in

$$a_p = (-1)^p \left(\frac{4k}{p^2} \right) \sqrt{\pi}, \qquad p = 1, 2, \ldots . \tag{2-268}$$

The special case of $p = 0$ gives

$$a_0 = \frac{1}{\sqrt{2\pi}} \int_{-\pi}^{+\pi} [V_0 + kx^2]\, dx, \tag{2-269}$$

or

$$a_0 = \sqrt{2\pi}\, V_0 + \left(\frac{k\pi^3}{3} \right) \sqrt{\frac{2}{\pi}} . \tag{2-270}$$

Multiplication of Eq. (2-265) by $(1/\sqrt{\pi}) \sin(px)$ and integrating gives the desired expression for b_p, i.e.,

$$b_p = \frac{1}{\sqrt{\pi}} \int_{-\pi}^{+\pi} V(x) \sin(px)\, dx, \qquad p = 1, 2, \ldots . \tag{2-271}$$

It is easily seen that

$$b_p = 0, \qquad p = 1, 2, \ldots . \tag{2-272}$$

This means that the expansion of the $V(x)$ of Eq. (2-265) in terms of a Fourier

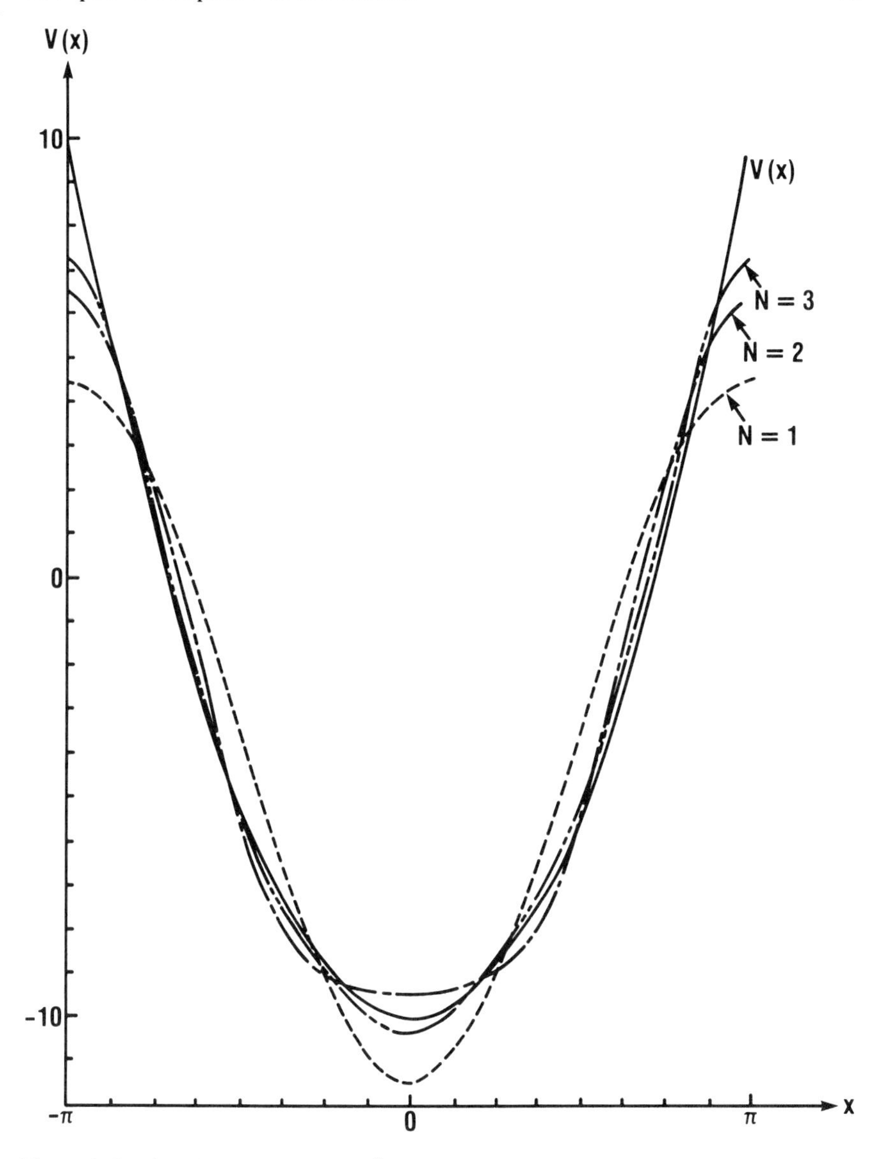

Figure 2.5. Plot of $V(x) = V_0 + kx^2$, along with one, two and three term Fourier series fits to $V(x)$.

series can be written as

$$V(x) = \left(V_0 + \frac{k\pi^2}{3} \right) + 4k \sum_{n=1}^{\infty} (-1)^n \left(\frac{1}{n^2} \right) \cos(nx). \qquad (2\text{-}273)$$

To illustrate how $V(x)$ is approximated better and better by taking more terms in the infinite series of Eq. (2-273), Fig. 2.5 shows the actual curve of $V(x)$ plus curves in which one, two, and three terms, respectively, were included from the infinite series.

Let us give one further example. If the arbitrary function is given as

$$W(x)=W_0+k'x^4, \tag{2-274}$$

then the desired Fourier coefficients c_n and d_n are found from

$$W(x)=\frac{c_0}{\sqrt{2\pi}}+\frac{1}{\sqrt{\pi}}\sum_{n=1}^{\infty}(c_n \cos nx+d_n \sin nx). \tag{2-275}$$

Using the same procedure as just illustrated for $V(x)$ gives

$$c_0=\sqrt{2\pi}\ W_0+\left(\frac{k'\pi^5}{5}\right)\sqrt{\frac{2}{\pi}}, \tag{2-276}$$

$$c_p=(-1)^p 8\sqrt{\pi}k' \cdot \left(\frac{\pi^2}{p^2}-\frac{6}{p^4}\right), \qquad p=1, 2, \ldots \tag{2-277}$$

$$d_p=0, \qquad p=1, 2, \ldots, \tag{2-278}$$

and thus

$$W(x)=\left(W_0+\frac{k'\pi^4}{5}\right)+8k'\sum_{n=1}^{\infty}(-1)^n\left[\frac{\pi^2}{n^2}-\frac{6}{n^4}\right]\cos nx. \tag{2-279}$$

These two examples are also convenient for illustrating, by means of a specific numerical example, the isomorphism between functions in function space and the vectors whose components are the infinite sequence of Fourier coefficients. Let **a** and **c** be the vectors formed from the Fourier coefficients of $V(x)$ and W(x), respectively, i.e.,

$$\mathbf{a}=(a_1, a_2, \ldots, a_n, \ldots), \tag{2-280}$$

$$\mathbf{c}=(c_1, c_2, \ldots, c_n, \ldots). \tag{2-281}$$

Then, let us choose $W_0 = V_0 = 0$, and $k = k' = 1$ for convenience, and compare the result obtained for the scalar product of **a** and **c** with the result obtained for the scalar product of $V(x)$ and $W(x)$. For the latter we obtain

$$(V, W)=\int_{-\pi}^{+\pi} x^6\, dx=862.940923.$$

For comparison purposes, Table 2.1 shows the results of taking increasing numbers of terms in the infinite series. The convergence to the same result is quite obvious, and emphasizes again that if an isomorphism between two spaces has been established, the question of which representation should be used is a question merely of which one is most convenient. The results will be the same in either representation.

Table 2.1. A Calculation using Eq. (2-218) of the Scalar Product of **a** with **c**, using Eqs. (2-268), (2-270), (2-276), and (2-277), with $V_0 = W_0 = 0$, $k = k' = 1$

N	$(\mathbf{a}, \mathbf{c}) = \sum_{i=0}^{N} a_i^* c_i$
10	862.657443
50	862.938356
100	862.940597
200	862.940882
300	862.940911
400	862.940918
500	862.940920

Thus, we see that it is quite easy to expand a function that is periodic in a Fourier series. In fact, there is no need to restrict the periodicity to be 2π, for it is easily verified that, for a function $f(x)$, which is periodic between $-l \leq x \leq +l$, the Fourier series expansion is

$$f(x) = \frac{a_0}{\sqrt{2l}} + \frac{1}{\sqrt{l}} \sum_{n=1}^{\infty} \left[a_n \cos\left(\frac{n\pi x}{l}\right) + b_n \sin\left(\frac{n\pi x}{l}\right) \right], \quad \text{(2-282)}$$

with

$$a_p = \left(\frac{1}{\sqrt{l}}\right) \int_{-l}^{+l} f(t) \cos\left(\frac{p\pi t}{l}\right) dt, \qquad p = 1, 2, \ldots \quad \text{(2-283)}$$

$$b_p = \left(\frac{1}{\sqrt{l}}\right) \int_{-l}^{+l} f(t) \sin\left(\frac{p\pi t}{l}\right) dt, \qquad p = 1, 2, \ldots . \quad \text{(2-284)}$$

Although it is obvious that a Fourier series expansion of a function is very useful if the function is periodic within some finite interval $[-l \leq x \leq +l]$, we shall often encounter functions in which the interval of its definition is $[-\infty \leq x \leq +\infty]$, and the function is not periodic within the interval. Even in this case, it is possible to perform an analysis similar to that involving the Fourier series, but where we must first consider the limiting process leading to the transition of $|l| \to \infty$. As the reader has perhaps noticed in this chapter, limiting processes such as these are nontrivial jobs, and we shall not attempt to treat this one in detail, but rather outline some of the more salient points in the development.[31]

In order to indicate how this generalization can be made plausible, let us first recast the Fourier series as expressed in Eqs. (2-282)–(2-284) in another form.

[31] For more details, the reader is referred to treatises such as R. Courant and D. Hilbert, *Methods of Mathematical Physics*, Vol. I, Interscience Publishers, New York, 1953, pp. 69–81.

By using the identity

$$e^{\pm ix}=\cos x\pm i\sin x, \tag{2-285}$$

we can rewrite the Fourier series expansion as

$$f(x)=\frac{a_0}{\sqrt{2l}}+\frac{1}{\sqrt{l}}\sum_{n=1}^{\infty}\left\{\frac{1}{2}[a_n-ib_n]e^{+in\pi x/l}+\frac{1}{2}[a_n+ib_n]e^{-in\pi x/l}\right\}$$

or

$$f(x)=\sum_{p=-\infty}^{+\infty}k_p e^{+ip\pi x/l}, \tag{2-286}$$

with

$$k_p=\begin{cases}\dfrac{1}{2\sqrt{l}}(a_n-ib_n), & p>0\\[2ex] \dfrac{a_0}{\sqrt{2l}}, & p=0\\[2ex] \dfrac{1}{2\sqrt{l}}(a_n+ib_n), & p<0\end{cases} \tag{2-287}$$

or

$$k_p=\left(\frac{1}{2l}\right)\int_{-l}^{+l}f(t)e^{-ip\pi t/l}\,dt, \qquad p=1,2,\ldots. \tag{2-288}$$

To generalize Eq. (2-286) properly, several properties must be present in $f(x)$. We shall assume that $f(x)$ is piecewise smooth, as discussed at the beginning of this section. Then, if

$$\int_{-\infty}^{+\infty}|f(x)|^2\,dx$$

exists, the desired result can be obtained.

Inserting Eq. (2-288) into Eq. (2-286), we obtain

$$f(x)=\sum_{p=-\infty}^{+\infty}\left(\frac{1}{2l}\right)\int_{-l}^{+l}f(t)\exp\left[-\frac{ip\pi}{l}(t-x)\right]dt. \tag{2-289}$$

Now let $\Delta\alpha = \pi/l$, which gives

$$f(x)=\left(\frac{1}{2\pi}\right)\sum_{p=-\infty}^{+\infty}\Delta\alpha\int_{-l}^{+l}f(t)\exp[-ip\Delta\alpha(t-x)]\,dt. \tag{2-290}$$

Thus far, we have done nothing new. What we now desire to do is to let the size of the region in which $f(x)$ is periodic become larger and larger, i.e., to let $|l| \to \infty$. Of course, if $|l| \to \infty$ then $\Delta\alpha \to 0$. One point needs to be noted here. The quantity $p\Delta\alpha$ in the exponential will have a value in the limit of $\Delta\alpha \to 0$ which depends upon p. Let us designate this quantity by $\alpha_p = p\Delta\alpha$. To make the limiting process more obvious (and perhaps making the difficulties less obvious), let us designate

$$g(\alpha_p, x) = \frac{1}{\sqrt{2\pi}} \int_{-l}^{+l} f(t) \exp[-i\alpha_p(t-x)]\, dt. \tag{2-291}$$

Taking the desired limit gives

$$f(x) = \frac{1}{\sqrt{2\pi}} \lim_{\Delta\alpha \to 0} \left\{ \sum_{p=-\infty}^{+\infty} \Delta\alpha g(\alpha_p, x) \right\}. \tag{2-292}$$

The quantity inside the braces in Eq. (2-292) is just the expression used for numerical integration of

$$\int_{-l}^{+l} g(\alpha_p, x)\, d\alpha, \tag{2-293}$$

where the step size $\Delta\alpha$ is kept constant throughout. Of course, this is really an oversimplification, since $|l|$ is going to infinity simultaneously, but the analogy is clear.

If this plausibility argument is accepted, we arrive at

$$f(x) = \left(\frac{1}{2\pi}\right) \int_{-\infty}^{+\infty} d\alpha \int_{-\infty}^{+\infty} f(t) \exp[-i\alpha(t-x)]\, dt, \tag{2-294}$$

which is known as the *Fourier integral formula.*[32]

The reason for inclusion of the factor $1/\sqrt{2\pi}$ in the definition of $g(\alpha)$ is now apparent, for after the limiting process just discussed is carried out, we can express the Fourier integral formula in a very compact form. By defining

$$g(\alpha) = \frac{1}{\sqrt{2\pi}} \int_{-\infty}^{+\infty} f(t) e^{-i\alpha t}\, dt, \tag{2-295}$$

we see that

$$f(t) = \frac{1}{\sqrt{2\pi}} \int_{-\infty}^{+\infty} g(\alpha) e^{+i\alpha t}\, d\alpha. \tag{2-296}$$

The functions $g(\alpha)$ and $f(t)$ as given by Eqs. (2-295) and (2-296) are said to be

[32] For additional discussion see, for example, W. Kaplan, *Advanced Calculus*, Addison-Wesley, Cambridge, MA, 1952, p. 434.

Fourier transforms of each other. Obviously, the knowledge of either function implies the other, and vice versa. This reciprocal relation is often very useful, for it allows more than one representation of the same function.

As an example of this alternate means of representation, let us consider the following function of x, y, and z:

$$f(r)=e^{-br}, \tag{2-297}$$

with $b \geq 0$, and b a real constant. To obtain the Fourier transform of $f(r)$, we need to evaluate the three-dimensional analog of Eq. (2-295), which is

$$g(\mathbf{k})=\frac{1}{(2\pi)^{3/2}}\int_{-\infty}^{+\infty}\int_{-\infty}^{+\infty}\int_{-\infty}^{+\infty} e^{-i(k_x x+k_y y+k_z z)}e^{-br}\,dx\,dy\,dz \tag{2-298}$$

$$=\frac{1}{(2\pi)^{3/2}}\int_{-\infty}^{+\infty}\int_{-\infty}^{+\infty}\int_{-\infty}^{+\infty} e^{-i(\mathbf{k},\mathbf{r})}e^{-br}\,dx\,dy\,dz. \tag{2-299}$$

Converting to spherical coordinates, we obtain

$$g(\mathbf{k})=\frac{1}{(2\pi)^{3/2}}\int_{0}^{\infty}\int_{0}^{\pi}\int_{0}^{2\pi} e^{-ikr\cos\vartheta}e^{-br}r^2\,dr\,\sin\vartheta\,d\vartheta\,d\varphi$$

$$=\frac{1}{\sqrt{2\pi}}\int_{0}^{\infty} e^{-br}r^2\,dr\int_{0}^{\pi} e^{-ikr\cos\vartheta}\sin\vartheta\,d\vartheta$$

$$=\frac{1}{ik\sqrt{2\pi}}\int_{0}^{\infty} re^{-br}(e^{ikr}-e^{-ikr})\,dr.$$

Integration by parts yields the result

$$g(\mathbf{k})=\left(\frac{1}{\sqrt{2\pi}}\right)\left(\frac{4b}{[b^2+k^2]^2}\right). \tag{2-300}$$

Let us examine another way of expressing the same result. If for some reason it would be inconvenient to work with $f(r)$ as expressed in Eq. (2-297), it may be easier to deal with an integral representation of $f(r)$, which we obtain by substitution of Eq. (2-300) into the three-dimensional analog of Eq. (2-296), i.e.,

$$f(r)=\frac{4b}{(2\pi)^2}\int_{-\infty}^{+\infty}\frac{1}{(b^2+k^2)^2}\,e^{i(\mathbf{k},\mathbf{r})}\,d\mathbf{k}. \tag{2-301}$$

Many transforms of the type introduced here are known, and an extensive compilation has been provided by the Bateman Manuscript Project.[33]

[33] *Tables of Integral Transforms*, Vols. I-II, edited by A. Erdelyi, McGraw-Hill Book Co., New York, 1954.

2-9. Extension to Continuum Functions

In the previous sections, we saw how one arbitrary function $F(x)$ in function space can be expanded in terms of a countable, complete set of orthonormal functions $\{f_1, f_2, \dots, f_n, \dots\}$. We also saw that with respect to this basis, $F(x)$ was uniquely specified by its Fourier coefficients. Furthermore, these Fourier coefficients formed a countably infinite sequence of complex numbers that are vectors in a Hilbert space $\mathcal{H}$.

However, this capability will not be sufficient to allow us to treat all of the problems of interest in later chapters. An indication of this has already been given in the discussion of the Fourier integral. In that case, an arbitrary function $f(x)$ is expanded in terms of the functions e^{ikx} over the interval $-\infty \le x \le +\infty$, i.e.,

$$f(x) = \frac{1}{\sqrt{2\pi}} \int_{-\infty}^{+\infty} a(k) e^{ikx}\, dk, \qquad (2\text{-}302)$$

where $a(k)$ are the expansion coefficients. Although this expansion is a very useful one, the basis functions e^{ikx} do not belong to $\mathcal{F}$, since they do not have a finite norm over the interval $-\infty \le x \le +\infty$. Thus, we shall expand the list of functions that is available for our use to include both those with a countable basis and finite norm, as well as those with a noncountable basis that behave at infinity like e^{ikx} [or $\sin(kx + \delta)$]. It is these latter functions that will be very useful in discussions of continuum problems.

It will be particularly convenient to define a special quantity for use in continuum problems, which is a generalization of the Kronecker delta δ_{ij} that was introduced in Eq. (2-120). This special quantity, called[34] the "Dirac δ-function," has several very peculiar properties when compared to ordinary functions, and cannot be considered to be a function in the usual sense.

We shall define the δ-function to be one in which

$$\int_{-\infty}^{+\infty} \delta(x)\, dx = 1, \qquad (2\text{-}303)$$

and where

$$\delta(x) = 0, \qquad x \neq 0. \qquad (2\text{-}304)$$

Thus, we see that the δ-function is strangely behaved indeed, for it is zero everywhere in space, except at $x = 0$. At that point, it goes to infinity in such a manner that the integral is unity. Clearly, its behavior at $x = 0$ is responsible for it being excluded from consideration as an ordinary function. Nevertheless, the δ-function and its properties will be quite useful to us.

---

[34] This "function" was introduced by P. A. M. Dirac, *Quantum Mechanics*, Oxford University Press, 1958, pp. 58–62. See also E. Meibacher, *Quantum Mechanics*, 2nd ed., Wiley, New York, 1970, pp. 82–85.

Several properties[35] of the δ-function can be seen directly from the formal definition in Eqs. (2-303) and (2-304). For example,

$$\int_{-\infty}^{+\infty} f(x)\delta(x)\,dx = f(0). \tag{2-305}$$

The establishment of Eq. (2-305) from the definition of $\delta(x)$ is obvious. Also,

$$\delta(x) = \delta(-x), \tag{2-306}$$

$$x\delta(x) = 0, \tag{2-307}$$

$$\delta(ax) = a^{-1}\delta(x), \tag{2-308}$$

$$\int \delta(a-x)\delta(x-b)\,dx = \delta(a-b). \tag{2-309}$$

It should be emphasized that the relations given in Eqs. (2-306)–(2-309) take on meaning only when both sides are multiplied by a conventional function $f(x)$ and integrated over dx. As an example of how these relations can be established, consider Eq. (2-309) when multiplied by $f(y)$ and integrated:

$$\int f(y)\,dy \int \delta(y-x)\delta(x-b)\,dx = \int \delta(x-b)\,dx \int f(y)\delta(y-x)\,dy, \tag{2-310}$$

$$= \int \delta(x-b)f(x)\,dx, \tag{2-311}$$

$$= f(b), \tag{2-312}$$

$$= \int f(y)\delta(y-b)\,dy, \tag{2-313}$$

where we have interchanged the order of integrations in Eq. (2-310) and used Eq. (2-307) when needed. The verifications of Eqs. (2-306)–(2-308) are left to the student. Thus, any of the quantities on the left-hand side of Eqs. (2-306)–(2-309) can always be replaced by the quantities on the right, as long as it is followed by multiplication by a conventional function and integration.

It is possible to obtain another representation of the δ-function by consideration of the Fourier integral formula of Eq. (2-294), which can be written as

$$f(x) = \int_{-\infty}^{+\infty} \left[\frac{1}{2\pi}\int_{-\infty}^{+\infty} e^{ik(x-x')}\,dk\right] f(x')\,dx'. \tag{2-314}$$

Thus, we see that the quantity in brackets has precisely the same property as the δ-function; and we write

$$\delta(x-x') = \frac{1}{2\pi}\int_{-\infty}^{+\infty} e^{ik(x-x')}\,dk, \tag{2-315}$$

which is a representation of the δ-function that will be useful later.

---

[35] Proper proofs of these properties can be given using Distribution Theory (cf. A. Messiah, *Quantum Mechanics*, Vol. 1, North-Holland Publishing Co., Amsterdam, 1965, Appendix A). The "proofs" given here are strictly formal proofs.

The above relation also leads us to the generalization of the notion of orthonormality to the case of continuous indices. If l and m are two (continuous) indices, then the functions $f(l, x)$ and $f(m, x)$ are said to be orthonormal if

$$\int_{-\infty}^{+\infty} f^*(l, x) f(m, x)\, dx = \delta(l-m). \tag{2-316}$$

We have already seen that the basis set $(1/\sqrt{2\pi})e^{ikx}$ is an orthonormal basis, since

$$\int_{-\infty}^{+\infty} e^{-ik'x} e^{+ik''x}\, dx = \delta(k'-k''), \tag{2-317}$$

and the expansion of an arbitrary function $f(x)$ in terms of this basis is given in Eq. (2-302). Also, it is apparent from Eq. (2-309) that the basis set $\delta(x - x')$, where x' denotes different members of the set, is orthonormal, since

$$\int \delta(x-x')\delta(x-x'')\, dx = \delta(x'-x''). \tag{2-318}$$

This basis set can also be used for the expansion of an arbitrary function $f(x)$, since

$$f(x) = \int_{-\infty}^{+\infty} f(x')\delta(x-x')\, dx'. \tag{2-319}$$

We have now extended the notion of orthonormality to the case of a continuous basis using the δ-function, and need only to carry out the same extension to the notion of a scalar product to complete the generalization.

In function space, we saw that all scalar products are defined and finite. In particular, the expansion of an arbitrary function $f(x)$ in the orthonormal basis $\phi_1(x)$, $\phi_2(x)$, ... is given by

$$f(x) \doteq \sum_{k=1}^{\infty} c_k \phi_k(x), \tag{2-320}$$

where the coefficient c_k is a generalized Fourier coefficient, i.e.,

$$c_k = (\phi_k, f). \tag{2-231}$$

For the case of a continuous basis $\phi(k, x)$, the expansion of an arbitrary function $f(x)$ is given by

$$f(x) = \int c(k)\phi(k, x)\, dk, \tag{2-322}$$

and the coefficients are obtained by forming the scalar product of $\phi^*(k, x)$ and $f(x)$, i.e.,

$$\begin{aligned} c(k) &= [\phi(k), f] \\ &= \int \phi^*(k, x) f(x)\, dx. \end{aligned} \tag{2-323}$$

We shall used the notation (f, g) for a scalar product throughout, and it can apply to functions and vectors both in function space and in the continuum.

2-10. Function Minimization with Constraints

In our study of quantum mechanics, especially in the determination of approximate solutions to the Schrodinger equation, we shall frequently be faced with the need to find the extremum of a function. Furthermore, one or more constraints on the function may need to be satisfied at the same time. Fortunately, there is a method that is conveniently applied in those cases, known as *Lagrange's Method of Undetermined Multipliers.*

To illustrate how this method works, we shall begin with a simple example. Suppose we wish to find the extremum of

$$f(x, y, z) = 3x + 4y^2 + 5(z-2)^2, \tag{2-324}$$

but subject to the following constraint:

$$g(k, y, z) = x - y + 1 = 0. \tag{2-325}$$

One way to solve this problem is to use Eq. (2-325) to solve for x in terms of y, and substitute the result back into Eq. (2-324), giving

$$f = 3(y-1) + 4y^2 + 5(z-2)^2. \tag{2-326}$$

Now we have a function of two variables instead of three which incorporates the constraint, and the extremum of f can be found from

$$\frac{\partial f}{\partial y} = 0, \qquad \frac{\partial f}{\partial z} = 0, \tag{2-327}$$

which yields

$$y = -3/8, \qquad z = 2, \qquad x = -7/8. \tag{2-328}$$

However, it is not always easy or even possible to eliminate variable(s) by use of the constraint equation(s) [Eq. (2-325) in the above example] and find the extremum via procedures such as given in Eq. (2-327). In such cases an alternate procedure is needed, and the Lagrange Method of Undetermined Multipliers provides a convenient and effective alternative way to accomplish the same goal.

To introduce this method, let us consider a slightly generalized version of the previous example. In particular, let us consider a function $f(x, y, z)$ and how the extrema of f can be found. The general condition for an extrema of f is

$$df = 0 = \left(\frac{\partial f}{\partial x}\right) dx + \left(\frac{\partial f}{\partial y}\right) dy + \left(\frac{\partial f}{\partial z}\right) dz. \tag{2-329}$$

If there are no further constraints, the necessary and sufficient conditions for determining extrema of f are

$$\frac{\partial f}{\partial x}=0, \qquad \frac{\partial f}{\partial y}=0, \qquad \frac{\partial f}{\partial z}=0. \tag{2-330}$$

If one or more constraints is to be applied while finding the extrema of f, then the variables will no longer be independent, but will be related in some manner. We can express this in general as

$$g(x, y, z)=0, \tag{2-331}$$

even though this equation may not be solvable explicitly for one variable in terms of the others.

However, if we form

$$dg=0=\left(\frac{\partial g}{\partial x}\right) dx+\left(\frac{\partial g}{\partial y}\right) dy+\left(\frac{\partial g}{\partial z}\right) dz. \tag{2-332}$$

followed by multiplication of Eq. (2-332) by an arbitrary constant (λ) and addition to Eq. (2-329), we obtain

$$0=\left[\frac{\partial f}{\partial x}+\lambda \frac{\partial g}{\partial x}\right] dx+\left[\frac{\partial f}{\partial y}+\lambda \frac{\partial g}{\partial y}\right] dy+\left[\frac{\partial f}{\partial z}+\lambda \frac{\partial g}{\partial z}\right] dz. \tag{2-333}$$

Thus, we have now incorporated the constraint into the process of finding the extrema of f as desired.

Let us now make use of the fact that λ is an arbitrary constant, and choose the value of λ such that[36]

$$\frac{\partial f}{\partial x}+\lambda \frac{\partial g}{\partial x}=0. \tag{2-334}$$

If such a value of L is found we have, in effect, eliminated one of the variables, and the extrema of f can be found from solving only the remaining two equations, i.e.,

$$\frac{\partial f}{\partial y}+\lambda \frac{\partial g}{\partial y}=0, \tag{2-335}$$

and

$$\frac{\partial f}{\partial z}+\lambda \frac{\partial g}{\partial z}=0. \tag{2-336}$$

For the particular example of Eqs. (2-324) and (2-325) given earlier, application of Eq. (2-334) gives

$$\lambda=-3,$$

---

[36] We assume here that $\frac{\partial g}{\partial x} \neq 0$.

and substitution of this result into Eqs. (2-335) and (2-336) gives the same results as obtained earlier [Eq. (2-238)].

This process is not restricted to the use of a single parameter (λ) to satisfy a single constraint. For example, if we have a function containing at least n variables, and if there are n constraints[37] that are to be applied while finding the extremum of f, i.e.,

$$\begin{aligned} g_1 &= 0 \\ g_2 &= 0 \\ &\vdots \\ g_n &= 0, \end{aligned} \tag{2-337}$$

then finding the extrema of f subject to these constraints can be written as

$$d[f + \lambda_1 g_1 + \lambda_2 g_2 + \cdots + \lambda_n g_n] = 0, \tag{2-338}$$

where $\lambda_1, \lambda_2, \ldots, \lambda_n$ are arbitrary constants (called *Lagrangian Multipliers*) that are chosen so as to guarantee satisfaction of the constraints.[38]

As we shall see in applications in later chapters,[39] it is not necessary in general to determine the actual values of the constants (λ_j). It is only necessary to know that there exists a value of the constant that will allow satisfaction of the equation in which it appears. Thus, in practice these constants are undetermined, giving the method its name.

2-11. Linear Operators

In the preceding sections we have discussed vector spaces at some length, and developed the idea of how an arbitrary vector in a vector space can always be expressed as a linear combination of a set of basis vectors of the space. We will now take up the question of how these vectors may be "changed" into new vectors. Our considerations will be done much more conveniently if we give a name to these operations that "change" vectors, and we shall define an *operator* as a set of instructions (which is written as a symbol) that changes or transforms the vector that appears to the right of it into another vector. This definition can be written symbolically as

$$\boldsymbol{\beta} = \mathcal{A}\boldsymbol{\alpha} \tag{2-339}$$

where $\mathcal{A}$ is an operator that will transform $\boldsymbol{\alpha}$ into another vector $\boldsymbol{\beta}$, that is in

---

[37] These constraints can take various forms. For example, an integral constraint on a component (f_i) of f might be written as

$$g_j = \int f_j \, d\tau - c_j = 0$$

where c_i is a constant.

[38] For additional discussion, see G. Arfken, *Mathematical Methods for Physicists*, Academic Press, New York, 1968, Sect. 17.6.

[39] See, for example, Section 12-2.

general not in the same space as $\boldsymbol{\alpha}$. We shall call the object on which $\mathcal{A}$ operates the *operand*.

In more precise mathematical terms, an operator $\mathcal{A}$ will be taken to be a rule (or set of instructions) by means of which the elements of one vector space V can be *mapped* or *transformed* into the elements of another vector space V'. This can be written symbolically as

$$\mathcal{A} : \{\alpha\} \rightarrow \{\beta\}, \tag{2-340}$$

where $\boldsymbol{\alpha}$ and $\boldsymbol{\beta}$ are vectors in V and V', respectively. Another way of stating this idea is by the use of an equation, i.e., as shown in Eq. (2-339). This can be thought of as a generalization of the "function" concept, $y = f(x)$, where the independent variable, x, is mapped into the dependent variable y.

The notion of an operator is certainly not a new one, for there are many common examples. For example, the symbol "log" means to change the function $f(x)$ into a new function $g(x) = \log f(x)$. Similarly, taking the square root of a function is an operator which is usually denoted by $\sqrt{\ }$. We can express this fact in the notation of Eq. (2-339) by letting "sqrt" denote the operation of taking the square root of the function that appears to the right of it. Thus,

$$g(x) = \text{sqrt}[f(x)] \tag{2-341}$$

is equivalent to the more familiar notation $g(x) = \sqrt{f(x)}$, but emphasizes the operator concept.

Although perhaps obvious, it should be pointed out that operators have meaning only for certain classes of operands. For example, d/dx denotes "take the derivative of," and is meaningful only if it operates on differentiable functions. Of course, an operator need not be as complicated as those just mentioned. For example, if b is a number, the relation

$$g(x) = bf(x)$$

illustrates the operation of transforming $f(x)$ into $g(x)$ by means of multiplication by b. Thus, "multiplication by b" is an operator, and is denoted simply by the symbol b.

An important class of operators to quantum mechanics is called *linear operators*, and they can be characterized in the following manner. An operator $\mathcal{L}$ is said to be a linear operator if the following two conditions are satisfied for any vectors in the space:

$$\mathcal{L}(\boldsymbol{\alpha}_i + \boldsymbol{\alpha}_j) = \mathcal{L}\boldsymbol{\alpha}_i + \mathcal{L}\boldsymbol{\alpha}_j, \tag{2-342}$$

$$\mathcal{L}(c\boldsymbol{\alpha}_i) = c(\mathcal{L}\boldsymbol{\alpha}_i), \tag{2-343}$$

where c is a complex scalar.

Now let us reconsider the examples of operators introduced earlier in the light of these ideas. If we are dealing with continuous functions $f(x)$ and $g(x)$, the linearity of the operator $\mathfrak{D} = d/dx$ is established as follows, i.e.,

$$\mathfrak{D}[f(x)+g(x)]=\mathfrak{D}f(x)+\mathfrak{D}g(x), \tag{2-344}$$

$$\mathfrak{D}[cf(x)]=c[\mathfrak{D}f(x)]. \tag{2-345}$$

This is in agreement with the results of elementary differential calculus.

For the example $\mathcal{S} = \sqrt{}$, the following shows that $\mathcal{S}$ is not linear, i.e.,

$$\mathcal{S}(a+b)\neq \mathcal{S}a+\mathcal{S}b, \tag{2-346}$$

since

$$\sqrt{a+b}\neq\sqrt{a}+\sqrt{b}. \tag{2-347}$$

In most of our studies to follow, we will confine our attention to linear mappings of a restricted type:

$$\mathfrak{L} : V\rightarrow V, \tag{2-348}$$

i.e., transformations which map a linear space onto itself or onto a subspace of itself.[40]

There are two elementary operators of particular importance, viz. the *identity* operator $\mathfrak{I}$, and the *null* (*or zero*) operator ϑ, defined by

$$\mathfrak{I}\boldsymbol{\alpha}=\boldsymbol{\alpha}, \tag{2-349}$$

and

$$\vartheta\boldsymbol{\alpha}=\mathbf{0}. \tag{2-350}$$

These operators are equivalent to multiplication by the scalars 1 and 0, respectively.

As another example, consider the space of two dimensional geometric vectors. The types of mappings $\mathfrak{L}:V \rightarrow V$ that are of interest in this case are summarized below:

1. The rotation of every vector through the same angle about the origin.
2. The multiplication of every vector by a real number c.
3. The reflection of every vector with respect to a fixed line l through the origin.
4. The projection of every vector on a fixed line through the origin.

Figure 2.6 shows one of the geometric vectors $\mathbf{v}$ and its images $\mathfrak{L}\mathbf{v}$ corresponding to the above operations (i.e., mappings).

Another concept of great importance is that of an *inverse operator* $\mathfrak{L}^{-1}$. To illustrate this notion first by means of an example, consider the following equation for ordinary functions:

$$yx=1. \tag{2-351}$$

---

[40] More general mappings of the form $\mathfrak{A}:V \rightarrow V'$ are discussed in most advanced books in linear algebra, see, e.g., K. Hoffman and R. Kunze, *Linear Algebra*, Prentice-Hall, Englewood Cliffs, NJ, 1961.

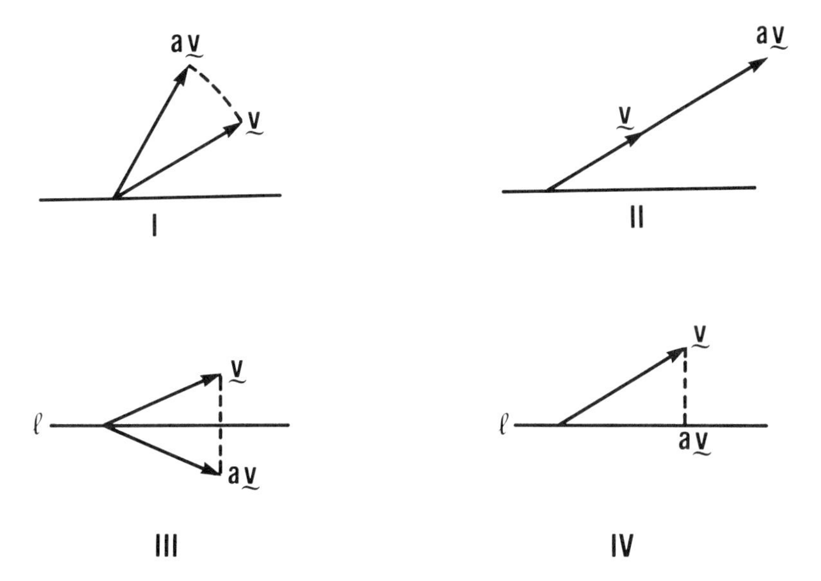

Figure 2.6. Depiction of a vector (**v**) and the result of rotation, multiplication by a constant, reflection, and projection operations.

The y that solves this equation is obviously

$$y = 1/x. \tag{2-352}$$

Thus, Eq. (2-352) has a solution whenever the function $(1/x)$ is defined, which is for all $x \neq 0$.

More formally, the inverse operator is connected with the inverse mapping:

$$\mathcal{L}^{-1}\colon\ \{\alpha\} \leftarrow \{\beta\} \tag{2-353}$$

where $\{\alpha\}$, $\{\beta\}$ are in V. This is usually denoted by

$$\alpha = \mathcal{L}^{-1}\beta. \tag{2-354}$$

It is left as an exercise to show that if $\mathcal{L}$ is a linear operator, then $\mathcal{L}^{-1}$ is also a linear operator. Proof of the existence of $\mathcal{L}^{-1}$ can be found in most textbooks on linear algebra.[41]

2-12. Algebra of Linear Operators

As we shall see, it is possible to manipulate linear operators themselves, apart from their operands. These manipulations lead to the development of an *algebra of linear operators*.

[41] See, for example, J. Indritz, *Methods in Analysis*, Macmillan, New York, 1963, p. 22.

We first define the *sum of two linear operators* $\mathcal{L}_1$ and $\mathcal{L}_2$ as

$$\mathcal{L}_1 + \mathcal{L}_2 = \mathcal{L}, \tag{2-355}$$

if for every operand Ω for which $\mathcal{L}_1$ and $\mathcal{L}_2$ are defined, the following relation holds:

$$(\mathcal{L}_1 + \mathcal{L}_2)\Omega = \mathcal{L}\Omega. \tag{2-356}$$

We can also define the *product of linear operators*. The relation $\mathcal{L}_1\mathcal{L}_2\Omega$ has the meaning that the operator $\mathcal{L}_2$ operates on Ω, and the operator $\mathcal{L}_1$ operates on $\mathcal{L}_2\Omega$, the result of the previous operation. Explicitly, we can write

$$\mathcal{L}_1\mathcal{L}_2\Omega = \mathcal{L}_1(\mathcal{L}_2\Omega), \tag{2-357}$$

if $\mathcal{L}_1\mathcal{L}_2\Omega$ is defined for all operands in the space. We may consider $\mathcal{L}_1\mathcal{L}_2$ to be a single operation instead of two separate operations. If we give this single operation to the symbol $\mathcal{L}$, then we can write the operator relation

$$\mathcal{L} = \mathcal{L}_1\mathcal{L}_2. \tag{2-358}$$

One point needs to be emphasized before continuing. Although we can manipulate operators in the algebraic fashion of Eqs. (2-355) and (2-358), an associated operand on which these operators can act is always implied. If there is doubt about a particular algebraic manipulation with operators, an arbitrary operand should be inserted and the equality verified. To illustrate this, consider the operators

$$\mathfrak{D} = \frac{d}{dx}, \qquad \mathfrak{A} = x.$$

Then

$$\mathfrak{A}(\mathfrak{D}\Omega) = x\left(\frac{d\Omega}{dx}\right). \tag{2-359}$$

If operator algebra were the same as ordinary algebra, we would expect that $\mathfrak{A}\mathfrak{D}\Omega = \mathfrak{D}\mathfrak{A}\Omega$. However,

$$\mathfrak{D}(\mathfrak{A}\Omega) = \frac{d}{dx}(x\Omega) = \Omega + x\left(\frac{d\Omega}{dx}\right) \neq \mathfrak{A}(\mathfrak{D}\Omega). \tag{2-360}$$

In operator form, the above equation can be written as

$$\frac{d}{dx}x = x\frac{d}{dx} + 1. \tag{2-361}$$

The pitfall to be avoided using operator equations is now apparent, for the term on the left of Eq. (2-361) means "multiply by x, and then take the derivative with respect to ." It does not mean "take the derivative of the function x," which would ignore the implied operand. Thus, we see that the order in which operators are written as exceedingly important, and in this case $\mathfrak{D}\mathfrak{A} \neq \mathfrak{A}\mathfrak{D}$.

If it does happen to turn out that

$$\mathcal{L}_1(\mathcal{L}_2\Omega)=\mathcal{L}_2(\mathcal{L}_1\Omega), \tag{2-362}$$

for all operands for which $\mathcal{L}_1$ and $\mathcal{L}_2$ have meaning, then the operators $\mathcal{L}_1$ and $\mathcal{L}_2$ are said to *commute*. In this connection, it is common to speak of the *commutator of two operators* (or *commutator bracket*) as

$$[\mathcal{L}_1,\ \mathcal{L}_2]=\mathcal{L}_1\mathcal{L}_2-\mathcal{L}_2\mathcal{L}_1. \tag{2-363}$$

Thus, in order to manipulate quantities using a noncommutative algebra, it is necessary to know the commutators for all pairs of operators. In the example given above, the commutator of d/dx and x is 1, i.e.,

$$\left[\frac{d}{dx},\ x\right]=\frac{d}{dx}x-x\frac{d}{dx}=1. \tag{2-364}$$

Now that we have a definition of the product of operators, we can speak of the *powers of an operator*. Thus, if we define

$$\mathcal{L}^2 \equiv \mathcal{L}\mathcal{L}, \tag{2-365}$$

then

$$\mathcal{L}^2\Omega=\mathcal{L}\mathcal{L}\Omega. \tag{2-366}$$

In a similar manner we can speak of higher powers of an opertor, such as

$$\mathcal{L}^n=\mathcal{L}\mathcal{L}\ \ldots\ \mathcal{L}, \tag{2-367}$$

where the product has been repeated n times.

Since both the sum and powers of an operator are acceptable, we can also conceive of combining these ideas to form the *power series of an operator*,

$$\mathcal{L}^0+\mathcal{L}^1+\mathcal{L}^2+\cdots+\mathcal{L}^n+\cdots. \tag{2-368}$$

By convention, we take

$$\mathcal{L}^0 \equiv \mathcal{I}. \tag{2-369}$$

Taking this idea a step further, we can multiply each quantity in Eq. (2-368) by a constant:

$$e^{\mathcal{L}} \equiv \mathcal{I}+\mathcal{L}+\left(\frac{1}{2!}\right)\mathcal{L}^2+\left(\frac{1}{3!}\right)\mathcal{L}^3+\cdots$$

where the symbol $e^{\mathcal{L}}$ has been given to represent the series of operators, by analogy to the power series expansion of the ordinary function e^X. By proceeding in this manner, we see that it appears reasonable to talk of functions of operators. However, whenever we write $f(\mathcal{L})$, it is understood that there must be some well-defined power series representation of it. This allows us to use the power series representation to define the operator $f'(\mathcal{L})$. For example, if

$$f(\mathcal{L})=\mathcal{L}^n \qquad (n\geq 0), \tag{2-370}$$

then we can define

$$f'(\mathfrak{L}) = n\mathfrak{L}^{n-1}. \tag{2-371}$$

In general, if

$$f(\mathfrak{L}) = \sum_{k=0}^{\infty} a_k \mathfrak{L}^k, \tag{2-372}$$

where the a_k are scalars, then

$$f'(\mathfrak{L}) = \sum_{k=0}^{\infty} a_k k \mathfrak{L}^{k-1}. \tag{2-373}$$

To illustrate these ideas of commutators and functions of operators, let us consider $\mathfrak{D} = d/dx$ and $\mathfrak{A} = x$ once again. As we have seen, $[\mathfrak{D}, \mathfrak{A}] = 1$. Suppose we now wish to find the commutator $[\mathfrak{D}, \mathfrak{A}^2]$. Using the definition of a commutator, we have

$$\begin{aligned}\mathfrak{D}\mathfrak{A}^2 &= (\mathfrak{D}\mathfrak{A})\mathfrak{A} = ([\mathfrak{D}, \mathfrak{A}] + \mathfrak{A}\mathfrak{D})\mathfrak{A} \\ &= (1 + \mathfrak{A}\mathfrak{D})\mathfrak{A} = \mathfrak{A} + \mathfrak{A}(\mathfrak{D}\mathfrak{A}) \\ &= \mathfrak{A} + \mathfrak{A}(1 + \mathfrak{A}\mathfrak{D}) \\ &= 2\mathfrak{A} + \mathfrak{A}^2\mathfrak{D}.\end{aligned} \tag{2-374}$$

Thus,

$$[\mathfrak{D}, \mathfrak{A}^2] \equiv \mathfrak{D}\mathfrak{A}^2 - \mathfrak{A}^2\mathfrak{D} = 2\mathfrak{A}. \tag{2-375}$$

In a similar manner, it can be easily verified that

$$[\mathfrak{D}, \mathfrak{A}^3] = 3\mathfrak{A}^2, \tag{2-376}$$

or, in general

$$[\mathfrak{D}, \mathfrak{A}^n] = n\mathfrak{A}^{n-1}. \tag{2-377}$$

Finally, using the commutator of Eq. (2-337) we obtain the more general result that

$$[\mathfrak{D}, f(\mathfrak{A})] = f'(\mathfrak{A}), \tag{2-378}$$

where $f(\mathfrak{A})$ is a power series in the operator $\mathfrak{A} \equiv x$. To verify this, we first establish that commutators themselves obey a *linear algebra*. For arbitrary linear operators, $\mathfrak{L}_1$, $\mathfrak{L}_2$, and $\mathfrak{L}_3$, it is easily seen that

$$[\mathfrak{L}_1, \mathfrak{L}_2 + \mathfrak{L}_3] = [\mathfrak{L}_1, \mathfrak{L}_2] + [\mathfrak{L}_1, \mathfrak{L}_3], \tag{2-379}$$

and

$$[\mathfrak{L}_1, c\mathfrak{L}_2] = c[\mathfrak{L}_1, \mathfrak{L}_2], \tag{2-380}$$

where c is an arbitrary scalar. Using this result, we can write

$$[\mathfrak{D}, f(\mathfrak{A})] = \left[\mathfrak{D}, \sum_{k=0}^{\infty} b_k \mathfrak{A}^k\right], \tag{2-381}$$

or

$$[\mathfrak{D}, f(\mathfrak{A})] = \sum_{n=0}^{\infty} b_n [\mathfrak{D}, \mathfrak{A}^n] = \sum_{n=0}^{\infty} b_n n \mathfrak{A}^{n-1}, \tag{2-382}$$

from which Eq. (2-378) follows.

2-13. Special Kinds of Linear Operators

We shall now discuss some linear operators that are defined on inner product spaces. Some of the properties of these operators will be examined here, although many important properties as well as several additional kinds of special operators will not be discussed until Chapter 3.

Suppose we consider the three vectors **u**, **v**, and **w**, which belong to an inner product space V'. If **v** and **w** are related by the linear operator $\mathfrak{L}$, i.e.

$$\mathbf{v} = \mathfrak{L}\mathbf{w} \tag{2-383}$$

then the scalar product of **u** and **v** can be written as

$$(\mathbf{u}, \mathbf{v}) = (\mathbf{u}, \mathfrak{L}\mathbf{w}). \tag{2-384}$$

As an example, suppose the vectors **u** and **w** are functions, $f(x)$ and $g(x)$. In this case, the scalar product would be written simply as

$$(f, \mathfrak{L}g) = \int_a^b f^*(x)[\mathfrak{L}g(x)]\, dx, \tag{2-385}$$

where the square brackets have been added only to emphasize that $\mathfrak{L}$ operates only on $g(x)$.

In connection with scalar products of this type, it is useful to define the *adjoint operator of* $\mathfrak{L}$, which is denoted by $\mathfrak{L}^\dagger$, by the following equation:

$$(\mathfrak{L}\mathbf{u}, \mathbf{v}) = (\mathbf{u}, \mathfrak{L}^\dagger \mathbf{v}) \tag{2-386}$$

$$= (\mathbf{v}, \mathfrak{L}\mathbf{u})^*. \tag{2-387}$$

In other words, $\mathfrak{L}$ and $\mathfrak{L}^\dagger$ are adjoint operators if their scalar products are complex conjugates of each other. Also, we see from Eq. (2-386) that the operator $\mathfrak{L}$ can operate on either operand of the scalar product, but the adjoint of $\mathfrak{L}$ is needed if we wish to change the operand on which the operator operates. Due to the form of Eq. (2-387), the definition of the adjoint is often referred to as the "*turnover rule.*" Also note that from the definition, the adjoint of $\mathfrak{L}^\dagger$ is $\mathfrak{L}$.

To illustrate the concept of an operator and its adjoint with a simple example, consider the following operator in one-dimension ($-\infty \le x \le +\infty$),

$$\mathcal{L} = \left(\frac{\hbar}{i}\right)\frac{\partial}{\partial x},$$

along with two functions

$$u_1 = e^{-a_1 x}, \qquad u_2 = e^{-a_2 x},$$

where a_1 and a_2 are constants. We construct

$$\begin{aligned} I &= (\mathcal{L}u_1, u_2) \\ &= \int_{-\infty}^{\infty} \left(\frac{\hbar}{i}\frac{\partial}{\partial x} e^{-a_1 x}\right)^* e^{-a_2 x}\, dx \\ &= a_1 \left(\frac{\hbar}{i}\right) \int_{-\infty}^{+\infty} e^{-(a_1+a_2)x}\, dx. \end{aligned}$$

On the other hand, let us construct

$$\begin{aligned} II &= (u_2, \mathcal{L}u_1) \\ &= \int_{-\infty}^{+\infty} (e^{-a_2 x})^* \left(\frac{\hbar}{i}\frac{\partial}{\partial x} e^{-a_1 x}\right) dx \\ &= -a_1 \left(\frac{\hbar}{i}\right) \int_{-\infty}^{\infty} e^{-(a_1+a_2)x}\, dx = I^*. \end{aligned}$$

Thus, we see that, for this simple example, $\mathcal{L} = \mathcal{L}^\dagger$, i.e., $\mathcal{L}^\dagger$ is equal to $\mathcal{L}$ itself.

To illustrate several general properties of adjoint operators, let us consider the following theorems. Let $\mathcal{L}_1$ and $\mathcal{L}_2$ be linear operators with scalar products defined with these operators and operands $\mathbf{u}$ and $\mathbf{v}$.

Theorem 2-10.

$$(\mathcal{L}_1 + \mathcal{L}_2)^\dagger = \mathcal{L}_1^\dagger + \mathcal{L}_2^\dagger \tag{2-388}$$

or

$$(\mathbf{u}, \{\mathcal{L}_1 + \mathcal{L}_2\}^\dagger \mathbf{v}) = (\mathbf{u}, \{\mathcal{L}_1^\dagger + \mathcal{L}_2^\dagger\}\mathbf{v}). \tag{2-389}$$

The proof of this theorem is left to the student.

Theorem 2-11.

$$(\mathcal{L}_1\mathcal{L}_2)^\dagger = \mathcal{L}_2^\dagger \mathcal{L}_1^\dagger. \tag{2-390}$$

Proof:

$$\begin{aligned} (\mathbf{u}, \{\mathcal{L}_1\mathcal{L}_2\}^\dagger \mathbf{v}) &= (\mathbf{v}, \mathcal{L}_1\mathcal{L}_2\mathbf{u})^* \\ &= (\mathcal{L}_1\mathcal{L}_2\mathbf{u}, \mathbf{v}). \end{aligned}$$

Using the turnover rule twice on the operators $\mathcal{L}_1$ and $\mathcal{L}_2$ separately, we obtain

$$\begin{aligned}(\mathcal{L}_1\mathcal{L}_2\mathbf{u},\ \mathbf{v}) &= (\mathcal{L}_2\mathbf{u},\ \mathcal{L}_1^\dagger\mathbf{v}) \\ &= (\mathbf{u},\ \mathcal{L}_2^\dagger\mathcal{L}_1^\dagger\mathbf{v}).\end{aligned} \tag{2-391}$$

which is the desired result.

A very important special operator is the one in which

$$\mathcal{A} = \mathcal{A}^\dagger. \tag{2-392}$$

In this case $\mathcal{A}$ is called a *self-adjoint* or *Hermitian operator*, and we have just seen a simple example of this kind of operator in the discussion following Eq. (2-387).

Thus, a Hermitian operator has the property that

$$(\mathbf{u},\ \mathcal{A}^\dagger\mathbf{v}) = (\mathbf{u},\ \mathcal{A}\mathbf{v}) \tag{2-393}$$

$$= (\mathcal{A}\mathbf{u},\ \mathbf{v}). \tag{2-394}$$

To show how an operator can be determined to be Hermitian or not, let us take the example:

$$\mathcal{L}_x = -i\hbar\left(y\frac{\partial}{\partial z} - z\frac{\partial}{\partial y}\right). \tag{2-395}$$

This operator will be encountered often in our study of quantum mechanics, and is the operator associated with the x-component of angular momentum. Then if we restrict our discussion to quadratically integrable functions $\Psi(x, y, z)$ only,[42] we have

$$\int_{-\infty}^{+\infty}\int_{-\infty}^{+\infty}\int_{-\infty}^{+\infty} \Psi^*(x, y, z)\Psi(x, y, z)\, dx\, dy\, dz < \infty.$$

Thus, it is necessary that $\Psi(x, y, z)$ vanish if any coordinate goes to infinity in order to guarantee quadratic integrability. Let us consider two such functions, Ψ_1 and Ψ_2, and write $dV = dx\, dy\, dz$. Then,

$$\begin{aligned}\int (\mathcal{L}_x\psi_1)^*\psi_2\, dV &= -\left(\frac{\hbar}{i}\right)\int\left(y\frac{\partial\psi_1^*}{\partial z} - z\frac{\partial\psi_1^*}{\partial y}\right)\psi_2\, dV \\ &= -\int(\mathcal{L}_x\psi_1^*)\psi_2\, dV.\end{aligned} \tag{2-396}$$

Also, since $\mathcal{L}_x$ is simply a sum of differentiable operators,

$$\mathcal{L}_x(\psi_1^*\psi_2) = \psi_2(\mathcal{L}_x\psi_1^*) + \psi_1^*(\mathcal{L}_x\psi_2). \tag{2-397}$$

[42] Such a constraint is frequently required in quantum mechanical applications, as discussed in Chapter 5, Section 5-2.

Using Eq. (2-397), we may rewrite Eq. (2-396) as

$$\int (\mathcal{L}_x\psi_1)^*\psi_2\, dV=\int \psi_1^*(\mathcal{L}_x\psi_2)\, dV-\int \mathcal{L}_x(\psi_1^*\psi_2)\, dV. \tag{2-398}$$

The last integral in the above equation is

$$\begin{aligned}
&\left(\frac{\hbar}{i}\right)\int_{-\infty}^{\infty}\int_{-\infty}^{\infty}\int_{-\infty}^{\infty} dx\, dy\, dz\left(y\frac{\partial}{\partial z}-z\frac{\partial}{\partial y}\right)(\psi_1^*\psi_2)\\
&=\left(\frac{\hbar}{i}\right)\int_{-\infty}^{\infty} dx\int_{-\infty}^{\infty} y\, dy\int_{-\infty}^{\infty}\frac{\partial}{\partial z}(\psi_1^*\psi_2)\, dz\\
&\quad-\left(\frac{\hbar}{i}\right)\int_{-\infty}^{\infty} dx\int_{-\infty}^{\infty} z\, dz\int_{-\infty}^{\infty}\frac{\partial}{\partial y}(\psi_1^*\psi_2)\, dy\\
&=\left(\frac{\hbar}{i}\right)\int_{-\infty}^{\infty} dx\int_{-\infty}^{\infty} dy\, y[\psi_1^*\psi_2]\Big|_{z=-\infty}^{z=\infty}\\
&\quad-\left(\frac{\hbar}{i}\right)\int_{-\infty}^{\infty} dx\int_{-\infty}^{\infty} z\, dz[\psi_1^*\psi_2]\Big|_{y=-\infty}^{y=+\infty}.
\end{aligned} \tag{2-399}$$

Since ψ_1 and ψ_2 are both quadratically integrable functions, both of the integrals in Eq. (2-399) must be zero since ψ_1 and ψ_2 must vanish as y or $z = \pm\infty$. Thus,

$$\int (\mathcal{L}_x\psi_1)^*\psi_2\, dV=\int \psi_1^*(\mathcal{L}_x\psi_2)\, dV, \tag{2-400}$$

and $\mathcal{L}_x$ is a Hermitian operator.

Theorem 2-12. Every operator $\mathcal{L}$ can be written in a unique way as

$$\mathcal{L}=\mathcal{L}_1+i\mathcal{L}_2 \tag{2-401}$$

where $\mathcal{L}_1$ and $\mathcal{L}_2$ are Hermitian operators.

Proof: Defining

$$\mathcal{L}_1=\frac{1}{2}(\mathcal{L}+\mathcal{L}^\dagger);\qquad \mathcal{L}_2=\frac{1}{2i}(\mathcal{L}-\mathcal{L}^\dagger), \tag{2-402}$$

Eq. (2-401) is clearly satisfied. It is left as an exercise for the student to show that $\mathcal{L}_1$ and $\mathcal{L}_2$ are Hermitian.

Theorem 2-13. A necessary and sufficient condition for the product of two Hermitian operators $\mathcal{L}_1$ and $\mathcal{L}_2$ to be Hermitian is that $\mathcal{L}_1$ and $\mathcal{L}_2$ commute.

Proof: Assume

$$[\mathcal{L}_1,\ \mathcal{L}_2]=0.$$

$$\mathcal{L}_1\mathcal{L}_2=\mathcal{L}_2\mathcal{L}_1=\mathcal{L}_2^\dagger\mathcal{L}_1^\dagger=(\mathcal{L}_1\mathcal{L}_2)^\dagger.$$

To prove the other half of the theorem, assume:

$$\mathcal{L}_1\mathcal{L}_2=(\mathcal{L}_1\mathcal{L}_2)^\dagger.$$

Then,

$$\mathcal{L}_1\mathcal{L}_2=(\mathcal{L}_1\mathcal{L}_2)^\dagger=\mathcal{L}_2^\dagger\mathcal{L}_1^\dagger=\mathcal{L}_2\mathcal{L}_1,$$

which completes the proof.

Another special kind of operator that arises frequently is called a unitary operator, and it is characterized by

$$\mathcal{U}^\dagger=\mathcal{U}^{-1}. \tag{2-403}$$

Thus, for this type of an operator

$$\mathcal{U}^\dagger\mathcal{U}=\mathcal{U}^{-1}\mathcal{U}=\mathcal{U}\mathcal{U}^{-1}=\mathcal{U}\mathcal{U}^\dagger=\mathcal{I}, \tag{2-404}$$

i.e.,

$$\mathcal{U}^\dagger\mathcal{U}=\mathcal{U}\mathcal{U}^\dagger=\mathcal{I}. \tag{2-405}$$

Unitary operators are very interesting, since they have the property that their action on operands leaves *norms* and *angles* (i.e., scalar products) *unchanged*. Consider the normalized operands $\mathbf{u}$ and $\mathbf{v}$ in V', and the unitary operator $\mathcal{U}$:

$$(\mathbf{u},\ \mathbf{u})=(\mathbf{u},\ \mathcal{U}^{-1}\mathcal{U}\mathbf{u}). \tag{2-406}$$

Using Eq. (2-403), we have

$$(\mathbf{u},\ \mathbf{u})=(\mathbf{u},\ \mathcal{U}^\dagger\mathcal{U}\mathbf{u}), \tag{2-407}$$

which, upon applying the turnover rule, yields

$$(\mathbf{u},\ \mathbf{u})=(\mathcal{U}\mathbf{u},\ \mathcal{U}\mathbf{u}) \tag{2-408}$$

and we see that the norm is preserved, i.e., left unchanged. Now consider

$$(\mathbf{u},\ \mathbf{v})=(\mathbf{u},\ \mathcal{U}^\dagger\mathcal{U}\mathbf{v}). \tag{2-409}$$

Again applying the turnover rule yields

$$(\mathbf{u},\ \mathbf{v})=(\mathcal{U}\mathbf{u},\ \mathcal{U}\mathbf{v}), \tag{2-410}$$

and we see that angles are also preserved. An example of such a unitary transformation is given by the rigid rotation of two-dimensional geometric vectors, discussed previously.[43]

Another type of special operator that will prove useful in later discussions is called a *projection operator*. To see how they are defined, consider the arbitrary vector $\mathbf{u}$ in an n-dimensional vector space V. If $\{\mathbf{v}_1,\ \mathbf{v}_2,\ \dots,\ \mathbf{v}_n\}$ is a basis for V, then we can write

$$\mathbf{u}=\sum_{i=1}^{n} c_i\mathbf{v}_i. \tag{2-411}$$

[43] See Fig. 2-6 and associated discussion in Section 2-11.

Now consider the operator defined by the following equation

$$\mathcal{P}_i \mathbf{u} = c_i \mathbf{v}_i, \qquad i = 1, 2, \ldots, n. \tag{2-412}$$

This operator maps the vector $\mathbf{u}$ in V onto a single "component", $c_i\mathbf{v}_i$. Such an operator is called a *projection operator*. It is left as an exercise for the student to show that the set of operators $\{\mathcal{P}_1, \mathcal{P}_2, \ldots, \mathcal{P}_n\}$ are linear.

If we apply $\mathcal{P}_i$ twice, i.e.,

$$\mathcal{P}_i^2 \mathbf{u} = \mathcal{P}_i(\mathcal{P}_i \mathbf{u}) = \mathcal{P}_i(c_i \mathbf{v}_i) = c_i \mathbf{v}_i, \tag{2-413}$$

we see that

$$\mathcal{P}^2 = \mathcal{P}. \tag{2-414}$$

An operator with the property shown in Eq. (2-414) is called an *idempotent* operator. If we apply $\mathcal{P}_i\mathcal{P}_j$ to $\mathbf{u}$, we see that

$$\mathcal{P}_i \mathcal{P}_j \mathbf{u} = \mathcal{P}_i(\mathcal{P}_j \mathbf{u}) = \mathcal{P}_i(c_i \mathbf{v}_j) = \mathbf{0}, \tag{2-415}$$

and thus

$$\mathcal{P}_i \mathcal{P}_j = 0; \qquad i \neq j. \tag{2-416}$$

Hence, the projection operators $\mathcal{P}_i$ and $\mathcal{P}_j$ are said to be *mutually exclusive*.

Since each $\mathcal{P}_i$ projects out the *ith component*, $c_i\mathbf{v}_i$, of $\mathbf{u}$, we can write $\mathbf{u}$ alternatively as

$$\mathbf{u} = \sum_{i=1}^{n} c_i \mathbf{v}_i = \sum_{i=1}^{n} \mathcal{P}_i \mathbf{u} = \left(\sum_{i=1}^{n} \mathcal{P}_i \right) \mathbf{u}, \tag{2-417}$$

or

$$\left(\mathcal{I} - \sum_{i=1}^{n} \mathcal{P}_i \right) \mathbf{u} = \mathbf{0}, \tag{2-418}$$

for every $\mathbf{u}$ in V. Hence the operator $\mathcal{I} - \sum_{i=1}^{n} \mathcal{P}_i$ must be a zero operator, since it maps each $\mathbf{u}$ in V into the zero element. Therefore, we can write

$$\mathcal{I} = \sum_{i=1}^{n} \mathcal{P}_i, \tag{2-419}$$

which is called the *resolution of the identity*.

As an example, consider the three-dimensional Cartesian space, where the position vector is

$$\mathbf{r} = \mathbf{i}x + \mathbf{j}y + \mathbf{k}z. \tag{2-420}$$

If the projection operators are given as $\mathcal{P}_x$, $\mathcal{P}_y$, and $\mathcal{P}_z$, where, for example,

$$\mathcal{P}_x \mathbf{r} = x\mathbf{i}, \tag{2-421}$$

then it is easy to verify that the resolution of the identity is

$$\mathcal{I} = \mathcal{P}_x + \mathcal{P}_y + \mathcal{P}_z. \tag{2-422}$$

Furthermore, it is easy to see that the component projected out by $(\mathcal{I} - \mathcal{P}_i)$, $i = x, y, z$, is always orthogonal to the component projected out by $\mathcal{P}_i$, and it is also idemponent. It is left as an exercise for the student to show that the $\mathcal{P}_i$ are also Hermitian.[44]

Finally, we conclude this section with the introduction of one other special kind of operator. These operators are called *normal operators*, and are defined as those operators that commute with their own adjoint. Thus, the operator $\mathfrak{N}$ is a normal operator if

$$[\mathfrak{N}, \mathfrak{N}^\dagger] = 0. \tag{2-423}$$

We have already seen two examples of normal operators, Hermitian and unitary operators, which are equal to their adjoint, and thus obviously obey Eq. (2-423).

Thus, we see that normal operators can be considered to be a generalization of Hermitian and unitary operators. The reader is referred to the excellent development of normal operators by P. O. Löwdin for further study.[45]

2-14. Eigenvalues and Eigenvectors

In Section 2-11, we observed that an operator is simply a prescription by which an element of a vector space is changed into a new element of that space. In cases that arise frequently, the effect that an operator has on an operand is nothing more than to change the length. In that case, the application of an operator $\mathcal{A}$ to a vector $\mathbf{v}$ produces $(a\mathbf{v})$ where a is a numerical constant, i.e.,

$$\mathcal{A}\mathbf{v} = a\mathbf{v}. \tag{2-424}$$

This particular situation is a very important one, and has been the subject of considerable study. When the operator has the effect on $\mathbf{v}$ as shown in Eq. (2-424), $\mathbf{v}$ is said to be the *eigenfunction* or *eigenvector*[46] of $\mathcal{A}$, and a is said to be the *eigenvalue* of $\mathcal{A}$.

Let us consider an example to clarify the notion. Let us take the operator $\mathcal{A} = d^2/dx^2$, and find its eigenfunctions and eigenvalues. This seemingly difficult

---

[44] For a development of the properties and uses of these operators, see, for example, P. O. Löwdin, *Phys. Rev.*, **97**, 1509 (1955); *Rev. Mod. Phys.*, **34**, 520 (1962); *J. Math Phys.*, **3**, 969 (1962). See also, J. V. Neumann, *Mathematical Foundations of Quantum Mechanics*, Princeton University Press, Princeton, 1955.

[45] P. O. Löwdin, *Rev. Modern Phys.*, **34**, 520 (1962).

[46] Some writers refer to eigenvectors and eigenvalues as characteristic vectors and values or proper vectors and values, respectively.

problem of solving the differential equation

$$\frac{d^2f(x)}{dx^2}=kf(x), \tag{2-425}$$

for the eigenfunction $f(x)$ and eigenvalue k is really quite simple. We are looking for a function that, when we take the second derivative of it, gives back the original function multiplied by the number k. This requirement clearly rules out most functions, and the only possibilities that are obvious are exponentials or sine and cosine functions of x. In fact, either choice will work in this case, and we shall choose exponentials for convenience. Let us try

$$f(x)=e^{-bx}. \tag{2-426}$$

Finding the necessary derivatives gives

$$\frac{d^2f(x)}{dx^2}=b^2e^{-bx}. \tag{2-427}$$

Thus, we see that $f(x) = e^{-bx}$ is an eigenfunction of the operator d^2/dx^2, and has an eigenvalue of $k = b^2$.

Although we have introduced the notion of eigenvalues and eigenvectors by solving a differential equation, the same problem is also found in matrix theory, which will be described in the next chapter. At that time we shall discuss the determination of eigenvalues and eigenfunctions in greater detail.

Problems

1. Show that the space of polynomial functions of degree $(n - 1)$ constitute a vector space, where addition and multiplication by a scalar are defined by Eqs. (2-19) and (2-20).
2. Show that the set of polynomials

   $f_0(x)=1,\ f_1(x)=x,\ f_2(x)=x^2,\ \ldots,\ f_{n-1}(x)=x^{n-1}, \quad (-\infty \leq x \leq +\infty)$

   is linearly independent.
3. If $\mathbf{u}_1$ is a vector belonging to the set of linearly independent vectors $\{\mathbf{u}_i\}$, show that $\mathbf{u}_1 \neq \mathbf{0}$.
4. Calculate the scalar product $(\mathbf{u}, \mathbf{v})$ and norms $[|\mathbf{u}|$ and $|\mathbf{v}|]$ of the following vectors:

   $\mathbf{u}=(2+3i,\ 4,\ 1+2i,\ 4i)$

   $\mathbf{v}=(3+i,\ 2+3i,\ 1,\ 3+3i)$

   which belong to a three-dimensional unitary vector space.
5. For any Euclidean or unitary space V, show that

   A. $|c\mathbf{v}|=|c|\ |\mathbf{v}|$

   where c is a real or complex scalar.

B. $|\mathbf{v}|>0$

unless $\mathbf{v} = \mathbf{0}$.

6. Show that the set of complex-valued functions of the real variable x, defined on the inverval $a \leq x \leq b$, and which are both integrable and square integrable, form a vector space. [Closure under addition and scalar multiplication is proven in Eqs. (2-94) through (2-104).]
7. Show by direct calculation that the following two functions are orthogonal over the range $(-\infty \leq x \leq +\infty)$:

$$f_1(x)=xe^{-x^2/2}$$

$$f_2(x)=(2x^2-1)e^{-x^2/2}$$

8. Use the Gram–Schmidt orthogonalization procedure to orthogonalize the following three functions $(0 \leq r \leq \infty)$:

$$g_1(r)=e^{-r}$$

$$g_2(r)=re^{-r} \qquad (dV=r^2dr)$$

$$g_3(r)=r^2e^{-r}.$$

9. Show that the following functions are mutually orthogonal,

$$\psi_1=\sin\vartheta e^{i\varphi} \qquad (0\leq\varphi\leq 2\pi)$$

$$\psi_2=\sin\vartheta e^{-i\varphi} \qquad (-\pi\leq\vartheta\leq\pi).$$

$$\psi_3=\cos\vartheta$$

It will be seen in Chapter 7 that these functions comprise the angular dependence of several low-lying excited states of the Hydrogen atom.

10. Given a set of vectors, in which each vector may contain an infinite number of components, show that these vectors constitute a vector space, if the norm is defined by Eq. (2-201) and addition and scalar multiplication are defined by Eqs. (2-203) and (2-205), respectively. [Closure under addition and scalar multiplication is proven in Eqs. (2-206) through (2-217).]
11. Show that the vectors in the preceding problem form an inner product space, if the inner product is defined in Eq. (2-220) converges.
12. Use Fourier expansions of length 2, 3, and 4 to fit the function

$$f(x)=(1-x)e^{-x}$$

over the range $-\pi \leq x \leq +\pi$.

13. An operator $\mathcal{A}$ that satisfies

$$|\mathcal{A}\mathbf{u}|\leq k|\dot{\mathbf{u}}|,$$

where $\mathcal{A}\mathbf{u}$ and $\mathbf{u}$ belong to V and k is a positive constant, is called a *bounded* operator. Show that the linear operator $\mathcal{L}:V \rightarrow V$ is always bounded if V is finite dimensional.

14. Show that the inverse operator $\mathcal{L}^{-1}$ is linear.

15. If $\mathcal{L}$ is linear operator and $e^{\mathcal{L}}$ is defined as

$$f(\mathcal{L}) = e^{\mathcal{L}} \equiv \mathcal{I} + \mathcal{L} + \left(\frac{1}{2!}\right)\mathcal{L}^2 + \left(\frac{1}{3!}\right)\mathcal{L}^3 + \cdots$$

show that

$$f'(\mathcal{L}) = f(\mathcal{L}).$$

16. Prove Theorem 2-10.
17. If $\mathcal{L}_1$ and $\mathcal{L}_2$ are Hermitian operators, and

$$\mathcal{L} = \mathcal{L}_1 + i\mathcal{L}_2,$$

show that $\mathcal{L}_1$ and $\mathcal{L}_2$ [as defined in Eq. (2-396)] are Hermitian.
18. If $\mathcal{U}$ is a unitary operator and

$$\mathcal{L}_2 = \mathcal{U}^\dagger \mathcal{L}_1 \mathcal{U}$$

where $\mathcal{L}_1$ and $\mathcal{L}_2$ are linear operators, show that

$$\mathcal{L}_1 = \mathcal{U}\mathcal{L}_2\mathcal{U}^\dagger.$$

19. Show that the set of projection operators $\{\mathcal{P}_1, \mathcal{P}_2, \ldots, \mathcal{P}_n\}$ defined by Eq. (2-412) are linear.
20. If $\mathcal{P}_i$ represents projection operators defined by Eq. (2-412), show that the $\mathcal{P}_i$ are Hermitian.
21. Determine whether the following operators ($\mathcal{A}$) are linear or not:

a. $\mathcal{A}\Psi = \Psi^2 \qquad (\Psi \text{ real})$

b. $\mathcal{A}\Psi = \dfrac{d^2}{dx^2}\Psi$

c. $\mathcal{A}\Psi = 0$

d. $\mathcal{A}\Psi = \Psi^*$

e. $\mathcal{A}\Psi = \int_a^b G(x, x')\Psi(x')\,dx'$, where $G(x, x')$ is the same given function of x, x' for all $\Psi(x)$.

22. If $\mathcal{A}$ and $\mathcal{B}$ are operators that each commutes with its commutator $[\mathcal{A}, \mathcal{B}]$, prove the following identity:

$$e^{\mathcal{A}}e^{\mathcal{B}} = e^{\mathcal{A} + \mathcal{B} + (1/2)[\mathcal{A}, \mathcal{B}]}$$

Hint: Consider $f(\lambda) = e^{\lambda\mathcal{A}}e^{\lambda\mathcal{B}}e^{-\lambda(\mathcal{A}+\mathcal{B})}$, establish the differential equation

$$\frac{df}{d\lambda} = \lambda[\mathcal{A}, \mathcal{B}]f$$

and integrate the differential equation.

23. Prove that $\mathcal{A} = x(d/dx)\ (-\infty \leq x \leq +\infty)$ is not Hermitian, and show how to construct a Hermitian operator from it.
24. Let Γ be a linear operator that operates on a set of four nonorthogonal functions according to the following rules:

$$\Gamma\varphi_1 = \varphi_2$$

$$\Gamma\varphi_2 = \varphi_1$$

$$\Gamma\varphi_3 = -\varphi_3$$

$$\Gamma\varphi_4 = \varphi_4.$$

For an arbitrary function

$$\Phi = \sum_{k=1}^{4} c_k \varphi_k,$$

a. Show that the operators

$$\mathcal{R}_1 = \frac{1}{2}(\mathcal{I} + \Gamma)$$

$$\mathcal{R}_2 = \frac{1}{2}(\mathcal{I} - \Gamma)$$

are idemponent.

b. Show that

$$x_1 = \mathcal{R}_1 \varphi$$

$$x_2 = \mathcal{R}_2 \varphi$$

are eigenfunctions of Γ, and determine their eigenvalues.

25. If $\mathcal{U}$ is a linear, unitary operator,
    a. Show that all eigenvalues of $\mathcal{U}$ have modulus unity.
    b. Show that eigenvectors corresponding to different eigenvalues of $\mathcal{U}$ are orthogonal.
26. Show that any scalar multiple of an eigenvector of an operator $\mathcal{A}$ is also an eigenvector of $\mathcal{A}$.

Chapter 3

Matrix Theory

3-1. Elements of Matrix Algebra

As we shall see presently, there is a mathematical technique called matrix theory that will allow us to represent many of the relations of quantum mechanics in a concise and easily manipulated form. In order to become familiar with the techniques involved in manipulating matrices, let us begin by considering some definitions and elementary properties.

For our purpose, we shall define a *matrix* as a collection of elements arranged in a two-dimensional array, where the position of each element in the array is determined by designating the row and column to which the element belongs. For example, the following arrangment of elements would be called a matrix $\mathbf{A}$:

$$\mathbf{A} = \begin{pmatrix} a_{11} & a_{12} & a_{13} & a_{14} \\ a_{21} & a_{22} & a_{23} & a_{24} \\ a_{31} & a_{32} & a_{33} & a_{34} \end{pmatrix}. \qquad (3\text{-}1)$$

The set of elements $\{a_{ij}\}$ within the matrix $\mathbf{A}$ are labeled with two subscripts, the first representing the *row* and the second denoting the *column* to which the element belongs. Note also that a matrix need not be square, but can be rectangular. The size of a matrix is indicated by stating the total number of rows and columns that are present. For example, the matrix $\mathbf{A}$ of Eq. (3-1) is referred to as a (3×4) matrix, with three rows and four columns.

Although a matrix is rectangular in general, we shall be dealing primarily with three particular kinds of matrices. A *square matrix* $(n \times n)$ will be denoted by a capital letter, as in Eq. (3-1). The other types of matrices of interest are called *row vectors* and *column vectors*, and will be denoted by small letters.

A row vector is a matrix having only one row, e.g.,

$$\mathbf{r}_i = (r_{i1}, r_{i2}, \ldots r_{in}), \tag{3-2}$$

and a column vector is a matrix having only one column:

$$\mathbf{c}_j = \begin{pmatrix} c_{1j} \\ c_{2j} \\ c_{3j} \\ \vdots \\ c_{nj} \end{pmatrix}. \tag{3-3}$$

The rationale behind the names of the matrices in Eqs. (3-2) and (3-3) is clear, due to the obvious analogy between the elements of a row or column vector and the elements of a vector as referred to earlier.

There are several particular matrices that occur that have special names. The *unit matrix* is an $(n \times n)$ square matrix with ones down the main diagonal,[1] and zeros elsewhere.

$$\mathbf{I} = \begin{pmatrix} 1 & 0 & 0 & \cdot & \cdot & \cdot & \cdot & 0 \\ 0 & 1 & 0 & \cdot & \cdot & \cdot & \cdot & 0 \\ 0 & 0 & 1 & 0 & & & & 0 \\ \cdot & \cdot & 0 & \cdot & & & & \cdot \\ \cdot & \cdot & \cdot & & \cdot & & & \cdot \\ \cdot & \cdot & \cdot & & & \cdot & & \cdot \\ \cdot & \cdot & \cdot & & & & \cdot & 0 \\ 0 & 0 & 0 & \cdot & \cdot & \cdot & 0 & 1 \end{pmatrix} \tag{3-4}$$

Another special matrix is the null matrix, $\mathbf{0}$, which is one in which each element is zero.

Using the above definitions, an algebra of matrices has been developed, in which the rules for combination of matrices (e.g., addition and multiplication) have many analogies in ordinary arithmetic. However, just as in the case of operators, we must be careful not to carry the analogies too far, as we shall see.

Since matrices involve more than one quantity, i.e., several elements, we expect that operations on matrices and properties of matrices will involve consideration of each of the elements. For example, two matrices, $\mathbf{A}$ and $\mathbf{B}$, are said to be *equal* if and only if

$$a_{ij} = b_{ij} \tag{3-5}$$

for all i and j.

Also, there are several ways in which matrices can be combined. We define the addition of two matrices, $\mathbf{A}$ and $\mathbf{B}$, as

$$\mathbf{C} = \mathbf{A} + \mathbf{B}, \tag{3-6}$$

[1] The main diagonal is the diagonal of a matrix having the same row and column index, e.g., d_{ii} would be the main diagonal of a square matrix $\mathbf{D}$.

where

$$c_{ij} = a_{ij} + b_{ij}, \tag{3-7}$$

for all i and j. It is important to note that the operation of addition is defined only if **A** and **B** are of the same size, i.e., if **A** is an $(m \times n)$ matrix **B** must also be an $(m \times n)$ matrix for addition of **A** and **B** to be defined. When this occurs, it is said that **A** and **B** *conform under addition*.

Multiplication of matrices is slightly more complicated, in the sense that there are two kinds of multiplication. The first kind, *multiplication by a scalar*, is defined as follows. If b is a scalar, then $b\mathbf{A}$ is defined as

$$\mathbf{C} = b\mathbf{A}, \tag{3-8}$$

where

$$c_{ij} = ba_{ij}, \tag{3-9}$$

for all i and j. Note that *every* element of **A** gets multiplied by b. It is left as an exercise for the student to show that the set of $(n \times m)$ matrices whose elements are complex forms a vector space if addition and scalar multiplication are defined as above (see Problem 3-1).

The other kind of multiplication, which is the *product of two matrices*, is defined as follows. If **A** is an $(m \times n)$ matrix and **B** is an $(n \times p)$ matrix, then the product of **A** and **B** is given by

$$\mathbf{C} = \mathbf{AB}, \tag{3-10}$$

where

$$c_{ij} = \sum_{k=1}^{n} a_{ik} b_{kj}, \tag{3-11}$$

for all i and j. This type of multiplication requires several other comments. First of all, the product indicated in Eq. (3-10) is defined only if the number of columns of **A** is precisely the same as the number of rows of **B**. When two matrices meet this requirement, they are said to *conform under multiplication*. Note also that the order in which the product is formed is very important, for the product **BA** is not defined in this example.

To make this concept clear, let us take a simple example. Let **A** be a (2×3) matrix, given by

$$\mathbf{A} = \begin{pmatrix} 1 & 2 & 3 \\ 4 & 5 & 6 \end{pmatrix},$$

and **B** be a (3×4) matrix, given by

$$\mathbf{B} = \begin{pmatrix} -1 & +1 & -2 & +2 \\ -3 & +3 & -4 & +4 \\ 0 & -5 & +5 & -1 \end{pmatrix}.$$

Then, the only acceptable product of **A** and **B** is **AB**, and not **BA**. The elements of the resulting matrix, **C**, are given by

$$\mathbf{C} = \mathbf{AB},$$

where

$$\begin{aligned}
c_{11} &= (1)(-1) + (2)(-3) + (3)(0) &&= -7 \\
c_{12} &= (1)(+1) + (2)(+3) + (3)(-5) &&= -8 \\
c_{13} &= (1)(-2) + (2)(-4) + (3)(+5) &&= +5 \\
c_{14} &= (1)(+2) + (2)(+4) + (3)(-1) &&= +7 \\
c_{21} &= (4)(-1) + (5)(-3) + (6)(0) &&= -15 \\
c_{22} &= (4)(+1) + (5)(+3) + (6)(-5) &&= -11 \\
c_{23} &= (4)(-2) + (5)(-4) + (6)(+5) &&= +2 \\
c_{24} &= (4)(+2) + (5)(+4) + (6)(-1) &&= +22.
\end{aligned}$$

Note that the dimensions of **C** are (2×4). This is an example of the general rule that, if **A** is an $(m \times n)$ matrix and **B** is an $(n \times p)$ matrix, then $\mathbf{C} = \mathbf{AB}$ will be an $(m \times p)$ matrix. Schematically, this can be viewed as

$$\underset{(m\times n)}{\mathbf{A}}\ \underset{(n\times p)}{\mathbf{B}} = \underset{(m\times p)}{\mathbf{C}}.$$

The student can easily verify that provided the matrices that are involved conform, the properties of *associativity* and *distributivity* apply to matrix multiplication. This means that

$$\mathbf{A}(\mathbf{BC}) = (\mathbf{AB})\mathbf{C} \tag{3-12}$$

and

$$\mathbf{A}(\mathbf{B} + \mathbf{C}) = \mathbf{AB} + \mathbf{AC}. \tag{3-13}$$

However, as indicated in the example given previously, a matrix product is not commutative in general, i.e.,

$$\mathbf{AB} \neq \mathbf{BA}, \tag{3-14}$$

even if the products **AB** and **BA** are both defined. That this is true is easily seen if one considers the two matrices

$$\mathbf{A} = \begin{pmatrix} 1 & 2 \\ 5 & 7 \end{pmatrix}; \quad \mathbf{B} = \begin{pmatrix} 5 & 0 \\ -3 & 1 \end{pmatrix}. \tag{3-15}$$

Forming the matrix products **AB** and **BA** we have

$$\mathbf{AB} = \begin{pmatrix} 1 & 2 \\ 5 & 7 \end{pmatrix}\begin{pmatrix} 5 & 0 \\ -3 & 1 \end{pmatrix} = \begin{pmatrix} -1 & 2 \\ 4 & 7 \end{pmatrix}, \tag{3-16}$$

and

$$\mathbf{BA} = \begin{pmatrix} 5 & 0 \\ -3 & 1 \end{pmatrix} \begin{pmatrix} 1 & 2 \\ 5 & 7 \end{pmatrix} = \begin{pmatrix} 5 & 10 \\ 2 & 1 \end{pmatrix}. \tag{3-17}$$

If we consider the $n \times n$ unit matrix $\mathbf{I}$ and any other $n \times n$ matrix $\mathbf{A}$, then it is clear that

$$\mathbf{IA} = \mathbf{A} = \mathbf{AI}. \tag{3-18}$$

Furthermore, if we consider the *scalar matrix*

$$\mathbf{C} = c\mathbf{I} = \begin{pmatrix} c & 0 & \cdots & 0 \\ 0 & c & \ddots & \vdots \\ \vdots & \ddots & \ddots & 0 \\ 0 & \cdots & 0 & c \end{pmatrix}, \tag{3-19}$$

it is left as an exercise for the student (see Problem 3-3) to show that

$$\mathbf{CA} = c\mathbf{A} = \mathbf{AC}. \tag{3-20}$$

Thus, for the special case of the multiplication of an $(n \times n)$ matrix with either the unit matrix or a scalar matrix, the order of multiplication does not matter.

One other simple example will serve to emphasize the importance of the order in which matrices are multiplied. Consider the following row and column vectors,

$$\mathbf{a}_i = (a_{i1}, a_{i2}, \ldots, a_{in}); \qquad \mathbf{b}_j = \begin{pmatrix} b_{1j} \\ b_{2j} \\ \vdots \\ b_{nj} \end{pmatrix}. \tag{3-21}$$

Since $\mathbf{a}_i$ is a $(1 \times n)$ matrix and $\mathbf{b}_j$ is an $(n \times 1)$ matrix, it is possible to form both $\mathbf{a}_i\mathbf{b}_j$ and $\mathbf{b}_j\mathbf{a}_i$. However,

$$\mathbf{a}_i\mathbf{b}_j = (a_{i1}, a_{i2}, \ldots, a_{in}) \begin{pmatrix} b_{1j} \\ b_{2j} \\ \vdots \\ b_{nj} \end{pmatrix} = \sum_{k=1}^{n} a_{ik} b_{kj} = c, \tag{3-22}$$

which is a scalar, i.e., c is a (1×1) matrix. Note that this is the same as the inner-product between two vectors $\mathbf{a}_i$ and $\mathbf{b}_j$ with real components, as shown in Eqs. (2-61), (2-62) and (2-63). On the other hand,

$$b_j a_i = \begin{pmatrix} b_{1j} \\ b_{2j} \\ \vdots \\ b_{nj} \end{pmatrix} (a_{i1} \; a_{i2} \; \cdots \; a_{in}) \tag{3-23}$$

$$= \begin{pmatrix} [b_{1j}a_{i1}] & [b_{1j}a_{i2}] & \cdots & [b_{1j}a_{in}] \\ \vdots & \vdots & & \vdots \\ [b_{nj}a_{i1}] & [b_{nj}a_{i2}] & \cdots & [b_{nj}a_{in}] \end{pmatrix} = \mathbf{C}, \tag{3-24}$$

which is an $(n \times n)$ matrix, which is sometimes called an *outer product*. It is clear that such column and row matrix multiplication is not commutative.

In anticipation of our study of determinants, we shall discuss one more technique that can be very useful in manipulating matrices. It is usually called *partitioning*, and involves blocking out a large matrix into smaller submatrices as illustrated in Eq. (3-25),

$$\mathbf{A} = \begin{pmatrix} a_{11} & a_{12} & a_{13} & \cdots & a_{1m} \\ a_{21} & a_{22} & a_{23} & \cdots & a_{2m} \\ a_{31} & a_{32} & a_{33} & \cdots & a_{3m} \\ \cdot & \cdot & \cdot & & \cdot \\ \cdot & \cdot & \cdot & & \cdot \\ \cdot & \cdot & \cdot & & \cdot \\ \cdot & \cdot & \cdot & & \cdot \\ a_{n1} & a_{n2} & a_{n3} & \cdots & a_{nm} \end{pmatrix} = \left(\begin{array}{c|c} \mathbf{A}_{11} & \mathbf{A}_{12} \\ \hline \mathbf{A}_{21} & \mathbf{A}_{22} \end{array}\right), \tag{3-25}$$

where, e.g.,

$$\begin{aligned} \mathbf{A}_{11} &= \begin{pmatrix} a_{11} & a_{12} & a_{13} \\ a_{21} & a_{22} & a_{23} \end{pmatrix}, \\ \mathbf{A}_{12} &= \begin{pmatrix} a_{14} & \cdots & a_{1m} \\ a_{24} & \cdots & a_{2m} \end{pmatrix}, \\ \mathbf{A}_{21} &= \begin{pmatrix} a_{31} & a_{32} & a_{33} \\ \vdots & \vdots & \vdots \\ a_{n1} & a_{n2} & a_{n3} \end{pmatrix}, \\ \mathbf{A}_{22} &= \begin{pmatrix} a_{44} & a_{45} & \cdots & a_{4m} \\ a_{54} & a_{55} & \cdots & a_{5m} \\ \vdots & \vdots & \ddots & \vdots \\ a_{n4} & a_{n5} & \cdots & a_{nm} \end{pmatrix}. \end{aligned} \tag{3-26}$$

Note that the submatrices may be either rectangular or square. The partitioning is usually carried out so that the resulting matrix can be easily manipulated, since there may be many submatrices.

Consider now a second matrix **B** that is partitioned in a similar fashion to **A**, i.e.

$$\mathbf{B} = \begin{pmatrix} b_{11} & b_{12} & \cdots & b_{1l} \\ b_{21} & b_{22} & \cdots & b_{2l} \\ \vdots & \vdots & & \vdots \\ b_{k1} & b_{k2} & \cdots & b_{kl} \end{pmatrix} = \left(\begin{array}{c|c} \mathbf{B}_{11} & \mathbf{B}_{12} \\ \hline \mathbf{B}_{21} & \mathbf{B}_{22} \end{array}\right). \tag{3-27}$$

If **A** and **B** are conformable under addition, then they can be added as

$$\mathbf{A} + \mathbf{B} = \left(\begin{array}{c|c} \mathbf{A}_{11} + \mathbf{B}_{11} & \mathbf{A}_{12} + \mathbf{B}_{12} \\ \hline \mathbf{A}_{21} + \mathbf{B}_{21} & \mathbf{A}_{22} + \mathbf{B}_{22} \end{array}\right), \tag{3-28}$$

provided the corresponding submatrices are also conformable under addition.

Similarly, for matrix multiplication, if **A** and **B** are conformable under matrix multiplication, we have

$$\mathbf{AB} = \left(\begin{array}{c|c} \mathbf{A}_{11}\mathbf{B}_{11} + \mathbf{A}_{12}\mathbf{B}_{21} & \mathbf{A}_{11}\mathbf{B}_{12} + \mathbf{A}_{12}\mathbf{B}_{22} \\ \hline \mathbf{A}_{21}\mathbf{B}_{11} + \mathbf{A}_{22}\mathbf{B}_{21} & \mathbf{A}_{21}\mathbf{B}_{12} + \mathbf{A}_{21}\mathbf{B}_{22} \end{array}\right), \tag{3-29}$$

provided also that the submatrices are conformable to matrix multiplications.

From the above, it is clear that addition and multiplication of partitioned matrices take the same form as that between unpartitioned matrices, except that the added or multiplied elements take the form of matrices and are, hence, subject to the rules of conformability and commutivity. If **A** and **B** are two matrices of the same dimension, then they are said to be *identically partitioned* if their block structure is the same.

Let us consider as an example the two matrices partitioned as indicated below:

$$\mathbf{A} = \left(\begin{array}{c|c} \mathbf{A}_{11} & \mathbf{A}_{12} \\ \hline \mathbf{A}_{21} & \mathbf{A}_{22} \end{array}\right) = \left(\begin{array}{c|cc} 1 & 3 & 1 \\ \hline 4 & 2 & 4 \\ 5 & 7 & 3 \end{array}\right), \tag{3-30}$$

and

$$\mathbf{B} = \left(\begin{array}{c|c} \mathbf{B}_{11} & \mathbf{B}_{12} \\ \hline \mathbf{B}_{21} & \mathbf{B}_{22} \end{array}\right) = \left(\begin{array}{c|cc} 3 & 2 & 6 \\ \hline 5 & 4 & 3 \\ 1 & 2 & 1 \end{array}\right). \tag{3-31}$$

The matrix multiplication **AB** is, using Eq. (3-29), given as,[2]

$$\mathbf{AB} = \left(\begin{array}{c|c} (1)(3) + (3\ \ 1)\begin{pmatrix} 5 \\ 1 \end{pmatrix} & (1)(2\ \ 6) + (3\ \ 1)\begin{pmatrix} 4 & 3 \\ 2 & 1 \end{pmatrix} \\ \hline \begin{pmatrix} 4 \\ 5 \end{pmatrix}(3) + \begin{pmatrix} 2 & 4 \\ 7 & 3 \end{pmatrix}\begin{pmatrix} 5 \\ 1 \end{pmatrix} & \begin{pmatrix} 4 \\ 5 \end{pmatrix}(2\ \ 6) + \begin{pmatrix} 2 & 4 \\ 7 & 3 \end{pmatrix}\begin{pmatrix} 4 & 3 \\ 2 & 1 \end{pmatrix} \end{array}\right) \tag{3-32}$$

Carrying out the indicated matrix multiplications within the large matrix yields

$$\mathbf{AB} = \left(\begin{array}{c|c} (3) + (16) & (2\ \ 6) + (14\ \ 10) \\ \hline \begin{pmatrix} 12 \\ 15 \end{pmatrix} + \begin{pmatrix} 14 \\ 38 \end{pmatrix} & \begin{pmatrix} 8 & 24 \\ 10 & 30 \end{pmatrix} + \begin{pmatrix} 16 & 10 \\ 34 & 24 \end{pmatrix} \end{array}\right). \tag{3-33}$$

Performing the matrix additions (note all matrices that need to be added do conform), we obtain

$$\mathbf{AB} = \left(\begin{array}{c|c} (19) & (16\ \ 16) \\ \hline \begin{pmatrix} 26 \\ 53 \end{pmatrix} & \begin{pmatrix} 24 & 34 \\ 44 & 54 \end{pmatrix} \end{array}\right) = \begin{pmatrix} 19 & 16 & 16 \\ 26 & 24 & 34 \\ 53 & 44 & 54 \end{pmatrix}, \tag{3-34}$$

---

[2] The terms in parentheses inside the matrix are also matrices.

which would also have been obtained by direct matrix multiplication between the unpartitioned matrices **A** and **B**.

It is also frequently useful to be able to describe the "magnitude" of a matrix using a single number. In that case the concept of a *matrix norm* is used, and a variety of definitions can be used. In particular, for an $(m \times n)$ matrix **A**, we shall define the norm $(\|\mathbf{A}\|)$ of **A** as follows:

$$\|\mathbf{A}\| = \left[\sum_{i=1}^{m} \sum_{j=1}^{n} a_{ij}^2 \right]^{1/2} \tag{3-35}$$

where $\|\mathbf{A}\|$ defined in this manner is known as the *Euclidean norm*[3] of **A**.

Such a definition is useful in a variety of ways. For example, when discussing binomial expansions using matrices, e.g.,

$$(\mathbf{I}+\mathbf{A})^{-1} = \mathbf{I} - \mathbf{A} + \mathbf{A}^2 - \mathbf{A}^3 + \cdots, \tag{3-36}$$

the criterion for convergence of the expansion that is typically used is that

$$\|\mathbf{A}\| < 1 \tag{3-37}$$

for convergence to be assured.

One other special notation will be used at times. We shall denote

$$\underbrace{\mathbf{AAA} \cdots \mathbf{A}}_{n \text{ products}} \equiv \mathbf{A}^n,$$

and refer to $\mathbf{A}^n$ as the nth *power of the square matrix* **A**.

3-2. Determinants

Before proceeding to a more detailed discussion of square matrices, we shall first consider another mathematical entity that takes the form of a square array of elements, and is called a *determinant*. Since a determinant can be formed from any square array of elements, one possible manner in which determinants can arise involves the determinant of a matrix. If a square matrix **A** is given by

$$\mathbf{A} = \begin{pmatrix} a_{11} & a_{12} & \cdots & a_{1n} \\ a_{21} & a_{22} & \cdots & a_{2n} \\ \vdots & \vdots & \ddots & \vdots \\ a_{n1} & a_{n2} & \cdots & a_{nn} \end{pmatrix} \tag{3-38}$$

where the elements of **A** are numbers, then the determinant of **A** can be thought of as an operator, written as $|\mathbf{A}|$ or det **A**, that acts on **A** to yield in general a

[3] Some authors refer to this definition as the Frobenius norm.

single number. As an illustration, consider the following matrix $\mathbf{A}$:

$$\mathbf{A} = \begin{pmatrix} a_{11} & a_{12} \\ a_{21} & a_{22} \end{pmatrix}. \tag{3-39}$$

The determinant of this matrix is defined in the following way

$$|\mathbf{A}| = \det |\mathbf{A}| = \begin{vmatrix} a_{11} & a_{12} \\ a_{21} & a_{22} \end{vmatrix} \equiv a_{11}a_{22} - a_{21}a_{12}. \tag{3-40}$$

Since the a_{ij} are numbers, the term $a_{11}a_{22} - a_{12}a_{21}$ is also a number. This example does not, however, show us how to treat the determinants of larger matrices, such as the one given by Eq. (3-38). The determinant of such an $(n \times n)$ matrix is said to be of *order n*.

There are two methods commonly employed to evaluate larger determinants, and we shall discuss each of them.

The first method involves the use of *minors* and *cofactors* of the original determinant. To illustrate this technique, let us consider the determinant of the matrix $\mathbf{A}$ given in Eq. (3-38), i.e.

$$|\mathbf{A}| = \begin{vmatrix} a_{11} & a_{12} & \cdots & a_{1j} & \cdots & a_{1n} \\ a_{21} & a_{22} & \cdots & a_{2j} & \cdots & a_{2n} \\ \vdots & \vdots & & \vdots & & \vdots \\ a_{i1} & a_{i2} & \cdots & a_{ij} & \cdots & a_{in} \\ \vdots & \vdots & & \vdots & & \vdots \\ a_{n1} & a_{n2} & \cdots & a_{nj} & \cdots & a_{nn} \end{vmatrix} \tag{3-41}$$

The *minor* (M_{ij}) of any element, a_{ij}, in $|\mathbf{A}|$ is found by forming a new determinant of order $(n - 1)$, consisting of the rows and columns that remain after removing the ith row and jth column from the original determinant. The *cofactor*, $|\mathbf{A}_{ij}|$, of any element a_{ij} of $|\mathbf{A}|$ is obtained by multiplying the minor of a a_{ij} by the factor $(-1)^{i+j}$. The evaluation of the determinant is then effected by summing up the products of the elements of any row or column and their corresponding cofactors, i.e.,

$$|\mathbf{A}| = \sum_{i=1}^{n} a_{ij}|\mathbf{A}_{ij}| = \sum_{i=1}^{n} a_{ij}(-1)^{i+j}M_{ij}, \tag{3-42}$$

where we have chosen to expand along the jth column. A similar expression exists for the expansion along any row.

Of course, if the determinant is larger than a 3×3 (three rows and three columns), each $|\mathbf{A}_{ij}|$ will be a determinant larger than 2×2, which can also be expanded in cofactors. This procedure is continued until only 2×2 determinants remain, and Eq. (3-40) can be used. This method of evaluating a determinant is usually called the *Laplace expansion of a determinant*.

To illustrate this procedure, consider the evaluation of the following determinant, which we shall expand in cofactors along the first row for convenience:

$$|\mathbf{A}| = \begin{vmatrix} a_{11} & a_{12} & a_{13} \\ a_{21} & a_{22} & a_{23} \\ a_{31} & a_{32} & a_{33} \end{vmatrix}$$

$$= a_{11} \begin{vmatrix} a_{22} & a_{23} \\ a_{32} & a_{33} \end{vmatrix} + (-1)a_{12} \begin{vmatrix} a_{21} & a_{23} \\ a_{31} & a_{33} \end{vmatrix} + a_{13} \begin{vmatrix} a_{21} & a_{22} \\ a_{31} & a_{32} \end{vmatrix}, \tag{3-43}$$

$$= a_{11}(a_{22}a_{33} - a_{23}a_{32}) - a_{12}(a_{21}a_{33} - a_{23}a_{31}) + a_{13}(a_{21}a_{32} - a_{22}a_{31}), \tag{3-44}$$

$$= a_{11}a_{22}a_{33} - a_{11}a_{23}a_{32} - a_{12}a_{21}a_{33} + a_{12}a_{23}a_{31} + a_{13}a_{21}a_{32} - a_{13}a_{22}a_{31}. \tag{3-45}$$

Before discussing the second method of determinant evaluation, let us examine some more usual uses of determinants. One important example involves the vector cross-product between two vectors $\mathbf{a} = (a_x, a_y, a_z)$ and $\mathbf{b} = (b_x, b_y, b_z)$. This can be represented using determinants as

$$\mathbf{a} \times \mathbf{b} = \begin{vmatrix} \mathbf{i} & \mathbf{j} & \mathbf{k} \\ a_x & a_y & a_z \\ b_x & b_y & b_z \end{vmatrix}, \tag{3-46}$$

where $\mathbf{i}$, $\mathbf{j}$, and $\mathbf{k}$ are the orthonormal Cartesian basis vectors (see Fig. 2.2). Using Eq. (3-44), we see that

$$\begin{vmatrix} \mathbf{i} & \mathbf{j} & \mathbf{k} \\ a_x & a_y & a_z \\ b_x & b_y & b_z \end{vmatrix} = \mathbf{i}(a_y b_z - a_z b_y) + \mathbf{j}(a_z b_x - a_x b_z) + \mathbf{k}(a_x b_y - a_y b_x). \tag{3-47}$$

Combining Eqs. (3-44) and (3-45), we obtain

$$\mathbf{a} \times \mathbf{b} = \mathbf{i}(a_y b_z - a_z b_y) + \mathbf{j}(a_z b_x - a_x b_z) + \mathbf{k}(a_x b_y - a_y b_x), \tag{3-48}$$

which is consistent with[4] the definition given in Eq. (2-8). It is important to note that $a \times b$ is a *vector*, and *not a scalar*. It should also be noted that the array of elements,

$$\begin{Bmatrix} \mathbf{i} & \mathbf{j} & \mathbf{k} \\ a_x & a_y & a_z \\ b_x & b_y & b_z \end{Bmatrix}$$

are not all numbers, so that we cannot speak of the matrix

$$\begin{pmatrix} \mathbf{i} & \mathbf{j} & \mathbf{k} \\ a_x & a_y & a_z \\ b_x & b_y & b_z \end{pmatrix}.$$

Nevertheless, this example serves to illustrate the fact that determinants can also be used as tools for formal manipulation of various mathematical objects, e.g.,

[4] See Problem 3-4.

scalars and vectors. However, the elements involved in the determinant must have the operation of multiplication defined among themselves.

As another example, consider the determinant made up of the four vectors **a**, **b**, **c**, **d**, i.e.,

$$\begin{vmatrix} \mathbf{a} & \mathbf{b} \\ \mathbf{c} & \mathbf{d} \end{vmatrix} = \mathbf{ad} - \mathbf{cb} \tag{3-49}$$

The multiplication of two vectors, e.g. **ad**, must be clearly defined, including the order of multiplication (i.e., **ad** or **da**), before the above determinant can be defined.

An alternate method of evaluating a determinant considers the indices of the elements in the determinant, and how they may be permuted. For example, we see that each of the products in Eq. (3-45) can be written in the form

$$a_{1j_1} \cdot a_{2j_2} \cdot a_{3j_3}$$

where the sign of the product has been ignored momentarily, and j_1, j_2, and j_3 represent some permutation of the indices 1, 2, 3. Note that there are 3! such products. To obtain the appropriate sign for each product, we determine the number of exchanges of j_1, j_2, and j_3 that are necessary to bring them to the order 1, 2, 3. For example, if $j_1 = 3, j_2 = 1, j_3 = 2$, there are two exchanges that are necessary, i.e.,

$$\overset{\frown}{31}2 \rightarrow 1\overset{\frown}{32} \rightarrow 123.$$

If the number of these exchanges is even, the sign of that term is taken to be positive; if the number of exchanges is odd, the sign is taken to be negative. It can be seen that the following series of exchanges

$$3\overset{\frown}{12} \rightarrow \overset{\frown}{32}1 \rightarrow \overset{\frown}{231} \rightarrow 1\overset{\frown}{32} \rightarrow 123$$

also brings 312 to the order 123. But again the *number* of exchanges is even. This evenness (or oddness) of the number of exchanges is known as the *parity* of the permutation. Parity is a characteristic of the permutation, and thus is not dependent on an expeditious choice of exchanges.

Then, the evaluation of a determinant consists of summing up the products for all possible permutations of the second index, giving the product the appropriate sign, as described above:

$$\begin{aligned} |A| &= \sum_{\substack{\text{all} \\ \text{permutations } (\mathcal{P})}}^{n!} (-1)^p \mathcal{P} a_{1j_1} a_{2j_2} a_{3j_3} \cdots a_{nj_n} \\ &= \det \{a_{1j_1}, a_{2j_2} \cdots a_{nj_n}\}, \end{aligned} \tag{3-50}$$

where $\mathcal{P}$ is a permutation operator that permutes the indices $j_1, \ldots, j_n$, and p is the parity of the permutation $j_1, j_2, \ldots, j_n$. The student should verify that the same result as in Eq. (3-42) is obtained using this method of evaluation.

We now proceed to prove several elementary theorems concerning the rows and columns of a determinant. The proofs are given for the manipulation of either rows or columns, but are applicable to both.

Theorem 3-1. Exchanging two rows or two columns in a determinant changes the sign of the determinant.

The proof of this theorem follows directly from the expression for the expansion of a determinant given in Eq. (3-50), since the exchange of two columns merely adds one to the number of permutations needed to bring the indices to the order 1, 2, ... , n. This multiplies each term by (-1), and hence the determinant is multiplied by (-1).

Theorem 3-2. If each of the elements of a given row or column is multiplied by a scalar k, the value of the determinant is multiplied by k.

To prove this theorem, we note from Eq. (3-42) that multiplication of each element of the jth column by k gives

$$\sum_{i=1}^{n} (ka_{ij})|A_{ij}| = k \sum_{i=1}^{n} a_{ij}|A_{ij}| = k|A|. \tag{3-51}$$

Theorem 3-3. If two or more rows or two or more columns are constant multiples of each other, then the value of the determinant is zero.

The use of Theorem 3-2 allows us to consider only the case when two columns are equal, without loss of generality. In that case, interchanging two columns that are alike must change the sign of the determinant, by Theorem 3-1. However, the determinant with two identical columns interchanged looks precisely like the original determinant. Therefore, the only way in which $|A| = -|A|$ can be fulfilled is for $|A| = 0$, which proves the theorem.

Theorem 3-4. The replacement of any (or column) of a determinant by a sum of that row (or column) with any other row (or column) leaves the value of the determinant unchanged.

To prove this theorem, suppose that the kth row of the determinant represented by Eq. (3-41) is replaced by $a_{kj} + \lambda a_{mj_m}$, where λ is an arbitrary scalar. Then Eq. (3-50) becomes

$$\sum_{\substack{\text{all}\\ \text{permutations}}}^{n!} (-1)^p a_{1j_1}a_{2j_2} \cdots a_{mj_m} \cdots (a_{kj_k} + \lambda a_{mj_m})a_{k+1,j_{k+1}} \cdots a_{nj_n}$$

$$= |A| + \lambda \sum_{\substack{\text{all}\\ \text{permutations}}}^{n!} (-1)^p a_{1j_1}a_{2j_2} \cdots a_{mj_m} \cdots a_{mj_m}a_{k+1,j_{k+1}} \cdots a_{nj_n}. \tag{3-52}$$

However, the second term in Eq. (3-52) is zero by Theorem 3-3, which provides the desired result.

To illustrate the use of determinants, let us now see how they can be used to facilitate the solution of simultaneous equations. The set of simultaneous, linear, nonhomogeneous equations given in Eq. (3-53) could easily be solved for x_1 and x_2 by the use of elementary algebra.

$$\begin{aligned} a_{11}x_1 + a_{12}x_2 &= b_1 \\ a_{21}x_1 + a_{22}x_2 &= b_2. \end{aligned} \tag{3-53}$$

The solutions are

$$x_1 = \frac{a_{22}b_1 - a_{12}b_2}{a_{11}a_{22} - a_{12}a_{21}}, \qquad x_2 = \frac{a_{11}b_2 - a_{21}b_1}{a_{11}a_{22} - a_{12}a_{21}} \tag{3-54}$$

However, we note that the terms in the denominator of the expressions for x_1 and x_2 are both identical, and can be expressed as the determinant of the coefficients:

$$|A| = \begin{vmatrix} a_{11} & a_{12} \\ a_{21} & a_{22} \end{vmatrix}.$$

Similarly, the numerators can be expressed as

$$|A_1| = \begin{vmatrix} b_1 & a_{12} \\ b_2 & a_{22} \end{vmatrix}, \qquad |A_2| = \begin{vmatrix} a_{11} & b_1 \\ a_{21} & b_2 \end{vmatrix}$$

This simple example is an illustration of *Cramer's Rule*, which states that for a system of n simultaneous, linear, inhomogeneous equations,

$$\begin{aligned} a_{11}x_1 + a_{12}x_2 + \cdots + a_{1n}x_n &= b_1 \\ a_{21}x_1 + a_{22}x_2 + \cdots + a_{2n}x_n &= b_2 \\ \vdots \qquad \vdots \qquad & \vdots \\ a_{n1}x_1 + a_{n2}x_2 + \cdots + a_{nn}x_n &= b_n, \end{aligned} \tag{3-55}$$

the desired solutions are given by

$$x_1 = \frac{|A_1|}{|A|}, \; x_2 = \frac{|A_2|}{|A|}, \; \ldots, \; x_n = \frac{|A_n|}{|A|}, \tag{3-56}$$

where $|A|$ is the determinant formed from the coefficients, and where $|A_i|$ is the determinant[5] obtained from $|A|$ by replacing the ith column by b_1, b_2, ..., b_n. It should be noted that the solution indicated in Eq. (3-56) exists only if $|A| \neq 0$. In the later discussion of matrices we shall see that the condition $|A| \neq 0$ is both necessary and sufficient to assure that the equations have a solution.

Two other cases of interest that will be stated, but not proved, are concerned with the case of homogeneous equations,[6] i.e., $b_1 = b_2 = \cdots = b_n = 0$. For

[5] Some authors refer to $|A_i|$ as the adjoint of $|A|$, but we shall use the term adjoint to refer to the complex-conjugate transpose of a matrix, which will be introducted later in this chapter.

[6] See, for example, F. B. Hildebrand, *Introduction to Numerical Analysis,* McGraw-Hill, New York, 1956, p. 427.

this situation, if $|A| \neq 0$, the only solution is the trivial one where $x_1 = x_2 = \cdots = x_n = 0$. Thus, we must have $|A| = 0$, for there to be a nontrivial solution in this case.

Before leaving the subject of determinants, let us consider a practical method for examining the linear independence of a set of vectors $\{\mathbf{v}_1, \mathbf{v}_2, \ldots, \mathbf{v}_n\}$ using determinants. As was discussed in Chapter 2, this set of vectors is linearly independent if and only if the linear combination

$$\sum_{k=1}^{n} c_k \mathbf{v}_k = 0. \tag{3-57}$$

implies that $c_1 = c_2 = \cdots = c_n = 0$.

Suppose we form the inner product of each vector in the set with Eq. (3-57), e.g.,

$$\left(\mathbf{v}_i, \sum_{k=1}^{n} c_k \mathbf{v}_k\right) = \sum_{k=1}^{n} c_k (\mathbf{v}_i, \mathbf{v}_k) = 0, \qquad i = 1, 2, \ldots, n. \tag{3-58}$$

We then obtain the following set of homogeneous linear equations,

$$\begin{aligned} c_1(\mathbf{v}_1, \mathbf{v}_1) + c_2(\mathbf{v}_1, \mathbf{v}_2) + \cdots + c_n(\mathbf{v}_1, \mathbf{v}_n) &= 0 \\ c_1(\mathbf{v}_2, \mathbf{v}_1) + c_2(\mathbf{v}_2, \mathbf{v}_2) + \cdots + c_n(\mathbf{v}_2, \mathbf{v}_n) &= 0 \\ \vdots \qquad\qquad \vdots \qquad\qquad\qquad \vdots \\ c_1(\mathbf{v}_n, \mathbf{v}_1) + c_2(\mathbf{v}_n, \mathbf{v}_2) + \cdots + c_n(\mathbf{v}_n, \mathbf{v}_n) &= 0. \end{aligned} \tag{3-59}$$

The c_i are the n unknowns, so that the determinant of the coefficients of the unknowns is given by

$$G = \begin{vmatrix} (\mathbf{v}_1, \mathbf{v}_1) & (\mathbf{v}_1, \mathbf{v}_2) & \cdots & (\mathbf{v}_1, \mathbf{v}_n) \\ (\mathbf{v}_2, \mathbf{v}_1) & (\mathbf{v}_2, \mathbf{v}_2) & \cdots & (\mathbf{v}_2, \mathbf{v}_n) \\ \vdots & \vdots & & \vdots \\ (\mathbf{v}_n, \mathbf{v}_1) & (\mathbf{v}_n, \mathbf{v}_2) & \cdots & (\mathbf{v}_n, \mathbf{v}_n) \end{vmatrix}. \tag{3-60}$$

This determinant is usually known as the *Gramian* or *Gram determinant*.

Referring to the discussion on the existence of trivial or nontrivial solutions for sets of linear homogeneous equations (see above), it follows that $G \neq 0$ if the set of vectors $\{\mathbf{v}_1, \mathbf{v}_2, \ldots, \mathbf{v}_n\}$ is linearly independent. This is the condition on the Gramian that gives the trivial solutions to Eq. (3-59), i.e., $c_1 = c_2 = \cdots = c_n = 0$. It also follows that if $G = 0$, that the set of vectors is linearly dependent.

If we are dealing with a set of functions $\{f_1, f_2, \cdots, f_n\}$ whose inner product is given by

$$(f_i, f_j) = \int_a^b f_i^*(x) f_j(x)\, dx, \tag{3-61}$$

then the Gramian becomes

$$G = \begin{vmatrix} (f_1, f_1) & (f_1, f_2) & \cdots & (f_1, f_n) \\ (f_2, f_1) & (f_2, f_2) & \cdots & (f_2, f_n) \\ \vdots & \vdots & & \vdots \\ (f_n, f_1) & (f_n, f_2) & \cdots & (f_n, f_n) \end{vmatrix}. \tag{3-62}$$

Note that elements of the Gram determinants are scalars, since they arise from inner products. Therefore, it is obvious that the value of the Gramian will also be a scalar.

As a concrete example of the use of Gram determinants, consider the two sets of vectors,

$$\begin{aligned} \mathbf{v}_1 &= (1, 1, 0, 0) \\ \mathbf{v}_2 &= (0, 1, 2, 0) \\ \mathbf{v}_3 &= (0, 0, 3, 4) \\ \mathbf{v}_4 &= (1, 1, 1, 1), \end{aligned} \tag{3-63}$$

and

$$\begin{aligned} \mathbf{v}_1' &= (1, 1, 0, 0) \\ \mathbf{v}_2' &= (0, 1, 2, 0) \\ \mathbf{v}_3' &= (0, 0, 3, 4) \\ \mathbf{v}_4' &= (1, 0, 1, 4), \end{aligned} \tag{3-64}$$

which differ by only one vector. It is not immediately obvious whether the sets are linearly independent or linearly dependent. To answer this question, we form the Gramians G and G' for the primed and unprimed sets of vectors, respectively, i.e.,

$$G = \begin{vmatrix} 2 & 1 & 0 & 2 \\ 1 & 5 & 6 & 3 \\ 0 & 6 & 25 & 7 \\ 2 & 3 & 7 & 4 \end{vmatrix}, \tag{3-65}$$

and

$$G' = \begin{vmatrix} 2 & 1 & 0 & 1 \\ 1 & 5 & 6 & 2 \\ 0 & 6 & 25 & 19 \\ 1 & 2 & 19 & 18 \end{vmatrix}. \tag{3-66}$$

Evaluation of the determinants shows that $G \neq 0$ and $G' = 0$. Hence, it follows that the set $\{\mathbf{v}_1, \mathbf{v}_2, \mathbf{v}_3, \mathbf{v}_4\}$ is linearly independent, while the set $\{\mathbf{v}_1', \mathbf{v}_2', \mathbf{v}_3', \mathbf{v}_4'\}$ is linearly dependent. It is left as an exercise for the student to show that $G = 0.0931$ and $G' = 0.000$.

3-3. Characterization of Square Matrices

Just as numbers are characterized by both their magnitude and sign, and vectors by their length and direction, square matrices are also characterized in several ways.

The first characteristic of a square matrix $\mathbf{A}$ of interest is the determinant of the matrix, which is written as $|\mathbf{A}|$. If $|\mathbf{A}| = 0$, the matrix $\mathbf{A}$ is said to be *singular*, and if $|\mathbf{A}| \neq 0$, the matrix $\mathbf{A}$ is said to be *nonsingular*.

There is an important theorem concerning determinants of matrices that will be used often in later discussions that we shall prove at this point.

Theorem 3-5. If $\mathbf{A}$ and $\mathbf{B}$ are $(n \times n)$ matrices, then

$$|\mathbf{AB}| = |\mathbf{A}| \cdot |\mathbf{B}| \tag{3-67}$$

Proof: Let us consider the determinant of size $(2n \times 2n)$

$$D = \begin{vmatrix} \mathbf{A} & \mathbf{0} \\ \mathbf{P} & \mathbf{B} \end{vmatrix}, \tag{3-68}$$

where $\mathbf{P}$ is a completely arbitrary $(n \times n)$ matrix. The Laplace expansion of D along its first row shows that

$$D = |\mathbf{A}| \cdot |\mathbf{B}| \tag{3-69}$$

for any choice of $\mathbf{P}$. Let us choose

$$\mathbf{P} = \begin{pmatrix} -1 & & & \\ & -1 & & 0 \\ 0 & & \ddots & \\ & & & -1 \end{pmatrix}. \tag{3-70}$$

We now multiply column 1 of D by b_{11}, and add it to the $(n+1)$st column of D, which does not change the value of the determinant (Theorem 3-4). This gives

$$D = \begin{vmatrix} a_{11} & a_{12} & \cdots & a_{1n} & a_{11}b_{11} & 0 & \cdots & 0 \\ a_{21} & a_{22} & \cdots & a_{2n} & a_{21}b_{11} & 0 & \cdots & 0 \\ \vdots & \vdots & \ddots & \vdots & \vdots & \vdots & \ddots & \vdots \\ a_{n1} & a_{n2} & \cdots & a_{nn} & a_{n1}b_{11} & 0 & \cdots & 0 \\ -1 & 0 & \cdots & 0 & 0 & b_{12}\,b_{13} & \cdots & b_{1n} \\ 0 & -1 & & \vdots & b_{21} & b_{22} & \cdots & b_{2n} \\ & & \ddots & \vdots & b_{31} & & \ddots & \\ \vdots & & & 0 & \vdots & & & \vdots \\ 0 & \cdots & 0 & -1 & b_{n1} & b_{n2} & \cdots & b_{nn} \end{vmatrix}. \tag{3-71}$$

In a similar manner, we now add the second column, multiplied by b_{21}, to the $(n + 1)$st column. This gives

$$D = \begin{vmatrix} a_{11} & a_{12} & \cdots & a_{1n} & (a_{11}b_{11} + a_{12}b_{21}) & 0 & \cdots & 0 \\ a_{21} & a_{22} & \cdots & a_{2n} & (a_{21}b_{11} + a_{22}b_{21}) & 0 & \cdots & 0 \\ \vdots & & \ddots & \vdots & \vdots & \vdots & \ddots & \vdots \\ a_{n1} & a_{n2} & \cdots & a_{nn} & (a_{n1}b_{11} + a_{n2}b_{21}) & 0 & \cdots & 0 \\ -1 & 0 & \cdots & 0 & 0 & b_{21} & \cdots & b_{1n} \\ 0 & -1 & 0 \cdots & 0 & 0 & b_{22} & \cdots & b_{n2} \\ \vdots & \ddots & \ddots & \vdots & b_{31} & & & \\ \vdots & & \ddots & 0 & \vdots & \vdots & \ddots & \vdots \\ 0 & \cdots & 0 & -1 & b_{n1} & b_{n2} & \cdots & b_{nn} \end{vmatrix}. \qquad (3\text{-}72)$$

By repeating this process n times, i.e., by adding the product of the ith column and b_{i1} to the $(n + 1)$st column $(i = 1, 2, \ldots, n)$, we obtain

$$D = \begin{vmatrix} a_{11} & a_{12} & \cdots & a_{1n} & \sum_{i=1}^{n} a_{1i}b_{i1} & 0 & \cdots & 0 \\ a_{21} & a_{22} & \cdots & a_{2n} & \sum_{i=1}^{n} a_{2i}b_{i2} & 0 & \cdots & 0 \\ \vdots & & \ddots & \vdots & \vdots & \vdots & \ddots & \vdots \\ a_{n1} & a_{n2} & \cdots & a_{nn} & \sum_{i=1}^{n} a_{ni}b_{in} & 0 & \cdots & 0 \\ -1 & 0 & \cdots & 0 & 0 & b_{12} & \cdots & b_{1n} \\ 0 & -1 & \ddots & \vdots & \vdots & b_{22} & \cdots & b_{n2} \\ \vdots & \ddots & \ddots & 0 & \vdots & \vdots & \ddots & \vdots \\ 0 & \cdots & 0 & -1 & 0 & b_{n2} & \cdots & b_{nn} \end{vmatrix}. \qquad (3\text{-}73)$$

Proceeding in a similar manner for the $(n + 2)$, $(n + 3)$, ..., $(2n)$th column of D, we obtain

$$D = \begin{vmatrix} & \mathbf{A} & & \mathbf{AB} \\ -1 & 0 & \cdots 0 & \\ 0 & -1 & \ddots \vdots & \\ \vdots & \ddots & \ddots 0 & \mathbf{0} \\ 0 & \cdots 0 & -1 & \end{vmatrix}. \qquad (3\text{-}74)$$

Finally, the Laplace expansion of the determinant in Eq. (3-74) along the $(2n)$th column shows that

$$D = |\mathbf{AB}|, \tag{3-75}$$

which proves the theorem. Note also that the determinant of a matrix product is commutative, since

$$|\mathbf{AB}| = |\mathbf{A}| \cdot |\mathbf{B}| = |\mathbf{B}| \cdot |\mathbf{A}| = |\mathbf{BA}|. \tag{3-76}$$

There are several other properties of matrices that will be useful to us in later discussions that are conveniently introduced at this point. The *rank* (r) of a matrix $\mathbf{A}$ is defined as the order of the largest square matrix with a determinant that does not vanish that can be taken from $\mathbf{A}$. Obviously $r \leqslant n$, where n is the order of the matrix $\mathbf{A}$. For example, the matrix

$$\mathbf{A} = \begin{pmatrix} 1 & 0 & 0 & 0 \\ 4 & 3 & 2 & 1 \\ 1 & 0 & 1 & 4 \\ 8 & 6 & 4 & 2 \end{pmatrix} \tag{3-76}$$

cannot be of rank four because $|\mathbf{A}| = 0$, since the second and fourth rows are multiples of each other. To see if $\mathbf{A}$ is of rank three, we must examine several possibilities, including

$$\begin{vmatrix} 1 & 0 & 0 \\ 4 & 3 & 2 \\ 1 & 0 & 1 \end{vmatrix} = 3, \quad \begin{vmatrix} 4 & 3 & 2 \\ 1 & 0 & 1 \\ 8 & 6 & 4 \end{vmatrix} = 0, \quad \begin{vmatrix} 3 & 2 & 1 \\ 0 & 1 & 4 \\ 6 & 4 & 2 \end{vmatrix} = 0, \quad \begin{vmatrix} 0 & 0 & 0 \\ 3 & 2 & 1 \\ 0 & 1 & 4 \end{vmatrix} = 0.$$

Since one of the above possibilities does have a nonzero determinant, $\mathbf{A}$ is of rank three. In practice we only need to examine the subdeterminants of $\mathbf{A}$ until we find one that is nonzero. (It is left as a student exercise to show that we can form 16 subdeterminants of order 3 for $\mathbf{A}$.)

If the rows or columns of a square matrix are treated as vectors, then we also note that the rank of the matrix is the same as the number of linearly independent rows or columns. This can be seen as follows. Let us consider the columns of some matrix $\mathbf{B}$,

$$\mathbf{B} = \begin{pmatrix} b_{11} & b_{12} & \cdots & b_{1n} \\ b_{21} & b_{22} & \cdots & b_{2n} \\ \vdots & \vdots & & \vdots \\ b_{n1} & b_{n2} & \cdots & b_{nn} \end{pmatrix} \tag{3-77}$$

as vectors, i.e.,

$$\mathbf{b}_1 = \begin{pmatrix} b_{11} \\ b_{21} \\ \vdots \\ b_{n1} \end{pmatrix}, \; \mathbf{b}_2 = \begin{pmatrix} b_{12} \\ b_{22} \\ \vdots \\ b_{n2} \end{pmatrix}, \; \ldots, \; \mathbf{b}_1 = \begin{pmatrix} b_{1n} \\ b_{2n} \\ \vdots \\ b_{nn} \end{pmatrix}. \tag{3-78}$$

This gives $\mathbf{B}$ in the following form

$$\mathbf{B} = (\mathbf{b}_1 \;\; \mathbf{b}_2, \; \ldots, \; \mathbf{b}_n). \tag{3-79}$$

For the $\mathbf{b}_i$ to be linearly independent, the equation

$$c_1\mathbf{b}_1 + c_2\mathbf{b}_2 + \cdots + c_n\mathbf{b}_n = \mathbf{0}, \tag{3-80}$$

must imply that $c_1 = c_2 = \cdots = c_n = 0$. If indicated multiplications in Eq. (3-80) are carried out, we obtain the following set of n homogeneous linear equations,

$$\begin{aligned} b_{11}c_1 + b_{12}c_2 + \cdots + b_{1n}c_n &= 0 \\ b_{21}c_1 + b_{22}c_2 + \cdots + b_{2n}c_n &= 0 \\ &\vdots \\ b_{n1}c_1 + b_{n2}c_2 + \cdots + b_{nn}c_n &= 0. \end{aligned} \tag{3-81}$$

From the discussion in Section 3-2 (see pages 114–115), it is clear that $c_1 = c_2 = \cdots = c_n = 0$ only if $|\mathbf{B}| \neq 0$. If $|\mathbf{B}| = 0$, we obtain the nontrivial solution (i.e., not all $c_i = 0$), and hence the $\mathbf{b}_i$ would be linearly dependent.

The same argument can be applied to the rows of **B** as was applied to the columns of **B**. This will lead to a determinant that is identical to **B**, except that its rows and columns have been interchanged. As was shown in Theorem 3-1, this determinant will have the same value as **B**.

Another useful idea is that of the *trace of a matrix*, which is defined as the sum of the diagonal elements of the matrix:

$$\text{tr}\,(\mathbf{A}) = \sum_{k=1}^{n} a_{kk}. \tag{3-82}$$

In the example of Eq. (3-76),

$$\text{tr}\,(\mathbf{A}) = 7.$$

An interesting property of the trace of a matrix that will be useful shortly is exhibited in the following theorem:

Theorem 3-6. If **A** and **B** are two $(n \times n)$ matrices, then tr $(\mathbf{AB}) =$ tr $(\mathbf{BA})$.

Proof: Since

$$(\mathbf{AB})_{ik} = \sum_{j=1}^{n} a_{ij}b_{jk}, \tag{3-83}$$

then

$$\text{tr}\,(\mathbf{AB}) = \sum_{i=1}^{n} \left[\sum_{j=1}^{n} a_{ij}b_{ji} \right]. \tag{3-84}$$

Also, since

$$(\mathbf{BA})_{jk} = \sum_{j=1}^{n} b_{ji}a_{ik}, \tag{3-85}$$

then

$$\operatorname{tr}(\mathbf{BA})=\sum_{j=1}^{n}\left[\sum_{i=1}^{n} b_{ji}a_{ij}\right]. \tag{3-86}$$

Since the summations in Eq. (3-86) are finite, we may interchange them, which gives,

$$\operatorname{tr}(\mathbf{BA})=\sum_{i=1}^{n}\sum_{j=1}^{n} b_{ji}a_{ij}=\sum_{i=1}^{n}\left[\sum_{j=1}^{n} a_{ij}b_{ji}\right]=\operatorname{tr}(\mathbf{AB}), \tag{3-87}$$

which completes the proof.

3-4. Matrix Inversion

An important concept concerning matrices is that of the *inverse of a matrix*. If upon multiplication of two matrices, **A** and **B**, we obtain

$$\mathbf{AB}=\mathbf{I}, \tag{3-88}$$

then we say that the matrix **B** is the inverse of **A**, and denote it by

$$\mathbf{B}=\mathbf{A}^{-1}. \tag{3-89}$$

We could also have said that **A** was the inverse of **B**, in which case we would denote **A** by

$$\mathbf{A}=\mathbf{B}^{-1}. \tag{3-90}$$

Note also that **A** and **B** must be square matrices. We can now rewrite Eq. (3-88), using Eq. (3-89), as

$$\mathbf{AA}^{-1}=I. \tag{3-91}$$

As in the case of operators, the inverse matrix does not always exist. We shall now see under what conditions an inverse is guaranteed.

Theorem 3-7. An $(n \times n)$ matrix **A** has an inverse if and only if $|\mathbf{A}| \neq 0$.

Proof: First suppose $|\mathbf{A}| \neq 0$. We now want to show that a matrix **B** exists such that Eq. (3-88) is satisfied. To do this we shall construct a matrix $\mathbf{B} = \mathbf{A}^{-1}$, which requires for its existence that $|\mathbf{A}| \neq 0$. We begin by choosing the ijth element of **B** as

$$(\mathbf{B})_{ij}=b_{ij}=\frac{|A_{ji}|}{|\mathbf{A}|}, \tag{3-91}$$

for all i and j, where $|A_{ji}|$ is the cofactor of $(\mathbf{A})_{ji}$. Let us now examine the elements of the product **AB**:

$$(\mathbf{AB})_{ij}=\sum_{k=1}^{n}(\mathbf{A})_{ik}(\mathbf{B})_{kj}=\sum_{k=1}^{n} a_{ik}b_{kj}. \tag{3-92}$$

Substituting Eq. (3-91) into Eq. (3-92) gives

$$(\mathbf{AB})_{ij} = \sum_{k=1}^{n} a_{ik} \frac{|A_{jk}|}{|\mathbf{A}|} = \frac{1}{|\mathbf{A}|} \left\{ \sum_{k=1}^{n} a_{ik} |A_{jk}| \right\} . \tag{3-93}$$

If $i = j$, Eq. (3-93) becomes

$$(\mathbf{AB})_{ii} = \frac{1}{|\mathbf{A}|} \left\{ \sum_{k=1}^{n} a_{ik} |A_{ik}| \right\} , \tag{3-94}$$

where the summation on the right-hand side of the equation is just the Laplace expansion of $|\mathbf{A}|$ along the ith row, as given by Eq. (3-40). Hence,

$$(\mathbf{AB})_{ii} = \frac{|\mathbf{A}|}{|\mathbf{A}|} = 1. \tag{3-95}$$

The summation for $i \neq j$, i.e.

$$\sum_{k=1}^{n} a_{ik} |A_{jk}|$$

can be shown to be the same as the Laplace expansion of $|\mathbf{A}|$ if the ith and jth rows are identical. From Theorem 3-3, this expansion must be zero, i.e.,

$$\sum_{k=1}^{n} a_{ik} |A_{jk}| = 0, \qquad i \neq j. \tag{3-96}$$

Combining Eqs. (3-96) and (3-93), we see that

$$(\mathbf{AB})_{ij} = 0, \qquad i \neq j. \tag{3-97}$$

Hence, we see that the matrix $\mathbf{B}$, defined as

$$\mathbf{B} = \frac{1}{|\mathbf{A}|} \begin{pmatrix} |A_{11}| & |A_{21}| & \cdots & |A_{n1}| \\ |A_{12}| & |A_{22}| & \cdots & |A_{2n}| \\ \vdots & \vdots & & \vdots \\ |A_{1n}| & |A_{2n}| & \cdots & |A_{nn}| \end{pmatrix} = \mathbf{A}^{-1} \tag{3-98}$$

is, in fact, $\mathbf{A}^{-1}$. The matrix in Eq. (3-98) is sometimes called the *adjoint* of $\mathbf{A}$ and is usually written as Adj($\mathbf{A}$).[7] This simplifies Eq. (3-98) to

$$\mathbf{B} = \frac{Adj(\mathbf{A})}{|\mathbf{A}|} = \mathbf{A}^{-1}. \tag{3-99}$$

To show that the existence of an inverse implies $|\mathbf{A}| \neq 0$, we write

$$\mathbf{I} = \mathbf{AA}^{-1}. \tag{3-100}$$

[7] See Footnote 5.

Taking the determinant of both sides of Eq. (3-100) yields

$$|\mathbf{I}|=1=|\mathbf{A}\mathbf{A}^{-1}|=|\mathbf{A}|\ |\mathbf{A}^{-1}|\neq 0. \tag{3-101}$$

This shows that neither $|\mathbf{A}|$ nor $|\mathbf{A}^{-1}|$ can equal zero, and also that

$$|\mathbf{A}^{-1}|=|\mathbf{A}|^{-1}, \tag{3-102}$$

which completes the proof.

Using Eq. (3-98), it can also be shown that

$$\mathbf{BA}=\mathbf{A}^{-1}\mathbf{A}=\mathbf{I}. \tag{3-103}$$

Combining Eqs. (3-91) and (3-103), we obtain

$$\mathbf{A}\mathbf{A}^{-1}=\mathbf{A}^{-1}\mathbf{A}=\mathbf{I}, \tag{3-104}$$

which shows that both a "right" and "left" inverse exist.

As an example, let us compute the inverse of the following matrix

$$\mathbf{A}=\begin{pmatrix} 1 & -1 & 0 \\ -1 & 0 & 1 \\ 2 & 1 & 1 \end{pmatrix}. \tag{3-105}$$

The matrix of cofactors is given by

$$\mathbf{A}'=\begin{pmatrix} -1 & +3 & -1 \\ 1 & 1 & -3 \\ -1 & -1 & -1 \end{pmatrix}, \tag{3-106}$$

where, e.g.,

$$|A_{12}|=(-1)\begin{vmatrix} -1 & 1 \\ 2 & 1 \end{vmatrix}=3. \tag{3-107}$$

The value of $|\mathbf{A}|$ is easily seen to be

$$|\mathbf{A}|=a_{11}|A_{11}|+a_{12}|A_{12}|+a_{13}|A_{13}| \tag{3-108}$$

$$=-4. \tag{3-109}$$

Next, we form

$$\mathrm{Adj}(\mathbf{A})=\begin{pmatrix} -1 & 1 & -1 \\ 3 & 1 & -1 \\ -1 & -3 & -1 \end{pmatrix}, \tag{3-110}$$

and obtain $\mathbf{A}^{-1}$ by dividing $\mathrm{Adj}(\mathbf{A})$ by $|\mathbf{A}|$, i.e.,

$$\mathbf{A}^{-1}=\begin{bmatrix} \frac{1}{4} & -\frac{1}{4} & \frac{1}{4} \\ -\frac{3}{4} & -\frac{1}{4} & +\frac{1}{4} \\ \frac{1}{4} & \frac{3}{4} & \frac{1}{4} \end{bmatrix} \tag{3-111}$$

Verification of the relation

$$\mathbf{A}^{-1}\mathbf{A} = \mathbf{A}\mathbf{A}^{-1} = \mathbf{I}$$

is left as an exercise for the student.

It is obvious that when matrices are greater than (4×4), this procedure is not convenient. There are, however, many suitable methods[8] for these cases that can be employed, which are well-suited for adaptation to digital computers.

To gain further familiarity with matrix manipulations and see directly how the use of them simplifies the manipulations, let us consider once again the now familiar problem of finding the solution to the following set of n simultaneous, linear, nonhomogeneous equations in n unknowns:

$$\begin{aligned} a_{11}x_1 + a_{12}x_2 + \cdots + a_{1n}x_n &= b_1 \\ a_{21}x_1 + a_{22}x_2 + \cdots + a_{2n}x_n &= b_2 \\ \vdots \quad\quad \vdots \quad\quad\quad\quad \vdots \quad &\quad \vdots \\ a_{n1}x_1 + a_{n2}x_2 + \cdots + a_{nn}x_n &= b_n, \end{aligned} \tag{3-112}$$

where a_{ij} and b_i are known, and we wish to find the various x_i. We first note that each of the n equations can be written in a somewhat more compact form:

$$\sum_{k=1}^{n} a_{ik}x_k = b_i. \tag{3-113}$$

Next, we note that, for each i, Eq. (3-113) has the same form as that of the product of a row vector and a column vector (in which we have suppressed the second subscript of the column vector),

$$(a_{i1} \;\; a_{i2} \;\; \cdots \;\; a_{in}) \begin{pmatrix} x_1 \\ x_2 \\ \vdots \\ x_n \end{pmatrix} = \mathbf{b}_i.$$

In other words, Eq. (3-113) has the form of a scalar product of two vectors [see, e.g., Eq. (3-22)]. Combining this result for all values of i, we see that the set of equations of Eq. (3-112) can be expressed neatly in matrix form as

$$\mathbf{A}\mathbf{x} = \mathbf{b}, \tag{3-114}$$

where

$$\mathbf{A} = \begin{pmatrix} a_{11} & a_{12} & \cdots & a_{1n} \\ a_{21} & a_{22} & \cdots & a_{2n} \\ \vdots & \vdots & & \vdots \\ a_{n1} & a_{n2} & \cdots & a_{nn} \end{pmatrix}, \mathbf{x}_1 = \begin{pmatrix} x_1 \\ x_2 \\ \vdots \\ x_n \end{pmatrix}, \quad \mathbf{b} = \begin{pmatrix} b_1 \\ b_2 \\ \vdots \\ b_n \end{pmatrix}. \tag{3-115}$$

[8] See, for example, A. Ralston and H. S. Wilf, *Mathematic Methods for Digital Computers,* Vol. 1, John Wiley, New York, 1960; see also G. W. Stewart, *Introduction to Matrix Computations,* Academic Press, New York, 1973.

From Eq. (3-114) it is easily seen that the desired solutions can be obtained by left multiplying both sides of the equation by A^{-1}, i.e.,

$$\mathbf{A}^{-1}\mathbf{A}\mathbf{x}=\mathbf{A}^{-1}\mathbf{b}. \tag{3-116}$$

Applying Eq. (3-104), we obtain finally

$$\mathbf{x}=\mathbf{A}^{-1}\mathbf{b}. \qquad (3\text{-}117)^{9}$$

Before discussing an example, it is instructive to see that solutions obtained in this manner are perfectly consistent with those obtained using Cramer's rule [see Eq. (3-56)]. Consider the ith solution, x_i. Using Eq. (3-117), this is given by

$$x_i=\sum_{k=1}^{n}(\mathbf{A}^{-1})_{ik}b_k. \tag{3-118}$$

Substituting the form of $\mathbf{A}^{-1}$ shown in Eq. (3-98) into Eq. (3-118), we arrive at

$$x_i=\frac{1}{|\mathbf{A}|}\sum_{k=1}^{n}|A_{ki}|b_k, \tag{3-119}$$

where, as before, $|A_{ki}|$ is the cofactor of a_{ki}.

Now consider Cramer's form of the solution

$$x_i=\frac{\begin{vmatrix} a_{11} & a_{12} & \cdots & a_{1,i-1} & b_1 & a_{1,i+1} & \cdots & a_{1n} \\ a_{21} & a_{22} & & a_{2,i-1} & b_2 & a_{2,i+1} & & a_{2n} \\ \vdots & \vdots & & \vdots & \vdots & \vdots & & \vdots \\ a_{n1} & a_{n2} & & a_{n,i-1} & b_n & a_{n,i+1} & \cdots & a_{nn} \end{vmatrix}}{|\mathbf{A}|}. \tag{3-120}$$

After using Laplace's expansion for the ith column, we obtain

$$x_i=\frac{1}{|\mathbf{A}|}\sum_{k=1}^{n}|A_{ki}|b_k,$$

which is identical to Eq. (3-119) above.

As an example of solving a system of linear nonhomogeneous equations, consider the following set

$$\begin{aligned} x_1 - x_2 \qquad &= 3 \\ -x_1 \qquad + x_3 &= 2. \\ 2x_1 + x_2 + x_3 &= 1 \end{aligned} \tag{3-121}$$

[9] For further information concerning the solutions of linear systems of equations such as $\mathbf{Ax} = \mathbf{b}$, including a discussion of the homogeneous case ($\mathbf{b} = \mathbf{0}$), see, e.g., P. Lancaster, *Theory of Matrices,* Academic Press, New York, 1969, pp. 45–49.

Writing these equations in matrix notation, we have

$$\begin{pmatrix} 1 & -1 & 0 \\ -1 & 0 & 1 \\ 2 & 1 & 1 \end{pmatrix}\begin{pmatrix} x_1 \\ x_2 \\ x_3 \end{pmatrix} = \begin{pmatrix} 3 \\ 2 \\ 1 \end{pmatrix}. \tag{3-122}$$

Note that the square matrix is the same as the one studied previously [see Eq. (3-105)]. The inverse of this matrix is given by Eq. (3-111). Therefore, the three solutions to the above equation are given by

$$\begin{pmatrix} x_1 \\ x_2 \\ x_3 \end{pmatrix} = \begin{pmatrix} \frac{1}{4} & -\frac{1}{4} & \frac{1}{4} \\ -\frac{3}{4} & -\frac{1}{4} & \frac{1}{4} \\ \frac{1}{4} & \frac{3}{4} & \frac{1}{4} \end{pmatrix}\begin{pmatrix} 3 \\ 2 \\ 1 \end{pmatrix}, \tag{3-123}$$

or

$$\begin{aligned} x_1 &= \left(\frac{1}{4}\right)(3) + \left(-\frac{1}{4}\right)(2) + \left(\frac{1}{4}\right)(1) &= \frac{1}{2} \\ x_2 &= \left(-\frac{3}{4}\right)(3) + \left(-\frac{1}{4}\right)(2) + \left(\frac{1}{4}\right)(1) &= \frac{5}{2}. \\ x_3 &= \left(\frac{1}{4}\right)(3) + \left(\frac{3}{4}\right)(2) + \left(\frac{1}{4}\right)(1) &= \frac{5}{2} \end{aligned} \tag{3-124}$$

The analogy between the form of Eq. (3-113) and an inner product can be taken even further. If two matrices **A** and **B** are multiplied together to form **C** = **AB**, where **A** is an $(m \times n)$ matrix and **B** is an $(n \times p)$ matrix, then each of the members of **C** can be thought of as resulting from the inner product of a row vector of **A** with a column vector of **B**, *i.e.*

$$c_{ij} = \sum_{k=1}^{n} a_{ik} b_{kj} = (a_{i1} \; a_{i2} \; \cdots \; a_{in}) \begin{pmatrix} b_{1j} \\ b_{2j} \\ \vdots \\ b_{nj} \end{pmatrix} = (\mathbf{a}_i, \mathbf{b}_j). \tag{3-125}$$

One of the questions left unanswered in the previous discussion is the uniqueness of $\mathbf{A}^{-1}$, to which we now turn.

Theorem 3-8. If **AB** = I exists, then **B** = $\mathbf{A}^{-1}$ is unique.

Proof: Assume **B** = $\mathbf{A}^{-1}$ is not unique, and that **C** is another inverse of **A**. Then, **CA** = **I**. Multiplying on the right of both sides of the above equation by **B** yields

$$\mathbf{C}(\mathbf{AB}) = \mathbf{IB} = \mathbf{B},$$

or

$$\mathbf{CI} = \mathbf{C} = \mathbf{B},$$

which is contrary to the initial assumption. Therefore, **B** is unique.

Finally, we shall prove one additional theorem of importance.

Theorem 3-9. If two matrices **A** and **B** each possess an inverse, then $(\mathbf{AB})^{-1} = \mathbf{B}^{-1}\mathbf{A}^{-1}$.

Proof: By the definition of an inverse,

$$(\mathbf{AB})(\mathbf{AB})^{-1} = \mathbf{I}.$$

Left multiply both sides of the equation by $(\mathbf{B}^{-1}\mathbf{A}^{-1})$, which yields

$$(\mathbf{B}^{-1}\mathbf{A}^{-1})(\mathbf{AB})(\mathbf{AB})^{-1} = (\mathbf{B}^{-1}\mathbf{A}^{-1})\mathbf{I}.$$

Applying the definition of inverse to the left-hand side of the above equation, we obtain

$$(\mathbf{AB})^{-1} = \mathbf{B}^{-1}\mathbf{A}^{-1}, \tag{3-126}$$

which completes the proof.

3-5. Matrices Having Special Properties

In this section we shall introduce several operations that can be applied to matrices, and introduce some terminology that is associated with matrices that have special properties.

There are three special operations that can be applied to matrices. The first of these is that of forming the *transpose* $\mathbf{A}^t$ of a matrix **A**, which is defined to be

$$(\mathbf{A}^t)_{ij} = a_{ji}, \tag{3-127}$$

for all i and j. This amounts to interchanging each of the rows and columns of the original matrix. Using the example of Eq. (3-76),

$$\mathbf{A}^t = \begin{pmatrix} 1 & 4 & 1 & 8 \\ 0 & 3 & 0 & 6 \\ 0 & 2 & 1 & 4 \\ 0 & 1 & 4 & 2 \end{pmatrix} \tag{3-128}$$

This operation also changes row and column vectors into column and row vectors, respectively.

The second operation of interest is that of *complex conjugation* of a matrix. The complex conjugate of $\mathbf{A}^*$ of a matrix **A** is formed by

$$(\mathbf{A}^*)_{ij} = a_{ij}^*, \tag{3-129}$$

for all i and j.

Finally, the *adjoint* $\mathbf{A}^\dagger$ of a matrix **A** is defined as

$$(\mathbf{A}^\dagger)_{ij} = a_{ji}^*, \tag{3-130}$$

for all i and j. Thus, $\mathbf{A}^\dagger$ can be thought of as the complex conjugate-transpose of **A**. Note also that if all of the elements of **A** are real, $\mathbf{A}^\dagger = \mathbf{A}^t$. As an example, the adjoint of a row or column vector

$$\mathbf{a} = (a_1 \ a_2 \ \cdots a_n), \qquad \mathbf{b} = \begin{pmatrix} b_1 \\ b_2 \\ \vdots \\ b_n \end{pmatrix}, \tag{3-131}$$

is

$$\mathbf{a}^\dagger = \begin{pmatrix} a_1^* \\ a_2^* \\ \vdots \\ a_n^* \end{pmatrix}, \qquad \mathbf{b}^\dagger = (b_1^* \; b_2^* \; \cdots \; b_n^*). \tag{3-132}$$

The following list gives the names and description of various types of matrices which are useful in many applications:

1. If $\mathbf{A}^* = \mathbf{A}$, then $\mathbf{A}$ is a *real matrix*.
2. If $\mathbf{A}^* = -\mathbf{A}$, then $\mathbf{A}$ is a *pure imaginary matrix*.
3. If $\mathbf{A}^t = \mathbf{A}$, then $\mathbf{A}$ is a *symmetric matrix*.
4. If $\mathbf{A}^t = -\mathbf{A}$, then $\mathbf{A}$ is a *skew-symmetric matrix*.
5. If $[\mathbf{A}, \mathbf{A}^\dagger] = 0$, then $\mathbf{A}$ is a *normal matrix*.
6. If $\mathbf{A}^\dagger = \mathbf{A}$, then $\mathbf{A}$ is a *Hermitian matrix*.
7. If $\mathbf{A}^\dagger = -\mathbf{A}$, then $\mathbf{A}$ is a *skew-Hermitian matrix*.
8. If $\mathbf{A}^{-1} = \mathbf{A}^t$, then $\mathbf{A}$ is an *orthogonal matrix*.
9. If $\mathbf{A}^{-1} = \mathbf{A}^\dagger$, then $\mathbf{A}$ is a *unitary matrix*.
10. If $\mathbf{A}^2 = \mathbf{A}$, then $\mathbf{A}$ is *idempotent*.
11. If $\mathbf{A}^p = \mathbf{0}$, for some positive integer p then $\mathbf{A}$ is *nilpotent*.
12. If $\mathbf{A}^2 = \mathbf{I}$, then $\mathbf{A}$ is *involutory*.
13. If $(\mathbf{A})_{ij} = 0$ for $i > j$, then $\mathbf{A}$ is an *upper triangular matrix*.
14. If $(\mathbf{A})_{ij} = 0$ for $i < j$, then $\mathbf{A}$ is a *lower triangular matrix*.
15. If $(\mathbf{A})_{ij} = 0$ for $|i-j| > 1$, then $\mathbf{A}$ is a *tridiagonal matrix*.
16. If $(\mathbf{A})_{ij} = 0$ for $|i-j| > l$, then $\mathbf{A}$ is a *banded matrix* with bandwidth $2l+1$

The choice of names such as Hermitian and unitary for matrices having the properties shown is not simply coincidental, for we shall see shortly that there is a close relationship between the matrices and the corresponding operators of the same name. In fact, there is an *isomorphorism* between them.

Finally, before leaving this section we shall prove a theorem that will be used frequently in later discussions.

Theorem 3-10. If $\mathbf{A}$ and $\mathbf{B}$ are two square matrices, then

$$(\mathbf{AB})^\dagger = \mathbf{B}^\dagger \mathbf{A}^\dagger. \tag{3-133}$$

Proof:

$$(\mathbf{AB})_{ij}^\dagger = \left(\sum_{k=1}^{n} a_{ik} b_{kj} \right)^\dagger. \tag{3-134}$$

Using Eq. (3-130) gives

$$(\mathbf{AB})^{\dagger}_{ij} = \left(\sum_{k=1}^{n} a_{jk} b_{ki} \right)^{*} = \sum_{k=1}^{n} a^{*}_{jk} b^{*}_{ki}. \qquad (3\text{-}135)$$

Applying Eq. (3-130) again yields

$$(\mathbf{AB})^{\dagger}_{ij} = \sum_{k=1}^{n} (\mathbf{A}^{\dagger})_{kj} (\mathbf{B}^{\dagger})_{ik} = \sum_{k=1}^{n} (\mathbf{B}^{\dagger})_{ik} (\mathbf{A}^{\dagger})_{kj}, \qquad (3\text{-}136)$$

for all i and j. Hence,

$$(\mathbf{AB})^{\dagger} = \mathbf{B}^{\dagger}\mathbf{A}^{\dagger}, \qquad (3\text{-}137)$$

which completes the proof.

3-6. Matrix Representations of Linear Operators and Matrix Transformations

In Chapter 2 we discussed the representation of a general vector in an n-dimensional vector space V. As discussed there, if $\{\boldsymbol{v}_1 \ \boldsymbol{v}_2 \ \cdots \ \boldsymbol{v}_n\}$ is a basis for V, then an arbitrary vector $\boldsymbol{u}$ can be represented as[10]

$$\boldsymbol{u} = \sum_{i=1}^{n} u_i \boldsymbol{v}_i. \qquad (3\text{-}138)$$

From Eq. (3-138), we also see that once the basis is chosen, the vector $\boldsymbol{u}$ is completely determined by the linear array of numbers:

$$\begin{pmatrix} u_1 \\ u_2 \\ \vdots \\ u_n \end{pmatrix}. \qquad (3\text{-}139)$$

This array can be considered to be a column vector, which is written as

$$\mathbf{u} = \begin{pmatrix} u_1 \\ u_2 \\ \vdots \\ u_n \end{pmatrix}. \qquad (3\text{-}140)$$

Now consider a second vector

$$\boldsymbol{w} = \sum_{i=1}^{n} w_i \boldsymbol{v}_i, \qquad (3\text{-}141)$$

[10] The script symbols $\boldsymbol{u}$ and $\boldsymbol{v}$ that are used in Eq. (3-138) and in Chapter 2 for general vectors in V should be distinguished from symbols such as $\mathbf{u}$ and $\mathbf{v}$, which are used in this chapter to represent column or row vectors (which are basis dependent).

represented by the column vector

$$\mathbf{w} = \begin{pmatrix} w_1 \\ w_2 \\ \vdots \\ w_n \end{pmatrix}, \tag{3-142}$$

and define addition and scalar multiplication of the column vectors by

$$\mathbf{u} + \mathbf{w} = \begin{pmatrix} u_1 + w_1 \\ u_2 + w_2 \\ \vdots \\ u_n + w_n \end{pmatrix}, \tag{3-143}$$

and

$$\mathbf{cu}_2 \begin{pmatrix} cu_1 \\ cu_2 \\ \vdots \\ cu_n \end{pmatrix}, \tag{3-144}$$

where c is a real or complex scalar. It can now be shown that the set of n column vectors $\{\mathbf{u}\}$ also forms an n-dimensional vector space V'. The proof of this is quite similar to that for the n-component vectors discussed in Section 2-1,[11] and is left as an exercise for the student. Furthermore, not only does the set of column vectors form a vector space, the discussion in Section 2-2 makes it clear that the space of vectors V is *isomorphic* to the space of column vectors V'. It is important to realize, however, that the column vector representations $\{\mathbf{u}\}$ of the vectors $\{\boldsymbol{u}\}$ are dependent on the particular basis that is chosen.

We also noted in our study in Chapter 2 that the type of linear transformation (or mapping) in which the vector space V is mapped onto itself, i.e., $\mathcal{L}: V \to V$, will be of particular interest to us. Let us now consider what relationships exist between the set of linear transformations $\mathcal{L}$ on vectors $\boldsymbol{u}$ in V, and the corresponding operations on column vectors $\mathbf{u}$ in the vector space V'.

Consider an n-dimensional vector space V with basis vectors $\{\boldsymbol{v}_1, \boldsymbol{v}_2, \cdots, \boldsymbol{v}_n\}$. From chapter 2, we know that any arbitrary vector $\boldsymbol{u}$ in this space can be expressed as a linear combination of these basis vectors, i.e.,

$$\boldsymbol{u} = \sum_{i=1}^{n} u_i \boldsymbol{v}_i, \tag{3-145}$$

where the u_i are scalars, known as the Fourier (or expansion) coefficients. For a linear operator $\mathcal{L}$ that maps V onto itself, the effect of $\mathcal{L}$ upon the arbitrary vector $\boldsymbol{u}$ can be written as

$$\mathcal{L}\boldsymbol{u} = \mathcal{L}\left[\sum_{i=1}^{n} u_i \boldsymbol{v}_i\right] = \sum_{i=1}^{n} u_i(\mathcal{L}\boldsymbol{v}_i), \tag{3-146}$$

[11] See, in particular, Eqs. (2-8) through (2-16).

by the definition of a linear operator. However, the effect of $\mathcal{L}$ on $\boldsymbol{v}_i$ is simply to produce another vector in V, which can also be uniquely represented in terms of the basis vectors, i.e.,

$$\mathcal{L}\boldsymbol{v}_i = \sum_{j=1}^{n} l_{ji}\boldsymbol{v}_j. \tag{3-147}$$

Substituting this result into Eq. (3-146) yields

$$\begin{aligned} \mathcal{L}\boldsymbol{u} &= \sum_{i=1}^{n} u_i \left[\sum_{j=1}^{n} l_{ji}\boldsymbol{v}_j\right] \\ &= \sum_{j=1}^{n} \left[\sum_{i=1}^{n} u_i l_{ji}\right] \boldsymbol{v}_i, \end{aligned} \tag{3-148}$$

where the summations have been rearranged for convenience.

However, since $\boldsymbol{u}$ is a vector in the vector space V, we know that operating upon $\boldsymbol{u}$ will simply produce another vector in the space, i.e.,

$$\mathcal{L}\boldsymbol{u} = \boldsymbol{w} \tag{3-149}$$

$$= \sum_{j=1}^{n} w_i \boldsymbol{v}_j. \tag{3-150}$$

Comparison of the coefficients of the vectors ($\boldsymbol{v}_j$) in Eqs. (3-148) and (3-150) shows that

$$w_j = \sum_{i=1}^{n} u_i l_{ji}, \qquad j = 1, 2, \dots, n. \tag{3-151}$$

This set of equations can be written compactly in matrix form as

$$\mathbf{Lu} = \mathbf{w}, \tag{3-152}$$

where

$$(\mathbf{L})_{ij} = l_{ij}, \tag{3-153}$$

for all i and j, and

$$\mathbf{u} = \begin{pmatrix} u_1 \\ u_2 \\ \vdots \\ u_n \end{pmatrix}, \qquad \mathbf{w} = \begin{pmatrix} w_1 \\ w_2 \\ \vdots \\ w_n \end{pmatrix}. \tag{3-154}$$

The matrix transformation of $\mathbf{u}$ into w by $\mathbf{L}$ is known as a *linear matrix transformation*.

The results indicated in Eqs. (3-152) and (3-149) are quite important, for they show that for every linear operator $\mathcal{L}$, there is, for a given basis $\{\boldsymbol{v}_i\}$, a unique matrix $\mathbf{L}$ corresponding to it, that maps column vectors in V' onto V'. In other words, for a given basis, there is a one-to-one correspondence between every linear operator $\mathcal{L}$ and matrix $\mathbf{L}$. The terminology associated with this relationship that is usually used is that $\mathbf{L}$ is *called a matrix representation of* $\mathcal{L}$ *in the basis* $\{\boldsymbol{v}\}$. This can be written symbolically as

$$\mathcal{L} \overset{\{\boldsymbol{v}_i\}}{\Leftrightarrow} \mathbf{L}. \tag{3-155}$$

The reader will recognize from the discussions in Chapter 2 that a one-to-one correspondence such as the one just established is the first step toward establishing an isomorphism between the linear operators $\mathcal{L}$ and their corresponding matrix representations $\mathbf{L}$. The remaining steps in the proof of this isomorphism are given in the following theorem.

Theorem 3-11. The set of matrices $\mathbf{L}: V' \to V'$ are isomorphic to the corresponding set of linear transformations $\mathcal{L}: V \to V$, with respect to addition, multiplication, and multiplication by scalars.

Proof: Consider first the case of addition. Let $\mathcal{L}'$ and $\mathcal{L}''$ be two linear transformations whose sum is given by

$$\mathcal{L} = \mathcal{L}' + \mathcal{L}''. \tag{3-156}$$

The effect of operating on a basis vector $\boldsymbol{v}_i$ by $\mathcal{L}' + \mathcal{L}''$ is given by

$$(\mathcal{L}' + \mathcal{L}'')\boldsymbol{v}_i = \mathcal{L}'\boldsymbol{v}_i + \mathcal{L}''\boldsymbol{v}_i, \qquad i = 1, 2, \ldots, n. \tag{3-157}$$

Applying Eq. (3-143) to both terms on the right-hand side of Eq. (3-157) gives

$$(\mathcal{L}' + \mathcal{L}'')\boldsymbol{v}_i = \sum_{j=1}^{n} l'_{ji}\boldsymbol{v}_j + \sum_{j=1}^{n} l''_{ji}\boldsymbol{v}_j, \tag{3-158}$$

$$= \sum_{j=1}^{n} (l'_{ji} + l''_{ji})\boldsymbol{v}_j, \qquad i = 1, 2, \ldots, n. \tag{3-159}$$

If the operator $\mathcal{L}$ of Eq. (3-156) operates on $\boldsymbol{v}_i$, we have, by Eq. (3-147),

$$\mathcal{L}\boldsymbol{v}_i = \sum_{j=1}^{n} l_{ji}\boldsymbol{v}_j, \qquad i = 1, 2, \ldots, n \tag{3-160}$$

Combining Eqs. (3-156), (3-159), and (3-160) gives

$$\sum_{j=1}^{n} l_{ji}\boldsymbol{v}_j = \sum_{j=1}^{n} (l'_{ji} + l''_{ji})\boldsymbol{v}_j, \qquad i = 1, 2, \ldots, n \tag{3-161}$$

which can be rearranged to

$$\sum_{j=1}^{n} [l_{ji} - (l'_{ji} + l''_{ji})]\boldsymbol{v}_j = 0, \qquad i = 1, 2, \ldots, n \tag{3-162}$$

Since the $\{\mathbf{v}_j\}$ are linearly independent, we obtain the result that

$$l_{ji} = l'_{ji} + l''_{ji}, \tag{3-163}$$

for all i and j, which is just the expression that defines matrix addition that was given in Eq. (3-7). Thus, we have shown that

$$\mathcal{L}' + \mathcal{L}'' \overset{\{\mathbf{v}_i\}}{\Leftrightarrow} \mathbf{L}' + \mathbf{L}''. \tag{3-164}$$

Next, we consider the case of multiplication. The product of linear transformations $\mathcal{L}'$ and $\mathcal{L}''$ is given by

$$\mathcal{L} = \mathcal{L}'\mathcal{L}''. \tag{3-165}$$

If $\mathcal{L}'\ \mathcal{L}''$ operates on $\mathbf{v}_i$, we have

$$(\mathcal{L}'\mathcal{L}'')\mathbf{v}_i = \mathcal{L}'(\mathcal{L}''\mathbf{v}_i), \qquad i = 1, 2, \ldots, n \tag{3-166}$$

which, upon applying Eq. (3-147) to the term in parentheses, yields

$$\mathcal{L}'(\mathcal{L}''\mathbf{v}_i) = \mathcal{L}'\left(\sum_{j=1}^{n} l''_{ji}\mathbf{v}_j\right), \qquad i = 1, 2, \ldots, n. \tag{3-167}$$

Since $\mathcal{L}'$ is a linear operator, Eq. (3-167) becomes

$$\mathcal{L}'(\mathcal{L}''\mathbf{v}_i) = \sum_{j=1}^{n} l''_{ji}(\mathcal{L}'\mathbf{v}_j), \qquad i = 1, 2, \ldots, n \tag{3-168}$$

Applying Eq. (3-146) again yields

$$\mathcal{L}'(\mathcal{L}''\mathbf{v}_i) = \sum_{j=1}^{n} l''_{ji}\left(\sum_{k=1}^{n} l'_{kj}\mathbf{v}_k\right), \tag{3-169}$$

$$= \sum_{k=1}^{n}\left(\sum_{j=1}^{n} l'_{kj}l''_{ji}\right)\mathbf{v}_k, \qquad i = 1, 2, \ldots, n. \tag{3-170}$$

As in the case of addition, combining Eqs. (3-165), (3-166), (3-170) and (3-160) and rearranging terms, we obtain

$$\sum_{k=1}^{n}\left[l_{ki} - \left(\sum_{j=1}^{n} l'_{kj}l''_{ji}\right)\right]\mathbf{v}_k = 0, \qquad i = 1, 2, \ldots, n \tag{3-171}$$

which yields

$$l_{ki} = \sum_{j=1}^{n} l'_{kj}l''_{ji}, \tag{3-172}$$

for all k and i. Examination of Eq. (3-172) shows that is is identical to the definition of matrix multiplication given in Eq. (3-11). Thus, we have shown

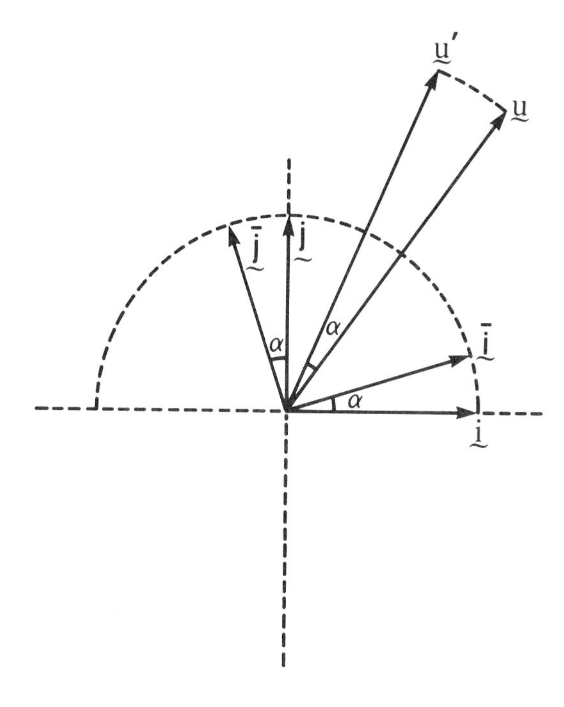

Figure 3.1. Rotation of a vector $\boldsymbol{u}$ (counterclockwise) into a vector $\boldsymbol{u}'$.

that

$$\mathcal{L}'\mathcal{L}'' \overset{\{\boldsymbol{v}_i\}}{\Leftrightarrow} \mathbf{L}'\mathbf{L}''. \tag{3-173}$$

It is left as an exercise for the student to show that

$$c\mathcal{L} \overset{\{\boldsymbol{v}_i\}}{\Leftrightarrow} c\mathbf{L}, \tag{3-174}$$

where c is a real or complex scalar, which completes the proof.

This is a very useful and important theorem, and we shall make extensive use of it in later chapters, when we discuss how quantities of physical interest in quantum mechanics can be represented in a computationally convenient form. Before proceeding with other aspects of matrix transformations, it will be useful to consider a specific example to illustrate how the isomorphism can be employed.

Let us consider a two-dimensional vector space with orthormal Cartesian basis vectors $\{\boldsymbol{i}, \boldsymbol{j}\}$ as shown in Fig. 3.1. Also, we shall be interested in two linear operators defined on this space. The first of these will be denoted by $\mathcal{R}_\alpha$, and represents a counterclockwise rotation of an arbitrary vector in the space by an angle α. The other operator will be denoted by $\mathcal{R}_l$, and represents a reflection of an arbitrary vector located at the origin through a line l, which also passes

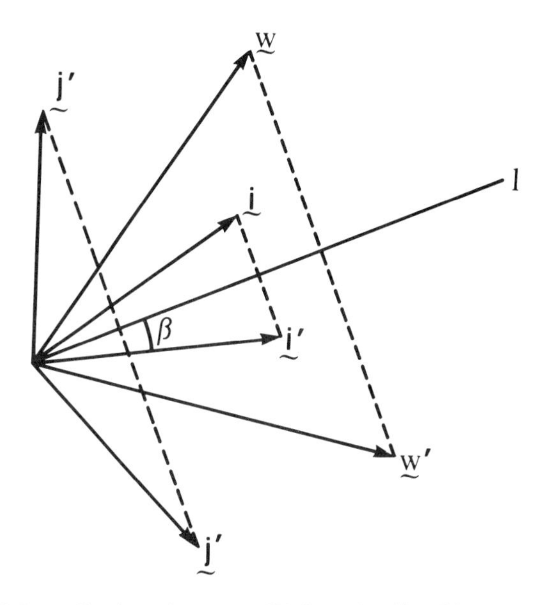

Figure 3.2. Reflection of a vector **W** through a line l into a vector **W**$'$.

through the origin (see Fig. 3.2). Clearly, such operators are examples of mappings that take geometric vectors in the plane onto themselves, as discussed earlier in Section 2-10.

Considering the rotation operator first, let us examine its effect on an arbitrary vector $\boldsymbol{u}$ in the space, i.e.,

$$\mathcal{R}_\alpha \boldsymbol{u} = \boldsymbol{u}'. \tag{3-175}$$

This mapping is illustrated pictorially in Fig. 3.1. In the Cartesian basis, the vectors $\boldsymbol{u}$ and $\boldsymbol{u}'$ are given by

$$\boldsymbol{u} = u_1 \boldsymbol{i} + u_2 \boldsymbol{j}, \tag{3-176}$$

and

$$\boldsymbol{u}' = u_1' \boldsymbol{i} + u_2' \boldsymbol{j}. \tag{3-177}$$

Operating on $\boldsymbol{u}$ in the Cartesian basis with the linear operator $\mathcal{R}_\alpha$, we obtain

$$\mathcal{R}_\alpha \boldsymbol{u} = u_1(\mathcal{R}_\alpha \boldsymbol{i}) + u_2(\mathcal{R}_\alpha \boldsymbol{j}). \tag{3-178}$$

From Fig. 3.1, we see that the effect of $\mathcal{R}_\alpha$ upon the basis vectors can be written as

$$\bar{\boldsymbol{i}} = \mathcal{R}_\alpha \boldsymbol{i} = \cos\alpha \cdot \boldsymbol{i} + \sin\alpha \cdot \boldsymbol{j}, \tag{3-179}$$

and

$$\bar{\boldsymbol{j}} = \mathcal{R}_\alpha \boldsymbol{j} = -\sin\alpha \cdot \boldsymbol{i} + \cos\alpha \cdot \boldsymbol{j}. \tag{3-180}$$

since the basis vectors are of unit length. Substituting Eqs. (3-179) and (3-180) into Eq. (3-178) and rearranging terms gives

$$\mathcal{R}_\alpha \boldsymbol{u} = [u_1 \cos\alpha - u_2 \sin\alpha]\boldsymbol{i} + [u_1 \sin\alpha + u_2 \cos\alpha]\boldsymbol{j}. \tag{3-181}$$

Combining Eqs. (3-175), (3-181), and equating coefficients yields

$$\begin{aligned} u_1' &= u_1 \cos\alpha - u_2 \sin\alpha \\ u_2' &= u_1 \sin\alpha + u_2 \cos\alpha. \end{aligned} \tag{3-182}$$

This can be written alternatively as

$$\begin{pmatrix} u_1' \\ u_2' \end{pmatrix} = \begin{pmatrix} \cos\alpha & -\sin\alpha \\ \sin\alpha & \cos\alpha \end{pmatrix} \begin{pmatrix} u_1 \\ u_2 \end{pmatrix} \tag{3-183}$$

or

$$\mathbf{u}' = \mathbf{R}_\alpha \mathbf{u}, \tag{3-184}$$

using matrix notation. Thus, we have constructed $\mathbf{R}_\alpha$, the matrix representation of the operator, for the particular basis set choice $\{\boldsymbol{i}, \boldsymbol{j}\}$. Using Theorem 3-11, the isomorphism between $\mathcal{R}_\alpha$ and $\mathbf{R}_\alpha$ can be written symbollically as

$$\mathcal{R}_\alpha \overset{\{\boldsymbol{i},\boldsymbol{j}\}}{\Leftrightarrow} \mathbf{R}_\alpha. \tag{3-185}$$

In words, Eq. (3-184) shows how the coordinates of the vector $\boldsymbol{u}$ taken in the particular basis system $\{\boldsymbol{i}, \boldsymbol{j}\}$ change when the vector is rotated counterclockwise by an angle α. Also, due to the isomorphism, we know that any properties we establish for $\mathbf{R}_\alpha$ will be true for $\mathcal{R}_\alpha$ as well.

It is important to note that the coordinates of the "new" vector $\boldsymbol{u}'$ are also taken with respect to the original basis vectors $\{\boldsymbol{i}, \boldsymbol{j}\}$. Such a coordinate transformation is said to be *active*, since the *vector itself* is transformed or changed with respect to the *same* basis system. We will shortly discuss the case of a *passive transformation*, where the vector stays fixed and the basis system is changed. It is left as an exercise for the student to show that the linear mapping

$$(\mathcal{R}_\beta \mathcal{R}_\alpha)\boldsymbol{u} = \boldsymbol{u}', \tag{3-186}$$

where $\mathcal{R}_\alpha$ and $\mathcal{R}_\beta$ represent counterclockwise rotations by angles α and β, respectively, can be represented by the matrix equation

$$\mathbf{R}_\beta \mathbf{R}_\alpha \mathbf{u} = \mathbf{u}' \tag{3-187}$$

in the $\{\boldsymbol{i}, \boldsymbol{j}\}$ basis.

Now let us consider the problem of applying the reflection operator $\mathcal{R}_l$ onto an arbitrary vector $\boldsymbol{w}$, i.e.,

$$\mathcal{R}_l \boldsymbol{w} = \boldsymbol{w}'. \tag{3-188}$$

This mapping is shown in Fig. 3.2. As before, we write the vectors $\boldsymbol{w}$ and $\boldsymbol{w}'$ in a Cartesian basis,

$$\boldsymbol{w} = w_1 \boldsymbol{i} + w_2 \boldsymbol{j}, \tag{3-189}$$

and

$$\boldsymbol{w}' = w_1' \boldsymbol{i} + w_2' \boldsymbol{j}. \tag{3-190}$$

Combining Eqs. (3-188) and (3-189) yields

$$\mathcal{R}_l \boldsymbol{w} = w_1 (\mathcal{R}_l \boldsymbol{i}) + w_2 (\mathcal{R}_l \boldsymbol{j}). \tag{3-191}$$

The new vectors $\mathcal{R}_l \boldsymbol{i}$ and $\mathcal{R}_l \boldsymbol{j}$ can be written, referring to Fig. 3.2, as

$$\boldsymbol{i}' = \mathcal{R}_l \boldsymbol{i} = \cos 2\beta \cdot \boldsymbol{i} + \sin 2\beta \cdot \boldsymbol{j}, \tag{3-192}$$

and

$$\begin{aligned} \boldsymbol{j} = \mathcal{R}_l \boldsymbol{j} &= \cos\left(\frac{\pi}{2} - 2\beta\right) \boldsymbol{i} - \sin\left(\frac{\pi}{2} - 2\beta\right) \boldsymbol{j} \\ &= \sin 2\beta \cdot \boldsymbol{i} - \cos 2\beta \cdot \boldsymbol{j}. \end{aligned} \tag{3-193}$$

Repeating the manipulations used in the case of rotation gives

$$\begin{pmatrix} w_1' \\ w_2' \end{pmatrix} = \begin{pmatrix} \cos 2\beta & \sin 2\beta \\ \sin 2\beta & -\cos 2\beta \end{pmatrix} \begin{pmatrix} w_1 \\ w_2 \end{pmatrix}, \tag{3-194}$$

or

$$\mathbf{w}' = \mathbf{R}_l \mathbf{w}. \tag{3-195}$$

Thus, similar to the previous case, $\mathbf{R}_l$ is a matrix representation of $\mathcal{R}_l$ with respect to the particular basis $\{\boldsymbol{i}, \boldsymbol{j}\}$, i.e.

$$\mathcal{R}_l \overset{\{\boldsymbol{i},\boldsymbol{j}\}}{\Leftrightarrow} \mathbf{R}_l.$$

Note that we are again dealing with an active transformation, which means that the coordinates of $\boldsymbol{w}$ and $\boldsymbol{w}'$ are both given with respect to the $\{\boldsymbol{i}, \boldsymbol{j}\}$ basis.

We have now discussed in detail the representation of linear operators with respect to a particular basis. This notion will be extremely useful in quantum mechanics, where the matrix representation of an operator is usually written in the following *inner product form*

$$(\boldsymbol{v}_i, \mathcal{L}\boldsymbol{v}_j) \equiv (\boldsymbol{\Lambda})_{ij} = \lambda_{ij}, \tag{3-196}$$

with respect to the $\{\boldsymbol{v}_1, \boldsymbol{v}_2, \ldots, \boldsymbol{v}_n\}$ basis, where $\mathcal{L}$ is a linear operator defined on this space. However, this matrix representation is *not* in general the same as the ones discussed previously, and we shall now see how it is related to the *true* matrix representation in the $\{\boldsymbol{v}_i\}$ basis.

As seen in Eq. (3-147), operation on the basis vectors by $\mathcal{L}$ gives

$$\mathcal{L}\boldsymbol{v}_j = \sum_{k=1}^{n} l_{kj} \boldsymbol{v}_k, \qquad j = 1, 2, \cdots, n. \tag{3-197}$$

Let us now form the matrix element λ_{ij}, i.e.,

$$\lambda_{ij} = \left(\mathbf{v}_i, \sum_{k=1}^{n} l_{kj} \mathbf{v}_k \right)$$

$$= \sum_{k=1}^{n} l_{kj} (\mathbf{v}_i, \mathbf{v}_k), \quad i, j = 1, 2, \cdots, n. \tag{3-198}$$

The inner products $(\mathbf{v}_i, \mathbf{v}_k)$ are simply elements of the Gramian matrix $\mathbf{G}$, which is the matrix corresponding to the Gram determinant shown in Eq. (3-60), viz.

$$\mathbf{G} = \begin{pmatrix} (\mathbf{v}_1, \mathbf{v}_1) & (\mathbf{v}_1, \mathbf{v}_2) & \cdots & (\mathbf{v}_1, \mathbf{v}_n) \\ (\mathbf{v}_2, \mathbf{v}_1) & (\mathbf{v}_2, \mathbf{v}_2) & \cdots & (\mathbf{v}_2, \mathbf{v}_n) \\ \vdots & \vdots & & \vdots \\ (\mathbf{v}_n, \mathbf{v}_1) & (\mathbf{v}_n, \mathbf{v}_2) & \cdots & (\mathbf{v}_n, \mathbf{v}_n) \end{pmatrix} \tag{3-199}$$

Equation (3-198) can now be written in matrix form as

$$\mathbf{\Lambda} = \mathbf{GL}, \tag{3-200}$$

where $\mathbf{L}$ is the *true* matrix representation of $\mathcal{L}$ in the $\{\mathbf{v}_i\}$ basis. Since the basis is linearly independent, the discussion on p. 115 showns that $|\mathbf{G}| \neq 0$, and hence $\mathbf{G}^{-1}$ exists. Therefore, we can also write the above equation as

$$\mathbf{L} = \mathbf{G}^{-1}\mathbf{\Lambda}. \tag{3-201}$$

It is important to emphasize that the matrices $\mathbf{\Lambda}_1, \mathbf{\Lambda}_2, \ldots$ corresponding to the set of operators $\mathcal{L}_1, \mathcal{L}_2, \ldots$ in the $\{\mathbf{v}_i\}$ basis are *not*, in general, isomorphic to each other.[12] Only the matrices $\mathbf{L}_1, \mathbf{L}_2, \ldots$ *are* isomorphic to $\mathcal{L}$. Hence, one must take care to employ *only* the $\mathbf{L}$'s, if correct results are to be obtained. Only for the special case where the basis that is chosen is orthonormal, i.e., where $\mathbf{G} = \mathbf{I}$, do we have $\mathbf{L} = \mathbf{\Lambda}$, and hence $\mathbf{\Lambda}$ is isomorphic to $\mathcal{L}$. The importance of this observation will become evident in the discussion of matrix eigenvalue problems in Section 3-8.

3-7. Changes of Basis and Similarity Transformations

Although the matrix representation of an operator with respect to a given basis is isomorphic to the operator itself, the actual *form* that the matrix representation takes on depends on the particular basis that has been chosen. We shall now be interested in examining the details of what happens to a matrix representation when the basis is changed.

Suppose we have two sets of basis vectors, $\{\mathbf{v}_1, \mathbf{v}_2, \ldots, \mathbf{v}_n\}$ and $\{\bar{\mathbf{v}}_1, \bar{\mathbf{v}}_2, \ldots, \bar{\mathbf{v}}_n\}$, in the vector space V. Then, an arbitrary vector $\boldsymbol{u}$ in V can be written

---

[12] The proof of this is left as an exercise for the student.

in terms of either basis as

$$\boldsymbol{u}=\sum_{i=1}^{n} u_i\boldsymbol{v}_i, \tag{3-202}$$

or

$$\boldsymbol{u}=\sum_{i=1}^{n} \bar{u}_i\bar{\boldsymbol{v}}_i. \tag{3-203}$$

The basis vectors themselves are also related, since each vector $\boldsymbol{v}_i$ can be mapped onto the corresponding vector $\bar{\boldsymbol{v}}_i$ by means of an operator $\mathcal{A}$, i.e.,

$$\mathcal{A}\boldsymbol{v}_i=\bar{\boldsymbol{v}}_i, \qquad i=1, 2, \ldots, n. \tag{3-204}$$

Since $\bar{\boldsymbol{v}}_i$ can be considered to be just another vector in V, it can be expressed in terms of the $\{\boldsymbol{v}_i\}$ basis. This means that Eq. (3-204) can be written alternatively as

$$\mathcal{A}\boldsymbol{v}_i=\sum_{j=1}^{n} a_{ji}\boldsymbol{v}_j, \qquad i=1, 2, \ldots, n, \tag{3-205}$$

where a_{ij} are the elements of the matrix representation $\mathbf{A}$ of $\mathcal{A}$ in the $\{\boldsymbol{v}_i\}$ basis. This set of equations [Eqs. (3-202)–(3-205)] can now be manipulated to show the effect of a change of basis. Combining Eqs. (3-203) and (3-204), we obtain

$$\boldsymbol{u}=\sum_{i=1}^{n} \bar{u}_i(\mathcal{A}\boldsymbol{v}_i), \tag{3-206}$$

which, on substitution of Eq. (3-205), gives

$$\begin{aligned}\boldsymbol{u}&=\sum_{i=1}^{n} \bar{u}_i\left(\sum_{j=1}^{n} a_{ji}\boldsymbol{v}_j\right)\\ &=\sum_{j=1}^{n}\left(\sum_{i=1}^{n} a_{ji}\bar{u}_i\right)\boldsymbol{v}_j\end{aligned} \tag{3-207}$$

Comparing coefficients of $\boldsymbol{v}_i$ in Eqs. (3-202) and (3-207) yields

$$\sum_{i=1}^{n} a_{ji}\bar{u}_i=u_j, \qquad j=1, 2, \ldots, n, \tag{3-208}$$

which can be written in matrix notation as

$$\mathbf{A}\bar{\mathbf{u}}=\mathbf{u}, \tag{3-209}$$

where $\bar{\mathbf{u}}$ and $\mathbf{u}$ are column vectors. This is the first relationship between the basis sets of interest. Before proceeding further, it is useful to summarize for future reference the isomorphic relationships that are present in our discussion thusfar. We have two isomorphisms with respect to the $\{\mathbf{v}_i\}$ basis:

$$\mathbf{A} \overset{\{\mathbf{v}_i\}}{\Leftrightarrow} \mathcal{A}, \tag{3-210}$$

$$\mathbf{u} \overset{\{\mathbf{v}_i\}}{\Leftrightarrow} \boldsymbol{u}, \tag{3-211}$$

and one with respect to the $\{\bar{\mathbf{v}}_i\}$ basis:

$$\bar{\mathbf{u}} \overset{\{\bar{\mathbf{v}}_i\}}{\Leftrightarrow} \boldsymbol{u}. \tag{3-212}$$

Now let us consider the linear mapping which takes the $\{\bar{\mathbf{v}}_i\}$ basis into the $\{\mathbf{v}_i\}$ basis [the inverse of the operation indicated in Eq. (3-204)]:

$$\mathcal{B}\bar{\mathbf{v}}_i = \mathbf{v}_i, \qquad i = 1, 2, \ldots, n. \tag{3-213}$$

The inverse relationship of $\mathcal{B}$ to $\mathcal{A}$ will be shown explicitly, shortly. Analogous to the previous considerations, we can write[13]

$$\mathcal{B}\bar{\mathbf{v}}_i = \sum_{j=1}^{n} \bar{b}_{ji}\bar{\mathbf{v}}_j. \tag{3-214}$$

Proceeding as before, we obtain in this case

$$\bar{\mathbf{B}}\mathbf{u} = \bar{\mathbf{u}}, \tag{3-215}$$

where

$$\bar{\mathbf{B}} \overset{\{\bar{\mathbf{v}}_i\}}{\Leftrightarrow} \mathcal{B}, \tag{3-216}$$

and Eqs. (3-211) and (3-212) also apply.

Equations (3-209) and (3-215) now contain the desired relationships, and we can combine them to give

$$(\mathbf{A}\bar{\mathbf{B}})\mathbf{u} = \mathbf{u} \tag{3-217}$$

or

$$\mathbf{A}\bar{\mathbf{B}} = \mathbf{I}. \tag{3-218}$$

Thus,

$$\bar{\mathbf{B}} = \mathbf{A}^{-1}, \tag{3-219}$$

which demonstrates that the inverse of $\mathbf{A}$ exists. Since $\mathbf{A}$ is isomorphic to $\mathcal{A}$, Eq.

---

[13] The "bar" on $\bar{b}_{ji}$ has been used to remind the reader that these matrix elements are formed using the $\{\bar{\mathbf{v}}_i\}$ basis.

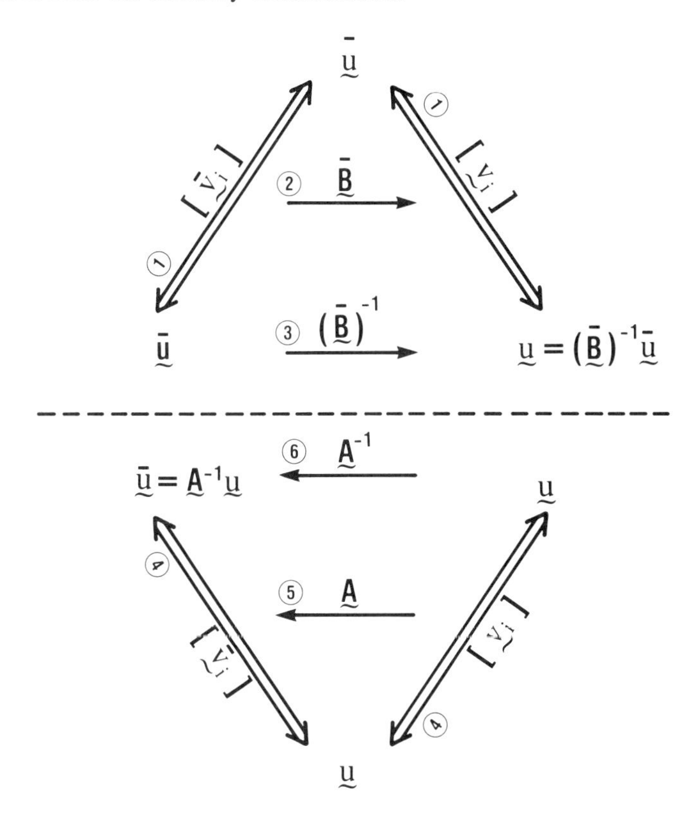

Figure 3.3. Relationships among representations of an arbitrary vector in different bases.

(3-219) also implies that $\mathcal{A}^{-1}$ exists.[14] Alternatively, we can write

$$\mathbf{A}=(\bar{\mathbf{B}})^{-1} \tag{3-220}$$

which assures that $\mathcal{B}^{-1}$ exists.

The result of primary interest can now be obtained by substitution of Eq. (3-219) into Eq. (3-215), which gives

$$\mathbf{A}^{-1}\mathbf{u}=\bar{\mathbf{u}}. \tag{3-221}$$

The analogous transformation of $\mathbf{u}$ from the $\{\bar{\mathbf{v}}_i\}$ basis to $\mathbf{u}$ in the $\{\mathbf{v}_i\}$ basis is easily seen to be

$$\mathbf{u}=(\bar{\mathbf{B}})^{-1}\bar{\mathbf{u}}. \tag{3-222}$$

Figure 3.3 summarizes the various relationships in pictorial form. The

[14] This could also be shown by combining Eqs. (3-204) and (3-213), which gives $(\mathcal{A}\mathcal{B})\bar{\mathbf{v}}_i = \bar{\mathbf{v}}_i$, or $\mathcal{A}\mathcal{B} = \mathcal{I}$ and hence $\mathcal{B} = \mathcal{A}^{-1}$.

following numbers refer to the numbers circled on the figure:

1. An arbitrary vector $\boldsymbol{u}$ can be represented isomorphically in many ways. The use of two different bases, $\{\bar{\mathbf{v}}_i\}$ and $\{\mathbf{v}_i\}$ are illustrated here.
2. By means of a matrix transformation $\mathbf{B}$ (which is a representation of the operator $\mathcal{B}$ in the $\{\bar{\mathbf{v}}_i\}$ basis), the basis vectors of $\{\bar{\mathbf{v}}_i\}$ can be transformed into the basis vectors $\{\mathbf{v}_i\}$.
3. The matrix transformation $(\bar{\mathbf{B}})^{-1}$ (which is a representation of the operator $\mathcal{B}^{-1}$ in the $\{\bar{\mathbf{v}}_i\}$ basis) takes the column vector representation of an arbitrary vector $\boldsymbol{u}$ in the $\{\bar{\mathbf{v}}_i\}$ basis to a new representation of $\boldsymbol{u}$ in the $\{\mathbf{v}_i\}$ basis.

The numbers 4–6 are analogous to 1–3, but starting from the representation of $\boldsymbol{u}$ in the $\{\mathbf{v}_i\}$ basis instead of in the $\{\bar{\mathbf{v}}_i\}$ basis. It is important to note that $\bar{\mathbf{B}}$ transforms the $\{\bar{\mathbf{v}}_i\}$ basis into the $\{\mathbf{v}_i\}$ basis, while the *inverse* of $\bar{\mathbf{B}}$, $(\bar{\mathbf{B}})^{-1}$, transforms the column vector representation of an arbitrary vector $\boldsymbol{u}$ from the $\{\bar{\mathbf{v}}_i\}$ to the $\{\mathbf{v}_i\}$ basis.

The type of transformation discussed here, where the *basis* of a vector description is changed, is known as a *passive transformation*. It should be carefully distinguished from the *active transformations* discussed earlier, in which the *vectors* themselves are transformed.

Before showing the explicit relationship between active and passive transformations, let us consider an example of a passive transformation. Consider once again the rotation of an arbitrary vector $\boldsymbol{u}$ about the origin of a two-dimensional vector space, as shown in Fig. 3-1. This arbitrary vector $\boldsymbol{u}$ can be described either by

$$\boldsymbol{u} = u_1\boldsymbol{i} + u_2\boldsymbol{j}, \tag{3-223}$$

or

$$\boldsymbol{u} = \bar{u}_1\bar{\boldsymbol{i}} + \bar{u}_2\bar{\boldsymbol{j}}, \tag{3-224}$$

using the two basis systems shown in Fig. 3-1.

From Eq. (3-204) and/or Fig. 3-1, we see that

$$\begin{aligned} \mathcal{R}_\alpha\boldsymbol{i} &= \bar{\boldsymbol{i}} \\ \mathcal{R}_\alpha\boldsymbol{j} &= \bar{\boldsymbol{j}}, \end{aligned} \tag{3-225}$$

where $\mathcal{R}_\alpha$ is the rotation operator described previously.[15] Substituting Eq. (3-225) into Eq. (3-224) gives

$$\boldsymbol{u} = \bar{u}_1(\mathcal{R}_\alpha\boldsymbol{i}) + \bar{u}_2(\mathcal{R}_\alpha\boldsymbol{j}). \tag{3-226}$$

Referring to Eqs. (3-179) and (3-180) or Fig. 3-1, we can write Eq. (3-226) as

$$\begin{aligned} \boldsymbol{u} &= \bar{u}_1[(\cos\alpha)\boldsymbol{i} + (\sin\alpha)\boldsymbol{j}] + \bar{u}_2[(-\sin\alpha)\boldsymbol{i} + (\cos\alpha)\boldsymbol{j}] \\ &= [(\cos\alpha)\bar{u}_1 + (-\sin\alpha)\bar{u}_2]\boldsymbol{i} + [(\sin\alpha)\bar{u}_1 + (\cos\alpha)\bar{u}_2]\boldsymbol{j}. \end{aligned} \tag{3-227}$$

---

[15] See Eq. (3-175) and following discussion.

Comparing Eqs. (3-227) and (3-223), we obtain the following matrix equation

$$\begin{pmatrix} u_1 \\ u_2 \end{pmatrix} = \begin{pmatrix} \cos\alpha & -\sin\alpha \\ \sin\alpha & \cos\alpha \end{pmatrix} \begin{pmatrix} \bar{u}_1 \\ \bar{u}_2 \end{pmatrix} \tag{3-228}$$

or

$$\mathbf{u} = \mathbf{R}_\alpha \bar{\mathbf{u}},$$

where $\mathbf{R}_\alpha$ is the matrix given in Eq. (3-184). This can also be written as

$$\bar{\mathbf{u}} = \mathbf{R}_\alpha^{-1}\mathbf{u}, \tag{3-229}$$

and since[16] $\mathbf{R}_\alpha$ is an orthogonal matrix, it follows that the above equation can be written as

$$\bar{\mathbf{u}} = \mathbf{R}_\alpha^{t}\mathbf{u}. \tag{3-230}$$

In expanded form, Eq. (2-230) becomes

$$\begin{pmatrix} \bar{u}_1 \\ \bar{u}_2 \end{pmatrix} = \begin{pmatrix} \cos\alpha & \sin\alpha \\ -\sin\alpha & \cos\alpha \end{pmatrix} \begin{pmatrix} u_1 \\ u_2 \end{pmatrix}. \tag{3-231}$$

Thus, $\bar{\mathbf{u}}$ is a second representation of $\boldsymbol{u}$, obtained from the representation $\mathbf{u}$ through a passive transformation involving a *change of basis* via Eq. (3-230).

Having seen an example of both active and passive transformations, we are now in a position to see how they are related. Suppose we had rotated the vector (an active transformation) $\boldsymbol{u}$ in the $\{\boldsymbol{i}, \boldsymbol{j}\}$ basis in a *clockwise* direction by an angle α (see Fig. 3.4a), instead of counterclockwise as considered earlier. For this case, Eq. (3-183) becomes

$$\begin{aligned} \begin{pmatrix} u_1' \\ u_2' \end{pmatrix} &= \begin{pmatrix} \cos(-\alpha) & -\sin(-\alpha) \\ \sin(-\alpha) & \cos(-\alpha) \end{pmatrix} \begin{pmatrix} u_1 \\ u_2 \end{pmatrix} \\ &= \begin{pmatrix} \cos\alpha & \sin\alpha \\ -\sin\alpha & \cos\alpha \end{pmatrix} \begin{pmatrix} u_1 \\ u_2 \end{pmatrix}. \end{aligned} \tag{3-232}$$

Thus, the matrix representing the clockwise rotation of $\boldsymbol{u}$ through an angle α is *identical* to the matrix representing a transformation of the basis (a *passive* transformation) from the $\{\boldsymbol{i}, \boldsymbol{j}\}$ to a new basis $\{\bar{\boldsymbol{i}}, \bar{\boldsymbol{j}}\}$, where each of the basis vectors of the latter basis is rotated counterclockwise by an angle α. In other words, active and passive transformations can be thought of as being equivalent, i.e., one can *either* transform the *vector or* transform the *basis*. The choice of which to use in practice is largely a matter of convenience. However, it must be remembered that the transformations involve inverse types of operations, e.g., if the *vector* is rotated *clockwise* by an angle α, the *basis* must be rotated *counterclockwise* by an angle α to achieve the same result.

Now that we have seen how the representation of a *vector* changes when the basis is changed, let us examine how the matrix representation of a *linear*

[16] Student exercise.

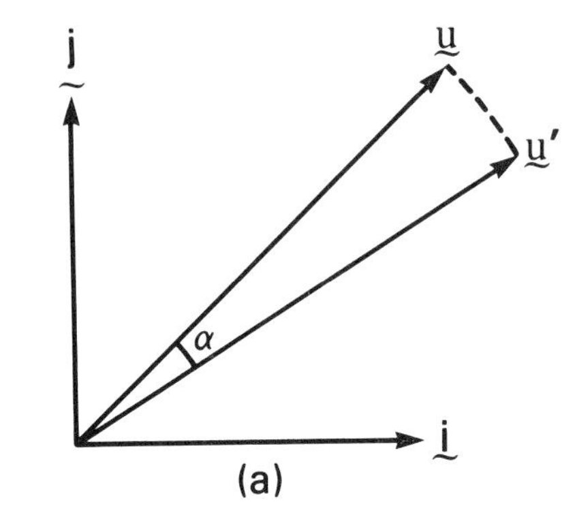

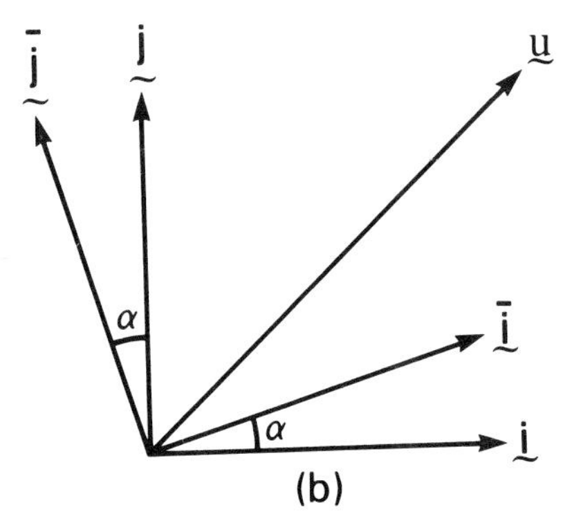

Figure 3.4. (a) Active transformation of a vector $\boldsymbol{u}$ into a vector $\boldsymbol{u}'$ with basis fixed. (b) Passive transformation of a vector $\boldsymbol{u}$ with basis rotated.

operator $\mathcal{L}$, defined on the finite-dimensional vector space V, changes when the basis for the space is changed, and how these representations are related. As shown in Eq. (3-143), the effect of $\mathcal{L}$ can be represented as

$$\mathcal{L}\boldsymbol{v}_i = \sum_{j=1}^{n} l_{ji}\boldsymbol{v}_j, \tag{3-233}$$

and

$$\mathcal{L}\bar{\boldsymbol{v}}_p = \sum_{q=1}^{n} \bar{l}_{qp}\bar{\boldsymbol{v}}_q, \tag{3-234}$$

in the $\{\boldsymbol{v}_i\}$ and $\{\bar{\boldsymbol{v}}_q\}$ bases, respectively. Combining Eqs. (3-234), (3-204), and (3-205) yields

$$\begin{aligned}\mathcal{L}\bar{\boldsymbol{v}}_p &= \mathcal{L}(\mathcal{A}\boldsymbol{v}_p)\\ &= \mathcal{L}\left(\sum_{r=1}^{n} a_{rp}\boldsymbol{v}_r\right).\end{aligned} \tag{3-235}$$

Since $\mathcal{L}$ is a linear operator, we can use Eq. (3-233) to write Eq. (3-235) as

$$\mathcal{L}\bar{\mathbf{v}}_p = \sum_{r=1}^{n} a_{rp}(\mathcal{L}\mathbf{v}_r)$$

$$= \sum_{r=1}^{n} a_{rp}\left(\sum_{s=1}^{n} l_{sr}\mathbf{v}_s\right) \tag{3-236}$$

Finally, on rearranging terms, we obtain

$$\mathcal{L}\bar{\mathbf{v}}_p = \sum_{s=1}^{n}\left(\sum_{r=1}^{n} l_{sr}a_{rp}\right)\mathbf{v}_s. \tag{3-237}$$

The effect of $\mathcal{L}$ operating on $\bar{\mathbf{v}}_p$ can also be expressed in a different manner using Eq. (3-234), by substitution of Eqs. (3-204) and (3-205), which gives

$$\mathcal{L}\bar{\mathbf{v}}_p = \sum_{q=1}^{n} \bar{l}_{qp}(\mathcal{A}\mathbf{v}_q)$$

$$= \sum_{q=1}^{n} \bar{l}_{qp}\left(\sum_{t=1}^{n} a_{tq}\mathbf{v}_t\right)$$

$$= \sum_{t=1}^{n}\left(\sum_{q=1}^{n} a_{tq}\bar{l}_{qp}\right)\mathbf{v}_t. \tag{3-238}$$

Equating Eqs. (3-237) and (3-238) and comparing coefficients of the $\mathbf{v}_i$ yields

$$\sum_{r=1}^{n} l_{sr}a_{rp} = \sum_{q=1}^{n} a_{sq}\bar{l}_{qp}, \qquad s, p = 1, 2, \ldots, n. \tag{3-239}$$

Eq. (3-239) can be written in matrix notation as

$$\mathbf{LA} = \mathbf{A\bar{L}}, \tag{3-240}$$

which, upon left multiplying by $\mathbf{A}^{-1}$, yields

$$\mathbf{A}^{-1}\mathbf{LA} = \mathbf{\bar{L}}. \tag{3-241}$$

Such a transformation is called a *similarity transformation*, and *any* two matrices related by such an equation are said to be *similar*. Hence, we see that the matrix representations of $\mathcal{L}$ in the two basis systems are similar.

Equation (3-241) can also be rearranged to give

$$\mathbf{L} = \mathbf{A\bar{L}A}^{-1}, \tag{3-242}$$

which, on substitution of Eq. (3-220), yields

$$\mathbf{L} = (\mathbf{\bar{B}})^{-1}\mathbf{\bar{L}\bar{B}}. \tag{3-243}$$

In a manner similar to that used in the discussion of the representation of an

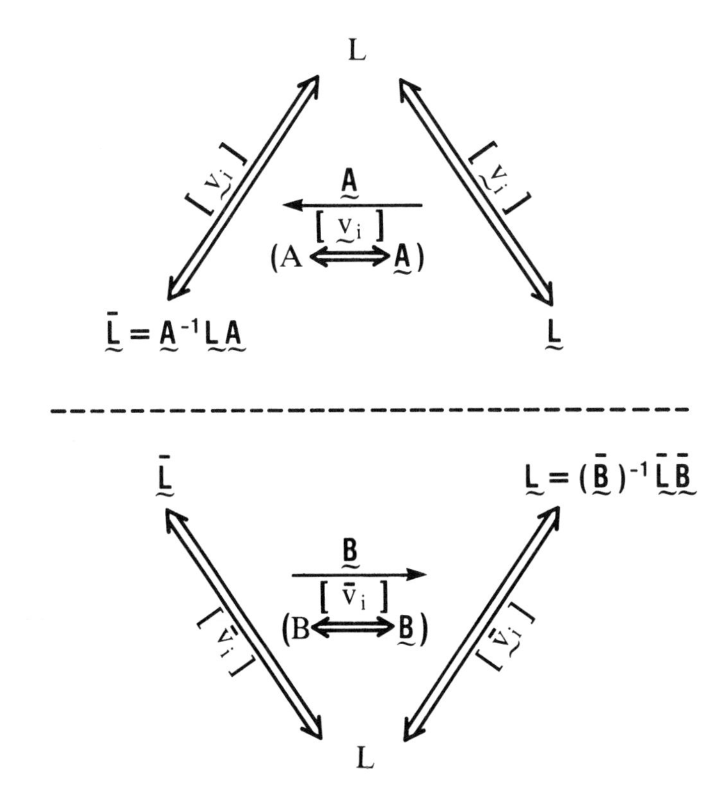

Figure 3.5. Relationships among representations of an operator $\mathcal{L}$ in different bases.

arbitrary vector in various bases (see Fig. 3.3), Fig. 3.5 indicates the relationships between the matrix representations $\mathbf{L}$ and $\bar{\mathbf{L}}$ of the *operator* $\mathcal{L}$ in two different bases. For example, suppose we start with a representation of $\mathcal{L}$ in the $\{\mathbf{v}_i\}$ basis, i.e., $\mathbf{L}$, and wish to effect a change of basis to the $\{\bar{\mathbf{v}}_i\}$ basis. The representation of $\mathcal{L}$ in the new basis, $\bar{\mathbf{L}}$, can be obtained from the representation in the old basis by application of Eq. (3-241), where $\mathbf{A}$ is the matrix representation of $\mathcal{A}$ in the $\{\mathbf{v}_i\}$ basis. It is important to remember that the transformation from one representation to another will always involve the use of transformation matrices (e.g., $\mathbf{A}$, $\bar{\mathbf{B}}^{-1}$) that are formed using the *old basis*, i.e., the basis from which the transformation is being made.

Similarity transformations will also be very useful to us in later discussions on how to transform a matrix to diagonal form. In preparation for that, let us prove several elementary theorems concerning the properties of such transformations.

Theorem 3-12. The trace of an $(n \times n)$ matrix $\mathbf{A}$ is invariant under a similarity transformation.

Proof: From the definition of a similarity transformation, we have

$$\mathbf{C} = \mathbf{B}^{-1}\mathbf{A}\mathbf{B}$$

and

$$\text{tr}(\mathbf{C}) = \text{tr}([\mathbf{B}^{-1}\mathbf{A}]\mathbf{B}).$$

From Theorem 3-6, we can write

$$\begin{aligned}\text{tr}(\mathbf{C}) &= \text{tr}(\mathbf{B}[\mathbf{B}^{-1}\mathbf{A}]) \\ &= \text{tr}(\mathbf{A}).\end{aligned}$$

Theorem 3-13. The determinant of an $(n \times n)$ matrix $\mathbf{A}$ is invariant under a similarity transformation.

Proof: From Theorem 3-5, we have

$$\begin{aligned}|\mathbf{C}| &= |\mathbf{B}^{-1}\mathbf{A}\mathbf{B}| \\ &= |\mathbf{B}^{-1}| \cdot |\mathbf{A}| \cdot |\mathbf{B}|.\end{aligned}$$

Since the order of multiplication of determinants is irrelevant, we have

$$\begin{aligned}|\mathbf{C}| &= |\mathbf{B}^{-1}| \cdot |\mathbf{B}| \cdot |\mathbf{A}| \\ &= |\mathbf{B}^{-1}\mathbf{B}| \cdot |\mathbf{A}| = |\mathbf{A}|.\end{aligned}$$

Before leaving this section, it is useful to point out one special kind of similarity transformation that will be of interest to us later. Specifically, we shall often be interested in the effect of a *unitary transformation* $\mathcal{U}$ on the basis, rather than the more general nonsingular transformation $\mathcal{A}$ discussed previously. In this case, the matrix representation of $\mathcal{U}$ satisfies

$$\mathbf{U}^{-1} = \mathbf{U}^{\dagger}, \tag{3-244}$$

or

$$\mathbf{U}\mathbf{U}^{\dagger} = \mathbf{U}^{\dagger}\mathbf{U} = \mathbf{I}. \tag{3-245}$$

For this special case, the similarity transformation of Eq. (3-241) can be written as

$$\bar{\mathbf{L}} = \mathbf{U}^{\dagger}\mathbf{L}\mathbf{U}, \tag{3-246}$$

and is called a *unitary similarity transformation*.

3-8. Matrix Eigenvalue Problems

In Chapter 2, it was pointed out that for a linear operator $\mathcal{A}$ defined on a vector space V, there may exist a special *subset* of the vectors in V such that

$$\mathcal{A}\boldsymbol{u} = \lambda\boldsymbol{u}, \tag{3-247}$$

where $\boldsymbol{u}$ is in V and λ is a scalar (real or complex). As introduced earlier, λ and $\boldsymbol{u}$ are said to be an *eigenvalue* and *eigenvector* of $\mathcal{A}$, respectively. The set of all eigenvalues $\{\lambda_1, \lambda_2, \ldots\}$, corresponding to the set of eigenvectors $\{\boldsymbol{v}_1, \boldsymbol{v}_2, \ldots\}$, is said to form the *spectrum* of $\mathcal{A}$.

Since we have found that matrix representations of operators can be constructed to be isomorphic to the operators themselves, it should not be surprising that the eigenvalue problem described earlier for operators also has a corresponding matrix form. To see this, suppose that $\{\mathbf{v}_1, \mathbf{v}_2, \ldots, \mathbf{v}_n\}$ is a basis for V. Then, $\mathcal{A}$ has a matrix representation $\mathbf{A}$, and the eigenvector $\boldsymbol{u}$ can be represented by the column vector $\mathbf{u}$. Hence, Eq. (3-247) becomes,

$$\mathbf{A}\mathbf{u} = \lambda\mathbf{u} \tag{3-248}$$

or, more explicitly,

$$\begin{pmatrix} a_{11} & a_{12} & \cdots & a_{1n} \\ a_{21} & a_{22} & \cdots & a_{2n} \\ \vdots & \vdots & \ddots & \vdots \\ a_{n1} & a_{n2} & \cdots & a_{nn} \end{pmatrix} \begin{pmatrix} u_1 \\ u_2 \\ \vdots \\ u_n \end{pmatrix} = \lambda \begin{pmatrix} u_1 \\ u_2 \\ \vdots \\ u_n \end{pmatrix}. \tag{3-249}$$

Rearranging Eq. (3-248), we obtain

$$(\mathbf{A} - \lambda\mathbf{I})\mathbf{u} = \mathbf{0}, \tag{3-250}$$

which has the form of a set of homogeneous equations,[17] i.e.,

$$\mathbf{B}(\lambda)\mathbf{u} = \mathbf{0}. \tag{3-251}$$

Equation (3-251) does not, however, represent exactly the same problem ordinarily encountered for homogeneous equations since not only must we determine the elements of the vector $\mathbf{u}$, we must also determine the unknown λ, of which $\mathbf{B}$ is a function.

From our previous discussion of homogeneous equations[17] it is clear that nontrivial solutions of Eq. (3-251) can be obtained only when

$$|\mathbf{B}(\lambda)| = |\mathbf{A} - \lambda\mathbf{I}| = 0. \tag{3-252}$$

$$= \begin{vmatrix} a_{11}-\lambda & a_{12} & \cdots & a_{1n} \\ a_{21} & a_{22}-\lambda & \cdots & a_{2n} \\ \vdots & \vdots & \ddots & \vdots \\ a_{n1} & a_{n2} & \cdots & a_{nn-\lambda} \end{vmatrix} = 0. \tag{3-253}$$

Equations (3-252) and (3-253) are called *secular equations*, and the determinant is usually referred to as the *secular determinant*.[18] Expansion of the determinant leads to an nth degree polynomial in λ, known as the *secular polynomial*,[19]

$$p_n(\lambda) = \lambda^n + \alpha_{n-1}\lambda^{n-1} + \cdots + \alpha_1\lambda + \alpha_0 = 0. \tag{3-254}$$

---

[17] See the discussion following Eq. (3-56).

[18] Mathematics texts often refer to these equations as characteristic equations and the determinant as the characteristic determinant.

[19] Sometimes referred to in mathematics texts as the characteristic polynomial.

The n, not necessarily distinct, roots of the secular polynomial obviously also satisfy the secular equations, and hence are the desired eigenvalues of Eq. (3-248). If the eigenvalues are all different, they are said to be *nondegenerate*. If two or more eigenvalues are equal, they are said to be *degenerate*, and the *degree of degeneracy* (or multiplicity) of the eigenvalue is given by the number of eigenvalues that are equal. Thus, a given eigenvalue spectrum could be totally nondegenerate, totally degenerate, or some combination of nondegenerate and degenerate eigenvalues.

It should be pointed out that since nth order polynomials may have complex roots, the eigenvalues of the matrix representation of a general linear operator are complex in general. As we shall prove shortly, the special case of Hermitian matrices is one in which only real eigenvalues arise. This has important implications in quantum mechanics, as we shall discuss later.

Since similar matrices are merely different matrix representations for the same operator, which respect to different bases, their eigenvalue spectra must be identical. This is just another way of saying that the eigenvalues and eigenvectors that correspond to a particular operator depend only on the form of the operator and the vector space on which it is defined, and not on the basis used to represent the operator.

Before proceeding let us consider a simple example. The matrix

$$\mathbf{B}=\begin{pmatrix} 1 & 2 \\ 2 & 1 \end{pmatrix} \tag{3-255}$$

gives rise to the secular polynomial

$$|\mathbf{B}-\lambda\mathbf{I}|=\lambda^2-2\lambda-3=0, \tag{3-256}$$

which roots $\lambda_1 = -1$, $\lambda_2 = 3$. To obtain the eigenvectors that correspond to λ_1 and λ_2, we must solve either Eq. (3-248) or (3-250) after substituting in the proper eigenvalue. We shall use Eq. (3-250), which gives

$$\begin{pmatrix} 1-\lambda_i & 2 \\ 2 & 1-\lambda_i \end{pmatrix}\begin{pmatrix} u_{1i} \\ u_{2i} \end{pmatrix}=\begin{pmatrix} 0 \\ 0 \end{pmatrix}. \tag{3-257}$$

If we consider first the eigenvalue $\lambda_1 = -1$, we have, on carrying out the required algebra,

$$2u_{11}+2u_{21}=0, \tag{3-258}$$

or

$$u_{11}=-u_{21}. \tag{3-259}$$

Note that a degree of freedom remains, in that we choose any value for either u_{11} or u_{21}, and still have a solution to Eq. (3-257). We can therefore write the solution as

$$\begin{pmatrix} u_{11} \\ u_{21} \end{pmatrix}=c_1\begin{pmatrix} 1 \\ -1 \end{pmatrix}, \tag{3-260}$$

where c_1 is an *arbitrary* constant. If we now substitute λ_2 into Eq. (3-257), we obtain, in an analogous fashion

$$\begin{pmatrix} u_{12} \\ u_{22} \end{pmatrix} = c_2 \begin{pmatrix} 1 \\ 1 \end{pmatrix}, \tag{3-261}$$

where c_2 is another *arbitrary* constant. It is left as an exercise for the student to verify that Eqs. (3-260) and (3-261) are indeed eigenvectors of $\mathbf{B}$, corresponding to eigenvalues $\lambda_1 = -1$ and $\lambda_2 = 3$, respectively.

The presence of the one arbitrary constant in each of the eigenvectors should not be surprising. For each eigenvalue, we have two equations and three unknowns (the eigenvalue and the two components of the eigenvector corresponding to it), and complete determination of all of the unknowns is not possible. In practice, the remaining parameter is usually assigned by requiring the eigenvector to be normalized.

This example also illustrates the procedure used in general for finding eigenvalues and eigenvectors of matrices.[20] First, the eigenvalues are obtained by finding the roots of the secular polynomial. The corresponding eigenvectors are found next by substitution of the eigenvalues one by one into the original matrix equations. Some precautions and modifications arise (which we shall discuss shortly) when degenerate eigenvalues are present, but the general procedure remains the same.

We shall now be concerned with several theorems that will establish the conditions under which the eigenvalues and eigenvectors of $n \times n$ matrices can be found, and some of the characteristics of the resulting eigenvalues and eigenvectors. In preparation for that study, we note first that it is possible to prove[21] that the secular polynomial of an $n \times n$ matrix *always* has *at least one root*, λ. We shall use this result frequently, either explicitly or implicitly, in several of the following discussions.

Theorem 3-14: Let $\mathbf{A}$ be an $n \times n$ matrix that is isomorphic to the linear operator $\mathcal{A}$, defined on the vector space V. If a particular eigenvalue of $\mathbf{A}$ is k-fold degenerate, then a linear combination of the k eigenvectors corresponding to λ is also an eigenfunction.

Proof: Let the column vectors $\mathbf{u}_1, \mathbf{u}_2, \ldots, \mathbf{u}_k$ each be eigenvectors of $\mathbf{A}$, corresponding to the same eigenvalue λ. Take the linear combination

$$\mathbf{u} = a_1\mathbf{u}_1 + a_2\mathbf{u}_2 + \cdots + a_k\mathbf{u}_k,$$

---

[20] It should be noted that as matrices become large, finding eigenvalues via solution of the secular polynomial becomes impractical and other techniques are employed. See, for example, J. H. Wilkinson, *The Algebraic Eigenvalue Problem,* Clarendon Press, Oxford, England, 1965, for a discussion of alternative approaches; see also, B. N. Parlett, *The Symmetric Eigenvalue Problem,* Prentice-Hall, Englewood Cliffs, NJ, 1980; E. R. Davidson, *J. Comp. Phys.,* **17**, 87–94 (1975); I. Shavitt, C. F. Bender, A. Pipano, and R. P. Hosteny, *J. Comp. Phys.,* **11**, 90–108 (1973); R. C. Raffenetti, *J. Comp. Phys.,* **32**, 403–419 (1979).

[21] For a proof of this see, for example, M. J. Weiss, *Higher Algebra,* John Wiley, NY, 1949, p. 80.

and left multiply it by **A**. This gives

$$\begin{aligned} A\mathbf{u} &= a_1(\mathbf{A}\mathbf{u}_1) + a_2(\mathbf{A}\mathbf{u}_2) + \cdots + a_k(\mathbf{A}\mathbf{u}_k) \\ &= a_1(\lambda\mathbf{u}_1) + a_2(\lambda\mathbf{u}_2) + \cdots + a_k(\lambda\mathbf{u}_k) \\ &= \lambda(a_1\mathbf{u}_1 + a_2\mathbf{u}_2 + \cdots + a_k\mathbf{u}_k) \\ &= \lambda\mathbf{u}_1 \end{aligned}$$

which completes the proof.

Before proceeding, it is useful to consider the implications of this theorem. The theorem states that *any* linear combination of eigenvectors that corresponds to the *same* eigenvalue is also an eigenvector. It says nothing about these eigenvectors being linearly independent, but merely that they span a subspace $V^{(\lambda)}$ of V, i.e., any vector belonging to $V^{(\lambda)}$ can be written as a linear combination of the eigenvectors corresponding to λ. However, as will be seen in the following example, the dimension of the subspace need not equal the multiplicity (or degeneracy) of the eigenvalue. This arises since[22] a set of vectors can span a space without being linearly independent, and we cannot guarantee that it will always be possible to find k linearly independent eigenvectors corresponding to λ.

To illustrate this point consider the matrix

$$\mathbf{C} = \begin{pmatrix} 1 & c \\ 0 & 1 \end{pmatrix}, \tag{3-262}$$

where c is a constant. The secular determinant of **C**

$$|\mathbf{C} - \lambda\mathbf{I}| = 0, \tag{3-263}$$

has roots $\lambda_1, \lambda_2 = 1$. As in the example discussed earlier, we obtain the eigenvectors from the following equation, i.e.,

$$\begin{pmatrix} 0 & c \\ 0 & 0 \end{pmatrix} \begin{pmatrix} u_{11} \\ u_{21} \end{pmatrix} = \begin{pmatrix} 0 \\ 0 \end{pmatrix}. \tag{3-264}$$

Carrying out the matrix multiplication, we obtain

$$0 \cdot u_{11} + cu_{21} = 0. \tag{3-265}$$

It is clear that $u_{11} = k$, where k is an arbitrary constant, and $u_{21} = 0$ satisfy the above equation, so that an acceptable eigenvector corresponding to $\lambda_1 = 1$ is

$$\begin{pmatrix} u_{11} \\ u_{21} \end{pmatrix} = k \begin{pmatrix} 1 \\ 0 \end{pmatrix}. \tag{3-266}$$

If we now attempt to find another eigenvector corresponding to $\lambda_2 = 1$, we find

[22] See the discussion associated with Eq. (2-36) and Theorem 2-2.

that

$$\begin{pmatrix} u_{12} \\ u_{22} \end{pmatrix} = K \begin{pmatrix} u_{11} \\ u_{21} \end{pmatrix}, \tag{3-267}$$

i.e., the second eigenvector will always be a multiple of the first (K is multiplicative factor) eigenvector, and hence the two eigenvectors will be linearly dependent and cannot span a two-dimensional vector space.[23]

However, for most cases of interest in quantum mechanics, we shall be able to find k linearly independent eigenvectors for a k-fold degenerate eigenvalue λ. This arises because only certain forms of matrices ($\mathbf{A}$) are acceptable, and these matrices allow the desired linearly independent eigenvectors to be found.

Returning now to the secular polynomial, we see that it can also be written in the following form, i.e.,

$$p_n^{(\lambda)} = \prod_{i=1}^{n} (\lambda_i - \lambda), \tag{3-268}$$

where the λ_i are the roots of the nth order secular polynomial. In the case of degeneracies, the above equation can be rewritten as

$$p_n^{(\lambda)} = \prod_{k=1}^{n'} (\lambda_k - \lambda)^{g_k}, \tag{3-269}$$

where g_k is the degeneracy of the root, and n' is the number of distinct roots.

If the λ_i are used as the elements of a diagonal matrix $\mathbf{\Lambda}$, i.e.,

$$\mathbf{\Lambda} = \begin{pmatrix} \lambda_1 & & & \\ & \lambda_2 & & 0 \\ & & \ddots & \\ & 0 & & \lambda_n \end{pmatrix}, \tag{3-270}$$

then the secular determinant corresponding to Eqs. (3-268) or (3-269) is just

$$|\mathbf{\Lambda} - \lambda \mathbf{I}| = 0. \tag{3-271}$$

Now the intriguing question arises whether, since similar matrices have identical eigenvalues, it is possible to find a similarity transformation which "diagonalizes" the matrix $\mathbf{A}$, i.e., brings it to the form of $\mathbf{\Lambda}$. If so, we shall have a powerful technique for finding eigenvalues. The next theorem sets down the conditions for this.

Theorem 3-15: for an $n \times n$ matrix $\mathbf{A}$, defined on the n-dimensional vector space V, with eigenvectors $\mathbf{u}_i$ and associated eigenvalues λ_i, i.e.,

$$\mathbf{A}\mathbf{u}_i = \lambda_i \mathbf{u}_i, \tag{3-272}$$

---

[23] To extend the above illustration to the general case is beyond the scope of this book. See, e.g., G. E. Shilov, *An Introduction to the Theory of Linear Spaces,* Prentice-Hall, Englewood Cliffs, NJ, 1961.

the matrix $\mathbf{A}$ can be digaonalized by the similarity transformation $\mathbf{U}^{-1}\mathbf{A}\mathbf{U} = \mathbf{\Lambda}$, if and only if the eigenvectors form a basis for V, where $\mathbf{U}$ is given by

$$\mathbf{U} = (\mathbf{u}_1 \ \mathbf{u}_2 \ \cdots \ \mathbf{u}_n) = \begin{pmatrix} u_{11} & u_{12} & \cdots & u_{1n} \\ u_{21} & u_{22} & \cdots & u_{2n} \\ \vdots & \vdots & \ddots & \vdots \\ u_{n1} & u_{n2} & \cdots & u_{nn} \end{pmatrix}. \tag{3-273}$$

Proof: We can write Eq. (3-272), using Eq. (3-273), in the following form,

$$\mathbf{A}\mathbf{U} = \mathbf{U}\mathbf{\Lambda}. \tag{3-274}$$

If the $\{\mathbf{u}_i\}$ form a basis, they are linearly independent, and hence the matrix $\mathbf{U}$ possesses linearly independent columns and can be inverted. Therefore, Eq. (3-274) becomes

$$\mathbf{U}^{-1}\mathbf{A}\mathbf{U} = \mathbf{\Lambda}. \tag{3-275}$$

The proof of the converse is left to the student (see Problem 3-20).

As an example of this theorem, consider the matrix given by Eq. (3-255), with eigenvectors given by Eqs. (3-260) and (3-261). The eigenvector matrix therefore becomes

$$\mathbf{U} = \begin{pmatrix} c_1 & c_2 \\ -c_1 & c_2 \end{pmatrix}. \tag{3-276}$$

Applying the similarity transformation, we have

$$\mathbf{U}^{-1}\mathbf{B}\mathbf{U} = \begin{pmatrix} \dfrac{1}{2c_1} & \dfrac{-1}{2c_1} \\ \dfrac{1}{2c_2} & \dfrac{1}{2c_2} \end{pmatrix} \begin{pmatrix} 1 & 2 \\ 2 & 1 \end{pmatrix} \begin{pmatrix} c_1 & c_2 \\ -c_1 & c_2 \end{pmatrix} \tag{3-277}$$

$$= \begin{pmatrix} -1 & 0 \\ 0 & 3 \end{pmatrix} \tag{3-278}$$

$$= \mathbf{\Lambda}, \tag{3-279}$$

as desired.

It should be noted that if the basis is orthonormal (this can always be accomplished using, e.g., the Gram–Schmidt procedure), then the matrix $\mathbf{U}$ becomes a unitary matrix. Hence, we have a unitary similarity transformation such that

$$\mathbf{U}^{\dagger}\mathbf{A}\mathbf{U} = \mathbf{\Lambda}. \tag{3-280}$$

Theorem 3-15 is of theoretical importance, but it is, unfortunately, not very specific, i.e., it does not provide any insight into the type of matrices that can be diagonalized by similarity transformations. For the case of normal matrices, i.e., matrices that satisfy:

$$\mathbf{A}\mathbf{A}^{\dagger}=\mathbf{A}^{\dagger}\mathbf{A}, \tag{3-281}$$

it is possible to prove a very general theorem concerning their diagonalizability. Before proving the theorem, however, we must first prove two other theorems.

Theorem 3-16: Every $n \times n$ matrix $\mathbf{A}$ can be reduced to upper triangular form, with its eigenvalues on the diagonal.

Proof: The proof will be carried out by induction. First, consider the case $n = 2$. Let $\mathbf{A}$ have eigenvalues λ_1 and λ_2, and associate with λ_2, the normalized eigenvector $\mathbf{u}_2$. (Note that we can guarantee the existence of only one independent eigenvector.) Now choose the normalized vector $\mathbf{u}_2$ such that

$$\mathbf{u}_2^{\dagger}\mathbf{u}_1=0. \tag{3-282}$$

Then, the matrix

$$\mathbf{U}=(\mathbf{u}_1,\ \mathbf{u}_2) \tag{3-283}$$

is unitary. Now let us apply the unitary similarity transformation

$$\begin{aligned}\mathbf{U}^{\dagger}\mathbf{A}\mathbf{U}&=\begin{pmatrix}\mathbf{u}_1^{\dagger}\\ \mathbf{u}_2^{\dagger}\end{pmatrix}\mathbf{A}(\mathbf{u}_1\ \ \mathbf{u}_2)\\ &=\begin{pmatrix}\mathbf{u}_1^{\dagger}\\ \mathbf{u}_2^{\dagger}\end{pmatrix}(\mathbf{A}\mathbf{u}_1,\ \mathbf{A}\mathbf{u}_2)\\ &=\begin{pmatrix}\mathbf{u}_1^{\dagger}\mathbf{A}\mathbf{u}_1 & \mathbf{u}_1^{\dagger}\mathbf{A}\mathbf{u}_2\\ \mathbf{u}_2^{\dagger}\mathbf{A}\mathbf{u}_1 & \mathbf{u}_2^{\dagger}\mathbf{A}\mathbf{u}_2\end{pmatrix}\end{aligned} \tag{3-284}$$

Since $\mathbf{A}\mathbf{u}_1 = \lambda_1\mathbf{u}_1$, we can rewrite Eq. (3-284) using Eq. (3-282), as

$$\mathbf{U}^{\dagger}\mathbf{A}\mathbf{U}=\begin{pmatrix}\lambda_1 & \mathbf{u}_1^{\dagger}\mathbf{A}\mathbf{u}_2\\ 0 & \mathbf{u}_2^{!}\mathbf{A}\mathbf{u}_2\end{pmatrix}. \tag{3-285}$$

Since $\mathbf{A}$ and $\mathbf{U}^{\dagger}\mathbf{A}\mathbf{U}$ are unitarily similar matrices, they possess identical eigenvalues. Therefore, the secular polynomial of Eq. (3-285) is given by

$$|\mathbf{A}-\lambda\mathbf{I}|=|\mathbf{U}^{\dagger}\mathbf{A}\mathbf{U}-\lambda\mathbf{I}|=(\lambda_1-\lambda)(\mathbf{u}_2^{\dagger}\mathbf{A}\mathbf{u}_2-\lambda)=0 \tag{3-286}$$

and thus

$$\lambda_2=\mathbf{u}_2^{\dagger}\mathbf{A}\mathbf{u}_2. \tag{3-287}$$

This proves the theorem for $n = 2$. Now we assume that the theorem holds for matrices of order $n - 1$. This implies that there exists a unitary matrix $\mathbf{V}$, of order $n - 1$, which triangularizes the $n - 1$ order matrix $\mathbf{B}$, and the diagonal elements of the transformed matrix are its eigenvalues, i.e.,

$$\mathbf{V}^{\dagger}\mathbf{B}\mathbf{V}=\mathbf{B}'=\begin{pmatrix}\lambda_1' & b_{12}' & \cdots & b_{1,n-1}'\\ & \lambda_2' & \ddots & \vdots\\ & 0 & \ddots & b_{n-2,n-4}\\ & & & \lambda_{n-1}'\end{pmatrix} \tag{3-288}$$

If $\mathbf{A}$ is of order n, we know that there exists at least one independent normalized eigenvector $\mathbf{u}_1$, corresponding to λ_1. We choose the vectors $\mathbf{u}_2, \mathbf{u}_3, \ldots, \mathbf{u}_n$ such that $\{\mathbf{u}_1, \mathbf{u}_2, \ldots, \mathbf{u}_n\}$ is an orthonormal set. Therefore, the matrix

$$\mathbf{U} = (\mathbf{u}_1 \;\; \mathbf{u}_2 \;\; \cdots \;\; \mathbf{u}_n) \tag{3-289}$$

is unitary. Again forming the unitary similarity transformation, we obtain

$$\mathbf{U}^\dagger\mathbf{A}\mathbf{U} = \left(\begin{array}{c|ccc} \lambda_1 & \mathbf{u}_1^\dagger\mathbf{A}\mathbf{u}_2 & \cdots & \mathbf{u}_1^\dagger\mathbf{A}\mathbf{u}_n \\ \hline & \mathbf{u}_2^\dagger\mathbf{A}\mathbf{u}_2 & \cdots & \mathbf{u}_2^\dagger\mathbf{A}\mathbf{u}_n \\ & \vdots & \ddots & \vdots \\ 0 & \vdots & & \vdots \\ & \mathbf{u}_n^\dagger\mathbf{A}\mathbf{u}_2 & \cdots & \mathbf{u}_n^\dagger\mathbf{A}\mathbf{u}_n \end{array}\right) \tag{3-290}$$

$$= \left(\begin{array}{c|c} \lambda_1 & \mathbf{C} \\ \hline \mathbf{0} & \mathbf{B} \end{array}\right)$$

From Eq. (3-288), we see that, by hypothesis, there is a unitary matrix $\mathbf{V}$ which triangularizes $\mathbf{B}$. Let us use this matrix to define a new unitary matrix:

$$\mathbf{T} = \left(\begin{array}{c|c} 1 & \mathbf{0} \\ \hline \mathbf{0} & \mathbf{V} \end{array}\right), \tag{3-291}$$

and apply a unitary similarity transformation using $\mathbf{T}$ to $\mathbf{U}^\dagger\mathbf{A}\mathbf{U}$, i.e.,

$$\begin{aligned} \mathbf{T}^\dagger(\mathbf{U}^\dagger\mathbf{A}\mathbf{U}) &= \left(\begin{array}{c|c} 1 & \mathbf{0} \\ \hline \mathbf{0} & \mathbf{V}^\dagger \end{array}\right)\left(\begin{array}{c|c} \lambda_1 & \mathbf{C} \\ \hline \mathbf{0} & \mathbf{B} \end{array}\right)\left(\begin{array}{c|c} 1 & \mathbf{0} \\ \hline \mathbf{0} & \mathbf{V} \end{array}\right) \\ &= \left(\begin{array}{c|c} \lambda_1 & \mathbf{C}\mathbf{V} \\ \hline \mathbf{0} & \mathbf{V}^\dagger\mathbf{B}\mathbf{V} \end{array}\right). \end{aligned} \tag{3-292}$$

Since $\mathbf{U}$ and $\mathbf{T}$ are unitary, $\mathbf{UT}$ is also (see Problem 3-7). Letting

$$\mathbf{W} = \mathbf{UT}, \tag{3-293}$$

and combining Eqs. (3-292), (3-293), and (3-288), we have

$$\mathbf{W}^\dagger\mathbf{A}\mathbf{W} = \left(\begin{array}{c|cccc} \lambda_1 & & \mathbf{CV} & & \\ \hline & \lambda_1' & b_{12}' & \cdots & b_{1,n-1}' \\ & & \lambda_2' & & \vdots \\ & 0 & & \ddots & \vdots \\ & & & & \lambda_{n-1}' \end{array}\right), \tag{3-294}$$

which completes the proof.

Theorem 3-17: If $\mathbf{U}$ is a unitary matrix, then $\mathbf{A}$ is a normal matrix if and only if $\mathbf{U}^\dagger\mathbf{A}\mathbf{U}$ is normal.

The proof of this theorem is left as an exercise for the student (see Problem 3-21).

Using Theorems 3-16 and 3-17, we can now prove the following very important theorem concerning normal matrices.

Theorem 3-18: An $n \times n$ matrix $\mathbf{A}$ can be diagonalized by a unitary similarity transformation if and only if $\mathbf{A}$ is normal.

Proof: Suppose $\mathbf{U}$ is a unitary matrix such that

$$\mathbf{U}^{\dagger}\mathbf{A}\mathbf{U}=\boldsymbol{\Lambda} \tag{3-295}$$

where

$$\boldsymbol{\Lambda}=\begin{pmatrix} \lambda_1 & & & \\ & \lambda_2 & & 0 \\ & & \ddots & \\ & 0 & & \lambda_n \end{pmatrix}. \tag{3-296}$$

From Eq. (3-296), it follows that

$$\boldsymbol{\Lambda}^{\dagger}\boldsymbol{\Lambda}=\sum_{i=1}^{n}|\lambda_i|^2=\boldsymbol{\Lambda}\boldsymbol{\Lambda}^{\dagger}, \tag{3-297}$$

so that, applying Eq. (3-295), we have

$$(\mathbf{U}^{\dagger}\mathbf{A}\mathbf{U})^{\dagger}(\mathbf{U}^{\dagger}\mathbf{A}\mathbf{U})=(\mathbf{U}^{\dagger}\mathbf{A}\mathbf{U})(\mathbf{U}^{\dagger}\mathbf{A}\mathbf{U})^{\dagger}. \tag{3-298}$$

On expanding the above equation and using the unitarity of $\mathbf{U}$, we obtain

$$\mathbf{U}^{\dagger}\mathbf{A}^{\dagger}\mathbf{A}\mathbf{U}=\mathbf{U}^{\dagger}\mathbf{A}\mathbf{A}^{\dagger}\mathbf{U}, \tag{3-299}$$

so that

$$\mathbf{A}^{\dagger}\mathbf{A}=\mathbf{A}\mathbf{A}^{\dagger}. \tag{3-300}$$

To prove the converse, assume $\mathbf{A}$ is normal. By Theorem 3-16, there exists an unitary matrix $\mathbf{U}$ such that

$$\mathbf{U}^{\dagger}\mathbf{A}\mathbf{U}=\mathbf{B}, \tag{3-301}$$

where

$$\mathbf{B}=\begin{pmatrix} \lambda_1 & b_{12} & \cdots & b_{1n} \\ & \lambda_2 & & \vdots \\ & 0 & \ddots & \vdots \\ & & & \lambda_n \end{pmatrix}. \tag{3-302}$$

From Theorem 3-17, it follows that $\mathbf{B}$ is also normal, i.e.,

$$\mathbf{B}\mathbf{B}^{\dagger}=\mathbf{B}^{\dagger}\mathbf{B}. \tag{3-303}$$

Combining Eqs. (3-302) and (3-303), we obtain the following equation for the (1,1)-elements of $\mathbf{B}^{\dagger}\mathbf{B}$ and $\mathbf{B}\mathbf{B}^{\dagger}$:

$$|\lambda_1|^2=|\lambda_1|^2+\sum_{i=2}^{n}|b_{1i}|^2 \tag{3-304}$$

or

$$\sum_{i=2}^{n} |b_{1i}|^2 = 0, \tag{3-305}$$

which implies

$$b_{1i} = 0, \; i = 2, 3, \ldots, n, \tag{3-306}$$

We can repeat this process for the 2,2-element, which gives

$$|\lambda_2|^2 = |\lambda_2|^2 + \sum_{i=2}^{n} |b_{2i}|^2, \tag{3-307}$$

and leads to the result

$$b_{2i} = 0, \qquad i = 3, 4, \ldots, n. \tag{3-308}$$

This process can be carried on in a similar fashion until all the (i,i)-elements are considered. We arrive at the conclusion that *all* off-diagonal elements must be zero, which completes the proof.

We shall now prove several theorems concerning eigenvalues and eigenvectors that will be very useful in the development of quantum chemistry. We shall state and prove these theorems using the matrix representation, but they obviously apply equally well to the operators themselves.

Theorem 3-19: The eigenvalues of a Hermitian matrix are real.

Proof: Let $\mathbf{A}$ be a Hermitian matrix with normalized eigenfunction $\mathbf{u}$ and eigenvalue λ. Then,

$$\mathbf{A}\mathbf{u} = \lambda\mathbf{u}$$

and

$$\mathbf{u}^\dagger\mathbf{A}\mathbf{u} = \lambda.$$

Since $\mathbf{A}$ is Hermitian and $\mathbf{u}^\dagger\mathbf{A}\mathbf{u}$ is a scalar, we also have

$$\begin{aligned}\mathbf{u}^\dagger\mathbf{A}\mathbf{u} &= \mathbf{u}^\dagger\mathbf{A}^\dagger\mathbf{u}\\ &= (\mathbf{u}^\dagger\mathbf{A}\mathbf{u})^\dagger\\ &= \lambda^*.\end{aligned}$$

Hence,

$$\lambda = \lambda^*,$$

which implies that λ must be real.

This theorem is of great importance in quantum mechanics since, as will be seen, there is a one-to-one correspondence between Hermitian matrices (or operators) and observables. Furthermore, the eigenvalues of these matrices correspond to the measured values of the observables, which must be real. In fact, the Schrödinger equation, which is the most important equation in molecular quantum mechanics, can be written as

$$\mathbf{Hu} = \epsilon \mathbf{u}, \tag{3-309}$$

where $\mathbf{H}$ is the matrix representation of the Hamiltonian operator and ϵ is the energy of the system. Since it can be shown that $\mathbf{H} = \mathbf{H}^\dagger$, it follows that ϵ will be real.

Theorem 3-20: The eigenvectors of a Hermitian matrix corresponding to different eigenvalues are orthogonal.

Proof: Let $\mathbf{A}$ be a Hermitian matrix having eigenvectors $\mathbf{u}_1$ and $\mathbf{u}_2$ and associated eigenvalues λ_1, and λ_2, with $\lambda_1 \neq \lambda_2$. Then

$$\mathbf{Au}_1 = \lambda_1 \mathbf{u}_1 \tag{3-310}$$

$$\mathbf{Au}_2 = \lambda_2 \mathbf{u}_2 \tag{3-311}$$

Left multiplying both sides of Eq. (3-310) by $\mathbf{u}_2^\dagger$ and Eq. (3-311) by $\mathbf{u}_1^\dagger$ gives

$$\mathbf{u}_2^\dagger \mathbf{Au}_1 = \lambda_1 \mathbf{u}_2^\dagger \mathbf{u}_1, \tag{3-312}$$

and

$$\mathbf{u}_1^\dagger \mathbf{Au}_2 = \lambda_2 \mathbf{u}_1^\dagger \mathbf{u}_2. \tag{3-313}$$

Taking the adjoint of both sides of Eq. (3-312) yields

$$\mathbf{u}_1^\dagger \mathbf{A}^\dagger \mathbf{u}_2 = \lambda_1^* \mathbf{u}_1^\dagger \mathbf{u}_2. \tag{3-314}$$

Since $\mathbf{A}$ is Hermitian and hence λ_1 is real, we have

$$\mathbf{u}_1^\dagger \mathbf{Au}_2 = \lambda_1 \mathbf{u}_1^\dagger \mathbf{u}_2. \tag{3-315}$$

Subtracting Eq. (3-313) from Eq. (3-315) gives

$$0 = (\lambda_1 - \lambda_2) \mathbf{u}_1^\dagger \mathbf{u}_2. \tag{3-316}$$

Since it has been assumed that $\lambda_1 \neq \lambda_2$, we observe that $\mathbf{u}_1$ and $\mathbf{u}_2$ must be orthogonal, i.e.,

$$\mathbf{u}_1^\dagger \mathbf{u}_2 = 0, \tag{3-317}$$

which completes the proof.

The above theorem does not say anything about the case of eigenvectors belonging to degenerate eigenvalues. The next theorem deals with this case.

Theorem 3-21: The eigenvectors $\{\mathbf{u}_i\}$ of the $n \times n$ Hermitian matrix $\mathbf{A}$, defined on the n-dimensional vector space V, form a basis for V.

Proof: From Theorem 3-18 it follows that there exists a unitary matrix $\mathbf{U}$ such that $\mathbf{A}$ can be diagonalized by a unitary similarity transformation, i.e.,

$$\mathbf{U}^\dagger \mathbf{AU} = \mathbf{\Lambda}.$$

Since $\mathbf{U}^\dagger = \mathbf{U}^{-1}$, it follows from Theorem 3-15 that the eigenvectors $\{\mathbf{u}_i\}$ that form the columns of $\mathbf{U}$ form a basis for V. This completes the proof.

From Theorem 3-21 it follows that the eigenvectors that belong to a particular degenerate eigenvalue are linearly independent, and form the basis of a subspace whose dimension is equal to the degeneracy of the eigenvalue. This is a useful point since, as was discussed previously [see the discussion associated with Eqs. (3-262)–(3-267)], some matrices possess degenerate eigenvalues whose eigenvectors do not form a basis of a subspace whose dimension is equal to the degeneracy of the eigenvalue.

Furthermore, applying the Gram–Schmidt procedure, it is possible to orthonormalize the eigenvectors belonging to a particular degenerate eigenvalue. From Theorem 3-14 it also follows that the orthonormalized vectors are eigenvectors belonging to the same eigenvalue. Finally, Theorem 3-20 shows that the above orthonormalized set is still orthogonal to all of the eigenvectors corresponding to different eigenvalues. In summary, the above discussion has shown that there exists a set of n orthonormal eigenvectors that belongs to every Hermitian matrix,[24] even when degeneracies are present.

Now let us consider another theorem of importance to the development of quantum mechanics. This theorem concerns itself with the simultaneous diagonalization of commuting matrices. As will be discussed in the next chapter, only those observables whose matrix representations or operators commute can be measured at the same time, and the theorem of direct interest.

Before proving this theorem, let us examine the following aspect of normal operators and their matrix representations. If $\mathbf{A}$ is an $(n \times n)$ normal matrix with eigenvectors $\mathbf{U}$, then

$$\mathbf{AU} = \mathbf{U\Lambda}, \tag{3-318}$$

where $\mathbf{\Lambda}$ is an $(n \times n)$ diagonal matrix containing the eigenvalues, and $\mathbf{U}$ is an $(n \times n)$ unitary matrix. Taking the adjoint of both sides of Eq. (3-318) gives

$$\mathbf{U}^\dagger\mathbf{A}^\dagger = \mathbf{\Lambda}^*\mathbf{U}^\dagger \tag{3-319}$$

since $\mathbf{\Lambda}$ is diagonal. Left and right multiplying both sides of the above equation by $\mathbf{U}$ gives

$$\mathbf{UU}^\dagger\mathbf{A}^\dagger\mathbf{U} = \mathbf{U\Lambda}^*\mathbf{U}^\dagger\mathbf{U}$$

or

$$\mathbf{A}^\dagger\mathbf{U} = \mathbf{U\Lambda}^*, \tag{3-320}$$

which will be useful shortly.

Theorem 3-22: It is possible to find a unitary matrix that simultaneously diagonalizes two Hermitian matrices if and only if the matrices commute.

Proof: To show the necessary condition, let us consider two Hermitian matrices, $\mathbf{A}$ and $\mathbf{B}$. By Theorem 3-18 we know that $\mathbf{A}$ can always be

[24] It is important to note that the above theorems and discussion also hold for matrices whose elements are real. In this case, the words Hermitian and unitary are replaced by symmetric and orthogonal, respectively.

diagonalized via a unitary similarity transformation:

$$\mathbf{V}^{\dagger}\mathbf{A}\mathbf{V}=\mathbf{d}$$

or

$$\mathbf{A}\mathbf{V}=\mathbf{V}\mathbf{d}, \tag{3-321}$$

where $\mathbf{d}$ is diagonal. Then, if

$$\mathbf{V}^{\dagger}\mathbf{B}\mathbf{V}=\mathbf{c},$$

or

$$\mathbf{B}\mathbf{V}=\mathbf{V}\mathbf{c}, \tag{3-322}$$

where $\mathbf{c}$ is diagonal, we can write

$$\mathbf{A}(\mathbf{B}\mathbf{V})=\mathbf{A}(\mathbf{V}\mathbf{c})=(\mathbf{A}\mathbf{V})\mathbf{c}=\mathbf{V}\mathbf{d}\mathbf{c}. \tag{3-323}$$

Also,

$$\mathbf{B}(\mathbf{A}\mathbf{V})=\mathbf{B}(\mathbf{V}\mathbf{d})=(\mathbf{B}\mathbf{V})\mathbf{d}=\mathbf{V}\mathbf{c}\mathbf{d}. \tag{3-324}$$

But $\mathbf{dc} = \mathbf{cd}$, since $\mathbf{c}$ and $\mathbf{d}$ are diagonal, and substraction of Eq. (3-324) from Eq. (3-323) yields

$$(\mathbf{A}\mathbf{B}-\mathbf{B}\mathbf{A})\mathbf{V}=\mathbf{0}.$$

Since $\mathbf{V}$ is unitary, right multiplication by $\mathbf{V}^{\dagger}$ gives the desired result, i.e.,

$$\mathbf{A}\mathbf{B}=\mathbf{B}\mathbf{A}. \tag{3-325}$$

To prove the sufficiency, we shall use the simple proof of P. O. Löwdin,[25] based on the use of normal matrices. If we assume that $\mathbf{A}$ and $\mathbf{B}$ are Hermitian matrices that commute, then $\mathbf{L}$ and $\mathbf{L}^{\dagger}$, obtained from

$$\mathbf{L}=\mathbf{A}+i\mathbf{B} \tag{3-326}$$

and

$$\mathbf{L}^{\dagger}=\mathbf{A}-i\mathbf{B} \tag{3-327}$$

are normal matrices. Then, from Eq. (3-318) and Eq. (3-320) we know that

$$\mathbf{L}\mathbf{V}=\mathbf{V}\boldsymbol{\Lambda} \tag{3-328}$$

$$\mathbf{L}^{\dagger}\mathbf{V}=\mathbf{V}\boldsymbol{\Lambda}^{*}, \tag{3-329}$$

where $\boldsymbol{\Lambda}$ is diagonal and $\mathbf{V}$ is unitary. If we write $\boldsymbol{\Lambda} = \mathbf{e} + i\mathbf{f}$, where $\mathbf{e}$ and $\mathbf{f}$ are real, diagonal matrices, then Eq. (3-328) and Eq. (3-329) can be written as

$$(\mathbf{A}+i\mathbf{B})\mathbf{V}=\mathbf{V}(\mathbf{e}+i\mathbf{f}) \tag{3-330}$$

$$(\mathbf{A}-i\mathbf{B})\mathbf{V}=\mathbf{V}(\mathbf{e}-i\mathbf{f}). \tag{3-331}$$

[25] P. O. Löwdin, *Rev. Mod. Phys.*, **34**, 520 (1962).

Adding Eqs. (3-330) and (3-331) gives

$$\mathbf{AV} = \mathbf{Ve},$$

and subtracting Eq. (3-330) from Eq. (3-331) gives

$$\mathbf{BV} = \mathbf{Vf}$$

and $\mathbf{V}$ is the matrix that diagonalizes both $\mathbf{A}$ and $\mathbf{B}$, and the proof is completed.

It should also be pointed out that the columns of the unitary matrix that diagonalize $\mathbf{A}$ and $\mathbf{B}$ also form a common set of eigenvectors of $\mathbf{A}$ and $\mathbf{B}$. Expressed in operator language, Theorem 3-22 would read as follows. It is possible to find simultaneous eigenfunctions of two Hermitian operators if and only if the operators commute.

This theorem does not guarantee that every unitary matrix that diagonalizes $\mathbf{A}$ will also diagonalize $\mathbf{B}$. It states only that it is possible to find a matrix that will simultaneously diagonalize both. That does not rule out the possibility that other unitary matrices may diagonalize either $\mathbf{A}$ or $\mathbf{B}$, but not both simultaneously.

One of the interesting characteristics of Hermitian or symmetric matrices that will be useful later is that it is possible to place a geometric interpretation on their eigenvalues and eigenvectors. In illustrating this we shall confine our attention to symmetric matrices for convenience. Let us consider the following symmetric matrix transformation $\mathbf{A}$ on $\mathbf{v}$,

$$\mathbf{Av} = \mathbf{u}, \tag{3-332}$$

where $\mathbf{v}$ and $\mathbf{u}$ are column vectors. Left multiplying by the transpose of $\mathbf{v}$ gives

$$\mathbf{v}^t\mathbf{Av} = \mathbf{v}^t\mathbf{u} = K, \tag{3-333}$$

or, in summation form,

$$\sum_{i,j=1}^{n} v_i a_{ij} v_j = K, \tag{3-334}$$

where K is a constant.

The above equation is called a *quadratic form*, and such forms appear in many problems in physics and chemistry, e.g., the theory of molecular vibrations, molecular polarizabilities, and nuclear and molecular quadrupole moments.

Consider first the special case when $\mathbf{A}$ is a (3×3) matrix. Since $\mathbf{A}$ is assumed to be symmetric, we know there exists an orthogonal transformation matrix $\mathbf{T}$ that diagonalizes $\mathbf{A}$, i.e.,

$$\mathbf{T}^t\mathbf{AT} = \mathbf{\Lambda} = \begin{pmatrix} \lambda_1 & 0 & 0 \\ 0 & \lambda_2 & 0 \\ 0 & 0 & \lambda_2 \end{pmatrix}. \tag{3-335}$$

Therefore, if we substitute Eq. (3-335) into Eq. (3-333) and use the orthogonality property of **T**, we obtain the following result

$$K=\bar{\mathbf{v}}'\mathbf{\Lambda}\bar{\mathbf{v}}, \tag{3-336}$$

or

$$K=\sum_{i=1}^{3}\lambda_i\bar{v}_1^2, \tag{3-337}$$

where

$$\bar{\mathbf{v}}=\mathbf{T}'\bar{\mathbf{v}}. \tag{3-338}$$

From Eq. (3-335) it is clear that the columns of **T** are the eigenvectors of **A**, where the matrix **T** is given by:

$$\mathbf{T}=\begin{pmatrix} \cos(\boldsymbol{i},\bar{\boldsymbol{i}}) & \cos(\boldsymbol{i},\bar{\boldsymbol{j}}) & \cos(\boldsymbol{i},\bar{\boldsymbol{k}}) \\ \cos(\boldsymbol{j},\bar{\boldsymbol{i}}) & \cos(\boldsymbol{j},\bar{\boldsymbol{j}}) & \cos(\boldsymbol{j},\bar{\boldsymbol{k}}) \\ \cos(\boldsymbol{k},\bar{\boldsymbol{i}}) & \cos(\boldsymbol{k},\bar{\boldsymbol{j}}) & \cos(\boldsymbol{k},\bar{\boldsymbol{k}}) \end{pmatrix}, \tag{3-339}$$

where $\cos(p, q)$ represents the direction cosine between the pth and qth basis vectors in the old and new coordinate systems, respectively. Hence, the components of the eigenvectors of **A** can also be thought of as the components of the new basis vectors, expressed in terms of the old basis. The new axis system in which **A** is diagonal is called the *principal axis system*, and the eigenvectors of **A** form the *principal axes*. In order to determine the relative orientation between the new and old axis systems, such as the relationship of $\bar{\mathbf{i}}$ to the old basis, we must determine the three quantities

$$\begin{aligned} (\boldsymbol{i},\bar{\boldsymbol{i}})&=\cos^{-1}[\cos(\boldsymbol{i},\bar{\boldsymbol{i}})] \\ (\boldsymbol{j},\bar{\boldsymbol{i}})&=\cos^{-1}[\cos(\boldsymbol{j},\bar{\boldsymbol{i}})] \\ (\boldsymbol{k},\bar{\boldsymbol{i}})&=\cos^{-1}[\cos(\boldsymbol{k},\bar{\boldsymbol{i}})] \end{aligned} \tag{3-340}$$

Let us illustrate these notions with a more concrete example. The equation representing the ellipse shown in Fig. 3.6 can be written in the quadratic form

$$5x^2-2\sqrt{3}xy+7y^2=8, \tag{3-341}$$

which can be written in matrix notation as

$$(x\ y)\begin{pmatrix} 1 & -\sqrt{3} \\ -\sqrt{3} & 7 \end{pmatrix}\begin{pmatrix} x \\ y \end{pmatrix}=8. \tag{3-342}$$

In order to find the equation of the ellipse in the barred (i.e., rotated) coordinate system where the cross-terms are eliminated, i.e., where the matrix is diagonal, we must first determine the eigenvalues and the corresponding eigenvectors of the above matrix. It is easily verified that these are given by

$$\lambda_1=4, \qquad \mathbf{t}_1=\begin{pmatrix} \sqrt{3}/2 \\ 1/2 \end{pmatrix}, \tag{3-343}$$

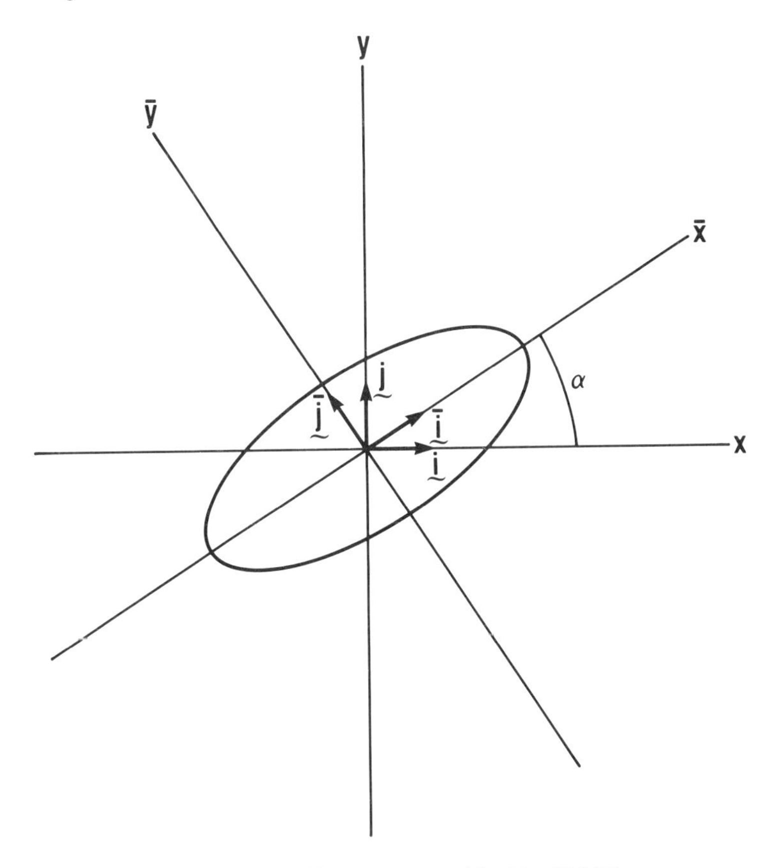

Figure 3.6. Ellipse represented by Eq. (3-341).

$$\lambda_2 = 8, \qquad \mathbf{t}_2 = \begin{pmatrix} -1/2 \\ \sqrt{3}/2 \end{pmatrix}.$$

Applying the inverse of the transformation shown in Eq. (3-338), we obtain

$$\begin{pmatrix} x \\ y \end{pmatrix} = (\mathbf{t}_1 \ \ \mathbf{t}_2) \begin{pmatrix} \bar{x} \\ \bar{y} \end{pmatrix} = \begin{pmatrix} \sqrt{3}/2 & -1/2 \\ 1/2 & \sqrt{3}/2 \end{pmatrix} \begin{pmatrix} \bar{x} \\ \bar{y} \end{pmatrix}. \tag{3-344}$$

Substitution into Eq. (3-342) yields

$$(\bar{x} \ \ \bar{y}) \begin{pmatrix} 4 & 0 \\ 0 & 8 \end{pmatrix} \begin{pmatrix} \bar{x} \\ \bar{y} \end{pmatrix} = 4\bar{x}^2 + 8\bar{y}^2 = 8, \tag{3-345}$$

which can be further simplified to

$$\frac{\bar{x}^2}{2} + \frac{\bar{y}^2}{1} = 1. \tag{3-346}$$

This is the equation of the ellipse shown in Fig. 3.6 with respect to the barred coordinate system. Finally, the angle α between the barred and unbarred coordinate systems is given, using Eq. (3-340), as

$$\alpha = \cos^{-1}(\sqrt{3}/2) = 30°. \tag{3-347}$$

Before leaving this section on matrix eigenvalue problems, we shall consider one additional problem of importance in quantum mechanics. As discussed in Section 3-6, a linear operator can be represented with respect to a nonorthogonal basis $\{\boldsymbol{\upsilon}_1, \boldsymbol{\upsilon}_2, \dots, \boldsymbol{\upsilon}_n\}$ as

$$\mathbf{A} = \mathbf{G}^{-1}\boldsymbol{\Omega}_a. \tag{3-348}$$

The elements of $\boldsymbol{\Omega}_a$ are given by

$$(\boldsymbol{\Omega}_a)_{ij} = (\boldsymbol{\upsilon}_i, \mathcal{A}\boldsymbol{\upsilon}_j), \tag{3-349}$$

and in quantum mechanics are usually called *matrix elements*. $\mathbf{G}$ is the Gramian matrix,[26] whose elements are given by Eq. (3-199). As previously discussed, $\mathbf{G} = \mathbf{I}$ when the basis is orthonormal, and $\boldsymbol{\Omega}_a$ corresponds to the true matrix representation of $\mathcal{A}$.

Now let us examine the form of the eigenvalue equation when written in terms of $\boldsymbol{\Omega}_a$, i.e.,

$$\mathbf{AU} = \mathbf{U\Lambda}. \tag{3-350}$$

Substituting Eq. (3-348) into the above equation yields

$$\mathbf{G}^{-1}\boldsymbol{\Omega}_a\mathbf{U} = \mathbf{U\Lambda}$$

or

$$\boldsymbol{\Omega}_a\mathbf{U} = \mathbf{GU\Lambda}. \tag{3-351}$$

Before actually solving Eq. (3-351), let us examine a few of the properties of $\mathbf{G}$ that are important for our further discussions, viz., that $\mathbf{G}$ is (1) Hermitian, and (2) positive definite.

To show that $\mathbf{G}$ is Hermitian, consider the form of the elements of $\mathbf{G}$ given by Eq. (3-199), i.e.,

$$\begin{aligned}(\mathbf{G})_{ij} &= (\boldsymbol{\upsilon}_i, \boldsymbol{\upsilon}_j)\\ &= (\boldsymbol{\upsilon}_j, \boldsymbol{\upsilon}_i)^* = (\mathbf{G}^\dagger)_{ij},\end{aligned} \tag{3-353}$$

which shows that $\mathbf{G}$ is Hermitian.

To show that $\mathbf{G}$ is a positive definite matrix, we first define a *positive definite Hermitian quadratic form* as

$$\mathbf{w}^\dagger\mathbf{Qw} > 0 \tag{3-354}$$

where $\mathbf{w} \neq \mathbf{0}$ and is an otherwise arbitrary column vector, and $\mathbf{Q}$ is Hermitian. Since $\mathbf{Q}$ is Hermitian, there exists a set of orthonormal eigenvectors $(\mathbf{v}_1, \mathbf{v}_2, \dots, \mathbf{v}_n)$

---

[26] $\mathbf{G}$ is also frequently called the *overlap* or *metric* matrix, and is denoted by $\mathbf{S}$ or $\boldsymbol{\Delta}$, respectively.

such that

$$\mathbf{v}_i^\dagger \mathbf{Q} \mathbf{v}_i = \lambda_i, \qquad i = 1, 2, \dots, n. \tag{3-355}$$

Then, since $\mathbf{w}$ is arbitrary, it follows from Eqs. (3-354) and (3-355) that

$$\lambda_i > 0, \qquad i = 1, 2, \dots, n. \tag{3-356}$$

Therefore, the matrix $\mathbf{Q}$ of a positive definite quadratic form possesses all positive eigenvalues, and is called a *positive definite matrix*.

We can now show that $\mathbf{G}$ is a Hermitian positive definite matrix in the following manner. Consider the arbitrary vector $\boldsymbol{u}$, given by

$$\boldsymbol{u} = \sum_{i=1}^{n} u_i \boldsymbol{v}_i. \tag{3-357}$$

If we form the inner product $(\boldsymbol{u}, \boldsymbol{u})$, we obtain

$$(\boldsymbol{u}, \boldsymbol{u}) = \sum_{i=1}^{n} \sum_{j=1}^{n} u_i (\boldsymbol{v}_i, \boldsymbol{v}_j) u_j > 0, \tag{3-358}$$

unless $\boldsymbol{u} = 0$. Equation (3-358) can also be written in matrix notation as

$$\mathbf{u}^\dagger \mathbf{G} \mathbf{u} > 0, \tag{3-359}$$

and hence $\mathbf{G}$ is a positive definite matrix.

We can now return to the task of solving Eq. (3-350). Since $\mathbf{G}$ is Hermitian it can be diagonalized by a unitary similarity transformation, i.e.,

$$\mathbf{S}^\dagger \mathbf{G} \mathbf{S} = \boldsymbol{\Gamma} = \begin{pmatrix} \gamma_1 & & & \\ & \gamma_2 & & 0 \\ & & \ddots & \\ & 0 & & \gamma_n \end{pmatrix}, \tag{3-360}$$

where $\gamma_i > 0$, $i = 1, 2, \dots, n$. We now construct the matrices

$$\boldsymbol{\Gamma}^{-1/2} \boldsymbol{\Gamma}^{+1/2} = \mathbf{I} = \begin{pmatrix} \gamma_1^{-1/2} & & & \\ & \gamma_2^{-1/2} & & 0 \\ & & \ddots & \\ & 0 & & \gamma_n^{-1/2} \end{pmatrix} \begin{pmatrix} \gamma_1^{+1/2} & & & \\ & \gamma_2^{+1/2} & & 0 \\ & & \ddots & \\ & 0 & & \gamma_n^{+/1/2} \end{pmatrix}, \tag{3-361}$$

which is possible due to the positive definiteness of $\mathbf{G}$. We note also that

$$\boldsymbol{\Gamma}^{-1/2} \boldsymbol{\Gamma} \boldsymbol{\Gamma}^{-1/2} = \mathbf{I}, \tag{3-362}$$

or, substituting Eq. (3-360) into Eq. (3-362),

$$\boldsymbol{\Gamma}^{-1/2} \mathbf{S}^t \mathbf{G} \mathbf{S} \boldsymbol{\Gamma}^{-1/2} = \mathbf{I}. \tag{3-363}$$

If we now define the matrix

$$\mathbf{T} = \mathbf{S} \boldsymbol{\Gamma}^{-1/2}, \tag{3-364}$$

then Eq. (3-363) becomes

$$\mathbf{T}^{\dagger}\mathbf{G}\mathbf{T} = \mathbf{I}. \tag{3-365}$$

$\mathbf{T}$ is not unitary, but its inverse is given by

$$\mathbf{T}^{-1} = (\mathbf{S}\boldsymbol{\Gamma}^{-1/2})^{-1} = \boldsymbol{\Gamma}^{1/2}\mathbf{S}^{\dagger}. \tag{3-366}$$

Having accomplished the task of diagonalizing $\mathbf{G}$, we can return to Eq. (3-351), and rewrite it as

$$\mathbf{T}^{\dagger}\boldsymbol{\Omega}_a(\mathbf{T}\mathbf{T}^{-1})\mathbf{U} = \mathbf{T}^{\dagger}\mathbf{G}(\mathbf{T}\mathbf{T}^{-1})\mathbf{U}\boldsymbol{\Lambda}, \tag{3-367}$$

which simplifies to

$$(\mathbf{T}^{\dagger}\boldsymbol{\Omega}_a\mathbf{T})(\mathbf{T}^{-1}\mathbf{U}) = (\mathbf{T}^{-1}\mathbf{U})\boldsymbol{\Lambda}. \tag{3-368}$$

Defining

$$\bar{\mathbf{U}} = \mathbf{T}^{-1}\mathbf{U}, \tag{3-369}$$

and

$$\bar{\boldsymbol{\Omega}}_a = \mathbf{T}^{\dagger}\boldsymbol{\Omega}_a\mathbf{T} \tag{3-370}$$

which is also Hermitian, Eq. (3-368) becomes

$$\bar{\boldsymbol{\Omega}}_a\bar{\mathbf{U}} = \bar{\mathbf{U}}\boldsymbol{\Lambda}. \tag{3-371}$$

Note that Eq. (3-371) is identical in form to Eq. (3-350). It is also important to note that the eigenvalues of Eqs. (3-350) and (3-371) are identical. This follows also from the fact that both $\mathbf{A}$ and $\boldsymbol{\Omega}_a$ are true matrix representations of $\mathcal{A}$, i.e., they differ only by the similarity transformation $\mathbf{T}^{-1}\,\mathbf{A}\mathbf{T} = \bar{\boldsymbol{\Omega}}_a$ (see Problem 3-23).

Since Eq. (3-371) is in standard eigenvalue equation form, it follows from our previous discussions that

$$\bar{\mathbf{U}}^{\dagger}\bar{\mathbf{U}} = \bar{\mathbf{U}}\bar{\mathbf{U}}^{\dagger} = \mathbf{I}. \tag{3-372}$$

Summarizing, the previous discussion has shown that Eq. (3-351) can be solved by transforming it first to "standard" eigenvalue form. Furthermore, we have described the prescription for forming the matrix $\mathbf{T}$ that accomplishes the transformation. After Eq. (3-371) is solved, the eigenvectors of Eq. (3-351) can then be found by rearranging Eq. (3-369), i.e.,

$$\mathbf{U} = \mathbf{T}\bar{\mathbf{U}}. \tag{3-373}$$

Referring to the general discussion of similarity transformations in Section 3-7, it follows that the transformation represented by $\mathbf{T}$ is a nonunitary transformation, which maps a nonorthogonal basis onto an orthogonal one. In fact, if $\mathbf{T}$ was unitary, we would not be able to obtain an orthogonal basis from a nonorthogonal basis, since the transformation would have preserved angles and distances.

Finally, before leaving the discussion of theorems that will be of importance in our study of quantum mechanics, it is useful to consider under what conditions the eigenvectors for different eigenvalues will not only be orthogonal, but also noninteracting. To see what is meant by this, we consider the following theorem.

Theorem 3-23: If **A** and **B** are Hermitian and commute, and if $\mathbf{V}_1$ and $\mathbf{V}_2$ are eigenvectors of **A** with nondegenerate eigenvalues a_1 and a_2, then

$$\mathbf{V}_1^\dagger \mathbf{B} \mathbf{V}_2 = 0. \tag{3-374}$$

Proof: We first consider

$$\mathbf{V}_1^\dagger \mathbf{B} \mathbf{A} \mathbf{V}_2 = a_2 \mathbf{V}_1^\dagger \mathbf{B} \mathbf{V}_2. \tag{3-375}$$

Alternatively, the expression on the left-hand side of Eq. (3-375) can be written as:

$$\begin{aligned} \mathbf{V}_1^\dagger \mathbf{B} \mathbf{A} \mathbf{V}_2 &= \mathbf{V}_1^\dagger \mathbf{A} \mathbf{B} \mathbf{V}_2 \\ &= \mathbf{V}_1^\dagger \mathbf{A}^\dagger \mathbf{B} \mathbf{V}_2 \\ &= (\mathbf{A}\mathbf{V}_1)^\dagger \mathbf{B} \mathbf{V}_2 \\ &= a_1 \mathbf{V}_1^\dagger \mathbf{B} \mathbf{V}_2. \end{aligned} \tag{3-376}$$

Combining Eqs. (3-375) and (3-376) gives

$$a_2 \mathbf{V}_1^\dagger \mathbf{B} \mathbf{V}_2 = a_1 \mathbf{V}_1^\dagger \mathbf{B} \mathbf{V}_2$$

or

$$(a_2 - a_1) \mathbf{V}_1^\dagger \mathbf{B} \mathbf{V}_2 = 0. \tag{3-377}$$

Thus, since $a_1 \neq a_2$, we have

$$\mathbf{V}_1^\dagger \mathbf{B} \mathbf{V}_2 = 0, \tag{3-378}$$

which proves the theorem.

3.9. Infinite Matrices and Linear Transformations on Hilbert Space

Although we shall not often encounter infinite matrices in the quantum chemical applications of later chapters, several comments are appropriate here within the mathematical discussions. As we shall see, the generalization of the techniques that were applied to matrices of finite dimension to those of infinite dimension is nontrivial. The following discussions are not intended to provide this generalization rigorously but, rather, to indicate some of the difficulties involved in the process. More detailed discussions of these difficulties can be found in other texts.[27]

---

[27] For a detailed account of infinite matrices, see Richard G. Cooke, *Infinite Matrices and Sequence Spaces,* Macmillan and Company, Ltd., London, 1950.

An example of the difficulties that are encountered is exhibited by the following discussion. Consider the infinite matrix

$$\mathbf{A} = \begin{pmatrix} a_{11} & a_{12} & \cdots & a_{1n} & \cdots \\ a_{21} & a_{22} & \cdots & a_{2n} & \cdots \\ \vdots & \vdots & & \vdots & \\ a_{n1} & a_{n2} & \cdots & a_{nn} & \cdots \\ \vdots & \vdots & & \vdots & \end{pmatrix}, \tag{3-379}$$

and the infinite column vectors[28] given by

$$\mathbf{u} = \begin{pmatrix} u_1 \\ u_2 \\ \vdots \\ u_n \\ \vdots \end{pmatrix}, \mathbf{w} = \begin{pmatrix} w_1 \\ w_2 \\ \vdots \\ w_n \\ \vdots \end{pmatrix}. \tag{3-380}$$

Next, consider the following matrix transformation,

$$\mathbf{Au} = \mathbf{w}, \tag{3-381}$$

whose elements are given by

$$w_i = \sum_{j=1}^{\infty} a_{ij} u_j, \qquad i = 1, 2, \cdots, n, \ldots . \tag{3-382}$$

In the case of finite matrix transformations, the above sum presents no problems. For infinite matrix transformations, however, it does not always follow that the above infinite sums are convergent. The necessary and sufficient condition that the sums converge for any vector $\mathbf{u}$ in $\mathcal{H}$ is given by

$$\sum_{j=1}^{\infty} |a_{ij}|^2 < \infty, \qquad i = 1, 2, \ldots, n, \ldots . \tag{3-383}$$

If some of the sums do not converge, then the norm of $\mathbf{w}$, $\Sigma_{j=1}^{\infty} |w_j|^2$, will not be finite. Hence, the vector $\mathbf{w}$ will not represent a vector in $\mathcal{H}$.

Thus, we see via the previous isolated example that difficulties are to be expected in the generalization to infinite matrices. Indeed, new concepts are frequently introduced to handle these difficulties, and the interested reader is referred to more extensive treatments[29] for further discussion.

---

[28] In an analogous fashion to the discussion in Section 3-6, it can be shown that $\mathbf{u}$ and $\mathbf{w}$ are isomorphic to the vectors $\boldsymbol{\mathcal{u}}$ and $\boldsymbol{\mathcal{w}}$ in Hilbert space.

[29] See, for example, V. I. Smirnov, *A Course of Higher Mathematics,* Vol. V, Addison-Wesley, Reading, MA, 1964.

3-10. Dirac Notation

As we have seen in our previous discussions, the cases of finite and infinite vectors often need to be considered separately, due to questions such as convergence in the infinite dimensional case. However, Dirac[30] has given a notation that considers generalized vectors (including both finite and infinite dimensional cases) that can be used to obtain many results of interest, regardless of the dimensionality of the space. Since the concepts and the notation introduced by Dirac will be useful in certain contexts later, we shall introduce them here.

Instead of dealing with the components of vectors, we shall refer only to the entire vector, which we shall call a *ket vector*, and give it the symbol $|\rangle$. Of course, this vector may refer either to a function or a column vector. A particular ket vector is distinguished by a label, such as $|v\rangle$. This vector may have either a finite or infinite number of components. Just as for the other vectors that we have considered, different kets can be added to give a new ket,

$$a_1|v\rangle + a_2|w\rangle = |z\rangle \tag{3-384}$$

where a_1 and a_2 are arbitrary complex constants.

In order to form scalar products using these kets, which we shall need in order to have a Hilbert space, we introduce another set of vectors, called *bra vectors*, and denote them by $\langle|$. These vectors must have the property that the formation of scalar products with ket vectors produces a linear vector space. Thus, picking an arbitrary bra vector $\langle y|$, we require that

$$\langle y|\{|v\rangle + |w\rangle\} = \langle y|v\rangle + \langle y|w\rangle, \tag{3-385}$$

and

$$\langle y|\{a|v\rangle\} = a\langle y|v\rangle, \tag{3-386}$$

where a is an arbitrary complex constant.

Thus far, we have seen no necessary connection between the space of the bra vectors and the space of the ket vectors, except that a scalar product can be formed. To assure the complete correspondence of scalar products using bra and ket vectors and the vectors of a Hermitian vector space that were previously discussed, an association of the bra and ket spaces is assumed. It is assumed that there is a one-to-one correspondence between every ket vector $|v\rangle$ and its corresponding bra vector $\langle v|$, such that the vector in the bra space corresponding to the vector $[|v\rangle + |w\rangle]$ in the ket space is the sum of the bras corresponding to $|v\rangle$ and $|w\rangle$, and that the bra vector corresponding to $a|v\rangle$ is $a^* \langle v|$. Thus,

$$\langle av|w\rangle = a^*\langle v|w\rangle \tag{3-387}$$

[30] P. A. M. Dirac, *Quantum Mechanics,* 4th ed., Oxford University Press, Oxford, 1958.

and

$$\langle v \mid aw \rangle = a\langle v \mid w \rangle.$$

The vectors in these two spaces are known as dual vectors. The advantage of such a notation is evident, for $\langle u \mid v \rangle$ can be used to represent the scalar product of two n-dimensional vectors,

$$\langle u \mid v \rangle = \sum_{i=1}^{n} u_i^* v_i, \tag{3-388}$$

or it could equally well represent the scalar product of two continuous functions

$$\langle u \mid v \rangle = \int u^*(x) v(x)\, dx. \tag{3-389}$$

Also, a linear operator $\mathcal{A}$ on a ket vector is one in which

$$\mathcal{A}\{\mid v \rangle + \mid w \rangle\} = \mathcal{A} \mid v \rangle + \mathcal{A} \mid w \rangle, \tag{3-390}$$

$$\mathcal{A}\{c \mid v \rangle\} = c\mathcal{A} \mid v \rangle, \tag{3-391}$$

where c is an arbitrary constant. In connection with operators, it should be noted that there is a vast difference between $\langle u \mid v \rangle$ and $\mid v \rangle\langle u \mid$. The scalar product $\langle u \mid v \rangle$ is simply a single quantity (e.g., a number), but $\mid v \rangle\langle u \mid$ is an *operator*, as can easily be seen. Forming the scalar product of $\mid v \rangle\langle u \mid$ with the ket $\mid w \rangle$ gives $\mid v \rangle\langle u \mid w \rangle$, which is another ket vector, multiplied by the number $\langle u \mid w \rangle$. Similarly, the scalar product of the bra $\langle w \mid$ with $\mid v \rangle\langle u \mid$ gives $\langle w \mid v \rangle\langle u \mid$, which is another bra vector, multiplied by the number $\langle w \mid v \rangle$. Thus, $\mid v \rangle\langle u \mid$ can be viewed as an operator that transforms a ket vector into another ket vector, or a bra vector into another bra vector.

We shall use both the bracket notation and the ordinary vector and scalar product notation throughout the text, and the student should be familiar with both.

Problems

1. Show that the set of $(n \times m)$ matrices whose elements are complex forms a vector space, if matrix addition and multiplication are defined as in Eqs. (3-6) through (3-9).
2. If $\mathbf{A}$ is an $(m \times n)$ matrix, $\mathbf{B}$ is an $(n \times p)$ and $\mathbf{C}$ is an $(m \times p)$ matrix, show that the associative and distributive properties apply to matrix multiplication, i.e., show that

$$\mathbf{A}(\mathbf{BC}) = (\mathbf{AB})\mathbf{C}$$

and

$$\mathbf{A}(\mathbf{B} + \mathbf{C}) = \mathbf{AB} + \mathbf{AC}.$$

3. If $\mathbf{A}$ is an $(n \times n)$ matrix and $\mathbf{C}$ is an $(n \times n)$ scalar matrix as defined in

Eq. (3-19), show that

$$\mathbf{CA} = \mathbf{AC}.$$

4. Show that Eqs. (3-46) and (2-8) are equivalent representations of a vector product in 3 dimensions.
5. The derivative with respect to a variable of an $(n \times n)$ determinant having differentiable elements is defined as the sum of n determinants, in each of which a different row (or column) in the original determinant is replaced by the derivative of each element of the row (or column). With this definition, show that

$$\frac{d}{dx}\begin{vmatrix} x^2 & 1 & 2x \\ 0 & 3x^2 & 3x \\ 1 & (2x-1) & 4 \end{vmatrix} = 60x^4 - 48x^3 + 9x^2 + 3$$

6. Prove that: $|\mathbf{A}| = |\mathbf{A}^t|$.
7. Show that the set of vectors represented by the columns of

$$\mathbf{A} = \begin{pmatrix} 3x & y^2 & 6x \\ 4 & x & 8 \\ 5y & z & 10y \end{pmatrix}$$

is not linearly independent.
8. Show that if the rows (or columns) of an $(n \times n)$ matrix $\mathbf{M}$ are treated as vectors, $\mathbf{M}^{-1}$ exists if and only if the rows and/or columns are linearly independent.
9. Find the inverse of

$$\mathbf{A} = \begin{pmatrix} 3 & 2 & 1 \\ 2 & 4 & 0 \\ 1 & 0 & 5 \end{pmatrix}.$$

10. If $\mathbf{A}$, $\mathbf{B}$, and $\mathbf{C}$ are each $(n \times n)$ matrices, show that

$$(\mathbf{ABC})^\dagger = \mathbf{C}^\dagger \mathbf{B}^\dagger \mathbf{A}^\dagger.$$

11. If $\mathbf{U}$ and $\mathbf{V}$ are $(n \times n)$ unitary matrices, show that

$$\mathbf{W} = \mathbf{UV}$$

is also unitary.
12. Prove that the rows (or columns) of a unitary matrix are orthonormal.
13. If $\mathbf{A}$ is Hermitian, show that $|\mathbf{A}|$ is a real number.
14. If $\mathbf{A}$ is unitary, show that $|\mathbf{A}|$ has unit modulus.
15. Given the passive transformation of the vector

$$\mathbf{v} = v_1 \mathbf{i} + v_2 \mathbf{j} + v_3 \mathbf{k}$$

in the i, j, k coordinate system to the vector

$$\boldsymbol{v} = \bar{v}_1 \bar{\boldsymbol{i}} + \bar{v}_2 \bar{j} + \bar{v}_3 \bar{k}$$

in the $\bar{\boldsymbol{i}}$, $\bar{\boldsymbol{j}}$, $\bar{\boldsymbol{k}}$ coordinate system, show that

$$\begin{pmatrix} v_1 \\ v_2 \\ v_3 \end{pmatrix} = \begin{pmatrix} \cos(\boldsymbol{i}, \bar{\boldsymbol{i}}) & \cos(\boldsymbol{i}, \bar{\boldsymbol{j}}) & \cos(\boldsymbol{i}, \bar{\boldsymbol{k}}) \\ \cos(\boldsymbol{j}, \bar{\boldsymbol{i}}) & \cos(\boldsymbol{j}, \bar{\boldsymbol{j}}) & \cos(\boldsymbol{j}, \bar{\boldsymbol{k}}) \\ \cos(\boldsymbol{k}, \bar{\boldsymbol{i}}) & \cos(\boldsymbol{k}, \bar{\boldsymbol{j}}) & \cos(\boldsymbol{k}, \bar{\boldsymbol{k}}) \end{pmatrix} \begin{pmatrix} \bar{v}_2 \\ \bar{v}_2 \\ \bar{v}_3 \end{pmatrix}$$

where $\cos(\mathbf{l}, \bar{\mathbf{m}})$ is the direction cosine between the $\mathbf{l}$ and $\bar{\mathbf{m}}$ basis vectors.

16. Show that the product of two commuting matrices $\mathbf{A}$ and $\mathbf{B}$ is invariant to a similarity transformation.
17. Prove that if two matrices are similar, they have the same secular polynomial, and therefore the same eigenvalues.
18. Prove that a similarity transformation leaves eigenvalues invariant.
19. Find the eigenvalues and normalized eigenvectors of the matrix

$$\mathbf{A} = \begin{pmatrix} 3 & 2 & 1 \\ 2 & 4 & 0 \\ 1 & 0 & 5 \end{pmatrix}.$$

20. Prove that, if an $(n \times n)$ matrix $\mathbf{A}$ can be diagonalized by a similarity transformation

$$\mathbf{U}^{-1}\mathbf{A}\mathbf{U} = \mathbf{\Lambda}$$

where $\mathbf{\Lambda}$ is diagonal, the eigenvectors form a basis for the n-dimensional vector space on which $\mathbf{A}$ is defined (converse of Theorem 3-15).

21. Prove Theorem 3-17.
22. In an n-dimensional space, if $\boldsymbol{u}_1, \boldsymbol{u}_2, \ldots, \boldsymbol{u}_m$ are eigenvectors $(m \leq n)$ having corresponding eigenvalues $\lambda_1, \lambda_2, \ldots, \lambda_m$ which are unique, show that the eigenvectors $\boldsymbol{u}_1, \boldsymbol{u}_2, \ldots, \boldsymbol{u}_m$ are linearly independent. (Hint: Proof by induction may be convenient.)
23. From the discussion involving Eqs. (3-348)–(3-371), show that

$$\mathbf{T}\mathbf{T}^{\dagger} = \mathbf{G}^{-1}$$

and that $\mathbf{A}$ and $\mathbf{\Omega}_a$ are true matrix representations of $\mathcal{A}$, i.e., that they differ only by the similarity transformation:

$$\bar{\mathbf{\Omega}}_a = \mathbf{T}^{-1}\mathbf{A}\mathbf{T}.$$

24. If $\mathbf{Q}$ is a symmetric positive definite matrix, show that

$$\mathbf{W}^t\mathbf{Q}\mathbf{W} > 0$$

for arbitrary $\mathbf{w}$, unless $\mathbf{W} \equiv \mathbf{0}$.

25. Let $\mathbf{A} = a\mathbf{BC} + b\mathbf{DEF}$, where a and b are scalars. Show that if a similarity transformation is applied to this equation, the algebraic relationships are maintained.

26. If **A** and **B** are Hermitian and **U** and **V** are unitary, show that
    a. $\mathbf{U}^{-1}\mathbf{AU}$ is Hermitian.
    b. $\mathbf{U}^{-1}\mathbf{VU}$ is unitary.
    c. $i[\mathbf{A}, \mathbf{B}]$ is Hermitian.
27. Show that Theorem (3-6) can be extended to

$$\mathrm{tr}(\mathbf{ABC}) = \mathrm{tr}(\mathbf{CAB}) = \mathrm{tr}(\mathbf{BCA}) \neq \mathrm{tr}(\mathbf{BAC}).$$

28. Let **A** be a real, symmetric $(n \times n)$ matrix, and **T** be a symmetric tridiagonal matrix of the form:

$$\mathbf{T} = \begin{pmatrix} \alpha_1 & \beta_1 & 0 & \cdots & 0 \\ \beta_1 & \alpha_2 & \beta_2 & \cdots & 0 \\ 0 & \beta_2 & \alpha_3 & \ddots & \vdots \\ \vdots & \vdots & \ddots & \alpha_{n-1} & \beta_{n-1} \\ 0 & 0 & \cdots & \beta_{n-1} & \alpha_n \end{pmatrix}$$

Also, let **Q** be an $(n \times n)$ orthogonal matrix that transforms **A** into **T**, i.e.,

$$\mathbf{Q}^{-1}\mathbf{AQ} = \mathbf{Q}^T\mathbf{AQ} = \mathbf{T}.$$

Then, show that the vectors $\mathbf{q}_j$ that are columns of **Q** are related by the following three-term recursion relation:

$$\beta_j\mathbf{q}_{j+1} = \mathbf{Aq}_j - \alpha_j\mathbf{q}_j - \beta_{j-1}\mathbf{q}_{j-1}, \qquad \beta_0 \equiv \beta_n \equiv 0,$$

where

$$\alpha_j = \mathbf{q}_j^T\mathbf{Aq}_j$$

$$\beta_j = \mathbf{q}_{j+1}^T\mathbf{Aq}_j.$$

This is known as Lanczos method of triangularization of a matrix.

29. Consider the matrix **A** defined as

$$\mathbf{A} = \mathbf{I} - \sigma\mathbf{x}_i\mathbf{x}_i^t,$$

where x_i is a normalized, real vector and σ is a real constant.
    a. Show that **A** is symmetric.
    b. Identify the conditions under which **A** is also orthogonal.
30. Let **A** be a real, symmetric matrix that satisfies the eigenvalue equation:

$$\mathbf{Ax}_i = \lambda_i\mathbf{x}_i, \qquad i = 1, 2, \ldots, n.$$

    a. Show that, if $\mathbf{A}'$ is defined as,

$$\mathbf{A}' = \mathbf{A} - \alpha\mathbf{x}_m\mathbf{x}_m^t,$$

    where α is constant, $\mathbf{A}'$ is real and symmetric.
    b. Determine the eigenvalues and eigenvectors of $\mathbf{A}'$ in terms of the eigenvalues of **A**.
31. Show that, for the infinite matrix transformation (cf. Section 3-9):

$$\mathbf{w} = \mathbf{Au} \qquad \mathbf{u}, \mathbf{w} \in \mathcal{H},$$

where

$$w_1 = \sum_{j=1}^{\infty} a_{1j} u_j$$

$$w_2 = \sum_{j=1}^{\infty} a_{2j} u_j$$

$$\vdots \qquad \vdots$$

$$w_i = \sum_{j=1}^{\infty} a_{ij} u_j$$

$$\vdots \qquad \vdots$$

the necessary and sufficient condition that the series converge is that

$$\sum_{j=1}^{\infty} |a_{ij}|^2 < \infty, \qquad i = 1, 2, \ldots, n, \ldots$$

for any $\mathbf{u}$ in $\mathcal{H}$.

Chapter 4

Postulates of Quantum Mechanics and Initial Considerations

In order to begin the actual study of quantum mechanics, we shall now state a set of five postulates (or axioms), which form the basic structure of the theory. These postulates provide a conceptual and mathematical framework that is strikingly different from classical mechanics in many ways. Within it, however, explanations for apparently contradictory results of early experiments such as quantization of energy and angular momentum can be achieved. In addition, it provides a framework for interpretation of future experiments, as well as additional development of the theory itself.

Although there are several formulations of quantum theory, we shall use the ''Copenhagen'' interpretation, which seems to have gained the widest acceptance at this time.[1] Also, there has been no attempt to use an absolute minimum number of postulates. Instead, we have chosen a sufficient set that is convenient for presentation of the concepts involved. The important concepts to be developed are those of *state, observables,* and *measurement*. In the following discussions it will become clear where the quantum mechanical interpretation attached to these concepts differs from their meaning in classical systems.

Postulate I concerned with a description of the state vector or state function, which defines completely the quantum state of the system.[2] Postulate II defines

---

[1] See A. Messiah, *Quantum Mechanics*, Vol. I, North Holland Publ. Co., Amsterdam, 1965, p. 8 for a more thorough discussion of this point. Also, see *Quantum Theory and Beyond*, edited by T. Bastin, Cambridge University Press, 1971, for a discussion of the difficulties associated with such an interpretation.

[2] A ''system'' for our purposes will be taken as an atom, a molecule, a collection of atoms and molecules, or, in general, any collection of nuclei and electrons.

the relationship between observable quantities and Hermitian operators, and Postulate III describes the time evolution of a quantum state. Finally, Postulates IV and V define the effect of measurements on a quantum system. Once the postulates have been formulated, several consequences of these postulates, such as the simultaneous measurability of observables and the Uncertainty Principle will be discussed.

4-1. Quantum Mechanical States and Observables

If we are dealing with a system having n degrees of freedom that are specified by the n *generalized position coordinates*[3] $q_1, q_2, \ldots, q_n$, then the quantum mechanical state of the system is given by Postulate I.

Postulate I. The state of a quantum mechanical system at time t, containing n degrees of freedom, is specified completely by a normalized *state vector* (*wavefunction* or *eigenfunction*):

$$|\Psi_t\rangle \equiv \Psi_t(q_1, q_2, \cdots, q_n).$$

Unless otherwise specified, normalization to unity is assumed, i.e.,

$$\langle\Psi_t|\Psi_t\rangle = \int \cdots \int \Psi_t^*(q_1, q_2, \cdots, q_n)\Psi_t(q_1, q_2, \cdots, q_n)\, d\tau_1 \cdots d\tau_n = 1,$$

where the integration is taken over all spatial and spin coordinates. Thus, each state is represented by a normalized vector[4] in Hilbert space.[5] Also, this state vector contains an arbitrary factor, e^{if}, of modulus unity, where f may be a function of the coordinates $q_1, q_2, \ldots, q_n$, or a multiplicative operator, which usually will be omitted from consideration.

Before discussing the implications of this postulate further, several points should be noted to clarify its meaning. First, the Hilbert space must be "extended" to include not only state vectors (wavefunctions) that correspond to operators having a discrete eigenvalue spectra, but also state vectors (wavefunctions) that correspond to operators having continuous eigenvalue spectra.[6]

Next, the time (t) has been singled out, and listed as a separate variable. This

---

[3] Since the coordinates to be discussed will be both spatial and spin coordinates, use of generalized position coordinates provides a convenient means of including both cases.

[4] Other constraints on eigenfunctions, e.g., those imposed via boundary conditions, will be discussed in Chapter 5.

[5] We shall use the term "Hilbert space" to represent the larger *abstract Hilbert space* that encompasses both the specialized Hilbert space and the function space that were discussed in Section 2-7, since both of these "specialized spaces" have been shown to be isomorphous to the abstract Hilbert space.

[6] See J. Von Neumann, *Mathematical Foundations of Quantum Mechanics*, Princeton University Press, Princeton, 1955, for a more detailed discussion of the mathematical techniques associated with these generalizations.

has been done for several reasons. First, it emphasizes a distinction that should be made between quantum mechanics and classical mechanics. In particular, the generalized spatial coordinates, $q_1, q_2 \ldots$, are taken to be *independent* of time in quantum mechanics, while they are thought of as functions of time in classical mechanics. Thus, we can think of the time dependence in quantum mechanics as occurring parametrically, i.e., as the quantum system evolves in time, it passes from one state to another, and hence is described by a different state vector in the Hilbert space for every instant of time. Also, the use of a special symbol for time implies that for the nonrelativistic discussions that will be the primary thrust of this text, the time variable is treated in a unique way, compared to that of the other variables, $q_1, q_2, \ldots$. This "special treatment" of time will be more apparent when the Schrödinger equation itself is introduced in Postulate III, where the time variable appears in quite a different manner from that of the other variables.[7] Having these distinctions in mind, it is, however, sometimes convenient to list time as one of the variables, i.e.,

$$\Psi_t(q_1 \, q_2 \cdots) \equiv \Psi(q_1, q_2 \cdots; t),$$

and the latter notation may be used in later discussions.

The most important implication that arises from Postulate I is that all possible information that can be known about a particular quantum system at time t must be contained in the wavefunction. However, Postulate I does not say anything about how information is to be extracted from the state vector.

Asking about how information is to be extracted from the state vector leads us to several questions, e.g., how is the state vector to be interpreted, and how can measurement of physical properties, such as the energy and momentum, be related to the state vector. We shall deal with the latter question in Postulate II; but before turning to that discussion, let us deal with interpretation of the state vector itself.

Since the state vector is a mathematical construct, it is not required (nor obvious) that it is measurable directly. Indeed, since it can have both positive and negative values, we would not expect the state vector itself (Ψ) to be the quantity that would represent, e.g., the likelihood of finding the particle corresponding to Ψ at a given point, since the latter requires a positive (or zero) number at all points. On the other hand, a particle would not be expected to be found where $\Psi = 0$, so that Ψ must be related in some fashion to the location of the particle corresponding to it. In order to accommodate these features, the interpretation of Ψ is taken via the following corollary to Postulate I:

Corollary: The probability P of measuring a particle associated with a state

---

[7] It is possible to formulate a relativistic theory of quantum mechanics in which all of the variables (including time) appear symmetrically, i.e., they are treated on an equal footing, but such an approach is not appropriate for the discussions in this text. For a more complete discussion see, for example, P. A. M. Dirac, *Quantum Mechanics*, 2nd ed., Oxford University Press, Oxford, 1958.

vector $\Psi_t(q_1 \ldots q_n)$ at a point between $q_1, q_2, \ldots, q_n$ and $q_1 + dq_1, q_2 + dq_2, \ldots, q_n + dq_n$ is given by

$$P(q_1 \cdots q_n)dq_1 \cdots dq_n$$
$$= \Psi_t^*(q_1 \cdots q_n)\Psi_t(q_1 \cdots q_n)dq_1 \cdots dq_n \qquad (4\text{-}1)$$
$$= |\Psi_t(q_1 \cdots q_n)|^2 dq_1 \cdots dq_n. \qquad (4\text{-}2)$$

In the above equation $|\Psi_t(q_1 \ldots q_n)|^2$ is known as the *probability density*. Such a definition is also consistent with the normalization condition introduced in Postulate I, i.e., the probability of finding the particle somewhere in the space is unity (i.e., certain).

However, we see that several unsettling (at least initially) features are incorporated into the interpretation of Ψ_t. First, Ψ_t itself is not a direct measure of the particle's position, and second, the position of the particle can be estimated only by means of a *probability* (not a certainty) of finding the particle at a particular point. However, as we shall see in later sections of this chapter and subsequent chapters, such an apparent lack of complete determinism will not preclude us from measuring[8] physical properties of interest. These physical properties, called *observables,* are the subject of Postulate II.

Postulate II. For each physical observable A, there exists a Hermitian operator $\mathcal{A}$ with a complete, orthornormal set of eigenfunctions. Furthermore, the only possible values that can be obtained for A by any measurement on the system are the eigenvalues of $\mathcal{A}$, i.e., $a_1, a_2, \ldots, a_n \ldots$.

The above postulate implies that there is an isomorphism of the form $A \Leftrightarrow \mathcal{A}$ between observables (corresponding to dynamic variables of classical mechanics) and Hermitian operators. However, it is important to realize that the isomorphism does not imply that they are identical. In fact, one of the ways in which quantum mechanics differs from classical mechanics is that the mathematical representation of an observable (an operator) and the actual numerical values that the observable can take on are to be carefully distinguished. This is an important distinction, since numbers (i.e., the values of an observable) commute, while this is not in general true for operators. The implications of this fact will be seen to have important consequences in subsequent discussions of the quantum theory of measurement.

The choice of Hermitian operators ($\mathcal{A}$) is not simply for convenience. Instead, since the eigenvalues of such operators are known to be real,[9] the use of Hermitian operators is required in order to guarantee measurability of the eigenvalues. Also, the assumption of a complete set of eigenfunctions and

---

[8] We shall assume that, in principle, such measurements can be carried out ideally, i.e., that the values obtained are infinitely precise.

[9] See Theorem 3-19 for a proof of this using the matrix representation of the operator.

associated eigenvalues assures us that all of the possible values of the observable can be described by the operator.[10]

Postulate II might be constructed to imply that all observables will possess only *discrete* spectra (i.e., eigenvalues). As we shall see shortly, some observables of importance to quantum mechanics possess *continuous* spectra, and others possess both discrete and continuous spectra.

First, however, let us consider observables with only discrete spectra. The operators corresponding to such observables satisfy the usual eigenvalue equation

$$\mathcal{A}|\alpha_i\rangle = a_i|\alpha_i\rangle, \tag{4-3}$$

where $i = 1, 2, \ldots, n, \ldots$. We shall assume for the moment that all eigenvalues are nondegenerate. The case of degenerate eigenvalues will be discussed in detail when the quantum theory of measurement is treated. Since the eigenvectors form a complete orthonormal set, it is possible to expand *any* function in the Hilbert space, Ψ, in terms of these eigenfunctions, i.e.,

$$\Psi \equiv |\Psi\rangle = \sum_{i=1}^{\infty} c_i|\alpha_i\rangle, \tag{4-4}$$

where

$$c_i = \langle\alpha_i|\Psi\rangle. \tag{4-5}$$

It is thus at this point where the connection is made between the state vector (eigenfunction) of a system that was introduced in Postulate I, and the information that can be learned from the state vector. Specifically, if the state of a system is written as an expansion in the eigenfunctions of a Hermitian operator $\mathcal{A}$ then the values of the observable A that can be observed are the eigenvalues (a_i) of $\mathcal{A}$. The details of the measurement process and what can be learned from it will be presented as a part of the discussion of Postulates IV and V.

An important example of an observable with a continuous spectrum is given by the *position operator* ($\mathcal{R}$) for a single particle, which satisfies the eigenvalue equation

$$\mathcal{R}|\rho(\mathbf{r}', \mathbf{r})\rangle = \mathbf{r}'|\rho(\mathbf{r}', \mathbf{r})\rangle, \tag{4-6}$$

where $\mathbf{r}'$ is continuous and $\rho(\mathbf{r}', \mathbf{r})$ is the eigenfunction. $\mathcal{R}$ represents simply multiplication by the position vector $\mathbf{r}$, so that the above equation can be written as

$$(\mathbf{r} - \mathbf{r}')|\rho(\mathbf{r}', \mathbf{r})\rangle = 0.$$

Since $\mathbf{r} - \mathbf{r}' = 0$ only when $r = r'$, the function $\rho(\mathbf{r}', \mathbf{r})$ must be zero

[10] There exist Hermitian operators that do not possess a complete set of eigenvectors. See, for example, A. Messiah, *Quantum Mechanics*, Vol. 1, John Wiley, New York, 1958, Vol. 1, p. 188. However, such operators do not correspond to observables, and hence will not be of interest to us.

everywhere *except* at $\mathbf{r} = \mathbf{r}'$. Dirac's delta function[11] is the one function which has this behavior, so that

$$|\rho(\mathbf{r}', \mathbf{r})\rangle = \delta(\mathbf{r}' - \mathbf{r}). \tag{4-7}$$

From Eq. (2-318), it follows that the set of functions $\delta(\mathbf{r}' - \mathbf{r})$ for all $\mathbf{r}'$ is also orthonormal, since

$$\int \delta(\mathbf{r}' - \mathbf{r})\delta(\mathbf{r}'' - \mathbf{r})\, d\mathbf{r} = \delta(\mathbf{r}' - \mathbf{r}''). \tag{4-8}$$

Furthermore, the functions $\delta(\mathbf{r}' - \mathbf{r})$ form a complete set, so that any arbitrary square integrable function may be expanded as

$$f(\mathbf{r}) = \int c(\mathbf{r}')\delta(\mathbf{r}' - \mathbf{r})\, d\mathbf{r}', \tag{4-9}$$

where the expansion coefficients $c(\mathbf{r}')$ are given, by analogy to Eq. (4-5) by

$$c(\mathbf{r}') = (\delta, f) = \int \delta(\mathbf{r}' - \mathbf{r})f(\mathbf{r})\, d\mathbf{r}. \tag{4-10}$$

Finally, suppose $f(\mathbf{r})$ is normalized to 1. Then we can write

$$1 = |f|^2 = \langle f|f\rangle = \int f^*(\mathbf{r})f(\mathbf{r})\, d\mathbf{r}. \tag{4-11}$$

Substituting the expression for $f(\mathbf{r})$ given by Eq. (4-9), we obtain

$$1 = |f|^2 = \int \left[\int c(\mathbf{r}')^*\delta^*(\mathbf{r}' - \mathbf{r})\, d\mathbf{r}'\right]\left[\int c(\mathbf{r}'')\delta(\mathbf{r}'' - \mathbf{r})\, d\mathbf{r}''\right] d\mathbf{r}.$$

Rearranging the integrations and simplifying yields

$$|f|^2 = \int c(\mathbf{r}')^*c(\mathbf{r}')\, d\mathbf{r}' = 1. \tag{4-12}$$

This relationship will be important in later considerations, and is an example of a *closure relation over continuous indices*.

There also exist important cases of operators with mixed spectra, i.e., both discrete and continuous eigenvalues. Perhaps the most important case of such an operator is the Hamiltonian operator, which is the main operator of interest throughout the text. Eigenfunctions in such a "mixed eigenbasis" will be written in the general form

$$\Psi(\mathbf{r}) = \sum_i c_i\phi_i(\mathbf{r}) + \int c(\mathbf{r}')\phi(\mathbf{r}', \mathbf{r})\, d\mathbf{r}'. \tag{4-13}$$

Such functions satisfy modified closure relations, i.e.,

$$|\Psi|^2 = \sum_i |c_i|^2 + \int |c(\mathbf{r})|^2\, d\mathbf{r}. \tag{4-14}$$

---

[11] See Chapter 2, Section 2-9.

In classical mechanics, if A is an observable, then any real function of A, say $g(A)$, is also an observable. However, from our previous discussion, we note that for this to be true in quantum mechanics, the function of an observable operator $g(A)$ must also possess an eigenvalue spectrum $\{g(a_i)\}$, where a_i is an eigenvalue of A. Examination of Eqs. (2-413)–(2-418) shows that this is indeed the case. It should also be noted that both A and $g(A)$ possess the *same* set of eigenfunctions.

An important difference between the classical and quantum mechanical description of a state is related to the mode of expressing its time dependence. In classical mechanics, the state of the system is defined by the *values* of certain observables, and thus *both* the state *and* the observables are functions of time. In quantum mechanics, however, the state function $|\Psi_t\rangle$ generally changes with time, but the observables are usually time independent. This is not surprising, since the eigenvalues and eigenfunctions of the observables do not contain time *explicitly*. This also implies that the operators corresponding to these observables are time independent.

We can explore this time dependence even further. Since the state function Ψ_t can be expanded in terms of the eigenfunctions of an observable operator as in Eq. (4-2), i.e.,

$$|\Psi_t\rangle = \sum_{i=1}^{\infty} c_i|\alpha_i\rangle = \sum_{i=1}^{\infty} \langle\alpha_i|\Psi_t\rangle|\alpha_i\rangle, \tag{4-15}$$

and since the eigenfunctions (α_i) are *not explicitly time dependent,* the time dependence of $\Psi_t(q_1, q_2 \ldots)$ must come from the coefficients

$$|c_i(t) = \langle\alpha_i|\Psi_t\rangle = \int \alpha_i^*(q_1, q_2 \cdots)\Psi(q_1, q_2 \cdots; t)\, dq_1 dq_2 \cdots. \tag{4-16}$$

It should be emphasized again that the time dependence of A and hence the time independence of its spectrum and eigenfunctions does *not* mean that measurements of the observable A are time independent, since their values depend on $|\Psi_t\rangle$. This point will be examined in greater detail when the quantum theory of measurement is discussed in Postulates IV and V.

4-2. Time Evolution of a Quantum State

Since the time evolution (or variation) of a quantum mechanical system depends on the statefunction $\Psi(\mathbf{q}, t)$, where $\mathbf{q} = (q_1, q_2 \ldots)$, we must now find out how this time evolution of the wavefunction occurs, which is the aim of Postulate III.

Postulate III. For every quantum system, there exists a linear, Hermitian operator $\mathcal{H}$ (the Hamiltonian operator), which determines the time variation of the state vector $\Psi(\mathbf{q}, t)$ through the time-dependent Schrödinger equation

$$\mathcal{H}(\mathbf{q}, t)\Psi(\mathbf{q}, t) = i\hbar \frac{\partial \Psi(\mathbf{q}, t)}{\partial t} = -\left(\frac{\hbar}{i}\right) \frac{\partial \Psi(\mathbf{q}, t)}{\partial t}, \tag{4-17}$$

during any time interval in which the system is not disturbed.[12]

Equation (4-17) is usually referred to as the *time-dependent Schrödinger equation*. In general, $\mathcal{H}$ may be time dependent although, for the majority of cases of interest to us, it will contain only time-independent terms. Furthermore, $\mathcal{H}$ is a linear, Hermitian operator that represents the observable associated with the total energy of the system. Note that none of the foregoing postulates describes the way in which one constructs a proper Hamiltonian operator. A detailed treatment of the formation of $\mathcal{H}$ in various circumstances is found in Chapter 8, and initial discussion of how to form Hamiltonian operators is given in Section 4-7. For the moment, let us assume that $\mathcal{H}$ is given.

For the particular case when $\mathcal{H}$ is independent of time, important simplifications occur in the Schrödinger equation that was introduced in Postulate III. For this situation, let us assume that the desired wavefunction can be written in "factored" form as the following product

$$\Psi(\mathbf{q}, t) = \chi(\mathbf{q})\Phi(t). \tag{4-18}$$

Substituting into Eq. (4-17) and subsequent division by $\Psi(\mathbf{q}, t)$ leads to[13]

$$\frac{1}{\chi(\mathbf{q})} \mathcal{H}(\mathbf{q})\chi(\mathbf{q}) = \frac{i\hbar}{\Phi(t)} \frac{\partial}{\partial t} \Phi(t). \tag{4-19}$$

Since the variables $\mathbf{q}$ and t are completely independent, the only way that Eq. (4-19) can be satisfied for arbitrary values of $\mathbf{q}$ and t is for *each* side of the equation to be *constant* for all values of the independent variables. Let us call that constant E. This means that Eq. (4-19) can be written as two equations, i.e.,

$$\mathcal{H}(\mathbf{q})\chi(\mathbf{q}) = E\chi(\mathbf{q}), \tag{4-20}$$

and

$$\frac{\partial}{\partial t} \Phi(t) = -\frac{iE}{\hbar} \Phi(t). \tag{4-21}$$

The first of these equations is known as the *time-independent Schrödinger equation*,[14] and it is the one whose solutions will comprise most of our attention

[12] A measurement represents such a disturbance. See also Section 4-3.

[13] Special consideration must be given to the possibility that $\Psi(\mathbf{q}, t) = 0$ for some t. Such considerations will not affect the desired results for cases of interest here and thus will be omitted.

[14] E. Schrödinger, *Ann. Phys.*, **79**, 361, 489 (1926); **80**, 437 (1926); **81**, 109 (1926).

in subsequent chapters. As we see, the eigenvalues of this equation (E) are simply the *allowed energy states*[15] that the system can have.

The other equation can be easily solved in general, with the result

$$\Phi(t) = e^{-iEt/\hbar}. \tag{4-22}$$

Thus, the total wavefunction that describes how a quantum system evolves in time when $\mathcal{H}$ is time independent is given by

$$\Psi(\mathbf{q}, t) = \chi(\mathbf{q}) e^{-iEt/\hbar}. \tag{4-23}$$

In order to specify Ψ completely in this case, we see that it is necessary only to obtain the solutions to the *time-independent* Schrödinger equation $[\chi(\mathbf{q})]$.

These solutions of the time-independent Schrödinger equation yield a set of time-independent eigenfunctions $\{\chi_i\}$, which correspond to the different allowed energies $\{E_i\}$ of the system. Such energy eigenstates $\{\chi_i\}$ of the system that are time independent are called *stationary states.* Furthermore, the stationary states of the system form a complete, orthonormal set, so that *any* wavefunction for the system can be expressed in terms of them (Postulate II). We shall see shortly that such an eigenbasis is very convenient for such expansions.

Let us now relate the current postulate to earlier discussions. In particular, Postulate I stated that the statefunction $\Psi_t(\mathbf{q})$ may be normalized (to 1) at some time $t = t_0$. It is important to show that the normalization of the wavefunction is maintained as the system evolves in time. Considering only a single generalized spatial coordinate (q) for convenience, this can be shown as follows:

$$\frac{d}{dt}\langle\Psi(q, t)|\Psi(q, t)\rangle = \left[\frac{\partial}{\partial q}\langle\Psi(q, t)|\Psi(q, t)\rangle\right]\frac{\partial q}{\partial t} + \frac{\partial}{\partial t}\langle\Psi(q, t)|\Psi(q, t)\rangle. \tag{4-24}$$

The first term on the right-hand side of Eq. (4-24) is zero, since q is *not* a function of t in quantum mechanics. This gives[16]

$$\frac{d}{dt}\langle\Psi(q, t)|\Psi(q, t)\rangle = \left\langle\frac{\partial}{\partial t}\Psi(q, t)\middle|\Psi(q, t)\right\rangle + \left\langle\Psi(q, t)\middle|\frac{\partial}{\partial t}\Psi(q, t)\right\rangle. \tag{4-25}$$

[15] The "energy states" referred to here include all types, e.g., electronic energy states, vibrational energy states, and rotational energy states.

[16] The derivative with respect to time can be taken into the brackets since the integration of

$$\langle\Psi(q, t)|\Psi(q, t)\rangle \equiv \int \Psi^*(q, t)\Psi(q, t)\, dq$$

is not carried out with respect to time.

Rearranging the time-dependent Schrödinger equation [Eq. (4-17)] as

$$\frac{\partial}{\partial t}\Psi(q, t) = -\frac{i}{\hbar}\mathcal{H}\Psi(q, t), \tag{4-26}$$

and substitution into Eq. (4-25) gives

$$\frac{d}{dt}\langle\Psi|\Psi\rangle = \frac{i}{\hbar}\{\langle\mathcal{H}\Psi|\Psi\rangle - \langle\Psi|\mathcal{H}\Psi\rangle\}. \tag{4-27}$$

Using Eq. (3-380) and the fact that $\mathcal{H}$ is Hermitian, we obtain

$$\frac{d}{dt}\langle\Psi|\Psi\rangle = 0. \tag{4-28}$$

Thus, the value of $\langle\Psi|\Psi\rangle$ does *not* change in time, and a wavefunction (Ψ) that was initially normalized to unity will remain normalized to unity for *all t*.

As mentioned in Section 4-1 in connection with the time dependence of state vectors, the evolution of a quantum system can be described as a passage between different states, each state being represented by a vector in the Hilbert space. This implies that the time evolution of a quantum system can be viewed as a mapping between vectors of a Hilbert space, i.e.,

$$|\Psi_t\rangle = \mathcal{U}(t, t_0)|\Psi_{t_0}\rangle, \tag{4-29}$$

where t_0 is an initial time, and $\mathcal{U}(t, t_0)$ is usually called the *time evolution* (or *time-development*) *operator*. The properties of this operator are discussed in greater detail elsewhere,[17] but it is useful at this point to see what form $\mathcal{U}(t, t_0)$ takes in the special case when the Hamiltonian is independent of time. Substitution of Eq. (4-29) into the time-dependent Schrödinger equation [Eq. (4-17)] gives

$$\mathcal{H}\mathcal{U}(t)\Psi_0 = i\hbar\frac{\partial}{\partial t}\mathcal{U}(t)\Psi_0 \tag{4-30}$$

where, for convenience, we have defined $\mathcal{U}(t) \equiv \mathcal{U}(t, t_0)$ and $\Psi_0 \equiv \Psi(\mathbf{q}, t_0)$. Rearranging Eq. (4-30) yields

$$\left[\mathcal{H}\mathcal{U}(t) - i\hbar\frac{\partial\mathcal{U}(t)}{\partial t}\right]\Psi_0 = 0, \tag{4-31}$$

for *all* Ψ_0. Hence, the time evolution operator satisfies the differential equation

$$\mathcal{H}\mathcal{U}(t) = i\hbar\frac{\partial}{\partial t}\mathcal{U}(t). \tag{4-32}$$

The solution of this equation for the special case when $\mathcal{H}$ is not a function of

[17] See P. O. Löwdin, *Adv. Quantum Chem.*, **3**, 324 (1967) for an extensive discussion of the properties of this operator.

time and, with the initial condition that $\mathcal{U}(t_0) = 1$ will give the desired explicit form of $\mathcal{U}(t)$. In particular, it is easily verified that $\mathcal{U}(t, t_0)$ has the form

$$\mathcal{U}(t, t_0) = \exp\left\{-\frac{i}{\hbar}\mathcal{H}(t_1 - t_0)\right\}. \tag{4-33}$$

Note that, since $\mathcal{U}$ is not Hermitian, it does *not* correspond to an observable. Also, since e^x (for x = operator) has a well-defined power series expansion [see Section 2-12], it follows that $\mathcal{U}(t)$ can be written as

$$\mathcal{U}(t) = \sum_{n=0}^{\infty} \frac{\left[-\frac{i}{\hbar}\mathcal{H}t\right]^n}{n!}, \tag{4-34}$$

where we have set $t_0 = 0$ for convenience. From Eqs. (4-33) and (4-34), it follows that $\mathcal{U}(t)$ is both unitary and linear [see Problem 4-1]. Equation (4-29) is now explicitly given by

$$\Psi(\mathbf{q}, t) = \exp\left[-\frac{i}{\hbar}\mathcal{H}t\right]\Psi_0(\mathbf{q}). \tag{4-35}$$

Thus, for the special case when $\mathcal{H}$ is not a function of time, the time evolvement of the system takes on a rather simple form, which requires knowledge only of the solutions to the time-independent Schrödinger equation $[\Psi(\mathbf{q})]$ in order to specify the system at any time t. This result is another way of stating a similar conclusion that was pointed out earlier [Eq. (4-23)] within the context of the eigenvalues (E) of $\mathcal{H}$ (instead of $\mathcal{H}$ itself).

We will return to some important ideas regarding the time dependence of various measured quantities after we examine several aspects of the quantum theory of measurement.

4-3. Quantum Theory of Measurement and Expectation Values

In the first three postulates the concepts of state vectors, observables, and time evolution of a quantum system were discussed independently. Now, the relationship between the observables of a quantum system and its wavefunction will be explained. This relationship is based on the quantum theory of measurement, of which the next two postulates form a basis.

Postulate IV. If the observable A, related to the Hermitian operator $\mathcal{A}$ with eigenvectors $\{\alpha\}$ and eigenvalues $\{a\}$ [discrete and/or continuous], is measured on a system in a quantum state $|\Psi\rangle$, the *probability* of finding a value of $A \Leftrightarrow$ equal to "a" at time t is given by

$$P_{(t)}(a) = \sum_k |\langle \alpha_k | \Psi_t \rangle|^2 = \sum_k |c_k(t)|^2, \tag{4-36}$$

for the discrete case, and

$$P_{(t)}(a) = \int_k |\langle \alpha(k) | \Psi_t \rangle|^2 \, dk, \tag{4-37}$$

for the continuous case. The summation or integration is taken over all values of k for which $a_k = a$ or $a(k) = a$.

Note first that the possibility of degeneracy is accounted for in this postulate, since the summation[18] is over *all* k. For example, consider the l-fold degenerate eigenvalue a_k. The probability of obtaining this value in a measurement is

$$P(a_k) = \sum_{j=k}^{k+l} |\langle \alpha_j | \Psi \rangle|^2 = \sum_{i=k}^{k+l} |c_j(t)|^2, \tag{4-38}$$

where $\alpha_j (j = k, k + 1, \ldots, k + l)$ are the l linearly independent eigenfunctions corresponding to the l-fold degenerate eigenvalue a_k.

Three additional points of importance are implied in Postulate IV. First, it is *not* possible generally to predict the precise outcome of a measurement on a quantum system, even though it is in a completely defined state, Ψ. This is due to the fact that Postulate IV only deals with the *probability* associated with a measurement,[19] and cannot predict the outcome with certainty. Second, two separate measurements (carried out identically) on the same state of a quantum system will not necessarily yield the same value. This point is related to the previous one, and also follows directly from the probabilistic nature of Postulate IV. It is obvious that these points represent dramatic departures from classical mechanics where, once the state of the system is known, the results of subsequent measurements of observables are known with certainty.

Finally, for the special case (and most important case for us) when the system is in an eigenstate of $\mathcal{A}$, i.e., $\Psi = \alpha_k$, then measurement of A will yield a_k with certainty. Not only is this important for measurement of a single eigenvalue (a_k) but, if Ψ is a simultaneous eigenfunction of several commuting operators, then measurement of the eigenvalue of each operator will field its eigenvalue with

[18] For convenience the following discussion will be given with respect to the discrete case although, with minor modifications, the continuous case can be treated likewise. Also, the subscript denoting time will be suppressed to clarify the notation.

[19] An important difference concerning measurements on quantum systems having discrete versus continuous eigenvalues should be noted here. As mentioned earlier, it is assumed that measurements on systems having discrete eigenvalues can be made with infinite precision. However, for operators having continuous spectra, an infinitely precise measurement would imply that the upper and lower limits in Eq. (4-38) are equal, which results in a zero probability of measuring that particular value. In this case, we take the limits to be as close as desired, but not exactly equal.

certainty (and simultaneously) if the system is in an eigenstate of each operator. This is of particular help in the study of states of atoms and molecules, as we shall see in later chapters.

Further justification for the definition of $P_t(a)$ as a probability in Postulate IV can be seen as follows. Since the wavefunction for the state Ψ is normalized, we have

$$\langle \Psi_t | \Psi_t \rangle = 1. \tag{4-39}$$

Using Eq. (4-2), we can rewrite the above equation as

$$1 = \sum_{i=1}^{\infty} \sum_{j=1}^{\infty} c_i^*(t) c_j(t) \langle \alpha_i | \alpha_j \rangle = \sum_{i=1}^{\infty} |c_i|^2, \tag{4-40}$$

or, from Eq. (4-3),

$$1 = \sum_{i=1}^{\infty} |\langle \alpha_i | \Psi_t \rangle|^2. \tag{4-41}$$

From Eq. (4-40) and the fact that

$$|c_i(t)|^2 \geq 0, \tag{4-42}$$

we see that

$$0 \leq |c_i(t)|^2 = |\langle \alpha_i | \Psi_t \rangle|^2 \leq 1, \tag{4-43}$$

which is what is expected for quantities that represent probabilities.

In order to deal with the possibility of making *repeated* measurements of an observable, we shall make use of the notions that are useful in the statistical discussion of multiple measurements. Suppose, for example, the set of values $x_1, x_2, \ldots, x_n$ is found in a series of measurements with frequencies $f_1, f_2, \ldots, f_n$. In statistics, the weighted average $(\bar{X})$ is then defined as

$$\bar{X} = \frac{\sum_{i=1}^{n} f_i x_i}{\sum_j f_j}. \tag{4-44}$$

The ratio $(f_i / \Sigma_j f_j)$ can be interpreted as the probability, p_i, that x_i will occur, i.e.

$$p_i = \frac{f_i}{\sum_j f_j} \tag{4-45}$$

Thus, Eq. (4-44) can be rewritten as

$$\bar{X} = \sum_{i=1}^{n} p_i x_i. \tag{4-46}$$

In this case $\bar{X}$ is called the *expectation value* of X.

Let us now use this concept to define expectation values in quantum mechanical systems. In particular, if a large number of repeated measurements of the observable A are made on a quantum system that is always in exactly the same quantum state Ψ, then the *average value* (or *expectation value*) that will be obtained from these measurements will be

$$\langle \mathcal{A} \rangle \equiv \bar{A} = \sum_{k=1}^{\infty} |c_k|^2 a_k \tag{4-47}$$

$$= \sum_{k=1}^{\infty} |\langle \alpha_k | \Psi \rangle|^2 a_k. \tag{4-48}$$

We shall now establish the relationship between the matrix element $\langle \Psi | \mathcal{A} \Psi \rangle$ and the expectation value of $\mathcal{A}$ as just defined. Substituting the expansion for Ψ in the matrix element yields

$$\langle \Psi | \mathcal{A} \Psi \rangle = \left\langle \sum_{k=1}^{\infty} \langle \alpha_k | \Psi \rangle \alpha_k \right| \mathcal{A} \left. \sum_{l=1}^{\infty} \langle \alpha_l | \Psi \rangle \alpha_l \right\rangle \tag{4-49}$$

$$= \sum_{k=1}^{\infty} |\langle \alpha_k | \Psi \rangle|^2 a_k. \tag{4-50}$$

Hence, comparing Eq. (4-50) with Eq. (4-48) shows that

$$\langle \mathcal{A} \rangle = \langle \Psi | \mathcal{A} \Psi \rangle. \tag{4-51}$$

When Ψ is not normalized to unity, $\langle \mathcal{A} \rangle$ is written as

$$\langle \mathcal{A} \rangle = \frac{\langle \Psi | \mathcal{A} \Psi \rangle}{\langle \Psi | \Psi \rangle}. \tag{4-52}$$

It is important to note that since the wavefunction Ψ is time dependent in general, $\langle \mathcal{A} \rangle$ is also a function of time, even though $\mathcal{A} \neq \mathcal{A}(t)$.

Another statistical concept which has important implications in quantum mechanics is that of the *mean square deviation*, $(\Delta x)^2$, associated with a series of measurements, that is defined as

$$(\Delta x)^2 = \sum_i p_i (x_i - \bar{x}_i)^2. \tag{4-53}$$

In an analogous fashion, we shall define the *mean square deviation* $(\Delta x)^2$ of an operator[20] in quantum mechanics as:

$$(\Delta \mathcal{A})^2 = \langle (\mathcal{A} - \langle \mathcal{A} \rangle)^2 \rangle. \tag{4-54}$$

$(\Delta \mathcal{A})^2$ can be written in a more transparent form by substituting the definitions

---

[20] This is sometimes called the "width" of an operator. Further important implications of $(\Delta \mathcal{A})$ will be found in the generalized uncertainty principle (see Section 4-4.A].

introduced earlier, i.e.,

$$(\Delta\mathcal{A})^2 = \langle\Psi|(\mathcal{A}^2 - 2\langle\mathcal{A}\rangle\mathcal{A} + \langle\mathcal{A}\rangle^2)\Psi\rangle \tag{4-55}$$

$$= \langle\Psi|\mathcal{A}^2|\Psi\rangle - 2\langle\mathcal{A}\rangle\langle\Psi|\mathcal{A}|\Psi\rangle + \langle\mathcal{A}\rangle^2\langle\Psi|\Psi\rangle, \tag{4-56}$$

where Ψ is taken here to be an *arbitrary* normalized function.[21] Collecting terms and simplifying yields

$$(\Delta\mathcal{A})^2 = \langle\mathcal{A}^2\rangle - \langle\mathcal{A}\rangle^2. \tag{4-57}$$

A special case of interest arises when the mean square deviation vanishes. In that case,

$$\langle\mathcal{A}^2\rangle = \langle\mathcal{A}\rangle^2, \tag{4-58}$$

which has an important consequence, as we shall see. Rewriting Eq. (4-58), and noting that $\mathcal{A}$ is Hermitian, we obtain

$$\langle\mathcal{A}\rangle\langle\Psi|\mathcal{A}\Psi\rangle = \langle\Psi|\mathcal{A}^2\Psi\rangle, \tag{4-59}$$

or

$$\langle\mathcal{A}\rangle\langle\mathcal{A}\Psi|\Psi\rangle = \langle\mathcal{A}\Psi|\mathcal{A}\Psi\rangle, \tag{4-60}$$

or

$$\langle\mathcal{A}\Psi|\langle\mathcal{A}\rangle\Psi\rangle = \langle\mathcal{A}\Psi|\mathcal{A}\Psi\rangle, \tag{4-61}$$

or

$$\mathcal{A}\Psi = \langle\mathcal{A}\rangle\Psi. \tag{4-62}$$

In other words, if $(\Delta\mathcal{A})^2 = 0$, then $\langle\mathcal{A}\rangle$ is an *eigenvalue* of $\mathcal{A}$ with eigenfunction Ψ. The converse can also be shown [see Problem 4-7]. This means that if the quantum state wavefunction is an eigenfunction of $\mathcal{A}$, then $(\Delta\mathcal{A})^2$ (the root mean square deviation) vanishes.

The last and probably most controversial postulate deals with what happens to the wavefunction of a quantum system after a measurement is made on that system. During the measuring process itself the quantum system is interacting with the measuring device and, based on Postulate III, any such disturbance destroys any possibility of describing the time evolution of the system in terms of its state vector. However, *after* the measuring process has stopped, the system will evolve in time again, according to the time dependent Schrödinger equation (or its equivalent, the time evolution operator). This forms the basis for Postulate V. In order to clarify the following discussion, the expansion of Ψ in terms of eigenfunctions of an observable operator will be written in a slightly

[21] It is assumed here that Ψ, $\mathcal{A}\Psi$, and $\mathcal{A}^2\Psi$ all exist, and lie in the Hilbert space. Operators such as $\mathbf{r}$ or $i\hbar\partial/\partial q$ sometimes cause problems in this regard, and must be treated with care. See A. Messiah, *Quantum Mechanics*, Vol. 1, John Wiley, 1958, p. 169 for a discussion of this point.

different form, i.e.,

$$\Psi = \sum_l \Psi_l, \tag{4-63}$$

where the summation index is taken over each of the different eigenvalues of $\mathfrak{A}$.

Some of the Ψ_l may represent degenerate eigenvalues,[22] so that, in general, the ψ_l are given by

$$\Psi_l = \sum_{r=1}^{n} c_l^{(r)} \alpha_l^{(r)}, \tag{4-64}$$

where the summation limit n represents the degeneracy of the eigenvalue a_l. By writing the expansion of Ψ in this particular form, another important point previously alluded to in Theorem 3-14 is illustrated. In particular, any linear combination of *degenerate* eigenfunctions is also an eigenfunction. Hence, an infinite number of possible eigenfunctions exist corresponding to a degenerate eigenfunction. In the case of a nondegenerate eigenvalue, however, the corresponding eigenfunction is *unique* to within a multiplicative constant.

Postulate V. If a measurement on a quantum system of the observable A, associated with operator $\mathfrak{A}$, yields a value a_l, then the state of the system *immediately* after the measurement is given by

$$\Psi_l = \frac{\sum_{r=1}^{n} c_l^{(r)} \alpha_l^{(r)}}{\left[\sum_{r=1}^{n} |c_l^{(r)}|^2\right]^{1/2}}, \tag{4-65}$$

where the denominator is added to assure normalization.

In order to understand the implications of Postulate V fully, consider first the case where the *initial* measurement of A yields the *nondegenerate* eigenvalue a_k. From Postulate II, the probability that a_k is found is $|c_k|^2$. If a second measurement of A is made before the system has evolved in time, then the second measurement should also yield the value a_k with complete certainty. This implies that before the second measurement, the system must be in an eigenstate corresponding to α_k that, from Eq. (4-65), is given by

$$\Psi_k = \frac{c_k}{|c_k|} \alpha_k. \tag{4-66}$$

Since $c_k/|c_k|$ is of modulus unity, Ψ_k and α_k are equivalent (see Postulate I).

[22] In this and the following section, the notation will be altered slightly to make the discussion more transparent. Instead of the subscript l in α_l referring to all eigenfunctions (including degeneracies), it will here refer only to the states having *distinct* eigenvalues, and the sum over different components in the case of degeneracies will be given as in Eq. (4-64).

Thus, the measurement can be considered to represent the selection of a "unique component" of the wavefunction of the system. In a sense, the measuring device acts as a "perfect filter," by taking out *all* the eigenbasis elements except α_k, which it passes on without distortion.

In the case of a degenerate eigenvalue a_k, the interpretation is more difficult. As shown in Postulate V, the system after the first measurement is in an eigenstate given by Eq. (4-65). However, the expansion coefficients $c_l^{(r)}$, are not in general known.[23] In fact, from the previous discussion based on Theorem 3-14, it follows that any linear combination of the degenerate eigenbasis $\{\alpha_l^{(1)}, \alpha_l^{(2)}, \ldots, \alpha_l^{(n)}\}$, e.g.,

$$\Psi_l' = \sum_{r=1}^{n} c_l^{(r)'} \alpha_l^{(r)}, \tag{4-67}$$

is also an appropriate eigenfunction corresponding to a_l. Thus, there remains an indeterminancy associated with the measurement that gives a degenerate eigenvalue of an observable.

Postulate V is of great importance, for it shows how information obtained from experiments can be related to the state vector. In general, the measurement of some observable can lead to a variety of results, each result having a certain probability of occurrence. However, once the measurement has been made and a particular result obtained, most of the state vector is superfluous. It is necessary to retain only that part of the wavefunction that would yield the same value of the observable as the preceding measurement.

Hence, Postulate V provides the basis for determining the state of a quantum system, i.e., its state vector. If the value measured for an observable corresponds to a nondegenerate eigenvalue, then the state of the system is uniquely determined (to within a phase factor). However, for the more general case of the measurement of a degenerate eigenvalue, further measurements of other observables are required to specify the state of the system completely, as we shall see in the following section.

While Postulate V deals with the case when the system is in a particular eigenstate, the concept of measurement introduced there is also applicable when the wavefunction is not an eigenfunction of a particular state. In that case, the concept of expectation (or average) value introduced in Eqs. (4-46) and (4-47) is used, and we speak of the average value $\langle A \rangle$ of the observable A as given by

$$\langle A \rangle = \langle \Psi | \mathcal{A} | \Psi \rangle,$$

where $\langle A \rangle$ is obtained as a result of a large (i.e., statistically significant) number of measurements.

---

[23] The expansion coefficients are indeterminant unless the state of the system, Ψ, *before*, the initial measurement is known. In that case, $c_l^{(r)} = \langle \alpha_l^{(r)} | \Psi \rangle$, as shown also by Eq. (4-3).

4-4. Compatible Observables and Commuting Operators

In classical mechanics the question of which observables can be simultaneously measured does not arise.[24] For example, in a classical mechanical system with n degrees of freedom, measurement of the n position coordinates $q_1, q_2, \ldots q_n$ does not affect the measurement of the n momentum coordinates $p_1, p_2, \ldots p_n$. This is not true in quantum mechanics. As we shall see, measurement of one set of observables causes loss of information of the conjugate set of variables.[25] In such cases the two sets of variables are said to be *incompatible*. The present section deals with variables that *can* be *simultaneously measured* and hence are called *compatible*. Compatibility and simultaneous measurability will thus be taken to be synonymous terms.

Before proceeding, a more precise definition of compatibility will be given. *The observables A and B are said to be compatible if and only if measurement, in rapid succession, of A and B with values a and b, again yields a on remeasurement of A*. This definition leads immediately to the following theorem.

Theorem 4-1. The observables A and B are compatible if and only if the corresponding operators $\mathcal{A}$ and $\mathcal{B}$ possess a complete, orthonormal set of *simultaneous* eigenfunctions.[26,27]

Proof: First we shall prove the necessary condition, i.e., if A and B are compatible, there exists a complete, orthonormal set of simultaneous eigenfunctions of $\mathcal{A}$ and $\mathcal{B}$.

Consider the system to be in an initial state given by

$$\Psi = \sum_{k=1}^{\infty} \sum_{r=1}^{n_k} c_{kr} \alpha_k^{(r)}, \tag{4-68}$$

where the $\alpha_k^{(r)}$ are the eigenfunctions of $\mathcal{A}$ corresponding to the kth eigenvalue a_k. The values of r represent the various components of the degenerate set $\alpha_k^{(1)}$, $\ldots$, $\alpha_k^{(n)}$.

If A is measured, yielding a value a_{k_0}, the corresponding eigenfunction is

---

[24] By simultaneous measurements we shall mean measurements that are performed in such rapid succession that the state of the system does not change between measurement.

[25] See Chapter 2 for a discussion of conjugate variables. In the above discussion, the set $q_1, q_2, \ldots$, q_n is conjugate to the set $p_1, p_2, \ldots , p_n$

[26] We shall confine ourselves here to operators with purely discrete spectra, but extension to the case of continuous of mixed spectra is straightforward.

[27] Since the operators corresponding to quantum mechanical observables are Hermitian, it is clear from the discussion in Section 5-1 that they possess at least separate, complete, orthonormal sets of eigenfunctions.

(see Postulate V) given by

$$\Psi_{k_0}=\frac{\sum_{r=1}^{n_{k_0}} c_{k_0}^{(r)}\alpha_{k_0}^{(r)}}{\left[\sum_{r=1}^{n_{k_0}}|c_{k_0}^{(r)}|^2\right]^{1/2}}. \tag{4-69}$$

Since Ψ_{k_0} is in general degenerate, its explicit form is not known, i.e., the $c_{k_0}^{(r)}$ are unknown. However, Ψ_{k_0} can now be expanded in terms of the eigenfunctions of $\mathcal{B}$, $\beta_l^{(s)}$, which is written as[28]

$$\Psi_{k_0}=\sum_{l=1}^{\infty}\sum_{s=1}^{n_l} d_l^{(s)}\beta_l^{(s)}. \tag{4-70}$$

Now B is measured, yielding an eigenvalue b_{l_0} with eigenfunction

$$\phi_{l_0}=\frac{\sum_{s=1}^{n_{l_0}} d_{l_0}^{(s)}\beta_{l_0}^{(s)}}{\left[\sum_{s=1}^{n_{l_0}}|d_{l_0}^{(s)}|^2\right]^{1/2}}. \tag{4-71}$$

Since A and B are assumed to be compatible, immediate remeasurement of A should yield the eigenvalue a_{k_0}, i.e.

$$\mathcal{A}\phi_{l_0}=a_{k_0}\phi_{l_0}. \tag{4-72}$$

Hence, ϕ_{l_0} must be a *simultaneous* eigenfunction of $\mathcal{A}$ and $\mathcal{B}$. b_{l_0} represents *any* eigenvalue of $\mathcal{B}$ for which $d_{l_0}^{(s)} \neq 0$, so that each eigenfunction of $\mathcal{B}$ appearing in Eq. (4-71) is also an eigenfunction of $\mathcal{A}$ with eigenvalue a_{k_0}.

Thus, any eigenfunction of $\mathcal{A}$ can be expanded in terms of a set of simultaneous eigenfunctions of $\mathcal{A}$ and $\mathcal{B}$. Therefore, the set of simultaneous eigenfunctions of $\mathcal{A}$ and $\mathcal{B}$ is *complete*. Finally, since they are eigenfunctions of Hermitian operators, they can be taken to be orthonormal without loss of generality. This completes the first part of the proof.

To prove the *sufficient condition*, assume that there exists a complete, orthonormal set of simultaneous eigenfunctions of $\mathcal{A}$ and $\mathcal{B}$, i.e., $\{\omega_{kl}^{(r)}\}$. Then

[28] Note that, in general, many of the $d_l^{(s)} = 0$. This arises due to the fact that once A is measured, the allowed values for B are restricted, and hence the eigenfunctions are also restricted. As an example, which will be discussed in more detail in Chapters 7 and 8, consider the hydrogen atom. If a measure of the energy of the system yields a certain value E_k, then the allowed values of the angular momentum are limited. Hence, the eigenfunctions corresponding to the now limited set of angular momentum eigenvalues constitute a subset of the total set of angular moment eigenfunctions. Expansion of the energy eigenfunction corresponding to E_k then requires only the subset angular momentum eigenfunctions. This is the same as saying that the expansion coefficients corresponding to eigenfunctions of unallowed angular momentum eigenvalues are equal to zero.

the state function, Ψ, can be expanded as

$$\Psi = \sum_{k=1}^{\infty} \sum_{l=1}^{l_k} \sum_{r=1}^{n_{kl}} c_{kl}^{(r)} \omega_{kl}^{(r)}. \tag{4-73}$$

The summation over l depends on the particular value for k and the n_{kl} represents the dimension of the subspace of simultaneous eigenfunctions corresponding to the eigenvalues a_k and b_l.

If measurement of A yields the eigenvalue a_{k_0}, then the associated eigenfunction is given by

$$\Psi_{k_0} = \frac{\displaystyle\sum_{l=1}^{l_{k_0}} \sum_{r=1}^{n_{k_0 l_0}} c_{k_0 l_0}^{(r)} \omega_{k_0 l_0}^{(r)}}{\left[\displaystyle\sum_{r}^{n_{k_0 l_0}} |c_{k_0 l_0}^{(r)}|^2\right]^{1/2}}. \tag{4-74}$$

Equation (4-74) illustrates the fact that eigenfunctions corresponding to a_{k_0}, i.e., $\omega_{k_0 l}^{(r)}$, may be associated with a number of eigenvalues of $\mathcal{B}$, i.e., b_l ($l = 1, 2, \ldots, l_{k_0}$), but necessarily *not all* possible b_l. That is to say that measurement of A restricts the number of possible values that measurement of B can produce.

Measurement of B then gives a value b_{l_0}, with eigenfunction

$$\phi_{l_0}^{(k_0)} = \frac{\displaystyle\sum_{r=1}^{n_{k_0 l_0}} c_{k_0 l_0}^{(r)} \omega_{k_0 l_0}^{(r)}}{\left[\displaystyle\sum_{r=1}^{n_{k_0 l_0}} |c_{k_0 l_0}^{(r)}|^2\right]^{1/2}}, \tag{4-75}$$

where the superscript indicates that $\phi_{l_0}^{(k_0)}$ is also an eigenfunction of $\mathcal{A}$ with eigenvalue a_{k_0}. It should be noted that the dimensions of the subspaces satisfy the following inequality,

$$n_{kl} \geq n_{k_0 l} \geq n_{k_0 l_0}. \tag{4-76}$$

Since $\phi_{l_0}^{(k_0)}$ is also an eigenfunction of $\mathcal{A}$ with eigenvalue a_{k_0}, remeasurement of A must give a_{k_0} with certainty, which assures that A and B are compatible, and completes the proof.

From Eq. (4-76) and the preceding discussion it follows that applying Postulate IV, the probability of simultaneously observing A and B with values a_{k_0} and b_{l_0} is

$$P(a_{k_0}, b_{l_0}) = \sum_{r=1}^{n_{k_0 l_0}} |c_{k_0 l_0}^{(r)}|^2. \tag{4-77}$$

An important link between the notion of *compatible observables* as previously described with respect to the process of measurement and the mathematical concept of *commuting operators* will now be developed. In

Theorem 3-22 and the discussion that followed, the following important result emerged, viz. *that it is possible to find a complete set of simultaneous eigenfunctions of two Hermitian operators, each of which also possesses a complete set of eigenfunctions, if and only if the operators commute.*

The above result can be combined with Theorem 4-1 into the following theorem (that requires no further proof):

Theorem 4-2. If, for two observables A and B with corresponding operators $\mathcal{A}$ and $\mathcal{B}$, any one of the three following conditions is satisfied:

1. A and B are compatible observables,
2. $\mathcal{A}$ and $\mathcal{B}$ possess a complete, orthonormal set of simultaneous eigenfunctions,
3. $[\mathcal{A}, \mathcal{B}] = 0$,

then the other two conditions are directly implied.

The preceding results can also be generalized to cases of continuous or mixed spectra. It should again be pointed out, however, that for the cases where continuous spectra are important, measurements of absolute accuracy are no longer possible.

Even though we can obtain definite values only for compatible observables, it is possible to obtain the probability of occurrence or the mean value of *any* observable, provided we know the wavefunction for the system. This has important implications as to which observables are *constant in time*, i.e., which observables are *constants of the motion* of a quantum mechanical system.

A. The Uncertainty Principle

Although the results of the previous section, in which we can be assured of the simultaneous measurability of observables to arbitrary accuracy if their operators commute, are of great importance and utility, they leave a significant question unanswered. In particular, it is important to consider if the same situation results if the operators for two observables do *not* commute. For this case (of incompatible observables), we shall see that fundamental limitations on the results of measurements arise, and provide another example of a major departure from classical mechanics. Furthermore, it has given rise to perhaps the most controversial aspects of the theory.

To establish the results of interest, we consider two Hermitian operators, $\mathcal{A}$ and $\mathcal{B}$, corresponding to observables A and B, that do not commute, i.e.,

$$[\mathcal{A}, \mathcal{B}] = i\mathcal{C} \tag{4-78}$$

where $i\mathcal{C}$ is a Hermitian operator.[29] It will be seen that this noncommuting property provides the basis for limitations on measurements of A and B.

To facilitate the analysis, it will be convenient to introduce several other

[29] See Problem 4-6.

operators related to $\mathcal{A}$ and $\mathcal{B}$, i.e.,

$$\mathcal{A}' = \mathcal{A} - \langle \mathcal{A} \rangle \mathcal{I}$$

$$\mathcal{B}' = \mathcal{B} - \langle \mathcal{B} \rangle \mathcal{I} \tag{4-79}$$

$$\mathcal{C}' = \mathcal{A}\mathcal{B} + \mathcal{B}\mathcal{A} \tag{4-80}$$

and

$$\mathcal{D}' = \mathcal{A}' + (a + ib)\mathcal{B}' \tag{4-81}$$

where $\mathcal{A}'$, $\mathcal{B}'$, and $\mathcal{C}'$ are each Hermitian[30] operators, and a and b are arbitrary real numbers. It can easily be seen that[31]

$$[\mathcal{A}', \mathcal{B}'] = i\mathcal{C}, \tag{4-82}$$

which means that properties derived for the operators $\mathcal{A}'$ and $\mathcal{B}'$ on the basis of Eq. (4-82) also apply to the operators $\mathcal{A}$ and $\mathcal{B}$.

If Ψ is an arbitrary, normalized function defined in the space of operators given above, we know that the norm of $\mathcal{D}'\Psi$ is always finite, i.e.,

$$\langle \mathcal{D}'\Psi | \mathcal{D}'\Psi \rangle \geq 0. \tag{4-83}$$

Inserting the definition of $\mathcal{D}'$ gives

$$\int [\mathcal{A}' + (a + ib)\mathcal{B}']^* \Psi^* [\mathcal{A}' + (a + ib)\mathcal{B}'] \Psi \, d\tau \geq 0, \tag{4-84}$$

or

$$\langle (\mathcal{A}')^2 \rangle + (a^2 + b^2)\langle (\mathcal{B}')^2 \rangle + a\langle \mathcal{C}' \rangle + b\langle \mathcal{C} \rangle \geq 0. \tag{4-85}$$

Since a and b are arbitrary, let us choose them to minimize the left-hand side of Eq. (4-85). This gives:

$$a = -\frac{\langle \mathcal{C}' \rangle}{2\langle (\mathcal{B}')^2 \rangle}, \qquad b = -\frac{\langle \mathcal{C} \rangle}{2\langle (\mathcal{B}')^2 \rangle}. \tag{4-86}$$

Substitution of a and b from Eq. (4-86) into Eq. (4-85) and multiplication of both sides by $\langle (\mathcal{B}')^2 \rangle$ gives

$$\langle (\mathcal{A}')^2 \rangle \langle (\mathcal{B}')^2 \rangle \geq \frac{\langle \mathcal{C}' \rangle^2}{4} + \frac{\langle \mathcal{C} \rangle}{4}. \tag{4-87}$$

To relate this result to the "width" $(\Delta\mathcal{A}')$ of the operator $\mathcal{A}'$, we note from Eq. (4-57) that

$$\langle (\mathcal{A}')^2 \rangle = (\Delta\mathcal{A}')^2 + \langle \mathcal{A}' \rangle^2. \tag{4-88}$$

---

[30] See Problem 4-8.

[31] This representation is based upon the analysis by A. Gamba, *Nuovo Cimento*, **7**, 378 (1950).

However, from the definition of $\mathcal{A}'$ [Eq. (4-79)], we note that

$$\langle \mathcal{A}' \rangle = 0 \tag{4-89}$$

and

$$\langle (\mathcal{A}')^2 \rangle = (\Delta\mathcal{A}')^2. \tag{4-90}$$

A similar result holds for $(\Delta\mathcal{B}')^2$. Substitution into Eq. (4-87) gives

$$(\Delta\mathcal{A}')^2(\Delta\mathcal{B}')^2 \geq \frac{\langle \mathcal{C} \rangle^2}{4} + \frac{\langle \mathcal{C}' \rangle^2}{4}. \tag{4-91}$$

It is also easily seen that

$$\begin{aligned} (\Delta\mathcal{A}')^2 &= (\Delta\mathcal{A})^2 \\ (\Delta\mathcal{B}')^2 &= (\Delta\mathcal{B})^2 \end{aligned} \tag{4-92}$$

which means that Eq. (4-91) can be written as

$$(\Delta\mathcal{A})^2(\Delta\mathcal{B})^2 \geq \frac{\langle \mathcal{C} \rangle^2}{4} + \frac{\langle \mathcal{C}' \rangle^2}{4}. \tag{4-93}$$

This is the general result of interest, known as the *Uncertainty Relation*. To see it in its more common form we neglect the last term, which results in:

$$(\Delta\mathcal{A})(\Delta\mathcal{B}) \geq \frac{\langle \mathcal{C} \rangle}{2}. \tag{4-94}$$

We now see that the noncommutivity of $\mathcal{A}$ and $\mathcal{B}$ has a significant impact on the ability to measure A and B. In particular, we see from Eq. (4-94) that there is a nonzero result arising from the expectation value of the commutator $(\mathcal{C})$ of $\mathcal{A}$ and $\mathcal{B}$, whose magnitude determines the minimum error that is associated with simultaneous measurements of A and B. Thus, in contrast to the case of commuting operators as well as to classical mechanics, we cannot expect *even in principle* to measure A and B simultaneously to arbitrary accuracy. That this should be controversial is now easy to understand, since in classical mechanics the accuracy to which we can measure observables is limited *only* by the quality of the measuring instrument.

To illustrate this general result with a specific example, consider the two operators (x-component),

$$\mathcal{A} = \frac{\hbar}{i}\frac{d}{dx}, \qquad \mathcal{B} = x. \tag{4-95}$$

Since

$$\left[x, \frac{\hbar}{i}\frac{d}{dx}\right] = i\hbar\mathcal{I}, \tag{4-96}$$

and since we shall see shortly[32] that the two operators ($\mathcal{A}$ and $\mathcal{B}$) as expressed in Eq. (4-95) represent the x-components of the momentum and position operators, respectively, the uncertainty relation of Eq. (4-94) gives

$$(\Delta p_x)(\Delta x) \geq \frac{\hbar}{2}. \tag{4-97}$$

Thus, the momentum and position of a particle cannot be measured simultaneously without incurring an uncertainty of at least $(\hbar/2)$ in the product of momentum times position.

$$(\Delta E)(\Delta t) \geq \frac{\hbar}{2}, \tag{4-98}$$

where Δt is the lifetime of the state whose energy is uncertain by ΔE.

This inherent uncertainty combined with the probabilistic interpretation of measurements of observables in quantum theory has caused substantial controversy among even the founders of the theory. Many have claimed[33] that the formulation of quantum theory is at fault, and that there are "hidden variables" that, if understood, would allow a more satisfactory formulation of quantum theory to be devised that does not contain such inherent uncertainties. However, as uncomfortable, unsettling, and annoying such concepts may be, a suitable alternative formulation has not yet appeared.

4-5. Constants of Motion and Transition Probabilities

As in classical mechanics, knowledge of the constants of motion of a quantum system also provides insight into the dynamics of the system. In order to study the constants of motion, let us consider the time dependence of an expectation value, i.e.,

$$\frac{d}{dt}\langle\mathcal{A}\rangle = \frac{d}{dt}\langle\Psi_t|\mathcal{A}|\Psi_t\rangle. \tag{4-99}$$

Carrying out the differentiation explicitly yields[34]

$$\frac{d}{dt}\langle\mathcal{A}\rangle = \left\langle\frac{\partial\Psi_t}{\partial t}\,\middle|\,\mathcal{A}\,\middle|\,\Psi_t\right\rangle + \left\langle\Psi_t\,\middle|\,\frac{\partial\mathcal{A}}{\partial t}\,\middle|\,\Psi_t\right\rangle + \left\langle\Psi_t\,\middle|\,\mathcal{A}\,\middle|\,\frac{\partial\Psi_t}{\partial t}\right\rangle, \tag{4-100}$$

---

[32] See Section 4-7.

[33] See, for example, Max Jammer, *The Conceptual Development of Quantum Mechanics*, McGraw Hill, New York, 1966, Chapter 7, for a discussion of the early debates that occurred on this subject.

[34] Note also that $\mathcal{A}$ is generally not a function of t, although there are some exceptions. For further discussion see Section 4-1.

where, as also given in Eq. (4-29),

$$\Psi_t = \mathcal{U}(t)\Psi_0. \tag{4-101}$$

Differentiation of Ψ_t can also be carried out explicitly, i.e.,

$$\frac{\partial \Psi_t}{\partial t} = \frac{\partial}{\partial t}[\mathcal{U}(t)\Psi_0], \tag{4-102}$$

$$= \frac{\partial}{\partial t}\left[\exp\left(-\frac{i}{\hbar}\mathcal{H}t\right)\right]\Psi_0, \tag{4-103}$$

$$= -\frac{i}{\hbar}\mathcal{H}\mathcal{U}(t)\Psi_0, \tag{4-104}$$

$$= -\frac{i}{\hbar}\mathcal{H}\Psi_t, \tag{4-105}$$

where the form of $\mathcal{U}(t)$ that has been used is appropriate only for the case when $\mathcal{H} \neq \mathcal{H}(t)$. Substitution of Eq. (4-105) into Eq. (4-100) yields

$$\frac{d}{dt}\langle \mathcal{A} \rangle = \left\langle -\frac{i}{\hbar}\mathcal{H}\Psi_t \middle| \mathcal{A} \middle| \Psi_t \right\rangle + \left\langle \Psi_t \middle| \mathcal{A} \middle| -\frac{i}{\hbar}\mathcal{H}\Psi_t \right\rangle + \left\langle \Psi_t \middle| \frac{\partial \mathcal{A}}{\partial t} \middle| \Psi_t \right\rangle. \tag{4-106}$$

Using the Hermiticity of $\mathcal{H}$, Eq. (4-106) can be written as

$$\frac{d}{dt}\langle \mathcal{A} \rangle = \left(\frac{i}{\hbar}\right)\langle \Psi_t | \mathcal{H}\mathcal{A} - \mathcal{A}\mathcal{H} | \Psi_t \rangle + \left\langle \Psi_t \middle| \frac{\partial \mathcal{A}}{\partial t} \middle| \Psi_t \right\rangle \tag{4-107}$$

$$= \left(\frac{i}{\hbar}\right)\langle [\mathcal{H}, \mathcal{A}] \rangle + \left\langle \frac{\partial \mathcal{A}}{\partial t} \right\rangle. \tag{4-108}$$

Since, in most cases $\partial A/\partial t = 0$, it follows that the observable $\mathcal{A}$ is compatible with the time-independent Hamiltonian, i.e., $[\mathcal{H}, \mathcal{A}] = 0$. In that case A is a constant of the motion, and Eq. (4-108) is given simply by

$$\frac{d}{dt}\langle A \rangle = \langle [\mathcal{H}, \mathcal{A}] \rangle. \tag{4-109}$$

Note that, for the particular case when $\mathcal{A} = \mathcal{H}$, the energy is a constant of the motion.

Furthermore, it can be shown that the uncertainty in $\mathcal{A}$, i.e., $\Delta\mathcal{A}$ [see Problem 4-10), is also constant in time if $[\mathcal{H}, \mathcal{A}] = 0$. Hence, if the system is in an eigenstate of $\mathcal{A}$ so that $\Delta\mathcal{A} = 0$, the system will *remain* in that state throughout time if the system is undisturbed. Another related point of importance is that if we specify a *set* of commuting observables that contains the (time-independent) Hamiltonian of the system, then the values of *each* of the observables after measurement will remain constant in time, i.e., the entire set of commuting observables is a constant of the motion.

Of course, quantum systems do not all remain in a given state, and it is of obvious importance to know the probability that a system in a specified (although not necessarily completely) quantum state of time t_0 will be in *another* quantum state at time t. The probability of such an occurrence is usually called a *transition probability*, and the basis for such transition probabilities will now be discussed.

Suppose a measurement of compatible observables A and B is made on a quantum system at time t_0, whose wavefunction is given by

$$\Psi_{t_0}=\sum_{k=1}^{\infty}\sum_{r=1}^{n_k} c_k^{(r)}\omega_k^{(r)}, \tag{4-110}$$

where $\{\omega_k^{(r)}\}$ is a complete set of orthonormal, simultaneous eigenfunctions of A and B.

It follows that immediately after the measurement, the system is described by the state function

$$\Psi_{k_0}(t_0)=\sum_{r=1}^{n_{k_0}} \tilde{c}_{k_0}^{(r)}(t)\omega_{k_0}^{(r)}, \tag{4-111}$$

where

$$\sum_{r=1}^{n_{k_0}} |\tilde{c}_{k_0}^{(r)}|^2=1, \tag{4-112}$$

as prescribed by Postulate V, where k_0 is identified with the eigenvalues a_i and b_j for the operators $\mathcal{A}$ and $\mathcal{B}$, respectively. From Postulate IV, Eq. (4-112) becomes

$$P(k_0;\, t_0)=\sum_{r=1}^{n_{k_0}} P^{(r)}(k_0;\, t_0)=1. \tag{4-113}$$

This shows that, after the state has been "prepared," its immediate remeasurement will yield the state again with certainty.

At a later time t, the state of the system is described by

$$\Psi_t=\mathcal{U}(t,\, t_0)\Psi_{t_0}. \tag{4-114}$$

The question now is[35]: What is the probability of measuring $a_{i'}$ and $b_{j'}$, simultaneously at time t, where $a_{i'} \neq a_i$, and $b_{j'} \neq b_j$? To answer this question, we expand Ψ_t in the complete, orthonormal set of simultaneous

[35] Throughout this discussion it should be remembered that the Hamiltonian has been assumed to be independent of time.

eigenfunctions $\{\omega_k^{(s)}\}$, i.e.,

$$\Psi_t = \sum_{l=1}^{\infty} \sum_{s=1}^{n_l} \langle \omega_l^{(s)} | \Psi_t \rangle \omega_l^{(s)}, \tag{4-115}$$

as shown in Eqs. (4-2) and (4-3). Hence, by Postulate IV,

$$P(l;\, t) = \sum_{s=1}^{n_l} |\langle \omega_l^{(s)} | \Psi_t \rangle|^2, \tag{4-116}$$

where l refers to one of the simultaneous eigenvalues $(a_{i'},\, b_{j'})$ of the system. Substituting Eqs. (4-111) and (4-114) into Eq. (4-116) yields

$$P(l;\, t) = \sum_{s=1}^{n_l} \left| \left\langle \omega_l^{(s)} \middle| \mathfrak{U}(t,\, t_0) \middle| \sum_{r=1}^{n_k} \tilde{c}_{k_0}^{(r)}(t_0) \omega_{k_0}^{(r)} \right\rangle \right|^2, \tag{4-117}$$

or

$$P(l;\, t) = \sum_{s=1}^{n_l} \left\{ \sum_{r,u} \tilde{c}_{k_0}^{(u)*}(t_0) \tilde{c}_{k_0}^{(r)}(t_0) \mathfrak{U}_{lk_0}^{(su)*}(t,\, t_0) \mathfrak{U}_{lk_0}^{(sr)}(t,\, t_0) \right\}, \tag{4-118}$$

where

$$\mathfrak{U}_{lk_0}^{(sr)}(t,\, t_0) \equiv \langle \omega_l^{(s)} | \mathfrak{U}(t,\, t_0) | \omega_{k_0}^{(r)} \rangle, \tag{4-119}$$

and is an element of a unitary matrix [see Eqs. (3-196)–(3-201)]. Rearranging the summations gives

$$P(l;\, t) = \sum_{s=1}^{n_l} \left\{ \sum_{r} |\tilde{c}_{k_0}^{(r)}|^2 \cdot |\mathfrak{U}_{lk_0}^{(sr)}|^2 + \sum_{r} \sum_{u \neq r} \tilde{c}_{k_0}^{(u)*} \tilde{c}_{k_0}^{(r)} \mathfrak{U}_{lk_0}^{(su)*} \mathfrak{U}_{lk_0}^{(sr)} \right\}. \tag{4-120}$$

Summing over *all* possible states l, and using Eqs. (4-112) and (4-113), gives

$$\sum_{l} P(l;\, t) = \sum_{r} P^{(r)}(k_0;\, t_0) \sum_{l,s} |\mathfrak{U}_{lk_0}^{(sr)}|^2 + \sum_{r} \sum_{u \neq r} \tilde{c}_{k_0}^{(u)*} \tilde{c}_{k_0}^{(r)} \left[\sum_{l,s} \mathfrak{U}_{lk_0}^{(su)*} \mathfrak{U}_{lk_0}^{(sr)} \right]. \tag{4-121}$$

Since the $\mathfrak{U}_{lk_0}^{(sr)}$ are elements of a unitary matrix, it follows that

$$\sum_{l,s} \mathfrak{U}_{lk_0}^{(su)*} \mathfrak{U}_{lk_0}^{(sr)} = \delta_{ru}, \tag{4-122}$$

so that

$$\sum_{l} P(l;\, t) = \sum_{r} P^{(r)}(k_0;\, t_0) = 1. \tag{4-123}$$

Hence we have the result that probability is "conserved" during a transition, i.e., the system is guaranteed to be in one of the possible states.

A more useful relation from the point of view of predicting to *which* state the system will evolve can be obtained from Eq. (4-121) by rearranging the summations in the first term on the right-hand side and using the fact that the second term vanishes, i.e.,

$$\sum_{l} P(l;\, t)=\sum_{l}\left\{\sum_{r,s} P^{(r)}(k_0;\, t_0)|\mathcal{U}_{lk_0}^{(sr)}(t,\, t_0)|^2\right\}, \tag{4-124}$$

which leads to the following identification of terms:

$$P(l;\, t)=\sum_{r,s} P^{(r)}(k_0;\, t_0)|\mathcal{U}_{lk_0}^{(sr)}(t,\, t_0)|^2. \tag{4-125}$$

The usual interpretation of this expression is as follows. Since $P(l;\, t)$ represents the probability that $a_{i'}$ and $b_{j'}$ will be measured at time t *after* the system has been prepared in the state k_0 [corresponding to the eigenvalues $\{a_i, b_j\}$], the $P(l;\, t)$ will be interpreted as *transition probabilities,* i.e., the probability that a transition will take place such that a system initially in a state k_0 at time t_0 will be found in state l [corresponding to the eigenvalue $\{a_{i'}, b_{j'}\}$], at time t. This is usually denoted by

$$S_{k_0\to l}\,(t_0,\, t) \;\equiv\; P(l;\, t). \tag{4-126}$$

Furthermore, since $P^{(r)}(k_0,\, t_0)$ is the probability that the system will be in the rth substate of k_0 at time t_0, the $|\mathcal{U}_{lk_0}^{(sr)}|^2$ can also be interpreted as transition probabilities. In this case, the notation that is usually used is

$$S_{k_0r\to ls}\,(t_0,\, t) \;\equiv\; |\mathcal{U}_{lk_0}^{(sr)}(t_0,\, t)|^2. \tag{4-127}$$

Thus, $S_{k_0r\to ls}\,(t_0,\, t)$ is said to be the probability that a system initially in the rth substate of k_0 at time t_0 makes a "transition" to the sth substate of l. Combining Eqs. (4-125), (4-126), and (4-127) gives

$$S_{k_0\to l}\,(t_0,\, t)=\sum_{r}\sum_{s} P^{(r)}(k_0;\, t_0)S_{k_0r\to ls}(t_0,\, t), \tag{4-128}$$

and hence the transition probability $S_{k_0\to l}\,(t_0,\, t)$ can be thought of as the sum of the transition probabilities between the substates of k_0 and l, *weighted* by the probability that the system is in a particular substate of k_0 at time t_0.

Combining Eqs. (4-122) and (4-127) shows that

$$\sum_{l,s} S_{k_0r\to ls}\,(t_0,\, t)=1, \tag{4-129}$$

and since it is clear from Eq. (4-127) that $S_{k_0 r \to ls}(t_0, t) \geq 0$, it follows that

$$0 \leq S_{k_0 r \to ls}(t_0, t) \leq 1. \tag{4-130}$$

Hence, the transition probabilities never should "blow-up."[36]

Examination of Eqs. (4-119) and (4-128) shows that the particular form of $\mathcal{U}(t_0, t)$ is important in determining the transition probability. Hence, before leaving this section, it will be useful to examine one possible form of $\mathcal{U}(t_0, t)$. In particular, for the case where $\mathcal{H} \neq \mathcal{H}(t)$, $\mathcal{U}(t_0, t)$ has been seen to have the following particularly simple form [see Eq. (4-33)]:

$$\mathcal{U}(t_0, t) = e^{-(i/\hbar)\mathcal{H}(t-t_0)} \tag{4-131}$$

Thus,

$$S_{lr \to ms}(t_0, t) = |\mathcal{U}_{lm}^{(rs)}|^2 = |\langle \omega_l^{(r)} | e^{-(i/\hbar)\mathcal{H}(t-t_0)} | \omega_m^{(s)} \rangle|^2 \neq 0, \tag{4-132}$$

where $\{\omega_l^{(r)}\}$ represent a complete, orthonormal set of simultaneous eigenfunctions of the commuting operators $\mathcal{A}$ and $\mathcal{B}$. Thus, Eq. (4-132) shows that transitions can occur between states l and m, even when $\mathcal{H}$ is time independent. However, as will now be shown, if $\mathcal{A}$ and $\mathcal{B}$ are *constants of the motion* of the system, no transitions will occur.

If $\mathcal{A}$ and $\mathcal{B}$ are constants of the motion, then $[\mathcal{H}, \mathcal{A}] = [\mathcal{H}, \mathcal{B}] = 0$. Since $\mathcal{U}(t_0, t)$ is given by Eq. (4-131), it is possible to write (see Section 2-12)

$$[\mathcal{A}, f(\mathcal{H})] = [\mathcal{A}, \mathcal{U}] = 0, \tag{4-133}$$

$$[\mathcal{B}, f(\mathcal{H})] = [\mathcal{B}, \mathcal{U}] = 0. \tag{4-134}$$

Since $\mathcal{A}$, $\mathcal{B}$, $\mathcal{H}$, and $\mathcal{U}$ all commute,[37] there exists a complete, orthornormal set of simultaneous eigenfunctions $\{\phi_k^{(q)}\}$ of these operators. The transition probability can then be written as

$$S_{lr \to ms}(t_0, t) = \| \langle \phi_l^{(r)} | \mathcal{U}(t_0, t) | \phi_m^{(s)} \rangle \|^2. \tag{4-135}$$

Since the $\phi_k^{(q)}$ are eigenstates of the energy, we can write

$$\mathcal{U}(t_0, t) | \phi_m^{(s)} \rangle = e^{-(i/\hbar)E_m(t-t_0)} | \phi_m^{(s)} \rangle, \tag{4-136}$$

which, on substitution into Eq. (4-135), yields

$$S_{lr \to ms}(t_0, t) = |\langle \phi_l^{(r)} | e^{-(i/\hbar)E_m(t-t_0)} | \phi_m^{(s)} \rangle|^2 \tag{4-137}$$

$$= \langle \phi_l^{(r)} | e^{-(i/\hbar)E_m(t-t_0)} | \phi_m^{(s)} \rangle \langle \phi_l^{(r)} | e^{-(i/\hbar)E_m(t-t_0)} | \phi_m^{(s)} \rangle^* \tag{4-138}$$

$$= |\langle \phi_l^{(r)} | \phi_m^{(s)} \rangle|^2 \tag{4-139}$$

---

[36] It will be seen later that the cases where $S_{k_0 r \to ls}(t_0, t) \to \infty$ occur only when approximate treatments are used, such as when perturbation theory is employed.

[37] The statements here apply even if $\mathcal{U}$ is not Hermitian, as long as it is unitary.

or

$$S_{lr\to ms}(t_0, t) = \delta_{lm}\delta_{rs}. \tag{4-140}$$

Thus, as anticipated, Eq. (4-140) shows that transitions between different states of the system do not occur if the states are stationary states.[38]

4-6. Different Pictures of Quantum Phenomena

Up to this point we have maintained a particular ''view'' of quantum mechanics, i.e., we have assumed that the time dependence of a quantum system is described by the time evolution of its state vector. The observable operators, except in special cases, were generally time independent. This particular view is usually called the *Schrödinger* picture.[39] It can be represented more explicitly by [Eq. (4-29)]

$$|\Psi_t^{(S)}\rangle = \mathcal{U}(t, t_0)|\Psi_{t_0}^{(S)}\rangle, \tag{4-141}$$

where the superscript S has been used here to associate the state vectors with the Schrödinger picture.

Another equivalent view of quantum mechanics that is of importance is known as the *Heisenberg picture.*[40] The Heisenberg state vector $|\Psi^{(H)}\rangle$ can be obtained from the Schrödinger picture by transforming the state vector $|\Psi_t^{(S)}\rangle$ with the *inverse* time development operator, i.e.,

$$|\Psi^{(H)}\rangle = \mathcal{U}(t, t_0)^{-1}|\Psi_t^{(S)}\rangle \tag{4-142}$$

$$= \mathcal{U}(t, t_0)^{\dagger}|\Psi_t^{(S)}\rangle, \tag{4-143}$$

since $\mathcal{U}$ is unitary. Combining Eqs. (4-141) and (4-142), and using the unitary property of $\mathcal{U}$, gives

$$|\Psi^{(H)}\rangle = |\Psi_{t_0}^{(S)}\rangle. \tag{4-144}$$

Hence, it is clear that the state vector $|\Psi^{(H)}\rangle$ in the Heisenberg representation does *not* evolve in time. More importantly, the time-dependent Schrödinger equation is no longer the equation of motion of a quantum system in the Heisenberg representation. In order to obtain the ''new'' equation of motion in

[38] Also, see Problem 4-15 for an alternative derivation of Eq. (4-140).

[39] This view of the theory is also called the Schrödinger ''representation.'' The use of the word representation in this context, however, is misleading, for it does not have the same significance as it does in the case of the representation of vectors and operators in a vector space by matrices.

[40] A postulational approach can also be used to obtain the Heisenberg representation. The particular approach used here is employed to bring out the connection with the Schrödinger representation more clearly.

the Heisenberg representation, we consider first the following operator relation

$$\mathcal{A}^{(S)}|\Psi^{(S)}\rangle = |\Psi^{(S)\prime}\rangle, \tag{4-145}$$

which maps vectors in the Schrödinger picture into new vectors. Substituting Eqs. (4-141) and (4-144) into the above equation yields

$$\mathcal{A}^{(S)}\mathcal{U}(t, t_0)|\Psi^{(H)}\rangle = \mathcal{U}(t, t_0)|\Psi^{(H)\prime}\rangle \tag{4-146}$$

Left multiplying by $\mathcal{U}(t, t_0)^\dagger$ gives

$$\mathcal{U}^\dagger(t, t_0)\mathcal{A}^{(S)}\mathcal{U}(t, t_0)|\Psi^{(H)}\rangle = \mathcal{U}^\dagger\mathcal{U}|\Psi^{(H)\prime}\rangle, \tag{4-147}$$

or

$$\mathcal{A}^{(H)}|\Psi^{(H)}\rangle = |\Psi^{(H)\prime}\rangle, \tag{4-148}$$

where we have set

$$\mathcal{A}^{(H)} = \mathcal{U}^\dagger\mathcal{A}^{(S)}\mathcal{U}. \tag{4-149}$$

It is useful to note that this equation is exactly analogous to the unitary similarity transformations discussed in Chapter 3, which assures that the spectrum of $\mathcal{A}^{(S)}$ and $\mathcal{A}^{(H)}$ is identical.

The quantum equation of motion in the Heisenberg picture can now be obtained as follows. Differentiating Eq. (4-149) with respect to time gives

$$\frac{d}{dt}\mathcal{A}^{(H)} = \frac{d}{dt}(\mathcal{U}^\dagger\mathcal{A}^{(S)}\mathcal{U}) \tag{4-150}$$

$$= \frac{\partial\mathcal{U}^\dagger}{\partial t}\mathcal{A}^{(S)}\mathcal{U} + \mathcal{U}^\dagger\frac{\partial\mathcal{A}^{(S)}}{\partial t}\mathcal{U} + \mathcal{U}^\dagger\mathcal{A}^{(S)}\frac{\partial\mathcal{U}}{\partial t}. \tag{4-151}$$

Using Eq. (4-32) and its adjoint, i.e.,

$$\frac{\partial\mathcal{U}}{\partial t} = -\frac{i}{\hbar}\mathcal{H}^{(S)}\mathcal{U}, \tag{4-152}$$

and

$$\frac{\partial\mathcal{U}^\dagger}{\partial t} = \frac{i}{\hbar}\mathcal{U}^\dagger\mathcal{H}^{(S)} \tag{4-153}$$

plus substitution into Eq. (4-151) yields

$$\frac{d}{dt}\mathcal{A}^{(H)} = \frac{i}{\hbar}\mathcal{U}^\dagger\mathcal{H}^{(S)}\mathcal{A}^{(S)}\mathcal{U} - \frac{i}{\hbar}\mathcal{U}^\dagger\mathcal{A}^{(S)}\mathcal{H}^{(S)}\mathcal{U} + \mathcal{U}^\dagger\left(\frac{\partial\mathcal{A}^{(S)}}{\partial t}\right)\mathcal{U} \tag{4-154}$$

$$= \mathcal{U}^\dagger\left\{\frac{i}{\hbar}(\mathcal{H}^{(S)}\mathcal{A}^{(S)} - \mathcal{A}^{(S)}\mathcal{H}^{(S)})\right\}\mathcal{U} + \mathcal{U}^\dagger\left(\frac{\partial\mathcal{A}^{(S)}}{\partial t}\right)\mathcal{U}. \tag{4-155}$$

Using Eq. (4-149) and the unitary nature of $\mathcal{U}$, Eq. (4-155) becomes[41]

$$\frac{d\mathcal{A}^{(\mathrm{H})}}{dt} = \left(\frac{i}{\hbar}\right)[\mathcal{H}^{(\mathrm{H})}, \mathcal{A}^{(\mathrm{H})}] + \frac{\partial \mathcal{A}^{(\mathrm{H})}}{\partial t}, \tag{4-156}$$

where $\mathcal{H}^{(\mathrm{H})} = \mathcal{U}^{\dagger}\mathcal{H}^{(\mathrm{S})}\mathcal{U}$. Equation (4-156) is *Heisenberg's equation of motion*. Note that it differs from the time-dependent Schrödinger equation in that it is an equation in terms of *operators only*. Note also that it is analogous in form to Eq. (4-108), which dealt with the time dependence of the expectation values of observable operators.

It should be observed that if $\mathcal{H}^{(\mathrm{S})}$ and $\mathcal{A}^{(\mathrm{S})}$ are time independent, it can be shown (see Problem 4-12) that $\mathcal{H}^{(\mathrm{H})}$ and $\mathcal{A}^{(\mathrm{H})}$ are also time independent. Since $[\mathcal{H}^{(\mathrm{H})}, \mathcal{A}^{(\mathrm{H})}] = \mathcal{U}^{\dagger}[\mathcal{H}^{(\mathrm{S})}, \mathcal{A}^{(\mathrm{S})}]\mathcal{U}$, then if $\mathcal{A}^{(\mathrm{S})}$ is a constant of the motion, $\mathcal{H}^{(\mathrm{H})}$ also must represent a constant of the motion.

An important aspect of Eq. (4-156) is that its form is analogous to the classical Hamiltonian equation of motion, i.e.,

$$\frac{dF}{dt} = -\{H, F\} + \frac{\partial F}{\partial t}, \tag{4-157}$$

where H is the classical Hamiltonian function, F is a function of the dynamic variables (i.e. generalized coordinates and momentum), and $\{H, F\}$ is the classical *Poisson bracket*.[42] If the implied correspondence,

$$\frac{[\mathcal{H}^{(\mathrm{H})}, \mathcal{A}^{(\mathrm{H})}]}{i\hbar} \Leftrightarrow -\{H, A\}, \tag{4-158}$$

is made, then Heisenberg's equation of motion is seen to be identical to the classical equation. (See Table 4-1 for a summary of comparisons between classical variables and quantum mechanical operators.)

Although the state vector is *time independent* in the Heisenberg picture, the full eigenfunctions of observable operators are now time dependent, as is seen in the following equations:

$$|\alpha_k^{(\mathrm{H})}(\mathrm{t})\rangle = \mathcal{U}^{\dagger}(t, t_0)|\alpha_k^{(\mathrm{S})}\rangle, \tag{4-159}$$

---

[41] We shall assume that

$$\mathcal{U}^{\dagger}\left(\frac{\partial \mathcal{A}^{(\mathrm{S})}}{\partial t}\right)\mathcal{U} = \frac{\partial}{\partial t}(\mathcal{U}^{\dagger}\mathcal{A}^{(\mathrm{S})}\mathcal{U}) = \frac{\partial}{\partial t}\mathcal{A}^{(\mathrm{H})}.$$

[42] The Poisson bracket of two functions of the dynamic variables $\{q_i\}$ and $\{p_i\}$ is given by

$$\{F, G\} = \sum_i \left[\frac{\partial F}{\partial q_i}\frac{\partial G}{\partial p_i} - \frac{\partial F}{\partial p_i}\frac{\partial G}{\partial q_i}\right].$$

See H. Goldstein, *Classical Mechanics*, Addison-Wesley, Reading, MA, 1950, pp. 255–258, for further details.

Table 4.1. Comparison of Classical Mechanical Variables and Quantum Mechanical Operators.

	Classical mechanical	Quantum mechanical
Hamiltonian of system	$H(q, p, t)$	$\mathcal{H}(\mathcal{Q}, \mathcal{P}, t)$
State of system	$\begin{Bmatrix} q=q(t) \\ p=p(t) \end{Bmatrix}$	$\Psi(q, t)$
Time evaluation of system	$\frac{\partial}{\partial p} H=\frac{dq}{dt}$ $-\frac{\partial H}{\partial q}=\frac{dp}{dt}$	$\mathcal{H}\Psi=i\hbar \frac{\partial \Psi}{\partial t}$
Initial conditions	$\begin{Bmatrix} q=q(0) \\ p=p(0) \end{Bmatrix}$	$\Psi=\Psi(q, 0)$

or

$$|\alpha_k^{(S)}\rangle=\mathcal{U}(t, t_0)|\alpha_k^{(H)}(t)\rangle. \tag{4-160}$$

The relationship between the two pictures can be further elucidated in the following manner. First, differentiate Eq. (4-160) with respect to time, noting that $|\alpha_k^{(S)}\rangle$ is time-independent, i.e.,

$$\frac{\partial}{\partial t}|\alpha_k^{(S)}\rangle=0=\left(\frac{\partial \mathcal{U}}{\partial t}\right)|\alpha_k^{(H)}(t)\rangle+\mathcal{U}\frac{\partial}{\partial t}|\alpha_k^{(H)}(t)\rangle. \tag{4-161}$$

Substituting Eq. (4-152) into the above equation gives

$$\frac{i}{\hbar}\mathcal{H}^{(S)}\mathcal{U}|\alpha_k^{(H)}(t)\rangle=\mathcal{U}\frac{\partial}{\partial t}|\alpha_k^{(H)}(t)\rangle. \tag{4-162}$$

If $\mathcal{H}^{(S)} \neq \mathcal{H}^{(S)}_{(t)}$, then $[\mathcal{H}^{(S)}, \mathcal{U}] = 0$, so that Eq. (4-162) can be rewritten as

$$\frac{i}{\hbar}\mathcal{U}\mathcal{H}^{(S)}|\alpha_k^{(H)}(t)\rangle=\mathcal{U}\frac{\partial}{\partial t}|\alpha_k^{(H)}(t)\rangle. \tag{4-163}$$

By left multiplying by $\mathcal{U}^\dagger$ and rearranging, we obtain

$$\mathcal{H}^{(S)}|\alpha_k^{(H)}(t)\rangle=-i\hbar\frac{\partial}{\partial t}|\alpha_k^{(H)}(t)\rangle, \tag{4-164}$$

which, except for the minus sign, is identical to the time-dependent Schrödinger equation. The interpretation of this *similarity* provides added insight into the two pictures of quantum theory. In the Schrödinger picture, the state vector undergoes a time-dependent "rotation" in Hilbert space, while the observable operators and the eigenvectors are time independent. In the Heisenberg picture, the state vector is stationary, while the observable operators and their associated eigenvectors "rotate" in the *opposite* direction.[43]

Finally, to show the equivalence between the two pictures, the probability of measuring a particular value of an observable, the expectation value of an observable, and the transition probability between two states must be shown to be the same. For the case of the measurement of an observable, we have, from Postulate IV,

$$P(a_k;\, t) = \sum_{k \in a_k} |\langle \alpha_k^{(S)} | \Psi_t^{(S)} \rangle|^2. \tag{4-165}$$

Substituting Eq. (4-159) into the above expression gives

$$P(a_k;\, t) = \sum_{k \in a_k} |\langle \alpha_k^{(H)}(t) | \mathcal{U}^\dagger | \Psi_t^{(S)} \rangle|^2 \tag{4-166}$$

$$= \sum_{k \in a_k} |\langle \alpha_k^{(H)}(t) | \Psi^{(H)} \rangle|^2, \tag{4-167}$$

where Eq. (4-143) was used. Hence, it is clear that Postulate IV is consistent in either picture. It is left to the reader to show that the expectation values of observables and the transition probabilities are also consistent with either picture.

Since the state vectors and operators are not directly measured quantities, it is therefore possible to derive an infinite number of pictures of quantum mechanics, all differing by unitary transformations. However, not all these new pictures would be advantageous in solving problems or clarifying theory. There is one picture, however, which has found frequent use in time-independent perturbation theory, the so-called *interaction picture*.

To develop this picture, consider first that the Hamiltonian defining the system can be separated into two parts, one that is time independent, and one that is time dependent,[44] i.e,

$$\mathcal{H}^{(S)}(t) = \mathcal{H}_0^{(S)} + \mathcal{V}^{(S)}(t). \tag{4-168}$$

Since $\mathcal{H}_0^{(S)}$ is time independent, its time evolution operator is given by Eqs. (4-33) or (4-34), where $\mathcal{H} \equiv \mathcal{H}_0^{(S)}$ and $\mathcal{U} \equiv \mathcal{U}_0$. Proceeding, the *total* time evolution operator is written as

$$\mathcal{U} = \mathcal{U}_0 \mathcal{U}_v, \tag{4-169}$$

---

[43] There is an analogy between the above discussion and the treatment of rigid body motion, i.e., the rigid body can be thought of as moving in a fixed coordinate system, or the body can remain stationary and the coordinate system can rotate in the opposite directions. Such transformations are examples of active and passive transformations, as discussed in Chapter 3, Section 3-7.

[44] Note that, in general, the Hamiltonian can be separated in any fashion that is convenient for the problem at hand. Since the interaction picture is especially important in time-dependent perturbation theory, we shall separate the Hamiltonian as in Eq. (4-147).

where $\mathcal{U}$, $\mathcal{U}_0$, and $\mathcal{U}_v$ are unitary, and substituting into Eq. (4-32), gives

$$(\mathcal{H}_0^{(S)} + \mathcal{V}^{(S)})\mathcal{U}_0\mathcal{U}_v = i\hbar \frac{\partial}{\partial t} \mathcal{U}_0\mathcal{U}_v, \tag{4-170}$$

or

$$(\mathcal{H}_0^{(S)}\mathcal{U}_0)\mathcal{U}_v + \mathcal{V}^{(S)}\mathcal{U}_0\mathcal{U}_v = i\hbar\left(\frac{\partial \mathcal{U}_0}{\partial t}\right)\mathcal{U}_v + i\hbar\mathcal{U}_0\left(\frac{\partial \mathcal{U}_v}{\partial t}\right). \tag{4-171}$$

Since $\mathcal{H}_0^{(S)}\mathcal{U}_0$ must satisfy Eq. (4-32), the above expression becomes

$$i\hbar\left(\frac{\partial \mathcal{U}_0}{\partial t}\right)\mathcal{U}_v + \mathcal{V}^{(S)}\mathcal{U}_0\mathcal{U}_v = i\hbar\left(\frac{\partial \mathcal{U}_0}{\partial t}\right)\mathcal{U}_v + i\hbar\mathcal{U}_0\left(\frac{\partial \mathcal{U}_v}{\partial t}\right), \tag{4-172}$$

or, cancelling terms,

$$\mathcal{V}^{(S)}\mathcal{U}_0\mathcal{U}_v = i\hbar\mathcal{U}_0 \frac{\partial \mathcal{U}_v}{\partial t}. \tag{4-173}$$

Finally, after left multiplying by $\mathcal{U}_0^\dagger$, we obtain

$$(\mathcal{U}_0^\dagger\mathcal{V}^{(S)}\mathcal{U}_0)\mathcal{U}_v = i\hbar \frac{\partial \mathcal{U}_v}{\partial t}. \tag{4-174}$$

The actual form of $\mathcal{U}_v$, which will be obtained later for use in time-dependent perturbation theory, can be found by solving the differential equation in the previous expression.

The interaction picture is then defined in a fashion analogous to the Heisenberg picture, i.e.,

$$|\Psi_t^{(I)}\rangle = \mathcal{U}_0^\dagger|\Psi_t^{(S)}\rangle, \tag{4-175}$$

and

$$\mathcal{A}^{(I)} = \mathcal{U}_0^\dagger\mathcal{A}^{(S)}\mathcal{U}_0, \tag{4-176}$$

so that the Heisenberg and interaction pictures are identical if $\mathcal{U}^{(S)} = 0$. It also follows from Eq. (4-176) that $\mathcal{H}_0^{(S)} = \mathcal{H}_0^{(I)}$. Note in this case that $|\Psi^{(I)}(t)\rangle$ in general depends on time. This is in contrast to the case of the Heisenberg picture, since substitution of Eq. (4-151) into Eq. (4-175) gives

$$|\Psi_t^{(I)}\rangle = \mathcal{U}_0^\dagger\mathcal{U}|\Psi_{t_0}^{(S)}\rangle. \tag{4-177}$$

Substitution of Eq. (4-169) into (4-177), and using the unitary property of $\mathcal{U}_0$ yields

$$|\Psi_t^{(I)}\rangle = \mathcal{U}_v(t, t_0)|\Psi_{t_0}^{(S)}\rangle, \tag{4-178}$$

which shows that the time dependence is maintained in $|\Psi_t^{(I)}\rangle$.

To see which equation of motion governs the state vectors in the interaction

Table 4.2. Comparison of Various Pictures of Quantum Mechanics.

	State vectors	Operators	Eigenvectors
Schrödinger picture	Moving	Stationary	Stationary
Heisenberg picture	Stationary	Moving	Moving
Interaction picture	Moving	Moving	Moving

picture, we differentiate both sides of Eq. (4-175) with respect to time, and use Eqs. (4-153) and (4-26) to obtain

$$\frac{\partial}{\partial t}|\Psi_t^{(\mathrm{I})}\rangle = \left(\frac{\partial \mathcal{U}_0^\dagger}{\partial t}\right)|\Psi_t^{(\mathrm{S})}\rangle + \mathcal{U}_0^\dagger \frac{\partial}{\partial t}|\Psi_t^{(\mathrm{S})}\rangle \tag{4-179}$$

$$= \frac{i}{\hbar}(\mathcal{U}_0^\dagger \mathcal{H}_0^{(\mathrm{S})})|\Psi_t^{(\mathrm{S})}\rangle + \mathcal{U}_0^\dagger\left(-\frac{i}{\hbar}\mathcal{H}^{(\mathrm{S})}|\Psi_t^{(\mathrm{S})}\rangle\right), \tag{4-180}$$

or

$$i\hbar\frac{\partial}{\partial t}|\Psi_t^{(\mathrm{I})}\rangle = \mathcal{U}_0^\dagger[\mathcal{H}^{(\mathrm{S})} - \mathcal{H}_0^{(\mathrm{S})}]\mathcal{U}_0\mathcal{U}_0^\dagger|\Psi_t^{(\mathrm{S})}\rangle \tag{4-181}$$

$$= (\mathcal{U}_0^\dagger \mathcal{V}^{(\mathrm{S})}\mathcal{U}_0)|\Psi_t^{(\mathrm{I})}\rangle \tag{4-182}$$

and finally

$$\mathcal{V}^{(\mathrm{I})}|\Psi_t^{(\mathrm{I})}\rangle = i\hbar\frac{\partial}{\partial t}|\Psi_t^{(\mathrm{I})}\rangle, \tag{4-183}$$

where we have defined $\mathcal{V}^{(\mathrm{I})} = \mathcal{U}_0^\dagger \mathcal{V}^{(\mathrm{S})}\mathcal{U}_0$. Equation (4-183) has the form of the time-dependent Schrödinger equation, except that the "Hamiltonian" is now the time-dependent component of the interaction picture.

In a fashion similar to that applied to the Heisenberg picture, differentiation of Eq. (4-186) leads to the following operator equation of motion

$$\frac{d\mathcal{A}^{(\mathrm{I})}}{dt} = \frac{i}{\hbar}[\mathcal{H}_0^{(\mathrm{I})}, \mathcal{A}^{(\mathrm{I})}] + \frac{\partial \mathcal{A}^{(\mathrm{I})}}{\partial t}, \tag{4-184}$$

in the interaction picture. Thus, the interaction picture determines the "motion" of the state vectors by $\mathcal{V}^{(\mathrm{I})}$, while the "motion" of the observable operators is determined by $\mathcal{H}_0^{(\mathrm{I})}$.

Table 4.2 presents a "capsule summary" of the various pictures discussed here. For the most part, the Schrödinger picture will prove most convenient for later discussions, and exceptions will be noted explicitly.

4-7. Hamiltonian Operator Construction: Initial Considerations

In Postulate III the time evolution of a quantum state was described, in which it was assumed that an appropriate quantum mechanical Hamiltonian operator existed and could be found. A detailed description of how such operators can be constructed and the various factors that need to be considered will be given in Chapter 8. However, to allow consideration of a number of important model problems, it is of interest to consider a special set of circumstances at this point.

The particular case of interest, and one that is frequently encountered, deals with the situation when we are interested in the *time-independent* solutions of the Schrödinger equation [Eq. (4-20)], when there are *no external fields present,* where the *potential energy is conservative,* and where *relativistic effects are ignored.*

For that particular set of circumstances, there are only two kinds of interactions that need to be described in the construction of an appropriate quantum mechanical Hamiltonian. To see this, we recall from classical mechanics (see Chapter 1) that, for the circumstances described above, consideration of the kinetic energy of each particle plus the potential energy that arises from each particle interacting with each other is sufficient to describe the system. Furthermore, for the current circumstances, only Coulombic interactions among the particles need to be considered.

Taking the potential energy terms first, the procedure that is used consists merely of writing down the Coulombic interactions that would contribute to the potential energy if the system of particles were to be considered as classical particles. Then, the quantum mechanical potential energy operator is taken simply as given by the corresponding classical expression. To illustrate the procedure, a system of N classical particles having charges $Z_1, Z_2, Z_3, \ldots, Z_N$ would have a potential energy given by

$$U=\sum_{j>k}^{N} \frac{Z_j Z_k}{r_{jk}}, \tag{4-185}$$

where r_{jk} is the distance between particles j and k. The corresponding quantum mechanical potential energy operator is given by

$$\mathcal{V}(r)=\sum_{j>k}^{N} \frac{Z_j Z_k}{r_{jk}} \tag{4-186}$$

where the distance between each pair of particles (r_{jk}) is interpreted as a multiplicative operator.

It is in the treatment of the kinetic energy portion where significant deviations from the corresponding classical situation occur. In particular, the kinetic energy of a system of N classical particles is given by

$$T=\frac{1}{2}\sum_{j=1}^{N} m_j v_j^2 \tag{4-187}$$

$$=\frac{1}{2}\sum_{j=1}^{N}\left(\frac{p_j^2}{m_j}\right) \tag{4-188}$$

where m_j is the mass of the jth particle, v_j is its velocity, and p_j is its linear momentum.

To convert this to a quantum mechanical operator, the momentum of each particle is replaced by an operator that differentiates the coordinates of the particle. Specifically, we have

$$(p_x)_j \Rightarrow \left(\frac{\hbar}{i}\right)\frac{\partial}{\partial x_j} \tag{4-189}$$

for the x-component of the momentum operator corresponding to particle j. We also note that in three dimensions,

$$[\mathbf{r}_j, \mathbf{p}_j]=i\hbar\mathcal{I}. \tag{4-190}$$

The quantity $(\hbar/i)$ in the definition of $\mathbf{p}_j$ is needed in part as a proportionality constant and in part to assume that the kinetic energy operator is Hermitian. For the kinetic energy operator itself, we have

$$\mathfrak{I}_j=\left(\frac{1}{2m_j}\right)[(p_x)_j^2+(p_y)_j^2+(p_z)_j^2]$$

$$=-\left(\frac{\hbar^2}{2m_j}\right)\left[\frac{\partial^2}{\partial x_j^2}+\frac{\partial^2}{\partial y_j^2}+\frac{\partial^2}{\partial z_j^2}\right]=-\left(\frac{\hbar^2}{2m_j}\right)\nabla_j^2 \tag{4-191}$$

for the jth particle. The total kinetic energy operator is thus given by

$$\mathfrak{I}=\sum_{j=1}^{N}\mathfrak{I}_j \tag{4-192}$$

$$=-\left(\frac{\hbar^2}{2}\right)\sum_{j=1}^{N}\left(\frac{1}{m_j}\right)\nabla_j^2. \tag{4-193}$$

The total quantum mechanical Hamiltonian operator is then given by

$$\mathcal{H}=\mathfrak{I}+\mathcal{V} \tag{4-194}$$

or

$$\mathcal{H}=-\frac{\hbar^2}{2}\sum_{j=1}^{N}\left(\frac{1}{m_j}\right)\nabla_j^2+\sum_{j>k}^{N}\frac{Z_jZ_k}{r_{jk}}. \tag{4-195}$$

Thus, the procedure for construction of an appropriate quantum mechanical Hamiltonian for the circumstances given above can be summarized as follows:

1. Construct classical mechanical expressions for the kinetic and potential energy of the particles in the system.
2. Replace p_j by $(\hbar/i)\nabla j$ for all j, and treat the distance terms in the potential energy expression as multiplicative operators.

Before leaving this section, it is appropriate to add several comments regarding coordinate transformations, and the form of the quantum mechanical Hamiltonian in various coordinate systems. In Eqs. (4-191) and (4-193), for example, the kinetic energy operator is expressed in cartesian coordinates. Such a coordinate system, in which we deal with three mutually perpendicular families of planes, i.e., x = constant, y = constant, and z = constant, is not always a convenient coordinate system for solution of the Schrödinger equation. In fact, it has been shown[45] that there are 11 different coordinate systems in which equations having the form of the Schrödinger equation are separable, i.e., that allow the use of a "separation of variables" approach to obtain solutions of the differential equation. Since the choice of coordinate system of interest is determined in general by the particular problem under consideration, it is therefore of interest to be able to express the Hamiltonian in various coordinate systems. Since the potential energy operators are typically multiplicative operators that are a function of interparticle distances only, and thus are relatively easy to transform from one coordinate system to another, we shall focus on expressing the kinetic energy operator in various coordinate systems.

If we wish to relate a point (x, y, z) in a cartesian coordinate system to a point (q_1, q_2, q_3) in another curvilinear coordinate system, it is necessary to specify the relationship between these coordinate systems for all points, i.e., we must find the relations that describe

$$\begin{aligned} x &= x(q_1, q_2, q_3) \\ y &= y(q_1, q_2, q_3) \\ z &= z(q_1, q_2, q_3), \end{aligned} \tag{4-196}$$

along with the inverse relationship

$$\begin{aligned} q_1 &= q_1(x, y, z) \\ q_2 &= q_2(x, y, z) \\ q_3 &= q_3(x, y, z) \end{aligned} \tag{4-197}$$

In our case, we shall be interested in the distance (dS) between any two points on a surface q_i = constant, which can be written in general as

$$dS^2 = dx^2 + dy^2 + dz^2 = \sum_{i,j} h_{ij}^2 dq_i dq_j, \tag{4-198}$$

[45] L. P. Eisenhart, *Phys. Rev.*, **45**, 427 (1934).

in which the coefficients h_{ij} (which are collectively known as the *metric*) define the relationship between the coordinate systems that we seek.

To determine the h_{ij}, we differentiate Eq. (4-196) which gives

$$dx = \frac{\partial x}{\partial q_1} dq_1 + \frac{\partial x}{\partial q_2} dq_2 + \frac{\partial x}{\partial q_3} dq_3, \tag{4-199}$$

with similar expressions for dy and dz. If Eq. (4-199) is substituted into Eq. (4-198), we see that

$$h_{ij}^2 = \frac{\partial x}{\partial q_i}\frac{\partial x}{\partial q_j} + \frac{\partial y}{\partial q_i}\frac{\partial y}{\partial q_j} + \frac{\partial z}{\partial q_i}\frac{\partial z}{\partial q_j}. \tag{4-200}$$

For the case of orthogonal coordinate systems, which is of interest to us, $h_{ij} = 0$ $(i \neq j)$. In such a case we may simplify the notation by writing $h_i \equiv h_{ii}$, and Eq. (4-198) can be rewritten as

$$dS^2 = (h_1 dq_1)^2 + (h_2 dq_2)^2 + (h_3 dq_3)^2. \tag{4-201}$$

This suggests that we can divide dS into components, i.e.,

$$ds_i = h_i dq_i. \tag{4-202}$$

This allows us to obtain the first result of interest, i.e., that the volume element can be expressed in general as

$$\begin{aligned} dV &= ds_1 ds_2 ds_3 \\ &= h_1 h_2 h_3 dq_1 dq_2 dq_3. \end{aligned} \tag{4-203}$$

The other general result that is needed is the effect of coordinate transformations on operators such as ∇ and ∇^2. Using analyses analogous to those used above,[46] the following results are obtained:

$$\boldsymbol{\nabla}\Psi(q_1, q_2, q_3) = \mathbf{i}\frac{\partial \Psi}{h_1 \partial q_1} + \mathbf{j}\frac{\partial \Psi}{h_2 \partial q_2} + \mathbf{k}\frac{\partial \Psi}{h_3 \partial q_3}, \tag{4-204}$$

$$\begin{aligned} \nabla^2\Psi(q_1, q_2, q_3) = \left(\frac{1}{h_1 h_2 h_3}\right)\left[\frac{\partial}{\partial q_1}\left(\frac{h_2 h_3}{h_1}\cdot\frac{\partial \Psi}{\partial q_1}\right) + \frac{\partial}{\partial q_2}\left(\frac{h_1 h_3}{h_2}\frac{\partial \Psi}{\partial q_2}\right)\right. \\ \left. + \frac{\partial}{\partial q_3}\left(\frac{h_1 h_2}{h_3}\frac{\partial \Psi}{\partial q_3}\right)\right]. \end{aligned} \tag{4-205}$$

To illustrate the application of these general ideas to a specific case, let us transform the quantum mechanical kinetic energy operator (taken to a single particle for convenience) of Eq. (4-191) from cartesian to spherical polar coordinates. To do this we note that the relationship between the cartesian coordinates (x, y, z) and the corresponding spherical polar coordinates (r, ϑ, φ)

---

[46] For a more detailed description of such an analysis see, for example, G. Arfken, *Mathematical Methods for Physicists*, Academic Press, New York, 1968, Section 2.2.

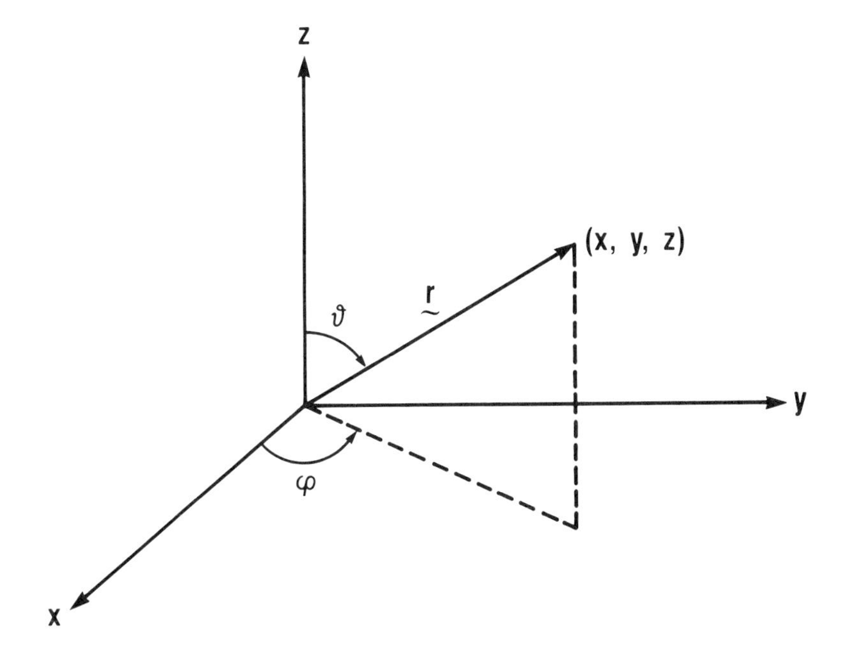

Figure 4.1. Spherical Polar Coordinate System.

is given by (see Figure 4-1):

$$x = r \sin \vartheta \cos \varphi$$
$$y = r \sin \vartheta \sin \varphi$$
$$z = r \cos \vartheta. \tag{4-206}$$

From Eqs. (4-200) and (4-206), it can easily be seen that

$$\begin{aligned} h_r &= \left[\left(\frac{\partial x}{\partial r} \right)^2 + \left(\frac{\partial y}{\partial r} \right)^2 + \left(\frac{\partial z}{\partial r} \right)^2 \right]^{1/2} \\ &= [\sin^2 \vartheta \cos^2 \varphi + \sin^2 \vartheta \sin^2 \varphi + \cos^2 \vartheta]^{1/2} \\ &= 1. \end{aligned} \tag{4-207}$$

Similarly,

$$h_\vartheta = r, \tag{4-208}$$

and

$$h_\varphi = r \sin \vartheta. \tag{4-209}$$

Using Eqs. (4-207)–(4-209), it is easily seen that Eqs. (4-203) and (4-205) can be written as follows for the case of spherical polar coordinates:

$$dV = r^2 \sin \vartheta \, dr \, d\vartheta \, d\varphi \tag{4-210}$$

$$\nabla^2 = \frac{1}{r^2} \frac{\partial}{\partial r} \left(r^2 \frac{\partial}{\partial r} \right) + \frac{1}{r^2 \sin \vartheta} \frac{\partial}{\partial \vartheta} \left(\sin \vartheta \frac{\partial}{\partial \vartheta} \right) + \frac{1}{r^2 \sin^2 \vartheta} \frac{\partial^2}{\partial \varphi^2}. \tag{4-211}$$

Although the use of the above procedure will provide appropriate quantum mechanical Hamiltonians in various coordinate systems, it is also clear that the procedure has simply been stated, and has not been derived. In a sense this should be expected, just as the postulates of quantum mechanics are not derived. As to why such a procedure is plausible and not totally unexpected, it should be remembered that the Schrödinger development of quantum mechanics came after deBroglie had showed that particles had wave character and vice versa.[47] Thus, it was to be expected that a "wave equation" should be found that determines the motion of particles, and which included particle mechanics as a limiting case. Thus, an extension of Hamilton's particle equation into a differential equation (for waves) was to be expected.

It should also be noted that many terms are omitted from the quantum mechanical Hamiltonian that has been constructed for the special case described above. These omitted terms are generally small in magnitude, but can be very important for interpretation of spectroscopic results (e.g., spin–orbit and spin–spin interactions). The origin and inclusion of these terms will be described in Chapter 8, but the terms introduced in Eq. (4-195) will suffice for discussion of electronic structural features in the next several chapters.

Problems

1. Show that the time evolution operator ($\mathcal{U}$) is both unitary and linear, where $\mathcal{U}$ is given by

$$\mathcal{U}(t) = \exp\left\{-\frac{i}{\hbar}\mathcal{H}t\right\} = \sum_{n=0}^{\infty} \frac{\{(i/\hbar)\mathcal{H}t\}^n}{n!},$$

   and $\mathcal{H} \neq \mathcal{H}(t)$.

2. If $\mathcal{H}\psi_k = E_k\psi_k$ holds, where $\mathcal{H} \neq \mathcal{H}(t)$, show that the time evolution operator ($\mathcal{U}$):

$$\mathcal{U}(t) = \exp\left[-\frac{i}{\hbar}\mathcal{H}t\right]$$

   satisfies the equation:

$$\mathcal{U}(t)\psi_k = \exp\left[-\frac{i}{\hbar}E_k t\right]\psi_k.$$

3. Using the time evolution operator ($\mathcal{U}$), show that the time-dependent state function $\Psi(q, t)$ maintains its normalization throughout time.
4. Assuming that a power series of the operator $\mathcal{A}$ exists, and hence $f(\mathcal{A})$ is

[47] For a more complete discussion of the historical events and factors that influenced Schrödinger's thinking, see, Max Jammer, *The Conceptual Development of Quantum Mechanics*, McGraw-Hill, New York, 1966, pp. 196–280.

defined, show that

$$\langle f(\mathcal{A})\rangle = \sum_{k=1}^{\infty} f(a_k)|\langle\alpha_k|\Psi\rangle|^2 = \sum_{k=1}^{\infty} f(a_k)|c_k|^2.$$

5. Show that expansion of the wavefunction in terms of stationary states, i.e.,

$$\Psi_i = \sum_k \psi_k c_k$$

where $\mathcal{H} \neq \mathcal{H}(t)$ and

$$\mathcal{H}\psi_i = E_i\psi_i$$

is also a solution of the time-dependent Schrödinger equation.

6. If $\mathcal{A}$ and $\mathcal{B}$ are Hermitian operators that do not commute, show that it is possible to write the commutator of $\mathcal{A}$ and $\mathcal{B}$ as

$$[\mathcal{A}, \mathcal{B}] = i\mathcal{C},$$

where $i\mathcal{C}$ is Hermitian.

7. Given an operator $\mathcal{A}$ such that

$$\mathcal{A}\Psi = \langle\mathcal{A}\rangle\Psi$$

show that

$$\Delta\mathcal{A} = 0,$$

where

$$(\Delta\mathcal{A})^2 = \langle\mathcal{A}^2\rangle - \langle\mathcal{A}\rangle^2.$$

8. If $\mathcal{A}$, $\mathcal{A}'$, $\mathcal{B}$, $\mathcal{B}'$ are each Hermitian operators, show that

$$[\mathcal{A}', \mathcal{B}'] = [\mathcal{A}, \mathcal{B}] = i\mathcal{C}$$

where $\mathcal{C}$ is also a Hermitian operator, and

$$\mathcal{A}' = \mathcal{A} - \langle\mathcal{A}\rangle\mathcal{I}$$

$$\mathcal{B}' = \mathcal{B} - \langle\mathcal{B}\rangle\mathcal{I}.$$

9. If $\mathcal{A}$ is a linear, Hermitian operator, show that

$$\langle\mathcal{A}^2\rangle \geq \langle\mathcal{A}\rangle^2.$$

10. If $\mathcal{A}$ is a linear, Hermitian operator, such that

$$[\mathcal{H}, \mathcal{A}] = 0,$$

show that

$$\frac{d}{dt}(\Delta\mathcal{A}) = 0.$$

11. Show that the expectation value of an observable is the same in the Schrödinger and Heisenberg pictures.
12. Show that, if $\mathcal{H}^{(S)} \neq \mathcal{H}^{(S)}(t)$ and $\mathcal{A}^{(S)} \neq \mathcal{A}^{(S)}(t)$,

$$\mathcal{H}^{(H)} \neq \mathcal{H}^{(H)}(t)$$

and

$$\mathcal{A}^{(H)} \neq \mathcal{A}^{(H)}(t).$$

13. Prove that in one dimension:

$$\frac{d}{dt}\langle x^2\rangle = \left(\frac{1}{m}\right)[\langle xp_x\rangle + \langle p_x x\rangle].$$

14. Prove that

$$\frac{d}{dt}\langle \mathbf{p}\rangle = -\langle \nabla V\rangle.$$

The above statement is known as Ehrenfest's Theorem, and is the quantum mechanical analog of Newton's second law. Hint: Use Green's Theorem, i.e,

$$\int_v (u\nabla^2 v - v\nabla^2 u)\, d\tau = \int (u\nabla v - v\nabla u) \cdot d\mathbf{S}$$

for continuously differential functions u and v.

15. Show that Eq. (4-150) can be derived in an alternative fashion from Eq. (4-145) by using the fact that the $\{\phi_k^{(q)}\}$ are eigenfunctions of $\mathcal{U}$ having eigenvalues of modulus unity.
16. In terms of distances r_A and r_B from the points $(0, 0, -a)$ and $(0, 0, +a)$, respectively, with coordinates ξ, η defined as

$$\xi = (r_A + r_B)/2, \qquad \eta = (r_A - r_B)/2,$$

and

$$x = a[(\xi^2 - 1)(1 - \eta^2)]^{1/2} \cos\varphi,$$
$$y = a[(\xi^2 - 1)(1 - \eta^2)]^{1/2} \sin\varphi, \qquad z = a\xi\eta,$$

show that ∇^2 in the (ξ, η, φ) coordinate system (confocal elliptic coordinates) is given by:

$$\nabla^2 = \left[\frac{\partial}{\partial\xi}\left\{(\xi^2 - 1)\frac{\partial}{\partial\xi}\right\} + \frac{\partial}{\partial\eta}\left\{(1 - \eta^2)\frac{\partial}{\partial\eta}\right\} + \frac{(\xi^2 - \eta^2)}{(\xi^2 - 1)(1 - \eta^2)}\frac{\partial^2}{\partial\varphi^2}\right] \Big/ a^2(\xi^2 - \eta^2).$$

Chapter 5

One-Dimensional Model Problems

5-1. General Comments

In the previous chapter, the basic ideas of quantum mechanics were introduced by means of several postulates. It was also seen that there are several possible equivalent ways of viewing quantum mechanics, commonly referred to as the Schrödinger picture, the Heisenberg picture, and the interaction picture. Since these are all equivalent views of the same phenomena, the choice of which one to employ when attempting to solve a problem of interest can be made on the basis of which one is most convenient to use.

However, it should be noted that the mathcmatical apparatus that is used differs, depending on the particular picture that is chosen. For example, if the Heisenberg picture is used, it is common to extract the eigenvalues of the system by examination of commutators involving the Hamiltonian and the other observables (e.g., position and/or momentum).[1] These relations can usually be used to establish a set of equations whose solutions give rise directly to the desired eigenvalues. On the other hand, if the Schrödinger picture is employed, the general solution of a second-order partial differential equation (the Schrödinger equation) is obtained first, and the energy levels are obtained on application of the boundary conditions. Since the Schrödinger picture has found widespread acceptance and use in chemistry, both in the past and currently, most of the applications given in this chapter will be couched in terms of this picture.[2]

---

[1] For a description of such procedures see, for example, L. I. Schiff, *Quantum Mechanics*, McGraw-Hill, New York, 1968, Chapter 6.

[2] For the solution of many of these problems using the Heisenberg representation, see H. S. Green, *Matrix Methods in Quantum Mechanics*, Barnes & Noble, New York, 1965.

In the following discussion we shall be concerned with the determination of the stationary state energy levels of several one-dimensional model problems. Although these problems cannot be considered to be direct illustrations of real systems, since they are one-dimensional, they can often be quite useful in facilitating the understanding of real problems. However, it is necessary to be careful, as we shall see, to consider the alterations that must be considered when the problem is extrapolated to higher dimensionality.

5-2. Wavefunction Criteria and Boundary Conditions

Before beginning the discussion of specific examples, it will be useful to discuss in greater detail the characteristics that are required for acceptable eigenfunctions of the Schrödinger equation for problems in which the particles exist in a bound state. In particular, we shall in general be interested in finding eigenfunctions $\Psi(x)$ of the Schrödinger equation having the following properties:

1. The quantity $\Psi^*(x)\Psi(x)$ must be single-valued over the entire range of x.
2. The wavefunction must be quadratically integrable, i.e.,

$$\int |\Psi(x)|^2 \, dx < \infty. \tag{5-1}$$

   This requirement will generally be taken to mean that $\Psi(x)$ must be *continuous everywhere*.
3. The integral of the derivative of $\Psi(x)$ must exist, i.e.,

$$\int \left| \frac{d\Psi(x)}{dx} \right|^2 dx < \infty. \tag{5-2}$$

Since a second derivative[3] is usually utilized in the Schrödinger equation, the constraint of Eq. (5-2) will generally be replaced by the more stringent requirement that the derivative of the wavefunction must be *continuous everywhere*.[4] However, we will sometimes encounter cases where this constraint is too strong. In those cases, we shall require that $\Psi'(x)$ be *piece-wise continuous*, and we shall find that the points at which singularities in $\Psi'(x)$ occur will usually be accompanied by an associated singularity in the potential. The implications of these singularities on measurable quantities cannot be ignored, and they will be explored in greater detail within the context of specific examples in later sections.

---

[3] The second derivative is usually employed in the kinetic energy operator, and the constraint in Eq. (5-2) can be shown to guarantee the existance of the expectation value of the kinetic energy. See J. C. Slater, *Phys. Rev.*, **51**, 846 (1937) and H. S. Silverstone, *Phys. Rev. (A)*, **5**, 1092 (1972).

[4] For examples of acceptable, albeit rather unconventional, wavefunctions that forego some of these continuity characteristics, see G. G. Hall, *Symposium of the Faraday Society*, No. 2, The University Press, Aberdeen, 1968, p. 69, and G. G. Hall, J. Hyslop, and D. Rees, *Int. J. Quantum Chem.*, **3**, 195 (1969).

In practice, these continuity conditions are generally translated into boundary conditions on the wavefunction and its first derivative at particular points. The particular constraints that are to be applied in every case are not, unfortunately, clear and straightforward to formulate, and depend on the particular problem being discussed. However, we can express the general procedure to be followed.

For a region of interest $l \leq x \leq m$, with $m > l$, we shall in general be interested in applying boundary conditions on the wavefunction and its first derivative at the points $x = l$ and $x = m$. If $\Psi(x)$ is the general solution within the region $l \leq x \leq m$, then the usual set of boundary conditions that we shall wish to impose[5] can be written as a set of homogeneous equations in $\Psi(x)$ and $\Psi'(x)$, i.e.,

$$\begin{aligned} a_{11}\Psi(l)+a_{12}\Psi'(l)&=0,\\ a_{21}\Psi(m)+a_{22}\Psi'(m)&=0, \end{aligned} \tag{5-3}$$

where a_{11}, a_{12}, a_{21}, and a_{22} are constants that are chosen to be suitable to the particular problem under consideration, and where a_{i1} and a_{i2} may not simultaneously be zero ($i = 1, 2$). The first of these equations corresponds to imposing a continuity condition on the wavefunction and an existence condition on its derivative at $x = l$, and the second corresponds to a similar set of conditions at $x = m$. Boundary conditions of this form are usually referred to as *one-point boundary conditions*. We shall encounter several examples of choices of a_{11}, a_{12}, a_{21}, and a_{22} in the specific problems that are investigated later in this chapter.

5-3. The Nondegeneracy Theorem

When eigenfunctions of one-dimensional problems are found that obey the boundary conditions of Eq. (5-3), they possess another property that can be expressed conveniently in the form of the following theorem.

Theorem 5-1: The eigenvalues of the one-dimensional Schrödinger equation with a continuous potential are nondegenerate, if the eigenfunctions satisfy the boundary conditions of Eq. (5-3).

Proof:[6] This theorem is most conveniently proved by contradiction. In particular, let us assume that the eigenvalue λ is doubly degenerate, with linearly independent eigenfunctions $\varphi_\lambda(x)$ and $\Psi_\lambda(x)$. From the definition of linear independence (Chapter 2, Section 2-2), it follows that for any x in the regions for

[5] See G. Hellwig, *Differential Operators of Mathematical Physics*, Addison-Wesley, Reading, MA, 1967, pp. 39–49 for a discussion of these conditions.

[6] For earlier proofs of this theorem, see, for example, G. Hellwig, *Differential Operators of Mathematical Physics*, Addison-Wesley, Reading, MA, 1967, p. 49. Also, see R. Courant and D. Hilbert, *Methods of Mathematical Physics*, Vol. 1, Interscience Publishers, New York, 1961, pp. 293–294.

which φ_λ and Ψ_λ are eigenfunctions,

$$\varphi_\lambda(x)\Psi_\lambda'(x)-\Psi_\lambda(x)\varphi_\lambda'(x)\neq 0. \tag{5-4}$$

On the other hand, if the solutions are to obey the boundary conditions of Eq. (5-3), we require, for example, that φ_λ and Ψ_λ both satisfy the first of the two equations in Eqn. (5-3), i.e.,

$$\begin{aligned} a_{11}\varphi_\lambda(l)+a_{12}\varphi_L'(l)&=0 \\ a_{11}\Psi_\lambda(l)+a_{12}\Psi_L'(l)&=0, \end{aligned} \tag{5-5}$$

with a_{11} and a_{12} not both identically zero. If we multiply the first of the above equations by $\Psi_\lambda'(l)$ and the second by $\varphi_\lambda'(l)$ and subtract the resulting equation, we obtain[7]

$$a_{11}[\varphi_\lambda(l)\Psi_\lambda'(l)-\Psi_\lambda(l)\varphi_\lambda'(l)]=0. \tag{5-6}$$

However, using Eq. (5-4), we see that a_{11} must be zero, in contradiction to the requirements of Eq. (5-5). A similar analysis leads to a contradiction involving a_{12}, which proves the theorem.

It is also of interest to note the assumptions that are necessary if the theorem is to be valid. First, it must be assumed that there are no discontinuities in the potential within the range of x. If discontinuities are introduced within the interior of the range of x, then degeneracies may occur.[8] We shall discuss the specific case of the effect on observables of discontinuities in the potential *at the boundaries* in greater detail in Section 5-6. The other assumption that is essential to the theorem is that only boundary conditions of the form specified by Eq. (5-3) be applied. If more general boundary conditions are employed, such as

$$\left.\begin{aligned} \alpha_{11}\Psi(l)+\alpha_{12}\Psi'(l)+\alpha_{13}\Psi(m)+\alpha_{14}\Psi'(m)=0, \\ \alpha_{21}\Psi(l)+\alpha_{22}\Psi'(l)+\alpha_{23}\Psi(m)+\alpha_{24}\Psi'(m)=0, \end{aligned}\right\} \tag{5-7}$$

where the rank of the matrix of coefficients is given by

$$\text{rank}\begin{Bmatrix} \alpha_{11} & \alpha_{12} & \alpha_{13} & \alpha_{14} \\ \alpha_{21} & \alpha_{22} & \alpha_{23} & \alpha_{24} \end{Bmatrix}=2,$$

then degeneracies can occur.[9] These more general boundary conditions correspond to placing two constraints on $\Psi(x)$ and $\Psi'(x)$ at both boundary points instead of one, and are usually called *two-point boundary conditions*. The two-point type of boundary condition arises in quantum mechanics

[7] The term in brackets in Eq. (5-6) is known as the *Wronskian* of φ and Ψ.

[8] An example of this notion is given by the one-dimensional hydrogen atom in which a two-fold degeneracy in the bound state energy levels is found. For a discussion of this example, see R. Loudon, *Am. J. Phys.*, **27**, 649 (1959).

[9] For a more detailed discussion of these boundary conditions, including examples, see G. Hellwig, *Differential Operators of Mathematical Physics*, Addison-Wesley, Reading, MA, 1967, pp. 39–49.

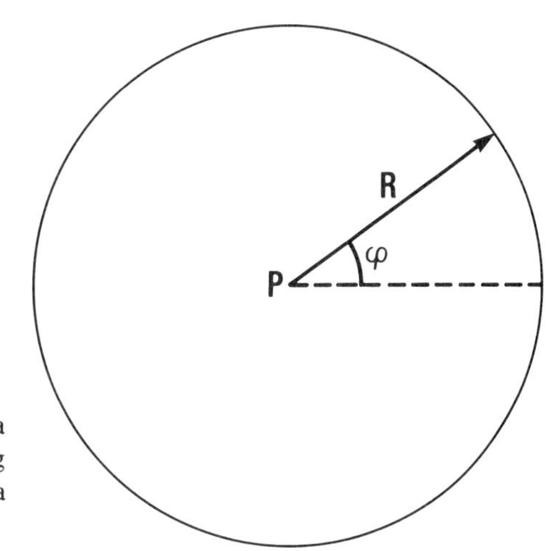

Figure 5.1. Depiction of the path of a one-dimensional "particle" moving about a point P, and constrained to a fixed radius R.

primarily when periodic boundary conditions are invoked, as we shall see in the discussion of the particle on a ring in the next section.

5-4. Particle on a Ring

As a first example of a one-dimensional model problem, let us consider the motion of a particle of mass m about a fixed point where the distance (r) of the particle from the point is held fixed at $r = R$. This motion is depicted in Fig. 5.1, and can be envisioned as the special case where r and ϑ are held fixed, e.g., with $\sin \vartheta = 1$, and the motion in the φ-variable ($0 \leq \varphi \leq 2\pi$) is examined. This one-dimensional model problem is quite useful as a first approximation (i.e., the "rigid rotor" approximation) to the rotational motion of a diatomic molecule, or in the "free electron" model[10] of cyclic conjugated systems.

Since there is no potential energy acting on the particle, the only factors determining the energy of the particle are its kinetic energy and the boundary conditions. As indicated in the previous chapter, the three-dimensional quantum mechanical kinetic energy operator for a single particle of mass m is given in spherical coordinates as

$$\mathfrak{T} = -\frac{\hbar^2}{2m} \nabla^2 \tag{5-8}$$

$$= -\frac{\hbar^2}{2m} \left\{ \frac{1}{r^2} \frac{\partial}{\partial r} \left(r^2 \frac{\partial}{\partial r} \right) + \frac{1}{r^2 \sin \vartheta} \frac{\partial}{\partial \vartheta} \left(\sin \vartheta \frac{\partial}{\partial \vartheta} \right) + \frac{1}{r^2 \sin^2 \vartheta} \frac{\partial^2}{\partial \varphi^2} \right\} . \tag{5-9}$$

[10] See, for example, L. Pauling, *J. Chem. Phys.*, **4**, 673 (1936), and Problem 5-5.

To create the one-dimensional case of interest here, let us fix[11] $r = R$, and $\vartheta = 90°$, which yields

$$\mathfrak{J} = -\frac{\hbar^2}{2mR^2}\left(\frac{\partial^2}{\partial\varphi^2}\right), \tag{5-10}$$

where the r and ϑ dependence has been eliminated by fixing r and ϑ at specific values.

The Schrödinger equation for the desired stationary state energy levels is then given by

$$\mathcal{H}\Psi(\varphi) = E\Psi(\varphi), \tag{5-11}$$

or

$$-\left(\frac{\hbar^2}{2mR^2}\right)\frac{d^2\Psi(\varphi)}{d\varphi^2} = E\Psi(\varphi), \tag{5-12}$$

where total derivatives can be employed since r and ϑ are not being considered. This equation can be rewritten as

$$\frac{d^2\Psi(\varphi)}{d\varphi^2} = -k^2\Psi(\varphi), \tag{5-13}$$

where

$$k^2 = \frac{2mR^2E}{\hbar^2}. \tag{5-14}$$

In this case the general solution to Eq. (5-13) is quite easy to establish on intuitive grounds, and we have already encountered equations of this form in earlier discussions.[12] In particular, we are seeking a function that, after being differentiated twice, gives itself back again (perhaps multiplied by a constant). The obvious possibilities are exponentials or sines and cosines, and we shall choose the former for convenience. Of course, more general techniques could also be used to establish the solutions.[13] In any case, the student can easily verify that a general solution to Eq. (5-14) can be written as

$$\Psi(\varphi) = Ae^{+ik\varphi} + Be^{-ik\varphi}, \tag{5-15}$$

where A and B are the two arbitrary constants that are expected in the general

---

[11] Arbitrarily fixing r and ϑ can be thought of as a kind of boundary condition that is applied to the motion of the particle. It differs from the "usual" boundary conditions that are required in order to produce acceptable wavefunctions, since it is not required, but introduced only to simplify the problem.

[12] See Chapter 2, Section 2-14.

[13] For a discussion of the general techniques that are useful in solving differential equations of this type see, for example, W. Kaplan, *Ordinary Differential Equations*, Addison-Wesley, Reading, MA, 1961, Chapter 4.

solution of a second order differential equation. These must be determined by application of the criteria listed earlier.

Regarding these boundary conditions, let us first examine whether $\Psi(\varphi)$ can be made continuous everywhere. Obviously, the only place where difficulties are anticipated is at the end point, i.e., when $\varphi = 0$ and $\varphi = 2\pi$. To gain continuity at these points, we require that

$$\Psi(0) = \Psi(2\pi). \tag{5-16}$$

We note that this boundary condition is of the more general two-point type (i.e., a periodic boundary condition) given in Eq. (5-5), with $\alpha_{11} = 1, \alpha_{12} = -1$, and $\alpha_{13} = \alpha_{14} = 0$. Also, it should be observed that satisfaction of this boundary condition will also satisfy the single-valued criterion.[14]

Applying the boundary conditions of Eq. (5-16) to the general solution of Eq. (5-15) gives

$$\begin{aligned} A + B &= Ae^{2\pi ik} + Be^{-2\pi ik} \\ &= A\cos(2\pi k) + iA\sin(2\pi k) \\ &\quad + B\cos(2\pi k) - iB\sin(2\pi k). \end{aligned} \tag{5-17}$$

For this equation to be satisfied, we see that two kinds of conditions need to be satisfied,[15] i.e.,

$$A = B \qquad \text{or} \qquad A = -B, \tag{5-18}$$

and

$$k = \text{integer}. \tag{5-19}$$

This latter condition is of particular interest, for it restricts the values of k that can be allowed[16] and, through Eq. (5-14), *quantizes the energy levels* of the particle. In particular, the allowed energy levels arc given by

$$E_k = \frac{\hbar^2 k^2}{2mR^2}, \qquad k = 0, \pm 1, \pm 2, \cdots. \tag{5-20}$$

The important point to note is that quantization of the energy levels did not have to imposed in an ad hoc manner, but arose naturally through the application of the boundary conditions. Also, since the energy is dependent on k^2, a double degeneracy is present, except for $k = 0$.

Summarizing the results obtained thusfar, the wavefunction for the case $A =$

[14] This kind of situation arises whenever an angular variable is involved.

[15] The choice $A = B = 0$ has been excluded, since it leads to the obviously uninteresting result that $\Psi \cong 0$ everywhere.

[16] The term $(e^{+ik\varphi})$ can be thought of as representing motion of the particle in the $+\varphi$ direction, and the term $(e^{-ik\varphi})$ as motion in the $-\varphi$ direction.

B can be written as

$$\begin{aligned}\Psi_k(\varphi) &= A(e^{ik\varphi} + l^{-ik\varphi}) \\ &= 2A\cos(k\varphi) \\ &= 2A\cos\left(\sqrt{\frac{2mR^2E_k}{\hbar^2}}\,\varphi\right).\end{aligned} \tag{5-21a}$$

where the positive square root has been taken for k for convenience. In a similar manner, the other wavefunction of the degenerate pair, i.e., $A = -B$, can be written

$$\Psi_k(\varphi) = 2iA\sin(k\varphi)$$

as

$$= 2iA\sin\left(\sqrt{\frac{2mR^2E_k}{\hbar^2}}\,\varphi\right). \tag{5-21b}$$

In addition to quantizing the energy levels, the continuity condition of $\Psi(\varphi)$ [Eq. (5-1)] can also be used to obtain the normalization constant[17] (A). In particular, by requiring the wavefunction to be normalized to unity, we obtain (for $A = B$)

$$\begin{aligned}1 &= \int_0^{2\pi} |\Psi(\varphi)|^2\,d\varphi = 4A^2\int_0^{2\pi}\cos^2(k\varphi)\,d\varphi \\ &= 4A^2\pi,\end{aligned} \tag{5-22}$$

or

$$A = \frac{1}{2\sqrt{\pi}}. \tag{5-23}$$

It is also easily established that an identical result is obtained for the case $A = -B$. Thus, the wavefunction is now completely determined, and is given by

$$\Psi_k(\varphi) = \frac{1}{\sqrt{\pi}}\cos(k\varphi), \qquad (A = B)$$

and (5-24)

$$\Psi_k(\varphi) = \frac{1}{\sqrt{\pi}}\sin(k\varphi), \qquad (A = -B).$$

[17] Is should be remembered that, although the continuity condition can be used to establish a normalization constant, the choice of any particular kind of normalization is *not* required by the continuity condition. In fact, the expectation value of any quantum mechanical operator is easily seen to be independent of the particular normalization chosen, as long as the wavefunction is square integrable [Eq. (5-1)].

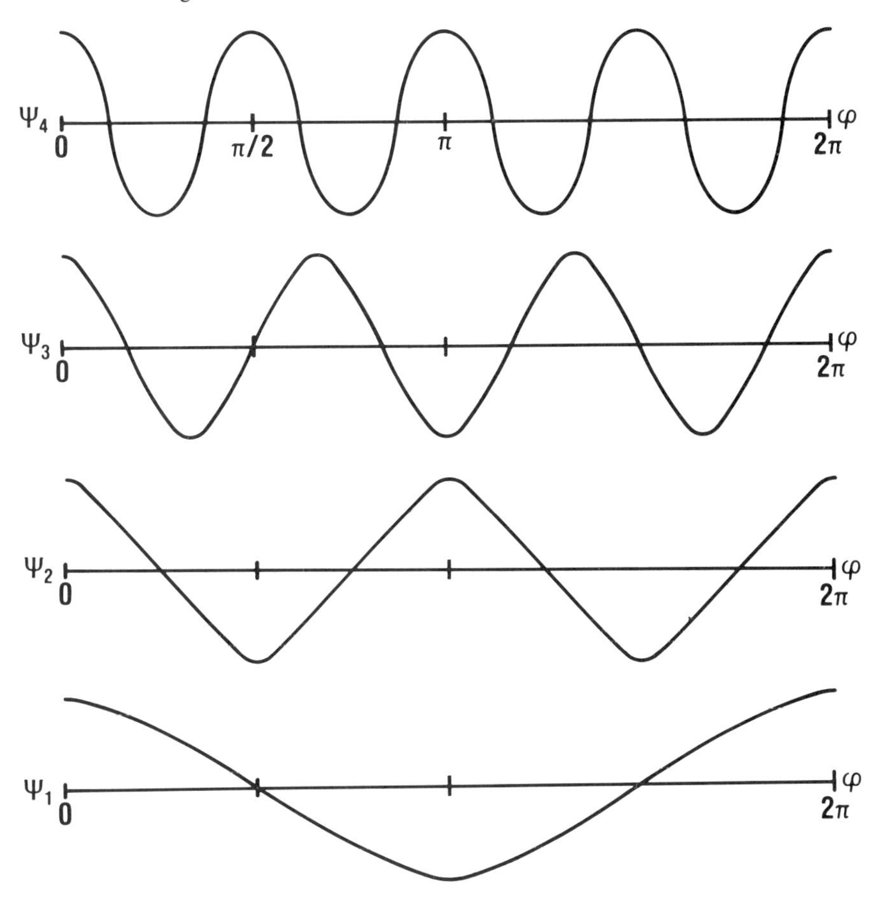

Figure 5.2. A graph of the four lowest eigenfunctions of a particle on a ring, corresponding to $k = 1, 2, 3$ and 4 for the case $A = B$. The corresponding plot for the case of $k = 0$ is simply a constant, and is not shown.

Now let us investigate the conditions that are required of $\Psi'(\varphi)$. In particular, the continuity condition on $\Psi'(\varphi)$ [Eq. (5-2)] needs to be examined at $x = 0$ and $x = 2\pi$, i.e., is the following condition satisfied?

$$\Psi'(0) = \Psi'(2\pi). \tag{5-25}$$

Since both $\Psi'(0)$ and $\Psi'(2\pi)$ are zero for the case $A = B$, and are equal to $k/\sqrt{\pi}$ for the case $A = -B$, we see that the condition that the derivative be continuous everywhere is also satisfied. This is indeed fortunate, since no additional parameters remain to be determined.

Several of the low-energy eigenfunctions ($k = 1, 2, 3, 4$) are shown in Fig. 5.2 for the case[18] $A = B$. Perhaps the most interesting feature of these graphs is

[18] The functions corresponding to the case $A = -B$ give rise to graphs having an odd number of nodes (not counting the end points) i.e., the $k = 0$ graph has one node, the $k = 1$ graph has 3 nodes, etc.

that the number of nodes increases as the energy of the particle increases. This observation, common to many one-dimensional eigenfunctions, is often carried over to multidimensional problems of practical interest that can be solved only approximately. In those cases, the energy of a given function is often estimated simply by counting its nodes. Although commonly used and often useful, the reader should be cautioned that no theorems guaranteeing the validity of such a generalization to more than one dimension are available. Finally, it should be noted [see Eq. (5-20)] that, consistent with chemical intuition, the energy of a "particle" constrained to move on a ring decreases if the mass is increased, or if the radius on which it moves is increased.

5-5. Particle Trapped in a Box

As a second example, let us consider the motion of a single particle of mass m in one dimension, constrained to remain within the confines of a "box," whose "walls" appear at $x = 0$ and $x = L$. This "box" is depicted in Fig. 5.3. The potential is taken to be infinite outside the box and zero inside it, i.e.,

$$V(x) = \begin{cases} \infty & x \leq 0,\ x \geq L \\ 0 & 0 < x < L. \end{cases} \tag{5-26}$$

Other constant values of $V(x)$ could be taken inside the box, without altering the results. The value of $V = 0$ within the box is chosen merely for convenience. Although this model problem obviously cannot represent a real problem, its

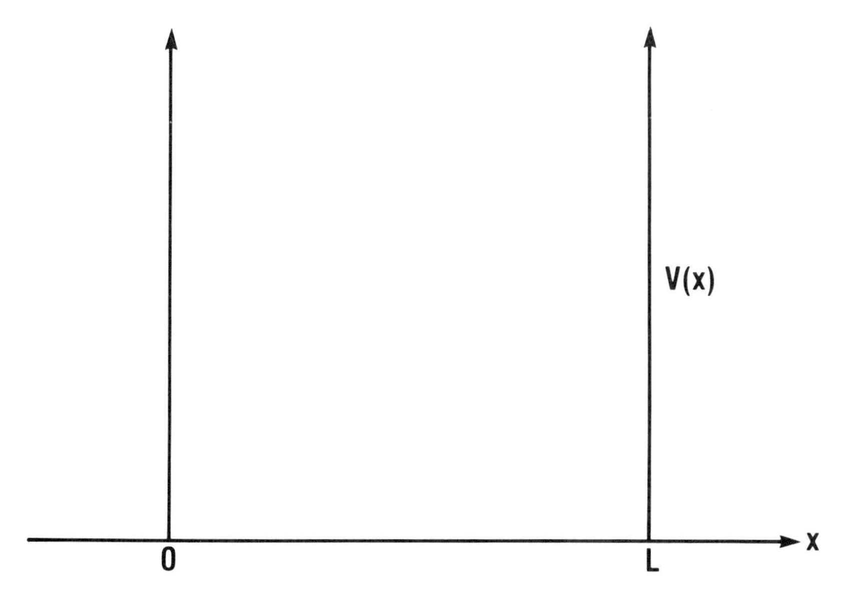

Figure 5.3. Depiction of a one-dimensional "box," within which $V(x) = 0$, and outside of which $V(x) = \infty$.

simplicity of discussion and interpretation have allowed it to be used often as a model of chemical phenomena. For example, the particle-in-a-box eigenfunctions have been used frequently as the simplest model of an electron in a molecule (e.g., the free-electron model of a π-electron in a conjugated molecule[19]), an electron in a metal (e.g., the electron gas model of a metal[20]), and a molecule in a gas (ideal gas[21]).

In searching for the eigenfunctions and eigenvalues associated with this particle we note first that the infinitely high potential at the walls and outside the box imply that the eigenfunction must be identically zero everywhere outside the box and at the walls. This arises because, if the wave function were nonzero at some point outside the box, it would imply that the particle would be required to have an infinite kinetic energy in order to satisfy the Schrödinger equation. Thus, we have

$$\Psi(x) \equiv 0 \qquad \text{for} \begin{cases} x \leq 0 \\ x \geq L. \end{cases} \tag{5-27}$$

This means that *boundary conditions* at two points have been identified which must be imposed upon the eigenfunctions for the region $0 \leq x \leq L$, if the complete eigenfunction is to satisfy all of the criteria listed earlier.

Considering the region inside the box, the Schrödinger equation that must be satisfied is

$$-\left(\frac{\hbar^2}{2m}\right)\frac{d^2\Psi(x)}{dx^2}=E\Psi(x) \tag{5-28}$$

or

$$\frac{d^2\Psi(x)}{dx^2}=-k^2\Psi(x), \tag{5-29}$$

where we have defined the constant k as

$$k^2=\frac{2mE}{\hbar^2}\ . \tag{5-30}$$

As comparison with Eq. (5-13) reveals, the differential equations of Eq. (5-29) and Eq. (5-13) are identical, except for the arbitrary labeling of the variable. Consequently, the general solution will be the same in both cases, and only the boundary conditions will give different final forms for the eigenfunctions.

For this case, let us choose the general solutions involving sines and cosines

---

[19] See, for example, J. R. Platt, K. Ruedenberg, C. W. Scherr, N. S. Ham, H. Labhart, and W. Licten, *Free Electron Theory of Conjugated Molecules*, John Wiley, New York, 1964.

[20] See, for example, J. C. Slater, *Quantum Theory of Molecules and Solids*, Vol. 3, McGraw-Hill, New York, 1967, p. 9.

[21] See, for example, R. Fowler and E. A. Guggenheim, *Statistical Thermodynamics*, Cambridge University Press, Cambridge, 1952, p. 72.

for variety, which gives

$$\Psi(x) = A \sin(kx) + B \cos(kx). \tag{5-31}$$

For this problem the boundary conditions are not periodic and, as Eq. (5-27) indicates, they can be written as

$$\Psi(0) = 0, \tag{5-32}$$

and

$$\Psi(L) = 0. \tag{5-33}$$

Incidentally, we note that these boundary conditions are consistent with one-point boundary conditions of the type given in Eq. (5-3), i.e., where $a_{11} = a_{12} = 1$.

However, this example has introduced a singularity in the potential at the points where the boundary conditions are to be examined, and careful investigation of the implications of the singularities on the properties of the system is needed.

Comparison of the expression for $\Psi(x)$ in Eq. (5-31) with the boundary condition in Eq. (5-32) requires that $B = 0$. Next, Eq. (5-33) requires that

$$0 = \Psi(L) = A \sin(kL). \tag{5-34}$$

Since we do not wish to choose $A = 0$ (which would result in $\Psi = 0$ everywhere), we must require that

$$kL = n\pi \tag{5-35}$$

where n is an integer, i.e.,

$$n = \pm 1, \ \pm 2, \ \pm 3, \ \cdots . \tag{5-36}$$

The case of $n = 0$ has been eliminated in this case in order to avoid the trivial result of $\Psi = 0$ everywhere. This requirement on n means that the values of the energy again are quantized, i.e., from Eqs. (5-30) and (5-35), we have

$$E_n = \frac{n^2 h^2}{8mL^2}, \qquad n = \pm 1, \ \pm 2, \ \cdots. \tag{5-37}$$

Thus, application of the boundary conditions has again given rise directly to quantization of the energy levels of the particle, without any kind of arbitrary or ad hoc postulation that quantization must occur.

In this manner similar to the discussion of the particle on a ring, we note that there is a suggested degeneracy corresponding to both positive and negative values of n giving the same energy. In this case, however, the degeneracy does not actually occur, since the functions corresponding to $\pm n$ differ by only a phase factor (± 1), and, hence, are not linearly independent. Therefore the nondegeneracy theorem is not violated, and we shall choose positive values of n in the following discussion for convenience.

The final constant can be obtained by requiring the eigenfunction to be normalized to unity, i.e.,

$$1=\int_{-\infty}^{\infty} \Psi^*(x)\Psi(x)\,dx \tag{5-38}$$

or

$$1=A^*A\int_0^L \sin^2\left(\frac{n\pi x}{L}\right)dx$$

$$=|A|^2(L/2),$$

or

$$A=\sqrt{\frac{2}{L}}\,. \tag{5-39}$$

Thus, the wavefunction has now been completely determined, and is given by

$$\Psi_n(x)=\sqrt{\frac{2}{L}}\sin\left(\frac{n\pi x}{L}\right). \tag{5-40}$$

Now let us investigate the implications of the discontinuities in the potential that are present at $x = 0$ and $x = L$. At $x = 0$, the value of $\Psi'(x)$ is easily seen to be

$$\Psi'(x)|_{x=0}=kA, \tag{5-41}$$

when the eigenfunction for the region inside the box is used. However, $\Psi'(x) \equiv 0$ for all $x < 0$, which implies that a discontinuity in $\psi'(x)$ exists at $x = 0$. A similar comment applies to $\psi'(x)$ at $x = L$. Thus, the discontinuity in the potential has given rise to a wavefunction whose derivative is only piece-wise continuous, and we need to investigate the implications of this property further. In particular, since $d\Psi(x)/dx$ is usually required to be continuous at all points to assure the correct behavior of the kinetic energy operator at all points [i.e., $(-\hbar 2/2m)(d^2/dx^2)$], we must now determine the contribution to the expectation value of the kinetic energy from the points $x = 0$ and $x = L$ where discontinuities occur.

A detailed treatment of this contribution from points at which discontinuities in $d\Psi(x)/dx$ occur is not appropriate here. However, such an analysis has been carried out by Hirschfelder and Nazaroff,[22] and the results of their analysis are of direct utility. They found that for a one-dimensional problem having a discontinuity at the point $x = x_0$, the contribution to the expectation value of the kinetic energy $[\delta \mathrm{T}(x_0)]$ is given by

$$\delta\bar{T}(x_0)=-\left(\frac{\hbar^2}{2m}\right)\left\{\Psi^*(x)\left[\frac{d\Psi_+(x)}{dx}-\frac{d\Psi_-(x)}{dx}\right]\right\}_{x=x_0}, \tag{5-42}$$

[22] J.O. Hirschfelder and G.V. Nazaroff, *J. Chem. Phys.*, **34**, 1666 (1961).

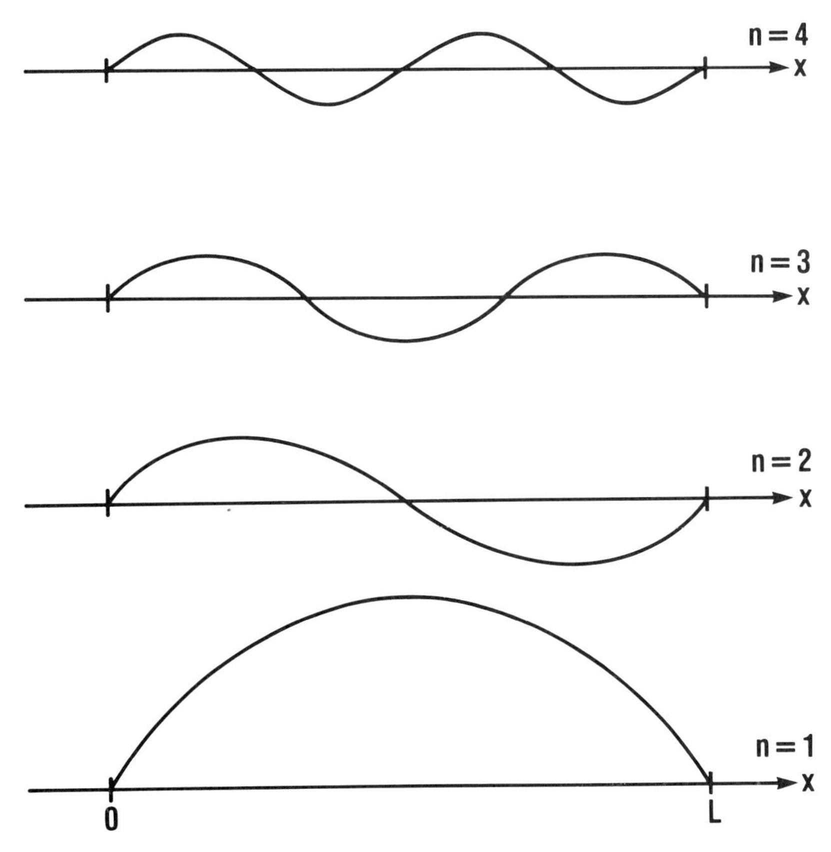

Figure 5.4. A qualitative plot of the eigenfunctions of a "particle" in a one-dimensional "box" for $n = 1, 2, 3$, and 4.

where Ψ_+ and Ψ_- are the eigenfunctions to the right and left of x_0, respectively. For the case of interest here, we note that $\Psi(x) = 0$ for $x = 0$ and $x = L$. Thus, we see from Eq. (5-42) that there is *zero* contribution to the expectation value of the kinetic energy from these points, even though we have not been able to require continuity of $\Psi'(x)$ at these points.

Figure (5.4) indicates the behavior of the wavefunctions [Eq. (5-40)] for the first several positive values of n. As in the case of the previous one-dimensional example, we note the increase in nodes as n increases (and consequently E increases). However, an increase in nodes does not necessarily change the values of observables. For example, let us compute the expectation value of x for this one-dimensional particle in the nth state. We have

$$X_n = \langle x \rangle_n = \int_{-\infty}^{+\infty} \Psi_n^*(x) x \Psi_n(x)\, dx$$

$$= \left(\frac{2}{L}\right) \int_0^L x \sin^2\left(\frac{n\pi x}{L}\right) dx,$$

or

$$\langle x \rangle_n = \frac{L}{2}. \tag{5-43}$$

Thus, a measurement of the position of the particle will result in an average value of $L/2$, *regardless* of the state in which the particle resides. However, it should be noted that Eq. (5-43) represents an *expectation value*, i.e., it is the value that would be obtained if the values obtained in a large number of individual measurements were averaged. Although this value is independent of n, it does not imply that the results of a *single* measurement will be independent of n. To see this explicitly, let us calculate the *most probable value* of x for the nth state, which is the value that would be expected if only a single measurement were made.

The probability of finding the particle in the nth state at a point x, within x and $x + dx$, is given by[23]

$$P_n(x)\, dx = \Psi_n^*(x)\Psi_n(x)\, dx. \tag{5-44}$$

Hence, to find the most probable value of x for a single measurement, we need to obtain the maximum values of $P_n(x)$. For $n = 1$, it is easily seen that the maximum in $P_1(x)$ occurs at $x = L/2$, in agreement with the average value of x given in Eq. (5-43). However, for the first excited state ($n = 2$), the maxima occur at $x = L/4$ and $3L/4$, thus emphasizing the difference between the most probable value of an observable and the average value of that observable.

There is one case where the most probable value and the average value of an observable will always be the same. This occurs when the energy eigenfunctions for the system are also eigenfunctions of the operator corresponding to the observable of interest. For this case, we know from Theorem 4-2 that this can occur only if the operator for the observable of interest commutes with the Hamiltonian (the energy operator). An interesting example, which will be developed in greater length in Chapter 6, is the z-component of angular momentum in atoms and diatomic molecules. For that case, it will be seen that the most probable value and average value are identical, since the Hamiltonian and the z-component of angular momentum commute.

5-6. Parity of Eigenfunctions

In addition to the straightforward calculation of expectation values of various operators using the eigenfunctions of a system, it is frequently possible to simplify the considerations considerably by the use of the concept of the *parity* of the eigenfunctions. To illustrate this concept, let $\Psi_i(x)$ be an eigenvector of a one-dimensional problem corresponding to the eigenvalue E_i. Then, if

$$\Psi_i(x) = +\Psi_i(-x), \tag{5-45}$$

[23] See Chapter 4, Section 4-1.

it is seen that $\Psi_i(x)$ is *symmetric* about the mid-point of the range of x ($x = 0$), and we say that the parity of Ψ_i is *even*. If, on the other hand,

$$\Psi_i(x) = -\Psi_i(-x), \tag{5-46}$$

i.e., $\Psi_i(x)$ is *antisymmetric* about $x = 0$, then the parity of Ψ_i is said to be *odd*. In other words, if Ψ_i has odd parity, then for every x, there is a corresponding point $(-x)$ at which Ψ_i has an exactly equal and opposite value, except for the special case of the mid-point ($x = 0$).

To see how the concept of parity can be utilized, consider two eigenvectors, Ψ_i and Ψ_j, where Ψ_i has even parity and Ψ_j has odd parity. Then, let us consider the integral

$$S_{ij} = \int_{-\infty}^{+\infty} \Psi_i^*(x)\Psi_j(x)\, dx. \tag{5-47}$$

Since $\Psi_i(x)$ has even parity, we know that for all $x \neq 0$, $\Psi(x) = \Psi(-x)$. However, since $\Psi_j(x)$ has odd parity, we know that $\Psi_j(x) = -\Psi_j(-x)$ for all $x \neq 0$. Thus, for every $x > 0$, the full integrand of Eq. (5-47) has a value that is exactly equal in magnitude but opposite in sign to the value of the integrand at $(-x)$. This means that the contribution to the integral for all $x > 0$ is exactly cancelled by the contributions from $x < 0$, and the total value of the integral depends only on the value of $\Psi_i^*(x)\Psi_j(x)$ at $x = 0$. Since Ψ_j has odd parity, we know that $\Psi_j(0) = 0$, which means that

$$S_{ij} = 0, \tag{5-48}$$

and we have proved that Ψ_i and Ψ_j are orthogonal, by using only the parity property of the eigenvectors. This result actually holds more generally than illustrated in this one-dimensional example, and will be useful frequently in later discussions. In particular, if the range of integration is symmetric about the mid-point and the parity of the integrand is odd over this range, then the value of the integral will be zero, since it depends only on the value of the integrand at the mid-point of the range.

The concept of parity of eigenfunctions can be developed even further. Suppose we have moved the potential describing the "box" of Eq. (5-26) along the x axis so that the boundaries occur at $x = \pm L/2$, i.e.,

$$V(x) = \begin{cases} \infty, & -\infty \leq x \leq -\dfrac{L}{2} \\ 0, & -\dfrac{L}{2} < x < +\dfrac{L}{2} \\ \infty, & +\dfrac{L}{2} \leq x \leq +\infty. \end{cases} \tag{5-49}$$

This procedure obviously cannot affect the physics and chemistry to be extracted from examination of this problem. However, it can have an effect on the mathematical manipulations required to obtain the results. In particular, we now have a potential that is symmetric about $x = 0$, i.e.,

$$V(x) = V(-x). \tag{5-50}$$

Let us now examine the implications of this on the Schrödinger equation. For $x \geq 0$, the Schrödinger equation to be solved is

$$-\left(\frac{\hbar^2}{2m}\right)\frac{d^2\Psi(x)}{dx^2} + V(x)\Psi(x) = E\Psi(x). \tag{5-51}$$

For the other range ($x \leq 0$), the Schrödinger equation is

$$-\left(\frac{\hbar^2}{2m}\right)\frac{d^2\Psi(-|x|)}{dx^2} + V(-|x|)\Psi(-|x|) = E\Psi(-|x|). \tag{5-52}$$

However, since the potential is symmetric about $x = 0$, Eq. (5-52) can be rewritten as

$$-\left(\frac{\hbar^2}{2m}\right)\frac{d^2\Psi(-|x|)}{dx^2} + V(x)\Psi(-|x|) = E\Psi(-|x|). \tag{5-53}$$

Hence, the terms (kinetic and potential energy operators) acting on $\Psi(x)$ have precisely the same form for *both* $\pm x$. This implies that the solution $[\Psi(x)]$ must also have the same form on both sides of the origin, except possibly for an arbitrary phase factor (± 1). Thus, it is necessary to search for solutions for (e.g.) only $x \geq 0$, and the full solution for all x can be constructed by taking either symmetric or antisymmetric combinations of such solutions, e.g.,

$$\Psi_{\text{sym}}(x) = \begin{cases} \Psi_{x\geq 0}(-|x|) & x \leq 0 \\ \Psi_{x\geq 0}(x) & x \geq 0, \end{cases} \tag{5-54}$$

with

$$\Psi_{\text{antisym}}(x) = \begin{cases} -\Psi_{x\geq 0}(-|x|) & x \leq 0 \\ \Psi_{x\geq 0}(x) & x \geq 0. \end{cases} \tag{5-55}$$

This ability to distinguish between symmetric and antisymmetric solutions (and, hence, to search for them separately) is frequently useful in simplifying the search for eigenfunctions of the Schrödinger equation. It is a technique that can be applied whenever the potential is symmetric, and we shall see a direct example of the utility of this observation in the next two sections.

5-7. Square Well Potential

Although the previous example is useful in illustrating several of the notions that can be extracted from examination of a one-dimensional model problem, there is one obvious deficiency that needs to be remedied. The deficiency arises from the

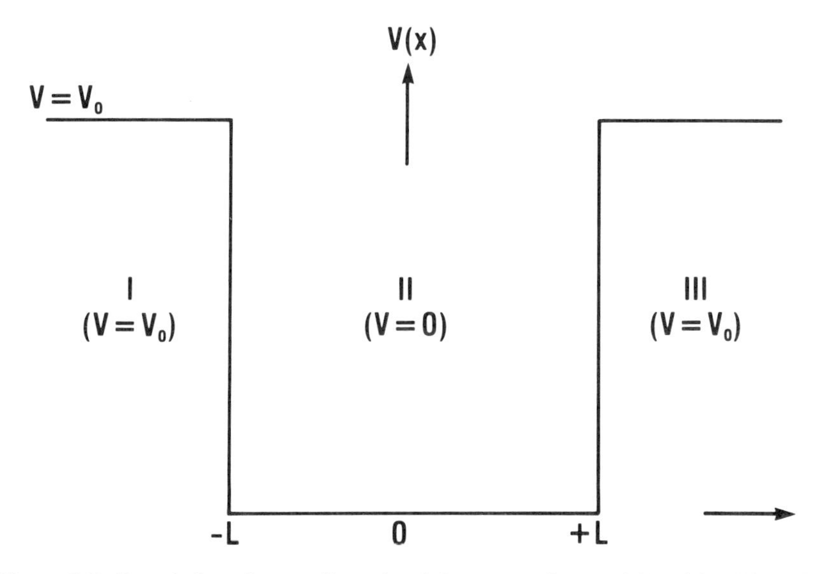

Figure 5.5. Description of a one-dimensional "square well potential" with $V(x) = V_0$ outside the well, and $V(x) = 0$ inside the well.

fact that infinitely high "walls" do not allow for the possibility of the particle to leave the box (e.g., ionization), which is of interest in chemical problems. It is clearly of importance to examine a model problem in which this possibility exists, which we shall now undertake. Such an example is a particle that is influenced by a square well potential, as depicted in Fig. 5.5. The potential $V(x)$ can be represented as

$$V(x)=\begin{cases} 0, & -L<x<+L \\ V_0, & -\infty\leq x\leq -L,\ +L\leq x\leq +\infty, \end{cases} \tag{5-56}$$

where V_0 is a positive constant. As for the other examples in this chapter, we shall be interested in the bound state solutions, which means that we shall be interested only in examining the solutions for which $E < V_0$.

Considering region I (see Fig. 5.5), the Schrödinger equation to be solved is

$$\left(-\frac{\hbar^2}{2m}\right)\frac{d^2\Psi_{\text{I}}(x)}{dx^2}+(V_0-E)\Psi_{\text{I}}(x)=0, \tag{5-57}$$

or

$$\frac{d^2\Psi_{\text{I}}(x)}{dx^2}=k^2\Psi_{\text{I}}(x), \tag{5-58}$$

where

$$k^2=\frac{2m(V_0-E)}{\hbar^2}\ . \tag{5-59}$$

Note that, since $V_0 > E$ by assumption, $k^2 \geq 0$. Equation (5-58) is very similar to the differential equations already encountered except for the sign of k^2. It is easily verified that the general solution of Eq. (5-58) is given by

$$\Psi_{\mathrm{I}}(x) = A_{\mathrm{I}}e^{+kx} + B_{\mathrm{I}}e^{-kx}, \tag{5-60}$$

where A and B are arbitrary constants that need to be determined.

It is also clear from the symmetry of the potential that the solution for region III must have the same form as the solution for region I, i.e.,

$$\Psi_{\mathrm{III}}(x) = A_{\mathrm{III}}e^{+kx} + B_{\mathrm{III}}e^{-kx}, \tag{5-61}$$

where A_{III} and B_{III} are arbitrary constants that need to be determined.

For the center region (region II), the Schrödinger equation to be solved is

$$-\left(\frac{\hbar^2}{2m}\right)\frac{d^2\Psi_{\mathrm{II}}(x)}{dx^2} = E\Psi_{\mathrm{II}}(x), \tag{5-62}$$

or

$$\frac{d^2\Psi_{\mathrm{II}}(x)}{dx^2} = -\kappa^2\Psi_{\mathrm{II}}(x), \tag{5-63}$$

where

$$\kappa^2 = \frac{2mE}{\hbar^2}. \tag{5-64}$$

As before, we note that $\kappa^2 \geq 0$. The differential equation in Eq. (5-63) is identical to the one encountered in Sections 5-4 and 5-5, and its general solution is

$$\Psi_{\mathrm{II}}(x) = A'_{\mathrm{II}}e^{i\kappa x} + B'_{\mathrm{II}}e^{-i\kappa x}, \tag{5-65}$$

or (equivalently),

$$\Psi_{\mathrm{II}}(x) = A_{\mathrm{II}}\cos(\kappa x) + B_{\mathrm{II}}\sin(\kappa x), \tag{5-66}$$

where A_{II} and B_{II} (or A'_{II} and B'_{II}) are arbitrary constants to be determined.

Since the potential is symmetric, we can make use of the observations of the previous section to simplify our considerations. In particular, let us consider only $x \geq 0$, and examine first those solutions of the Schrödinger equation that are symmetric with respect to $x = 0$. This requires that $B_{\mathrm{II}} = 0$ in Eq. (5-66), and the wavefunction to be examined further can be written[24] as

$$^{(s)}\Psi_{x\geq 0}(x) = \begin{cases} A_{\mathrm{II}}\cos(\kappa x), & 0 \leq x < +L \\ B_{\mathrm{III}}e^{-kx}, & +L \leq x \leq +\infty, \end{cases} \tag{5-67}$$

where the choice $A_{\mathrm{III}} = 0$ is required in $\Psi_{\mathrm{III}}(x)$ [Eq. (5-61)] to assure that

$$\Psi_{\mathrm{III}}(x) \to 0 \text{ as } x \to +\infty, \tag{5-68}$$

[24] See Problem 5-2 for determination of A_{II} and B_{III}.

which satisfies the boundary condition associated with $x \to +\infty$. From Eq. (5-67), we also note that

$$^{(s)}\Psi'_{x\geq 0}(x) = \begin{cases} -\kappa A_{\text{II}} \sin(\kappa x) & 0 \leq x < +L \\ -kB_{\text{III}} e^{-kx} & +L \leq x \leq +\infty. \end{cases} \tag{5-69}$$

In order to fix the remaining parameters, boundary conditions at $x = +L$ need to be applied. In particular, we desire that both $\Psi(x)$ and $\Psi'(x)$ be continuous at $x = +L$, i.e.,

$$\begin{aligned} \Psi_{\text{II}}(L) &= \Psi_{\text{III}}(L) \\ \Psi'_{\text{II}}(L) &= \Psi'_{\text{III}}(L). \end{aligned} \tag{5-70}$$

Using Eqs. (5-67) and (5-69), the above equation gives rise to

$$A_{\text{II}} \cos(\kappa L) = B_{\text{III}} e^{-kL}, \tag{5-71}$$

and

$$\kappa A_{\text{II}} \sin(\kappa L) = kB_{\text{III}} e^{-kL}. \tag{5-72}$$

Note that due to the symmetry in the potential, the boundary conditions for negative x do not need to be considered explicitly, since their effect on the arbitrary parameters in the wavefunction will be the same as the boundary conditions for $x > 0$. In other words, the boundary conditions for $x < 0$ will not lead to any new information, and need not be considered further.

Assuming that the allowed energy levels of the system are such that

$$\kappa L \neq \frac{m\pi}{2}, \qquad m = 1, 2, \cdots \tag{5-73}$$

we may divide Eq. (5-72) by Eq. (5-71) to give

$$\tan(\kappa L) = k/\kappa. \tag{5-74}$$

This equation obviously cannot be satisfied for all values of κ and k. Hence, the only values of the energy (E) that are allowed are those that satisfy Eq. (5-74), and the application of the boundary conditions has once again led to quantization of the energy levels.

Before considering a specific numerical example, let us return to the general solutions and examine the *antisymmetric* solutions in greater detail. For these solutions, it is necessary to choose $A_{\text{II}} = 0$ in Eq. (5-66), which gives

$$^{(a)}\Psi_{x\geq 0}(x) = \begin{cases} B_{\text{II}} \sin(\kappa x), & 0 \leq x < +L \\ B_{\text{III}} e^{-kx}, & +L \leq x \leq +\infty \end{cases} \tag{5-75}$$

and

$$^{(a)}\Psi'_{x\geq 0}(x) = \begin{cases} \kappa B_{\text{II}} \cos(\kappa x), & 0 \leq x < +L \\ -kB_{\text{III}} e^{-kx}, & +L < x < +\infty. \end{cases} \tag{5-76}$$

Imposing the boundary conditions of Eq. (5-70) on $\Psi(x)$ and $\Psi'(x)$ for $x = +L$

gives rise to

$$B_{\text{II}} \sin(\kappa L) = B_{\text{III}} e^{-kL}, \tag{5-77}$$

and

$$\kappa B_{\text{II}} \cos(\kappa L) = -kB_{\text{III}} e^{-kL}. \tag{5-78}$$

Assuming the allowed energy levels are such that

$$\kappa L \neq n\pi, \quad n = 0, 1, 2, \cdots, \tag{5-79}$$

we may divide Eq. (5-78) by Eq. (5-77) to give

$$\cot(\kappa L) = -k/\kappa, \tag{5-80}$$

which again restricts the energies that are allowed for the system.

In order to see what energies might be allowed from Eqs. (5-74) and (5-80), let us see how well this model potential works as a model of the hydrogen atom. In particular, let us choose the length of the "box" to be the diameter of the first Bohr orbit, the mass to be that of the electron rest mass, and V_0 to be the ground state ionization potential. Thus, in cgs units, the various parameters that are needed have the values

$$\begin{aligned} \hbar &= 1.05459 \times 10^{-27} \text{ erg} \cdot \text{sec} \\ m &= 9.10956 \times 10^{-28} \text{ g} \\ L &= 5.29177 \times 10^{-9} \text{ cm} \\ V_0 &= 13.605 \text{ eV} = 2.17978 \times 10^{-11} \text{ erg}. \end{aligned} \tag{5-81}$$

Examining the symmetric case first, introduction of the definition of x and k into Eq. (5-74) gives

$$\tan\left[\left(\frac{L}{\hbar}\right)\sqrt{2mE}\right] = \sqrt{\frac{V_0 - E}{E}}, \tag{5-82}$$

or

$$\tan(\xi) = \left[\left(\frac{2mL^2V_0}{\hbar^2\xi^2}\right) - 1\right]^{1/2}, \tag{5-83}$$

where

$$\xi^2 = \left(\frac{L^2}{\hbar^2}\right)(2mE). \tag{5-84}$$

Introduction of the particular choice of parameters in Eq. (5-81) allows Eq. (5-83) to be rewritten as

$$\tan(\xi) = \frac{(1 - \xi^2)^{1/2}}{\xi}. \tag{5-85}$$

In order to find the allowed values of the energy, we must find the values of ξ

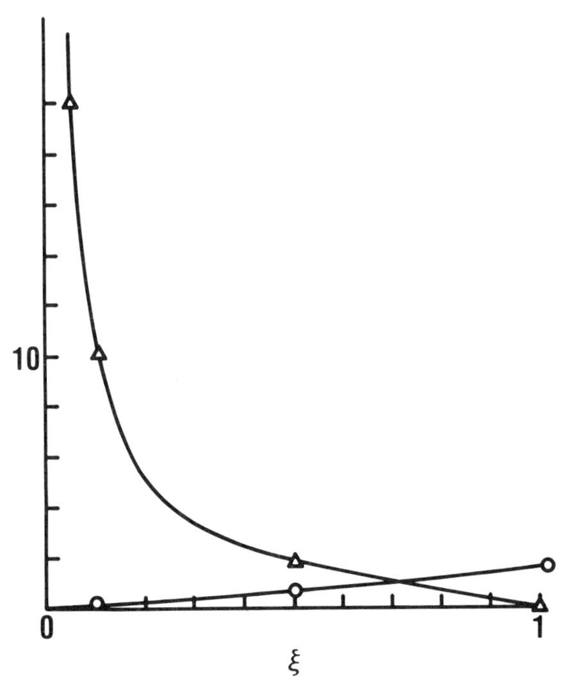

Figure 5.6. Plot of tan(ζ) (circled points) and the right-hand side of Eq. (5-85) (points in triangles) as a function of ξ.

that satisfy Eq. (5-85). In Fig. 5.6 the left- and right-hand sides of Eq. (5-85) have been plotted, and it is seen that the two functions intersect at only one value of $\xi > 0$, i.e., $\xi = 0.739$, which corresponds to

$$E = 1.19 \times 10^{-11} \text{ erg}$$
$$= 7.43 \text{ eV}.$$

A similar plot for the antisymmetric case Eq. (5-80) shows that no allowed energy levels occur for this choice of parameters[25] (i.e., there is no crossing of the $\cot(\xi)$ vs. $-(1 - \xi^2)^{1/2}/\xi$ curve for the range $0 \leq \xi \leq 1$). It should also be noted that there is no need to consider $\xi > 1$, since these values correspond to energies greater than V_0, and the particle would not be bound.

Thus, this simplified model of the hydrogen atom produces only one bound state with an energy of 7.43 eV. Since the top of the well is located at 13.605 eV, this means that the ionization potential of the hydrogen atom is predicted to be 6.17 eV from this model potential, substantially smaller than observed experimentally (13.605 eV). Another deficiency that this simple model of the hydrogen atom has is that only one bound state is predicted. Since it is well

[25] See D. Bohm, *Quantum Theory*, Prentice-Hall, Englewood Cliffs, NJ, 1957, pp. 247–251, for a discussion of the allowed energy levels for other choices of V_0.

known that there are an infinite number of bound states for the hydrogen atom, we see that a description of excited states from this potential is entirely absent.[26] However, we shall be able to utilize the ground state information further in the next example.

Before leaving this example, it is of interest to observe the behavior of the energy levels as the depth of the well is increased. As $V_0 \to \infty$, we know that the probability of finding the particle outside the box must go to zero.

Considering the symmetric case for convenience, we note that $V_0 \to \infty$ corresponds to $k \to \infty$. From Eq. (5-71), we see that the requirement of continuity of $\Psi(x)$ at $x = L$ leads in this case to

$$\cos(\kappa L) = 0$$

or

$$2\kappa L = n\pi, \qquad n = 1, 2, \cdots,$$

or

$$E = \frac{n^2 h^2}{8m(4L^2)}$$

which is the same result that was obtained earlier in Section 5-5 [See Eq. (5-37)], if we take into account that the "box" under consideration here is twice as large as the one represented in Section 5-5.

To illustrate how the eigenfunctions are modified by the introduction of finite walls, Fig. 5.7 shows a plot of the ground state eigenfunction of the particle in a box with infinite walls, as well as a plot of the one bound state eigenfunction (the symmetric one), using the parameters of Eq. (5-81) in both cases.

5-8. Double Wells and Tunneling

In addition to the quantization of possible energy levels, the example of the previous section can be seen to exhibit another major departure from the results that would be obtained by treating the problem as one in a classical mechanics. In particular, we see from Eqs. (5-67) and (5-75) that the probability of finding the particle *outside* the region $-L \leq X \leq +L$ is *not* zero, even though the energy of the "particle" treated in a classical interpretation is below that needed to escape from the box. This implies that we might expect to find a drastic departure from the expected classical results, if a suitable experimental arrangement could be devised. In Fig. 5.8 such a situation is depicted.[27] Since

[26] Of course, other choices for V_0 will give rise to different numbers of bound states.

[27] Examination of this problem for cases of interest to applications in physics has been carried out by P. A. Deutchman, *Am. J. Phys.*, **39**, 952 (1971).

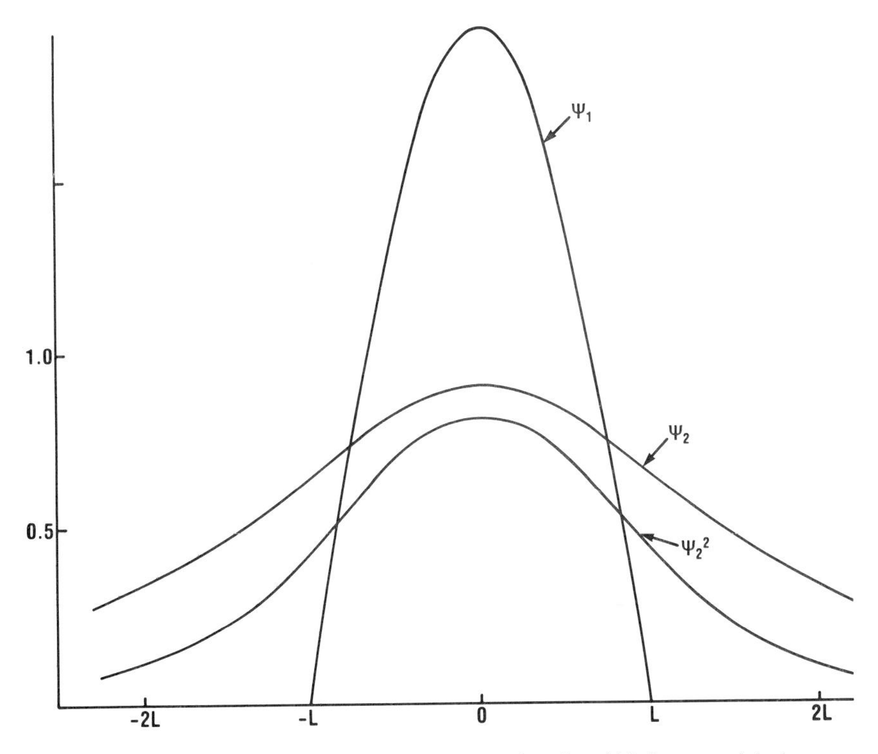

Figure 5.7. Comparison of the ground state wavefunction (ψ_1) for a particle in a one-dimensional box having infinite walls with the corresponding wavefunction (ψ_2) for a box having finite walls.

there are now two "boxes," and since there is a finite probability of finding the particle between the "boxes", it should be possible for the particle to "tunnel" from one box to the other, even if it does *not* possess sufficient energy to get out of the box in which it initially resides. Let us now determine the solutions to this model problem, to see how these notions become quantified.

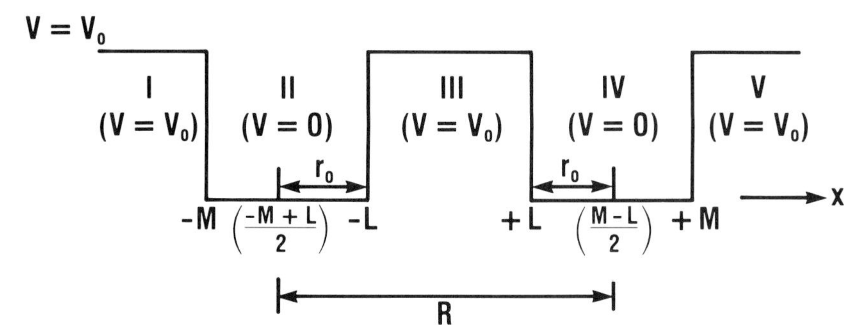

Figure 5.8. A one-dimensional double well potential, with a well depth of V_0. ($V_0 > 0$).

By comparison to the previous example, it is easily seen that the general solutions for the various regions are

$$\Psi(x)=\begin{cases} A_{\mathrm{I}}e^{+kx}+B_{\mathrm{I}}e^{-kx} & -\infty\leq x\leq -M \\ A_{\mathrm{II}}\cos(\kappa x)+B_{\mathrm{II}}\sin(\kappa x) & -M<x\leq -L \\ A_{\mathrm{III}}e^{+kx}+B_{\mathrm{III}}e^{-kx}, & -L<x<+L \\ A_{\mathrm{IV}}\cos(\kappa x)+B_{\mathrm{IV}}\sin(\kappa x) & +L\leq x<+M \\ A_{\mathrm{V}}e^{+kx}+B_{\mathrm{V}}e^{-kx} & +M\leq x\leq +\infty, \end{cases} \tag{5-86}$$

where

$$k^2=\frac{2m(V_0-E)}{\hbar^2} \tag{5-87}$$

and

$$\kappa^2=\frac{2mE}{\hbar^2}\ . \tag{5-88}$$

Since the potential is symmetric about the origin, as in the previous example, it is necessary to consider only solutions for $x \geq 0$, and the symmetric and antisymmetric solutions may be considered separately.

To obtain solutions which are symmetric about the origin, it is necessary that $A_{\mathrm{III}} = B_{\mathrm{III}}$ in Eq. (5-86). Also, in order to satisfy the boundary condition

$$\Psi(x)\rightarrow 0 \text{ as } x\rightarrow \pm\infty, \tag{5-89}$$

it is necessary to choose $A_{\mathrm{V}} = 0$. This gives

$${}^{(s)}\Psi_{x\geq 0}(x)=\begin{cases} A_{\mathrm{III}}(e^{+kx}+e^{-kx}) & 0\leq x<+L \\ A_{\mathrm{IV}}\cos(\kappa x)+B_{\mathrm{IV}}\sin(\kappa x) & +L\leq x<+M \\ B_{\mathrm{V}}e^{-kx} & +M\leq x\leq +\infty. \end{cases} \tag{5-90}$$

The derivative of $\Psi(x)$ is also needed, and is easily seen to be

$${}^{(s)}\Psi_{x\geq 0}(x)=\begin{cases} kA_{\mathrm{III}}(e^{+kx}-e^{-kx}) & 0\leq x<+L \\ \kappa[-A_{\mathrm{IV}}\sin(\kappa x)+B_{\mathrm{IV}}\cos(\kappa x)] & +L\leq x<+M \\ -kB_{\mathrm{V}}e^{-kx} & +M\leq x\leq +\infty. \end{cases} \tag{5-91}$$

To determine the arbitrary constants in the wavefunction, it is necessary to apply boundary conditions at $x = +L$ and $x = +M$, which are that $\Psi(X)$ and $\Psi'(x)$

must be continuous at those points, i.e., we require that

$$\Psi_{\mathrm{III}}(L) = \Psi_{\mathrm{IV}}(L)$$
$$\Psi'_{\mathrm{III}}(L) = \Psi'_{\mathrm{IV}}(L), \tag{5-92}$$

and a corresponding set of conditions on Ψ and Ψ' at $X = M$.

Continuity of the wavefunction and its first derivative $X = +L$ gives rise to

$$A_{\mathrm{III}}(e^{+kL} + e^{-kL}) = A_{\mathrm{IV}} \cos(\kappa L) + B_{\mathrm{IV}} \sin(\kappa L) \tag{5-93}$$

and

$$kA_{\mathrm{III}}(e^{+kL} - e^{-kL}) = \kappa[-A_{\mathrm{IV}} \sin(\kappa L) + B_{\mathrm{IV}} \cos(\kappa L)]. \tag{5-94}$$

These two expressions can be solved for the constant A_{III}, and subsequently combined to give the following relationship between A_{IV} and B_{IV} that must be satisfied if the boundary conditions at $X = L$ are to be fulfilled:

$$A_{\mathrm{IV}} = -\left\{\frac{k \sin(\kappa L) - \kappa \coth(kL) \cos(\kappa L)}{k \cos(\kappa L) + \kappa \coth(kL) \sin(\kappa L)}\right\} B_{\mathrm{IV}}. \tag{5-95}$$

It is possible to obtain another relationship between A_{IV} and B_{IV} from the continuity conditions on the wavefunction and its derivative at $X = M$, which give rise to

$$A_{\mathrm{IV}} \cos(\kappa M) + B_{\mathrm{IV}} \sin(\kappa M) = B_{\mathrm{V}} e^{-kM}, \tag{5-96}$$

and

$$\kappa[-A_{\mathrm{IV}} \sin(\kappa M) + B_{\mathrm{IV}} \cos(\kappa M)] = -kB_{\mathrm{V}} e^{-kM}. \tag{5-97}$$

Elimination of B_{V} and subsequent combination of Eqs. (5-96) and (5-97) yields

$$A_{\mathrm{IV}} = -\left\{\frac{k \sin(\kappa M) + \kappa \cos(\kappa M)}{k \cos(\kappa M) - \kappa \sin(\kappa M)}\right\} B_{\mathrm{IV}}. \tag{5-98}$$

Thus, Eqs. (5-95) and (5-98) give us two expressions for the relationship between the constants A_{IV} and B_{IV} that must be satisfied simultaneously, if the boundary conditions at both $X = L$ and $X = M$ are to be satisfied. This can be achieved only if

$$\frac{k \sin(\kappa L) - \kappa \coth(\kappa L) \cos(\kappa L)}{k \cos(\kappa L) + \kappa \coth(\kappa L) \sin(\kappa \lambda)}$$
$$= \frac{k \sin(\kappa M) + \kappa \cos(\kappa M)}{k \cos(\kappa M) - \kappa \sin(\kappa M)}. \tag{5-99}$$

By rearranging the above equation and making use of multiple angle formulas, it is possible to rewrite it in the following form, which is better suited for further examination:

$$\cot[\kappa(L-M)] = \frac{k^2 - \kappa^2 \coth(kL)}{k\kappa[1 + \coth(\mathrm{kL})]} \tag{5-100}$$

Introduction of the definition of k and κ and subsequent rearrangement gives

$$\cot\left[\left(\frac{L-M}{\hbar}\right)\sqrt{2mE}\right]=\frac{-1}{\sqrt{E(V_0-E)}}\cdot\left[E-\frac{V_0}{[1+\coth\{(L/\hbar)\sqrt{2m(V_0-E)}\}]}\right]. \tag{5-101}$$

Hence, the allowed energy levels (for the symmetric eigenfunctions) are those that satisfy the above equation. We shall return to this equation shortly.

For the eigenfunctions that are *antisymmetric* about $\chi = 0$, it is necessary that $A_{\mathrm{III}} = -B_{\mathrm{III}}$. Also, choosing $A_V = 0$, we have

$$\begin{array}{ll} A_{\mathrm{III}}(e^{+kx}-e^{-kx}) & 0\leq x<+L \\ A_{\mathrm{IV}}\cos(\kappa x)+B_{\mathrm{IV}}\sin(\kappa x) & +L\leq x<+M \\ B_{\mathrm{V}}e^{-kx} & +M\leq x\leq+\infty \end{array} \tag{5-102}$$

and

$$\Psi'_{x\geq0}(x)=\begin{cases} kA_{\mathrm{III}}(e^{+kx}-e^{-kx}) & 0\leq x<+L \\ \kappa[-A_{\mathrm{IV}}\sin(\kappa x)+B_{\mathrm{IV}}\cos(\kappa x)] & +L\leq x<+M \\ -kB_{\mathrm{V}}e^{-kx} & +M\leq x\leq+\infty. \end{cases} \tag{5-103}$$

Following a procedure similar to that for the symmetric solutions, it can be seen that the continuity conditions at $X = +L$ require that

$$A_{\mathrm{IV}}=-\left\{\frac{k\sin(\kappa L)-\kappa\tanh(kL)\cos(\kappa L)}{k\cos(\kappa L)+\kappa\tanh(kL)\sin(\kappa L)}\right\}B_{\mathrm{IV}} \tag{5-104}$$

Similarly, the continuity conditions at $X = +M$ give rise to

$$A_{\mathrm{IV}}=-\left\{\frac{k\sin(\kappa M)+\kappa\cos(\kappa M)}{k\cos(\kappa M)-\kappa\sin(\kappa M)}\right\}B_{\mathrm{IV}}, \tag{5-105}$$

and the above two equations can be satisfied only if

$$\frac{k\sin(\kappa L)-\kappa\tanh(kL)\cos(\kappa L)}{k\cos(\kappa L)+\kappa\tanh(kL)\sin(\kappa L)}=\frac{k\sin(\kappa M)+\kappa\cos(\kappa M)}{k\cos(\kappa M)-\kappa\sin(\kappa M)}. \tag{5-106}$$

Simplification of the above equation leads to

$$\cot[\kappa(L-M)]=\frac{k^2-\kappa^2\tanh(kL)}{k\kappa[1+\tanh(kL)]}, \tag{5-107}$$

and introduction of the definition of k and κ gives

$$\cot\left[\left(\frac{L-M}{\hbar}\right)\sqrt{2mE}\right]=\frac{-1}{\sqrt{E(V_0-E)}}\cdot\left\{E-\frac{V_0}{\left[1+\tanh\left\{\left(\frac{L}{\hbar}\right)\sqrt{2m(V_0-E)}\right\}\right]}\right\}. \tag{5-108}$$

Now let us see how the allowed energy levels of our model problem [that satisfy Eqs. (5-101) and (5-108)] can be used to investigate the notions of tunneling. Continuing the analogy to the hydrogen atom that was begun in the previous section, let us choose an electron as the "particle," and the other parameters (see Fig. 5-8) as in the previous example of a single well, i.e.,

$$\begin{aligned}\hbar&=1.05459\times10^{-27}\ \text{erg}\cdot\text{sec}\\ m&=9.109558\times10^{-28}\ \text{g}\\ r_0&=5.29117715\times10^{-9}\ \text{cm}\\ V_0&=13.605\ \text{eV}=2.17978\times10^{-11}\ \text{erg},\end{aligned} \tag{5-109}$$

and investigate what happens as the two wells are brought together. From Fig. 5.8, it is easily seen that this corresponds to investigation of the energy levels as a function of R, with

$$\begin{aligned}L&=\frac{R}{2}-r_0\\ M&=\frac{R}{2}+r_0.\end{aligned} \tag{5-110}$$

For a suitable choice of R, this potential can be thought of as a simple model of the H_2^+ molecule.

As a first example, let us consider $R = 10.0$ Å, which corresponds to a large separation between the boxes. In Fig. 5.9, plots of cot(x) are given with

$$\kappa=\left(\frac{L-M}{\hbar}\right)\sqrt{2mE},$$

as well as the right-hand side of Eqs. (5-101) and (5-108) as a function of E. For this particular choice of R, the right-hand sides of Eqs. (5-101) and (5-108) give rise to identical curves, and the symmetric and antisymmetric energy levels are degenerate. The particular value of E that results is seen to be $E \cong 7.4319$ eV, which is identical to the value found for the single box. This means that the potential that acts on a particle in either of the boxes is exactly the same as it would be if the other box were not there.

It also implies that tunneling will not be likely to take place. To see this, let us

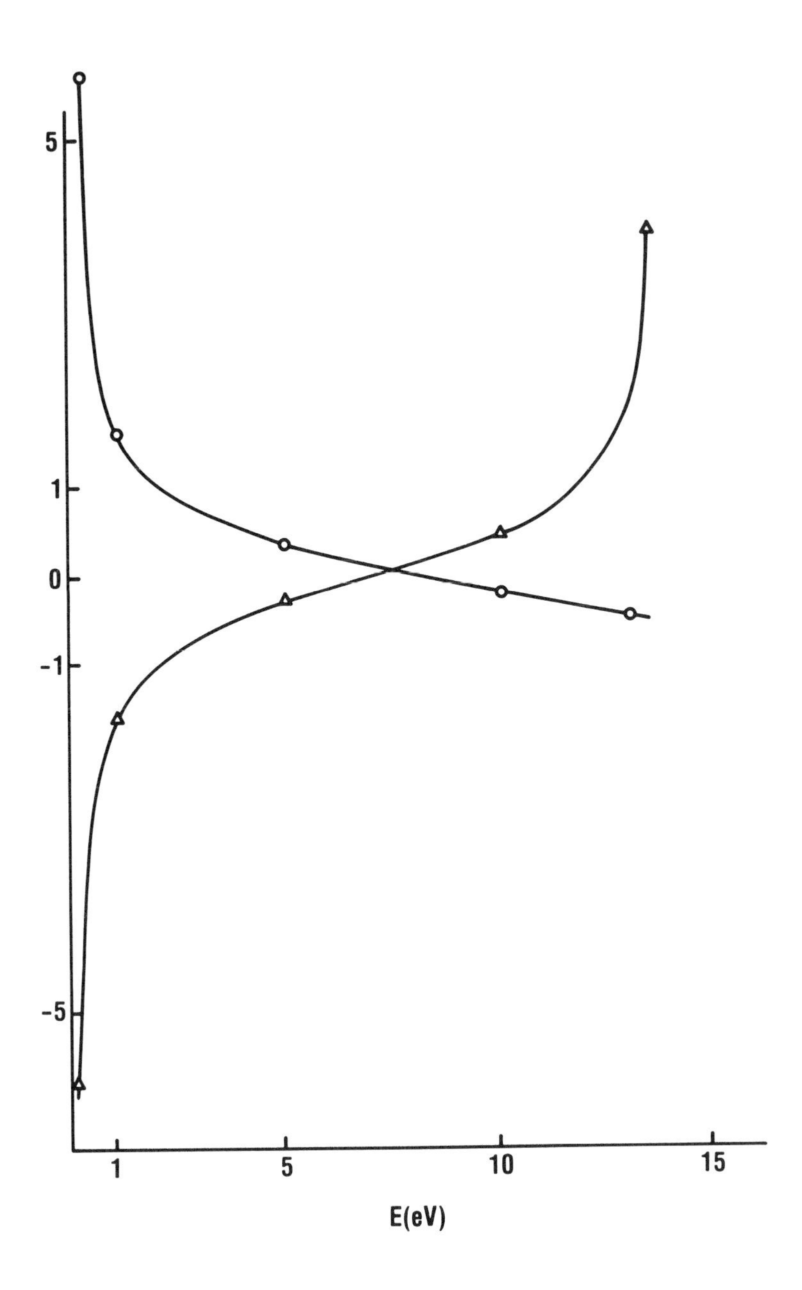

Figure 5.9. Plot of $\cot\{[(L - M)/\hbar]\sqrt{2mE}\}$ (points in circles), the right hand side of Eq. (5-101) (points in triangles), and the right hand side of Eq. (5-108) (points in squares) versus E, with $R = 10.0$ Å. For this particular choice of R, the points in triangles and the points in circles give rise to identical curves.

Table 5.1. Allowed Energy Levels for a Double Square Well Potential as a Function of the Distance between the Wells

R(Å)	Barrier width $2L$ (Å)	E_{sym} (eV)	$E_{antisym.}$ (eV)	$\Delta E = E_{sym} - E_{antisym}$ (eV)	Inversion frequency $\nu(sec^{-1})$	Inversion period $(1/\nu)$ (sec)
10.000	8.942	7.4319	7.4319	0.0000	0	∞
2.000	0.9416	7.4319	7.4319	0.0500	0	∞
1.500	0.4416	7.4174	7.4466	0.0292	7.060×10^{12}	0.1416×12^{-12}
1.200	0.1416	6.8194	8.1876	1.3682	3.308×10^{14}	0.3023×10^{-14}
1.100	0.0416	5.322	10.422	5.100	1.233×10^{15}	0.8110×10^{-15}

assume that the symmetric and antisymmetric curves are *exactly* identical.[28] In that extreme case, $E = E_{antisym.} - E_{sym.} = 0$. If we interpret this in a "semi-classical" manner, then the frequency of tunneling, i.e., the number of times per second that the "particle" placed in one of these boxes tunnels from one box into the other box is found from the DeBroglie relations

$$\nu(\sec^{-1}) = \frac{\Delta E}{h}, \tag{5-111}$$

which is zero for $R = 10.0$ Å. In other words, the particle will act as a "classical" particle in this case, and will stay in whichever box it is placed, and will not "tunnel" into the other box.

Other values of R are considered in Table 5.1. As we see from the table, the energy levels remain degenerate until the distance between the center of the boxes (R) becomes smaller than approximately 2.0 Å. At points closer than this (see Fig. 5.10 for the plots corresponding to $R = 1.10$ Å), a very interesting phenomenon occurs that is completely absent in a classical description of this system. At these distances the energy levels no longer remain degenerate, but the symmetric level is lowered and the antisymmetric level is raised from the value when the boxes are separated by large distances. Also, the energy level separation increases as R decreases. This means again using the "semiclassical" picture of a particle that "tunnels" between boxes, the frequency of tunneling is no longer zero, and increases as R becomes smaller. This also implies that the inversion period (i.e., the time it would take a "classical particle" to go from one box to the other and back again) decreases as R decreases.

Since the observed inversion frequency for a molecule like NH_3, which can be considered to invert approximately as shown in Fig. 5.11, is found[29] to be 0.7934 cm^{-1} (2.279×10^{10} sec^{-1}) (which corresponds to a period of

[28] This implies that the two curves should be the same to whatever numerical precision is being considered "exact" for this problem. For the purpose of this discussion, the curves will be assumed to be identical if they agree to five significant figures.

[29] C. C. Costain, *Phys. Rev.*, **82**, 108 (1951).

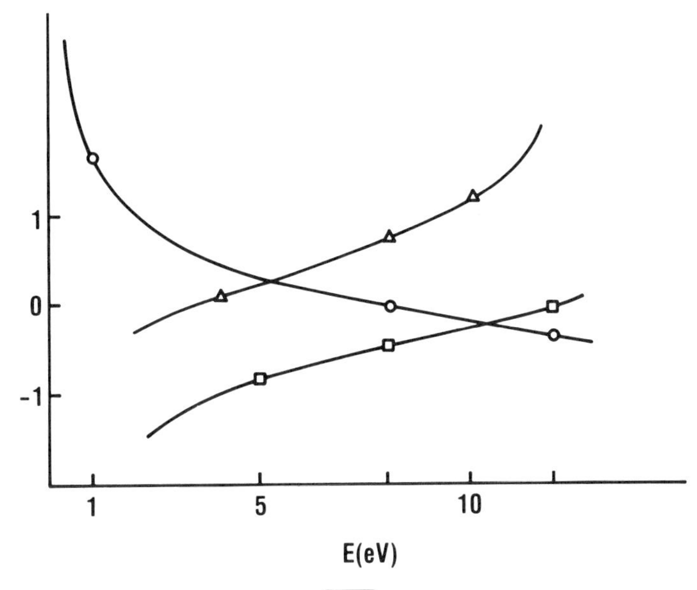

Figure 5.10. Plot of $\cot\{[(L - M)/\hbar]\sqrt{2mE}\}$ (points in circles), the right-hand side of Eq. (5-111) (points in triangles), and the right-hand side of Eq. (5-108) (points in squares) versus E, with $R = 1.10$ Å.

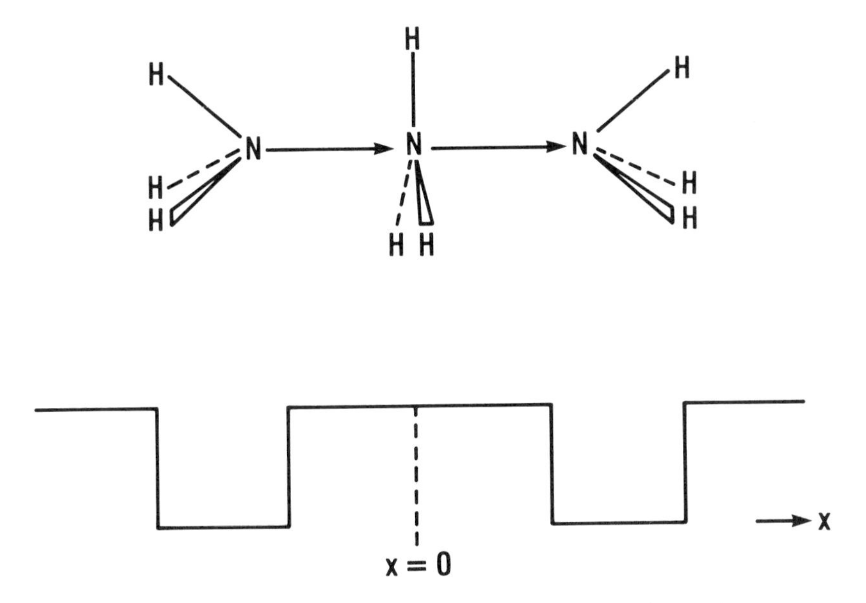

Figure 5.11. Schematic representation of the NH_3 inversion motion.

4.203×10^{-11} sec), we see that the tunneling frequency of the particle between the two square wells is greater than the inversion frequency of NH_3 by the point at which $R < 2.00$ Å. Of perhaps greater interest in the behavior of the particle when $R \cong 1.10$ Å, which is approaching the equilibrium internuclear distance observed for H_2^+ (1.058 Å). Using this double well model of H_2^+, we see that the "electron" (the one-dimensional particle whose mass is that of the electron rest mass) acts effectively as if the barrier were not there at all, having a frequency of tunneling that is five orders of magnitude larger than observed in NH_3. This simply emphasizes further the ease of motion of electrons compared to nuclei (with mass even as small as hydrogen). This delocalized nature of electrons within molecular environments to encompass more than one nucleus will play an important role in later discussions of chemical bonding and molecular orbital theory. Regardless of the adequacy of treating the electron within the H_2^+ molecule as an electron that tunnels between two potential wells, this section has brought out several important general points that will be useful in discussing the quantum behavior of real systems.

First, particles whose mass ranges from that of the electron to that of light nuclei can undergo tunneling under appropriate circumstances, which is a phenomenon that would be entirely absent in any classical description of the same system. The second point of general interest concerns the splitting of energy levels when the two boxes were brought close together. We shall observe this phenomenon frequently in later discussions, and within the context of many different examples. The general principle exhibited in these cases is that, whenever systems whose eigenfunctions possess nearly the same energy levels are allowed to interact,[30] a new system of eigenfunctions will be formed that are a linear combination of the previous ones, and whose eigenvalues will no longer be degenerate. In general half of the eigenvalues will be above the original degenerate eigenvalue, and half will be below.[31]

5-9. The Harmonic Oscillator

As a final example of a one-dimensional model problem that finds great utility in both qualitative and quantitative discussions of molecular quantum mechanics, let us examine the motion of a one-dimensional "particle" under the influence of a potential $[V(x)]$ that is parabolic, i.e.,

$$V(x) = \frac{k}{2}(x - x_0)^2. \tag{5-112}$$

[30] In the absence of external fields, this will occur whenever the degeneracy is "accidental," i.e., is not required by the symmetry of the system.

[31] If an odd number of accidentally degenerate eigenfunctions interact, some eigenvalues will remain unchanged from the original value.

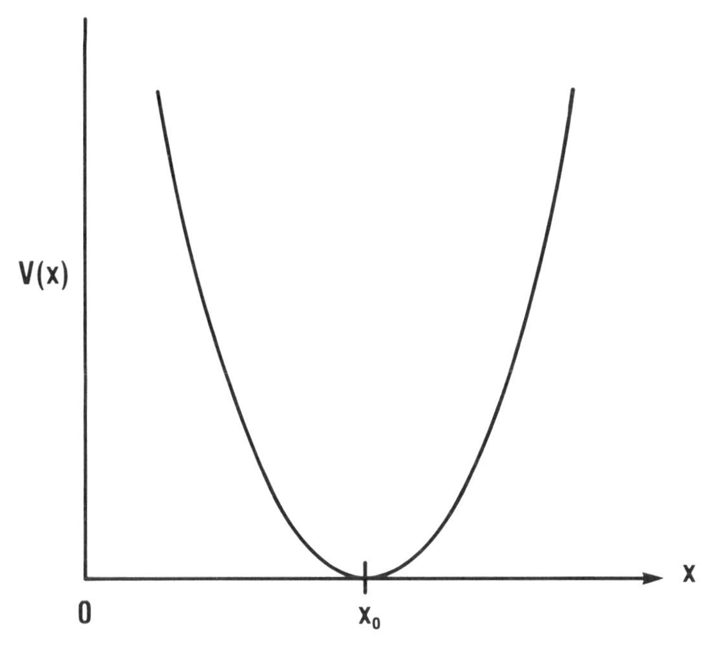

Figure 5.12. Harmonic oscillator potential.

Such a potential is depicted pictorially in Fig. 5.12. This type of potential is familiar in classical mechanics,[32] and corresponds to assuming that the force acting to restore the particle to its equilibrium position is proportional to the distance from equilibrium, i.e.,

$$F(x) = -\frac{dV(x)}{dx} = -k(x - x_0), \tag{5-113}$$

where k is known as the *force constant*, and x_0 is the equilibrium position. For the classical case, the frequency with which the particle oscillates is given by

$$\nu = \sqrt{\frac{k}{m}}, \tag{5-114}$$

where m is the mass of the particle. One of the interesting aspects of the following discussion will be to determine in what manner the description of the quantum mechanical "particle" differs from its corresponding classical analog.

Also, in keeping with the observation that matrix methods are of great importance in contemporary quantum chemistry, this example will be solved using matrix methods, as opposed to developing techniques for the solution of

[32] See, for example, W. Hauser, *Introduction to The Principles of Mechanics*, Addison-Wesley, Reading, MA, 1965, Section 4-9, p. 99.

differential equations that will be of limited utility later. Of course, due to isomorphism between operators and their matrix representations,[33] the results will be the same using either approach. However, to gain familiarity with matrix manipulations, as well as to provide results for later comparisons using approximations to the exact solutions (Chapter 9), the use of matrix techniques is particularly appropriate.

To allow analogies between the two approaches to be pointed out as the development proceeds, let us begin by writing the quantum mechanical Hamiltonian in operator form, i.e.,

$$\mathcal{H} = -\left(\frac{\hbar^2}{2m}\right)\frac{d^2}{dx^2} + V(x), \tag{5-115}$$

or

$$\mathcal{H} = -\left(\frac{\hbar^2}{2m}\right)\frac{d^2}{dx^2} + \left(\frac{k}{2}\right)x^2, \tag{5-116}$$

where total derivatives have been employed instead of partial derivations, since the problem is only one-dimensional. The origin has been chosen to coincide with x_0 for convenience, i.e., $x_0 = 0$.

If a complete and orthonormal basis is chosen that suitably spans the space $(-\infty \leq x \leq +\infty)$, then it is possible to form a matrix representation $(\mathbf{H})$ of $\mathcal{H}$ in the basis, whose elements are given by

$$(\mathbf{H})_{rs} = \langle r|\mathcal{H}|s\rangle, \tag{5-117}$$

where r and s are members of the basis. Of course, the size of the basis that is needed to span the space may be infinite,[34] and thus the matrix $\mathbf{H}$ may contain an infinite number of rows and columns. However, for this particular example, we shall see that the actual manipulation of the infinite matrices that arise can be carried out quite conveniently.

The matrix analog of Eq. (5-116) is easily seen to be

$$\mathbf{H} = \left(\frac{1}{2m}\right)\mathbf{P}^2 + \left(\frac{k}{2}\right)\mathbf{X}^2, \tag{5-118}$$

where

$$(\mathbf{P})_{rs} = \left\langle r\left|\left(\frac{\hbar}{i}\right)\frac{d}{dx}\right|s\right\rangle, \tag{5-119}$$

[33] Chapter 3, Section 3-6. (Theorem 3-11).

[34] Also, the basis that is chosen may not necessarily consist only of discrete members, which adds considerably to the complications. For the particular example under consideration, however, an entirely discrete (but infinite) basis can be chosen, and will be assumed in the following discussion.

and

$$(\mathbf{X})_{rs} = \langle r|x|s\rangle. \tag{5-120}$$

An important point to note is that since the momentum operator is not simply multiplicative, its matrix representation may not commute with other matrices. In particular, we note that

$$[\mathbf{X}, \mathbf{P}] = \mathbf{XP} - \mathbf{PX} = i\hbar\mathbf{I}, \tag{5-121}$$

which follows directly from the corresponding operator relation.[35]

In the operator form, the equation to be solved for the stationary states is given by

$$\mathcal{H}\Psi = E\Psi, \tag{5-122}$$

where both the allowed values of the energy eigenvalue (E) and the associated eigenvectors (Ψ) are desired. In matrix form, the equation to be solved is

$$\mathbf{H\Psi} = \mathbf{\Psi E}, \tag{5-123}$$

where $\mathbf{E}$ is a diagonal matrix containing the desired eigenvalues and $\mathbf{\Psi}$ is the matrix of eigenvectors corresponding to the eigenvalues in $\mathbf{E}$. Since the Hamiltonian matrix is Hermitian, we know that the eigenvalues in $\mathbf{E}$ must be real,[36] and that $\mathbf{H}$ can be diagonalized via a unitary transformation,[37] i.e.,

$$\mathbf{U}^\dagger\mathbf{HU} = \boldsymbol{\lambda}, \tag{5-124}$$

where $\boldsymbol{\lambda}$ is diagonal, and

$$\mathbf{U}^\dagger\mathbf{U} = \mathbf{UU}^\dagger = \mathbf{I}. \tag{5-125}$$

Multiplication of Eq. (5-124) on the left by $\mathbf{U}$ and comparison with Eq. (5-123) indicate that the form of the two equations is identical. This means merely that the desired matrix of eigenvectors ($\mathbf{\Psi}$) is unitary, i.e.,

$$\mathbf{\Psi}^\dagger\mathbf{\Psi} = \mathbf{I}, \tag{5-126}$$

and that the Hamiltonian matrix is diagonal in this basis. In the following discussion, let us *assume* that these eigenvectors are known. This does *not* mean that we shall be able to write down the eigenvectors. Rather, it implies only that we know that they *exist*. Since it will *not* be necessary to know them explicitly to carry out the following discussion, the knowledge that they exist will be all that is needed.[38]

---

[35] Chapter 4, Section 4-7.

[36] Chapter 3, Theorem 3-19.

[37] Chapter 3, Theorem 3-18.

[38] This is nothing more than finding the eigenvalues of a matrix, without knowing the eigenvectors explicitly. See Chapter 3, Section 3-8, for an example using a 2 × 2 matrix.

To facilitate the discussion, let us define the following matrix[39]:

$$\mathbf{A}=\left(\frac{i}{\sqrt{2}}\right)\left(\frac{1}{\sqrt{m}}\mathbf{P}-i\sqrt{k}\mathbf{X}\right), \tag{5-127}$$

and thus,

$$\mathbf{A}^{\dagger}=\left(\frac{1}{\sqrt{2}i}\right)\left(\frac{1}{\sqrt{m}}\mathbf{P}^{\dagger}+i\sqrt{k}\mathbf{X}^{\dagger}\right). \tag{5-128}$$

Hence $\mathbf{A}$ is a matrix representation, in some basis, of an operator[40] $\mathcal{A}$, which is the following linear combination of operators:

$$\mathcal{A}=\frac{1}{\sqrt{2}}\left(\frac{1}{\sqrt{m}}P-i\sqrt{k}x\right).$$

From these definitions, it is easily seen that

$$\mathbf{A}\mathbf{A}^{\dagger}=\mathbf{H}+\frac{\hbar}{2}\sqrt{\frac{k}{m}}\,\mathbf{I}, \tag{5-129}$$

and

$$\mathbf{A}^{\dagger}\mathbf{A}=\mathbf{H}-\frac{\hbar}{2}\sqrt{\frac{k}{m}}\,\mathbf{I}, \tag{5-130}$$

and Eq. (5-118) can be rewritten as

$$\mathbf{H}=\frac{1}{2}\left(\mathbf{A}\mathbf{A}^{\dagger}+\mathbf{A}^{\dagger}\mathbf{A}\right). \tag{5-131}$$

For later reference, we also note[41] that

$$[\mathbf{A},\,\mathbf{A}^{\dagger}]=\left(\hbar\sqrt{\frac{k}{m}}\right)\mathbf{I}, \tag{5-132}$$

---

[39] If an analog to the solution of this problem in the operator representation is desired, this step might be thought of an analogous to choosing the correct technique to solve the differential equation. Hence, in either approach, this is the step that is more difficult, and requires the greatest experience and intuition. Also there is an arbitrary phase ($e^{+i\gamma}$) that is omitted from Eqs. (5-121) and (5-122) for convenience. This corresponds to choosing $\gamma = 0$. See G. P. Alldredge, *Am. J. Phys.*, **38**, 1357 (1970) for a discussion of this point.

[40] It is appropriate to point out that the operators $\mathcal{A}$ and $\mathcal{A}^{\dagger}$ (or their matrix representatives $\mathbf{A}$ and $\mathbf{A}^{\dagger}$) are also useful in other theoretical discussions. In particular, these operators play an important role as "*annihilation*" ($\mathcal{A}$) and "*creation*" ($\mathcal{A}^{\dagger}$) *operators* in the quantum theory of the electromagnetic field. For a discussion, see, for example, J. M. Ziman, *Elements of Advanced Quantum Theory*, Cambridge University Press, Cambridge, 1969, Chapter 2.

[41] See Problem 5-8.

$$[\mathbf{A}, \mathbf{H}] = \left(\hbar \sqrt{\frac{k}{m}}\right) \mathbf{A}, \tag{5-133}$$

and

$$[\mathbf{A}^{\dagger}, \mathbf{H}] = -\hbar \sqrt{\frac{k}{m}}\, \mathbf{H}^{\dagger}, \tag{5-134}$$

which can be established easily using the relations found above.

Now let us use this information to find the allowed eigenvalues of the harmonic oscillator. Let $\mathbf{\Psi}_E$ be one of the columns of $\mathbf{\Psi}$ (that we have assumed exists), corresponding to energy E. Then, we consider the following scalar product:

$$(\mathbf{A\Psi}_E)^{\dagger}(\mathbf{A\Psi}_E) = \mathbf{\Psi}_E^{\dagger}\mathbf{A}^{\dagger}\mathbf{A\Psi}_E \tag{5-135}$$

$$= \mathbf{\Psi}_E^{\dagger}\left(\mathbf{H} - \frac{\hbar}{2}\sqrt{\frac{k}{m}}\,\mathbf{I}\right)\mathbf{\Psi}_E$$

$$= \mathbf{\Psi}_E^{\dagger}\left(E - \frac{\hbar}{2}\sqrt{\frac{k}{m}}\right)\mathbf{\Psi}_E \tag{5-136}$$

$$= \left(E - \frac{\hbar}{2}\sqrt{\frac{k}{m}}\right)\mathbf{\Psi}_E^{\dagger}\mathbf{\Psi}_E$$

$$= \left(E - \frac{\hbar}{2}\sqrt{\frac{k}{m}}\right).$$

Since the above is an expression for the scalar product of a matrix $(\mathbf{A\Psi}_E)$ with itself, it follows that[42]

$$E - \left(\frac{\hbar}{2}\right)\sqrt{\frac{k}{m}} \geq 0,$$

or

$$E \geq \left(\frac{\hbar}{2}\right)\sqrt{\frac{k}{m}}. \tag{5-137}$$

This is the first of a series of important results. Namely, we see that the particle *cannot* have zero energy (which would give difficulty with the Uncertainty

[42] Chapter 2, Eq. (2-71a).

Principle), as it can in classical mechanics, but must be greater than or equal to $(\hbar/2)\sqrt{k/m}$.

The matrices $\mathbf{A}$ and $\mathbf{A}^\dagger$ can be used to obtain even more information concerning the allowed energy states of the system. Choosing one particular vector of $\mathbf{\Psi}$ again, i.e., $\mathbf{\Psi}_E$, application of $\mathbf{A}$ to the left of both sides of Eq. (5-123) gives

$$\mathbf{AH\Psi}_E = \mathbf{A\Psi}_E E. \tag{5-138}$$

From Eq. (5-133), this equation can be rewritten as

$$\mathbf{H}(\mathbf{A\Psi}_E) = \left(E - \hbar\sqrt{\frac{k}{m}}\right)(\mathbf{A\Psi}_E). \tag{5-139}$$

Thus, we see that $(\mathbf{A\Psi}_E)$ is *also* an eigenfunction of $\mathbf{H}$ (as well as $\mathbf{\Psi}_E$), but its eigenvalue is given by $(E - \hbar\sqrt{k/m})$, instead of E. In other words, the application of $\mathbf{A}$ onto $\mathbf{\Psi}_E$ produces a *new* eigenfunction $(\mathbf{A\Psi}_E)$ of $\mathbf{H}$, whose eigenvalue is lowered by $\hbar\sqrt{k/m}$ from the original eigenvalue, E. Consequently, it is customary to refer to the operator $\mathcal{A}$ (or its matrix representation $\mathbf{A}$) as a *lowering operator*.

If we carry out this procedure again, we obtain

$$\mathbf{AAH\Psi}_E = (\mathbf{AA\Psi}_E)\mathbf{E}, \tag{5-140}$$

or

$$\mathbf{A}\left(\mathbf{HA} + \hbar\sqrt{\frac{k}{m}}\,\mathbf{A}\right)\mathbf{\Psi}_E = \mathbf{AA\Psi}_E\mathbf{E},$$

which can be written as

$$\mathbf{AHA\Psi}_E = \mathbf{AA\Psi}_E\left(E - \hbar\sqrt{\frac{k}{m}}\right),$$

or

$$\left(\mathbf{HA} + \hbar\sqrt{\frac{k}{m}}\,\mathbf{A}\right)\mathbf{A\Psi}_E = \mathbf{AA\Psi}_E\left(E - \hbar\sqrt{\frac{k}{m}}\right),$$

which gives finally

$$\mathbf{H}(\mathbf{AA\Psi}_E) = \left(E - 2\hbar\sqrt{\frac{k}{m}}\right)(\mathbf{AA\Psi}_E). \tag{5-141}$$

Hence $\mathbf{A}^2\mathbf{\Psi}_E$ is also an eigenfunction of $\mathbf{H}$ with eigenvalue $(E - 2\hbar\sqrt{k/m})$. From these examples, it is easily seen that, in general, $(\mathbf{A}^n\mathbf{\Psi}_E)$ is an eigenfunction of $\mathbf{H}$, with eigenvalue $(E - n\hbar\sqrt{k/m})$. However, we know that this process cannot be carried out indefinitely, since a contradiction to Eq. (5-137) would ultimately occur. Let us suppose that

$$\mathbf{\Psi}_0 = \mathbf{\Psi}_{E-n} = \mathbf{A}^n \mathbf{\Psi}_E \tag{5-142}$$

is the last eigenvector whose eigenvalue does not violate Eq. (5-137). Then, we require that

$$\mathbf{A}\mathbf{\Psi}_0 = \mathbf{0}, \tag{5-143}$$

which also implies that

$$\mathbf{A}^\dagger \mathbf{A}\mathbf{\Psi}_0 = \mathbf{0}. \tag{5-144}$$

But, using Eq. (5-130), the above equation can be written as

$$\left(\mathbf{H} - \frac{\hbar}{2}\sqrt{\frac{k}{m}}\,\mathbf{I}\right)\mathbf{\Psi}_0 = \mathbf{0},$$

or

$$\mathbf{H}\mathbf{\Psi}_0 = \left(\frac{\hbar}{2}\sqrt{\frac{k}{m}}\right)\mathbf{\Psi}_0. \tag{5-145}$$

Hence, we arrive at the important result that the *lowest* allowed energy level of the quantum mechanical harmonic oscillator is

$$E_0 = \frac{\hbar}{2}\sqrt{\frac{k}{m}}, \tag{5-146}$$

and that, from the previous discussion, all of the allowed energy eigenvalues can be described by the equation

$$E_n = \left(n + \frac{1}{2}\right)\hbar\sqrt{\frac{k}{m}}, \qquad n = 0, 1, 2, \cdots. \tag{5-147}$$

Hence, although the customary approach is to obtain the stationary state eigenvalues of the Schrödinger equation by application of appropriate boundary conditions to the general solution of the Schrödinger differential equation, we see that the result can also be obtained using matrix methods, without introduction of differential equation solution techniques.

However, not only the energy eigenvalues, but the associated eigenvectors as well as expectation values of other operators are of interest. To obtain these, we again use the matrix representation ($\mathbf{A}$) of the lowering operator ($\mathcal{A}$) and its properties to deduce the desired results. First we note that, while $\mathbf{A}$ is the matrix representation of a lowering operator, the matrix $\mathbf{A}^\dagger$ has the following opposite property. From Eq. (5-123), we have

$$\mathbf{A}^{\dagger}\mathbf{H}\boldsymbol{\Psi}_E=\mathbf{A}^{\dagger}\boldsymbol{\Psi}_E E, \tag{5-148}$$

and use of the results of Problem 5-8 gives

$$\left(\mathbf{H}\mathbf{A}^{\dagger}-\hbar\sqrt{\frac{k}{m}}\,\mathbf{A}^{\dagger}\right)\boldsymbol{\Psi}_E=(\mathbf{A}^{\dagger}\boldsymbol{\Psi}_E)E,$$

which can be rearranged to

$$\mathbf{H}(\mathbf{A}^{\dagger}\boldsymbol{\Psi}_E)=\left(E+\hbar\sqrt{\frac{k}{m}}\right)(\mathbf{A}^{\dagger}\boldsymbol{\Psi}_E), \tag{5-149}$$

which shows that $\mathbf{A}^{\dagger}$ is a matrix representation of an operator $\mathfrak{A}^{\dagger}$ that can be considered to be a *raising* operator, since application of $\mathbf{A}^{\dagger}$ to $\boldsymbol{\Psi}_E$ produces another eigenfunction of $\mathbf{H}$, whose eigenvalue is raised from E to $(E+\hbar\sqrt{k/m})$.

To obtain the eigenfunctions, we first need to consider the difference in normalization constants that might arise when the matrices representing the raising or lowering operators are applied, i.e., we need to evaluate the constants a_n and b_n in

$$\mathbf{A}\boldsymbol{\Psi}_n=a_n\boldsymbol{\Psi}_{n-1}, \tag{5-150}$$

$$\mathbf{A}^{\dagger}\boldsymbol{\Psi}_n=b_n\boldsymbol{\Psi}_{n+1}. \tag{5-151}$$

In this discussion, we shall assume that the eigenfunctions are each normalized to unity,[43] i.e.,

$$\boldsymbol{\Psi}^{\dagger}_{n'}\boldsymbol{\Psi}_n=\delta_{n',n}. \tag{5-152}$$

Multiplication on the left of both sides of Eq. (5-150) by $\boldsymbol{\Psi}^{\dagger}_{n'}$ yields

$$\begin{aligned}\boldsymbol{\Psi}^{\dagger}_{n'}\mathbf{A}\boldsymbol{\Psi}_n&=a_n\boldsymbol{\Psi}^{\dagger}_{n'}\boldsymbol{\Psi}_{n-1}\\&=a_n\delta_{n',n-1}.\end{aligned} \tag{5-153}$$

A similar procedure using Eq. (5-151) yields

$$\begin{aligned}\boldsymbol{\Psi}^{\dagger}_{n'}\mathbf{A}^{\dagger}\boldsymbol{\Psi}_n&=b_n\boldsymbol{\Psi}^{\dagger}_{n'}\boldsymbol{\Psi}_{n+1}\\&=b_n\delta_{n',n+1}.\end{aligned} \tag{5-154}$$

However, we also note that [from Eq. (5-151)],

$$\begin{aligned}b_n^*&=(\boldsymbol{\Psi}^{\dagger}_{n+1}\mathbf{A}^{\dagger}\boldsymbol{\Psi}_n)^{\dagger}=\boldsymbol{\Psi}^{\dagger}_n\mathbf{A}\boldsymbol{\Psi}_{n+1}\\&=a_n\delta_{n,n+1}.\end{aligned} \tag{5-155}$$

---

[43] It should be remembered that, in matrix products like Eq. (5-152), integration is implied in the formation of the product. For example, in function notation, Eq. (5-152) would be written as $\int \Psi^*_{n'}\Psi_n\,d\tau=\delta_{n',n}$.

Comparison of Eqs. (5-154) and (5-155) reveals a relationship of interest, i.e.,

$$a_{n+1} = b_n^*. \tag{5-156}$$

To obtain the other relation that will allow evaluation of a_n and b_n, we examine

$$\begin{aligned}\boldsymbol{\Psi}_{n'}^{\dagger}\mathbf{A}\mathbf{A}^{\dagger}\boldsymbol{\Psi}_n &= \boldsymbol{\Psi}_{n'}\left[\mathbf{H} + \left(\frac{\hbar}{2}\sqrt{\frac{k}{m}}\right)\mathbf{I}\right]\boldsymbol{\Psi}_n \\ &= \hbar\sqrt{\frac{k}{m}}\,(n+1)\delta_{n',n},\end{aligned} \tag{5-157}$$

where Eqs. (5-129) and (5-147) have been used to obtain the result. However, this equation can also be written as

$$\begin{aligned}\boldsymbol{\Psi}_{n'}^{\dagger}\mathbf{A}\mathbf{A}^{\dagger}\boldsymbol{\Psi}_n &= b_n\boldsymbol{\Psi}_{n'}^{\dagger}\mathbf{A}\boldsymbol{\Psi}_{n+1} \\ &= a_{n+1}b_n\delta_{n',n},\end{aligned} \tag{5-158}$$

which means that

$$a_{n+1}b_n = \hbar\sqrt{\frac{k}{m}}\,(n+1). \tag{5-159}$$

Using Eq. (5-156), we see that[44]

$$a_{n+1}b_n = a_{n+1}a_{n+1}^* = |a_{n+1}|^2 = b_n^*b_n = |b_n|^2 = \hbar\sqrt{\frac{k}{m}}\,(n+1),$$

$$a_n = \left[\sqrt{\frac{k}{m}}\,\hbar n\right]^{1/2}, \tag{5-160}$$

$$b_n = \left[\sqrt{\frac{k}{m}}\,\hbar\,(n+1)\right]^{1/2}. \tag{5-161}$$

With this information, we can now obtain the desired eigenfunctions. We form

$$\begin{aligned}(\mathbf{A} + \mathbf{A}^{\dagger})\boldsymbol{\Psi}_n &= \mathbf{A}\boldsymbol{\Psi}_n + \mathbf{A}^{\dagger}\boldsymbol{\Psi}_n \\ &= \left[\sqrt{\frac{k}{m}}\,\hbar\right]^{1/2}\{\sqrt{n}\,\boldsymbol{\Psi}_{n-1} + \sqrt{n+1}\,\boldsymbol{\Psi}_{n+1}\},\end{aligned} \tag{5-162}$$

[44] Equations (5-160) and (5-161) yield expressions for the magnitude of a_a and b_n only. However, the phase choice that remains arbitrary will not affect the stationary state eigenvalues under discussion here.

where Eqs. (5-160) and (5-161) have been used. However, another expression for $(\mathbf{A}+\mathbf{A}^{\dagger})\boldsymbol{\Psi}_n$ can be obtained from the definitions given in Eqs. (5-127) and (5-128), i.e.,

$$(\mathbf{A}+\mathbf{A}^{\dagger})\boldsymbol{\Psi}_n=\left\{\frac{1}{\sqrt{2}}\left(\frac{1}{\sqrt{m}}\mathbf{P}-i\sqrt{k}\mathbf{X}\right)+\left(\frac{1}{\sqrt{2}i}\right)\right.$$

$$\left.\cdot\left(\frac{1}{\sqrt{m}}\mathbf{P}^{\dagger}+i\sqrt{k}\mathbf{X}\right)\right\}\boldsymbol{\Psi}_n$$

$$=\sqrt{2k}\mathbf{X}\boldsymbol{\Psi}_n. \tag{5-163}$$

Comparison of Eqs. (5-162) and (5-163) yields

$$\sqrt{n}\boldsymbol{\Psi}_{n-1}-\left(\frac{2\sqrt{km}}{\hbar}\right)^{1/2}\mathbf{X}\boldsymbol{\Psi}_n+\sqrt{n+1}\boldsymbol{\Psi}_{n+1}=\mathbf{0}. \tag{5-164}$$

At this point, we have established a relationship among the eigenvectors corresponding to three different eigenvalues, and this completes the information that we shall need to determine the eigenvectors. The remaining steps will only make comparison with other forms of the solution easier. Relationships between different eigenvectors such as the above are known as *recurrence relations*.

From the isomorphism between functions, operators, and their matrix representations that were shown earlier,[45] we know that any relation that is derived from the matrix representation will also hold for the functions and operators themselves. Hence Eq. (5-164) can be written alternatively in function form as

$$\sqrt{n}\Psi_{n-1}(\xi)-\sqrt{2}\xi\Psi_n(\xi)+\sqrt{n+1}\Psi_{n+1}(\xi)=0, \tag{5-165}$$

where

$$\xi=\left(\frac{\sqrt{km}}{\hbar}\right)^{1/2}x. \tag{5-166}$$

Equation (5-165) is the same recurrence relation (except possibly for normalization constants) that would be obtained if the problem had been approached via the solution of the Schrödinger differential equation itself. The only remaining step is to explore whether any functions that have been characterized previously[46] also correspond to such a recurrence relation. Examination of

[45] Chapter 3, Section 3-6 and Chapter 2, Section 2-7.

[46] For an extensive discussion of such functions, see *Higher Transcendental Functions*, Vols. 1-2, condensed by A. Erdelyi, (editor), McGraw-Hill, New York, 1953.

available tabulations of such relations[47] reveals that except possibly for normalization constants, an equation of the above form is satisfied by the *Hermite orthogonal polynomials*. Furthermore, these functions, usually designated $H_n(\xi)$, are orthogonal over the range $(-\infty \leq \xi \leq +\infty)$ with weight factor $e^{-\xi^2}$ to assure quadratic integrability, i.e.,

$$\int_{-\infty}^{\infty} e^{-\xi^2} H_n(\xi) H_m(\xi)\, d\xi = 0 \qquad (m \neq n). \tag{5-167}$$

This suggests that the eigenfunctions of the harmonic oscillator can be written as

$$\Psi_n(\xi) = N_n e^{-\xi^2/2} H_n(\xi), \tag{5-168}$$

where N_n is a normalization constant.[48] Use of Eq. (5-167) with $m = n$ gives

$$N_n = \frac{\beta^{1/4}}{[2^n n! \sqrt{\pi}]^{1/2}}, \qquad \beta = \frac{\sqrt{mk}}{\hbar}, \tag{5-169}$$

and substitution of Eqs. (5-168) and (5-169) into (5-165) gives

$$H_{n+1}(\xi) - 2\xi H_n(\xi) + 2nH_{n-1}(\xi) = 0, \tag{5-170}$$

which is the usual recurrence relation[49] among Hermite polynomials. Other relations that are useful include

$$\begin{aligned} H_n'(\xi) &= 2nH_{n-1}(\xi) \\ H_{n+1}(\xi) &= 2\xi H_n - H_n'(\xi) \end{aligned} \tag{5-171}$$

and

$$H_n(\xi) = (-1)^n e^{\xi^2} \frac{d^n e^{-\xi^2}}{d\xi^n}, \tag{5-172}$$

where $H_n'(\xi) = dH_n(\xi)/d\xi$. When expressed in terms of the harmonic oscillator eigenfunctions, Eqs. (5-171) and (5-172) become

$$\xi \Psi_n(\xi) + \Psi_n'(\xi) = \sqrt{2n}\, \Psi_{n-1}(\xi), \tag{5-173}$$

and

$$\xi \Psi_n(\xi) - \Psi_n'(\xi) = \sqrt{2(n+1)}\, \Psi_{n+1}(\xi). \tag{5-174}$$

[47] For tabulation of recurrence relations and function values, see M. Abramowitz and I. Stegun, *Handbook of Mathematical Functions*, Dover, New York, 1965. See Chapter 22 for a discussion of the functions needed here.

[48] See Problem 5-12.

[49] See M. Abramowitz and I. Stegun, *Handbook of Mathematical Functions*, Dover, New York, 1965, p. 782.

In order to visualize the eigenfunctions of the harmonic oscillator (Eq. (5-168), Fig. 5.13 shows plots of the eigenfunctions for the three lowest states as a function of ξ. Among the wavefunction features that are illustrated is the fact that Ψ_0 is symmetric about $\xi = 0$, while Ψ_1 is antisymmetric and Ψ_2 is symmetric. Such a pattern also continues for higher states, i.e., they are alternatively symmetric and antisymmetric. In addition, the number of nodes is seen to increase linearly with energy state, i.e., Ψ_0 has zero nodes (excluding the nodes at $\xi = \pm\infty$), Ψ_1 has one node, Ψ_2 has two nodes, etc. Such an increase in nodal structure with energy is frequently found, not only in model problems but in actual chemical systems as well.

In addition to the energy eigenvalues, it is frequently of interest to obtain expectation values of other operators, e.g.,

$$\langle\bar{x}\rangle_n = \langle\Psi_n|x|\Psi_n\rangle = (\mathbf{x})_{nn}. \tag{5-175}$$

This can be accomplished either by direct integration using the eigenfunctions in Eq. (5-168) or by using the matrix representation of the eigenvectors. In the discussion here, we shall first obtain the matrix elements of the raising and lowering matrices, $\mathbf{A}^\dagger$ and $\mathbf{A}$, and use them to obtain several other expectation values of interest.

From the matrix representation–function isomorphism property that we have been using frequently, we see that Eq. (5-150) can also be written as

$$\mathcal{A}\Psi_n = \left[\sqrt{\frac{k}{m}}\,\hbar n\right]^{1/2} \Psi_{n-1}, \tag{5-176}$$

where the explicit expression for a_n [Eq. (5-160)] has been inserted. Formation of the matrix element with $\Psi_m^\dagger$ yields

$$(\mathbf{A})_{mn} = \langle\Psi_m|\mathcal{A}|\Psi_n\rangle = \left[\sqrt{\frac{k}{m}}\,\hbar n\right]^{1/2} \langle\Psi_m|\Psi_{n-1}\rangle$$

$$= \left[\sqrt{\frac{k}{m}}\,\hbar n\right]^{1/2} \delta_{m,n-1}. \tag{5-177}$$

A similar treatment for the matrix elements of $\mathbf{A}^\dagger$ yields

$$(\mathbf{A}^\dagger)_{mn} = \left[\sqrt{\frac{k}{m}}\,\hbar(n+1)\right]^{1/2} \delta_{m,n+1}. \tag{5-178}$$

With this information, expectation values can be easily calculated. For example,

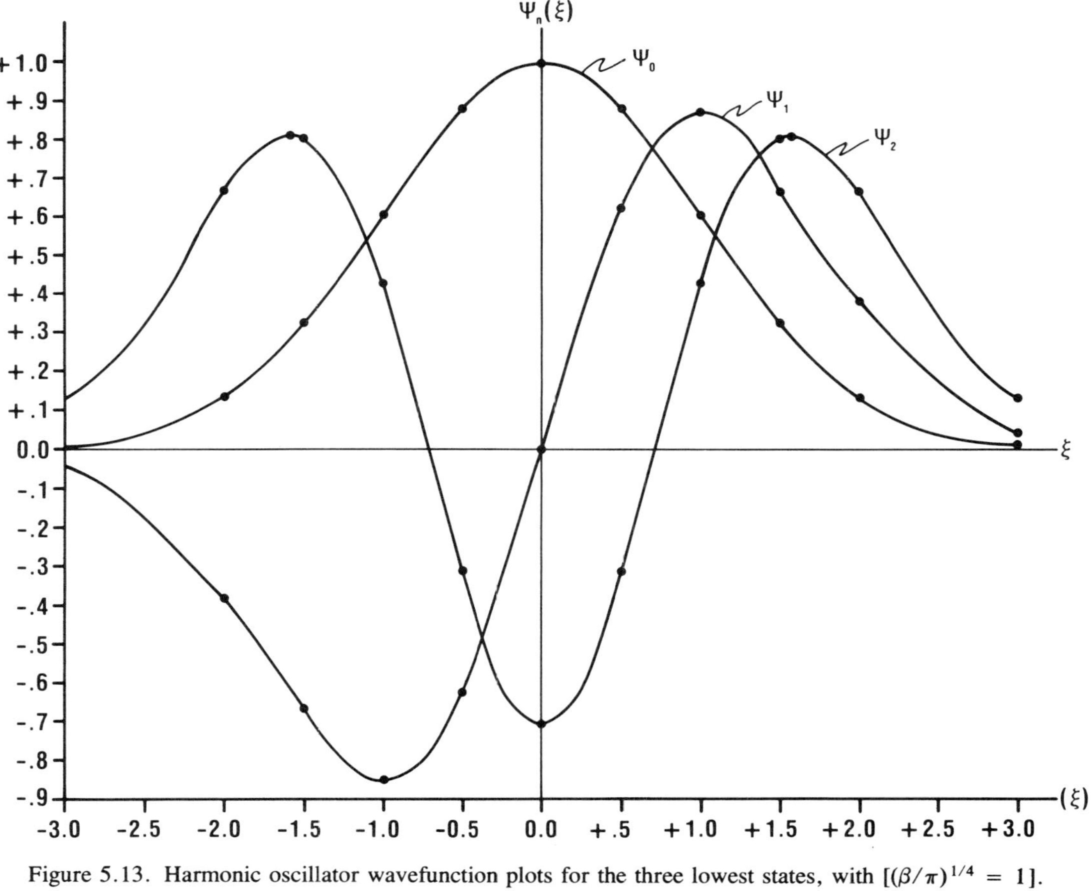

Figure 5.13. Harmonic oscillator wavefunction plots for the three lowest states, with $[(\beta/\pi)^{1/4} = 1]$.

Eq. (5-163) can be written as[50]

$$\mathbf{X} = \frac{1}{\sqrt{2k}}(\mathbf{A} + \mathbf{A}^+)$$

$$= \frac{1}{\sqrt{2}}\begin{pmatrix} 0 & \left[\frac{\hbar}{\sqrt{km}}\right]^{1/2} & 0 & & \\ \left[\frac{\hbar}{\sqrt{km}}\right]^{1/2} & 0 & \left[\frac{2\hbar}{\sqrt{km}}\right]^{1/2} & 0 & \\ 0 & \left[\frac{2\hbar}{\sqrt{km}}\right]^{1/2} & 0 & \left[\frac{3\hbar}{\sqrt{km}}\right]^{1/2} & 0 \quad \cdots \\ 0 & 0 & \left[\frac{3\hbar}{\sqrt{km}}\right]^{1/2} & 0 & \ddots \\ \vdots & \vdots & 0 & \ddots & \ddots \\ & & \vdots & & \end{pmatrix}. \tag{5-179}$$

Note that, in *every* state of the harmonic oscillator, its average position is at the origin, i.e.,

$$\langle \Psi_n | x | \Psi_n \rangle = 0. \tag{5-180}$$

for all n. In a similar manner, it can be seen by formation of $(\mathbf{A} - \mathbf{A}^\dagger)$ that

$$\mathbf{P} = -i\sqrt{\frac{m}{2}}(\mathbf{A} - \mathbf{A}^+)$$

$$= \frac{1}{\sqrt{2}}\begin{pmatrix} 0 & -i[\hbar\sqrt{km}]^{1/2} & 0 & 0 & \cdots \\ +i[\hbar\sqrt{km}]^{1/2} & 0 & -i[2\hbar\sqrt{km}]^{1/2} & 0 & \cdots \\ 0 & +i[2\hbar\sqrt{km}]^{1/2} & 0 & -i[3\hbar\sqrt{km}]^{1/2} & \\ 0 & 0 & +i[3\hbar\sqrt{km}]^{1/2} & 0 & \ddots \\ \vdots & \vdots & \ddots & & \ddots \end{pmatrix}, \tag{5-181}$$

and we see that the average value of the momentum in any state is also zero, i.e,

$$\langle \Psi_n | p | \Psi_n \rangle = 0, \tag{5-182}$$

for all n. However, not all powers of the average values of x and p are zero.[51]

Hence, we see that matrix representations can often provide very useful

[50] The labeling on the rows (and columns) of the matrix in Eq. (5-179) is as follows: row 1 corresponds to $n = 0$, row 2 corresponds to $n = 1$, etc.

[51] See Problem 5-9.

alternatives to the use of the functions themselves. We shall see many more examples of the use of matrix representations in later chapters, where exact solutions are not in general possible, and matrix techniques using finite basis sets provide a powerful tool for obtaining approximate solutions.

5-10. Zero Point Energy and the Uncertainty Principle

The lack of commutation of the position and momentum operators (or their matrix representatives) that was observed explicitly in the preceding discussion of the harmonic oscillator implies a more general characteristic of quantum mechanics that should be noted. In particular, the Heisenberg uncertainty relations that were introduced in Chapter 4 indicate that there are fundamental limitations on the accuracy with which simultaneous measurements of (e.g.) position and momentum can be made. These limitations also have implications on the energy states that are allowed for atomic and molecular systems, and the discussion in this section will illustrate how these restrictions apply to the examples just discussed.

For the particle on a ring, we found the allowed energy states [Eq. (5-20)] to be given as

$$E_k = \frac{\hbar^2 k^2}{2mR^2}, \qquad k = 0,\ \pm 1,\ \pm 2,\ \cdots. \tag{5-183}$$

The uncertainty relation for position and momentum is[52]

$$\Delta p_x \Delta x \geqslant \frac{\hbar}{2}. \tag{5-184}$$

For the particle on a ring, there is no potential energy term, and the total energy is just the kinetic energy of the particle, which can be written as

$$E_p = \frac{\overline{p_\varphi^2}}{2m} \tag{5-185}$$

where $\bar{p}_\varphi$ is the average value of the angular momentum. Thus, Eq. (5-184) can be written as

$$\sqrt{2m\Delta E_p}\, \Delta\varphi > \hbar. \tag{5-186}$$

Hence, $E = 0$ implies that *no* uncertainty in the energy exists, which might seem to be a contradiction of the uncertainty principle. However, the angle φ

[52] Chapter 4, Section 4-4A.

(i.e., the position of the particle on the ring) is completely *undetermined* at this point, implying an arbitrarily large uncertainty $\Delta\varphi$ in Eq. (5-186) when $k = 0$. Hence, the uncertainly principle is still satisfied, even for $k = 0$.

For the particle in a box, we found [Eq. (5-37)] that

$$E_n = \frac{n^2h^2}{8mL^2}, \qquad n = \pm 1, \pm 2, \cdots. \tag{5-187}$$

Since the potential was chosen to be zero inside the box, the total energy is related to the momentum in a manner similar to the case of a particle on a ring, i.e.,

$$E_n = \frac{(\overline{p_x^2})_n}{2m}. \tag{5-188}$$

Using the eigenfunctions of Eq. (5-40), the average value of the momentum in the nth energy state is given by

$$\begin{aligned}
(\bar{p}_x)_n = \langle p_x \rangle_n &= \int_{-\infty}^{+\infty} \Psi_n^*(x) p_x \Psi_n(x)\, dx \\
&= \left(\frac{2k}{iL}\right) \int_0^L \sin\left(\frac{n\pi x}{L}\right) \frac{d}{dx}\left[\sin\left(\frac{n\pi x}{L}\right)\right] dx \\
&= \left(\frac{2\hbar}{iL}\right) \int_0^{n\pi} \sin y \cos y\, dy = 0.
\end{aligned} \tag{5-189}$$

However, this does not necessarily imply that there is zero uncertainty in the momentum. In fact, using Eqs. (5-188) and (5-189), the average value of the square of the momentum is seen to be

$$(\overline{p_x^2})_n = \langle p_x^2 \rangle_n = \frac{n^2h^2}{4L^2}, \tag{5-190}$$

which implies that the momentum itself has values of $\pm nh/2L$. Thus, the uncertainty is

$$\Delta p_x = +\frac{nh}{2L} - \left(-\frac{nh}{2L}\right) = \frac{nh}{L}. \tag{5-191}$$

As for the position, it is easily established that

$$\langle x \rangle_n = \frac{L}{2} \tag{5-192}$$

is the uncertainty in x, so that $\Delta x = L$. This result indicates nothing more than that we know only that the particle is *in* the box, but do not know anything about its position inside the box.

Thus, for the particle in a box, we have

$$(\Delta p_x)_n(\Delta x)_n = \left(\frac{nh}{L}\right)(L) = nh, \tag{5-193}$$

which shows that the uncertainty principle is satisfied for all states ($n > 0$). Note that $n = 0$ is *not* allowed, since the uncertainty principle would be violated, which is a result already obtained via different reasoning in the earlier discussion.

For the case of the harmonic oscillator, we found [Eq. (5-147)] that

$$E_n = \left(n + \frac{1}{2}\right)\hbar\sqrt{\frac{k}{m}}, \qquad n = 0,\ 1,\ 2,\ \cdots. \tag{5-194}$$

As in the previous cases, we see that even in the lowest allowed state, there is a residual energy that is present that could be translated into uncertainties in position and momentum (see Problem 5-5). Thus, this residual energy, which is called the *zero-point energy* when the ground state is being discussed, does not simply arise accidentally. Instead, *every* quantum system must possess such a zero-point energy, in order to be consistent with the uncertainty principle.[53] We shall see other illustrations of this principle in later chapters, when potential energy curves are discussed.

Problems

1. Calculate the expectation value of the square of the position of a particle in energy state n, i.e., $\langle x^2\rangle_n$, assuming the particle is trapped in the box with a potential given by Eq. (5-26).
2. For the symmetric wavefunction for a particle in a finite box [Eq. (5-67)] and the parameters of Eq. (5-81), show that

$$\frac{B_{\text{III}}}{A_{\text{II}}} = 1.4502.$$

3. Show that the normalization constant for the symmetric wavefunction for a particle in a finite box is given by

$$N = \frac{1}{[L + (1/2\kappa)\sin(2\kappa L) + (1/k)e^{-2kL}]^{1/2}}.$$

[53] Apparent violations of this principle may occur in model problems of less than three dimensions. For those cases, the zero-point energy will be found by consideration of the problems in which the eigenvalues are actually observable, i.e., the three-dimensional problem.

4. Consider a potential of the form (in one dimension):

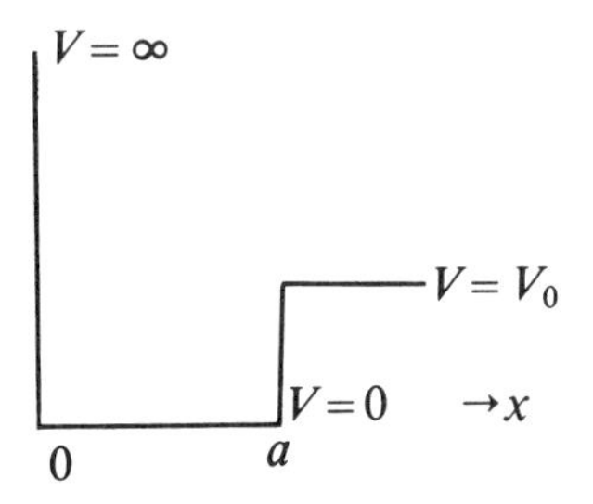

with a particle incident from the right with energy E_0 such that

$$0<E_0<V_0.$$

Find the lowest energy level and wavefunction. What is the least depth of the well for which a bound state exists?

5. Calculate the product of the uncertainties, $\Delta p_x \Delta x$, for a one-dimensional harmonic oscillator in its first excited state ($n = 2$), and compare to the limiting value indicated by the Uncertainty Principle.

6. Perhaps the simplest type of particle motion is that of a "free particle," i.e., one in which (in one dimension):

$$V(x)=0, \qquad -\infty \leq x \leq +\infty.$$

a. Write the time-independent Schrödinger equation for such a particle, and verify that the eigenfunctions are given by

$$\Psi(x)=Ae^{ikx}+Be^{-ikx},$$

where A and B are constants, and

$$k=\left[\frac{2mE}{\hbar^2}\right]^{1/2}.$$

b. Verify that the above wavefunction is also an eigenfunction of the momentum operator. Further, show that the choice of $B = 0$ corresponds to a free particle moving to the right, and that the alternative choice of $A = 0$ corresponds to a free particle moving to the left.

7. The "free electron" model of a molecule consists of representing the molecule by a "box" whose dimensions are the same as the molecule, and whose potential is taken as zero inside the box and infinite at the walls. For a description of the π-electrons in a molecule like butadiene, the free electron model considers a one-dimensional "tube" of length L. The Schrödinger equation for the system is given by

$$-\frac{\hbar^2}{2m}\frac{d^2\Psi(x)}{dx^2}=E\Psi(x).$$

Find the energy levels and wavefunctions for this system.

8. A "generating function" $G(x, t)$ for a set of orthogonal polynomials, $p_n(x)$, is a function such that a McLaurin series expansion of $G(x, t)$ in powers of t gives $p_n(x)$ as coefficients in the series, i.e.,

$$G(x, t)=\sum_{n=0}^{\infty} \frac{\partial^n}{\partial t^n}[G(x, t)]_{t=0}\left(\frac{t^n}{n!}\right)$$

$$=\sum_{n=0}^{\infty} p_n(x)\left(\frac{t^n}{n!}\right).$$

a. Show that

$$G(x, t)=e^{(2tx-t^2)}$$

is a generating function for Hermite polynomials.

b. Use $G(x, t)$ to generate explicit expressions for the first four hermite polynomials.

c. Use

$$e^{(2tx-t^2)}=\sum_{n=0}^{\infty} H_n(x)\left(\frac{t^n}{n!}\right)$$

to derive Eq. (5-165) and expressions for the first and second derivatives of $H_n(x)$.

9. Prove Eqs. (5-126)–(5-128) using Eqs. (5-121)–(5-125).
10. Determine which of $\langle x^2\rangle$, $\langle x^3\rangle$, $\langle p^2\rangle$, and $\langle p^3\rangle$ are non-zero for the harmonic oscillator.
11. Prove Eq. (5-165) using the definition of Hermite polynomials given in Eq. (5-166) and other recursion relations as needed.
12. For $n = 0$ and 1, show that the harmonic oscillator wavefunction of Eq. (5-168) is an eigenfunction of the quantum mechanical Hamiltonian of Eq. (5-116), with eigenvalues given by Eq. (5-147).
13. A potential that provides an improvement over the harmonic oscillator potential for the description of the vibrational motion of diatomic molecules is known as the Morse potential, and is given by

$$V(x)=D\{e^{-2a(x-x_0)}-2e^{-a(x-x_0)}\},$$

where D is the depth of the potential well, x_0 is the equilibrium distance, and a determines the width of the potential well. The energies and wavefunctions corresponding to this potential are given by

$$E_n=-D+\left(n+\frac{1}{2}\right)\left(\frac{4D}{K}\right)-\left(n+\frac{1}{2}\right)^2\left(\frac{4D}{K^2}\right),$$

$$\Psi_n=A_n e^{-t/2}t^{\alpha}\xi_n(t),$$

where A_n is a normalization constant,

$$K = \left(\frac{8mD}{\hbar^2 a^2}\right)^{1/2},$$

$$t = Ke^{-a(x-x_0)},$$

and $\xi_n(t)$ is a polynomial of degree n, whose first few members are given by

$$\xi_0(t) = 1$$

$$\xi_1(t) = 1 + (K-2)^{-1}t$$

$$\xi_2(t) = 1 + 2(K-4)^{-1}t + (K-4)^{-2}t^2$$

a. Show that Ψ_1, Ψ_2, and Ψ_3 are mutually orthogonal.

b. Show that if the harmonic oscillator force constant (k) is related to the Morse potential (V) via

$$k = \left(\frac{\partial^2 V}{Dx^2}\right)_{x=x_0},$$

then the vibration frequencies of the Morse oscillator are given by

$$\nu = \frac{1}{2\pi}\sqrt{\frac{2a^2D}{m}}.$$

c. Show that the energy expression can be rewritten in terms of ν as

$$E_n = -D + \left(n + \frac{1}{2}\right)h\nu - \left(\frac{1}{4D}\right)\left(n + \frac{1}{2}\right)^2 h^2\nu^2,$$

and calculate the correction to the transition energy ($E_n - E_0$) for HCl using the Morse potential compared to the harmonic oscillator potential.

Chapter 6

Angular Momentum

6-1. Introduction

In Chapter 1 we saw that one of the areas where strikingly different results are found, compared to the results expected from classical considerations, is associated with the measurement of angular momentum. In particular, the allowed states of orbital angular momentum are quantized at the microscopic level, in sharp contrast to the corresponding classical analog. Also, a new kind of angular momentum, associated with the intrinsic spin of particles, is observed at the microscopic level, which has no classical analog and is also quantized. In this chapter we shall investigate the properties of angular momentum in general, and show how both spin and orbital angular momentum can be considered to be different examples of a general quantum theory of angular momentum.

Although the analogies to classical mechanics are expected to be limited due to the differences just mentioned, it will nevertheless be useful to employ classical mechanical ideas to aid in the formation of the appropriate quantum mechanical operators of interest. In particular, it was observed in Chapter 1 [Eq. (1-5)] that the angular momentum of a particle about a point Q (*the orbital angular momentum*) is given in classical mechanics by

$$\mathbf{L}_Q = (\mathbf{r} - \mathbf{r}_Q) x \mathbf{p}. \tag{6-1}$$

Choosing the origin to be at the point $\mathbf{r}_Q$, and using the rules for formation of quantum mechanical operators from corresponding classical expressions that were introduced in Chapter 4 (Section 4-7), the quantum mechanical operator for orbital angular momentum is seen to be given by

$$\mathfrak{L} = \left(\frac{\hbar}{i}\right)(\mathbf{r} x \boldsymbol{\nabla}) = \left(\frac{\hbar}{i}\right) \begin{vmatrix} \mathbf{i} & \mathbf{j} & \mathbf{k} \\ x & y & z \\ \frac{\partial}{\partial x} & \frac{\partial}{\partial y} & \frac{\partial}{\partial z} \end{vmatrix}. \tag{6-2}$$

Writing out the components explicitly, we have

$$\mathcal{L}_x = \left(\frac{\hbar}{i}\right)\left(y\frac{\partial}{\partial z} - z\frac{\partial}{\partial y}\right), \tag{6-3}$$

$$\mathcal{L}_y = \left(\frac{\hbar}{i}\right)\left(z\frac{\partial}{\partial x} - x\frac{\partial}{\partial z}\right), \tag{6-4}$$

and

$$\mathcal{L}_z = \left(\frac{\hbar}{i}\right)\left(x\frac{\partial}{\partial y} - y\frac{\partial}{\partial x}\right). \tag{6-5}$$

One of the interesting properties of these operators is exhibited by examination of the commutators of the various components of $\mathcal{L}$. For example, the commutator of $\mathcal{L}_x$ and $\mathcal{L}_y$ is

$$\begin{aligned}[\mathcal{L}_x, \mathcal{L}_y] &= -\hbar^2 \left\{\left(y\frac{\partial}{\partial z} - z\frac{\partial}{\partial y}\right)\left(z\frac{\partial}{\partial x} - x\frac{\partial}{\partial z}\right)\right. \\ &\quad \left. - \left(z\frac{\partial}{\partial x} - x\frac{\partial}{\partial z}\right)\left(y\frac{\partial}{\partial z} - z\frac{\partial}{\partial y}\right)\right\} \\ &= \hbar^2\left(x\frac{\partial}{\partial y} - y\frac{\partial}{\partial x}\right),\end{aligned}$$

or

$$[\mathcal{L}_x, \mathcal{L}_y] = i\hbar\mathcal{L}_z. \tag{6-6}$$

In a similar manner, it is easily established (see Problem 6-1) that

$$[\mathcal{L}_y, \mathcal{L}_z] = i\hbar\mathcal{L}_x, \tag{6-7}$$

and

$$[\mathcal{L}_z, \mathcal{L}_x] = i\hbar\mathcal{L}_y. \tag{6-8}$$

Thus, the various components of the orbital angular momentum operator do not commute. Expressed in terms of measurable quantities, this lack of commutation indicates that it is not possible to construct a function that is a simultaneous eigenfunction of each of the components.[1] This in turn implies that it is not possible to make a simultaneous measurement of any two components of $\mathcal{L}$, in contrast to the classical result where such measurements are possible. Further details of this measurement process will be discussed shortly, but the point of interest here is that introduction of quantum mechanical operators in place of the corresponding classical quantities has far reaching effects, that become apparent at even the very early stages of analysis.

---

[1] Theorem 3-22, Chapter 3.

6-2. General Angular Momentum Considerations

Having seen that the quantum mechanical operators for orbital angular momentum can be expected to lead to results that are substantially different from their classical analogs, and that spin angular momentum must yet be accounted for, it should not be particularly surprising that more general considerations are appropriate in the quantum mechanical discussion.

To carry out this more general treatment, we shall assume that the basic operator *form* that was introduced specifically for orbital angular momentum is applicable to *all* kinds of angular momentum. If this is the case, then the eigenvalues that are found for this operator should include those of orbital angular momentum and spin angular momentum as special cases.

Thus, by analogy to Eqs. (6-2) through to (6-5), we define a *general angular momentum operator* ($\mathfrak{M}$), to be one with components $\mathfrak{M}_x$, $\mathfrak{M}_y$, and $\mathfrak{M}_z$, that obey the following commutation relation:

$$[\mathfrak{M}_p, \mathfrak{M}_q] = i\hbar\mathfrak{M}_r \tag{6-9}$$

where p, q, r represent x, y, z or any cyclic permutation of x, y, z. We shall also be interested in the square of the *general angular momentum operator* ($\mathfrak{M}^2$), which is defined as

$$\mathfrak{M}^2 = \mathfrak{M}_x^2 + \mathfrak{M}_y^2 + \mathfrak{M}_z^2, \tag{6-10}$$

and the so-called "*step-up*" and "*step-down*" operators[2] ($\mathfrak{M}_+$ and $\mathfrak{M}_-$, respectively), which are defined as:

$$\mathfrak{M}_+ = \mathfrak{M}_x + i\mathfrak{M}_y \tag{6-11}$$

and

$$\mathfrak{M}_- = \mathfrak{M}_x - i\mathfrak{M}_y. \tag{6-12}$$

It should be noted that $\mathfrak{M}_+$ and $\mathfrak{M}_\varphi$ are not Hermitian operators, while $\mathfrak{M}^2$ and the components of $\mathfrak{M}$ are Hermitian operators.[3]

$\mathfrak{M}^2$ can also be related to $\mathfrak{M}_+$ and $\mathfrak{M}_-$ using Eq. (6-10) as follows:

$$\begin{aligned}\mathfrak{M}^2 &= \mathfrak{M}_z^2 + \mathfrak{M}_x^2 + \mathfrak{M}_y^2 \\ &= \mathfrak{M}_z^2 + \frac{1}{2}[\mathfrak{M}_x^2 - i\mathfrak{M}_x\mathfrak{M}_y + i\mathfrak{M}_y\mathfrak{M}_x + \mathfrak{M}_y^2 + \mathfrak{M}_x^2 + i\mathfrak{M}_x\mathfrak{M}_y - i\mathfrak{M}_y\mathfrak{M}_x \\ &\quad + \mathfrak{M}_y^2] \\ &= \mathfrak{M}_z^2 + \frac{1}{2}\{\mathfrak{M}_+\mathfrak{M}_- - \mathfrak{M}_-\mathfrak{M}_+\}.\end{aligned} \tag{6-13}$$

Now let us explore some commutation relation among these various

[2] These operators are also known as "*raising*" and "*lowering*" operators, or "*ladder*" *operators*.

[3] See Problem 6-2.

operators. For example, using the previous definitions, we have

$$
\begin{aligned}
[\mathfrak{M}^2, \mathfrak{M}_z] &= [\mathfrak{M}_x^2 + \mathfrak{M}_y^2 + \mathfrak{M}_z^2, \mathfrak{M}_z] \\
&= \mathfrak{M}_x(\mathfrak{M}_x\mathfrak{M}_z) + \mathfrak{M}_y(\mathfrak{M}_y\mathfrak{M}_z) - (\mathfrak{M}_z\mathfrak{M}_x)\mathfrak{M}_x - (\mathfrak{M}_z\mathfrak{M}_y)\mathfrak{M}_y \\
&= \mathfrak{M}_x(\mathfrak{M}_z\mathfrak{M}_x - i\hbar\mathfrak{M}_y) + \mathfrak{M}_y(\mathfrak{M}_z\mathfrak{M}_y + i\hbar\mathfrak{M}_x) \\
&\quad - (\mathfrak{M}_x\mathfrak{M}_z + i\hbar\mathfrak{M}_y)\mathfrak{M}_x - (\mathfrak{M}_y\mathfrak{M}_z - i\hbar\mathfrak{M}_x)\mathfrak{M}_y
\end{aligned}
$$

or

$$[\mathfrak{M}^2, \mathfrak{M}_z] = 0. \tag{6-14}$$

In a similar manner, it can be shown that

$$[\mathfrak{M}^2, \mathfrak{M}_x] = 0 \tag{6-15}$$

and

$$[\mathfrak{M}^2, \mathfrak{M}_y] = 0. \tag{6-16}$$

These are interesting results since, in contrast to the results for any two components of $\boldsymbol{\mathfrak{M}}$, we see that it *is* possible to make a simultaneous measurement of $\mathfrak{M}^2$ and any one of the components of $\boldsymbol{\mathfrak{M}}$. Expressed another way,[4] the commutation of $\mathfrak{M}^2$ with any of the components of $\boldsymbol{\mathfrak{M}}$ assures that it is possible to find a function that is simultaneously an eigenfunction of both $\mathfrak{M}^2$ and *one* of the components of $\boldsymbol{\mathfrak{M}}$, e.g., $\mathfrak{M}_z$. If $(\mu^2\hbar^2)$ is the eigenvalue of $\mathfrak{M}^2$ that is observed, and $(\mu_z\hbar)$ is the eigenvalue of $\mathfrak{M}_z$ that is observed, then we can summarize this part of the discussion by the equations

$$\mathfrak{M}^2 Y = (\mu^2\hbar^2) Y \tag{6-17}$$

and

$$\mathfrak{M}_z Y = (\mu_z\hbar) Y, \tag{6-18}$$

where Y is a simultaneous eigenfunction of $\mathfrak{M}^2$ and $\mathfrak{M}_z$.

Next, let us consider the quantity $(\mathfrak{M}_+ Y)$, and the effect that $\mathfrak{M}_z$ has on it:

$$
\begin{aligned}
\mathfrak{M}_z(\mathfrak{M}_+ Y) &= \mathfrak{M}_z(\mathfrak{M}_x + i\mathfrak{M}_y) Y \\
&= (\mathfrak{M}_x\mathfrak{M}_z + i\hbar\mathfrak{M}_y) Y + i(\mathfrak{M}_y\mathfrak{M}_z - i\hbar\mathfrak{M}_x) Y \\
&= (\mathfrak{M}_x + i\mathfrak{M}_y)\mathfrak{M}_z Y + \hbar(\mathfrak{M}_x + i\mathfrak{M}_y) Y
\end{aligned}
$$

or

$$\mathfrak{M}_z(\mathfrak{M}_+ Y) = \mathfrak{M}_+(\mathfrak{M}_z + \hbar) Y.$$

Since Y is an eigenfunction of $\mathfrak{M}_z$, the previous equation can be rewritten as

$$
\begin{aligned}
\mathfrak{M}_z(\mathfrak{M}_+ Y) &= \mathfrak{M}_+(\mu_z\hbar + \hbar) Y \\
&= \hbar(\mu_z + 1) Y.
\end{aligned} \tag{6-19}
$$

---

[4] Theorem 3-22, Chapter 3.

Hence, we have the interesting result that not only is Y an eigenfunction of $\mathcal{M}_z$, but $(\mathcal{M}_+ Y)$ is also an eigenfunction of $\mathcal{M}_z$, with eigenvalue $(\mu_z + 1)\hbar$, which is the reason for referring to $\mathcal{M}_+$ as a "step-up" operator. In a similar manner, it is easily established that $(M_- Y)$ is also an eigenfunction of $\mathcal{M}_z$, with eigenvalue $(\mu_z - 1)\hbar$, i.e.,

$$\mathcal{M}_z(\mathcal{M}_- Y) = \hbar(\mu_z - 1)(\mathcal{M}_- Y). \tag{6-20}$$

Next, we note that

$$\begin{aligned}
\mathcal{M}_z(\mathcal{M}_+ \mathcal{M}_+ Y) &= \mathcal{M}_z(\mathcal{M}_x + i\mathcal{M}_y)(\mathcal{M}_x + i\mathcal{M}_y) Y \\
&= [(\mathcal{M}_x + i\mathcal{M}_y)\mathcal{M}_z + \hbar(\mathcal{M}_x + i\mathcal{M}_y)](\mathcal{M}_x + i\mathcal{M}_y) Y \\
&= \mathcal{M}_+(\mathcal{M}_z + \hbar)(\mathcal{M}_x + i\mathcal{M}_y) Y \\
&= \mathcal{M}_+[(\mathcal{M}_x + i\mathcal{M}_y)\mathcal{M}_z Y + 2\hbar(\mathcal{M}_x + i\mathcal{M}_y) Y] \\
&= \mathcal{M}_+\mathcal{M}_+(\mathcal{M}_z + 2\hbar) Y
\end{aligned}$$

or

$$\mathcal{M}_z(\mathcal{M}_+ \mathcal{M}_+ Y) = (\mu_z + 2)\hbar(\mathcal{M}_+ \mathcal{M}_+ Y). \tag{6-21}$$

Hence, repeated application of $\mathcal{M}_+$ is possible, with a raising of the eigenvalue of $\mathcal{M}_z$ by 1 (in units of $\hbar$) for each application of $\mathcal{M}_+$. Analogous results are obtained in the case of repeated application of $\mathcal{M}_-$.

Finally, it can easily be established[5] that both $\mathcal{M}_+$ and $\mathcal{M}_-$ act as identity operators with respect to the eigenvalues of $\mathcal{M}^2$, i.e.,

$$\mathcal{M}^2(\mathcal{M}_\pm Y) = \mu^2\hbar^2(\mathcal{M}_\pm Y). \tag{6-22}$$

These results indicate that it is possible to construct a whole series of eigenfunctions of $\mathcal{M}_z$ by appropriate application of $\mathcal{M}_-$ or $\mathcal{M}_+$ on Y, having eigenvalues

$$\cdots,\ (\mu_z - 2)\hbar,\ (\mu_z - 1)\hbar,\ \mu_z\hbar,\ (\mu_z + 1)\hbar,\ (\mu_z + 2)\hbar,\ \cdots . \tag{6-23}$$

At the same time, each of these eigenfunctions is also an eigenfunction of $\mathcal{M}^2$, having the *same* eigenvalue $(\mu^2\hbar^2)$. We also note that since all of the eigenvalues given in Eq. (6-23) are different, the various eigenfunctions associated with these eigenvalues [$\ldots$, $\mathcal{M}_-\mathcal{M}_- Y$, $\mathcal{M}_- Y$, Y, $\mathcal{M}_+ Y$, $\mathcal{M}_+\mathcal{M}_+ Y$, $\ldots$] are orthogonal, by Theorem 3-20.

The next step in the investigation of general angular momentum operators is to determine whether the number of eigenvalues listed in Eq. (6-23) is infinite, or whether the sequence terminates at some point. To explore this question, we return to the definition of $\mathcal{M}^2$ [Eq. (6-10)], and examine its expectation value,

---

[5] See Problem 6-3.

i.e.,

$$\begin{aligned}\langle \mathfrak{M}^2 \rangle &= \langle Y | \mathfrak{M}^2 | Y \rangle \\ &= \langle Y | \mathfrak{M}_x^2 + \mathfrak{M}_y^2 + \mathfrak{M}_z^2 | Y \rangle \\ &= \langle Y | \mathfrak{M}_x^2 | Y \rangle + \langle Y | \mathfrak{M}_y^2 | Y \rangle + \langle Y | \mathfrak{M}_z^2 | Y \rangle,\end{aligned}$$

or

$$\langle \mathfrak{M}^2 \rangle \geq \langle \mathfrak{M}_z^2 \rangle. \tag{6-24}$$

Thus, the expectation values of $\mathfrak{M}^2$ and $\mathfrak{M}_z$, computed using a simultaneous eigenfunction of both operators (which we have seen can be found), are such that

$$\mu^2 \geq \mu_z^2. \tag{6-25}$$

This is a very useful result, since we have already observed that the eigenvalue of $\mathfrak{M}^2$ is *not* affected by application of $\mathfrak{M}_+$ or $\mathfrak{M}_-$ to Y, while the eigenvalue of $\mathfrak{M}_z$ *is* affected. Thus, we see that the sequence of eigenvalues cannot be infinite, but must terminate at some point at each end of the spectrum, in order for μ_z to remain consistent with Eq. (6-25). In other words, at some point, the operation of $\mathfrak{M}_+$ (or $\mathfrak{M}_-$ at the other end of the eigenvalue spectrum) on the previous eigenfunction must produce zero. This characteristic can be used to deduce precisely how many and what kind of eigenvalues can be allowed for $\mathfrak{M}_z$, as we shall see next.

Suppose that the largest eigenvalue of $\mathfrak{M}_z$ that is consistent with the eigenvalue $(\mu^2 \hbar^2)$ for $\mathfrak{M}^2$ is called $(\hbar k_{\max})$, i.e.,

$$\hbar^2 k_{\max}^2 \leq \mu^2 \hbar^2. \tag{6-26}$$

Then, application of $\mathfrak{M}_+$ onto the eigenfunction of $\mathfrak{M}_z$ having eigenvalue $hk_{\max}$ must result in zero, i.e.,

$$\mathfrak{M}_+ Y(\hbar k_{\max}, \mu^2 \hbar^2) = 0. \tag{6-27}$$

Next, we apply $\mathfrak{M}_-$ to Eq. (6-27), resulting in

$$\begin{aligned}\mathfrak{M}_- \mathfrak{M}_+ Y(\hbar k_{\max}, \mu^2 \hbar^2) &= 0 \\ &= (\mathfrak{M}_x - i\mathfrak{M}_y)(\mathfrak{M}_x + iM_y) Y(\hbar k_{\max}, \mu^2 \hbar^2)\end{aligned}$$

or

$$0 = (\mathfrak{M}^2 - \mathfrak{M}_z^2 - \hbar \mathfrak{M}_z) Y(\hbar k_{\max}, \mu^2 \hbar^2). \tag{6-28}$$

However, $Y(\hbar k_{\max}, \mu^2 \hbar^2)$ is an eigenfunction of each of the operators in Eq. (6-28), which results in

$$0 = \hbar^2 (\mu^2 - k_{\max}^2 - k_{\max}) Y(\hbar k_{\max}, \mu^2 \hbar^2) \tag{6-29}$$

or

$$k_{\max} = \frac{1}{2} [-1 + \sqrt{1 + 4\mu^2}], \tag{6-30}$$

where the positive square root has been chosen, since we are interested in the largest eigenvalue.

To find the lowest allowed eigenvalue of $\mathcal{M}_z$ that is consistent with Eq. (6-26), we assume that $(\hbar k_{\min})$ is the lowest allowed eigenvalue. Application of $\mathcal{M}_-$ to the eigenfunction corresponding to that eigenvalue gives

$$\mathcal{M}_- Y(\hbar k_{\min}, \mu^2\hbar^2)=0, \tag{6-31}$$

and application of $\mathcal{M}_+$ to Eq. (6-31) followed by arguments analogous to those used in the previous discussion of the maximum allowed eigenvalue for $\mathcal{M}_z$ leads to

$$k_{\min}=\frac{1}{2}[1-\sqrt{1+4\mu^2}]. \tag{6-32}$$

Thus, we have

$$k_{\max}=-k_{\min}. \tag{6-33}$$

Since the spectrum of possible eigenvalues of $\mathcal{M}_z$ has the form

$$k_{\min}\hbar,\ (k_{\min}+1)_c,\ (k_{\min}+2)\hbar,\ \cdots,\ (k_{\max}-2)\hbar,\ (k_{\max}-1)\hbar,\ k_{\max}\hbar, \tag{6-34}$$

we see that

$$k_{\max}=k_{\min}+n, \tag{6-35}$$

where n is a positive integer. Using Eq. (6-33), we obtain

$$2k_{\max}=n. \tag{6-36}$$

This result allows several observations of interest to be made regarding the eigenvalues of a general angular momentum operator. First, we see that the eigenvalues of $\mathcal{M}_z$ are *quantized*, since n can take on only certain (integer) values. Next, we see that the eigenvalue spectrum of $\mathcal{M}_z$ is entirely discrete, i.e., there are no continuum contributions, and each of the eigenvalues is distinct, i.e., they are all nondegenerate. Finally, we see that Eq. (6-36) can be satisfied by either of two possible choices of $k_{\max}$. In particular, it can be chosen to be *either* an integer or a half-integer. This is entirely consistent with the notion that there should be two kinds of angular momentum for a single particle in quantum systems, and we shall see shortly how one choice corresponds to the allowed eigenvalues of orbital angular momentum, while another corresponds to the allowed eigenvalues of spin angular momentum.

Considering first the case when $k_{\max}$ = integer, the general form that the allowed eigenvalues can have can be seen by examination of Eq. (6-29), which can be written as

$$\mu^2=k_{\max}^2+k_{\max}=k_{\max}(k_{\max}+1). \tag{6-37}$$

This shows that μ^2 has the form

$$\mu^2=l(l+1) \tag{6-38}$$

with l a nonnegative integer. Combining this result with Eq. (6-25), we have

$$l(l+1) \geq m_z^2, \tag{6-39}$$

where m_z has been used to label the eigenvalue of $\mathfrak{M}_z$ for this particular case. Using this equation, consideration of the various allowed values for l (e.g., l = 0, 1, 2, ...) shows that the eigenvalues of $\mathfrak{M}_z$ are restricted to

$$-l \leq m_z \leq +l. \tag{6-40}$$

Thus, summarizing the results for the case when k_{max} is an integer, we have

$$\mathfrak{M}^2 Y(l, \mathfrak{M}_z) = l(l+1)\hbar^2 Y(l, \mathfrak{M}_z) \tag{6-41}$$

and

$$\mathfrak{M}_z Y(l, \mathfrak{M}_z) = m_z \hbar Y(l, \mathfrak{M}_z), \tag{6-42}$$

where l and m_z are integers (and l is nonnegative). As comparison with the experimental results[6] indicates, this corresponds to the allowed states of *orbital* angular momentum for electrons in quantum systems. Thus, the operators $\mathfrak{M}^2$ and $\mathfrak{M}_z$ correspond to $\mathfrak{L}^2$ and $\mathfrak{L}_z$, respectively, for this case.

The other case allowed in Eq. (6-36) is also of interest, namely, when k_{max} = half-integer. For this case, the allowed eigenvalues of $\mathfrak{M}_z$ will take the form [see Eq. (6-34)]

$$-|k_{\text{max}}|\hbar,\ -(|k_{\text{max}}|+1)\hbar,\ \cdots,\ -\frac{\hbar}{2},\ +\frac{\hbar}{2},\ \cdots,\ (|k_{\text{max}}|-1)\hbar,\ |k_{\text{max}}|\hbar, \tag{6-43}$$

where k_{max} is a half-integer that satisfies Eq. (6-37). If μ^2 is taken for this case[7] as

$$\mu^2 = s(s+1) \tag{6-44}$$

then we have

$$s(s+1) \geq s_z^2, \tag{6-45}$$

where s and s_z are half-integer (and s is nonnegative). Thus, we have

$$\mathfrak{M}^2 Y(s, s_z) = s(s+1)\hbar^2 Y(s, s_z) \tag{6-46}$$

and

$$\mathfrak{M}_z Y(s, s_z) = s_z \hbar Y(s, s_z). \tag{6-47}$$

---

[6] Chapter 1, Section 1-6. Also, see Chapter 7, Section 7-2, for illustration of orbital angular momentum eigenvalues and eigenvalues for the hydrogen atom.

[7] This case is treated entirely analogously to the case of k_{max} = integer, except that s (instead of l) will be used to denote the eigenvalue of $\mathfrak{M}^2$, and s_z (instead of m_z) will be used to denote the eigenvalue of $\mathfrak{M}_z$.

Consequently, it should not come as a great surprise that there is more than one kind of angular momentum that can be observed in quantum systems. Indeed, for the second case (k_{max} = half-integer), it is found experimentally[8] that electrons can have two spin states,[9] corresponding to[10] $s_z\hbar = \pm\hbar/2$, and $s(s+1)\hbar^2 = 3\hbar^2/4$, but this case does not have a suitable classical analog. However, we see that two kinds of angular momentum are to be expected for quantum systems (which we shall subsequently refer to as *orbital* and *spin* angular momentum), and both of them will need to be considered in discussions of atomic and molecular systems.

Thus far, we have concentrated on the *eigenvalues* of angular momentum in quantum systems. Before leaving the general discussion of angular momentum, let us discuss how the *eigenfunctions* of the two kinds of angular momentum can be represented. To do this, we shall find a matrix representation most convenient, since we shall see that both orbital and spin angular momentum can then be discussed simultaneously, without requiring a specific representation for the eigenfunctions.

The results for orbital and spin angular momentum that have been obtained thus far [Eqs. (6-41), (6-42), (6-46), and (6-47)] can be summarized as

$$\mathfrak{M}^2 Y(\mu, \mu_z) = (\mu)(\mu+1) Y(\mu, \mu_z), \tag{6-48}$$

and

$$\mathfrak{M}_z Y(\mu, \mu_z) = \mu_z Y(\mu, \mu_z), \tag{6-49}$$

where $\mathfrak{M}^2 = \mathcal{L}^2$ or $\mathcal{S}^2$, $\mathfrak{M}_z = \mathcal{L}_z$ or $\mathcal{S}_z$, $\mu = l$ or s, μ_z, m_z, or s_z, $|\mu| \geqslant \mu_z$, and where we have chosen to work in units of $\hbar$. Also, μ_z and μ may take on either integer or half-integer values.

Since the eigenvalues of $\mathfrak{M}^2$ and $\mathfrak{M}_z$ are all distinct, it is easily seen from Eq. (6-48) that

$$\langle Y(\mu', \mu_z')|\mathfrak{M}_z|Y(\mu, \mu_z)\rangle = \mu_z \delta_{\mu,\mu'} \delta_{\mu_z',\mu_z}. \tag{6-50}$$

In order to facilitate the discussion, we shall also be interested in the matrix elements of the raising and lowering operators. To obtain these, we note first that

$$\mathfrak{M}_+ Y(\mu, \mu_z) = a_{\mu_z} Y(\mu, \mu_z+1) \tag{6-51}$$

and

$$\mathfrak{M}_- Y(\mu, \mu_z) = b_{\mu_z} Y(\mu, \mu_z-1), \tag{6-52}$$

where the constants a_{μ_z} and b_{μ_z} arise because of the difference in normalization

[8] See Chapter 1, Section 1-6.

[9] Other particles (e.g., nuclei) are observed with spin states that are either intergral or half-integral, but electrons are apparently restricted to $s_z = \pm\hbar/2$.

[10] The specific operators having eigenvalues $s_z\hbar$ and $s(s+1)\hbar^2$ will be designated $\mathcal{S}_z$ and $\mathcal{S}^2$, respectively.

of $Y(\mu, \mu_z)$, $Y(\mu, \mu_z + 1)$, and $Y(\mu, \mu_z - 1)$. Multiplication of Eq. (6-51) by $Y(\mu, \mu_z + 1)$ gives

$$a_{\mu_z} = \langle Y(\mu, \mu_z + 1) | \mathfrak{M}_+ | Y(\mu, \mu_z) \rangle. \tag{6-53}$$

However, since [see Eqs. (6-11) and (6-12)]

$$\mathfrak{M}_-^\dagger = \mathfrak{M}_+, \tag{6-54}$$

Eq. (6-53) can be rewritten as

$$\begin{aligned} a_{\mu_z} &= \langle Y(\mu, \mu_z + 1) | \mathfrak{M}_-^\dagger | Y(\mu, \mu_z) \rangle \\ &= \langle \mathfrak{M}_- Y(\mu, \mu_z + 1) | Y(\mu, \mu_z) \rangle \end{aligned}$$

or

$$a_{\mu_z} = b^*_{\mu_z + 1}. \tag{6-55}$$

To obtain an explicit expression of these coefficients, we note first that

$$\begin{aligned} \mathfrak{M}_- \mathfrak{M}_+ Y(\mu, \mu_z) &= \mathfrak{M}_- [a_{\mu_z} Y(\mu, \mu_z + 1)] \\ &= a_{\mu_z} b_{\mu_z + 1} Y(\mu, \mu_z) \\ &= |a_{\mu_z}|^2 Y(\mu, \mu_z). \end{aligned} \tag{6-56}$$

Also, we note that an alternative expression for the above can be obtained using Eq. (6-28), i.e.,

$$\begin{aligned} \mathfrak{M}_- \mathfrak{M}_+ Y(\mu, \mu_z) &= (\mathfrak{M}^2 - \mathfrak{M}_z^2 - \mathfrak{M}_z) Y(\mu, \mu_z) \\ &= [\mu(\mu + 1) - \mu_z^2 - \mu_z] Y(\mu, \mu_z) \\ &= [\mu(\mu + 1) - \mu_z(\mu_z + 1)] Y(\mu, \mu_z). \end{aligned} \tag{6-57}$$

Thus, combining Eqs. (6-56) and (6-57), we obtain

$$|a_{\mu_z}| = [\mu(\mu + 1) - \mu_z(\mu_z + 1)]^{1/2} \tag{6-58}$$

or

$$|a_{\mu_z}| = [(\mu - \mu_z)(\mu + \mu_z + 1)]^{1/2}. \tag{6-59}$$

This also means that

$$b_{\mu_z} = |a^*_{\mu_z - 1}| = [(\mu + \mu_z)(\mu - \mu_z + 1)]^{1/2}. \tag{6-60}$$

The desired matrix elements of $\mathfrak{M}_+$ and $\mathfrak{M}_-$ are now easily seen to be given by

$$\langle Y(\mu', \mu_z') | \mathfrak{M}_+ | Y(\mu, \mu_z) \rangle = [(\mu - \mu_z)(\mu + \mu_z + 1)]^{1/2} \delta_{\mu', \mu} \delta_{\mu_z', \mu_z + 1}, \tag{6-61}$$

and

$$\langle Y(\mu', \mu_z') | \mathfrak{M}_- | Y(\mu, \mu_z) \rangle = [(\mu + \mu_z)(\mu - \mu_z + 1)]^{1/2} \delta_{\mu, \mu'} \delta_{\mu_z', \mu_z - 1}. \tag{6-62}$$

Finally, the matrix elements of $\mathfrak{M}^2$ are simply

$$\langle Y(\mu', \mu_z')|\mathfrak{M}^2| Y(\mu, \mu_z)\rangle = \mu(\mu+1)\delta_{\mu',\mu}\delta_{\mu_z',\mu_z}. \tag{6-63}$$

Thus, the matrix elements that are given in Eqs. (6-50), (6-61), (6-62), and (6-63) provide all of the information that is needed to represent the eigenvectors of angular momentum. To illustrate the actual form that these matrix representations take, let us consider a few specific examples.

For the simplest case of $\mu = 0$, (and thus $\mu_z = 0$) we see from Eqs. (6-59) and (6-60) that $a_{\mu_z} = b_{\mu_z} = 0$. Also, the dimension of the representation is one, since only one μ_z eigenvalue is present. Thus, the matrix representation of the operators $\mathfrak{M}^2$, $\mathfrak{M}_z$, $\mathfrak{M}_+$, and $\mathfrak{M}_-$ for this case is seen from Eqs. (6-50) and (6-61) through (6-63) to be simply

$$\mathbf{M}^2 = \mathbf{0} = \mathbf{M}_z = \mathbf{M}_+ = \mathbf{M}_-. \tag{6-64}$$

In other words, the matrix representation of each of the operators for this case consists of a null matrix, with dimension 1×1.

Next, for $\mu = \frac{1}{2}$, we have $\mu_z = \pm\frac{1}{2}$, and thus we expect a two dimensional representation for each of the operators. Using Eqs. (6-50) and (6-61) through (6-63), we see that

$$\mathbf{M}^2 = \begin{pmatrix} 3/4 & 0 \\ 0 & 3/4 \end{pmatrix}, \qquad \mathbf{M}_z = \begin{pmatrix} -1/2 & 0 \\ 0 & +1/2 \end{pmatrix} \tag{6-65}$$

$$\mathbf{M}_+ = \begin{pmatrix} 0 & 1 \\ 0 & 0 \end{pmatrix}, \qquad \mathbf{M}_- = \begin{pmatrix} 0 & 0 \\ 1 & 0 \end{pmatrix} \tag{6-66}$$

where the first row corresponds to $\mu_z = -\frac{1}{2}$ and the second row corresponds to $\mu_z = +\frac{1}{2}$.

Finally, for the case $\mu = 1$, we have $\mu_z = +1, 0, -1$, which results in

$$\mathbf{M}^2 = \begin{pmatrix} 2 & 0 & 0 \\ 0 & 2 & 0 \\ 0 & 0 & 2 \end{pmatrix}, \qquad \mathbf{M}_z = \begin{pmatrix} 1 & 0 & 0 \\ 0 & 0 & 0 \\ 0 & 0 & -1 \end{pmatrix} \tag{6-67}$$

and

$$\mathbf{M}_+ = \begin{pmatrix} 0 & \sqrt{2} & 0 \\ 0 & 0 & \sqrt{2} \\ 0 & 0 & 0 \end{pmatrix}, \qquad \mathbf{M}_- = \begin{pmatrix} 0 & 0 & 0 \\ \sqrt{2} & 0 & 0 \\ 0 & \sqrt{2} & 0 \end{pmatrix}, \tag{6-68}$$

where the first, second, and third rows correspond to $\mu_z = +1, 0 -1$, respectively. Matrix representations for other eigenvalues of μ and μ_z' are equally straightforward to create.[11]

[11] See Problem 6-5.

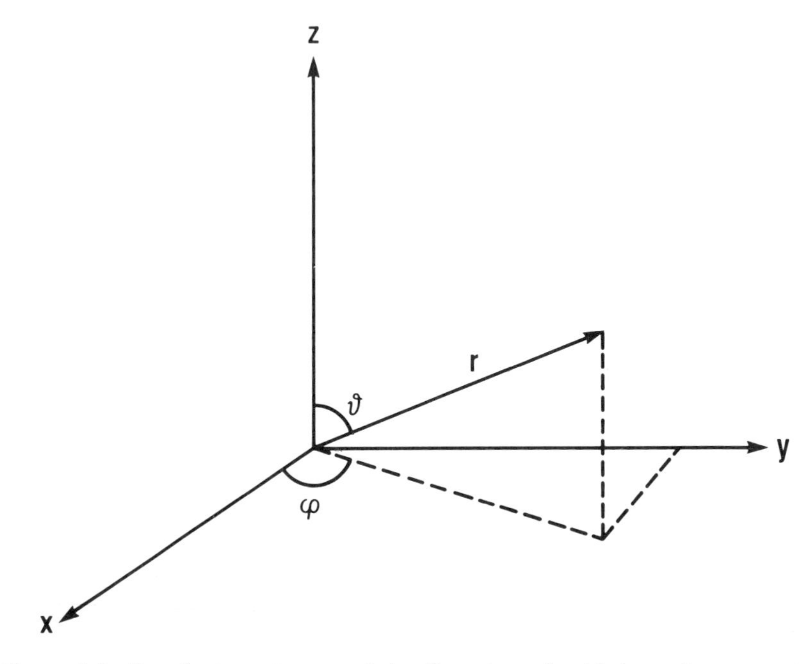

Figure 6.1. Coordinate systems used in discussion of orbital angular momentum operators.

6-3. Orbital Angular Momentum

For the case of integer values of μ (and μ_z), which we have seen correspond to the allowed eigenvalues of *orbital* angular momentum, we might expect that since there is a classical analog to this kind of angular momentum, there will be another representation of the eigenfunctions that uses the ordinary three-dimensional [e.g., (x, y, z)] representation that is found in classical mechanics. This is indeed the case, and we shall now investigate that representation in greater detail.

Since the treatment of this representation is carried out conveniently in spherical polar coordinates, we begin by transforming the operators of interest to spherical polar coordinates. In particular, using Fig. 6.1, we see that

$$\begin{aligned} r^2 &= x^2 + y^2 + z^2 \\ \vartheta &= \cos^{-1}(z/r) \\ \varphi &= \tan^{-1}(y/x) \end{aligned} \tag{6-69}$$

and

$$\begin{aligned} z &= r \cos \vartheta \\ y &= r \sin \vartheta \sin \varphi \\ x &= r \sin \vartheta \cos \varphi . \end{aligned} \tag{6-70}$$

Using these relations, Eqs. (6-3), (6-4), (6-5), (6-10), and the discussion in Chapter 4 (Section 4-7), it can be shown[12] that

$$\mathfrak{L}_z = \left(\frac{\hbar}{i}\right)\frac{\partial}{\partial\varphi}$$

$$\mathfrak{L}_y = \left(\frac{\hbar}{i}\right)\left\{\cos\varphi\frac{\partial}{\partial\vartheta} - \cot\vartheta\sin\varphi\frac{\partial}{\partial\varphi}\right\} \tag{6-71}$$

$$\mathfrak{L}_x = -\left(\frac{\hbar}{i}\right)\left\{\sin\varphi\frac{\partial}{\partial\vartheta} + \cot\vartheta\cos\varphi\frac{\partial}{\partial\varphi}\right\}$$

and

$$\mathfrak{L}^2 = -\hbar^2\left\{\frac{1}{\sin\vartheta}\frac{\partial}{\partial\vartheta}\left(\sin\vartheta\frac{\partial}{\partial\vartheta}\right) + \frac{1}{\sin^2\vartheta}\frac{\partial^2}{\partial\varphi^2}\right\}. \tag{6-72}$$

One of the advantages of transforming to spherical coordinates is now revealed, i.e., it is apparent that $\mathfrak{L}^2$ and the components of $\mathfrak{L}$ can be written in a form that depends only upon two of the three independent variables in this coordinate system. Thus, the equations to be solved for the eigenfunctions in this representation can be written as

$$-\left\{\frac{1}{\sin\vartheta}\frac{\partial}{\partial\vartheta}\left(\sin\vartheta\frac{\partial}{\partial\vartheta}\right) + \frac{1}{\sin^2\vartheta}\frac{\partial^2}{\partial\varphi^2}\right\} Y_{l,m_z}(\vartheta,\varphi) = l(l+1)\,Y_{l,m_z}(\vartheta,\varphi) \tag{6-73}$$

and

$$\left(\frac{1}{i}\right)\frac{\partial}{\partial\varphi} Y_{l,m_z}(\vartheta,\varphi) = m_z Y_{l,m_z}(\vartheta,\varphi), \qquad (-l \le m_z \le +l), \tag{6-74}$$

where we have again chosen to work in units of $\hbar$, and the form of the eigenvalues that was found in the previous section [Eqs. (6-41) and (6-42)] was inserted.

To obtain the explicit form for $Y_{l,m_z}(\vartheta,\varphi)$, let us assume that the dependence on the variables can be separated, i.e.,

$$Y_{l,m_z}(\vartheta,\varphi) = \Theta(\vartheta)\Phi(\varphi), \tag{6-75}$$

and see if Eqs. (6-73) and (6-74) are consistent with this assumption. Insertion of Eq. (6-75) into (6-73) and (6-74) and division by $Y_{l,m_z}(\vartheta,\varphi)$, gives[13]

$$-\frac{\sin\vartheta}{\Theta(\vartheta)}\frac{\partial}{\partial\vartheta}\left(\sin\vartheta\frac{\partial\Theta(\vartheta)}{\partial\vartheta}\right) - l(l+1)\sin^2\vartheta = \frac{1}{\Phi(\varphi)}\frac{\partial^2\Phi(\varphi)}{\partial\varphi^2} \tag{6-76}$$

[12] See Problem 6-6.

[13] This division requires special attention if $Y_{l,m_z}(\vartheta,\varphi) = 0$ for some ϑ and φ. However, examination of the eigenfunctions that are obtained using the assumption that $Y_{l,m_z}(\vartheta,\varphi) \neq 0$ will show that Eqs. (6-76) and (6-77) are satisfied even when $Y_{l,m_z}(\vartheta,\varphi) = 0$.

and

$$\frac{\partial \Phi(\varphi)}{\partial \varphi} = i m_z \Phi(\varphi), \tag{6-77}$$

and the variables have been separated, thus justifying the assumption.

The solution to Eq. (6-77) is easily seen to be

$$\Phi(\varphi) = A e^{i m_z \varphi} \qquad (0 \le \varphi \le 2\pi) \tag{6-78}$$

where the choice $A = 1/\sqrt{2\pi}$ is seen to normalize $\Phi(\varphi)$ to unity. Turning to Eq. (6-76), we see that if it is to hold for all ϑ and φ, it is necessary that

$$\frac{1}{\Theta(\vartheta)}\left[\sin\vartheta \frac{\partial}{\partial\vartheta}\left(\sin\vartheta \frac{\partial\Theta(\vartheta)}{\partial\vartheta}\right) + l(l+1)\sin^2\vartheta\,\Theta(\vartheta)\right] = \text{constant} = K, \tag{6-79}$$

and

$$\frac{1}{\Phi(\varphi)}\frac{\partial^2\Phi(\varphi)}{\partial\varphi^2} = K \tag{6-80}$$

be satisfied simultaneously. However, we note that the eigenfunction of Eq. (6-77) that is given in Eq. (6-78) also satisfies Eq. (6-80), if

$$K = -m_z^2. \tag{6-81}$$

Thus, we have that $\Phi(\varphi)$ is simultaneously an eigenfunction of both $\mathcal{L}_z$ and $\mathcal{L}^2$, which is consistent with

$$[\mathcal{L}_z, \mathcal{L}^2] = 0,$$

and we are left only with the problem of solving

$$\left\{\left(\frac{1}{\sin\vartheta}\right)\frac{\partial}{\partial\vartheta}\left(\sin\vartheta\frac{\partial}{\partial\vartheta}\right) - \left(\frac{m_z^2}{\sin^2\vartheta}\right)\right\}\Theta(\vartheta) = -l(l+1)\Theta(\vartheta). \tag{6-82}$$

To express this equation in a more "familiar" form, it is useful to let

$$\xi = \cos\vartheta \qquad (-1 \le \xi \le +1), \tag{6-83}$$

which allows Eq. (6-82) to be rewritten as

$$\frac{\partial}{\partial\xi}\left\{(1-\xi^2)\frac{\partial P(\xi)}{\partial\xi}\right\} - \frac{m_z^2}{(1-\xi^2)}P(\xi) = -l(l+1)P(\xi), \tag{6-84}$$

where the desired solution is referred to as $P(\xi)$ instead of $\Theta(\vartheta)$.

Equations of this form and their solutions have been studied in great detail, and the form of differential equation is known as the *associated Legendre equation,*[14] and the eigenfunctions are known as the *associated Legendre*

[14] See, for example, G. Sansone, *Orthogonal Functions*, Interscience Publishers, New York, 1959, Chapter 3.

functions. Since the procedures for obtaining these solutions will not be of general utility in later discussions, they will not be introduced here. Instead, the reader can easily establish[15] that the following functions *are* eigenfunctions of the operator represented in Eq. (6-84), and further details of the procedures that can be used to obtain such solutions can be found elsewhere[14]:

$$P_l^{m_z}(\xi)=(1-\xi^2)^{|m_z|/2}\frac{d^{|m_z|}}{d\xi^{|m_z|}}P_l(\xi), \tag{6-85}$$

where $P_l^{m_z}(\xi)$ are the associated Legendre functions, and $P_l(\xi)$ are the ("ordinary") Legendre polynomials that are given by

$$P_l(\xi)=\frac{1}{2^l l!}\frac{d^l}{d\xi^l}(\xi^2-1)^l. \tag{6-86}$$

Formulas such as given in Eqs. (6-85) and (6-86) that define the function of interest in terms of derivatives of other functions are frequently known as *generating formulas*, and the particular formula in Eq. (6-86) is known as the *Rodrigues generating formula*. Some examples of the first several Legendre polynomials and associated Legendre functions are as follows:

$$\begin{aligned}
&P_0(\xi)=1 && P_0^0(\xi)=1\\
& && P_1^1(\xi)=P_1^{-1}(\xi)=(1-\xi^2)^{1/2}\\
&P_1(\xi)=\xi && P_1^0(\xi)=\xi\\
& && P_2^2(\xi)=P_2^{-2}(\xi)=3(1-\xi^2)\\
&P_2(\xi)=\frac{1}{2}(3\xi^2-1) && P_2^1(\xi)=P_2^{-1}(\xi)=3\xi(1-\xi^2)^{1/2}\\
& && P_2^0(\xi)=\frac{1}{2}(3\xi^2-1).
\end{aligned} \tag{6-87}$$

Several points of interest should be noted about these functions. First, since the eigenvalues $l(l+1)$ are all different, we are assured that each of the $P_l(\xi)$ are orthogonal,[16] i.e.,

$$\int_{-1}^{+1} P_l(\xi)P_{l'}(\xi)\,d\xi=0 \qquad (l\neq l'). \tag{6-88}$$

Next, it will be of interest to normalize each function to unity. To do this, we use

---

[15] See Problem 6-7.

[16] Theorem 3-20.

the Rodrigues formula to evaluate

$$I_l = \int_{-1}^{+1} P_l(\xi)P_l(\xi)\, d\xi = \frac{1}{2^{2l}(l!)^2} \int_{-1}^{+1} \left[\frac{d^l}{d\xi^l}(\xi^2-1)^l\right]\left[\frac{d^l}{d\xi^l}(\xi^2-1)^l\right] d\xi. \tag{6-89}$$

Integrating by parts gives

$$I_l = \frac{1}{2^{2l}(l!)^2}\left\{\left[\frac{d^l}{d\xi^l}(\xi^2-1)^l\right]\left[\frac{d^{l-1}}{d\xi^{l-1}}(\xi^2-1)^l\right]\Bigg|_{-1}^{+1} - \int_{-1}^{+1}\left[\frac{d^{l-1}}{d\xi^{l-1}}(\xi^2-1)^l\right]\left[\frac{d^{l+1}}{d\xi^{l+1}}(\xi^2-1)^l\right] d\xi. \right. \tag{6-90}$$

Since the first term in Eq. (6-90) contains only the $(l-1)$st derivative of $(\xi^2-1)^l$, there will be a factor of (ξ^2-1) in each term. Thus, the first term in Eq. (6-91) will vanish at each end point, and we have

$$I_l = \left[\frac{-1}{2^{2l}(l!)^2}\right] \int_{-1}^{+1}\left[\frac{d^{l-1}}{d\xi^{l-1}}(\xi^2-1)^l\right]\left[\frac{d^{l+1}}{d\xi^{l+1}}(\xi^2-1)^l\right] d\xi. \tag{6-91}$$

Repeated integration by parts until it has been done l times yields

$$I_l = \left[\frac{(-1)^l}{2^{2l}(l!)^2}\right] \int_{-1}^{+1} (\xi^2-1)^l \left[\frac{d^{2l}}{d\xi^{2l}}(\xi^2-1)^l\right] d\xi. \tag{6-92}$$

However, we note that

$$\frac{d^{2l}}{d\xi^{2l}}(\xi^2-1)^l = (2l)! \tag{6-93}$$

and

$$\int_{-1}^{+1} (1-x)^n(1+x)^n\, dx = \left(\frac{(n!)^2}{(2n)!}\right) \cdot \frac{2^{2n+1}}{(2n+1)}, \tag{6-94}$$

which allows Eq. (6-92) to be written as

$$I_l = \left(\frac{2}{2l+1}\right). \tag{6-95}$$

Thus, the *normalized Legendre polynomials* are given by

$$\mathcal{P}_l(\xi) = \left[\frac{2l+1}{2}\right]^{1/2} P_l(\xi) \tag{6-96}$$

A similar procedure can be used for the associated Legendre functions[17] to

[17] See Problem 6-8.

show that they are orthogonal, and that, when normalized to unity, are given by

$$\mathcal{P}_l^{m_z}(\xi)=\left[\left(\frac{2l+1}{2}\right)\left(\frac{(l-|m_z|)!}{(l+|m_z|)!}\right)\right]^{1/2} P_l(\xi). \tag{6-97}$$

In addition to generating Legendre polynomials and associated Legendre functions by means of formulas such as the generating formula of Eq. (6-86), there are other relationships that are frequently used. These relationships, called *recursion relations*, connect a given Legendre polynomial or associated Legendre function to another (or the derivative of another).

Extensive tabulations and derivations of such relations exist,[18] and several of these are listed below to illustrate the idea.

$$(l+1)P_l(x)=P'_{l+1}(x)-xP'_l(x), \tag{6-98}$$

$$(2l+1)P_l(x)=P'_{l+1}(x)-P'_l(x), \tag{6-99}$$

$$(l+1)P_{l+1}(x)+lP_{l-1}(x)=(2l+1)xP_l(x), \tag{6-100}$$

$$lP_n(x)=xP'_l(x)-P'_{l-1}(x), \tag{6-101}$$

$$(x^2-1)P'_l(x)=lxP_l(x)-lP_{l-1}(x), \tag{6-102}$$

where

$$P'_l(x)=\frac{d}{dx}P_l(x). \tag{6-103}$$

For the associated Legendre functions, examples include

$$P^m_{l-1}(x)-P^m_{l+1}(x)=-(2l+1)(x^2-1)^{1/2}P_l^{m-1}(x), \tag{6-104}$$

$$P_l^{m+2}(x)+2(m+1)x(x^2-1)^{-1/2}P_l^{m+1}(x)=(l-m)(l+m+1)P_l^m(x), \tag{6-105}$$

$$(2l+1)xP_l^m(x)=(l-m+1)P^m_{l+1}(x) + (l+m)P^m_{l-1}(x). \tag{6-106}$$

These relationships are useful not only in generating new Legendre polynomials or associated Legendre functions, but also in evaluating integrals over these functions.

Finally, there is a relationship of interest that connects different functions with different arguments that should be noted, known as the *addition*

[18] See, for example, H. Bateman, *Higher Transcendental Functions*, Vol. I, Bateman Manuscript Project, A. Erdelyi, ed., McGraw Hill, New York, 1953, Section 3.8; G. Szegö, *Orthogonal Polynomials*, Vol. 23, Amer. Math. Soc. Colloq. Publications, Amer. Math. Soc., Providence, RI, 1939; E. A. Hylleraas, *Mathematical and Theoretical Physics,* Vol. 1, Wiley-Interscience, New York, 1970, Section 38.

theorem,[19] i.e.,

$$P_l(\cos\Theta) = P_l(\cos\vartheta_1)P_l(\cos\vartheta_2)$$

$$+2\sum_{m=-l}^{+l}\frac{(l-m)!}{(l+m)!}\cdot P_l^m(\cos\vartheta_1)P_l^m(\cos\vartheta_2)\cos m(\varphi_1-\varphi_2), \quad (6\text{-}107)$$

where Θ is the angle formed between two lines passing through the origin having angles ϑ_1, φ_1, and ϑ_2, φ_2.

If we combine the results obtained, we can obtain another result of interest. In particular, combination of Eqs. (6-78), (6-83), and (6-97) allows (6-75) to be rewritten as

$$Y_{l,m_z}(\vartheta, \varphi) = \left[\left(\frac{2l+1}{4\pi}\right)\left(\frac{(l-|m_z|)!}{(l+|m_z|)!}\right)\right]^{1/2} P_l^{m_z}(\cos\vartheta)e^{im_z\varphi}. \quad (6\text{-}108)$$

These $Y_{l,m_z}(\vartheta, \varphi)$ are orthonormal, i.e.,

$$\int_0^{2\pi} d\varphi \int_{-\pi}^{+\pi} \sin\vartheta\, d\vartheta\, Y^*_{l',m_z'}(\vartheta, \varphi)\, Y_{l,m_z}(\vartheta, \varphi) = \delta_{l,l'}\delta_{m_z',m_z}, \quad (6\text{-}109)$$

and are known as *spherical harmonics*. They will be used frequently in later discussions. Several examples of the normalized spherical harmonics for selected l and m_z are given in Table 6.1. It should be noted that there are other ways in which the spherical harmonics are frequently written, that maintain the same normalization to unity. The difference arises because of the inclusion (or lack of it) of an additional factor of $(-1)^{m_z}$ in the definition of the normalization constant in Eq. (6-108). Such an additional factor does not change any of the properties of these functions, but does require caution in making comparisons of various treatments.

6-4. Spin Angular Momentum

From the Stern–Gerlach experiments, we saw earlier[20] that only two values of $\mathcal{S}_z$ are observed experimentally for electron spin, i.e.,

$$s_z = \pm\frac{1}{2}, \quad (6\text{-}110)$$

where we have chosen to work in units of $\hbar$. This also implies that the only allowed eigenvalue of $\mathcal{S}^2$ is

$$s(s+1) = \frac{1}{2}\left(\frac{1}{2}+1\right) = \frac{3}{4}. \quad (6\text{-}111)$$

---

[19] See, for example, H. Margenau and G. M. Murphy, *The Mathematics of Physics and Chemistry,* Vol. I, Van Nostrand, New York, 1961, Section 3.7.

[20] See Chapter 1, Section 1-6.

Table 6.1. Tabulation of the First Several Normalized Spherical Harmonics.

l	m_z	$Y_{lm_z}(\vartheta, \varphi)$
0	0	$Y_{00}=\left(\frac{1}{4\pi}\right)^{1/2}$
1	0	$Y_{10}=\left(\frac{3}{4\pi}\right)^{1/2} \cos\vartheta$
	± 1	$Y_{1,\pm 1}=\left(\frac{3}{8\pi}\right)^{1/2} \sin\vartheta e^{\pm i\varphi}$
2	0	$Y_{20}=\frac{1}{2}\left(\frac{5}{4\pi}\right)^{1/2} (3\cos^2\vartheta-1)$
	± 1	$Y_{2,\pm 1}=\left(\frac{15}{8\pi}\right)^{1/2} \sin\vartheta \cos\vartheta e^{\pm i\varphi}$
	± 2	$Y_{2,\pm 2}=\left(\frac{1}{4}\right)\left(\frac{15}{2\pi}\right)^{1/2} \sin^2\vartheta e^{\pm 2i\varphi}$
3	0	$Y_{30}=\left(\frac{1}{2}\right)\left(\frac{7}{4\pi}\right)^{1/2} (5\cos^3\vartheta-3\cos^3\vartheta)$
	± 1	$Y_{3,\pm 1}=\left(\frac{1}{4}\right)\left(\frac{21}{4\pi}\right)^{1/2} \sin\vartheta(5\cos^2\vartheta-1)e^{\pm i\varphi}$
	± 2	$Y_{3,\pm 2}=\left(\frac{1}{4}\right)\left(\frac{105}{2\pi}\right)^{1/2} \sin^2\vartheta \cos\vartheta e^{\pm 2i\varphi}$
	± 3	$Y_{3,\pm 3}=\left(\frac{1}{4}\right)\left(\frac{35}{4\pi}\right) \sin^3\vartheta e^{\pm 3i\varphi}$

Earlier in this chapter we saw how to construct a matrix representation of the square of the general angular momentum operator and its z-component [Eqs. (6-65)] for the eigenvalues found for electron spin. In order to complete the matrix description of the eigenfunctions, a representation for the x- and y-components will now be obtained. From the definition of $\mathfrak{M}_+$ and $\mathfrak{M}_-$ [see Eqs. (6-11) and (6-12)], we have (using "$\mathcal{S}$" instead of "$\mathfrak{M}$" to represent electron spin operators)

$$\frac{1}{2}(\mathcal{S}_+ + \mathcal{S}_-) = \mathcal{S}_x$$
$$\frac{1}{2}(\mathcal{S}_+ - \mathcal{S}_-) = i\mathcal{S}_y. \tag{6-112}$$

The matrix analog of Eq. (6-112) is thus

$$\mathbf{S}_x = \frac{1}{2}[\mathbf{S}_+ + \mathbf{S}_-]$$
$$\mathbf{S}_y = \left(\frac{i}{2}\right)[\mathbf{S}_- - \mathbf{S}_+]. \tag{6-113}$$

Using the specific representations of $\mathbf{S}_+$ and $\mathbf{S}_-$ that have already been obtained for the case of interest [Eq. (6-66)], we obtain the desired result:

$$\mathbf{S}_x = \begin{pmatrix} 0 & 1/2 \\ 1/2 & 0 \end{pmatrix} \tag{6-114}$$

and

$$\mathbf{S}_y = \begin{pmatrix} 0 & -i/2 \\ i/2 & 0 \end{pmatrix}. \tag{6-115}$$

Therefore, we now have a matrix representation of $\mathfrak{M}^2$ and all of the components of $\mathfrak{M}$ for the case of electron spin angular momentum. The factors of $\frac{1}{2}$ are usually omitted during calculations (and inserted at the end), and the matrices ($\boldsymbol{\sigma} = 2\mathbf{S}$), consisting of

$$\boldsymbol{\sigma}_1 = \boldsymbol{\sigma}_x = \begin{pmatrix} 0 & 1 \\ 1 & 0 \end{pmatrix}, \qquad \boldsymbol{\sigma}_2 = \boldsymbol{\sigma}_y = \begin{pmatrix} 0 & -i \\ i & 0 \end{pmatrix},$$
$$\boldsymbol{\sigma}_3 = \boldsymbol{\sigma}_z = \begin{pmatrix} 1 & 0 \\ 0 & -1 \end{pmatrix}, \qquad \boldsymbol{\sigma}^2 = \frac{1}{2}\begin{pmatrix} 3 & 0 \\ 0 & 3 \end{pmatrix}, \tag{6-116}$$

which are known as the *Pauli spin matrices*, are used.[21]

In the case of electron spin it is not only *convenient* to use a matrix representation as we have done (the Pauli matrices), but it is *necessary*. This arises from the fact that although *orbital* angular momentum has a classical analog in a particle revolving about a point, there is *no satisfactory direct classical analog* for spin angular momentum. Therefore, we do not expect to be able to find a representation of the eigenfunctions of spin in a (r, ϑ, φ) coordinate system, since there is no classical mechanical analog from which the corresponding quantum mechanical spin operator can be formed. Rather, *only* the matrix representation is possible.

This lack of an ordinary three-dimensional representation of electron spin is why it is usually referred to as an *intrinsic property of the electron*. This means that, just like the charge or the mass of the electron (also intrinsic properties), we do not expect to obtain an (r, ϑ, φ) representation of them. However, since electron spin has more than one possible value, we must identify the spin

---

[21] See Section 8-3A for illustration of how Pauli spin matrices can be seen to arise directly from the solution of an appropriate relativistic form of the Schrödinger equation.

quantum numbers (i.e., the "spin state") as well as other quantum numbers in order to specify the state of an electron completely. In later chapters we shall see how this can be done for multielectron systems. Finally, we should also label the state of the electron by the eigenvalue of $\mathcal{S}^2(\mathcal{M}^2)$, but since the only value that is allowed is $s(s+1) = 3/4$, it is usually omitted for convenience.

Problems

1. From the definitions of $\mathcal{L}_x$, $\mathcal{L}_y$, and $\mathcal{L}_z$ given in Eqs. (6-3)–(6-5), show that

$$[\mathcal{L}_y, \mathcal{L}_z] = i\hbar\mathcal{L}_x$$

and

$$[\mathcal{L}_z, \mathcal{L}_x] = i\hbar\mathcal{L}_y.$$

2. From the definitions of Eqs. (6-9)–(6-13),
   a. Show that $\mathcal{M}^2$, $\mathcal{M}_x$, $\mathcal{M}_y$ and $\mathcal{M}_z$ are Hermitian.
   b. Show that $\mathcal{M}_+$ and $\mathcal{M}_-$ are not Hermitian.
3. Prove that

$$\mathcal{M}^2(\mathcal{M}_+ Y) = \mu^2\hbar^2(\mathcal{M}_\pm Y).$$

4. If Y is normalized to unity, show that $(\mathcal{M}_+ Y)$ can be normalized to unity with normalization constant:

$$N = \frac{1}{[m^2 - m_z^2 - m_z]^{1/2} \cdot \hbar}.$$

5. Construct matrix representations of $\mathcal{M}^2$, $\mathcal{M}_z$, $\mathcal{M}_+$ and $\mathcal{M}_-$ for the case $\mu = 3/2$.
6. Verify Eqs. (6-71) and (6-72), using the relations

$$\frac{\partial}{\partial x} = \sin\vartheta\cos\varphi\frac{\partial}{\partial r} + \cos\vartheta\cos\varphi\left(\frac{1}{r}\right)\frac{\partial}{\partial\vartheta} - \frac{\sin\varphi}{r\sin\vartheta}\frac{\partial}{\partial\varphi},$$

$$\frac{\partial}{\partial y} = \sin\vartheta\sin\varphi\frac{\partial}{\partial r} + \cos\vartheta\sin\varphi\left(\frac{1}{r}\right)\frac{\partial}{\partial\vartheta} + \frac{\cos\varphi}{r\sin\vartheta}\frac{\partial}{\partial\varphi},$$

$$\frac{\partial}{\partial z} = \cos\vartheta\frac{\partial}{\partial r} - \sin\vartheta\left(\frac{1}{r}\right)\frac{\partial}{\partial\vartheta},$$

   along with Eqs. (6-3), (6-4), (6-5), (6-10) and Eq. (4-205).
7. Verify by direct substitution that $\Phi(\varphi)$ given by Eq. (6-78) is a solution of Eq. (6-77), and that $P_l^{m_z}$ given by Eq. (6-85) is a solution to Eq. (6-84).
8. Show that the $P_l^m(\xi)$ given by Eq. (6-97) are orthonormal.
9. Show that, for a particle acted upon by a potential $V(r)$ that depends only on the distance,

$$[\mathcal{H}, \mathcal{L}_t] = 0 \qquad t = x, y, z$$

and

$$[\mathcal{H}, \mathcal{L}^2] = 0.$$

10. Using the generating formula of Eq. (6-86), drive the following recursion relation [Eq. (6-98)]:

$$(l+1)P_l(x) = P'_{l+1}(x) - xP'_l(x).$$

11. Show that $(1 - 2xt + t^2)^{-1/2}$ is a generating function for Legendre polynomials. (See Problem 5-8 for a definition of "generating function.")
12. Prove the following identities (in units of $\hbar$):

a. $\mathfrak{M}^2 = \mathfrak{M}_+\mathfrak{M}_- + \mathfrak{M}_z^2 - \mathfrak{M}_z$

b. $\mathfrak{M}^2 = \mathfrak{M}_-\mathfrak{M}_+ + \mathfrak{M}_z^2 + \mathfrak{M}_z$

c. $[\mathfrak{M}_+, \mathfrak{M}_z] = -\mathfrak{M}_+$

d. $[\mathfrak{M}_-, \mathfrak{M}_z] = \mathfrak{M}_-$

e. $[\mathfrak{M}_+, \mathfrak{M}_-] = 2\mathfrak{M}_z$.

Chapter 7

The Hydrogen Atom, Rigid Rotor, and the H_2^+ Molecule

After developing the notions of angular momentum in the previous chapter, we are now ready to use these ideas to help solve some important applications. As a first example, we shall examine the hydrogen atom and hydrogen-like atoms. It should be noted that the hydrogen atom is not only an important historical contribution to theoretical chemistry. As we shall see later, a substantial number of the qualitative and quantitative concepts that are used concerning complex atoms and molecules are couched in terms of hydrogenic orbitals. Consequently, it is of importance to study the eigenfunctions and associated eigenvalues of the hydrogen atom in some detail.[1]

7-1. Separation of Motion of Center of Mass

Throughout the following discussion and most of the remainder of this text, we shall assume that relativistic and external field effects will be small enough to be ignored. This is not always a valid assumption, and the reader should examine each problem individually before adopting this approximation.[2] Within this approximation, we shall now be interested in obtaining the solutions to the time-independent Schrödinger equation for the hydrogen atom. Since there are only two bodies in the problem, it is easily seen that the quantum mechanical

[1] Because of the historical importance of the hydrogen atom and its quantum mechanical treatment, the approach used here will consider the differential equations involved. However, this problem can also be solved using a matrix approach, cf. M. Bauder and C. Itzykson, *Rev. Mod. Phys.*, **838**, 330 (1966).

[2] For an excellent discussion of relativistic and external field effects, see H. A. Bethe and E. E. Saltpeter, *Quantum Mechanics of One- and Two-Electron Atoms*, Springer-Verlag, Berlin, 1957. See also Chapter 8 in this text for discussion of these effects.

Hamiltonian for the system is given by

$$\mathcal{H} = -\frac{\hbar^2}{2m_N}\nabla_N^2 - \frac{\hbar^2}{2m_e}\nabla_e^2 - \frac{Ze^2}{|\mathbf{r}_N - \mathbf{r}_e|}, \tag{7-1}$$

where the first two terms represent the kinetic energy of the nucleus and electron, respectively. The final term represents the electrostatic interaction between the electron and the nucleus, i.e., it is the quantum mechanical operator corresponding to the classical Coulombic attraction of two particles with charge $-Ze$ and $+e$. For the particular case of the hydrogen atom, $Z = 1$. However, we shall also be interested in examining the Schrödinger equation for arbitrary values of Z. These latter cases are usually called hydrogen-like atoms, since they also contain only one electron.

The time-independent Schrödinger equation to be solved has the form

$$\mathcal{H}\Psi(\mathbf{r}_N, \mathbf{r}_e) = E\Psi(\mathbf{r}_N, \mathbf{r}_e). \tag{7-2}$$

Since $\Psi(\mathbf{r}_N, \mathbf{r}_e)$ contains information concerning both the nuclear and electron motion, it must be a function of six coordinates, but we are free to measure these coordinates from any arbitrary origin. It will be easier to solve this problem by choosing an alternate set of six coordinates that is related to the original x, y, z coordinates of the nucleus and electron in the following manner. The alternate coordinates will be chosen to be those corresponding to the motion of the center of mass of the nucleus-electron combination, and the coordinates describing the motion of the electron relative to the nucleus. These coordinates are indicated in Fig. 7.1.

From classical mechanics, the coordinates of the center of mass of the system are given by

$$\mathbf{R} = \frac{m_N\mathbf{r}_N + m_e\mathbf{r}_e}{m_N + m_e}. \tag{7-3}$$

The other coordinates can be easily seen from the figure to be

$$\mathbf{r} = \mathbf{r}_N - \mathbf{r}_e. \tag{7-4}$$

Thus, we desire to solve the Schrödinger equation using the new set of coordinates $\mathbf{R}$ and $\mathbf{r}$, instead of $\mathbf{r}_N$ and $\mathbf{r}_e$. In order to express the Hamiltonian in this new coordinate system, we note that

$$(m_N + m_e)\mathbf{R} = m_N\mathbf{r}_N + m_e(\mathbf{r}_N - \mathbf{r}), \tag{7-5}$$

or

$$\mathbf{r}_N = \mathbf{R} + \left(\frac{\mu}{m_N}\right)\mathbf{r}, \tag{7-6}$$

where μ is the reduced mass of the system, i.e.,

$$\mu = \frac{m_N m_e}{m_N + m_e}. \tag{7-7}$$

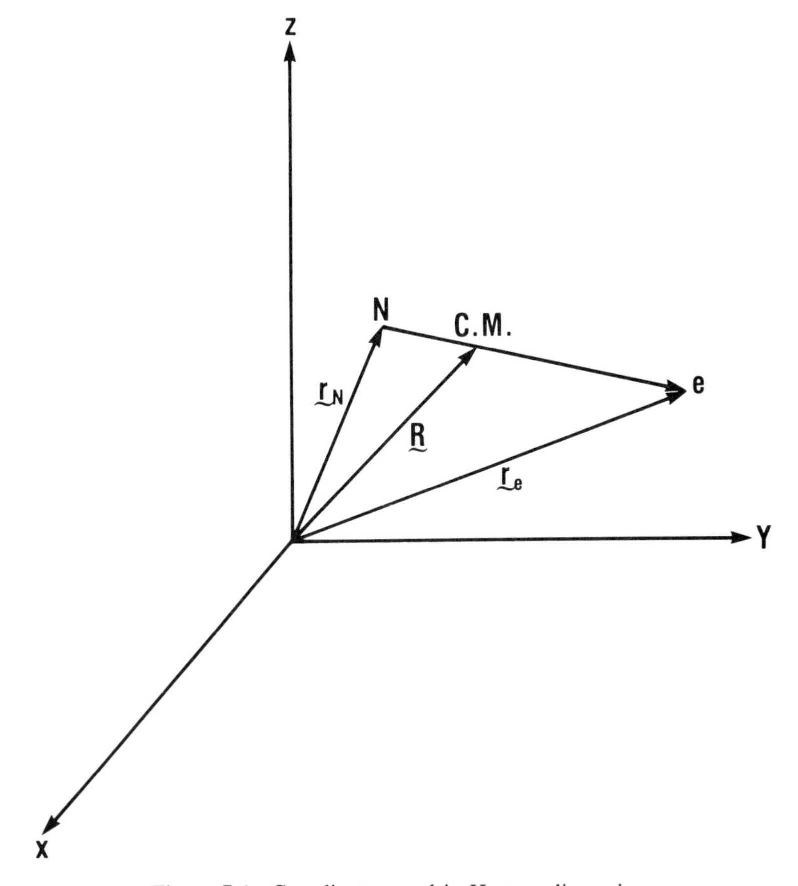

Figure 7.1. Coordinates used in H-atom discussion.

In a similar manner, it is easily seen that

$$\mathbf{r}_e = \mathbf{R} - \left(\frac{\mu}{m_e}\right)\mathbf{r}. \tag{7-8}$$

The terms in the Laplacian are now easily evaluated, e.g.,

$$\frac{\partial}{\partial x_N} = \frac{\partial X}{\partial x_N}\frac{\partial}{\partial X} + \frac{\partial x}{\partial x_N}\frac{\partial}{\partial x}, \tag{7-9}$$

where X is the x-component of $\mathbf{R}$, and

$$\frac{\partial X}{\partial x_N} = \frac{m_N}{m_N + m_e}, \tag{7-10}$$

and

$$\frac{\partial x}{\partial x_N} = +1. \tag{7-11}$$

This gives

$$\nabla_N = \left(\frac{\mu}{m_e}\right)\nabla_R + \nabla. \tag{7-12}$$

In a similar manner, it can be shown[3] that

$$\nabla_e = \left(\frac{\mu}{m_N}\right)\nabla_R - \nabla. \tag{7-13}$$

Thus, the Laplacians can be expressed as

$$\begin{aligned}\nabla_N^2 &= \nabla_N \nabla_N \\ &= \left(\frac{\mu}{m_e}\right)^2 \nabla_R^2 + \left(\frac{\mu}{m_e}\right)(\nabla \cdot \nabla_R + \nabla_R \cdot \nabla) + \nabla^2,\end{aligned} \tag{7-14}$$

and

$$\nabla_e^2 = \left(\frac{\mu^2}{m_N^2}\right)\nabla_R^2 - \left(\frac{\mu}{m_N}\right)(\nabla \cdot \nabla_R + \nabla_R \cdot \nabla) + \nabla^2. \tag{7-15}$$

Also, it is obvious that the potential term in the new coordinate system is

$$V = \frac{-Ze^2}{|\mathbf{r}_N - \mathbf{r}_e|} = \frac{-Ze^2}{|\mathbf{r}|} = \frac{-Ze^2}{r}. \tag{7-16}$$

Thus, the quantum mechanical Hamiltonian in the new coordinate system is

$$\begin{aligned}\mathcal{H} = &-\left(\frac{\hbar^2}{2m_N}\right)\left[\left(\frac{\mu}{m_e}\right)^2 \nabla_R^2 + \left(\frac{\mu}{m_e}\right)(\nabla \cdot \nabla_R + \nabla_R \cdot \nabla) + \nabla^2\right] \\ &-\left(\frac{\hbar^2}{2m_e}\right)\left[\left(\frac{\mu}{m_N}\right)^2 \nabla_R^2 - \left(\frac{\mu}{m_N}\right)(\nabla \cdot \nabla_R + \nabla_R \cdot \nabla) + \nabla^2\right] \\ &-\frac{Ze^2}{r},\end{aligned} \tag{7-17}$$

or

$$\mathcal{H} = -\frac{\hbar^2}{2(m_N + m_e)}\nabla_R^2 - \frac{\hbar^2}{2\mu}\nabla^2 - \frac{Ze^2}{r}. \tag{7-18}$$

The motivation for this change of coordinates will become more obvious momentarily.

We now write the Schrödinger equation as

$$-\frac{\hbar^2}{2(m_N + m_e)}\nabla_R^2 \Psi(\mathbf{R}, \mathbf{r}) = \left[\left(\frac{\hbar^2}{2\mu}\right)\nabla^2 + \frac{Ze^2}{r} + E\right]\Psi(\mathbf{R}, \mathbf{r}), \tag{7-19}$$

[3] See Problem 7-1.

which suggests that it might be profitable to search for solution $\Psi(\mathbf{R}, \mathbf{r})$ of the form

$$\Psi(\mathbf{R}, \mathbf{r}) = \Phi(\mathbf{R})\psi(\mathbf{r}) \tag{7-20}$$

where we have assumed that the variables can be separated. Insertion of Eq. (7-20) into Eq. (7-19) leads to

$$-E_t - \left[\frac{\hbar^2}{2(m_N + m_e)\Phi(\mathbf{R})}\right] \nabla_R^2 \Phi(\mathbf{R}) = \left[\frac{1}{\psi(\mathbf{r})}\right]\left[\frac{\hbar^2}{2\mu}\nabla^2 + \frac{Ze^2}{r} + E_e\right]\psi(\mathbf{r}), \tag{7-21}$$

where

$$E = E_e + E_t, \tag{7-22}$$

and the assumption that the variables can be separated is justified. Thus, the solution to the Schrödinger equation can be obtained by solving the following two equations separately:

$$-\frac{\hbar^2}{2(m_N + m_e)} \nabla_R^2 \Phi(\mathbf{R}) = E_t \Phi(\mathbf{R}) \tag{7-23}$$

$$\left[-\left(\frac{\hbar^2}{2\mu}\right)\nabla^2 - \frac{Ze^2}{r}\right]\psi(\mathbf{r}) = E_e \psi(\mathbf{r}). \tag{7-24}$$

The solution of Eq. (7-23) has already been discussed earlier, i.e., the lack of a potential involving $\mathbf{R}$ causes $\Phi(\mathbf{R})$ to behave as a plane wave. Explicitly, we have

$$\Phi(\mathbf{R}) = e^{i\mathbf{k}\mathbf{R}} \tag{7-25}$$

where $\mathbf{k}$ is a vector of arbitrary direction, but constant magnitude:

$$|\mathbf{k}| = \left[\frac{2(m_N + m_e)E_t}{\hbar^2}\right]^{1/2}. \tag{7-26}$$

Also, $\mathbf{k}$ is not quantized, since E_t can take on any value. In other words, the motion of the center of mass is that of a free particle of mass $= m_N + m_e$, and its energy is given by Eq. (7-26).

7-2. Solution of Equation for Relative Electron Motion of the Hydrogen Atom and Hydrogen-Like Atoms

Before discussing the solution to the equation of motion for the electron movement relative to the nucleus, it will be convenient first to introduce a system of units that will be found to be very useful in the current example and in later applications. This system of units, usually called *Hartree atomic units*,[4] is the one in which the choice

$$\hbar = e = m_e = 1 \tag{7-27}$$

is made.[5] This choice has the advantage that if calculations are carried out using these units, the results are independent of the currently accepted best values of constants. Thus, comparisons between calculations done in this manner are made in an easy manner, and are not plagued with the problem knowing which value was used for each of the constants. In this system of units, the energy is given in units of $\mu e^4/\hbar^2$, which is called a *Hartree* atomic unit. The unit of length is the Bohr radius (a_0 = 0.529172 Å), and is called a *Bohr*. The motivation for the choice of energy unit is obvious, since the constants $\hbar$, m_e and e all disappear from the Schrödinger equation. In the case of distance, the convenience of this particular choice will be apparent shortly.

For the particular case we are dealing with at the moment, we also choose $\mu = 1$ for convenience, since $\mu \cong m_e$. Thus, in atomic units, the Schrödinger equation to be solved is

$$\left(-\frac{1}{2}\nabla^2 - \frac{Z}{r}\right)\psi(\mathbf{r}) = E_e\psi(\mathbf{r}), \tag{7-28}$$

where $Z = 1$ for the particular case of the Hydrogen atom, and the origin is taken to be at the proton.

To facilitate the solution of Eq. (7-28), we express it in terms of spherical coordinates, i.e.,

$$\left\{-\frac{1}{2}\left[\frac{1}{r^2}\frac{\partial}{\partial r}\left(r^2\frac{\partial}{\partial r}\right) + \frac{1}{r^2 \sin\vartheta}\frac{\partial}{\partial\vartheta}\left(\sin\vartheta\frac{\partial}{\partial\vartheta}\right)\right.\right.$$
$$\left.\left. + \frac{1}{r^2\sin^2\vartheta}\frac{\partial^2}{\partial\varphi^2}\right] - \frac{Z}{r}\right\}\psi(\mathbf{r}) = E_e\psi(\mathbf{r}) \tag{7-29}$$

or

$$\left\{-\frac{1}{2}\left[\frac{1}{r^2}\frac{\partial}{\partial r}\left(r^2\frac{\partial}{\partial r}\right) + \left(\frac{1}{r^2}\right)\mathcal{L}^2\right] - \frac{Z}{r}\right\}\psi(\mathbf{r}) = E_e\psi(\mathbf{r}), \tag{7-30}$$

where we have used the result of the previous chapter, which shows that the angular dependence of ∇^2 is just the square of the orbital angular momentum operator. This observation also suggests that the "separation of variables" approach may again be worthwhile. Thus, we assume that $\psi(\mathbf{r})$ can be written as

$$\psi(\mathbf{r}) = R(r)\,Y_{lm_l}(\vartheta, \varphi), \tag{7-31}$$

[4] H. Shull and G. G. Hall, *Nature (London)*, **184**, 1559 (1959).

[5] Other systems of units are also employed, such as "Rydbergs" (which are double Hartree atomic units) for energy. Consequently, care should be exercised when comparing literature values so that the same units are used consistently.

where $Y_{lm_l}(\vartheta, \varphi)$ is the spherical harmonic that is known to be an eigenfunction of $\mathcal{L}^2$. Using this observation, Eq. (7-30) can be rewritten as

$$\left\{-\frac{1}{2}\left[\frac{1}{r^2}\frac{\partial}{\partial r}\left(r^2\frac{\partial}{\partial r}\right)+\frac{l(l+1)}{r^2}\right]-\frac{Z}{r}\right\}R(r)=E_eR(r), \tag{7-32}$$

where $l(l+1)$ is the eigenvalue of $\mathcal{L}^2$ in atomic units, and is restricted to the values

$$l=0,\ 1,\ 2,\ \cdots. \tag{7-33}$$

Of course, we also know from our previous studies of orbital angular momentum that

$$m_l=0,\ \pm 1,\ \pm 2,$$
$$-l\leq m_l\leq +l.$$

The solution to the r-dependence of Eq. (7-32) can be obtained in a manner entirely similar to that used earlier.[6] As in previous cases, the introduction of boundary conditions, i.e., that the resulting function be a quantum mechanical operand that is acceptable by the conditions set down earlier, causes quantization of E_e. The only allowed values that E_e can take on are

$$E_e(n)=-\frac{Z^2}{2n^2} \tag{7-34}$$

where

$$n=1,\ 2,\ \cdots. \tag{7-35}$$

This quantization also results in upper limits on the allowed values of l, i.e.,

$$l\leq n-1. \tag{7-36}$$

In the original unit system, E_n would be given as

$$E_e(n)=-\frac{Z^2\mu e^4}{2n^2\hbar^2}\,. \tag{7-37}$$

Returning to Hartree atomic units, several observations are appropriate at this point. The *ground state* of the hydrogen atom, which is the lowest allowed energy level, is that state in which $n = 1$, $l = 0$, $m_z = 0$, and

$$E_e(1)=-\frac{1}{2}\text{ Hartree}. \tag{7-38}$$

We note further that E_e depends only on the value of n, and is independent of l and m_z. Also, the energy does not depend upon the particular spin state that the

[6] See H. Eyring, J. Walter, and G. E. Kimball, *Quantum Chemistry*, John Wiley, New York, 1944, Chapters 4 and 6, for the details of the solution of the differential equation in Eq. (7-32).

Table 7.1. Degeneracy of Hydrogen Atom Energy Levels

n	l	m_l	m_s	Degree of degeneracy of nth energy level
1	0	0	$\pm\frac{1}{2}$	2
2	0	0	$\pm\frac{1}{2}$	8
	1	$+1, 0, -1$	$\pm\frac{1}{2}$	
3	0	0	$\pm\frac{1}{2}$	18
	1	$+1, 0, -1$	$\pm\frac{1}{2}$	
	2	$+2, +1, 0, -1, -2$	$\pm\frac{1}{2}$	

electron possesses. Consequently, the energy levels are degenerate, and the degree of degeneracy varies with n. Several examples of the allowed quantum numbers and associated energy degeneracy for low-lying states of the hydrogen atom are given in Table 7.1. In this table, the eigenvalue of the z-component of orbital angular momentum is labled by m_l, and the eigenvalue of spin is labeled m_s.

As verified by examination of the Table 7.1, the general formula that expresses the degeneracy of the nth energy level is

$$\text{Degree of degeneracy} = 2n^2. \tag{7-39}$$

The explicit eigenfunctions of the radial equation are given by

$$R_{nl}(r) = N_r \rho^l e^{-\rho/2} L_{n+l}^{2l+1}(\rho), \tag{7-40}$$

where

$$\rho = \frac{2Zr}{n}. \tag{7-41}$$

N_r is a normalization constant, given by

$$N_r = -\left[\left(\frac{2Z}{n}\right)^3 \frac{(n-l-1)!}{2n\{(n+l)!\}^3}\right]^{1/2}, \tag{7-42}$$

and $L_{n+l}^{2l+1}(\rho)$ is known as an *Associated Laguerre Polynomial.*

As seen in texts[7] that discuss solutions to the differential equation of Eq. (7-32), a general formula that can be used to generate any desired $L_s^t(\rho)$ is

$$L_s^t(\rho) = \frac{d^t}{d\rho^t} L_s(\rho) \tag{7-43}$$

[7] See, for example, G. Sansone, *Orthogonal Functions*, Interscience Publishers, New York, 1959, Chapter 4.

Table 7.2. Explicit Expressions for Several Associated Laguerre Polynomials

n	l	$L_{n+l}^{2l+1}(\rho)$
1	0	$L_1^1(\rho) = -1$
2	0	$L_2^1(\rho) = 2\rho - 4$
2	1	$L_3^3(\rho) = -6$
3	0	$L_3^1(\rho) = -3\rho^2 + 18\rho - 18$
3	1	$L_4^3(\rho) = 24\rho - 96$
3	2	$L_5^5(\rho) = -120$

where

$$L_s(\rho) = e^{\rho} \frac{d^s}{d\rho^s} (e^{-\rho}\rho^s). \tag{7-44}$$

For the first several values of n and l, the explicit expression for $L_{n+l}^{2l+1}(\rho)$ is given in Table 7.2.

One of the interesting properties of the functions $R_{nl}(r)$ is that they form an orthonormal set, i.e.,

$$\int_0^\infty R_{nl}^*(r) R_{n',l'}(r) r^2 \, dr = \delta_{n,n'} \delta_{l,l'}. \tag{7-45}$$

When this result is combined with the orthonormality properties of the spherical harmonics, we see that the set of eigenfunctions for the hydrogen-like atom form an orthonormal set:

$$\int_0^\infty \int_0^\pi \int_0^{2\pi} \psi_{nlm_l}^*(r, \vartheta, \varphi) \psi_{n',l',m_l'}(r, \vartheta, \varphi) r^2 \, dr \sin\vartheta \, d\vartheta \, d\varphi$$

$$= \delta_{n,n'} \cdot \delta_{l,l'} \cdot \delta_{m_l,m_l'}. \tag{7-46}$$

By combining the results obtained thusfar, we find the following expression for the normalized total wavefunctions for the hydrogen-like atom:

$$\psi_{nlm_l}(r, \vartheta, \varphi) = N_r \rho^l e^{-\rho/2} L_{n+l}^{2l+1}(\rho) Y_{lm_l}(\vartheta, \varphi). \tag{7-47}$$

Examples of the explicit wavefunction for several low-lying states of the hydrogen-like atom are given in Table 7.3.

The quantum numbers and designations are used in Table 7.3 have historically been given names, i.e.,

n = principal quantum number

l = azimuthal quantum number

m_l = magnetic quantum number,

Table 7.3. Normalized Hydrogen-like Atomic Wavefunctions

n	l	m_l	$\Psi_{nlm_l}(r, \vartheta, \varphi)$
1	0	0	$\Psi_{100}=\left(\frac{Z^3}{\pi}\right)^{1/2} e^{-Zr}$
2	0	0	$\Psi_{200}=\left(\frac{Z^3}{8\pi}\right)^{1/2}\left(1-\frac{Zr}{2}\right) e^{-Zr/2}$
2	1	0	$\Psi_{210}=\left(\frac{1}{4}\right)\left(\frac{Z^5}{2\pi}\right)^{1/2} r\cos\vartheta e^{-Zr/2}$
2	1	1	$\Psi_{211}=\left(\frac{1}{8}\right)\left(\frac{Z^5}{\pi}\right)^{1/2} r\sin\vartheta e^{-Zr/2}\cdot e^{+i\varphi}$
2	1	−1	$\Psi_{21-1}=\left(\frac{1}{8}\right)\left(\frac{Z^5}{\pi}\right)^{1/2} r\sin\vartheta e^{-Zr/2}\cdot e^{-i\varphi}$
3	0	0	$\Psi_{300}=\left(\frac{1}{81}\right)\left(\frac{Z^3}{3\pi}\right)^{1/2}(2Z^2r^2-18Zr+27)e^{-Zr/3}$
3	1	0	$\Psi_{310}=\left(\frac{1}{27}\right)\left(\frac{2Z^5}{\pi}\right)^{1/2}\left(2-\frac{Zr}{3}\right) r\cos\vartheta e^{-Zr/3}$
3	1	1	$\Psi_{311}=\left(\frac{1}{27}\right)\left(\frac{2Z^5}{\pi}\right)^{1/2}\left(2-\frac{Zr}{3}\right) r\sin\vartheta e^{+i\varphi}\cdot e^{-Zr/3}$
3	1	−1	$\Psi_{31-1}=\left(\frac{1}{27}\right)\left(\frac{2Z^5}{\pi}\right)^{1/2}\left(2-\frac{Zr}{3}\right) r\sin\vartheta e^{-i\varphi}\cdot e^{-Zr/3}$
3	2	0	$\Psi_{320}=\left(\frac{1}{81}\right)\left(\frac{Z^5}{6\pi}\right)^{1/2} r^2(3\cos^2\vartheta-1)e^{-Zr/3}$
3	2	1	$\Psi_{321}=\left(\frac{1}{81}\right)\left(\frac{Z^5}{\pi}\right)^{1/2} r^2\sin\vartheta\cos\vartheta e^{-Zr/3}\cdot e^{+i\varphi}$
3	2	−1	$\Psi_{32-1}=\left(\frac{1}{81}\right)\left(\frac{Z^5}{\pi}\right)^{1/2} r^2\sin\vartheta\cos\vartheta e^{-Zr/3}\cdot e^{-i\varphi}$
3	2	2	$\Psi_{322}=\left(\frac{1}{162}\right)\left(\frac{Z^5}{\pi}\right)^{1/2} r^2\sin\vartheta e^{-Zr/3}\cdot e^{+2i\varphi}$
3	2	−2	$\Psi_{32-2}=\left(\frac{1}{162}\right)\left(\frac{Z^5}{\pi}\right)^{1/2} r^2\sin\vartheta e^{-Zr/3}\cdot e^{-2i\varphi}$

and the various values of l are referred to as follows:

1	Label	Symbol
0	sharp	s
1	principal	p
2	diffuse	d
3	fundamental	f
4	—	g
etc.	etc.	etc.

These labels have arisen because of the appearance of the corresponding lines in the atomic spectra, and are used extensively.

For this specific case of the Hydrogen atom, the energy level associated with the various states is indicated in Fig. 7.2. From the figure, we see that $n = \infty$ corresponds to ionization of the electron. Hence, the ionization potential of the Hydrogen atom is $+\frac{1}{2}$ Hartree, or 13.6 eV.

7-3. Wavefunction Shapes

Let us now examine the shape of several of these wavefunctions. Considering the ground state first, we have

$$\begin{aligned} \psi_{100}(r, \vartheta, \varphi) &= Ne^{-\rho/2} \\ &= Ne^{-Zr}, \end{aligned} \tag{7-48}$$

where N is constant. Hence, the lowest state is one that does not depend on angular variables, and is spherically symmetric. Since the dependence is only on the radial variable, we can observe the shape of this wavefunction by a simple plot of $\psi_{100}(r)$ vs. r, as given in Fig. 7.3. As Z increases, the wavefunction will fall off faster, thus decreasing the distance at which the orbital has essentially a zero value.

However, the wavefunction itself does not provide an observable quantity. As we have seen earlier, only the probability of observing the particle at a given point can be given, and this probability is proportional to the square of the wavefunction. In particular, the probability of observing the electron at a position in space (r, ϑ, φ), within r and $r + dr$, $\vartheta + d\vartheta$, and $\varphi + d\varphi$, is given by

$$\begin{aligned} P(r, \vartheta, \varphi)\, dV &= \psi^*_{100}(\mathrm{r}, \vartheta, \varphi)\psi_{100}(r, \vartheta, \varphi)\, dV \\ &= |\psi_{100}|^2 r^2\, dr \sin\vartheta\, d\vartheta\, d\varphi. \end{aligned} \tag{7-49}$$

However, since $\psi_{100}(r)$ depends only on r, this means that all possible angular variables are equally likely, and we should average over all possible values. Since the angular variables are continuous, this averaging takes the form of

$E_\infty = 0$ — $n = \infty$
$n = 6$
$n = 5$
$n = 4$
$n = 3$

$E_2 = -1/8$ — $n = 2$

$E_1 = -1/2$ — $n = 1$

Figure 7.2. Energy levels of the hydrogen atom (in atomic units).

integration, and the resulting probability function, $p(r)$, is given by

$$p(r)\,dr = \int_0^{2\pi} \int_0^{\pi} d\varphi \sin\vartheta\, d\vartheta\, |\psi_{100}|^2 r^2\, dr = r^2 |R_{10}(r)|^2\, dr. \qquad (7\text{-}50)$$

In Eq. (7-50), $p(r)$ is known as the *radial distribution function*, and a plot of $p(r)$ for the ground state of the hydrogen atom is given in Fig. 7.4.

Since $p(r)$ represents a probability, it means that we can determine the point

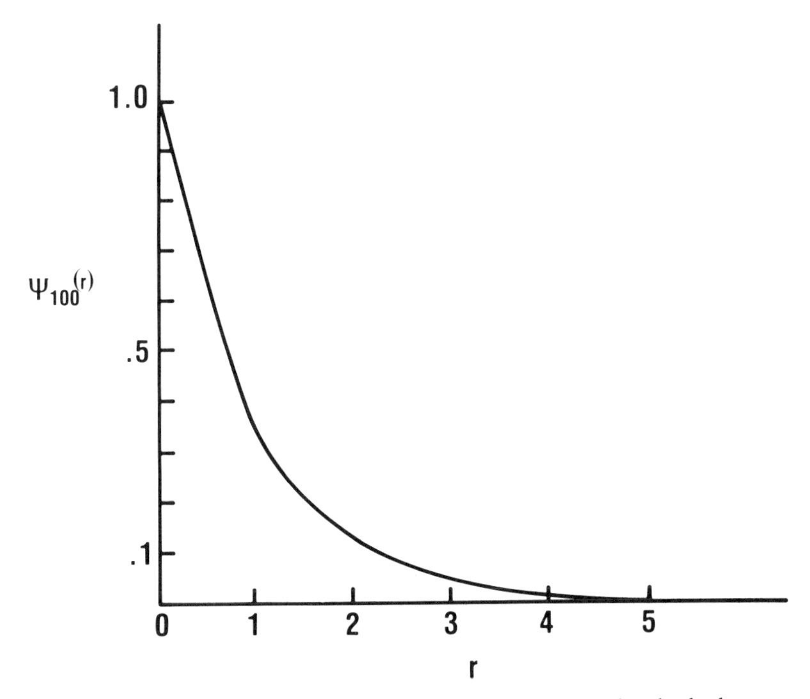

Figure 7.3. Shape of unnormalized ground state orbital (ψ_{100}) for the hydrogen atom ($Z = 1$).

of maximum probability by examination of $p(r)$, i.e., at what distance is the electron most likely to be found. From Eq. (7-50) we have

$$\frac{dp(r)}{dr} = N^2 \frac{d}{dr}(r^2 e^{-2Zr}),$$

or

$$\frac{dp(r)}{dr} = 8\pi N^2 r(1 - Zr)e^{-2Zr}. \tag{7-51}$$

From the shape of $p(r)$ as given in Fig. 7.4, we are assured that the derivative of $p(r)$ will be zero only at the *maximum* of $p(r)$, and at infinity. Thus we obtain the maximum value of $p(r)$ from $dp(r)/dr = 0$ as

$$\begin{aligned} r &= 1/Z \\ &= 1 \text{ (when } Z = 1). \end{aligned} \tag{7-52}$$

The other solutions, $r = 0$ and $r = \infty$, correspond to minima of $p(r)$. We now see the rationale for the choice of units for distance measurement. For the case of the hydrogen atom, the most likely distance from the proton at which the electron will be found is $r = 1$ Bohr. In the old system of units, this corresponds

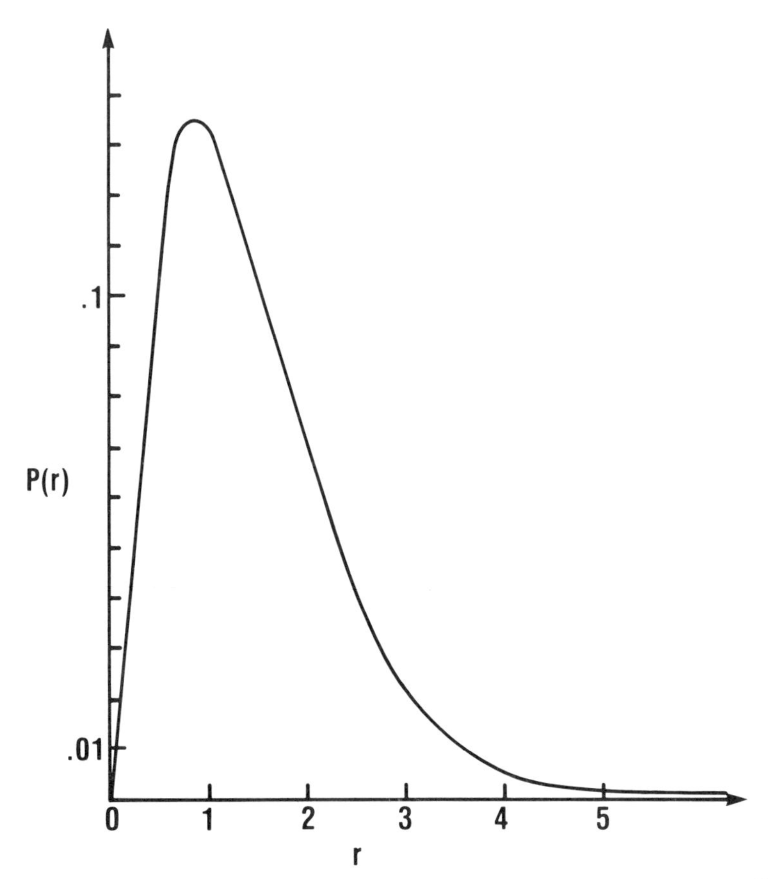

Figure 7.4. Radial distribution function $[p(r)]$ for the ground state of (unnormalized) hydrogen atom ($Z = 1$).

to $r = a_0$, where a_0 is the radius of the first Bohr orbit. Also, the most likely distance decreases as the nuclear charge increases, as anticipated.

One other point concerning the shape of the ground state wavefunction is worth mentioning. In a Cartesian coordinate system, the ground state wavefunction has the form

$$\psi_{100}(x, y, z) = N \exp\{-Z(x^2 + y^2 + z^2)^{1/2}\}. \tag{7-53}$$

It is interesting to examine the behavior of this orbital in the region near the origin. In particular, we shall examine the behavior along a given axis, e.g., the x-axis, in which case $y = z = 0$. First, we note that the function itself is continuous at $x = 0$, since

$$\lim_{x \to 0_+} [\psi_{100}(x, y, z)]_{\substack{y=0 \\ z=0}} = \lim_{x \to 0_-} [\psi_{100}(x, y, z)]_{\substack{y=0 \\ z=0}} \tag{7-54}$$

Now let us examine the first derivative of $\psi_{100}(x)$ for the hydrogen atom ($Z = 1$).

We have

$$\left[\frac{\partial}{\partial x}\psi_{100}(x, y, z)\right]_{\substack{y=0\\z=0}} = \left\{\frac{\partial}{\partial x}[N\exp\{-(x^2+y^2+z^2)^{1/2}\}]\right\}_{\substack{y=0\\z=0}} \tag{7-55}$$

$$= \left\{\frac{-xN}{(x^2+y^2+z^2)^{1/2}}\exp[-(x^2+y^2+z^2)^{1/2}]\right\}_{\substack{y=0\\z=0}}$$

$$= -N\frac{x}{|x|}\exp(-|x|), \tag{7-56}$$

since $(x^2)^{1/2} = |x|$. From this result, we see that

$$\lim_{x\to 0_+}\left\{\left[\frac{\partial}{\partial x}\psi_{100}(x, y, z)\right]\right\}_{\substack{y=0\\z=0}} = -N \tag{7-57}$$

and

$$\lim_{x\to 0_-}\left\{\left[\frac{\partial}{\partial x}\psi_{100}(x, y, z)\right]\right\}_{\substack{y=0\\z=0}} = +N. \tag{7-58}$$

Thus, although the function itself is continuous (and normalizable), its first derivative is discontinuous at the origin. This behavior of the derivative is indicated in Fig. 7.5. This peculiar behavior of the first derivative of the ground state wavefunction at the origin is called a *cusp* at the origin. We shall see in later chapters that the satisfaction of the appropriate cusp behavior for an approximate solution is a generally difficult problem. In fact, much effort has been devoted to investigation of this question, and any approximation to the true wavefunction of a given system that desires to obtain a total energy close to the true energy must satisfy the appropriate cusp conditions.

For the case of higher values of n, with $l = m_1 = 0$, the visualization is again straightforward, since ψ_{n00} has spherical symmetry for all n. Figure 7.6 gives plots of both ψ_{n00} and $p(r)$ for several higher values of n. We note that the number of nodes in ψ_{n00} increases linearly with n. In particular, the number of nodes of ψ_{n00} is precisely equal to n, if the node at infinity is included.

Next, let us discuss the shape of the various wavefunctions and their associated probability densities when l and m_l are nonzero. The interpretation of these wavefunctions is considerably more complicated, due to the dependence on *several* variables, which makes visualization difficult. In addition, for $m_l \neq 0$, the φ dependence of these wavefunctions is a complex exponential, further complicating the interpretation. As a result of these difficulties, several steps are usually taken to aid in interpretation.

First, we treat the question of the appearance of the complex exponential. Since the energy depends on the value of the principal quantum number n, we know that for a given n and l, there are $2l + 1$ values of m_l that are possible. Associated with these $2l + 1$ values of m_l are $2l + 1$ eigenfunctions of $\mathfrak{L}_z$, that

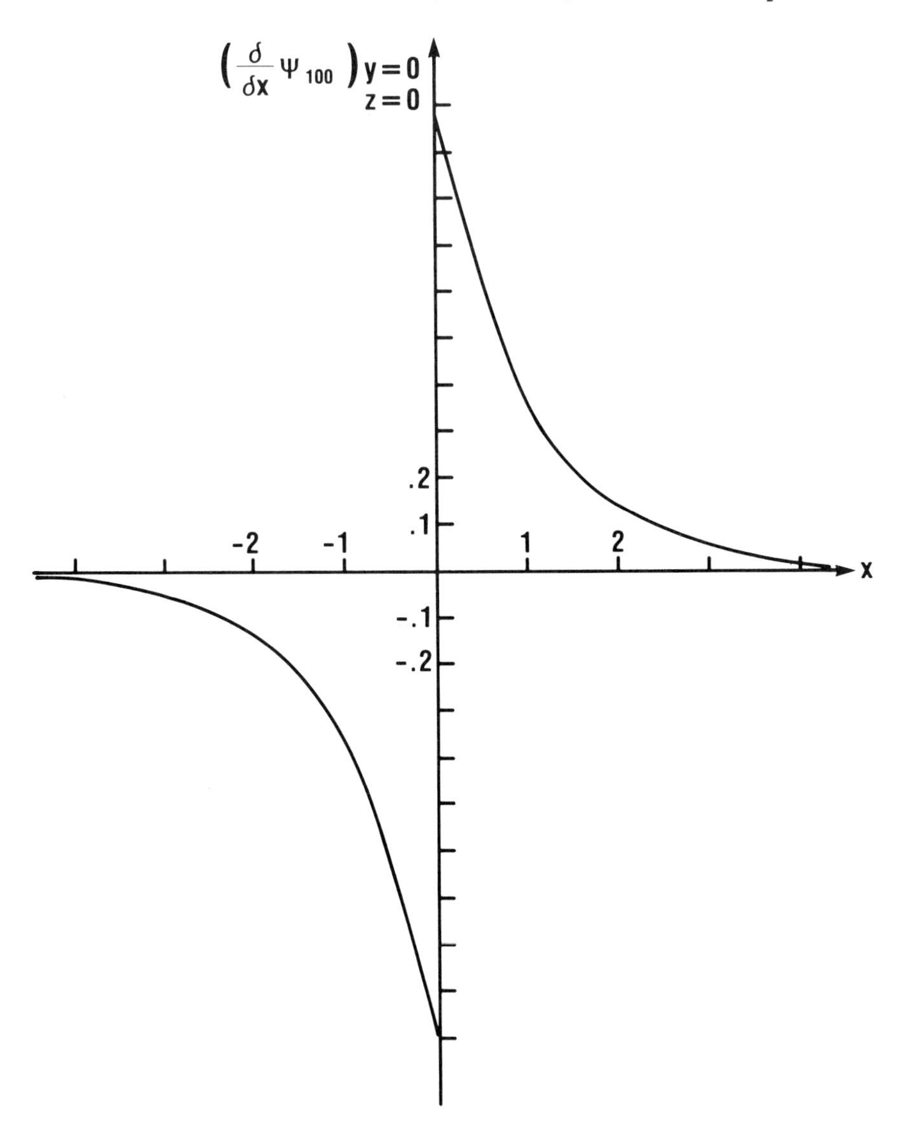

Figure 7.5. Behavior of the first derivative (unnormalized) of $\psi_{100}(x, y, z)$.

we found to be

$$\Phi_{m_l}(\varphi) = \frac{1}{\sqrt{2\pi}} e^{im_l\varphi} \qquad -l \leq m_l \leq +l. \tag{7-59}$$

We also note that, for every eigenfunction corresponding to $+|m_l|$, there is another eigenfunction that corresponds to the eigenvalue $-|m_l|$. The case of $m_l = 0$ is an exception to this observation, but it causes no problem, since $\Phi_0(\varphi)$ is just a constant in this case. Considering one of the pairs of eigenfunctions for

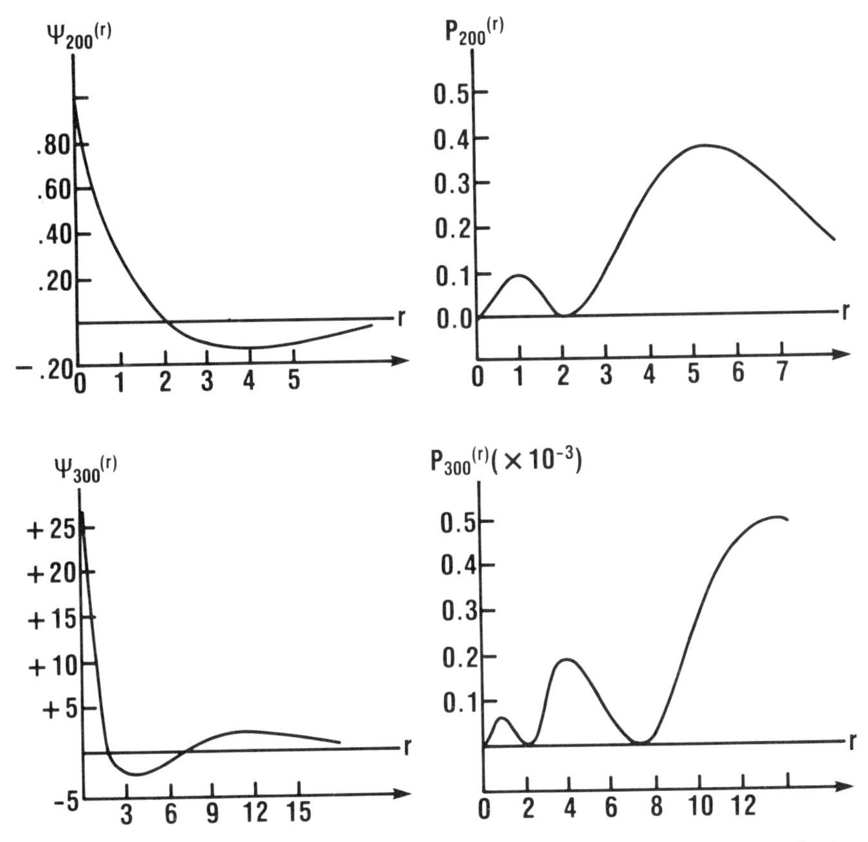

Figure 7.6. Plots of unnormalized ψ_{n00} and their corresponding radial distribution functions for $n = 2, 3$.

$m_l = 0$, we have

$$\Phi_{|m_l|}(\varphi) = \frac{1}{\sqrt{2\pi}} e^{i|m_l|\varphi} = \frac{1}{\sqrt{2\pi}} (\cos|m_l|\varphi + i \sin|m_l|\varphi) \tag{7-60}$$

and

$$\Phi_{-|m_l|}(\varphi) = \frac{1}{\sqrt{2\pi}} e^{-i|m_l|\varphi} = \frac{1}{\sqrt{2\pi}} (\cos|m_l|\varphi - i \sin|m_l|\varphi). \tag{7-61}$$

Since a linear combination of eigenfunctions is always another eigenfunction, we see from the above two equations that

$$\Phi'_{|m_l|}(\varphi) = \Phi_{|m_l|}(\varphi) + \Phi_{-|m_l|}(\varphi) \tag{7-62}$$

$$= \frac{2}{\sqrt{2\pi}} \cos|m_l|\varphi \tag{7-63}$$

and

$$\Phi''_{|m_l|}(\varphi)=\Phi_{|m_l|}(\varphi)-\Phi_{-|m_l|}(\varphi) \tag{7-64}$$

$$=\left(\frac{2i}{\sqrt{2\pi}}\right)\sin|m_l|\varphi. \tag{7-65}$$

Thus, we see that, except possibly for a normalization constant, $\Phi'_{|m_l|}$ and $\Phi''_{|m_l|}(\varphi)$ form equally as acceptable eigenfunctions as do $\Phi_{|m_l|}(\varphi)$ and $\Phi_{-|m_l|}(\varphi)$. In other words, it is possible to choose a set of eigenfunctions of $\mathfrak{L}_z$ that is all *real*, if we use the above relations for every pair of eigenvalues $\pm|m_l|$.

As an example, let us consider the various possibilities from $n = 2$. For $l = 0$, we have only $m_l = 0$, which is already real, i.e.,

$$\Phi_0(\varphi)=\frac{1}{\sqrt{2\pi}} \tag{7-66}$$

for $l = 1$, we can have $m_l = +1, 0, -1$, which gives rise to

$$\Phi_1(\varphi)=\frac{1}{\sqrt{2\pi}}e^{i\varphi}, \tag{7-67}$$

$$\Phi_0(\varphi)=\frac{1}{\sqrt{2\pi}}, \tag{7-68}$$

and

$$\Phi_{-1}(\varphi)=\frac{1}{\sqrt{2\pi}}e^{-i\varphi}. \tag{7-69}$$

Using the preceding analysis, we see that an equally acceptable pair of eigenfunctions corresponding to $|m_l| = 1$ that can be used instead of $\Phi_1(\varphi)$ and $\Phi_{-1}(\varphi)$ is

$$\Phi'(\varphi)=N\cos\varphi \tag{7-70}$$

and

$$\Phi''(\varphi)=N\sin\varphi. \tag{7-71}$$

Notice, however, that the label $|m_l| = +1$ or -1 is no longer applicable to either $\Phi'(\varphi)$ or $\Phi''(\varphi)$, since each latter eigenfunction is a linear combination of Φ_1 and Φ_{-1}. Instead, an alternate notation has been developed to label these eigenfunctions. Combining $\Phi'(\varphi)$ and $\Phi''(\varphi)$ with the appropriate eigenfunctions of ϑ and r, we use the notation

$$\psi_{2p_x}=N\rho e^{-\rho/2}L_3^3(\rho)\sin\vartheta\sin\varphi=N'xe^{-\rho/2}L_3^3(\rho) \tag{7-72}$$

$$\psi_{2p_y}=N\rho e^{-\rho/2}L_3^3(\rho)\sin\vartheta\cos\varphi=N'ye^{-\rho/2}L_3^3(\rho). \tag{7-73}$$

Table 7.4. Real Representation of Hydrogen-like Atomic Wavefunctions

n	l	Orbital	Designation and Form
1	0	$1s$	$= N_{11}e^{-Zr}$
2	0	$2s$	$= N_{2s}(2 - Zr)e^{-Zr/2}$
2	1	$2p_z$	$= N_{2p_z}(r \cos \vartheta)e^{-Zr/2}$
		$2p_x$	$= N_{2p_x}(r \sin \vartheta \cos \varphi)e^{-Zr/2}$
		$2p_y$	$= N_{2p_y}(r \sin \vartheta \sin \varphi)e^{-Zr/2}$
3	0	$3s$	$= N_{3s}(27 - 18Zr + 2Z^2r^2)e^{-sr/3}$
3	1	$3p_z$	$= N_{3p_z}(6 - Zr)e^{-Zr/3} \cos \vartheta$
		$3p_x$	$= N_{3p_x}(6 - Zr)e^{-Zr/3} \sin \vartheta \cos \varphi$
		$3p_y$	$= N_{3p_y}(6 - Zr)e^{-Zr/3} \sin \vartheta \sin \varphi$
3	2	$3d_{z^2}$	$= N_{3d_{z^2}}r^2e^{-Zr/3}(3 \cos^2 \vartheta - 1)$
		$3d_{xy}$	$= N_{3d_{xy}}r^2e^{-Zr/3} \sin^2 \vartheta \sin 2\varphi$
		$3d_{xz}$	$= N_{3d_{xz}}r^2e^{-Zr/3} \sin \vartheta \cos \vartheta \cos \varphi$
		$3d_{yz}$	$= N_{3d_{yz}}r^2e^{-Zr/3} \sin \vartheta \cos \vartheta \sin \varphi$
		$3d_{x^2-y^2}$	$= n_{3d_{x^2-y^2}}r^2e^{-Zr/3} \sin^2 \vartheta \cos 2\varphi$

For the case $m_l = 0$, we have

$$\psi_{2p_z} = N\rho e^{-\rho/2}L_3^3(\rho) \cos \vartheta = N' z e^{-\rho/2}L_3^3(\rho). \tag{7-74}$$

The rationale for the above notation is clear, since $\cos \vartheta$ represents the projection of the function onto the z-axis, while ψ_{2px} and ψ_{2py} represent projections onto the x and y axes, respectively. A summary of the real form of hydrogen-like atom eigenfunctions and associated designation for several low-lying states is given in Table 7.4. Thus, we see that a real representation of the eigenfunctions is always possible, which solves one of the problems encountered in interpretation of the eigenfunctions.

The other problem of representing graphically the shape of a function of as many as three variables (r, ϑ, φ) is not so unambiguously solved. One way of approaching the problem is simply to examine the dependence on one variable at a time. An example of such an approach has already been given in the examination of the radial dependence of hydrogen-like atom wavefunctions above. Another example, which is frequenly found in elementary chemistry textbooks, involves plotting the ϑ dependence of orbitals as a function of ϑ. In particular, "polar plots" are typically shown in which the magnitude of the ϑ dependence is plotted as a function of the angle ϑ. For the case of the $2p_z$ wavefunction, the ϑ dependence is given simply by $\cos \vartheta$, and the "polar plot" of $\cos \vartheta$ vs. ϑ is given in Fig. 7.7. Since there is no φ dependence, i.e., all possible values of φ are allowed, the tangent circles in the "polar plot" that have been given for one dimension in Fig. 7.7 will create two spheres along the z-axis when both the ϑ and φ dependence are considered. It should be emphasized, however, that such a plot represents only the angular dependence of the wavefunction, and should include consideration of the radial dependence of the wavefunction at the same time if the true "shape" is to be understood.

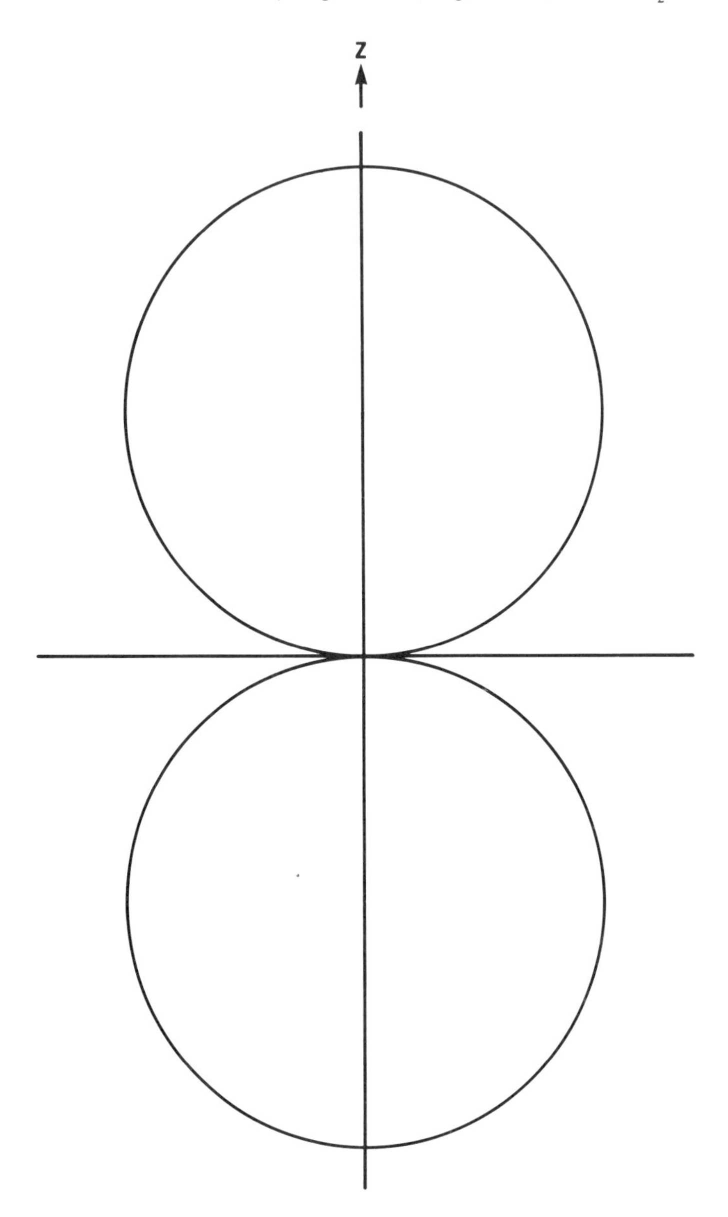

Figure 7.7. "Polar plot" of the ϑ dependence of the $2p_z$ orbital.

Furthermore, it is the *square* of the wavefunction (times the volume element) that is related to the probability of finding an electron at a given point. This usually is taken into account by plotting *contours* of $|\psi|^2$, i.e., finding the values of r, ϑ and φ for which the probability density $|\psi|^2$ is constant, and connecting those points to create a surface. For example, Fig. 7.8 shows the two-dimensional contour surface corresponding to $|\psi|^2 \cong 0.0024$ for a

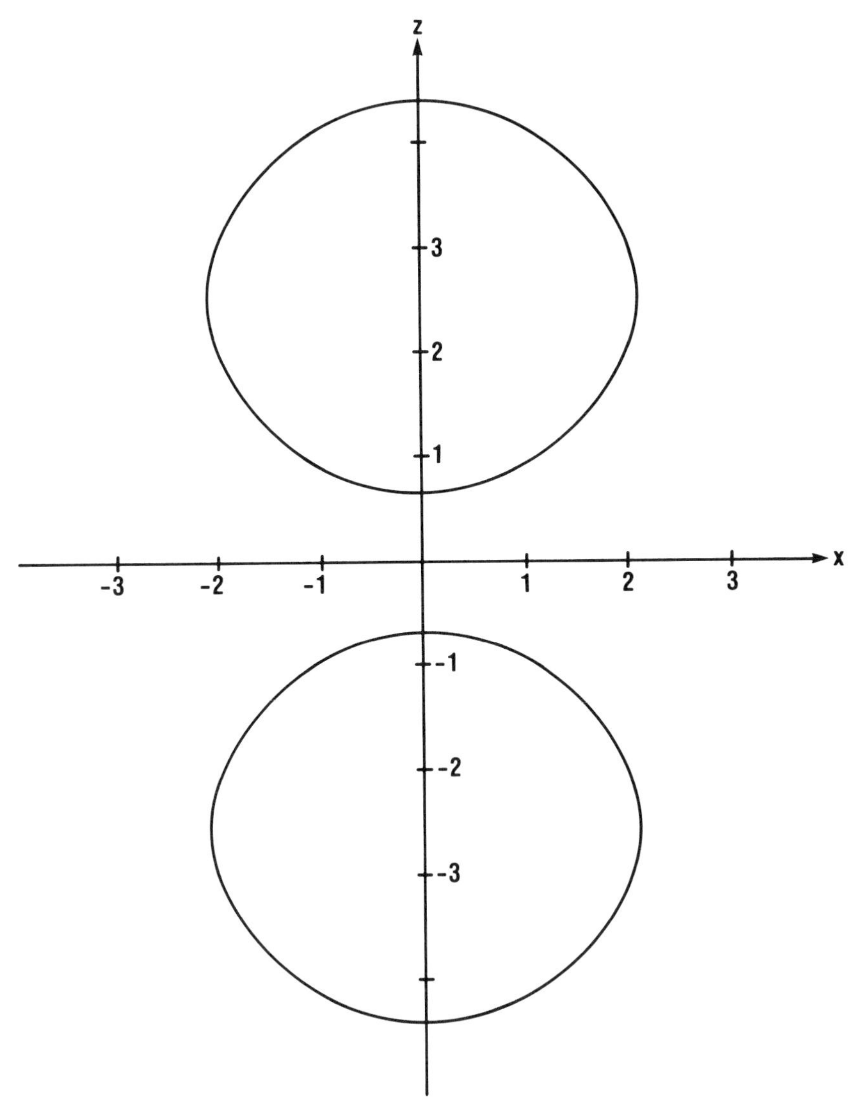

Figure 7.8. Contour diagram for the square of a normalized $2p_z$ orbital, with $|\Psi_{2p_z}|^2 \simeq$ 0.0024 electrons/(Bohr)^3.

normalized $2p_z$ wavefunction.[8] In these dimensions, this figure becomes two three-dimensional ellipsoids, instead of spheres (as found in Fig. 7.7 for the angular dependence contour diagram). Since depiction of $|\psi|^2$ is related to observables, it is a better method of visualizing hydrogen-like atom wavefunctions than looking only at their components.

[8] Shapes of wavefunctions and how they relate to observables have been discussed frequently, e.g., see R. E. Powell, *J. Chem. Ed.*, **45** (1968); E. A. Ogryzlo, *J. Chem. Ed.*, **42**, 150 (1965); I. Cohen, *J. Chem. Ed.*, **40**, 256 (1963).

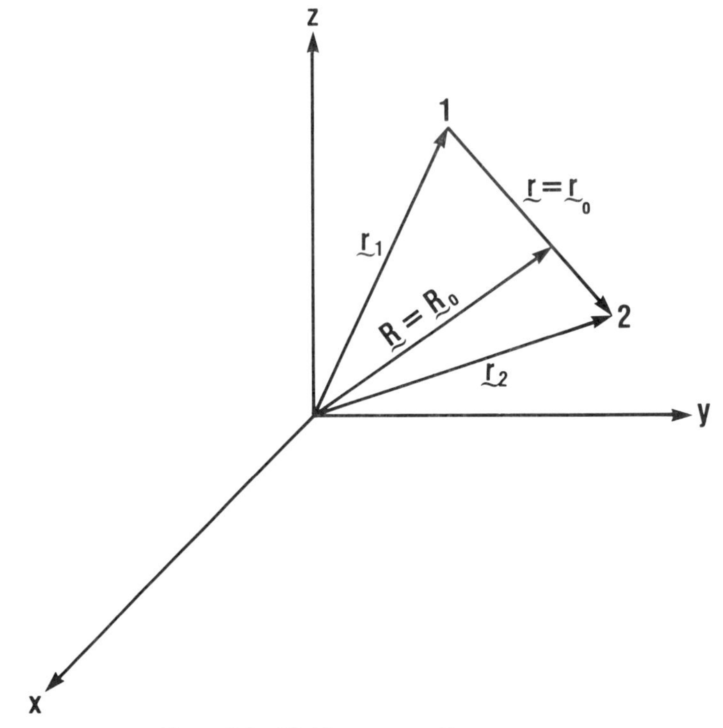

Figure 7.9. Rigid rotor coordinate systems.

7-4. Rigid Rotor

As a second example, it is useful to consider one that is easily related to the discussions concerning the hydrogen atom that have just been completed. This example is usually referred to as the *rigid rotor*. This rather simple model of the rotation of a diatomic molecule is quite useful in providing a useful approximation to spectral features associated with the rotational spectrum of diatomic molecules.

The basic model consists of two particles, whose motion is restricted in two different ways. These particles, with masses m_1, and m_2, are restricted to motion under the constraints that the distance of their center of mass ($\mathbf{R}$) from an arbitrary origin is held fixed and, second, that the relative distance ($\mathbf{r}$) of the two particles with respect to each other is also held fixed. This model is depicted in Fig. 7.9, in which $\mathbf{R} = \mathbf{R}_0$ and $\mathbf{r} = \mathbf{r}_0$. As we shall see shortly, this model considers the motion of the pair of particles to be restricted to rotation about the center of mass.

The analogy to the hydrogen atom is now clear, but with two important differences: the motion of *both* particles is considered (instead of just considering the motion of the electron relative to the nucleus), and the distance between the particles is held fixed. However, even in this more constrained environment, we shall see the quantization of the energy levels again occurs, and that the resulting model is quite useful as a first approximation in the discussion of diatomic molecule rotational spectra.

To begin the discussion, we note that the quantum mechanical Hamiltonian for use in describing the motion of these particles (in atomic units) is given by:

$$\mathcal{H} = -\frac{1}{2m_1}\nabla_1^2 - \frac{1}{2m_2}\nabla_2^2. \tag{7-75}$$

It should be noted that, while each of the particles possesses a kinetic energy term, there is no potential energy term. This results from the assumption that neither particle possesses a charge, and thus there is no potential energy term corresponding to the Coulombic interaction of the two particles. Thus, if quantization of energy levels is to occur, it is expected to be due to the constraints of fixed distances. This is analoguous to the imposition of boundary conditions in one-dimensional examples considered earlier.

The time-independent Schrödinger equation to be solved is thus written as:

$$\mathcal{H}\Psi(\mathbf{r}_1, \mathbf{r}_2) = E\Psi(\mathbf{r}_1, \mathbf{r}_2). \tag{7-76}$$

Just as in the case of the hydrogen atom, the center of mass motion can be separated using the following definitions:

$$\mu = \frac{m_1 m_2}{m_1 + m_2}, \tag{7-77}$$

$$\mathbf{R} = \mathbf{R}_0 = \frac{m_1\mathbf{r}_1 + m_2\mathbf{r}_2}{m_1 + m_2}, \tag{7-78}$$

$$\mathbf{r} = \mathbf{r}_0 = \mathbf{r}_2 - \mathbf{r}_1, \tag{7-79}$$

where $\mathbf{R}$ and $\mathbf{r}$ have been fixed at $\mathbf{R}_0$ and $\mathbf{r}_0$, respectively. When the Hamiltonian and Schrödinger equation are expressed in the $(\mathbf{R}, \mathbf{r})$ coordinate system, the variables can be separated, i.e., a wavefunction of the form:

$$\Psi(\mathbf{R}, \mathbf{r}) = \Phi(\mathbf{R})\psi(\mathbf{r}) \tag{7-80}$$

can be chosen, which results in

$$\Phi(\mathbf{R}) = e^{i\mathbf{k}\cdot\mathbf{R}} \tag{7-81}$$

with

$$|\mathbf{k}| = [2(m_1 + m_2)E_t]^{1/2}, \tag{7-82}$$

where E_t is the translational (unquantized) energy of the motion of the center of mass.

The more interesting part of the problem is associated with $\psi(\mathbf{r})$, which is to be determined via solution of:

$$-\left(\frac{1}{2\mu}\right)\nabla^2\psi(r_0) = E_r\psi(r_0), \tag{7-83}$$

where

$$E = E_t + E_r. \tag{7-84}$$

Writing out the above equation in detail gives

$$-\left(\frac{1}{2\mu r_0^2}\right)\mathcal{L}^2\psi(r_0, \vartheta, \varphi) = E_r\psi(r_0, \vartheta, \varphi), \tag{7-85}$$

which is analogous to Eq. (7-30) with $r = r_0$. The rigid rotor therefore is simply another example of angular motion, and the allowed energy levels of the above equation are seen to be

$$E_r = \frac{J(J+1)}{2\mu r_0^2}, \qquad J = 0, 1, 2, \cdots \tag{7-86}$$

corresponding to Eq. (7-32), in which the symbol J has been used instead of l to distinguish the example under consideration here from general orbital angular momentum considerations.

As in the case of earlier one-dimensional examples, we see that imposition of boundary conditions (which in this case corresponds to fixing the distance between the particles) leads to quantization of energy levels. In this case, determination of the specifically allowed values was simplified considerably due to the earlier discussion of the generalized angular momentum operator and its eigenvalues. It is also of interest to note (by referring to the earlier angular momentum discussion) that the energy levels are $(2J + 1)$-fold degenerate.

To illustrate the utility of this model, let us consider a diatomic molecule like HCl. From the discussion in Chapter 1, the moment of inertia can be seen to be given for the rigid rotor example as

$$I = \mu r_0^2 \tag{7-87}$$

which allows the energy expression to be rewritten as

$$E_r = \frac{J(J+1)}{2I}. \tag{7-88}$$

Of course, only *differences* between energy levels are observed, and the frequency associated with transitions between adjacent energy levels can be seen to be

$$\nu = \Delta E = \left(\frac{J+1}{I}\right).$$

If we define the *rotational constant* B (in Hartree atomic units) as

$$B = \frac{1}{2I} \tag{7-90}$$

Figure 7.10. Calculated HCl rigid rotor rotational spectrum (above) compared to observed (far-infrared) spectrum (below). $B = 10.34\ \text{cm}^{-1}$ (calc.).

the transition frequencies are seen to be given by

$$\nu = 2B(J+1). \tag{7-91}$$

As seen in Fig. 7.10, the frequencies observed experimentally[9] for HCl correspond very closely to the frequencies given by Eq. (7-91). This illustrates not only how useful the rigid rotor model can be expected to be, but also that, at least for diatomic molecules like HCl, the only transitions that are apparently allowed are those between *adjacent* energy levels. This latter observation is an example of the concept of *selection rules*, in which not all possible transitions between energy levels are expected to be observed experimentally.

7-5. The H_2^+ Molecule

Thus far we have seen that the hydrogen atom is a system appearing in nature for which an exact solution to the Schrödinger equation can be determined in closed form. Unfortunately, it (and the hydrogen-like, one-electon atoms) represents the only such system that allows such a solution. However, as we shall see below, the hydrogen molecule-ion H_2^+, is another one-electron system that can be completely solved, although the solution can be given in this case only in tabular form instead of as an analytical expression for the eigenfunctions. Nevertheless, the H_2^+ system has played an important role in the development of molecular quantum mechanics. This arises because it is the simplest (and only) *molecular* system for which the exact solution can be obtained (numerically).

To see how this solution can be obtained, we begin by noting that the non-relativistic Hamiltonian for this system of one electron and two protons can be written in atomic units as:

$$\mathcal{H} = -\frac{1}{2}\nabla^2 - \frac{1}{r_A} - \frac{1}{r_B} + \frac{1}{R}, \tag{7-92}$$

where the first term represents the kinetic energy operator for the electron, the next two terms represent the attraction of the electron to each of the nuclei, and

[9] See M. Czerny, *Z. Phys.*, **34**, 227 (1925); see also G. Herzberg, *Molecular Spectra and Molecular Structure*, I. *Spectra of Diatomic Molecules*, Van Nostrand, Princeton, NJ, 1950, p. 58.

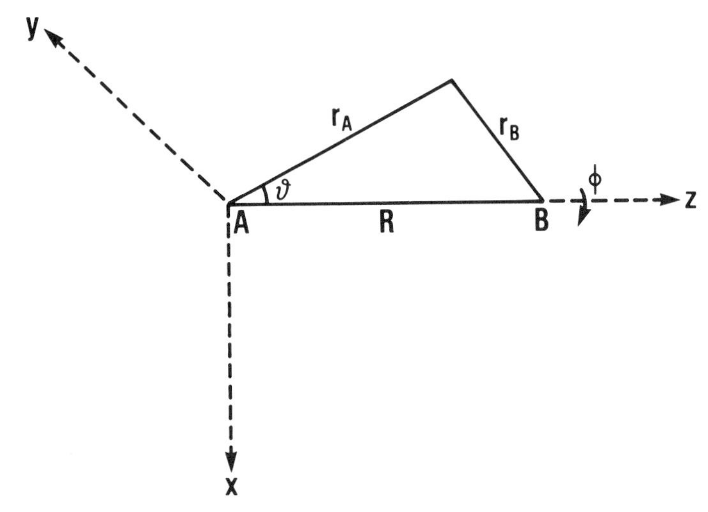

Figure 7.11. Coordinate system for the H_2^+ molecule. The origin is chosen at nucleus A, along with a left-handed coordinate system.

the last term represents the repulsion of the two protons. This system is depicted in Fig. 7.11.

It should be noted that the two nuclei will be assumed to be fixed with respect to each other[10] (i.e., R is a constant), and the solution to the Schrödinger equation that is sought is one that describes the motion of the electron about the two fixed nuclei.

Instead of using spherical coordinates, this problem will be more conveniently attacked if a coordinate system is chosen that more closely represents the symmetry of the system. Such a coordinate system is available in this case, and is known as a *confocal elliptical coordinate system.*[11] In such a coordinate system, the coordinates r, ϑ, ϕ are replaced by μ, ν, ϕ, where μ and ν are defined as

$$\mu = \frac{r_A + r_B}{R} \qquad (0 \leq \mu \leq \infty) \tag{7-93}$$

$$\nu = \frac{r_A - r_B}{R} \qquad (-1 \leq \nu \leq +1) \tag{7-94}$$

and the azimuthal coordinate (ϕ) remains the same in both coordinate systems.

---

[10] Such a restriction on the nuclear motion is known as the Born–Oppenheimer approximation, and will be discussed in greater detail in the next chapter (Section 8-1).

[11] Such a coordinate system was applied to H_2^+ first by O. Burran, *Kgl. Danske Videnskab. Selskab. Mat. Fys. Medd.*, **7**, (14), 1 (1927).

If the elliptical coordinate system is introduced, it can be seen[12] that the nonrelativistic Schrödinger equation to be solved can be written as:

$$\left\{\frac{\partial}{\partial\mu}\left[(\mu^2-1)\frac{\partial}{\partial\mu}\right]+\frac{\partial}{\partial\nu}\left[(1-\nu^2)\frac{\partial}{\partial\nu}\right]+\left[\frac{1}{(\mu^2-1)}+\frac{1}{(1-\nu^2)}\right]\frac{\partial^2}{\partial\phi^2}\right.$$
$$\left.+\frac{R^2}{2}\left[(\mu^2-\nu^2)\left(E-\frac{1}{R}\right)+\frac{4\mu}{R}\right]\right\}\Psi(\mu,\nu,\phi)=0. \tag{7-95}$$

The rationale for the choice of coordinate system is now clear, since the variables in Eq. (7-95) appear to be separable. In fact, if we assume that the desired wavefunction can be factored into a product of functions, i.e.,

$$\Psi(\mu,\nu,\phi)=M(\mu)N(\nu)\Phi(\phi), \tag{7-96}$$

then the Schrödinger equation of Eq. (7-95) can be written as three separate ordinary differential equations,[13]

$$\left\{\frac{d}{d\mu}\left[(\mu^2-1)\frac{d}{d\mu}\right]+A+2R\mu+\frac{R^2E'\mu^2}{2}-\frac{m^2}{(\mu^2-1)}\right\}M(\mu)=0, \tag{7-97}$$

$$\left\{\frac{d}{d\nu}\left[(1-\nu^2)\frac{d}{d\nu}\right]-A-\frac{R^2E'\nu^2}{2}-\frac{m^2}{(1-\nu^2)}\right\}N(\nu)=0, \tag{7-98}$$

and

$$\frac{d^2\Phi(\phi)}{d\phi^2}=-m^2\Phi(\phi), \tag{7-99}$$

where E' is known as the electronic energy, and is given by

$$E'=E-\frac{1}{R}. \tag{7-100}$$

Also, A, m^2, and E' are constants to be determined. As we shall see shortly, the allowed values for these constants will be restricted (i.e., quantized) in order for acceptable solutions to Eqs. (7-97)–(7-99) to be found.

Equation (7-99) has been encountered before,[14] and the solutions were seen to be

$$\Phi(\phi)=\frac{1}{\sqrt{2\pi}}e^{im\phi}, \tag{7-101}$$

[12] See Problem 7-3, as well as Problem 4-16.

[13] See Problem 7-4.

[14] See Section 6-3.

with allowed values of m being given by

$$m = 0, \pm 1, \pm 2, \cdots, \tag{7-102}$$

which implies that eigenfunctions with $m > 0$ are doubly degenerate. Also, $\Phi(\phi)$ is independent of R, indicating that the Φ-equation is always separable, regardless of the value of R. In such a case, the quantum number (m in this case) is said to be a "*good quantum number*," since its value is independent of R.

Equations (7-97) and (7-98) turn out not to have solutions that have convenient closed-form expressions[15] as we found for the hydrogen atom. In particular, only numerical solutions to these two equations have been found in practice,[16] and these solutions are a function of R. However, very accurate numerical solutions have been obtained, meaning that effectively "exact" solutions have been found for various R values within the constraint of the Born–Oppenheimer approximation at each R value. The total energy (E) of several of the low-lying states of H_2^+ as a function of R is depicted in Fig. 7.12, as calculated by Bates et al.[16]

Several points of interest are present in the data of Fig. 7.12. First, we see that with the exception of the ground state, no states exhibit a minimum (if at all) deep enough to expect a stable excited state to be formed. This means that electronic spectral studies of H_2^+ are not in general possible, and most information about this molecule and the structure of its states has come from theoretical studies. Next, the ground state binding energy (the energy difference between $R_e = 2.0$ and $R = \infty$) is 0.1026 Hartree, or 64.4 kcal/mol. Thus, even the single electron in H_2^+ is sufficient to form a stable chemical bond.

Also, we note that the notation used to label the states is similar but not identical to the labeling used for atomic states of the hydrogen atom. To see the rationale for this, we turn to the electronic energies [E' from Eq. (7-100)] of H_2^+, which are shown for several states calculated (using data from Bates et al.[16]) as a function of R in Fig. 7.13. At $R = 0$, we see that the corresponding states of the He^+ atom are found ($E = -2/n^2$). This extreme is known as the "*united atom*" limit, and labeling the H_2^+ state relative to its corresponding He^+ atom state ($1s$, $2s$, $2p$ ...) is therefore appropriate.

In addition, however, the H_2^+ states have been labeled according to their behavior as $R \to \infty$, which is known as the "*separated atom*" limit.[17] In this

[15] For discussion of possible solutions to Eq. (7-98) see, for example, J. A. Stratton, P. M. Morse, L. J. Chu, and R. A. Hunter, *Elliptic, Cylinder, and Spheroidal Wavefunctions*, John Wiley, New York, 1941 and E. A. Hylleraas, *Z. Phys.*, **71**, 739 (1931).

[16] See G. Jaffe, *Z. Phys.*, **87**, 535 (1934) and D. R. Bates, K. Ledsham, and A. L. Stewart, *Phil. Trans. R. Soc. (London)*, **A246**, 215 (1954), for examples of numerical solutions to Eqs. (7-97) and (7-98); see also E. M. Roberts, M. R. Foster, and F. F. Selig, *J. Chem. Phys.*, **37**, 485 (1962); H. Wind, *J. Chem. Phys.*, **42**, 2371 (1965); and J. L. Schaad and W. V. Hicks, *J. Chem. Phys.*, **53**, 851 (1970).

[17] In this case a proton plus an H atom are obtained, and the states of the H atom are given by $E = -1/2n^2$.

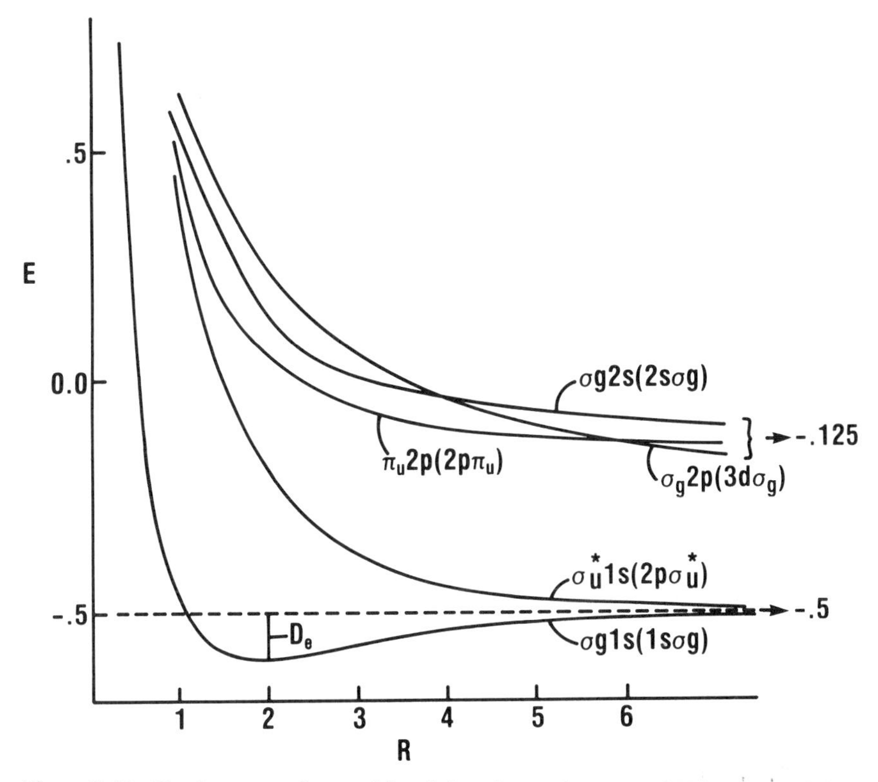

Figure 7.12. Total energy of several low-lying electronic states of H_2^+ calculated as a function of R. ($R_e = 1.997$ for the ground state.) The separated atom designation is used to label each state, while the united atom designation is listed in parentheses.

part of the labeling, two additional features are included. First, states of increasing energy are labeled according to m-value as in the hydrogen atom, but using Greek letters to distinguish the molecular case from the atomic case, i.e., σ, π, δ ... are used instead of s, p, d ... . The second feature in the labeling has to do with whether the wavefunction for the state of interest is symmetric or antisymmetric with respect to inversion of the electron coordinates through the origin.[18] When the wavefunction is symmetric under inversion, the wavefunction is labeled "g" (for "gerade"), and when the wavefunction is antisymmetric it is labeled "u" (for "ungerade").

The net result is a set of labels for states that combine the united and separated atom labels as follows near $R = 0$: $1s\sigma_g$, $1s\sigma_u$, $2s\sigma_g$, $2s\sigma_u$, $2p\sigma_g$, $2p\sigma_u$, $2p\pi_g$, ... etc. On the other hand, as $R \to \infty$ the states are labeled first by the separated atom notation, e.g., $\sigma_g 1s$, $\sigma_u 1s$, $\sigma_g 2s$, ..., etc.

Finally, the states are labeled as to whether the wavefunctions are symmetric

[18] See Problem 7-6 for justification of the use of such labels for H_2^+ states.

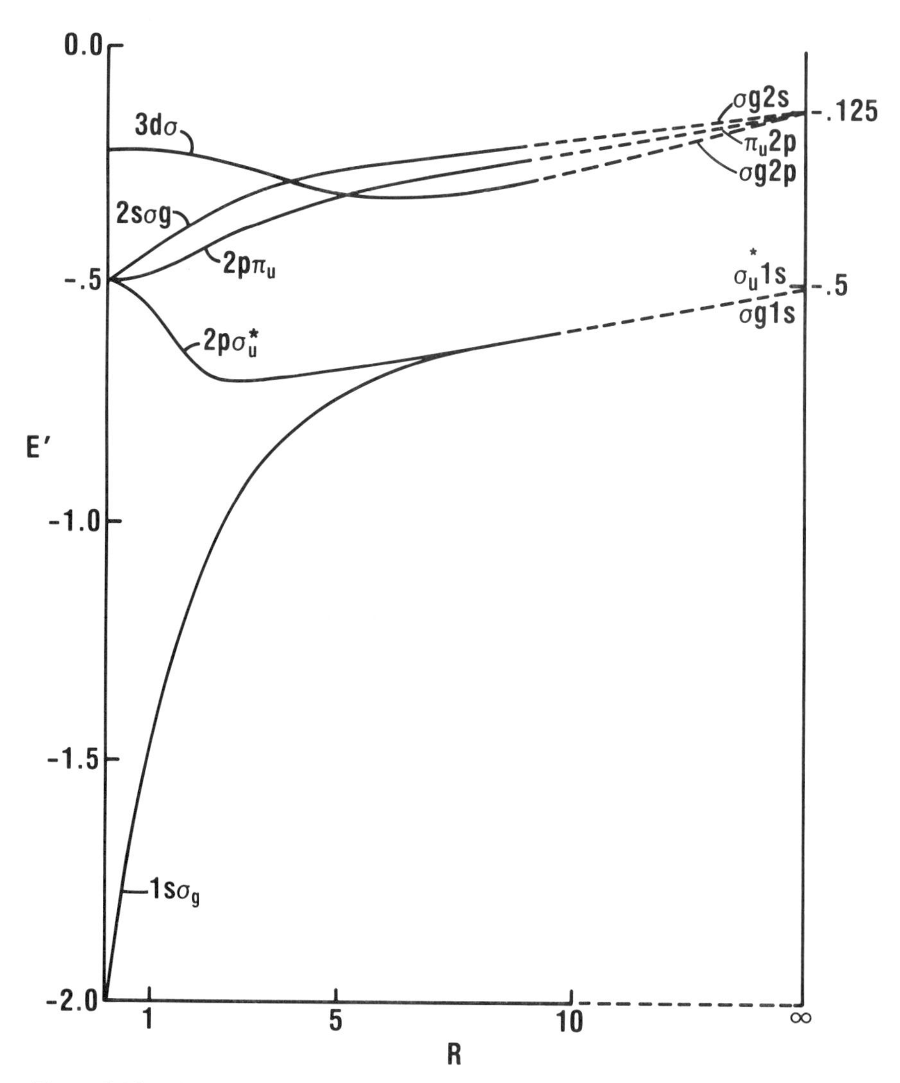

Figure 7.13. Electronic energy (in Hartrees) of H_2^+ states plotted as a function of R.

or antisymmetric with respect to reflection of the electron coordinates in a plane perpendicular to the molecular axis and midway between the two protons. States that are antisymmetric are labeled with an asterisk. Examples that illustrate these states are given in Fig. 7-14, where wavefunction contour diagrams are shown for the $1s\sigma_g(\sigma_g 1s)$ and $2p\sigma_u^*(\sigma_u^* 1s)$ states. The symmetry and antisymmetry of the two states, respectively, are apparent from the figure since, in order to be antisymmetric with respect to reflection, there must be a node at the mid-point between the nuclei.

It should also be noted that when discussing non-hydrogen separated (or

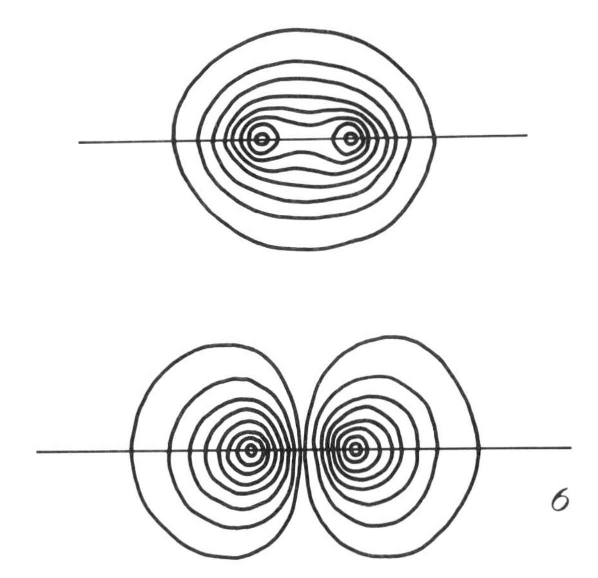

Figure 7.14. Wavefunction contour diagrams for the $1s\sigma_g$ (upper diagram) and $2p\sigma_u^*$ (lower diagram) states of H_2^+ at $R = 2.0$ Bohrs.

united) atoms, the hydrogen atom wavefunctions are not exact solutions. However, they (or other functions having similar shapes) are frequently useful in providing a qualitative description of more complicated atoms and molecules. In such cases it is important to distinguish them from exact eigenfunctions, and the term *orbitals* is frequently used. Indeed, such terminology will be utilized essentially exclusively for all systems beyond the H atom and H_2^+.

As we shall discover in greater detail later, "orbitals" that have electron density nodes at the mid-point (i.e., are labeled with an asterisk) lead to unstable bonds, and those leading to increased electron density in the region between nuclei lead to stable bonds. For these reasons the orbitals with asterisks are called *antibonding orbitals* and the orbitals without asterisks are called *bonding* orbitals.

To help understand why the states in Fig. 7.13 behave as shown, as well as to introduce a concept that will be useful in constructing a process for understanding and predicting the structure of diatonic molecules in general, it is useful to introduce the "*noncrossing rule*."[19] To understand this "rule," we consider a hypothetical diatomic molecule having only two states with two orthogonal basis functions, Ψ_1 and Ψ_2. The total wavefunction for the system for a given value of the internuclear distance R can thus be written as a linear combination of these two functions, i.e.,

$$\Psi = c_1\Psi_1 + c_2\Psi_2. \tag{7-103}$$

[19] See J. von Neumann and E. Wigner, *Z. Phys.*, **30**, 467 (1929) and E. Teller, *J. Chem. Phys.*, **41**, 109 (1937).

By using the linear variation method that is described later,[20] it can be shown that the optimum energy levels for such a system are found by solving the following determinant:

$$\begin{vmatrix} H_{11}-E & H_{12} \\ H_{21} & H_{22}-E \end{vmatrix} = 0, \tag{7-104}$$

with

$$H_{ij} = \int \Psi_i^* \mathcal{H} \Psi_j \, d\tau, \tag{7-105}$$

where $\mathcal{H}$ is the Hamiltonian for the system. Expanding Eq. (7-104) gives

$$E = \frac{(H_{11}+H_{22}) \pm [(H_{11}+H_{22})^2 - 4(H_{11}H_{22} - H_{12}H_{21})]^{1/2}}{2}. \tag{7-106}$$

Since the calculated energy will depend upon the value of R that is chosen, one of the questions of interest is whether the energy curves as a function of R for each of the two states can ever cross for some particular choice of R. Since the answer to this question is most conveniently proved via the use of group theory, which will be covered later, we shall use the results of that analysis without repeating it here.[21]

What Teller[21] showed was that if the symmetry properties of Ψ_1 and Ψ_2 were different (e.g., if Ψ_1 was symmetric under inversion through the origin and Ψ_2 was antisymmetric[22]), then

$$H_{12} = H_{21} = 0 \tag{7-107}$$

and

$$E = \begin{cases} H_{11} & \text{or} \\ H_{22} \end{cases} \tag{7-108}$$

Since H_{11} and H_{22} are functions of R and have no necessary relationship to each other, it is possible that for some choice of $R = R_0$,

$$H_{11} = H_{22} \tag{7-109}$$

For that value of R, the energy for both states will be the same, and the curves of E_1 vs. R and E_2 vs. R will cross at $R = R_0$.

If the two functions have the *same symmetry*, there is no constraint on H_{12}

---

[20] See Section 9-5.

[21] See E. Teller, *J. Chem. Phys.*, **41**, 109 (1936) for a more complete discussion of the proof. Also, see Chapter 10, Section 2.

[22] Another example in which Eq. (7-107) applies is when Ψ_1 is an even function and Ψ_2 is odd, or vice versa.

and H_{21} as given in Eq. (7-107), and Eq. (7-106) must be used to calculate E_1 and E_2. Since the term within the brackets in Eq. (7-106) will be nonzero, there will always be a splitting of the E_1 and E_2 levels, and the case $E_1 = E_2$ cannot arise. In other words, the E_1 vs. R for E_2 vs. R curves for the two states having the same symmetry cannot cross for any value of R if Ψ_1 and Ψ_2 have the same symmetry.[23]

Such a diagram is given in Figure 7.14 for a hypothetical homonuclear diatomic molecule. It not only indicates how united atom orbitals correlate with separated atom orbitals, but also gives a rough idea about the possible energy level ordering for orbitals in the diatomic molecule. Of course, since the energy level spacings cannot be drawn to scale for a hypothetical molecule, the figure cannot be used in any quantitative sense. However, as we shall see in Chapter 10 (Section 10-1), a correlation diagram of this type is quite useful in constructing an "aufbau principle" for diatomic molecules analogous to that used for atoms.

Problems

1. Prove Eq. (7-13).
2. Find the probability that an electron in the ground state of the H-atom will be found at a distance from the nucleus that is greater than 1 Bohr. (The latter is the distance that is predicted by the earliest quantum theory of Bohr.)
3. For the elliptical coordinate system defined in Eqs. (7-93) and (7-94) plus Fig. 7.12, show that the quantum mechanical Hamiltonian for H_2^+ can be written as in Eq. (7-95), with

$$dV = \frac{R^3}{8} (\mu^2 - \nu^2)\, d\mu\, d\nu\, d\phi$$

and

$$1 \leq \mu \leq \infty$$
$$-1 \leq \nu \leq +1$$
$$0 \leq \phi \leq 2\pi.$$

4. Show that, by the use of Eq. (7-96), Eq. (7-95) can be written as Eqs. (7-97), (7-98), and (7-99), with E' as defined in Eq. (7-100).
5. Verify by direct calculation that for the ground state of the hydrogen atom,

$$E = \frac{1}{2} \langle \mathcal{V} \rangle = \langle \mathfrak{T} \rangle.$$

[23] While this statement is generally true, there are limitations on its applicability due to the way in which the noncrossing rule was proved. See F. L. Pilar, *Elementary Quantum Chemistry*, McGraw-Hill, New York, 1968, p. 463 for examples of the limitations.

This is a specific example of the virial theorem, which will be discussed in Section 10-6.

6. Show that the "inversion" operator, i.e., an operator that changes

$$\varphi \to \varphi + \pi$$

$$\vartheta \to \vartheta + \pi$$

$$r \to r$$

commutes with the Hamiltonian operator for H_2^+. This means that simultaneous eigenfunctions of the inversion and Hamiltonian operators can be found, and that the wavefunctions H_2^+ can be labeled as to their "inversion state" (i.e., symmetric or antisymmetric with respect to inversion) as well as energy.

Chapter 8

The Molecular Hamiltonian

8-1. General Principles and Discussion

In Chapter 4 a set of postulates for the study of quantum mechanics was introduced, including the Schrödinger equation plus a prescription for forming the quantum mechanical Hamiltonian for a time-independent, field-free, conservative system in which relativistic effects are ignored. Although such a Hamiltonian is indeed a restricted one compared to the circumstances found in many experiments, use of it has allowed a number of important principles to be established using one-dimensional and other examples, as well as the introduction of angular momentum from a quantum mechanical point of view.

In this chapter we shall examine the formation of Hamiltonian operators more fully, in order to see what changes occur when the restrictions mentioned above are relaxed. In particular, we shall examine the effects of introducing external fields as well as relativistic effects on the formation of quantum mechanical Hamiltonians, as well as indicating some of the deficiencies of the theory that still remain even when external fields and relativistic effects are considered.

As we begin these considerations, it is useful to make more explicit the processes and constraints to be used. In particular, since quantum mechanics has been *postulated* (and *not derived* from classical mechanics), and since the relationship to classical mechanics is not determined explicitly by the postulates, it is important to understand how and in what ways the results of quantum mechanics are to be expected to correspond to the results of classical mechanical experiments. Unfortunately, the guidelines that are currently known are relatively few in number, and leave significant areas in the theory in which further developments are needed.

A. The Correspondence Principle

Although we have not referred to it explicitly in previous discussions, one of the guidelines that played a very important role in the development of quantum theory is known at *The Correspondence Principle*. This principle, initially stated and subsequently developed by Bohr[1] as a part of what is now known as the "old quantum theory," was first stated from the point of view that quantum theory should contain classical mechanics as its limiting case. Restated for use here, the Correspondence Principle can be taken to require that, however the laws of quantum theory are chosen, the limit in which many quanta are involved and appropriately averaged must lead to the results of classical mechanics. Thus, while not considered explicitly in the Postulates of Chapter 4, the Correspondence Principle provides a kind of "overarching" consistency constraint for the formulation of quantum theory.[2]

This principle was employed and generalized by Dirac and others in the conceptualization and construction of quantum mechanical operators. In particular, by considering classical Hamiltonian concepts of momentum and position plus the expectation value concept associated with a series of measurements, a successful prescription was devised by which the observable properties of a wavefunction may be deduced. We discussed this as a part of Postulate IV in Chapter 4, and illustrated how the original Correspondence Principle was utilized. In the current context, we shall discover in the following sections that consideration of external fields and relativistic effects is not as unambiguously treated as in the discussions of Section 4-7. Thus, the Correspondence Principle remains as a "guiding light" for future as well as past formulations of quantum theory.

8-2. Introduction of External Fields

We shall now begin the generalization of quantum mechanical Hamiltonian formation, starting with consideration of external electric and magnetic fields. In other words, consideration of the forces acting on a particular microscopic particle will now be expanded to include not only interactions with other microscopic particles ("internal interactions") but also with *externally applied* electric and magnetic fields. To do this, we begin by returning to the corresponding model from classical electromagnetic theory.

In particular, if a classical particle of mass m at point $\mathbf{r}$ with velocity $\mathbf{v}$ and charge q is acted on by a homogeneous external electric field ($\mathbf{E}$) and magnetic field ($\mathbf{B}$), then the force acting on the particle (separate from and in addition to the "internal" forces exerted on the particle by other microscopic particles)

---

[1] N. Bohr, *Z. Phys.* **6**, 1–9 (1921); *Proc. Cambridge Philos. Soc. (Suppl.)* 22 (1924).

[2] For a more extended discussion of the relationship between quantum and classical concepts and the role of the Correspondence Principle, see D. Bohm, *Quantum Theory*, Prentice-Hall, NY, 1951, Chapter 23; see also J. L. Powell and B. Crasemann, *Quantum Mechanics*, Addison-Wesley, Reading, MA, 1961, Section 6-13 for additional discussion and examples.

from these fields is given[3] by

$$\mathbf{F}(\mathbf{r}) = q\left\{\mathbf{E}(\mathbf{r}) + \left(\frac{1}{c}\right)(\mathbf{v}\times\mathbf{B}(\mathbf{r})\right\}, \tag{8-1}$$

where c is the speed of light. A number of comments about Eq. (8-1) are appropriate before continuing. First, the electric field ($\mathbf{E}$) at $\mathbf{r}$ is unaffected by the motion of the charge, and the effect of the electric field is dependent only on the magnitude and direction of the electric field. However, the magnetic field effect is clearly dependent on the fact that the charge is moving (with velocity $\mathbf{v}$), and the net effect depends not only on the strength of $\mathbf{B}$ but also on the velocity of the particle ($\mathbf{v}$). In other words, if the particle is at rest, the contribution of the magnetic field to the force acting on the particle disappears.

Conversely, there are effects on the external fields due to charges that are added to the system. In particular, the response of external fields to charges is given by *Maxwell's equations*, i.e.,

$$\begin{aligned}
\boldsymbol{\nabla}\cdot\mathbf{E} &= 4\pi\rho,\\
(\boldsymbol{\nabla}\times\mathbf{E}) + \left(\frac{1}{c}\right)\frac{\partial\mathbf{B}}{\partial t} &= 0,\\
\boldsymbol{\nabla}\cdot\mathbf{B} &= 0,\\
(\boldsymbol{\nabla}\times\mathbf{B}) - \left(\frac{1}{c}\right)\frac{\partial\mathbf{E}}{\partial t} &= \left(\frac{4\pi}{c}\right)\mathbf{j},
\end{aligned} \tag{8-2}$$

where ρ and $\mathbf{j}$ are the charge density and current density, respectively.

As to the effect on the movement of the particle that we expect to see as a result of introducing these external fields, we know that the force acting on a particle is related to its momentum, and we thus expect to see a change in the classical (as well as quantum mechanical) Hamiltonian, since the latter depends on $(p^2/2m)$ or its quantum mechanical analog.

Continuing from a classical mechanical point of view, we note that, since Newton's laws give

$$\mathbf{F} = m\mathbf{a}, \tag{8-3}$$

we can combine Eqs. (8-1) and (8-3) to give

$$\mathbf{F} = m\mathbf{a} = q\mathbf{E} + \left(\frac{q}{c}\right)(\mathbf{v}\times\mathbf{B})$$

[3] See, for example, E. R. Peck, *Electricity and Magnetism*, McGraw-Hill, NY, 1953, p. 201, for additional discussion and background. The units employed here are "mixed Gaussian" units, i.e., electric quantities are expressed in esu's, and magnetic quantities are in emu's. See, for example, C. A. Coulson, *Electricity*, Interscience/Oliver and Boyd, Edinburgh, 1953, Chapter 14.

or

$$m\,\frac{d^2x}{dt^2}=qE_x+\left(\frac{q}{c}\right)\left[\left(\frac{\partial y}{\partial t}\right)B_z-\left(\frac{\partial z}{\partial t}\right)B_y\right], \tag{8-4}$$

with similar expressions for the y- and z-components. It should be noted that while the order of terms within the brackets of Eq. (8-4) is immaterial from a classical mechanical point of view, a significant difference may result when each of the terms is converted to the corresponding quantum mechanical operator. Thus, a potential ambiguity has already been introduced, whose resolution awaits appropriate comparison to experiment.

In practice, dealing with the fields per se turns out to be typically less convenient than dealing with the vector (**A**) and scalar (ϕ) potentials that can be used to derive[4] **E** and **B**, i.e.,

$$\mathbf{E}=-\nabla\phi-\left(\frac{1}{c}\right)\frac{\partial\mathbf{A}}{\partial t}, \tag{8-5}$$

and

$$\mathbf{B}=\nabla\times\mathbf{A}, \tag{8-6}$$

where it should be noted that both **A** and ϕ are functions of x, y, z, and t in general. Insertion of Eqs. (8-5) and (8-6) into Eq. (8-4) gives the following equation of motion (x-component):

$$\frac{d}{dt}(m\dot{x})=\left(\frac{q}{c}\right)\left[-c\,\frac{\partial\phi}{\partial x}-\frac{\partial A_x}{\partial t}+\dot{y}\,\frac{\partial A_y}{\partial x}-\dot{y}\,\frac{\partial A_x}{\partial y}+\dot{z}\,\frac{\partial A_z}{\partial x}-\dot{z}\,\frac{\partial A_x}{\partial z}\right]$$
$$+\left(\frac{q}{c}\right)\left[\dot{x}\,\frac{\partial A_x}{\partial x}-\dot{x}\,\frac{\partial A_x}{\partial x}\right], \tag{8-7}$$

where the last term on the right-hand side of the above equation (which equals zero) has been added to make the following analysis more transparent. Similar expressions exist for the y- and z-components. By noting that

$$\frac{dA_x}{dt}=\frac{\partial A_x}{\partial t}+\dot{x}\,\frac{\partial A_x}{\partial x}+\dot{y}\,\frac{\partial A_x}{\partial y}+\dot{z}\,\frac{\partial A_x}{\partial z}, \tag{8-8}$$

Eq. (8-7) can be rewritten as

$$\frac{d}{dt}\left[m\dot{x}+\left(\frac{q}{c}\right)A_x\right]=-q\,\frac{\partial\phi}{\partial x}+\left(\frac{q}{c}\right)\left[\dot{x}\,\frac{\partial A_x}{\partial x}+\dot{y}\,\frac{\partial A_y}{\partial x}+\dot{z}\,\frac{\partial A_z}{\partial x}\right]. \tag{8-9}$$

In three dimensions, Eq. (8-9) becomes

$$\frac{d}{dt}\left[m\mathbf{v}+\left(\frac{q}{c}\right)\mathbf{A}\right]=-q\nabla\phi+\left(\frac{q}{c}\right)[\dot{x}(\nabla A_x)+\dot{y}(\nabla A_y)+\dot{z}(\nabla A_z)]. \tag{8-10}$$

---

[4] See, for example, W. K. H. Panofsky and M. Phillips, *Electricity and Magnetism*, Addison-Wesley, Reading, MA, 1955, Chapter 13. See also J. D. Jackson, *Classical Electrodynamics*, Wiley, New York 1975.

While Eq. (8-10) is perfectly acceptable as a classical expression of the equation of motion of a particle in the presence of an external electric and magnetic field, it will be convenient to construct a corresponding classical Hamiltonian before converting it to a quantum mechanical operator.

From earlier discussions,[5] we know that the classical Hamiltonian can be derived from knowledge of the Lagrangian (L), i.e.,

$$H=\sum_i p_i \dot{r}_i - L. \tag{8-11}$$

If we choose,

$$L=\left(\frac{1}{2m}\right)(m\mathbf{v})^2 - q\phi + \left(\frac{q}{c}\right)(\mathbf{v}\cdot\mathbf{A}), \tag{8-12}$$

the corresponding Lagrangian equations of motion [Eq. (1-28)] are given by (x-component):

$$\frac{d}{dt}\frac{\partial L}{\partial \dot{x}} - \frac{\partial L}{\partial x} = 0. \tag{8-13}$$

Inserting Eq. (8-12) into (8-13) gives

$$\frac{d}{dt}\left[m\dot{x}+\left(\frac{q}{c}\right)A_x\right] = -q\frac{\partial\phi}{\partial x} + \left(\frac{q}{c}\right)\left(\dot{x}\frac{\partial A_x}{\partial x} + \dot{y}\frac{\partial A_y}{\partial x} + \dot{z}\frac{\partial A_z}{\partial x}\right), \tag{8-14}$$

where the rationale for the particular choice of L in Eq. (8-12) is now apparent, i.e., this choice of L gives the same equation of motion as found in Eq. (8-9).

Insertion of Eq. (8-12) into Eq. (8-11) and use of Eq. (1-30) gives the classical Hamiltonian that corresponds to the equations of motion of Eq. (8-10), i.e.,

$$H=\left(\frac{1}{2m}\right)\left[\left(p_x - \left(\frac{q}{c}\right)A_x\right)^2\right] + q\phi_x, \tag{8-15}$$

with similar expressions for the y- and z-components.

At this point we utilize the procedure for constructing operators that was introduced in Section 4-7 to convert the classical Hamiltonian in Eq. (8-15) into a quantum mechanical operator,[6] which gives

$$\mathcal{H}=\left(\frac{1}{2m}\right)\left\{\left[\left(\frac{\hbar}{i}\right)\nabla - \left(\frac{q}{c}\right)\mathbf{A}\right]\cdot\left[\left(\frac{\hbar}{i}\right)\nabla - \left(\frac{q}{c}\right)\mathbf{A}\right]\right\} + q\phi. \tag{8-16}$$

Since

$$(\nabla\cdot\mathbf{A})f = \mathbf{A}\cdot(\nabla f) + f(\nabla\cdot\mathbf{A}) \tag{8-17}$$

---

[5] See Section 1-2.

[6] Problems 8-1 and 8-2 illustrate important invariance properties of the Hamiltonian operator of Eq. (8-16). Also, see Problem 8-3 for a one-dimensional analog of Eqs. (8-16)–(8-18).

for a function f, we can rewrite Eq. (8-16) as

$$\mathcal{H}=\left(\frac{1}{2m}\right)\left\{-\hbar^2\nabla^2+\left(\frac{i\hbar q}{c}\right)[(\nabla\cdot\mathbf{A})+2\mathbf{A}\cdot\nabla]+\left(\frac{q^2}{c^2}\right)|\mathbf{A}|^2\right\}+q\phi. \tag{8-18}$$

From the results of Problem 8-3, we see that, without loss of generality, we can take

$$(\nabla\cdot\mathbf{A})=0, \tag{8-19}$$

which allows Eq. (8-18) to be written as

$$\mathcal{H}=\left(\frac{1}{2m}\right)\left\{-\hbar^2\nabla^2+\left(\frac{2i\hbar q}{c}\right)\mathbf{A}\cdot\nabla+\left(\frac{q^2}{c^2}\right)|\mathbf{A}|^2\right\}+q\phi. \tag{8-20}$$

If there are N particles instead of one,[7] Eq. (8-20) is written as

$$\mathcal{H}=-\left(\frac{\hbar^2}{2}\right)\sum_{j=1}^{N}\left\{\left(\frac{1}{m_j}\right)\nabla_j^2+\left(\frac{2i\hbar q_j}{c}\right)\mathbf{A}_j\cdot\nabla_j+\left(\frac{q_j}{c}\right)^2|\mathbf{A}_j|^2\right\}+\sum_{j=1}^{N}q_j\phi_j. \tag{8-21}$$

If interparticle Coulombic interactions are also included, we obtain the following Hamiltonian operator, which includes all internal and external effects introduced thus far:

$$\mathcal{H}=-\left(\frac{\hbar^2}{2}\right)\sum_{j=1}^{N}\left\{\left(\frac{1}{m_j}\right)\nabla_j^2+\left(\frac{2i\hbar q_j}{c}\right)\mathbf{A}_j\cdot\nabla_j+\left(\frac{q_j}{c}\right)^2|\mathbf{A}_j|^2\right\}$$
$$+\sum_{i<j}^{N}\frac{q_iq_j}{r_{ij}}+\sum_{j=1}^{N}q_j\phi_j. \tag{8-22}$$

As a simple illustration of external fields, application of an electromagnetic wave in direction ($\mathbf{k}$) gives a vector potential

$$\mathbf{A}(\mathbf{r},t)=\mathbf{A}_0\cos(\mathbf{k}\cdot\mathbf{r}-\omega t),$$

where $\mathbf{A}_0$ is the amplitude and $\hbar\omega$ is the energy associated with $\mathbf{A}$. Such an applied field can, for example, remove an electron from an atom if the energy ($\hbar\omega$) is sufficiently large.

It should be noted that the above Hamiltonian operator treats all particles on an equal footing, e.g., it does not require an assumption that nuclei are held fixed relative to each other. Later in this chapter we shall return to this point (when the Born–Oppenheimer approximation is introduced), but for the present,

[7] The notation in Eq. (8-21) and beyond regarding the vector potential ($\mathbf{A}_j$) is not intended to imply a different potential for every particle (j). Instead, $\mathbf{A}_j$ refers to the value of the potential at the location of particle j.

all particles are considered to be moving. Also, it should be noted that "spin effects," i.e., the intrinsic electromagnetic interactions between particles, are not revealed by the "quasiclassical" approach to Hamiltonian operator formation that we have previously used. While it is possible to "add on" terms[8] corresponding to spin interactions, we shall eschew such an approach since these terms are truly "nonclassical" in origin, and will arise naturally when relativistic effects are considered.

8-3. Introduction of Relativistic Effects

The final set of effects to be considered are those arising from special relativity.[9] In particular, it is found that, in order to assure that identical results are obtained regardless of the coordinate system that is chosen, Newton's equations of motion must be written as (x-component)

$$F_x = \frac{d}{dt}(m\dot{x}) = \frac{d}{dt}(mv_x), \tag{8-23}$$

where the mass of the particle is allowed to vary with its velocity according to the relation

$$m = \frac{m_0}{[1-(v^2/c^2)]^{1/2}}, \tag{8-24}$$

where m_0 is the *rest mass* of the particle. As the above equation indicates, the mass of the particle can usually be taken as m_0, as long as the velocity (v) of the particle is small relative to the speed of light (c). While the modifications indicated above appear to be minor, and apparently arise only to assure invariance to coordinate system transformations[10] (and not because of, e.g., new kinds of external fields), we shall see shortly that the presence of the velocity term within the square root will cause major differences in the results that are obtained.

If we introduce the modifications of Eqs. (8-23) and (8-24), we see that the equations of motion that were given by Eq. (8-4) are now given by (x-component)

$$\frac{d}{dt}\left\{\frac{m_0 v_x}{[1-(v^2/c^2)]^{1/2}}\right\} = qE_x + \left(\frac{q}{c}\right)\left[\left(\frac{\partial y}{\partial t}\right)B_z - \left(\frac{\partial z}{\partial t}\right)B_y\right]. \tag{8-25}$$

[8] See, for example, J. C. Slater, *Quantum Theory of Atomic Structure*, Vol. 2, McGraw-Hill, New York, 1960, Chapter 24.

[9] For a discussion of special relativity in classical mechanics see, for example, *Principles of Mathematical Physics*, by W. V. Houston, McGraw-Hill, New York, 1948, Chapter 16.

[10] The general postulates of relativity require invariance of any physical law to what are known as Lorentz transformations, and the modifications introduced in Eqs. (8-23) and (8-24) assure Lorentz invariance for Newton's equations. For additional discussion see, for example, R. C. Tolman, *Relativity, Thermodynamics, and Cosmology*, Oxford University Press, New York, 1934.

By proceeding in a manner similar to the analysis involving Eqs. (8-5) through (8-10), it can be seen[11] that the equations of motion can be written as

$$\frac{d}{dt}\left\{\frac{m_0\mathbf{v}}{[1-(v^2/c^2)]^{1/2}}+\left(\frac{q}{c}\right)\mathbf{A}\right\}=-q\nabla\phi+\left(\frac{q}{c}\right)(\mathbf{v}\cdot\nabla\mathbf{A}). \quad (8\text{-}26)$$

It can also be seen[12] that choice of the classical Lagrangian as follows:

$$L=-m_0c^2\left[1-\frac{v^2}{c^2}\right]^{1/2}+\left(\frac{q}{c}\right)(\mathbf{v}\cdot\mathbf{A})-q\phi, \quad (8\text{-}27)$$

will give rise to Lagrangian equations of motion [Eq. (8-13)], which are identical to the Newtonian equations of motion given in Eq. (8-26).

Finally, we utilize Eq. (1-30), i.e.,

$$p_x=\frac{\partial L}{\partial v_x}=\frac{m_0v_x}{[1-(v_x^2/c^2)]^{1/2}}+\left(\frac{q}{c}\right)A_x, \quad (8\text{-}28)$$

and Eq. (8-27) to form the classical Hamiltonian for the system using Eq. (8-11), i.e., (for the x-component):

$$\begin{aligned}H_x&=p_xv_x-L_x\\&=\frac{m_0c^2}{[1-(v_x^2/c^2)]^{1/2}}+q\phi,\end{aligned} \quad (8\text{-}29)$$

or

$$H_x=\left\{m_0^2c^4+\left[p_x-\left(\frac{q}{c}\right)A_x\right]^2c^2\right\}^{1/2}+q\phi. \quad (8\text{-}30)$$

In three dimensions, when interparticle interactions (V) are included, we have

$$H=c\left\{\left[\mathbf{p}-\left(\frac{q}{c}\right)\mathbf{A}\right]^2+m_0^2c^2\right\}^{1/2}+q\phi+V. \quad (8\text{-}31)$$

This classical Hamiltonian includes all internal, external, and relativistic effects for a single particle, and thus should allow us to convert it to a quantum mechanical Hamiltonian that is correspondingly comprehensive in its description of internal, external, and relativistic effects at the microscopic level. However, the mathematical form of Eq. (8-31) is not nearly as convenient as we had earlier [Eq. (8-15)]. In particular, the presence of the square root in Eq. (8-31) complicates things considerably. Formulation of a satisfactory approach, at least for the case of a single electron, was given first by Dirac, and it is to a discussion of this formulation to which we now proceed.

---

[11] See Problem 8-5.

[12] See Problem 8-6.

A. The Dirac Equation

In formulating an alternative to direct use of a quantum mechanical Hamiltonian based on Eq. (8-31), we return to the case of a single particle, in which case Eq. (8-31) can be written as

$$H=\left\{m_0^2c^4+\left[\mathbf{p}-\left(\frac{q}{c}\right)\mathbf{A}\right]^2c^2\right\}^{1/2}+q\phi. \tag{8-32}$$

By rearranging and squaring, we obtain

$$(H-q\phi)^2=m_0^2c^4+c^2\left[\mathbf{p}-\left(\frac{q}{c}\right)\mathbf{A}\right]^2. \tag{8-33}$$

Defining

$$\pi_0=(H-q\phi)/c, \tag{8-34}$$

$$\pi_j=p_j-\left(\frac{q}{c}\right)A_j, \qquad j=x,\ y,\ z, \tag{8-35}$$

Eq. (8-33) can be rewritten as

$$\pi_0^2-m_0^2c^2=\pi_1^2+\pi_2^2+\pi_3^2. \tag{8-36}$$

The advantages that are achieved by such manipulations are that the square root has been eliminated and each of the variables (t, x, y, z) that will appear in a quantum mechanical operator formed from Eq. (8-36) appear in the *same* form in the equation. This latter point is particularly relevant to questions of Lorentz invariance. However, as appealing as Eq. (8-36) is,[13] it does not give rise to solutions that describe electrons, and further work is needed. In particular, it does not require the existence of the intrinsic spin of the electron. Also, since each term in Eq. (8-36) is squared, it implies that the time dependence will not be linear, as it is in the nonrelativistic case.

In order to alleviate these difficulties, Dirac took the approach of "factoring" Eq. (8-36) by writing it in the form

$$\left(\pi_0+\sum_{j=1}^{3}\alpha_j\pi_j+\alpha_4m_0c\right)\left(\pi_0-\sum_{j=1}^{3}\alpha_j\pi_j-\alpha_4m_0c\right)=0, \tag{8-37}$$

where α_1, α_2, α_3 and α_4 are chosen to be independent of $\mathbf{p}$ and $\mathbf{r}$, and so that Eq. (8-36) is satisfied. The particular choice

$$\begin{aligned}\alpha_j^2&=1 & j&=1,\ 2,\ 3\\ \alpha_j\alpha_k+\alpha_k\alpha_j&=0 & (j&\neq k)\end{aligned} \tag{8-38}$$

can be seen to make Eqs. (8-37) and (8-36) alike.[14]

---

[13] When converted to operator form, Eq. (8-36) is known as the *Klein–Gordon equation*. See, for example, H. A. Bethe and E. E. Saltpeter, *Quantum Mechanics of One- and Two-Electron Atoms*, Springer-Verlag, Berlin, 1957, Section 45, for additional discussion. See also Problem 8-7.

[14] See Problem 8-8.

In several senses, the approach taken by Dirac appears to have made the situation substantially more complicated instead of simpler. For example, conversion of Eq. (8-37) to quantum mechanical operator form implies that the $[\alpha_j]$ may be operators, for which no obvious classical mechanical analog is available to help in choosing appropriately. However, it has a number of advantages, especially from a conceptual point of view. First, we note that each of the factors in Eq. (8-37) is linear in its time dependence, as desired. Next, the fact that the αs may be operators for which there are no classical analogs should not necessarily be disturbing. In particular, we have already seen[15] in the general discussion of angular momentum that two kinds of angular momentum are possible, one of which does not have a classical mechanical analog. Thus, even though the "factoring" of Eq. (8-37) to produce the correct time dependence requires the introduction of new quantities (αs), we shall see that these quantities are just what is needed to provide an understanding of electron spin properties. It should also be noted that the choice of values for the αs as defined in Eq. (8-38) is not unique, and therefore provide only one representation of the α-operators of Eq. (8-37). Finally, we note that, while either factor in Eq. (8-37) can be used (with identical results), there exists the possibility that there may be positive as well as negative energy states that are acceptable solutions. We shall concern ourselves, however, only with those energy values corresponding to particles having positive kinetic energy[16] (e.g., electrons). In particular, we shall consider the equation

$$\left(\pi_0-\sum_{j=1}^{3}\alpha_j\pi_j-\alpha_4 m_0 c\right)\Psi=0. \tag{8-39}$$

Using the "traditional" definition

$$\beta=\alpha_4, \tag{8-40}$$

and a scalar product notation, we can rewrite Eq. (8-39) as

$$(\pi_0-\boldsymbol{\alpha}\cdot\boldsymbol{\pi}-\beta m_0 c)\Psi=0, \tag{8-41}$$

where

$$\boldsymbol{\alpha}\cdot\boldsymbol{\pi}=\sum_{j=1}^{3}\alpha_j\pi_j. \tag{8-42}$$

If we now convert the quantities in Eq. (8-41) to operator form, we have

[15] Chapter 6, Section 2.

[16] The states with negative kinetic energy correspond to positrons, which are particles having the same mass and opposite charge from that of electrons, and which were predicted from theory by Dirac for the first time.

[using Eqs. (8-24) and (8-35)]

$$\pi_0 = \left(i\hbar \frac{\partial}{\partial t} - q\phi \right) \Big/ c, \tag{8-43}$$

$$\pi_j = \left[\left(\frac{\hbar}{i} \right) \frac{\partial}{\partial r_j} - \left(\frac{q}{c} \right) A_j \right], \qquad j = x, y, z, \tag{8-44}$$

which allows Eq. (8-41) to be rewritten as

$$i\hbar \left(\frac{\partial \Psi}{\partial t} \right) = (c\boldsymbol{\alpha} \cdot \boldsymbol{\pi} + q\phi + \beta m_0 c^2)\Psi, \tag{8-45}$$

where the resemblance to the Schrödinger equation is made more apparent. However, we must remember that β and the components of α may now be operators (which are independent of, and therefore commute with, the momentum operators). This equation [Eq. (8-45)] is known as the relativistic *Dirac equation of state* for a single particle in the presence of an external electromagnetic field.

Turning next to the operators α_i, it is not expected that an explicit representation in terms of **r** and t will be possible, since there is no classical analog. However, a matrix representation is easily constructed. Since four components (corresponding to **r** and t) will be needed, a four-dimensional matrix representation of the operators α_i is expected. It can be easily verified[17] that the following matrices provide one representation of the α_i that is consistent with the anticommutation relations in Eq. (8-38):

$$\boldsymbol{\alpha}_1 = \begin{pmatrix} 0 & 0 & 0 & 1 \\ 0 & 0 & 1 & 0 \\ 0 & 1 & 0 & 0 \\ 1 & 0 & 0 & 0 \end{pmatrix} \tag{8-46}$$

$$\boldsymbol{\alpha}_2 = \begin{pmatrix} 0 & 0 & 0 & -i \\ 0 & 0 & +i & 0 \\ 0 & -i & 0 & 0 \\ +i & 0 & 0 & 0 \end{pmatrix} \tag{8-47}$$

$$\boldsymbol{\alpha}_3 = \begin{pmatrix} 0 & 0 & 1 & 0 \\ 0 & 0 & 0 & -1 \\ 1 & 0 & 0 & 0 \\ 0 & -1 & 0 & 0 \end{pmatrix} \tag{8-48}$$

$$\boldsymbol{\alpha}_4 = \begin{pmatrix} 1 & 0 & 0 & 0 \\ 0 & 1 & 0 & 0 \\ 0 & 0 & -1 & 0 \\ 0 & 0 & 0 & -1 \end{pmatrix} = \boldsymbol{\beta} \tag{8-49}$$

---

[17] See Problem 8-9.

In terms of the Pauli spin matrices (σ_i) introduced earlier [Eq. (6-116)], the above matrices can be written as

$$\alpha_1 = \begin{pmatrix} 0 & \sigma_1 \\ \sigma_1 & 0 \end{pmatrix}, \quad \alpha_2 = \begin{pmatrix} 0 & \sigma_2 \\ \sigma_2 & 0 \end{pmatrix}$$
$$\alpha_3 = \begin{pmatrix} 0 & \sigma_3 \\ \sigma_3 & 0 \end{pmatrix}, \quad \alpha_4 = \begin{pmatrix} \mathbf{I} & 0 \\ 0 & -\mathbf{I} \end{pmatrix} \tag{8-50}$$

where each of the submatrices defining the σs are 2×2 matrices. This connection to the Pauli spin matrices introduced earlier in the discussion of general angular momentum characteristics is another indication that the introduction of the αs in Eq. (8-37) may have physical as well as mathematical relevance. Specifically, the αs are referred to as *spin operators*.

Of course, such a representation implies that the wavefunctions that are the solution of Eq. (8-45) also contain four components. If we label these components Ψ_1, Ψ_2, Ψ_3, Ψ_4, then the Dirac equation of Eq. (8-45) can be written using the results of Problem 8-10 in explicit component form as

$$\pi_0\Psi_1 - \pi_1\Psi_4 + i\pi_2\Psi_4 - \pi_3\Psi_3 - (m_0c)\Psi_1 = 0, \tag{8-51}$$

$$\pi_0\Psi_2 - \pi_1\Psi_3 - i\pi_2\Psi_3 + \pi_3\Psi_4 - (m_0c)\Psi_2 = 0, \tag{8-52}$$

$$\pi_0\Psi_3 - \pi_1\Psi_2 + i\pi_2\Psi_2 - \pi_3\Psi_1 + (m_0c)\Psi_3 = 0, \tag{8-53}$$

$$\pi_0\Psi_4 - \pi_1\Psi_1 - i\pi_2\Psi_1 + \pi_3\Psi_2 + (m_0c)\Psi_4 = 0. \tag{8-54}$$

Illustrating the above equations explicitly, Eq. (8-51) can be written as

$$\left[\frac{i\hbar}{c}\frac{\partial}{\partial t} - \left(\frac{q}{c}\right)\phi - (m_0c)\right]\Psi_1 - \left[\hbar\left(\frac{1}{i}\frac{\partial}{\partial x} - \frac{\partial}{\partial y}\right) + \left(\frac{q}{c}\right)(A_x - iA_y)\right]\Psi_4$$
$$-\left[\left(\frac{\hbar}{i}\right)\frac{\partial}{\partial z} - \left(\frac{q}{c}\right)A_z\right]\Psi_3 = 0. \tag{8-55}$$

Similar explicit equations are found for Eqs. (8-52)–(8-54). Of course, *each* of the four components of Ψ is a function of x, y, z, and t, and these components are usually referred to as *spinor components*.

Solutions of Eq. (8-45) or the component form thereof [Eqs. (8-51)–(8-54)] give results that do agree with experiment.[18] The intuitive genius of Dirac in arriving at Eq. (8-45) is thus revealed, and is indeed remarkable since Eq. (8-45) was not derived, yet it is fully Lorentz-invariant and provides a satisfactory explanation for the existence of both electron spin and the positron.

In addition, we can, with a bit more analysis and manipulation, indicate the origin of specific spin effects (e.g., spin-orbit coupling). To do this, let us consider the equations for the hydrogen atom in the Dirac formulation.[19] The

[18] See, for example, P. A. M. Dirac, *Quantum Mechanics*, Oxford University Press, Oxford England, 1958, Sections 72 and 73.

[19] For an expanded discussion of hydrogen atom fine structure, see R. Shankar, *Principles of Quantum Mechanics*, Plenum Press, New York, 1985, Chapter 20.

potential term is given by

$$V = q\phi = -\frac{e^2}{r}, \tag{8-56}$$

and we shall be interested in steady-state solutions of the Dirac equation, which means that Eq. (8-45) can be written as

$$(E - V - \beta m_0 c^2 - c\boldsymbol{\alpha} \cdot \boldsymbol{\pi})\Psi = 0. \tag{8-57}$$

To facilitate further analysis, we shall write the wavefunction without loss of generality as

$$\Psi = \begin{pmatrix} \mathrm{X} \\ \Phi \end{pmatrix}, \tag{8-58}$$

where X and Φ each have two spinor components. Insertion of Eq. (8-58) into Eq. (8-57) gives

$$(E - V - m_0 c^2)\mathrm{X} - c(\boldsymbol{\sigma} \cdot \boldsymbol{\pi})\Phi = 0 \tag{8-59}$$

and

$$(E - V + m_0 c^2)\Phi - c(\boldsymbol{\sigma} \cdot \boldsymbol{\pi})\mathrm{X} = 0. \tag{8-60}$$

For the case where there is no external vector potential, i.e., $\mathbf{A} = 0$, we have

$$\boldsymbol{\pi} = \mathbf{P},$$

and Eq. (8-60) can be solved[20] for Φ as

$$\Phi = c(E - V + m_0 c^2)^{-1}(\boldsymbol{\sigma} \cdot \mathbf{P})\mathrm{X}. \tag{8-61}$$

Insertion of this expression into Eq. (8-50) gives

$$(E - V - m_0 c^2)\mathrm{X} - c^2(\boldsymbol{\sigma} \cdot \mathbf{P}) \left[\frac{1}{(E - V + m_0 c^2)}\right] (\boldsymbol{\sigma} \cdot \mathbf{P})\mathrm{X}. \tag{8-62}$$

Using the operator version of the expansion

$$\left(\frac{1}{1 + \mathrm{x}}\right) = 1 - \mathrm{x} + \mathrm{x}^2 - \mathrm{x}^3 + \cdots, \tag{8-63}$$

we can write Eq. (8-62) as

$$(E - V - m_0 c^2)\mathrm{X} = \left(\frac{1}{m_0^2}\right) (\boldsymbol{\sigma} \cdot \mathbf{P})^2 \mathrm{X} - \left(\frac{1}{m_0^2 c^2}\right) (\boldsymbol{\sigma} \cdot \mathbf{P})(E - V)(\boldsymbol{\sigma} \cdot \mathbf{P})\mathrm{X}, \tag{8-64}$$

where the expansion in Eq. (8-63) has been terminated after the second term on

[20] Throughout this discussion we assume that the inverse of operators (and operator expressions) exist as needed.

the right-hand side. By defining

$$E_s = E - m_0 c^2 \tag{8-65}$$

and noting the identity,

$$(\mathbf{C} \cdot \mathbf{A})(\mathbf{C} \cdot \mathbf{B}) = (\mathbf{A} \cdot \mathbf{B}) + i\mathbf{C} \cdot (\mathbf{A} \times \mathbf{B}) \tag{8-66}$$

we can write Eq. (8-64) as

$$E_s \mathrm{X} = \left[\left(\frac{P^2}{m_0} \right) + V \right] \mathrm{X} - \left[\frac{(\sigma \cdot \mathbf{P})(E - V)(\sigma \cdot \mathbf{P})}{m_0^2 c^2} \right] \mathrm{X}. \tag{8-67}$$

But

$$(E - V)(\sigma \cdot \mathbf{P}) = (\sigma \cdot \mathbf{P})(E - V) + \sigma \cdot [(E - V), \mathbf{P}], \tag{8-68}$$

which allows Eq. (8-67) to be written [after using Eq. (8-66) several times] as

$$\begin{aligned} E_s \mathrm{X} &= \left\{ \frac{P^2}{m_0} + V - \frac{P^2(E - V)}{m_0^2 c^2} - \frac{(\sigma \cdot \mathbf{P})[E - V, \sigma \cdot \mathbf{P}]}{m_0^2 c^2} \right\} \mathrm{X} \\ &= \left\{ \frac{P^2}{m_0} + V - \frac{P^4}{m_0^3 c^2} - \frac{(\sigma \cdot \mathbf{P})\sigma \cdot [V, \mathbf{P}]}{m_0^2 c^2} \right\} \mathrm{X} \\ &= \left\{ \frac{P^2}{m_0} + V - \frac{P^4}{m_0^3 c^2} - \frac{i\sigma \cdot \mathbf{P} \times [\mathbf{P}, V]}{m_0^2 c^2} - \frac{\mathbf{P} \cdot [\mathbf{P}, V]}{m_0^2 c^2} \right\} \mathrm{X}. \end{aligned} \tag{8-69}$$

But

$$[\mathbf{P}, V] = -i\hbar \nabla V = -i\hbar \mathbf{r}, \tag{8-70}$$

which allows Eq. (8-69) to be written as

$$E_s \mathrm{X} = \left\{ \frac{P^2}{m_0} + V - \frac{P^4}{m_0^3 c^2} + \frac{i\mathbf{L} \cdot \sigma}{m_0^2 c^2} - \frac{\mathbf{P} \cdot [\mathbf{P}, V]}{m_0^2 c^2} \right\}. \tag{8-71}$$

The main points of this analysis are now clear, i.e., the right-hand side of the above equation has terms whose form corresponds to the nonrelativistic terms $[(P^2/m_0) + V]$, a relativistic correction to the kinetic energy $(p^4/m_0^3 c^2)$, and a spin–orbit term $(\mathbf{L} \cdot \boldsymbol{\sigma})$. Including higher order terms from the expansion in Eq. (8-63) and analysis similar to that above[21] gives rise to other relativistic corrections. Thus, the Dirac equation contains the necessary generality to allow description of terms to include such terms that are presented usually as ad hoc additions to the nonrelativistic Schrödinger equation.

[21] See, for example, R. Shankar, *Principles of Quantum Mechanics*, Plenum Press, New York, 1985, Chapter 20.

B. The Breit Equation

As ingenious and remarkable is Dirac's equation of state for a single electron, the rigorous generalization of that equation to multielectron systems has not been possible to date. Numerous useful approximate generalizations have been given, however, and we shall see several examples in the current section.[22]

One of the most useful approximations for multiparticle systems has been given by Breit,[23] and can be considered to be an approximate generalization of the Dirac equation [Eq. (8-45)] to multiple particle systems that takes into account the variation of mass with velocity in a manner analogous to the discussions at the beginning of Section 8-3. In particular, if we write Eq. (8-45) for a stationary state as

$$\mathcal{H}(1)\Psi = E\Psi, \tag{8-72}$$

where

$$\mathcal{H}(1) = c\boldsymbol{\alpha}_1 \cdot \boldsymbol{\pi}_1 + q_1\phi + \beta m_0 c^2, \tag{8-73}$$

then the Breit equation for *two electrons* having mass m and charge q in a stationary state is written as

$$\left(E - \mathcal{H}(1) - \mathcal{H}(2) - \frac{q_1 q_2}{r_{12}}\right)\Psi = -\frac{1}{2r_{12}}\left[\boldsymbol{\alpha}_1 \cdot \boldsymbol{\alpha}_2 + \frac{(\boldsymbol{\alpha}_1 \cdot \mathbf{r}_{12})(\boldsymbol{\alpha}_2 \cdot \mathbf{r}_{12})}{r_{12}^2}\right]\Psi. \tag{8-74}$$

Several important changes and additions are present in Eq. (8-74) compared to Eq. (8-72). First, the wavefunction in Eq. (8-74) is a function of both electrons, and therefore contains $4 \times 4 = 16$ spinor components. Next, we see that the $(q_1 q_2/r_{12})$ term is the quantum mechanical operator analog of the usual classical electrostatic term that describes the interaction of two particles with charges q_1 and q_2 separated by a distance r_{12}. The terms on the right-hand side of Eq. (8-74) provide an approximation to the relativistic interaction between the two electrons, and require separate spin operators (α_1 and α_2) for each electron. Clearly, a classical electrostatic analog for the right-hand side of Eq. (8-74) is not expected, although subsequent manipulation will allow some identification with "classical-like" interactions.

In particular, if Eq. (8-74) is divided by c^2 and expanded in powers of (p/mc), then by ignoring terms higher than $(1/c^2)$ and carrying out manipulations similar to those in the previous section, an equation is obtained that is known as

[22] Since the equations of state to be discussed in this section are only approximate, no rigorous derivations will be given, and only heuristic descriptions that illustrate the concepts involved will be utilized. For additional discussion see, for example, H. A. Bethe and E. E. Saltpeter, *Quantum Mechanics of One- and Two-Electron Atoms*, Springer-Verlag, Berlin, 1957, Chapter 2, Sect. 38.

[23] G. Breit, *Phys. Rev.* **34**, 553 (1929); **36**, 383 (1930); **39**, 616 (1932); see also J. R. Oppenheimer, *Phys. Rev.* **35**, 461 (1930).

the Pauli approximation[24] to the Breit equation:

$$E\Psi = (\mathcal{H}_0 + \mathcal{H}_1 + \mathcal{H}_2 + \mathcal{H}_3 + \mathcal{H}_4 + \mathcal{H}_5 + \mathcal{H}_6)\Psi, \tag{8-75}$$

where

$$\mathcal{H}_0 = -\frac{\hbar^2}{2m}(\nabla_1^2 + \nabla_2^2) - \frac{q}{r_1} - \frac{q}{r_2} + \frac{q^2}{r_{12}} + q\phi(r_1) + q\phi(r_2), \tag{8-76}$$

$$\mathcal{H}_1 = \left(\frac{\hbar^4}{8m^3c^2}\right)(\nabla_1^4 + \nabla_2^4), \tag{8-77}$$

$$\mathcal{H}_2 = \left(\frac{\hbar^2}{2m^2c^2}\right)\left[\left(\frac{1}{r_{12}}\right)(\nabla_1 \cdot \nabla_2) + \frac{\mathbf{r}_{12} \cdot (\mathbf{r}_{12} \cdot \nabla_1)\nabla_2}{r_{12}^3}\right], \tag{8-78}$$

$$\mathcal{H}_3 = \left(\frac{\mu\hbar}{imc}\right)\left\{\left[(\mathcal{E}_1 \times \nabla_1) + 2\,\frac{(\mathbf{r}_{12} \times \nabla_2)}{r_{12}^3}\right] \cdot \mathbf{s}_1 + \left[(\mathcal{E}_2 \times \nabla_2) + 2\,\frac{(\mathbf{r}_{21} \cdot \nabla_1)}{r_{12}^3}\right] \cdot \mathbf{s}_2, \right. \tag{8-79}$$

$$\mathcal{H}_4 = \left(\frac{\hbar^2}{4m^2c^2}\right)(\nabla_1 \cdot \mathcal{E}_1 + \nabla_2 \cdot \mathcal{E}_2), \tag{8-80}$$

$$\mathcal{H}_5 = 4\mu^2\left\{-\frac{8\pi}{3}(\mathbf{s}_1 \cdot \mathbf{s}_2)\delta^{(3)}(\mathbf{r}_{12}) + \left(\frac{1}{r_{12}^3}\right) \cdot \left[(\mathbf{s}_1 \cdot \mathbf{s}_2) - \frac{3(\mathbf{s}_1 \cdot \mathbf{r}_{12})(\mathbf{s}_2 \cdot \mathbf{r}_{12})}{r_{12}^2}\right]'\right\}, \tag{8-81}$$

and

$$\mathcal{H}_6 = 2\mu[\mathbf{H}_1 \cdot \mathbf{s}_1 + \mathbf{H}_2 \cdot \mathbf{s}_2] + \left(\frac{\hbar}{imc}\right)[\mathbf{A}_1 \cdot \nabla_1 + \mathbf{A}_2 \cdot \nabla_2]. \tag{8-82}$$

In the above equations,

$$\mu = \frac{\hbar}{2mc}, \tag{8-83}$$

$\mathcal{E}_1$ is the coulomb field due to the nucleus plus electron two plus any external field, $\mathbf{s}_1$ and $\mathbf{s}_2$ are spin operators for electrons 1 and 2, respectively, and $\mathbf{H}_1$ and $\mathbf{H}_2$ are external magnetic fields.

Interpretations of the various terms in Eq. (8-75) using "classical" concepts are possible in most cases. For example, $\mathcal{H}_0$ is the ordinary nonrelativistic Hamiltonian and $\mathcal{H}_6$ represents the interaction of each electron with an external magnetic field. In addition, $\mathcal{H}_1$ and $\mathcal{H}_2$ are relativistic corrections, with $\mathcal{H}_1$

---

[24] W. Pauli, *Z. Phys.* **43**, 601 (1927); for the particular form of the Pauli approximation cited here and the discussion leading to it, see, H. A. Bethe and E. E. Saltpeter, *Quantum Mechanics of One- and Two-Electron Atoms*, Springer-Verlag, Berlin, 1957, Sect. IIb.

representing the variation of mass with velocity already seen for one electron in the previous section, and $\mathcal{H}_2$ representing the electron–electron relativistic correction. $\mathcal{H}_3$ and $\mathcal{H}_5$ represent "nonclassical" terms, in that they represent interactions between electron spin and orbital or other spin angular momentum. $\mathcal{H}_4$ is a term characteristic of Dirac theory, and represents the interaction of the electrons with electric fields.

The main point of indicating how the Pauli approximation to the Breit equation arises is not to proceed to solve Eq. (8–75) directly. Instead, it is intended to show that at least two major types of approximations are needed in order to arrive at Eq. (8-75), i.e., the relativistic interactions among particles are included only approximately in the Breit equation, and the Pauli approximation includes only terms through $(1/c^2)$ in the Breit equation. Furthermore, it should be remembered that Eq. (8-75) considers only two electrons, and additional terms in the overall Hamiltonian are expected for $N > 2$.

Nevertheless, the process shows how several important kinds of interactions of a quantum mechanical and/or relativistic nature can be identified, and the basis from which they arise. For example, the "spin–orbit" terms in $\mathcal{H}_3$ [similar to the spin–orbit term identified for a single electron in Eq. (8-71)] and the "spin–spin" terms in Eq. (8-81) that are important to NMR and ESR experiments are frequently introduced as "add-on" terms to the nonrelativistic Hamiltonian.[25] However, they can be seen as a result of the current discussion to be a natural consequence of including relativistic effects in quantum theory. Also, since these terms are higher order terms in the expansion of the various terms in the Hamiltonian in powers of $(1/mc)$, it is expected that their magnitude will be small (but not necessarily their importance) relative to other terms in the Hamiltonian, and that perturbation theory techniques[26] can be used to estimate the effects and compare them to experiments. This process has been used quite successfully in the application of these concepts to NMR and ESR problems of interest.[27]

8-4. The Born-Oppenheimer Approximation

In our discussions of quantum mechanical Hamiltonians thus far, we have not considered the difference in mass between particles, e.g., between protons and electrons, and the simplifications in concepts and computations that may be

[25] See, for example, A. Carrington and A. D. McLachlan, *Introduction to Magnetic Resonance*, Harper & Row, New York, 1967, Sect. 2.2; see also D. P. Slichter, *Principles of Magnetic Resonance*, Harper & Row, New York, 1963, Chapter 8.

[26] See Chapter 9, Section 8.

[27] See, for example, the selected bibliography in C. P. Slichter, *Principles of Magnetic Resonance*, Harper & Row, New York, 1963, pp. 238–243, and the bibliography in J. A. Pople, W. G. Schneider and H. I. Bernstein, *High-Resolution Nuclear Magnetic Resonance*, McGraw-Hill, NY, 1959.

possible if this difference is considered. In this section the most important of these simplifications will be considered that will have significant utility in later discussions, i.e., the Born–Oppenheimer approximation.[28]

As we shall see, the relatively large difference in mass between electrons and most nuclei will allow us to consider the nuclei as fixed in space relative to each other and to the motion of the electrons. This approximation, sometimes referred to as the "clamped-nuclei" approximation, is very often utilized and results in negligible error, at least for nonhydrogen nuclei.

To see why such an approximation is so important and so widely utilized, let us consider a molecule containing N electrons and P nuclei with charge Z_p and mass M_p, in a nonrelativistic environment without external fields, and seek to determine the stationary states of the system. From Chapter 4 we know that the quantum-mechanical Hamiltonian for this system of particles is given (in atomic units) by

$$\mathcal{H} = -\frac{1}{2}\sum_{i=1}^{N} \nabla_i^2 - \frac{1}{2}\sum_{\alpha=1}^{P}\left(\frac{1}{M_\alpha}\right)\nabla_\alpha^2 - \sum_{\alpha,i}^{P,N}\frac{Z_\alpha}{r_{\alpha i}} + \sum_{\alpha>\beta}^{P}\frac{Z_\alpha Z_\beta}{R_{\alpha\beta}} + \sum_{i>j}^{N}\frac{1}{r_{ij}}, \tag{8-84}$$

where $r_{\alpha i}$ is the distance between nucleus α and electron i, $R_{\alpha\beta}$ is the distance between nuclei α and β, r_{ij} is the distance between electrons i and j, and both nuclei and electrons are considered to be moving particles relative to each other. To make clear the assumptions to follow, we rewrite Eq. (8-84) in the form

$$\mathcal{H} = \left[-\frac{1}{2}\sum_{i=1}^{N}\nabla_1^2 - \sum_{\alpha,i}^{P,N}\frac{Z_\alpha}{r_{\alpha i}} + \sum_{i>j}^{N}\frac{1}{r_{ij}}\right] + \left[-\frac{1}{2}\sum_{\alpha=1}^{P}\left(\frac{1}{M_\alpha}\right)\nabla_\alpha^2 + \sum_{\alpha>\beta}^{P}\frac{Z_\alpha Z_\beta}{R_{\alpha\beta}}\right]. \tag{8-85}$$

In the above expression we see that the second term on the right-hand side is the one that involves both electron and nuclear motion in a nonseparable way, and thus prevents a rigorous separation of nuclear and electronic motion.

However, if we assume (as did Born and Oppenheimer) that the nuclear motion is so slow relative to electron motion that the nuclei can be considered to be fixed in space relative to electron motion, considerable simplification results. In particular, only the terms in the first bracket of Eq. (8-85) now involve variables that change, and we can write the wavefunction for the system as a product of two separate factors, i.e.,

$$\Psi(\mathbf{r}_i, \mathbf{R}_\alpha) = \Phi(\mathbf{R}_\alpha) \cdot \varphi(\mathbf{r}_i, \mathbf{R}_\alpha). \tag{8-86}$$

The term $\varphi(\mathbf{r}_i, \mathbf{R}_\alpha)$ in the above wavefunction is known as the electronic eigenfunction of $\mathcal{H}$ that describes the motion of the electrons ($\mathbf{r}_i$) for a particular (fixed) choice of nuclear positions ($\mathbf{R}_\alpha$). Since we may choose different (fixed)

---

[28] M. Born and J. R. Oppenheimer, *Ann. Phys.* **84**, 457 (1927); M. Born and K. Huang, *Dynamical Theory of Crystal Lattices*, Oxford University Press, London, 1954, Appendix 8; M. Born, *Nachr. Akad. Wiss. Göttigen*, 1 (1951).

values of the nuclear positions, the electronic wavefunction will in general be different for different choices of the nuclear positions. Hence, the electronic eigenfunction is listed as being a function of the nuclear positions ($\mathbf{R}_\alpha$), but it depends only *parametrically* on them (i.e., once the nuclear positions are chosen, they enter the Schrödinger equation only as fixed parameters thereafter). The other term [$\Phi(\mathbf{R}_\alpha)$] is the part of the wavefunction that depends *only* on the nuclear coordinates, and hence is a constant if the nuclei are assumed to remain fixed.

If Eqs. (8-85) and (8-86) are utilized, the Schrödinger equation for the system becomes

$$\mathcal{H}\Psi = E\Psi, \tag{8-87}$$

or

$$\begin{aligned}&\left[-\frac{1}{2}\sum_{i=1}^{N}\nabla_i^2 - \sum_{\alpha,i}^{P,N}\frac{Z_\alpha}{r_{\alpha i}} + \sum_{i>j}^{N}\frac{1}{r_{ij}}\right]\Phi(\mathbf{R}_\alpha)\varphi(\mathbf{r}_i, \mathbf{R}_\alpha)\\ &\qquad + \left[-\frac{1}{2}\sum_{\alpha=1}^{P}\left(\frac{1}{M_\alpha}\right)\nabla_\alpha^2 + \sum_{\alpha>\beta}^{P}\frac{Z_\alpha Z_\beta}{R_{\alpha\beta}}\right]\Phi(\mathbf{R}_\alpha)\cdot\varphi(\mathbf{r}_i, \mathbf{R}_\alpha)\\ &= E\Phi(\mathbf{R}_\alpha)\varphi(\mathbf{r}_i, \mathbf{R}_\alpha).\end{aligned} \tag{8-88}$$

We now note that, since ∇_i acts only on electron coordinates ($\mathbf{r}_i$) and ∇_α acts only on nuclear coordinates ($\mathbf{R}_\alpha$), we have

$$\nabla_i^2\Phi(\mathbf{R}_\alpha)\varphi(\mathbf{r}_i, \mathbf{R}_\alpha) = \Phi(\mathbf{R}_\alpha)[\nabla_i^2\varphi(\mathbf{r}_i, \mathbf{R}_\alpha)], \tag{8-89}$$

and

$$\begin{aligned}\nabla_\alpha^2\Phi(\mathbf{R}_\alpha)\varphi(\mathbf{r}_i, \mathbf{R}_\alpha) &= \varphi(\mathbf{r}_i, \mathbf{R}_\alpha)[\nabla_\alpha^2\Phi(\mathbf{R}_\alpha)] + 2[\nabla_\alpha\varphi(\mathbf{r}_i, \mathbf{R}_\alpha)]\cdot[\nabla_\alpha\Phi(\mathbf{R}_\alpha)]\\ &\quad + \Phi(\mathbf{R}_\alpha)\nabla_\alpha^2\varphi(\mathbf{r}_i, \mathbf{R}_\alpha).\end{aligned} \tag{8-90}$$

Inserting Eqs. (8-89) and (8-90) into Eq. (8-88) gives

$$\begin{aligned}&\Phi(\mathbf{R}_\alpha)\left\{-\frac{1}{2}\sum_{i=1}^{N}\nabla_i^2 - \sum_{\alpha,i}^{P,N}\frac{Z_\alpha}{r_{\alpha i}} + \sum_{i>j}^{N}\frac{1}{r_{ij}}\right]\varphi(\mathbf{r}_i, \mathbf{R}_\alpha)\\ &\quad + \left\{-\sum_{\alpha=1}^{P}\left(\frac{1}{M_\alpha}\right)[\nabla_\alpha\varphi(\mathbf{r}_i, \mathbf{R}_\alpha)][\nabla_\alpha\Phi(\mathbf{R}_\alpha)]\right.\\ &\quad \left. - \Phi(\mathbf{R}_\alpha)\sum_{\alpha=1}^{P}\left(\frac{1}{2M_\alpha}\right)\nabla_\alpha^2\varphi(\mathbf{r}_i, \mathbf{R}_\alpha)\right\}\\ &\quad + \left[-\frac{1}{2}\sum_{\alpha}^{P}\left(\frac{1}{M_\alpha}\right)\varphi(\mathbf{r}_i, \mathbf{R}_\alpha)\nabla_\alpha^2 + \varphi(\mathbf{r}_i, \mathbf{R}_\alpha)\sum_{\alpha>\beta}^{P}\frac{Z_\alpha Z_\beta}{R_{\alpha\beta}}\right]\Phi(\mathbf{R}_\alpha)\\ &= E\Phi(\mathbf{R}_\alpha)\varphi(\mathbf{r}_i, \mathbf{R}_\alpha).\end{aligned} \tag{8-91}$$

If we now neglect the terms in the second set of braces in Eq. (8-91), we obtain

$$\Phi(\mathbf{R}_\alpha)\left\{-\frac{1}{2}\sum_{i=1}^{N}\nabla_i^2-\sum_{\alpha,i}^{P,N}\frac{Z_\alpha}{r_{\alpha i}}+\sum_{i>j}^{N}\frac{1}{r_{ij}}\right]\varphi(\mathbf{r}_i,\mathbf{R})$$

$$+\left[-\frac{1}{2}\sum_{\alpha=1}^{P}\left(\frac{1}{M_\alpha}\right)\varphi(\mathbf{r}_i,\mathbf{R}_\alpha)\nabla_\alpha^2+\varphi(\mathbf{r}_i,\mathbf{R}_\alpha)\sum_{\alpha>\beta}^{P}\frac{Z_\alpha Z_\beta}{R_{\alpha\beta}}\right]\Phi(\mathbf{R}_\alpha)$$

$$=(E_e+E_n)\Phi(\mathbf{R}_\alpha)\varphi(\mathbf{r}_i,\mathbf{R}_\alpha), \tag{8-92}$$

where we have defined the electronic energy (E_e) and nuclear energy (E_n) by

$$E=E_e+E_n. \tag{8-93}$$

Rearranging Eq. (8-92) gives

$$\frac{1}{\varphi(\mathbf{r}_i,\mathbf{R}_\alpha)}\left\{-\frac{1}{2}\sum_{i=1}^{N}\nabla_i^2-\sum_{\alpha,i}^{P,N}\frac{Z_\alpha}{r_{\alpha i}}+\sum_{i>j}^{N}\frac{1}{r_{ij}}\right\}\varphi(\mathbf{r}_i,\mathbf{R}_\alpha)-E_e$$

$$=\frac{1}{\Phi(\mathbf{R}_\alpha)}\left\{\frac{1}{2}\sum_{\alpha}^{P}\left(\frac{1}{M_\alpha}\right)\nabla_\alpha^2-\sum_{\alpha>\beta}^{P}\frac{Z_\alpha Z_\beta}{R_{\alpha\beta}}\right\}\Phi(\mathbf{R}_\alpha)+E_n. \tag{8-94}$$

Thus, as long as the approximation leading to Eq. (8-92) is justified, we see that the variables in the Schrödinger equation can be separated, as shown in Eq. (8-94). This means that we can solve each of the sides of Eq. (8-94) separately, i.e.,

$$\left[-\frac{1}{2}\sum_{i=1}^{N}\nabla_i^2-\sum_{\alpha,i}^{P,N}\frac{Z_\alpha}{r_{\alpha i}}+\sum_{i>j}^{N}\frac{1}{r_{ij}}\right]\varphi(\mathbf{r}_i,\mathbf{R}_\alpha)=E_e\varphi(\mathbf{r}_i,\mathbf{R}_\alpha), \tag{8-95}$$

and

$$\left[-\frac{1}{2}\sum_{\alpha=1}^{P}\left(\frac{1}{M_\alpha}\right)\nabla_\alpha^2+\sum_{\alpha>\beta}^{P}\frac{Z_\alpha Z_\beta}{R_{\alpha\beta}}\right]\Phi(\mathbf{R}_\alpha)=E_n\Phi(\mathbf{R}_\alpha). \tag{8-96}$$

We can rewrite Eq. (8-96) using Eq. (8-93) as

$$\left[-\frac{1}{2}\sum_{\alpha=1}^{P}\left(\frac{1}{M_\alpha}\right)\nabla_\alpha^2+\sum_{\alpha>\beta}^{P}\frac{Z_\alpha Z_\beta}{R_{\alpha\beta}}+E_e\right]\Phi(\mathbf{R}_\alpha)=E\Phi(\mathbf{R}_\alpha). \tag{8-97}$$

At this point both the conceptual as well as mathematical aspects of the Born–Oppenheimer approximation are evident. In particular, the electronic Schrödinger equation [Eq. (8-95)] is the familiar equation from Chapter 4. In addition, however, we now see that the equation for nuclear motion [Eq. (8-97)] involves a Hamiltonian for nuclear motion in which the electronic energy (E_e) must be known for all values of (i.e., as a function of) the nuclear coordinates ($\mathbf{R}_\alpha$), since it appears as a part of the potential energy term for nuclear motion. In mathematical terms, the separation of nuclear and electronic motion into two equations [Eqs. (8-95) and (8-97)] is valid as long as the contribution of

$[-1/2\ \Sigma_{\alpha}^{P}\ (1/M_{\alpha})\nabla_{\alpha}^{2}\Phi]$ is much larger than $\{-\Sigma_{\alpha=1}^{P}\ [\nabla_{\alpha}\varphi]\cdot[\nabla_{\alpha}\Phi] + \Phi\ \Sigma_{\alpha}^{P}\ \nabla_{\alpha}^{2}\varphi\}$. Since φ is typically a slowly varying function of the nuclear coordinates, the approximation is usually well justified, and will be utilized extensively in later chapters.

However, there are cases in which this approximation is not adequate. These cases, i.e., when the coupling of nuclear and electronic motion is important, require that the energy be corrected for the coupling. Such cases can be expected in a variety of circumstances, and some of these will be discussed in later chapters. These circumstances include (1) when nuclear velocities are high (e.g., in hot atom reactions), (2) when the potential energy surfaces for two electronic states of the same symmetry cross, (3) when an electronic state is degenerate and the nuclear configuration is symmetric, and (4) when the molecule has a nonzero electronic momentum and is rotating rapidly.

These physical situations will be reflected mathematically in increased contributions from the nuclear kinetic energy terms. Specifically, when the coordinate system is transformed to use of relative coordinates[29] that are convenient for calculation,[30] it is found that coupling terms between nuclear and electronic kinetic energy arise. In addition to these terms, there are also contributions from relativistic effects that need to be considered.

These various correction terms have also been classified into two main types in the literature. The first of these are those that shift all levels. They are called *diagonal corrections*, and the inclusion of these corrections to the Born–Oppenheimer energy is referred to as the *adiabatic approximation*.[31] The second kinds of correction are those that selectively modify electronic energy levels and potential energy surfaces that, e.g., can allow new transitions to occur. These are referred to as *nondiagonal* corrections, and studies in which both diagonal and nondiagonal corrections to the Born–Oppenheimer energy are included are referred to as *nonadiabatic approximations*.

Problems

1. Show that the form of the equations representing the electric and magnetic fields as given in Eqs. (8-5) and (8-6) are invariant to transformations of the

[29] Relative coordinates in this context are those that relate distances of electrons and nuclei to each other, rather than to a fixed origin (as we have done in the discussions thus far).

[30] Many choices of relative coordinates can be made depending on the number of nuclei in the molecular and its overall complexity. For a discussion of the various options, see for example, A. Fröman, *J. Chem. Phys.* **36**, 1490 (1962); D. W. Jepsen and J. O. Hirschfelder, *J. Chem. Phys.* **32**, 1323 (1960).

[31] For examples of the various corrections described here for the case of a diatomic molecule, see Problems 8-11 and 8-12. It should also be noted that treatment of Born–Oppenheimer corrections represents a large and important area of work that is not treated in this text. For an excellent review, see J. O. Hirschfelder and W. J. Meath, *Adv. Chem. Phys.* **12**, 3–106 (1967).

type

$$\mathbf{A}' = \mathbf{A} - \nabla f$$

$$\Phi' = \Phi + \left(\frac{1}{c}\right)\frac{\partial f}{\partial t},$$

where f is an arbitrary function of position and time. Such a property for the electric and magnetic fields is referred to as *gauge invariance*, and transformations of the type shown above are known as gauge transformations. Such invariance is important because it assures that effects resulting from the potentials (**A** and Φ) are due to interaction of the charges and currents in the system, and are not artifacts of the particular form of **A** and Φ that is chosen.

2. When constructing quantum mechanical operators, it is important to assure that expectation values calculated from eigenfunctions of the operator are gauge invariant, e.g., invariant to the choice of origin. For the momentum portion of the quantum mechanical operator in Eq. (8-16), i.e.,

$$\boldsymbol{\pi} = \left[\frac{\hbar}{i}\nabla - \left(\frac{q}{c}\right)\mathbf{A}\right],$$

show that expectation values of π are invariant to a gauge change as introduced in Problem 1 as long as the wavefunction is written as

$$\Psi' = \exp(i\lambda)\Psi,$$

where λ is an arbitrary constant.

3. Expand the terms in the one-dimensional operator

$$\mathcal{H}_x = \frac{1}{2m}\left(\frac{\hbar}{i}\frac{\partial}{\partial x} - \frac{q}{c}A_x\right)^2$$

and show that it can be written as

$$\mathcal{H}_x = \left(\frac{1}{2m}\right)\left\{-\hbar^2\frac{\partial^2}{\partial x^2} + \left(\frac{i\hbar q}{c}\right)\left[2A_x\frac{\partial}{\partial x} + \frac{\partial A_x}{\partial x}\right] + \left(\frac{q^2}{c^2}\right)A_x^2\right\}.$$

4. Show that it is always possible to choose the function f in Problem 1 such that

$$\nabla \cdot \mathbf{A} = 0.$$

5. Verify Eq. (8-26), starting from Eq. (8-25) and using an analysis similar to that used in Eqs. (8-5) through (8-10).
6. With the Lagrangian of Eq. (8-27), verify that the Lagrangian equations of motion [Eq. (8-13)] are the same as the Newtonian equations of motion [Eq. (8-26)] for a relativistic particle.
7. With the definitions

$$\pi_1 = -i\hbar\frac{\partial}{\partial x}, \quad \pi_2 = -i\hbar\frac{\partial}{\partial y}, \quad \pi_3 = -i\hbar\frac{\partial}{\partial z}, \quad \pi_4 = -i\hbar\frac{\partial}{\partial t}$$

$$A_1 = A_x, \; A_2 = A_y, \; A_3 = A_z, \; A_4 = i\phi, \; x_4 = ict,$$

show that the Klein–Gordon equation corresponding to Eq. (8-36) can be written as

$$\left\{ \sum_{j=1}^{4} \left(\pi_j + \frac{q}{c} A_j \right)^2 + m_0 c_0^2 \right\} \Psi = 0.$$

Also show that the general solution for the above equation can be written as

$$\Psi(\mathbf{r}, t) = u(\mathbf{r}) \exp \left\{ -\frac{i}{\hbar} (E + m_0^2 c^2) t \right\},$$

where $u(\mathbf{r})$ is the solution of

$$\left\{ \left[\frac{\hbar^2}{2m_0} \nabla^2 + E - V \right] - \left(\frac{q}{m_0 c} \right) \sum_{j=1}^{3} (A_j \pi_j) \right.$$
$$\left. + \left(\frac{1}{2m_0 c} \right) \left[(E - V)^2 - q^2 \sum_{j=1}^{3} A_j^2 \right] \right\} u = 0.$$

8. Using the relations in Eq. (8-38) and treating the $[\alpha_j]$ as dimensionless operators, show that Eq. (8-37) can be reduced to Eq. (8-36).
9. Verify that the matrix representation of the α-matrices given by Eqs. (8-46)–(8.49) satisfies Eq. (8-38) when the latter is considered a matrix equation, and that each of the α-matrices is Hermitian with eigenvalues of ± 1.
10. Using the definitions of π_0 through π_3 in Eqs. (8-43) and (8-44) plus the α matrices of Eqs. (8-46) through (8-49), show that a wavefunction having four components

$$\boldsymbol{\Psi} = \begin{pmatrix} \Psi_1 \\ \Psi_2 \\ \Psi_3 \\ \Psi_4 \end{pmatrix}$$

results in the following:

$$-\boldsymbol{\alpha}_1 \pi_1 \boldsymbol{\Psi} = -\pi_1 \begin{pmatrix} \Psi_4 \\ \Psi_3 \\ \Psi_2 \\ \Psi_1 \end{pmatrix}, \qquad -\boldsymbol{\alpha}_2 \pi_2 \boldsymbol{\Psi} = -i\pi_2 \begin{pmatrix} \Psi_4 \\ -\Psi_3 \\ \Psi_2 \\ -\Psi_1 \end{pmatrix},$$

$$-\boldsymbol{\alpha}_3 \pi_3 \boldsymbol{\Psi} = -\pi_3 \begin{pmatrix} \Psi_3 \\ -\Psi_4 \\ \Psi_1 \\ -\Psi_2 \end{pmatrix}, \qquad -\boldsymbol{\alpha}_4 (m_0 c) \boldsymbol{\Psi} = (m_0 c) \begin{pmatrix} \Psi_1 \\ \Psi_2 \\ -\Psi_3 \\ -\Psi_4 \end{pmatrix}.$$

11. For a diatomic molecule with N electrons and nuclei A and B (with masses m_A and m_B), and with coordinates relative to a central origin designated as

$\mathbf{r}_i'$ $(i' = 1, 2, \ldots, N)$, $\mathbf{R}_A$ and $\mathbf{R}_B$, we define relative coordinates as

$$\mathbf{R}=\mathbf{R}_B-\mathbf{R}_A,$$

$$\mathbf{r}_i=\mathbf{r}_i'-\left(\frac{m_A\mathbf{R}_A+m_B\mathbf{R}_B}{m_A+m_B}\right),$$

and the reduced mass (μ) as

$$\mu=\left(\frac{m_Am_B}{m_A+m_B}\right).$$

The quantum mechanical Hamiltonian $(\mathcal{H})$, which takes into account nuclear as well as electronic motion, can be written as

$$\mathcal{H}=\mathcal{H}_e-\left(\frac{1}{\mu}\right)\nabla_R^2-\frac{1}{2(m_A+m_B)}\left[\sum_{i=1}^{N}\nabla_i^2+\sum_{i<j}^{N}\nabla_i\cdot\nabla_j\right],$$

where $\mathcal{H}_e$ is the quantum mechanical Hamiltonian in the "clamped nuclei" approximation, i.e.,

$$\mathcal{H}_e=-\frac{1}{2}\sum_{i=1}^{N}\nabla_i^2-\sum_{i=1}^{N}\left[\frac{Z_A}{r_{Ai}}+\frac{Z_B}{r_{Bi}}\right]+\sum_{i<j}^{N}\frac{1}{r_{ij}}+\frac{Z_AZ_B}{R}.$$

Assuming that a complete set of solutions [i.e., $\psi_k(\mathbf{r}, \mathbf{R})$] for $\mathcal{H}_e$ can be found having energies $E_k(\mathbf{R})$ for all values of $\mathbf{R}$, show that expansion of the total wavefunction $[\Psi(\mathbf{r}, \mathbf{R})$ in the form

$$\Psi(\mathbf{r}, \mathbf{R})=\sum_k \chi_k(\mathbf{R})\Psi_k(\mathbf{r}, \mathbf{R})$$

where $\{\chi_k(\mathbf{R})\}$ is the set of functions describing the nuclear motion, gives rise to equations that must be solved in order to determine $\chi_k(\mathbf{R})$ as follows:

$$\left[-\left(\frac{1}{2\mu}\right)\nabla_R^2+E_l(\mathbf{R})+E_{ll}'(\mathbf{R})+\mathbf{E}_{ll}''(\mathbf{R})\cdot\nabla_R-E\right]\chi_k(\mathbf{R})$$
$$=-\sum_{k\neq l}\left[E_{lk}'(\mathbf{R})+\mathbf{E}_{lk}'(\mathbf{R})\cdot\nabla_R\right]\chi_k(\mathbf{R}),$$

where E is the total energy associated with $\mathcal{H}$, and

$$E_{lk}'(\mathbf{R})=-\left(\frac{1}{2\mu}\right)\int\psi_l^*(\mathbf{r}, \mathbf{R})\nabla_R^2\psi_R(\mathbf{r}, \mathbf{R})\,d\mathbf{r}$$
$$-\left[\frac{1}{2(m_A+m_B)}\right]\sum_i^N\int\psi_l^*(\mathbf{r}, \mathbf{R})\nabla_i^2\psi_k(\mathbf{r}, \mathbf{R})\,d\mathbf{r}$$
$$-\frac{1}{(m_A+m_B)}\sum_{i<j}^N\int\psi_l^*(\mathbf{r}, \mathbf{R})\nabla_i\cdot\nabla_j\psi_k(\mathbf{r}, \mathbf{R})\,d\mathbf{r},$$

and

$$\mathbf{E}''_{lk}(\mathbf{R}) = -\left(\frac{1}{\mu}\right) \int \psi_l^*(\mathbf{r}, \mathbf{R}) \nabla_R \psi_R(\mathbf{r}, \mathbf{R})\, d\mathbf{r}.$$

12. Using the results of Problem 8-11,
    a. Show that, if the $\{\psi_l\}$ are taken to be real, the diagonal term vanishes, i.e.,

$$E'_{ll}(\mathbf{R}) = 0.$$

    b. Show that in the Born–Oppenheimer approximation, i.e., when all coupling terms $[E'_{lk}(\mathbf{R})]$ are ignored, the equation for the nuclear motion is given by

$$\left[-\left(\frac{1}{2\mu}\right)\nabla_R^2 + E_l(\mathbf{R}) - E\right] \chi_l(\mathbf{R}) = 0.$$

    c. Show that in the adiabatic approximation, i.e., where all nondiagonal coupling terms are ignored, the equation for the nuclear motion is given by

$$\left[-\left(\frac{1}{2\mu}\right)\nabla_R^2 + V_l(\mathbf{R}) - E\right] \chi_l(\mathbf{R}) = 0,$$

    where

$$V_l(\mathbf{R}) = E_l(\mathbf{R}) + E'_{ll}(\mathbf{R}).$$

Chapter 9

Approximation Methods for Stationary States

In several of the preceding chapters we have discussed application of the postulates of quantum mechanics to several important types of examples. Although these applications were varied, there is one common characteristic shared by all of them. In each case, the Schrödinger equation could be solved exactly for the eigenvalues and associated eigenfunctions. If this were possible for all problems of interest to chemists, our insight into the nature of chemical phenomena would certainly be increased enormously. However, the fact is that only a very few problems are solvable exactly. In particular, it has not yet been possible to obtain an exact solution to the Schrödinger equation for any system containing more than one electron. Consequently, our knowledge of chemistry as revealed through quantum mechanics has been restricted primarily to that obtainable from an examination of *approximate* solutions to the Schrödinger equation. However, the advent of large-scale computers has greatly expanded both the accuracy and the scope of approximate solutions that can be obtained. We shall now discuss some of the methods that are available for obtaining approximate solutions to the Schrödinger equation for stationary states, and that form the basis for most applications of quantum mechanics to chemical problems.

9-1. The Variation Principle

One of the most powerful techniques for obtaining approximate solutions of the Schrödinger equation of high accuracy is based on the use of the Variation Theorem. In the following discussion, it will be assumed that there are no relativistic effects or external fields present.

To derive the Variation Theorem, we consider a system of particles with a quantum mechanical Hamiltonian, $\mathcal{H}$, for which the Schrödinger equation cannot be solved in closed form. This does not imply that exact eigenvalues and associated eigenvectors do not exist for this system. Rather, it implies only that we cannot find an analytic expression for them as we found, e.g., in the case of the hydrogen atom. We shall now describe one of the energy eigenstates that we assume to exist by a function Φ that is only an approximation to the exact eigenfunction for that state. Specifically, we shall be interested in the ground state, and in establishing some measure of how close this approximation is to the exact eigenfunction.

It should be pointed out that the above assumption concerning the existence of exact solutions, even though they cannot apparently be found, is one that cannot be proved in general. However, there is considerable evidence that supports the usefulness of that assumption. For example, transitions between energy levels in the He atom can be measured experimentally, thus confirming the existence of energy levels. However, an exact expression for the eigenfunctions corresponding to these observed levels has not been found. To assume that these exact eigenfunctions do not exist leads quickly to a nonproductive viewpoint in which no progress can be made, and which ultimately would force abandonment of the theory. Consequently, we shall assume that there exist eigenvalues E_0, E_1, E_2, ... for the system, with associated normalized eigenvectors Ψ_0, Ψ_1, Ψ_2, ... For convenience, let us order these eigenstates such that $E_0 \leq E_1 \leq E_2 \ldots$. Without loss of generality, we can assume that the eigenfunctions Ψ_k are mutually orthogonal (even if degeneracies exist), by the use of Theorems 3-20 and 3-21.

Since the eigenfunctions form a complete set (by Postulate I), we can express Φ in terms of them, i.e.,

$$\Phi = \sum_k c_k \Psi_k, \qquad c_k = \int \Phi^* \Psi_k \, d\tau. \tag{9-1}$$

In the language of vector spaces, Eq. (9-1) simply states that an arbitrary vector in the space can always be expressed as a linear combination of the complete set of functions that spans the space. Incidentally, this implies that Φ is not entirely arbitrary, but that it must be constructed so that it is an acceptable vector in the space, i.e., that it obeys the appropriate boundary conditions.

Now let us examine the expectation value of the energy associated with this approximate function:

$$\mathcal{E} = \frac{\int \Phi^* \mathcal{H} \Phi \, d\tau}{\int \Phi^* \Phi \, d\tau}, \tag{9-2}$$

where the integral in the denominator is present to assure normalization of Φ, and the integration is taken over the space of all the particles. Insertion of Eq. (9-1) into Eq. (9-2) gives

$$\mathcal{E} = \frac{\sum_k \sum_l c_k^* c_l \int \Psi_k^* \mathcal{H} \Psi_l \, d\tau}{\sum_k \sum_l c_k^* c_l \int \Psi_k^* \Psi_l \, d\tau}. \tag{9-3}$$

However, we have assumed that the Ψ_i are the exact eigenfunctions of the Hamiltonian, i.e.,

$$\mathcal{H}\Psi_i = E_i \Psi_i, \tag{9-4}$$

which allows Eq. (9-3) to be rewritten as

$$\mathcal{E} = \frac{\sum_k |c_k|^2 E_k}{\sum_k |c_k|^2}, \tag{9-5}$$

where the mutual orthogonality of the eigenvectors has been employed. Writing out the above equation in detail, we have

$$\mathcal{E} = \frac{1}{\left(\sum_k |c_k|^2\right)} \{|c_0|^2 E_0 + |c_1|^2 E_1 + |c_2|^2 E_2 + \cdots\}. \tag{9-6}$$

If we replace each E_k by E_0 in Eq. (9-6), we obtain an expression whose value is clearly smaller than Eq. (9-6), since each term will be replaced by another that is smaller (or equal to it). Consequently, we have

$$\mathcal{E} = \frac{\sum_k |c_k|^2 E_k}{\sum_k |c_k|^2} \geq \frac{E_0\left(\sum_k |c_k|^2\right)}{\sum_k |c_k|^2} = E_0, \tag{9-7}$$

which is the desired result. It tells us that any approximate wavefunction will always have an energy expectation value that is above that of the true ground state energy,[1] except in the special case when $\phi \equiv \psi_0$. In other words, the energy expectation value of any appropriate wavefunction[2] will provide an upper bound to the exact energy of the lowest state, although it will not tell us how high above E_0 the approximate energy E is.

This result is still of great practical utility, for it allows us to obtain a measure of the "closeness" of an approximation to the exact eigenfunction by examination of its energy expectation value. If more than one approximate eigenfunction is proposed as being useful, the one having the lowest energy will be the one that most closely represents the exact eigenfunction for that property. Incidentally, it should be pointed out the choice of approximate wavefunction for the energy does not imply automatically that the approximate eigenfunction having the lowest energy is also the best approximation for describing other molecular properties, such as dipole moment, quadrupole coupling constants,

[1] This result was shown by C. E. Eckert, *Phys. Rev.*, **36**, 878 (1930).

[2] It should be remembered that, in addition to considering the energy expectation value, the approximate wavefunction will also have to satisfy other boundary conditions, e.g., be single-valued and quadradically integrable. See Chapter 5, Section 5-2.

etc. We shall return to the discussion of this matter later. However, since energy levels play such a central role in quantum chemistry and spectroscopy, it is appropriate that this discussion be continued with the assumption that an optimum description of the energy is the desired result.

The Variation Principle, as expressed in Eq. (9-7), appears to be applicable only to the ground state of a system. However, as we shall see, there are also circumstances in which it also has application to excited states. To see this, let us return to Eq. (9-5) and rewrite it with emphasis on the first excited state (with exact energy E_1), i.e.,

$$\begin{aligned}\mathcal{E}_1 - E_1 &= \sum_k |c_k|^2 (E_k - E_1) \\ &= |c_0|^2(E_0 - E_1) + |c_2|^2(E_2 - E_1) + |c_3|^2(E_3 - E_1) + \cdots,\end{aligned} \tag{9-8}$$

where we have used the normalization condition $[\Sigma_k |c_k|^2 = 1]$ and have subtracted E_1 from each side of Eq. (9-5). Also, since we are now interested in describing the first excited state by Φ (which we shall label Φ_1), it should be noted that the coefficients in Eq. (9-8) will differ from those in Eq. (9-6).

If we have the case that $c_0 = 0$, then Eq. (9-8) becomes

$$\mathcal{E}_1 - E_1 = |c_2|^2(E_2 - E_1) + |c_3|^2(E_3 - E_1) + \cdots. \tag{9-9}$$

Using the earlier assumption that $E_1 < E_2 < E_3, \ldots$, Eq. (9-9) can be written as

$$\mathcal{E}_1 - E_1 \geq 0 \tag{9-10}$$

or

$$\mathcal{E}_1 \geq E_1. \tag{9-11}$$

Thus, under the conditions just described, the Variation Principle has application to excited states as well. However, the number of times that $c_0 = 0$ is limited. For example, we shall see in later discussions[3] of symmetry that it sometimes occurs that the first excited state has a different symmetry than the ground state, which means that the wavefunctions for the ground and first excited states also have different symmetry. In that case we do find that

$$c_0 = \int \Phi_1^* \psi_0 \, d\tau = 0,$$

since Φ_1 must have the same symmetry properties as ψ_1, and we see that the Variation Principle can be applied.

9-2. Accuracy Considerations

Let us now see if there is a possibility of determining a more quantitative measure of the "goodness" of the approximate function Φ.

[3] See also Problems 9-5, 9-6, and 9-7.

Assuming that Φ has been normalized, Eq. (9-5) can be rewritten as

$$E = \sum_k |c_k|^2 E_k$$
$$= |c_0|^2 E_0 + \sum_{k \geq 1} |c_k|^2 E_k, \tag{9-12}$$

or

$$E \geq |c_0|^2 E_0 + E_1 \sum_{k \geq 1} |c_k|^2, \tag{9-13}$$

where E_k ($k \geq 1$) has been replaced by E_1 in each term. From the normalization condition, this equation can be rewritten as

$$E \geq |c_0|^2 E_0 + E_1(1 - |c_0|^2),$$

or

$$E \geq E_1 + |c_0|^2 (E_0 - E_1). \tag{9-14}$$

We shall return to this result in a moment.

What is desired in this discussion is a means of showing that, as $E \to E_0$, $\Phi \to \Psi_0$. One of the common measures of how close Φ is to Ψ_0 is the mean square deviation, ϵ^2, defined as

$$\epsilon^2 = \int |\Phi - \Psi_0|^2 \, d\tau. \tag{9-15}$$

Clearly, as $\Phi \to \Psi_0$, $\epsilon^2 \to 0$. We shall now relate this to the energy expectation value of Φ. Expanding Eq. (9-15), we have

$$\epsilon^2 = 2 - \int \Phi^* \Psi_0 \, d\tau - \int \Phi \Psi_0^* \, d\tau. \tag{9-16}$$

But,

$$\int \Phi^* \Psi_0 \, d\tau = \sum_k c_k^* \int \Psi_k^* \Psi_0 \, d\tau = c_0^*. \tag{9-17}$$

Similarly,

$$\int \Phi \Psi_0^* \, d\tau = c_0. \tag{9-18}$$

This allows Eq. (9-16) to be rewritten as

$$\epsilon^2 = 2 - (c_0^* + c_0)$$
$$= 2[1 - Re(c_0)].$$

If we restrict our consideration to real wavefunctions for the moment, then

$$\epsilon^2 = 2(1 - c_0),$$

or

$$c_0 = 1 - \frac{\epsilon^2}{2}. \tag{9-19}$$

Thus,

$$c_0^2 = \left(1 - \frac{\epsilon^2}{2}\right)^2$$
$$= 1 - \epsilon^2 + \frac{\epsilon^4}{4}. \tag{9-20}$$

Insertion of Eq. (9-20) into Eq. (9-14) yields

$$E \geq E_1 + \left(1 - \epsilon^2 + \frac{\epsilon^4}{4}\right)(E_0 - E_1), \tag{9-21}$$

or, substracting E_0 from both sides (which does not change the inequality, since $E > E_0$), we have

$$E - E_0 \geq \left(\epsilon^2 - \frac{\epsilon^4}{4}\right)(E_1 - E_0). \tag{9-22}$$

If we now assume that the energy levels E_0 and E_1 are nondegenerate, and recall that $E_1 > E_0$, we can write Eq. (9-22) as

$$\left(\frac{E - E_0}{E_1 - E_0}\right) \geq \epsilon^2 - \frac{\epsilon^4}{4}. \tag{9-23}$$

Finally, if the trial wavefunction Φ is close enough to Ψ_0 so that $\epsilon^4 \ll \epsilon^2$, then Eq. (9-23) can be written as

$$\epsilon^2 \leq \left(\frac{E - E_0}{E_1 - E_0}\right). \tag{9-24}$$

Thus, E is more than simply an upper bound to the exact energy. We see from Eq. (9-24) that $(E - E_0)$ is directly proportional to the mean square error in the trail wavefunction.

9-3. Example: The Hydrogen Atom

As an illustration of the Variation Principle, let us return to the hydrogen atom ground state problem, and show how it might have been approached if the exact solution were not available. In particular, let us assume that the angular dependence has been satisfactorily treated, but that the radial equation has not been solved. This means that the wavefunction for the hydrogen atom can be written as

$$\Psi(r, \vartheta, \varphi) = \eta(r) Y_{l,m_l}(\vartheta, \varphi), \tag{9-25}$$

where $\eta(r)$ is the desired solution to the radial equation.

If the exact solution for $\eta(r)$ is not available, we need to employ some sort of intuition coupled with mathematical requirements[4] in order to choose various trial forms for $\eta(r)$. Of course, once the trial functions have been chosen, their effectiveness (or lack of it) can be judged by examination of the energy associated with each trial function.

In this case, imposition of the requirements of acceptable operands as outlined in earlier chapters will help to limit drastically the kinds of functions that are considered. For example, if quadratic integrability is demanded, then at least two kinds of possibilities are appropriate for consideration. These are

$$\eta_1(r) = N_1 e^{-\zeta r} \qquad (N_1 = 2\zeta^{3/2}) \tag{9-26}$$

and

$$\eta_2(r) = N_2 e^{-\rho r^2} \tag{9-27}$$

where N_2 is a normalization constant and ζ and ρ are parameters (called *orbital exponents*) that need to be specified. Other types of functions such as polynomials in r need not be considered, due to their lack of quadratic integrability. Of course, more complicated trial functions are also possible, e.g.,

$$\eta_3(r) = N_3 p(r) e^{-\rho' r^2}, \tag{9-28}$$

where $p(r)$ is a polynomial in r, since $\eta_3(r)$ for this choice of $p(r)$ is also quadratically integrable. This also points to one of the difficulties associated with the use of Variation Principle in practice; namely, the choice of acceptable possibilities for trial functions is typically far larger than either our patience or available computer time will allow us to investigate. However, let us focus our attention on $\eta_1(r)$ for the moment, in order to illustrate the procedure.

In the case of $\eta_1(r)$, we need to find the optimum value of ζ, i.e., the value of ζ that gives the lowest energy. The most obvious procedure for finding the optimum ζ is simply to calculate E as a function of ζ. Then, one either plots the results or differentiates E with respect to ζ analytically, and sets the resulting equation equal to zero to find the optimum value of ζ.

In this case, the problem is simple enough to allow the differentiation to be carried out explicitly. To obtain the dependence of E on ζ, we form

$$E = \int \Psi^* \mathcal{H} \Psi \, d\tau. \tag{9-29}$$

Since $l = 0$ for the ground state, we have

$$\begin{aligned} E &= \int_0^\infty r^2\,dr \int_0^\pi \sin\vartheta\,d\vartheta \int_0^{2\pi} d\varphi\,\eta_1^*(r)\,Y_{00}^*(\vartheta,\varphi) \\ &\quad \cdot \left\{ -\frac{1}{2r^2}\frac{\partial}{\partial r}\left(r^2\frac{\partial}{\partial r}\right) - \frac{1}{r} \right\} \eta_1(r)\,Y_{00}(\vartheta,\varphi) \\ &= N_1^2 \int_0^\infty r^2\,dr\,e^{-\zeta r} \left\{ -\frac{1}{2r^2}\frac{\partial}{\partial r}\left(r^2\frac{\partial}{\partial r}\right) - \frac{1}{r} \right\} e^{-\zeta r}. \end{aligned} \tag{9-30}$$

[4] In general, these requirements consist of the criteria discussed earlier in Chapter 5, Section 5-2, i.e., quadratic integrability, single-valued, satisfying boundary conditions, etc.

But, it is easily seen that

$$-\frac{1}{2r^2}\frac{\partial}{\partial r}\left(r^2\frac{\partial}{\partial r}\right)e^{-\zeta r}=\left(\frac{\zeta}{r}-\frac{\zeta^2}{2}\right)e^{-\zeta r}, \tag{9-31}$$

which allows Eq. (9-30) to be rewritten as

$$E=N_1^2\int_0^\infty dre^{-2\zeta r}\left[(\zeta-1)r-\left(\frac{\zeta^2}{2}\right)r^2\right]. \tag{9-32}$$

However,

$$\int_0^\infty x^p e^{-qx}\,dx=\frac{p!}{(q)^{p+1}}, \tag{9-33}$$

which gives

$$\begin{aligned} E&=N_1^2\left\{(\zeta-1)\left(\frac{1}{4\zeta^2}\right)-\left(\frac{\zeta^2}{2}\right)\left(\frac{2}{8\zeta^3}\right)\right\}\\ &=N_1^2\left(\frac{\zeta^2}{2}-\zeta\right). \end{aligned} \tag{9-34}$$

In this case, Eq. (9-34) can also be used to obtain the optimum ζ by differentiation, i.e.,

$$0=\frac{\partial E}{\partial \zeta}=\zeta-1,$$

which gives

$$\zeta=1,$$

which we already knew was the correct answer from our previous study of the hydrogen atom.

9-4. Example: Variational Treatment of the Helium Atom

Now let us turn to a problem where the correct choice of a trial wavefunction is not so obvious, since the exact eigenfunction is not known. Let us consider the ground state of the helium atom, whose quantum mechanical Hamiltonian (in the absence of external fields and relativistic effects) can be written in Hartree atomic units as

$$\mathcal{H}=-\frac{1}{2}\nabla_1^2-\frac{1}{2}\nabla_2^2-\frac{2}{r_1}-\frac{2}{r_2}+\frac{1}{r_{12}}. \tag{9-35}$$

The reason that this problem is much more difficult than the hydrogen atom is the presence of the electron–electron repulsion term $(1/r_{12})$. This term inextricably connects the coordinates of the two electrons in a manner that does not allow separation of the variables as in the case of the hydrogen atom. Hence, we must seek approximate solutions to the problem. To obtain a rationale for the choice of a trial wavefunction, let us rewrite the Hamiltonian in Eq. (9-35) as

$$\mathcal{H}=\left(-\frac{1}{2}\nabla_1^2-\frac{\zeta}{r_1}\right)+\left(-\frac{1}{2}\nabla_2^2-\frac{\zeta}{r_2}\right)+(\zeta-2)\left(\frac{1}{r_1}+\frac{1}{r_2}\right)+\frac{1}{r_{12}}. \quad (9\text{-}36)$$

The reason for doing this is that the term $[-(1/2)\nabla_1^2-(\zeta/r_1)]$ has as its normalized eigenfunction the $1s$ orbital:

$$1s(1)=N_1e^{-\zeta r_1}, \quad (9\text{-}37)$$

with eigenvalue

$$E=-\frac{\zeta^2}{2}. \quad (9\text{-}38)$$

A similar comment applies to the term $[-(1/2)\nabla_2^2-(\zeta/r_2)]$ for the second electron. Thus, we have found the eigenfunctions for part of the Hamiltonian, and suggests that perhaps a trial wavefunction of the following form might be a useful approximation to the true solution:[5]

$$\Phi(1,\,2)=1s(1)\cdot 1s(2). \quad (9\text{-}39)$$

When Eq. (9-39) is used, the expression for the energy expectation value is given by

$$\begin{aligned}E=&\int\int 1s^*(1)1s^*(2)\left\{\left(-\frac{1}{2}\nabla_1^2-\frac{\zeta}{r_1}\right)+\left(-\frac{1}{2}\nabla_2^2-\frac{\zeta}{r_2}\right)\right.\\&\left.+(\zeta-2)\left(\frac{1}{r_1}+\frac{1}{r_2}\right)+\frac{1}{r_{12}}\right\}1s(1)1s(2)\,dV_1\,dV_2\\=&\int 1s^*(2)1s(2)\,dV_2\int 1s^*(1)\left[-\frac{1}{2}\nabla_1^2-\frac{\zeta}{r_1}\right]1s(1)\,dV_1\\&+\int 1s^*(1)1s(1)\,dV_1\int 1s^*(2)\left[-\frac{1}{2}\nabla_2^2-\frac{\zeta}{r_2}\right]1s(2)\,dV_2\\&+(\zeta-2)\int 1s^*(2)1s(2)\,dV_2\int 1s^*(1)\frac{1}{r_1}1s(1)\,dV_1\\&+(\zeta-2)\int 1s^*(1)1s(1)\,dV_1\int 1s^*(2)\frac{1}{r_2}1s(2)\,dV_2\\&+\int\int 1s^*(1)1s^*(2)\frac{1}{r_{12}}1s(1)1s(2)\,dV_1\,dV_2,\end{aligned} \quad (9\text{-}40)$$

---

[5] We shall see in the next chapter (Chapter 10, Section 10-2) that an acceptable trial wavefunction must be antisymmetric with respect to exchange of electrons. The function in Eq. (9-39) is easily seen to be *symmetric* with respect to electron exchange, but the antisymmetry property can be achieved through the electron spin component. Since the spin component can be factored from the space component (as will be seen in Chapter 10, Section 10-2), the analysis here can proceed without loss of generality.

but

$$\int 1s^*(i)1s(i)\, dV_i = 1 \qquad i = 1, 2$$

and

$$\left\{ -\frac{1}{2}\nabla_i^2 - \frac{\zeta}{r_i} \right\} 1s(i) = -\left(\frac{\zeta^2}{2}\right) 1s(i), \qquad i = 1, 2, \tag{9-41}$$

so we obtain

$$E = -\zeta^2 + 2(\zeta - 2) \int 1s^*(1) \frac{1}{r_1} 1s(1)\, dV_1$$
$$+ \int\int 1s^*(1)1s^*(2) \frac{1}{r_{12}} 1s(1)1s(2)\, dV_1\, dV_2. \tag{9-42}$$

The evaluation of the remaining integrals is not difficult and can be found in several places.[6] The result is

$$\int 1s^*(1) \frac{1}{r_1} 1s(1)\, dV_1 = \zeta, \tag{9-43}$$

and

$$\int\int 1s^*(1)1s^*(2) \frac{1}{r_{12}} 1s(1)1s(2)\, dV_1\, dV_2 = \left(\frac{5}{8}\right)\zeta, \tag{9-44}$$

so we have

$$E = -\zeta^2 + 2(\zeta - 2)\zeta + \left(\frac{5}{8}\right)\zeta$$

or

$$E = \zeta^2 - \left(\frac{27}{8}\right)\zeta. \tag{9-45}$$

To find the ζ that will minimize E, we form

$$\frac{dE}{d\zeta} = 0 = 2\zeta - \frac{27}{8}$$

or

$$\zeta = \frac{27}{16} = 1.6875.$$

---

[6] See, for example, H. Eyring, J. Walter, and G. E. Kimball, *Quantum Chemistry*, John Wiley, New York, 1944, pp. 103–106.

Substituting this result into the energy expression [Eq. (9-45)], we obtain $E = -2.848$ Hartree.

This is the best energy that we can obtain using only *one* basis orbital of the 1*s* type. However, the experimental energy is $E = -2.9037$ Hartree. Thus, the ability to choose the "best" ζ-value has brought us 98% of the way to the experimental energy value.

We can also say intuitively that a reasonable description of the He atom has been obtained by considering that either one of the electrons does not "see" a nucleus of charge $Z = +2$, but rather $Z \cong 1.7$, i.e., the presence of the other electron "screens" the nuclear charge by an amount 0.3.

However, let us look more carefully at the error involved in the "best" calculated energy. This error is 0.056 Hartree, which is

$$(0.056 \text{ Hartrees})\left(627\,\frac{\text{kcal/mol}}{\text{Hartree}}\right) = 35 \text{ kcal/mol}.$$

Since the changes involved in molecular systems are usually of the order of magnitude of 1–10 kcal/mol, we see that, unless it can be shown that errors cancel, we are on dangerous ground if we try to draw conclusions of a general nature when the errors in the calculation are three times as large as the effects that we are attempting to describe.

Thusfar, we have taken our trial functions to be single terms, with nonlinear parameters to be optimized. However, the Variation Principle applies to *all* types of parameters in a trial function, not only to nonlinear parameters. In fact, in many cases the variation of nonlinear parameters is such an arduous task that the nonlinear parameters are *fixed* at the outset of the calculation, and a trial wavefunction is constructed in which only linear parameters appear which are variable. This latter type of procedure is quite common, and deserves special attention.

9-5. The Linear Variation Method

Let us assume that a set of basis functions $\{\phi_i\}$ has been chosen, in which the ϕ_i are linearly independent, but not necessarily orthogonal or normalized. Furthermore, any nonlinear parameters in the ϕ_i will be assumed to be given and fixed. If only a finite number (N) of the ϕ_i are chosen to represent the trial wavefunction, then we can write

$$\Phi = \sum_{k}^{N} c_k \phi_k, \tag{9-46}$$

or, in matrix form,

$$\Phi = \boldsymbol{\varphi}\mathbf{c}, \tag{9-47}$$

where $\mathbf{c}$ and $\boldsymbol{\phi}$ are $(N \times 1)$ column and row $(1 \times N)$ vectors respectively, i.e.,

$$\mathbf{c} = \begin{pmatrix} c_1 \\ c_2 \\ \vdots \\ c_N \end{pmatrix}, \qquad \boldsymbol{\varphi} = \widetilde{\phi_1 \ \phi_2 \ \cdots \ \phi_N}. \tag{9-48}$$

If Φ is to be the best approximation possible for these N basis functions that have been chosen, then we must choose the c_k so that the energy expectation value is as low as possible. We are assured from the Variation Principle that this energy expectation value will always be above the exact energy.

The energy expectation value of Φ is easily seen to be given by

$$E = \frac{\int \Phi^* \mathcal{H} \Phi \, d\tau}{\int \Phi^* \Phi \, d\tau} \tag{9-49}$$

or

$$E = \frac{\mathbf{c}^\dagger \mathbf{H} \mathbf{c}}{\mathbf{c}^\dagger \mathbf{S} \mathbf{c}} \tag{9-50}$$

where

$$H_{kl} = \int \phi_k^* \mathcal{H} \phi_l \, d\tau \tag{9-51}$$

and

$$S_{kl} = \int \phi_k^* \phi_l \, d\tau. \tag{9-52}$$

$\mathbf{S}$ is usually known as the *overlap matrix*.

In order to minimize E, we wish to ensure that

$$\frac{\partial E}{\partial c_p^*} = 0. \tag{9-53}$$

A similar comment applies to c_p itself, but the resulting set of equations has exactly the same solution as those resulting from Eq. (9-53), so that either set can be used to minimize E.

Carrying out in Eq. (9-50) the operations indicated in Eq. (9-53) gives

$$\left(\frac{\partial E}{\partial c_p^*}\right)(\mathbf{c}^\dagger \mathbf{S} \mathbf{c}) + E \frac{\partial}{\partial c_p^*}(\mathbf{c}^\dagger \mathbf{S} \mathbf{c}) = \frac{\partial}{\partial c_p^*}(\mathbf{c}^\dagger \mathbf{H} \mathbf{c}), \qquad p = 1, 2, \cdots, N. \tag{9-54}$$

However,

$$\begin{aligned} \frac{\partial}{\partial c_p^*}(\mathbf{c}^\dagger \mathbf{S} \mathbf{c}) &= \frac{\partial}{\partial c_p^*}\left(\sum_k^N \sum_l^N c_k^* S_{kl} c_l\right) \\ &= \sum_l^N S_{pl} c_l, \end{aligned} \tag{9-55}$$

and similarly

$$\frac{\partial}{\partial c_p^*}(\mathbf{c}^\dagger\mathbf{H}\mathbf{c}) = \sum_l^N H_{pl}c_l. \tag{9-56}$$

If E is a minimum with respect to variations in c_p^*, then

$$\frac{\partial E}{\partial c_p^*} = 0, \tag{9-57}$$

and Eq. (9-54) can be written as

$$E\left(\sum_l^N S_{pl}c_l\right) = \sum_l^N H_{pl}c_l, \qquad p = 1, 2, \cdots, N,$$

or

$$\sum_l^N (H_{pl} - ES_{pl})c_l = 0, \qquad p = 1, 2, \cdots, N. \tag{9-58}$$

This set of N homogeneous equations in $N + 1$ unknowns is similar to those already discussed in Section 3-8. From that discussion, we know that a non-trivial solution to Eq. (9-58) exists only if

$$\det(\mathbf{H} - E\mathbf{S}) = 0, \tag{9-59}$$

i.e.,

$$0 = \begin{vmatrix} H_{11} - ES_{11} & H_{12} - ES_{12} & \cdots & H_{1n} - ES_{1n} \\ H_{21} - ES_{21} & H_{22} - ES_{22} & \cdots & H_{2n} - ES_{2n} \\ \vdots & \vdots & \ddots & \vdots \\ H_{n1} - ES_{n1} & H_{n2} - ES_{n2} & \cdots & H_{nn} - ES_{nn} \end{vmatrix}. \tag{9-60}$$

It should be noted at this point that the determinant of Eq. (9-60) differs from that of the secular equation of Eq. (3-253) in the introduction of the desired eigenvalues into every element. This arises because of the lack of any assumption concerning the orthogonality of the basis functions. However, it is easy to see how the two problems are actually equivalent, and we shall now demonstrate.

Consider the overlap matrix $\mathbf{S}$. Since $\mathbf{S}$ is obviously Hermitian, we know that, from Theorem 3-18, $\mathbf{S}$ can always be brought to diagonal form by means of a unitary similarity transformation,[7] i.e.,

---

[7] The orthogonalization procedure discussed here is usually referred to as *symmetric orthogonalization*, since it deals with each function on an equal footing. It is an alternative procedure to the Gram–Schmidt orthogonalization procedure that was introduced in Section 2-4.

$$\mathbf{U}^{\dagger}\mathbf{S}\mathbf{U} = \mathbf{d} = \begin{pmatrix} d_{11} & & & \\ & d_{22} & & 0 \\ & & \ddots & \\ 0 & & & d_{nn} \end{pmatrix}, \tag{9-61}$$

where $\mathbf{U}$ is a unitary matrix and $\mathbf{d}$ is diagonal. We now define

$$\mathbf{V} = \begin{bmatrix} \frac{1}{\sqrt{d_{11}}} & & & \\ & \frac{1}{\sqrt{d_{22}}} & & 0 \\ & & \ddots & \\ & 0 & & \frac{1}{\sqrt{d_{nn}}} \end{bmatrix}. \tag{9-62}$$

This definition will always be possible as long as

$$d_{ii} \neq 0, \qquad i = 1, 2, \cdots, n. \tag{9-63}$$

However, if any $d_{ii} = 0$, it would imply that the basis set that has been chosen is linearly dependent, and this basis should be altered to remove the linear dependence before proceeding. In fact, one of the ways in which the linear dependence of a given set is checked in practice is through examination of the eigenvalues of the overlap matrix ($\mathbf{S}$).

Assuming that all linear dependence problems have now been removed, we can apply $\mathbf{V}$ (and $\mathbf{V}^{\dagger}$) to Eq. (9-61) with the following result:

$$\mathbf{V}^{\dagger}\mathbf{U}^{\dagger}\mathbf{S}\mathbf{U}\mathbf{V} = \mathbf{I}. \tag{9-64}$$

Returning to Eq. (9-59), let us use the result just obtained in the following way:

$$\det(\mathbf{V}^{\dagger}\mathbf{U}^{\dagger}) \cdot \det(\mathbf{H} - E\mathbf{S}) \cdot \det(\mathbf{U}\mathbf{V}) = 0,$$

or[8]

$$\det[(\mathbf{V}^{\dagger}\mathbf{U}^{\dagger}\mathbf{H}\mathbf{U}\mathbf{V}) - E(\mathbf{V}^{\dagger}\mathbf{U}^{\dagger}\mathbf{S}\mathbf{U}\mathbf{V})] = 0,$$

or

$$\det(\mathbf{H}' - E\mathbf{I}) = 0, \tag{9-65}$$

where

$$\mathbf{H}' = \mathbf{V}^{\dagger}\mathbf{U}^{\dagger}\mathbf{H}\mathbf{U}\mathbf{V}. \tag{9-66}$$

[8] Here we have used the theorem from Chapter 3 (Theorem 3-5) that $\det(\mathbf{AB}) = \det(\mathbf{A}) \cdot \det(\mathbf{B})$.

Thus, we see that Eq. (9-65) has the same form as Eq. (3-253), and the current formulation using nonorthogonal basis functions can be transformed to the one treated in Chapter 3 by means of the procedure just outlined. We also note that the eigenvalues (E_i) are *unaffected* by the transformation described in Eq. (9-64), and the desired eigenfunctions of the original Hamiltonian matrix ($\mathbf{H}$) can be obtained from

$$\mathbf{c} = \mathbf{U}\mathbf{V}\mathbf{c}' \tag{9-67}$$

where $\mathbf{c}'$ is the matrix of eigenvectors of $\mathbf{H}'$ that results from the solution of Eq. (9-65).

Once the eigenvalues have been obtained, the eigenvectors can be found by substitution into Eq. (9-58) plus the use of the normalization condition. Optimization of a wavefunction of this matter is usually referred to as the *Raleigh–Ritz Variation Method* or the *Linear Variation Method.*

As the number of terms in the basis set is increased, the basis becomes more and more complete, and the approximate wavefunction Φ approaches closer and closer to Ψ_1, the desired eigenfunction for the ground state. Obviously, the larger the basis set, the better the results in general. It is also easy to understand why the introduction of digital computers has had such a tremendous impact on the field of quantum chemistry. It has transformed literally impossible problems into routine problems by allowing, for example, the diagonalization of a (100×100) matrix in a matter of seconds.

Thus far our discussion has focused on the ground state eigenvalue and eigenfunction. However, as noted earlier, expansion of the determinant in Eq. (9-60) gives rise to a *polynomial* of degree N in E, which means that N allowed values of E will result from the solution of Eq. (9-60). This suggests that the other values of E might be approximations to excited state energies. This is indeed the case, as we shall now see by deriving a result known as *MacDonald's Theorem.*[9]

We begin by assuming that we have already solved the secular determinant of Eq. (9-60) using a set of N orthonormal basis functions (χ_k), i.e., we have solved

$$\det(H - EI). \tag{9-68}$$

This gives rise to a set of N eigenvalues ($E_k^{(N)}$) and eigenfunctions ($\Phi_k^{(N)}$), i.e.,

$$\Phi_k^{(N)} = \sum_{l=1}^{N} d_{kl}^{(N)} \chi_l. \tag{9-69}$$

For simplicity of analysis, we shall assume that the energies associated with the

[9] See, for example, E. A. Hylleraas and B. Undheim, *Z. Phys.*, **65**, 759 (1930); J. K. L. MacDonald, *Phys. Rev.*, **43**, 830 (1933); H. Shull and P. O. Löwdin, *Phys. Rev.*, **110**, 1466 (1958).

$\Phi_k^{(N)}$ are nondegenerate, and that they have been placed in the order

$$E_1^{(N)} < E_2^{(N)} < \cdots < E_N^{(N)}. \tag{9-70}$$

We also note for later use that Eq. (9-70) implies that

$$\langle \Phi_k^{(N)} | \mathcal{H} | \Phi_l^{(N)} \rangle = E_k \delta_{kl}. \tag{9-71}$$

To prove MacDonald's Theorem we now construct an $(N + 1)$ term wavefunction, starting with the set of N functions $\{\Phi_k^{(N)}\}$ and adding one more function (Φ_{N+1}). This means that the approximate wavefunction now has the form

$$\Psi = \sum_{k=1}^{N} a_k \Phi_k^{(N)} + a_{N+1} \Phi_{N+1}. \tag{9-72}$$

For convenience, we shall choose Φ_{N+1} such that

$$\langle \Phi_{N+1} | \Phi_j \rangle = \delta_{N+1,j}, \qquad j = 1, 2, \cdots, N+1. \tag{9-73}$$

If we now construct the secular determinant associated with the wavefunction of Eq. (9-72), it is easily seen that it takes the form

$$\begin{vmatrix} E_1^{(N)} - \mathcal{E} & 0 & \cdots & 0 & H_{1,N+1} \\ 0 & E_2^{(N)} - \mathcal{E} & \cdots & 0 & H_{2,N+2} \\ \vdots & \vdots & \ddots & \vdots & \vdots \\ 0 & 0 & \cdots & E_N^{(N)} - \mathcal{E} & H_{N,N+1} \\ H_{N+1,1}^* & H_{N+1,2}^* & \cdots & H_{N+1,N}^* & H_{N+1,N+1} - \mathcal{E} \end{vmatrix} = 0. \tag{9-74}$$

Expansion of the determinant gives:

$$(H_{N+1,N+1} - \mathcal{E}) \cdot \prod_{k=1}^{N} (E_k^{(N)} - \mathcal{E}) - \sum_{l=1}^{N} |H_{l,N+1}|^2 \cdot \prod_{\substack{k=1 \\ (\neq l)}}^{N} (E_k^{(N)} - \mathcal{E}) = 0 = f(\mathcal{E}). \tag{9-75}$$

The above equation is an $(N = 1)$st-order polynomial in $\mathcal{E}$ [which we have called $f(\mathcal{E})$], which has several properties of interest,[10] i.e.,

$$\left. \begin{array}{ll} f(\mathcal{E}) \to \infty & \text{as } \mathcal{E} \to -\infty \\ f(\mathcal{E}) \to (-1)^{N+1} \cdot \infty & \text{as } \mathcal{E} \to +\infty. \end{array} \right\} \tag{9-76}$$

Other properties of $f(\mathcal{E})$ can be ascertained by looking at $f(\mathcal{E})$ at the points that are the roots of the Nth-order polynomial, i.e., examine $f(\mathcal{E} = E_k^{(N)})$. In those cases, Eq. (9-75) becomes:

$$f(E_k^{(N)}) = -|H_{k,N+1}|^2 \prod_{\substack{m=1 \\ (\neq k)}}^{N} (E_m^{(N)} - E_k^{(N)}). \tag{9-77}$$

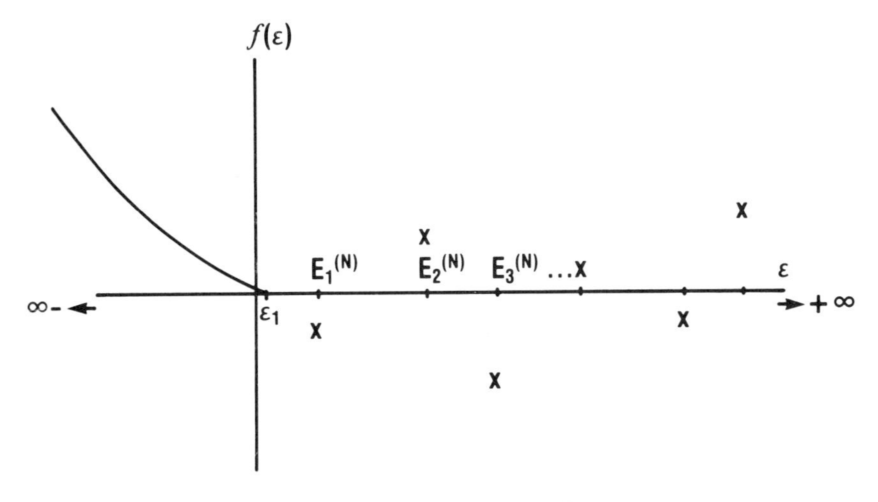

Figure 9.1. Schematic description of $f(\epsilon)$ behavior.

Because of the ordering of eigenvalues we have chosen [see Eq. (9-70)], Eq. (9-77) can be seen to have the following properties:

$$f(E_k^{(N)}) = \begin{cases} \geq 0, & k \text{ even} \\ \leq 0, & k \text{ odd} \end{cases}. \tag{9-78}$$

These properties of $f(\mathcal{E})$ are illustrated schematically in Fig. 9-1. From the figure we can see, for example, that $f(\mathcal{E})$ must be zero (at least) once between $E_1^{(N)}$ and $E_2^{(N)}$, and similarly between each $E_k^{(N)}$ and $E_{k+1}^{(N)}$. Since there are $(N - 1)$ such pairs [thus giving $N - 1$ roots of $f(\mathcal{E})$] plus two additional points where $f(\mathcal{E}) = 0$ that are associated with $f(\mathcal{E})$ behavior as $\mathcal{E} \rightarrow \pm\infty$, we conclude that $f(\mathcal{E})$ can be zero only once between each pair $E_k^{(N)}$ and $E_{k+1}^{(N)}$. This means that at any level of approximation, each root $(\mathcal{E}_k)$ is an upper bound to the exact energy of state, i.e.,

$$\mathcal{E}_k \geq E_k, \qquad k = 1, 2, \cdots. \tag{9-79}$$

This result is known as *MacDonald's Theorem*, and illustrates that the Linear Variation Method has applicability to excited states as well.[11] In addition, results of the calculations continue to improve as more terms are added to the approximate wavefunction. For example, if N, $N + 1$, and $N + 2$ terms are used to describe the pth excited state, then

$$\mathcal{E}_p^{(N)} \geq \mathcal{E}_p^{(N+1)} \geq \mathcal{E}_p^{(N+2)} \geq E_p.$$

[10] See Problem 9-9.

[11] It should be noted that MacDonald's Theorem has been derived with the assumption of non-degenerate eigenvalues. When degeneracies are present, significant complications can occur.

9-6. Example: The Hydrogen Atom Revisited

To illustrate the linear variation method, let us consider the ground state of the hydrogen atom once again. However, even though we now know that the true ground state eigenfunction is a simple exponential, let us choose a different kind of trial function so see if a satisfactory result can be obtained. In anticipation of their use in later applications, let us choose a basis set of Gaussian functions for this example.

Using the notation for the previous section, the trial wavefunction will be chosen as

$$\Phi = \boldsymbol{\varphi}\mathbf{c}, \tag{9-80}$$

where

$$\boldsymbol{\varphi} = (\phi_1 \;\; \phi_2 \;\; \cdots \;\; \phi_N), \tag{9-81}$$

and

$$\phi_i = \exp(-a_i r^2) \tag{9-82}$$

and $\mathbf{c}$ is a column vector of coefficients whose values we wish to determine so as to minimize the total energy E. Since we are dealing with the linear variation method, we shall not treat the a_i as variable parameters. Instead, the a_i will be chosen in advance (for a given N).

The matrix elements to be evaluated are

$$H_{kl} = \int \phi_k^* \mathcal{H} \phi_l \, d\tau \tag{9-83}$$

and

$$S_{kl} = \int \phi_k^* \phi_l \, d\tau. \tag{9-84}$$

It is at this point at which one of the reasons why spherical Gaussians have been chosen as basis orbitals becomes apparent. In particular, each of the integrals that arises in Eqs. (9-83) and (9-84) can be evaluated in closed form, i.e., no approximations need to be made in their evaluation.[12] When the hydrogen atom Hamiltonian is inserted into Eq. (9-83), it can be seen that the integrals to be evaluated are given as follows:

$$\langle \phi_k | \phi_l \rangle = \int \exp\{-(a_k + a_l) r^2\} \, dV = \left(\frac{\pi}{a_k + a_l} \right)^{3/2}, \tag{9-85}$$

$$\begin{aligned} \left\langle \phi_k \middle| -\frac{1}{2} \nabla^2 \middle| \phi_l \right\rangle &= -\frac{1}{2} \int \exp(-a_k r^2)[\nabla^2 \exp(-a_l r^2)] \, dV \\ &= \left(\frac{3 a_k a_l}{a_k + a_l} \right) \left(\frac{\pi}{a_k + a_l} \right)^{3/2}, \end{aligned} \tag{9-86}$$

[12] See S. F. Boys, *Proc. R. Soc. (London)*, **A200** 542 (1950) for the details of the integral evaluation. See also I. Shavitt, *Meth. Computat. Phys.*, **2**, 1 (1963).

Table 9.1. Description of the Hydrogen Atom Using Gaussians

Number of terms	c_k (unnormalized)	a_k (optimized)	E (Hartrees)
3	$c_1 = 1.0$	$a_1 = 0.159$	-0.49689
	$c_2 = 1.843$	$a_2 = 0.775$	
	$c_3 = 1.168$	$a_3 = 5.74$	
8	$c_1 = 1.0$	$a_1 = 0.09190$	-0.49992
	$c_2 = 0.9632$	$a_2 = 0.2013$	
	$c_3 = 1.9605$	$a_3 = 0.2716$	
	$c_4 = 2.1531$	$a_4 = 0.4795$	
	$c_5 = 3.0079$	$a_5 = 1.0139$	
	$c_6 = 2.4455$	$a_6 = 2.4474$	
	$c_7 = 2.2566$	$a_7 = 8.2450$	
	$c_8 = 1.3673$	$a_8 = 52.666$	

$$\left\langle \phi_k \left| \frac{1}{r} \right| \phi_l \right\rangle = \int \exp\{-(a_k + a_l)r^2\} \left(\frac{1}{r}\right) dV = \left(\frac{2\pi}{a_k + a_l}\right). \tag{9-87}$$

Thus, even though we know that our Gaussian basis orbitals do not have the correct functional form to give the exact answer, the evaluation of the integrals is simple enough to allow use of a large number of Gaussians if needed in order to give a good approximation to the exact answer.

In Table 9.1, two examples of different basis set sizes are given. The first of these consists of a basis set of three Gaussians, and the second basis set includes eight Gaussians. Since this problem is not particularly difficult from a computational point of view, the values of a_k were also varied in the original work of Longstaff and Singer.[13] However, for our purposes the values of the a_k in Table 9-1 can be considered to be fixed in advance.

Several points of interest emerge in this example. First, the energy converges rather rapidly to the exact answer (-0.5 Hartrees), with even the three-term wavefunction giving an energy that is 99.4% of the exact energy. In chemical terms, the three-term wavefunction has an error of 2 kcal/mol, while the error in the eight-term wavefunction is only 0.05 kcal/mol. Hence, at least from an energetic point of view, either of the trial functions ought to be acceptable in describing hydrogen atom behavior.

As for the wavefunction itself, we would expect that the Gaussian basis orbitals would be deficient in two different ways, compared to the exact solution

[13] See J. V. L. Longstaff and K. Singer, *Proc. R. Soc. (London)*, **A258**, 421 (1960) for the original report and discussion of this example.

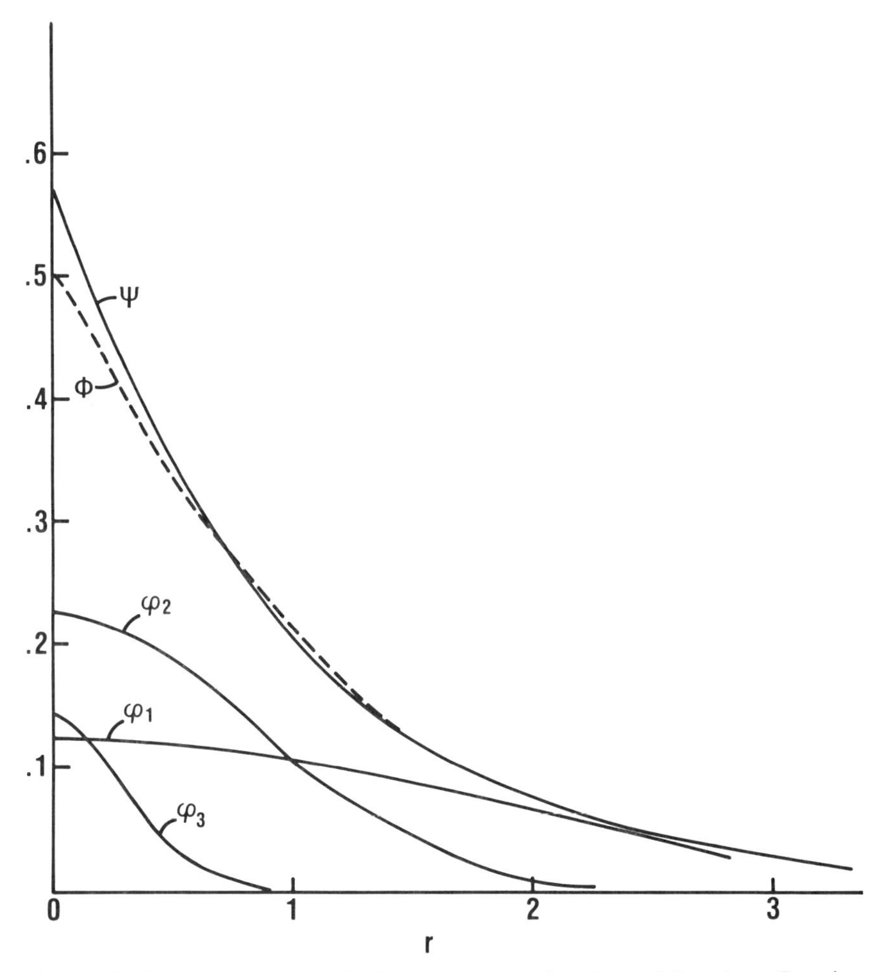

Figure 9.2. Comparison of exact hydrogen atom wavefunction and three-term Gaussian wavefunction.

(a single exponential). Specifically, the r^2 dependence in the exponential of Gaussian basis orbitals will fall off faster as $r \rightarrow \infty$ than the r dependence in the exact wavefunction. Also, the Gaussian basis functions do not possess the cusp behavior[14] at the origin that is present in the exact solution. For each of these reasons we expect to have to employ more than one Gaussian in order to emulate the behavior of the exact wavefunction. To see how this emulation takes place, Fig. 9.2 shows the behavior of each of the three Gaussian basis functions (φ_1, φ_2, φ_3) as well as the total wavefunction (Φ) as a function of r. Also shown is the exact hydrogen atom wavefunction (Ψ).

The deficiency in cusp behavior of the approximate Gaussian wavefunction is

[14] See Section 7-3.

clearly seen in Fig. 9-2, where the largest deviation occurs between the origin and approximately 0.5 Bohr. The role of each of the three terms in the Gaussian approximation is also clearly discernible. The first component (φ_1) primarily describes the long-range behavior of Φ, while the second component (φ_2) describes the mid-range behavior and the third component (φ_3) attempts to reproduce the appropriate shape near the origin. While a perfect fit has not been obtained, it is seen that, overall, a quite good approximation to the exact wavefunction has resulted. In later chapters we shall see how this type of approximation, plus the ease of integral evaluation, will be quite useful in obtaining approximate wavefunctions for molecular systems.

9-7. Lower Bounds

Given the power and flexibility of the Variation Principle and its application through methodology such as the Linear Variation Method to the calculation of *upper bounds* to energies, it is not surprising to find that efforts to formulate methods for calculation of corresponding *lower bounds* to energies have been attempted. As we shall see, however, a number of practical and other problems have prevented analogous developments for lower bounds.

To see how the calculation of lower bounds to energies can be approached, we again consider a system having eigenvalues

$$E_0 \le E_1 \le E_2 \le \cdots$$

with associated eigenfunctions Ψ_0, Ψ_1, Ψ_2 ... . We shall also use the expansion introduced earlier [Eq. (9-1)], i.e., we shall construct a normalized function Φ as

$$\Phi = \sum_k c_k \Psi_k. \tag{9-88}$$

Now let us examine the quantity

$$J = \langle \Phi | (\mathcal{H} - E_0)(\mathcal{H} - E_1) | \Phi \rangle. \tag{9-89}$$

Inserting Eq. (9-88) into Eq. (9-89) gives

$$\begin{aligned} J &= \sum_{k,l} c_k^* c_l \langle \Psi_k | (\mathcal{H} - E_0)(\mathcal{H} - E_1) | \Psi_l \rangle \\ &= \sum_{k,l} c_k^* c_l \langle \Psi_k | (E_l - E_0)(E_l - E_1) | \Psi_l \rangle \\ &= \sum_l c_l^* c_l (E_l - E_0)(E_l - E_1). \end{aligned} \tag{9-90}$$

We note that, since E_0 and E_1 are, by assumption, the two lowest states,

$$E_l \ge E_0, \ E_l \ge E_1. \tag{9-91}$$

This means that

$$J = \langle \Phi | (\mathcal{H} - E_0)(\mathcal{H} - E_1) | \Phi \rangle \geq 0. \tag{9-92}$$

Expanding Eq. (9-92) gives

$$\langle \Phi | \mathcal{H}^2 | \Phi \rangle - E_1 \langle \Phi | \mathcal{H} | \Phi \rangle \geq E_0 [\langle \Phi | \mathcal{H} | \Phi \rangle - E_1]. \tag{9-93}$$

Under the condition that

$$E_0 \leq \langle \Phi | \mathcal{H} | \Phi \rangle \leq E_1, \tag{9-94}$$

which is expected if Φ is a reasonably good approximation to the exact (lowest) eigenfunction and E_0 and E_1 are well separated, we can rewrite Eq. (9-93) as

$$E_0 \geq \frac{E_1 \langle \Phi | \mathcal{H} | \Phi \rangle - \langle \Phi | \mathcal{H}^2 | \Phi \rangle}{E_1 - \langle \Phi | \mathcal{H} | \Phi \rangle}. \tag{9-95}$$

This formula, which provides a lower bound estimate to E_0, is known as the *Temple formula.*[15] Other forms of the lower bound estimate have also been given,[16] e.g.,

$$E_0 \geq \langle \Phi | \mathcal{H} | \Phi \rangle - [\langle \Phi | \mathcal{H}^2 | \Phi \rangle - \langle \Phi | \mathcal{H} | \Phi \rangle^2]^{1/2}. \tag{9-96}$$

Unfortunately, several difficulties arise in the calculation of lower bound estimates that have limited the application of formulas such as given in Eqs. (9-95) and (9-96). For example, evaluation of the integrals involved using the $\mathcal{H}^2$ operator is quite difficult in general, and analytical expressions have been obtained only for the case when Gaussian basis orbitals are used.[17] In addition, optimization of lower bounds produces wavefunctions that do not satisfy the virial theorem,[18] while use of wavefunctions optimized for upper bounds produce lower bounds that in general are too far below the desired energy to be useful measures of wavefunction accuracy. To illustrate this latter point, calculations on H_3^+ that minimize the upper bound produced[19]

$$E_{\text{upper}} = -1.33764 \text{ Hartrees},$$

compared to an estimated exact energy of

$$E_0 = -1.34 \text{ Hartrees}.$$

---

[15] G. Temple, *Proc. R. Soc. (London)*, **A119**, 276 (1928).

[16] The form presented in Eq. (9-96) was given by D. H. Weinstein, *Phys. Rev.*, **40**, 797; **41**, 839 (1932); *Proc. Nat. Acad. Sci. U.S.A.*, **20**, 529 (1934), and J. K. L. MacDonald, *Phys. Rev.*, **46**, 828 (1934). See also A. F. Stevenson and M. F. Crawford, *Phys. Rev.*, **54**, 375 (1938). For a review of work in this area, see F. Weinhold, *Adv. Quantum Chem.*, **6**, 299–331 (1972).

[17] I. T. Keaveny and R. E. Christoffersen, *J. Chem. Phys.*, **50**, 80–85 (1969).

[18] The "virial theorem" will be discussed in Chapter 10, Section 10-6. For its application to lower bounds, see G. L. Caldow and C. A. Coulson, *Proc. Cambridge Philos. Soc.*, **57**, 341 (1961).

[19] I. T. Keaveny and R. E. Christoffersen, *J. Chem. Phys.*, **50**, 80–85 (1969).

However, use of that wavefunction to calculate $\mathcal{H}^2$ integrals gave

$$\langle \Phi | \mathcal{H}^2 | \Phi \rangle = 2.195881 \text{ Hartrees},$$

and use of Eq. (9-84) gives a lower bound estimate of

$$E_{\text{lower}} = -1.975292 \text{ Hartrees},$$

which is too far below the true energy to be useful as a test of wavefunction quality. Other approaches have been used with some success,[20] but full exploitation of the bracketing potential of upper and lower bounds as a means of improving approximate wavefunctions and assessing wavefunction quality remains to be accomplished.

9-8. Rayleigh–Schrödinger Perturbation Theory

While use of the variational principle in its various forms is widespread, an alternative approach using the techniques of *perturbation theory* is often found to be useful. The way in which perturbation theory is formulated and the systems to which it is applied vary somewhat, and the development given below [known as *Rayleigh–Schrödinger perturbation theory* (RSPT)] will describe at least some of the major areas of interest.

In general, perturbation theory begins with the assumption that it will be useful to take a given Hamiltonian operator ($\mathcal{H}$) and split it into two (or more parts, e.g.,

$$\mathcal{H} = \mathcal{H}_0 + \lambda \mathcal{H}'. \tag{9-97}$$

In the above equation there are several characteristics of the division of $\mathcal{H}$ that must be met in general. In particular, $\mathcal{H}_0$ is referred to as the "unperturbed" Hamiltonian, and it must be such that the *exact* solutions to the time-independent Schrödinger equation with $\mathcal{H}_0$ can be found. Also, λ is a real number, usually referred to as a "perturbation parameter." Both $\mathcal{H}_0$ and $\mathcal{H}'$ are assumed to be Hermitian, and λ is assumed to be "small." This latter assumption is interpreted to mean that the perturbation term has small effects on the eigenfunctions of the unperturbed operator ($\mathcal{H}_0$). Quantification of this concept is an important aspect of the theory, and will be considered within the context of the discussions to follow.

A. Nondegenerate States

The case that is usually used to introduce the concepts of perturbation theory is where the eigenvalues of the unperturbed operator are each nondegenerate. For that particular case, we assume that the exact eigenfunctions and eigenvalues of

[20] See, for example, N. W. Bazley, *Proc. Natl. Acad. Sci. U.S.A.*, **45**, 850 (1959); *Phys. Rev.*, **120**, 44 (1960). See also N. W. Bazley and D. W. Fox, *J. Res. Nat. Bus. Stand.*, **B65**, 105 (1961); *Phys. Rev.*, **124**, 483 (1961); *J. Math. Phys.*, **3**, 469 (1962).

$\mathcal{H}_0$ can be found, i.e., that

$$\mathcal{H}_0 \psi_n^{(0)} = E_n^{(0)} \psi_n^{(0)} \qquad n = 0, 1, 2, \cdots \tag{9-98}$$

can be solved exactly for all values of n, giving energy levels $E_0^{(0)}, E_1^{(0)}, E_2^{(0)} \ldots$ that are each nondegenerate, plus associated eigenfunctions $\Psi_0^{(0)}, \Psi_1^{(0)}, \Psi_2^{(0)} \ldots$. The superscript has been added in order to indicate that we are dealing at this point with eigenvalues and eigenvectors of the unperturbed problem.

A perturbation of the form $(\lambda \mathcal{H}')$ is now added to the system, and we want to determine what effect it will have on the unperturbed eigenfunctions and eigenvalues. The basic approach of perturbation theory is to assume that, since the Hamiltonian depends on λ, the eigenvectors and eigenvalues will also depend on λ, and the effect of the perturbation on both the eigenvalues and eigenvectors can be described by a power series expansion in λ, i.e.,

$$\Psi_n = \psi_n^{(0)} + \lambda \psi_n^{(1)} + \lambda^2 \psi_n^{(2)} + \cdots \tag{9-99}$$

$$E_n = E_n^{(0)} + \lambda E_n^{(1)} + \lambda^2 E_n^{(2)} + \cdots \tag{9-100}$$

for all n. In the above equations, Ψ_n and E_n represent the eigenfunction and eigenvalue of the nth state of the *perturbed* system, while $\psi_n^{(1)}, \psi_n^{(2)} \ldots$ represent the first-order correction, second-order correction, etc., to the unperturbed wavefunction $(\psi_n^{(0)})$ that results from the introduction of the perturbation. Similarly, $E_n^{(1)}, E_n^{(2)}, \ldots$ represent the first-order correction, second-order correction, etc., the unperturbed energy $(E_n^{(0)})$. All correction terms are assumed to be independent of λ, and the wavefunction correction terms are assumed to be orthonormal, i.e.,

$$\int \psi_n^{(j)*} \psi_n^{(k)} \, d\tau = \delta_{jk}. \tag{9-101}$$

The basic assumption of importance underlying this approach is that the power series represented in Eqs. (9-99) and (9-100) are convergent. Such an assumption *cannot* be justified in general without examination of the particulars of the problem being considered, and needs to be considered explicitly for each case.

For the purposes of our discussion at the moment, let us assume that the power series expansions converge appropriately. Then, substitution of Eqs. (9-99) and (9-100) into the Schrödinger equation for the perturbed system

$$(\mathcal{H}_0 + \lambda \mathcal{H}')\Psi_n = E_n \Psi_n \tag{9-102}$$

gives

$$\begin{aligned}(\mathcal{H}_0 + \lambda \mathcal{H}')&(\psi_n^{(0)} + \lambda \psi_n^{(1)} + \lambda^2 \psi_n^{(2)} + \cdots) \\ &= (E_n^{(0)} + \lambda E_n^{(1)} + \lambda^2 E_n^{(2)} + \cdots)(\psi_n^{(0)} + \lambda \psi_n^{(1)} + \lambda^2 \psi_n^{(2)} + \cdots)\end{aligned} \tag{9-103}$$

for all n. Since λ is an arbitrary parameter, the only way to assure that Eq. (9-103) will hold for all values of λ is for the coefficient of each power of λ on opposite sides of Eq. (9-103) to be equal.

Carrying out this process, we obtain from Eq. (9-103) the following equation for the terms independent of λ:

$$\mathcal{H}_0 \psi_n^{(0)} = E_n^{(0)} \psi_n^{(0)}. \tag{9-104}$$

This equation gives us no new information, since it simply represents the equation for the unperturbed system whose solutions we have assumed are known.

Considering next the terms linear in λ, we obtain

$$\mathcal{H}_0 \psi_n^{(1)} + \mathcal{H}' \psi_n^{(0)} = E_n^{(0)} \psi_n^{(1)} + E_n^{(1)} \psi_n^{(0)}, \tag{9-105}$$

which is known as the *first order equation*. To obtain the first order energy correction ($E_n^{(1)}$), we multiply Eq. (9-105) by $\Psi_n^{(0)*}$ and integrate, which gives

$$E_n^{(1)} = \int \psi_n^{(0)*} \mathcal{H}_0 \psi_n^{(1)} \, d\tau + \int \psi_n^{(0)*} \mathcal{H}' \psi_n^{(0)} \, d\tau - E_n^{(0)} \int \psi_n^{(0)*} \psi_n^{(1)} \, d\tau. \tag{9-106}$$

The Hermitian property of $\mathcal{H}_0$ and the orthonormality of the functions can be used to show that the first and third terms on the right-hand side of the above equation are zero, which gives

$$E_n^{(1)} = \int \psi_n^{(0)*} \mathcal{H}' \psi_n^{(0)} \, d\tau. \tag{9-107}$$

This is the first result of interest. It indicates that the first-order correction to the energy due to the presence of the perturbation can be evaluated from only a knowledge of the *unperturbed* wavefunction ($\Psi_n^{(0)}$) and evaluation of the integral in Eq. (9-107).

This type of analysis can be carried further. For the case of the λ^2 terms and the λ^3 terms, following a process as described for the terms linear in λ gives[21]

$$E_n^{(2)} = \int \psi_n^{(0)*} \mathcal{H}' \psi_n^{(1)} \, d\tau, \tag{9-108}$$

and

$$E_n^{(3)} = \int \psi_n^{(0)*} \mathcal{H}' \psi_n^{(2)} \, d\tau. \tag{9-109}$$

Through some additional manipulations, the results just obtained can be simplified further. For example, multiplication of the second-order equation (i.e., the equation representing the coefficients of λ^2) by $\Psi_n^{(1)}*$ and integration gives

$$\int \psi_n^{(1)*} \mathcal{H}' \psi_n^{(1)} \, d\tau = E_n^{(1)} - \int \psi_n^{(1)*} \mathcal{H}_0 \psi_n^{(2)} \, d\tau. \tag{9-110}$$

Multiplication of the first-order equation by $\Psi_n^{(2)*}$ and integration gives

$$\int \psi_n^{(1)*} \mathcal{H}_0 \psi_n^{(2)} \, d\tau = -\int \psi_n^{(0)*} \mathcal{H}' \psi_n^{(2)} \, d\tau. \tag{9-111}$$

Combining Eqs. (9-110) and (9-111) and substitution into Eq. (9-109) gives

$$E_n^{(3)} = \int \psi_n^{(1)*} \mathcal{H}' \psi_n^{(1)} \, d\tau - E_n^{(1)}. \tag{9-112}$$

This is the second important result. Stated in its generalized form, Eq. (9-112) indicates that the nth-order energetic correction can be obtained from a

[21] See Problem 9-3.

knowledge of the $(n - 2)$nd-order correction to the unperturbed wavefunction. For this process to be useful, it is essential to be able to calculate the various correction terms to the wavefunction, to which we now turn our attention.

Since the eigenfunctions of the unperturbed system form a basis for description of the system prior to initiation of the perturbation, these same eigenfunctions will provide an acceptable basis after the perturbation has been applied. However, the coefficients can be expected to change, due to the change in the system as a result of the perturbation. Hence, our next task is to determine the coefficients in the various correction terms to the unperturbed system.

For the first-order correction, we can write

$$\psi_n^{(1)} = \sum_l \psi_l^{(0)} a_{ln}, \tag{9-113}$$

where the a_{ln} are to be determined, and the summation may include continuum contributions, depending on the system under consideration. Substitution of Eq. (9-113) into the first-order equation [Eq. (9-105)] gives

$$\sum_l a_{ln}(\mathcal{H}_0 - E_n^{(0)})\psi_l^{(0)} = (E_n^{(1)} - \mathcal{H}')\psi_n^{(0)}$$

or

$$\sum_l a_{ln}(E_l^{(0)} - E_n^{(0)})\psi_l^{(0)} = (E_n^{(1)} - \mathcal{H}')\psi_n^{(0)}. \tag{9-114}$$

Multiplication by $\psi_r^{(0)*}$ (with $r \neq n$) and integration gives

$$a_{rn} = \frac{\int \psi_r^{(0)*} \mathcal{H}' \psi_n^{(0)}\, d\tau}{E_n^{(0)} - E_r^{(0)}}, \tag{9-115}$$

or

$$a_{rn} = \frac{H'_{rn}}{E_n^{(0)} - E_r^{(0)}}, \tag{9-116}$$

where we have defined

$$H'_{rn} = \int \psi_r^{(0)*} \mathcal{H}' \psi_n^{(0)}\, d\tau. \tag{9-117}$$

Substitution of Eqs. (9-116) and (9-117) into the expressions for the first-, second-, and third-order energetic corrections [Eqs. (9-107), (9-108), and (9-109)] gives the following expression for the energy of the perturbed system [with $\lambda = 1$ in Eq. (9-100)]:

$$E_n = E_n^{(0)} + H'_{nn} + \sum_{r \neq n} \frac{|H'_{nr}|^2}{E_n^{(0)} - E_r^{(0)}} + \cdots \tag{9-118}$$

where each of the terms listed explicitly on the right-hand side of the above equation can be obtained by use of the energies and eigenfunctions of only the unperturbed system.

Collecting the results for the correction of the wavefunction, we have

$$\Psi_n=\psi_n^{(0)}+\sum_{r\neq n}\left(\frac{H'_{rn}}{E_n^{(0)}-E_r^{(0)}}\right)\psi_r^{(0)}+\cdots. \tag{9-119}$$

B. Example: The Anharmonic Oscillator[22]

To illustrate the use of perturbation theory, let us consider the harmonic oscillator once again. The potential for the harmonic oscillator that was discussed in Section 5-9 was

$$V(x)=\frac{k}{2}(x-x_0)^2. \tag{9-120}$$

While such a potential forms a good first approximation for understanding the vibrational spectral features of diatomic molecules, there are several limitations inherent in its use. First, and most important, the harmonic oscillator potential is a parabola, and cannot describe the process of ionization that is known to occur. This drawback to the harmonic oscillator potential is depicted qualitatively in Fig. 9.3. The basic reason for this deficiency is due to the fact that $V(x)$ in Eq. (9-120) is only one term of an infinite series, i.e.,

$$V(x)=V_0+x\left(\frac{\partial V}{\partial x}\right)_{x=x_0}+\frac{x^2}{2!}\left(\frac{\partial^2 V}{\partial x^2}\right)_{x=x_0}$$
$$+\frac{x^3}{3!}\left(\frac{\partial^3 V}{\partial x^3}\right)_{x=x_0}+\frac{x^4}{4!}\left(\frac{\partial^4 V}{\partial x^4}\right)_{x=x_0}+\cdots \tag{9-121}$$

in which only the quadratic term is considered in the case of the harmonic oscillator. Of course, the choice of V_0 is irrelevant since it simply moves all levels up or down. Also, since x_0 has been chosen as the minimum in the $V(x)$ curve (see Fig. 9-3), the term linear in x in Eq. (9-121) will be zero. Hence, while the harmonic oscillator model includes the most important first term, higher order terms must be considered if a more adequate description of spectral features is to be obtained.

Let us consider the correction to Eq. (9-121) due to the cubic and quartic terms that can be deduced by the use of perturbation theory. For the cubic term, we shall consider a perturbation Hamiltonian of the form

$$\mathcal{H}'=ax^3. \tag{9-122}$$

The zero-order Hamiltonian ($\mathcal{H}_0$) plus associated eigenvalues and eigenfunctions are given[23] by

$$\mathcal{H}_0=-\left(\frac{\hbar^2}{2m}\right)\frac{d^2}{dx^2}+\left(\frac{k}{2}\right)x^2, \tag{9-123}$$

[22] Additional applications of RSPT to molecular problems can be found in Chapter 12.

[23] See Eqs. (5-109), (5-141), (5-162), (5-163), and (5-166).

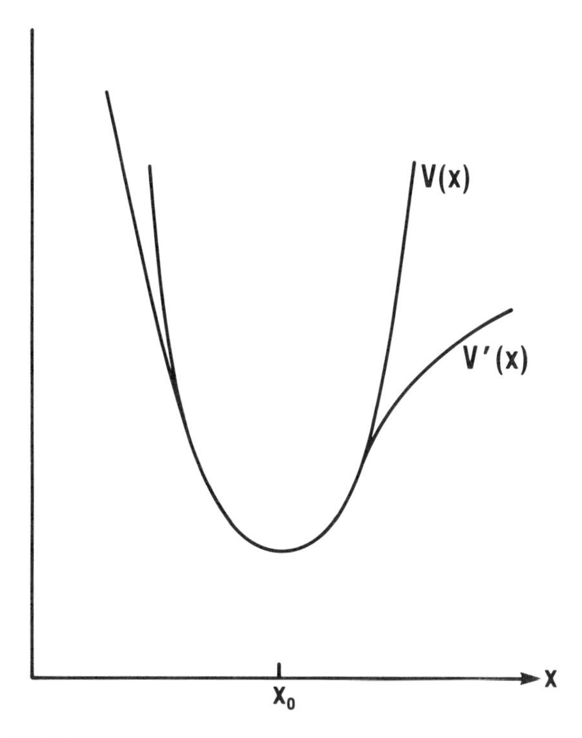

Figure 9.3. Qualitative comparison in one dimension of the harmonic oscillator potential [$V(x)$] with a typical diatomic molecule potential [$V'(x)$].

$$E_n^{(0)} = \left(n + \frac{1}{2}\right)\hbar\sqrt{\frac{k}{m}}, \qquad n = 0, 1, 2, \cdots \tag{9-124}$$

$$\Psi_n^{(0)} = N_n e^{-x^2/2} H_n(x), \tag{9-125}$$

where $H_n(x)$ are the Hermite polynomials and N_n is a normalization constant.

The first order energetic correction due to the cubic perturbation is given by

$$E_{n,3}^{(1)} = \int \Psi_n^{(0)*} \mathcal{H}' \Psi_n^{(0)}\, d\tau, \tag{9-126}$$

where the second subscript on $E^{(1)}$ has been added to distinguish the cubic term from others to be considered below. However, it is easily seen that $\mathcal{H}'$ is an odd function of x, while $\Psi_n^{(0)}$ is an even function of x. Thus, we conclude that

$$E_{n,3}^{(1)} = 0. \tag{9-127}$$

Now let us examine the first-order correction due to the quartic term, i.e., let us consider a perturbation of the form

$$\mathcal{H}'' = bx^4. \tag{9-128}$$

The first-order energetic correction for this term is given by

$$E_{n,4}^{(1)} = \int \Psi_n^{(0)*} \mathcal{H}'' \Psi_n^{(0)} \, d\tau. \tag{9-129}$$

In this case the integrand is an even function of x, and $E_{n,4}^{(1)}$ cannot be evaluated from symmetry considerations alone. However, the matrix elements that were evaluated in Section 5-9 can be of direct use here. In particular, it was seen [Eq. (5-173)] that the matrix elements of the position operator of the unperturbed harmonic oscillator were given by[24]

$$\mathbf{X} = \begin{pmatrix} 0 & \left(\frac{\hbar}{\sqrt{km}}\right)^{1/2} & 0 & 0 & \cdots \\ \left(\frac{\hbar}{\sqrt{km}}\right)^{1/2} & 0 & \left(\frac{2\hbar}{\sqrt{km}}\right)^{1/2} & 0 & \cdots \\ 0 & \left(\frac{2\hbar}{\sqrt{km}}\right)^{1/2} & 0 & \left(\frac{3\hbar}{\sqrt{km}}\right)^{1/2} & 0 \quad \cdots \\ \vdots & 0 & \left(\frac{3\hbar}{\sqrt{km}}\right)^{1/2} & 0 & \ddots \\ & \vdots & & \ddots & \ddots \end{pmatrix} \tag{9-130}$$

From this result the matrix elements of $\mathbf{X}^2$ can be calculated directly, i.e.,

$$(\mathbf{X}^2)_{n,n+2} = \sum_l (\mathbf{X})_{n,l}(\mathbf{X})_{l,n+2} \tag{9-131}$$

$$= (\mathbf{X})_{n,n+1}(\mathbf{X})_{n+1,n+2} \tag{9-132}$$

or

$$(\mathbf{X}^2)_{n,n+2} = \frac{\sqrt{(n+1)(n+2)}\,\hbar}{km}. \tag{9-133}$$

In a similar manner it can be seen that

$$(\mathbf{X}^2)_{n,n-2} = (\mathbf{X})_{n,n-1}(\mathbf{X})_{n-1,n-2}$$

$$= \frac{\sqrt{n(n-1)}\,\hbar}{km} \tag{9-134}$$

---

[24] It should be remembered that, in Eq. (9-130), the labeling of the rows as follows: Row 1 corresponds to $n = 0$, Row 2 corresponds to $n = 1$, etc.

and

$$(\mathbf{X}^2)_{n,n} = (\mathbf{X})_{n,n+1}(\mathbf{X})_{n+1,n} + (\mathbf{X})_{n,n-1}(\mathbf{X})_{n-1,n}$$

or

$$(\mathbf{X}^2)_{n,n} = \frac{(2n+1)\hbar}{km}. \tag{9-135}$$

All other matrix elements of $\mathbf{X}^2$ can be seen to be zero.

To obtain the desired result for the perturbation represented in Eq. (9-128), we form the diagonal elements of $(\mathbf{X}^4)$, i.e.,

$$\begin{aligned}(\mathbf{X}^4)_{n,n} &= \sum_l (\mathbf{X}^2)_{nl}(\mathbf{X}^2)_{ln} \\ &= (\mathbf{X}^2)_{n,n-2}(\mathbf{X}^2)_{n-2,n} + (\mathbf{X}^2)_{n,n}(\mathbf{X}^2)_{n,n} + (\mathbf{X}^2)_{n,n+2}(\mathbf{X}^2)_{n+2,n} \\ &= \left(\frac{\hbar}{km}\right)^2 [n(n-1) + (2n+1)^2 + (n+1)(n+2)] \end{aligned} \tag{9-136}$$

or

$$(\mathbf{X}^4)_{n,n} = 6\left(\frac{\hbar}{km}\right)^2 \left[n^2 + n + \frac{1}{2}\right], \tag{9-137}$$

and the first-order correction to the nth energy level for the quartic perturbation is seen to be nonzero, and given by

$$E_{n,4}^{(1)} = 6b\left(\frac{\hbar}{km}\right)^2 \left(n^2 + n + \frac{1}{2}\right), \qquad n = 0, 1, 2, \cdots. \tag{9-138}$$

Thus, while a portion of this correction simply moves all levels up or down (depending on the sign of b), the portion of Eq. (9-138) that depends on n^2 adds a new kind of potential term and nonzero energy dependence[25] on n that will fix substantially the incorrect shape of the original harmonic oscillator potential and its energy levels.

C. Degenerate States

Thusfar our discussion has assumed that all of the states of the unperturbed system are nondegenerate. However, the analysis can be modified appropriately to include the case when degeneracies are present, as shown below.

Suppose that the kth energy level is q-fold degenerate.[26] This means that

$$\mathcal{H}_0 \psi_{jk}^{(0)} = E_k^{(0)} \psi_{jk}^{(0)} \qquad j = 1, 2, \cdots, q, \tag{9-139}$$

---

[25] It should be noted that the second-order correction to the cubic perturbation also adds an energy dependence on n^2, which has not been considered here.

[26] While the analysis given here assumes for convenience that only one level is degenerate, the case of more than one level having a degeneracy can be treated using the same techniques as described here.

i.e., there are q different eigenfunctions corresponding to the energy level $E_k^{(0)}$. From our earlier discussions,[27] we know that the set of unperturbed functions form a basis for description of the system, and any linear combination of the $\psi_{jk}^{(0)}$ is also an eigenfunction of $\mathcal{H}_0$ with eigenvalue $E_k^{(0)}$, i.e.,

$$\Psi_k^{(0)} = \sum_{j=1}^{q} a_{jk}\psi_{jk}^{(0)}. \tag{9-140}$$

In the presence of a perturbation, the time-independent Schrödinger equation to be solved is

$$(\mathcal{H}_0 + \lambda\mathcal{H}')\Psi_k = E_k\Psi_k. \tag{9-141}$$

As before, we expand Ψ_k and E_k as a power series in λ, i.e.,

$$\Psi_k = \psi_k^{(0)} + \lambda\psi_k^{(1)} + \lambda^2\psi_k^{(2)} + \cdots \tag{9-142}$$

and

$$E_k = E_k^{(0)} + \lambda E_k^{(1)} + \lambda^2 E_k^{(2)} + \cdots. \tag{9-143}$$

Of course, the difference between the current case and the nondegenerate case is the more complicated form of $\Psi_k^{(0)}$ that, when inserted in Eq. (9-142), gives

$$\Psi_k = \left(\sum_{j=1}^{q} a_{jk}\psi_{jk}^{(0)}\right) + \lambda\psi_k^{(1)} + \lambda^2\psi_k^{(2)} + \cdots. \tag{9-144}$$

Substituting Eqs. (9-144) and (9-143) into Eq. (9-141) and examination of terms in various orders of λ can now be carried out as in the case of nondegenerate energy levels. In a manner similar to the nondegenerate case, we shall assume that each of the functions on the right-hand side of Eq. (9-144) is orthonormal.

For the terms independent of λ, we have

$$\mathcal{H}_0 = \left(\sum_{j=1}^{q} a_{jk}\psi_{jk}^{(0)}\right) = E_k^{(0)}\left(\sum_{j=1}^{q} a_{jk}\psi_{jk}^{(0)}\right). \tag{9-145}$$

As before, this is simply a restatement of the unperturbed problem that we have assumed we know how to solve, and so no new information is obtained.

For the terms linear in λ, we obtain:

$$\mathcal{H}_0\psi_k^{(1)} + \mathcal{H}'\left(\sum_{j=1}^{q} a_{jk}\psi_{jk}^{(0)}\right) = E_k^{(0)}\psi_k^{(1)} + \left(\sum_{j=1}^{q} a_{jk}\psi_{jk}^{(0)}\right)E_k^{(1)}. \tag{9-146}$$

We now see how the presence of the degeneracy has complicated the problem. In particular, the presence of the perturbation may affect the zero-order terms as well as higher order terms, and we need to determine the coefficients (a_{jk}) in Eq.

[27] See Chapter 2, Sections 2-4 to 2-8.

(9-146) in order to determine the first-order energy correction. Toward this end, we multiply Eq. (9-146) by $\psi_{pk}^{(0)*}$ (where $\psi_{pk}^{(0)}$ is any of the q-fold degenerate unperturbed eigenfunctions) and integrate, which gives

$$\int \psi_{pk}^{(0)*}\mathcal{H}_0\psi_k^{(1)}\,d\tau+\sum_{j=1}^{q} a_{jk}\int \psi_{pk}^{(0)*}\mathcal{H}'\psi_{jk}^{(0)}\,d\tau$$

$$=E_k^{(0)}\int \psi_{pk}^{(0)*}\psi_k^{(1)}\,d\tau+E_k^{(1)}\sum_{j=1}^{q} a_{jk}\int \psi_{pk}^{(0)*}\psi_{jk}^{(0)}\,d\tau. \tag{9-147}$$

Using the orthonormality of the functions and the Hermitian property of $\mathcal{H}_0$ gives

$$\sum_{j=1}^{q} a_{jk}H'_{pj}-E_k^{(1)}a_{pk}=0, \qquad p=1,\,2,\,\cdots,\,q, \tag{9-148}$$

where

$$H'_{pj}=\int \psi_{pk}^{(0)*}\mathcal{H}'\psi_{jk}^{(0)}\,d\tau. \tag{9-149}$$

Equation (9-148) can be rewritten as

$$\sum_{j=1}^{q}\left[H'_{pj}-E_k^{(1)}\delta_{j,p}\right]a_{jk}=0, \qquad p=1,\,2,\,\cdots,\,q. \tag{9-150}$$

Writing out these equations in detail gives

$$\begin{array}{lllll}
(H'_{11}-E_k^{(1)})a_{1k}+H'_{12}a_{2k} & & +\ \cdots & & +H'_{1q}a_{qk}=0\\
H'_{21}a_{1k}+(H'_{22}-E_k^{(1)})a_{2k} & & +\ \cdots & & +H'_{2q}a_{qk}=0\\
\vdots & & \ddots & & \vdots\\
H'_{q1}a_{1k}+ & & \cdots & & +(H'_{qq}-E_k^{(1)})a_{qk}=0
\end{array} \tag{9-151}$$

This set of equations is familiar from earlier discussions, i.e., it is a set of q homogeneous, linear equations in $q+1$ unknowns. As we saw earlier,[28] this set of equations has a nontrivial solution only if the determinant of the coefficients is identically zero, i.e.,

$$\begin{bmatrix}
(H'_{11}-E_k^{(1)}) & H'_{12} & \cdots & H'_{1q}\\
H'_{21} & (H'_{22}-E_k^{(1)}) & \vdots & H'_{2q}\\
\vdots & & \ddots & \vdots\\
H'_{q1} & & \cdots & (H'_{qq}-E_k^{(1)})
\end{bmatrix}=0. \tag{9-152}$$

[28] See Chapter 3, Section 3-2.

Thus, the solutions of Eq. (9-152) will give us the q first-order energetic corrections to the energy due to the perturbation. If all of these are different, the original q-fold degeneracy will be completely removed by the introduction of the perturbation. This is not always the case, however, and the introduction of a perturbation frequently removes only a portion of the original degeneracy. Once the values of $E_k^{(1)}$ have been found from a solution of Eq. (9-152), the values of the a_{jk} can be obtained by substitution of the $E_k^{(1)}$ into Eq. (9-151) and solving[29] for the a_{jk}.

D. Example: The Stark Effect

To illustrate the principles first discussed, let us consider an example of substantial interest in spectroscopy. In particular, let us consider the effect on the energy levels of the hydrogen atom that is caused by the introduction of a constant electric field as an external perturbation. This experiment and its result is called the *Stark Effect*.

If a constant electric field of magnitude $\mathcal{E}$ is applied along an axis (e.g., the z axis), the perturbation term is given (in atomic units) by

$$\mathcal{H}' = -\mathcal{E}z = -\mathcal{E}r \cos \vartheta. \tag{9-153}$$

For the hydrogen atom, we know that the unperturbed eigenfunctions and eigenvalues are given by[30]

$$\Psi_{nlm}(r, \vartheta, \varphi) = R_{nl}(r)\, Y_{lm}(\vartheta, \varphi) \tag{9-154}$$

$$E_n = \frac{-1}{2n^2}, \qquad n = 1, 2, \cdots. \tag{9-155}$$

As the above equation indicates, except for the ground state, the energy has a degeneracy of n^2.

To see how this degeneracy is affected by the introduction of the external electric field, let us examine the first excited state of the hydrogen atom ($Z = 1$). The four eigenfunctions, each of which corresponds to the unperturbed energy of

$$E_1^{(0)} = -\left(\frac{1}{8}\right) \text{ Hartrees} \tag{9-156}$$

are given by

$$\psi_{11}^{(0)} = \left(\frac{1}{8\pi}\right)^{1/2} \left(1 - \frac{r}{2}\right) e^{-r/2}$$

$$\psi_{21}^{(0)} = \frac{1}{2}\left(\frac{1}{8\pi}\right)^{1/2} re^{-r/2} \cos \vartheta \tag{9-157}$$

---

[29] This process will leave one parameter undetermined, which can be found by requiring that the wavefunction be normalized.

[30] See Section 7-2.

$$\psi_{31}^{(0)}=\frac{1}{8}\left(\frac{1}{\pi}\right)^{1/2} re^{-r/2} \sin \vartheta e^{i\varphi}$$

$$\psi_{41}^{(0)}=\frac{1}{8}\left(\frac{1}{\pi}\right)^{1/2} re^{-r/2} \sin \vartheta e^{-i\varphi}.$$

As noted earlier, the group of functions $[\psi_{j1}^{(0)}]$ span the space of E_1, and the zero-order function in the perturbed system can be written as a linear combination of the $n = 2$ eigenfunctions, i.e.,

$$\Psi_1^{(0)}=\sum_{j=1}^{4} a_{j1}\psi_{j1}^{(0)}. \tag{9-158}$$

The first-order energy correction due to the external electric field can be obtained by solving Eq. (9-152) for the current example, i.e.,

$$\begin{vmatrix} (H'_{11}-E_2^{(1)}) & H'_{12} & H'_{13} & H'_{14} \\ H'_{21} & (H'_{22}-E_2^{(1)}) & H'_{23} & H'_{24} \\ H'_{31} & H'_{32} & (H'_{33}-E_2^{(1)}) & H'_{34} \\ H'_{41} & H'_{42} & H'_{43} & (H'_{44}-E_2^{(1)}) \end{vmatrix}=0. \tag{9-159}$$

Since the operator representing the perturbation is simply a multiplicative operator, it is easily seen that

$$H'_{ij}=H'_{ji}. \tag{9-160}$$

Considering the remaining elements one at a time, we have

$$\begin{aligned} H'_{11} &= -\mathcal{E}\int \psi_{11}^{(0)*}(r\cos\vartheta)\psi_{11}^{(0)}\,dV \\ &= -\mathcal{E}N^2\int_0^\infty r^3e^{-r}\left(1-\frac{r}{2}\right)^2 dr\int_0^\pi \sin\vartheta\, d\vartheta\int_0^{2\pi} d\varphi\cos\vartheta\cdot Y_{00}^2 \\ &= 0, \end{aligned} \tag{9-161}$$

due to the integration over ϑ.

Similar reasoning can be used to show that

$$H'_{22}=H'_{33}=H'_{44}=H'_{13}=H'_{14}=H'_{23}=H'_{24}=H'_{34}=0. \tag{9-162}$$

The remaining element can be evaluated as follows:

$$\begin{aligned} H'_{12} &= \left(\frac{-\mathcal{E}}{16\pi}\right)\int_0^\infty r^4\left(1-\frac{r}{2}\right)e^{-r}\,dr\int_0^\pi \sin\vartheta\cos^2\vartheta\,d\vartheta\int_0^{2\pi}d\varphi \\ &= \left(\frac{-\mathcal{E}}{12}\right)\left\{\int_0^\infty r^4e^{-r}\,dr-\frac{1}{2}\int_0^\infty r^5e^{-r}\,dr\right\}, \end{aligned}$$

or

$$H'_{12}=+3\mathcal{E}. \tag{9-163}$$

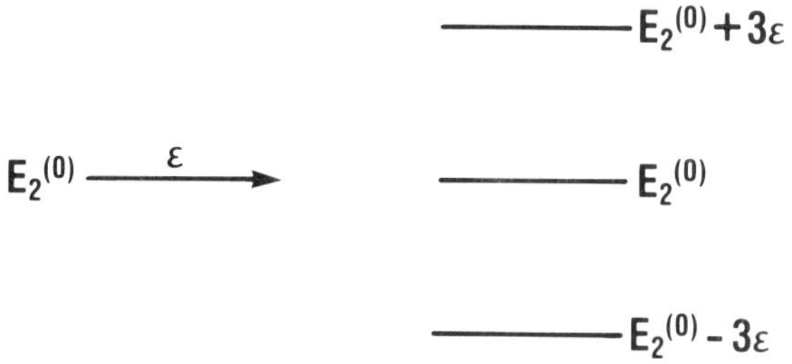

$E_2^{(0)}+3\varepsilon$

$E_2^{(0)} \xrightarrow{\varepsilon}$ $E_2^{(0)}$

$E_2^{(0)}-3\varepsilon$

Figure 9.4. Schematic description of *H*-atom energy level splitting due to an external electric field.

Assembling these results, we obtain

$$\begin{vmatrix} -E_2^{(1)} & 3\mathcal{E} & 0 & 0 \\ 3\mathcal{E} & -E_2^{(1)} & 0 & 0 \\ 0 & 0 & -E_2^{(1)} & 0 \\ 0 & 0 & 0 & -E_2^{(1)} \end{vmatrix} = 0. \qquad (9\text{-}164)$$

Expansion of the determinant gives

$$(E_2^{(1)})^2[(E_2^{(1)})^2 - 9\mathcal{E}^2] = 0. \qquad (9\text{-}165)$$

Thus, two of the energy corrections are zero, and the other two are given by:

$$E_2^{(1)} = +3\mathcal{E}. \qquad (9\text{-}166)$$

The result is that a portion of the original degeneracy has been removed by the introduction of the external electric field. As depicted schematically in Fig. 9.4, the 4-fold degeneracy has been split into three levels, two of which are nondegenerate and one of which is 2-fold degenerate. Furthermore, the splitting between the perturbed energy levels is dependent linearly on the external electric field. Thus, the size of the splitting can be adjusted by the magnitude of the external field in a manner that optimizes experimental conditions for observation of transitions between levels.

9-9. Brillouin–Wigner Perturbation Theory

Thus far our discussion of approximation methods has utilized either a variational approach or a perturbation theory approach. However, it is frequently the case that the exact solutions to a suitable unperturbed problem are not known, suggesting that a variational approach to even the unperturbed problem is required. Even in that case where the unperturbed wavefunction is approximate, it is possible to use perturbation theory to obtain estimates of the effect of the perturbation, as we shall see below.

The approach to be described is usually known as *Brillouin–Wigner perturbation theory* (BWPT). However, it is actually a combination of variational and perturbation theory methods, and is also known as *variation–perturbation theory.*[31]

We begin by assuming that we have a set of N orthonormal functions ($\psi_n^{(0)}$) that have been used to construct approximate wavefunctions (Ψ_i) of the Hamiltonian ($\mathcal{H}$), having corresponding approximate energies E_i. Thus, we have

$$\Psi_i = \sum_{n=1}^{N} c_{ni}\psi_n^{(0)}, \tag{9-167}$$

$$E_i = \frac{\langle \Psi_i | \mathcal{H} | \Psi_i \rangle}{\langle \Psi_i | \Psi_i \rangle} = \frac{\sum_{m=1}^{N}\sum_{n=1}^{N} c_{mi}^* c_{ni} \langle \psi_m^{(0)} | \mathcal{H} | \psi_n^{(0)} \rangle}{\sum_{m=1}^{N}\sum_{n=1}^{N} c_{mi}^* c_{ni} \langle \psi_m^{(0)} | \psi_n^{(0)} \rangle}, \tag{9-168}$$

where the approximate energy values (E_i) and wavefunction coefficients (c_{mi}) have been obtained via solution of the secular equation

$$\sum_{n=1}^{N} (H_{mn} - E\delta_{mn})c_n = 0, \qquad m = 1, 2, \cdots, N, \tag{9-169}$$

or

$$\begin{pmatrix} H_{11} - E & H_{12} & \cdots & H_{1N} \\ H_{21} & H_{22} - E & \cdots & H_{2N} \\ \vdots & \vdots & \ddots & \vdots \\ H_{N1} & H_{N2} & \cdots & H_{NN} - E \end{pmatrix} \begin{pmatrix} c_1 \\ c_2 \\ \vdots \\ c_n \end{pmatrix} = \begin{pmatrix} 0 \\ 0 \\ \vdots \\ 0 \end{pmatrix} \tag{9-170}$$

where

$$H_{mn} = \langle \psi_m^{(0)} | \mathcal{H} | \psi_n^{(0)} \rangle. \tag{9-171}$$

Thus, the N eigenvalues (E_i) and associated eigenfunctions (Ψ_i) resulting from the solution of Eq. (9-170) are variationally determined and, while not exact solutions, the E_i represent upper bounds to the lowest N states of the system.

There are two cases where additional analysis is desirable, both of which we shall discuss below. The first case occurs when the secular determinant arising from Eq. (9-170) is very large and there is interest in only one or a few states (e.g., the ground state). In that case it is sometimes advantageous to seek

[31] L. Brillouin, *J. Phys. Radium*, **3**, 373 (1935); E. Wigner, *Math. Naturw. Anz. ungar. Akad. Wiss.*, **53**, 477 (1935). The presentation used here is based on the analysis using partitioning techniques of P. O. Löwdin, *J. Chem Phys.*, **19**, 1396 (1951).

alternative ways of estimating the energy levels of interest than by solving the entire secular determinant.

Considering the ground state (whose energy we shall call E_1 for convenience) as an example of this case, we assume that $\psi_1^{(0)}$ is a good approximation, so that c_{11} dominates the expansion, i.e.,

$$\Psi_1 = \sum_{n=1}^{N} c_{n1}\psi_n^{(0)} \tag{9-172}$$

$$= c_{11}\psi_1^{(0)} + \sum_{n=2}^{N} \text{(small corrections)}. \tag{9-173}$$

In order to obtain estimates of the energetic corrections to

$$E_1^{(0)} = \langle \psi_1^{(0)} | \mathcal{H} | \psi_1^{(0)} \rangle \tag{9-174}$$

due to the correction terms in Eq. (9-161) without solving the secular determinant, we partition the matrices in Eq. (9-170) as follows:

$$\left(\begin{array}{c|cccc} H_{11} - E & H_{12} & \cdots & H_{1N} \\ \hline H_{21} & H_{22} - E & \cdots & H_{2N} \\ \vdots & \vdots & \ddots & \vdots \\ H_{N1} & H_{N2} & \cdots & H_{NN} - E_1 \end{array}\right)\left(\begin{array}{c} c_1 \\ \hline c_2 \\ \vdots \\ c_n \end{array}\right) = \left(\begin{array}{c} 0 \\ \hline 0 \\ \vdots \\ 0 \end{array}\right). \tag{9-175}$$

The individual equations corresponding to the above can be written as:

$$(E_1 - H_{11})c_1 = H_{12}c_2 + H_{13}c_3 + \cdots H_{1N}c_{1N} \tag{9-176}$$

$$(E_1 - H_{22})c_2 = H_{21}c_1 + H_{23}c_3 + \cdots H_{2N}c_{2N}$$

$$\vdots \qquad\qquad \vdots$$

$$(E_1 - H_{NN})c_N = H_{N1}c_1 + H_{N2}c_2 + \cdots H_{N-1}c_{N-1}. \tag{9-177}$$

If expressions for $c_2, c_3, \cdots, c_N$ are obtained from Eq. (9-177) and substituted into Eq. (9-176), we obtain

$$\begin{aligned}(E_1 - H_{11})c_1 = {} & H_{12}\left\{\frac{1}{(E_1 - H_{22})}\sum_{n\neq 2}^{N} H_{2n}c_n\right\} \\ & + H_{13}\left\{\frac{1}{(E_1 - H_{33})}\sum_{n\neq 3}^{N} H_{3n}c_n\right\} \\ & + \cdots \\ & + H_{1N}\left\{\frac{1}{(E_1 - H_{NN})}\sum_{n\neq N}^{N} H_{Nn}c_n\right\}.\end{aligned} \tag{9-178}$$

Rearrangement of terms in the above equation gives

$$
\begin{aligned}
(E_1 - H_{11})c_1 = \sum_{n\neq 1}^{N} \left[\frac{H_{1n}H_{n1}}{(E_1 - H_{nn})} \right] c_1 + \frac{H_{12}}{(E_1 - H_{22})} \sum_{\substack{n\neq 1 \\ n\neq 2}}^{N} H_{2n}c_n \\
+ \frac{H_{13}}{(E_1 - H_{33})} \sum_{\substack{n\neq 1 \\ n\neq 3}}^{N} H_{3n}c_n + \cdots + \frac{H_{1N}}{(E_1 - H_{NN})} \sum_{\substack{n\neq 1 \\ n\neq N}}^{N} H_{Nn}c_n.
\end{aligned}
\tag{9-179}
$$

If expressions for c_2, c_3, $\cdots$, c_N from Eq. (9-177) are substituted again, this time into Eq. (9-179), we obtain[32]

$$
\begin{aligned}
(E_1 - H_{11})c_1 = \sum_{n\neq 1}^{N} \left[\frac{H_{1n}H_{n1}}{(E_1 - H_{nn})} \right] c_1 \\
+ \sum_{m\neq 1}^{N} \sum_{\substack{n\neq 1 \\ n\neq m}}^{N} \left[\frac{H_{1m}H_{mn}H_{n1}}{(E_1 - H_{mm})(E_1 - H_{nn})} \right] c_1 \\
+ (\text{terms involving } c_2, \cdots, c_N).
\end{aligned}
\tag{9-180}
$$

A third substitution for c_2, $\cdots$, c_N into the above equation gives

$$
\begin{aligned}
(E_1 - H_{11})c_1 = \sum_{n\neq 1}^{N} \left[\frac{H_{1n}H_{n1}}{(E_1 - H_{nn})} \right] c_1 \\
+ \sum_{m\neq 1}^{N} \sum_{\substack{n\neq 1 \\ n\neq m}}^{N} \left[\frac{H_{1m}H_{mn}H_{n}}{(E_1 - H_{mm})(E_1 - H_{nn})} \right] c_1 \\
+ \sum_{l\neq 1}^{N} \sum_{\substack{m\neq 1 \\ m\neq l}}^{N} \sum_{\substack{n\neq 1 \\ n\neq m \\ n\neq l}}^{N} \left[\frac{H_{1l}H_{lm}H_{mn}H_{n1}}{(E_1 - H_{ll})(E_1 - H_{mm})(E_1 - H_{nn})} \right] c_1 \\
+ (\text{terms involving } c_2, \cdots, c_N).
\end{aligned}
\tag{9-181}
$$

If we truncate the above equation beyond terms in c_1, we obtain

$$
\begin{aligned}
E_1 = E_1^{(0)} + \sum_{n\neq 1}^{N} \left[\frac{H_{1n}H_{n1}}{(E_1 - H_{nn})} \right] \\
+ \sum_{m\neq 1}^{N} \sum_{\substack{n\neq 1 \\ n\neq m}}^{N} \left[\frac{H_{1m}H_{mn}H_{n1}}{(E_1 - H_{mm})(E_1 - H_{nn})} \right] \\
+ \sum_{l\neq 1}^{N} \sum_{\substack{m\neq 1 \\ m\neq l}}^{N} \sum_{\substack{n\neq 1 \\ n\neq m \\ n\neq l}}^{N} \left[\frac{H_{1l}H_{lm}H_{mn}H_{n1}}{(E_1 - H_{ll})(E_1 - H_{mm})(E_1 - H_{nn})} \right],
\end{aligned}
\tag{9-182}
$$

[32] See Problem 9-14.

where c_1 has been canceled from each term (and can be determined from the normalization condition), and the second, third, and fourth terms on the right-hand side can be thought of as second-, third- and fourth-order corrections, respectively, to the zeroth-order estimate ($E_1^{(0)}$). Of course, higher order corrections can be obtained by direct extension of the procedures used above.

Thus, we have obtained a kind of perturbation expansion for the ground state energy (E_1) in terms of an "unperturbed" contribution ($E_1^{(0)}$) plus correction terms. The obvious advantage to the expression in Eq. (9-182) is that higher order "corrections" to the energy can be found using elements (H_{mn}) but without solving the secular determinant. The equally obvious complication is the appearance of the desired energy (E_1) in each of the correction terms. In practice this complication can be solved by solving Eq. (9-182) iteratively, e.g., using $E_1^{(0)}$ as an estimate for E_1 as the right-hand side of Eq. (9-182), calculating a new value of E_1 from Eq. (9-182), and iterating until convergence is reached. The eigenfunctions corresponding to the converged value of E_1 can then be determined using Eqs. (9-176) and (9-177), assuming $c_1 = 1$, followed by determination of c_1 from the overall normalization condition.

The second case of interest corresponds to generalization of the previous discussion to the case where the "unperturbed" system is not represented by a single term, but a series of terms. In particular, we partition the Hamiltonian matrix as follows:

$$\mathbf{H} = \left(\begin{array}{c|c} \mathbf{H}_{aa} & \mathbf{H}_{ab} \\ \hline \mathbf{H}_{ba} & \mathbf{H}_{bb} \end{array}\right) \qquad \begin{array}{l} a = 1, \cdots, N \\ b = 1, \cdots, M \end{array}, \tag{9-183}$$

where $\mathbf{H}_{aa}$ is a matrix usually referred to as the "major" or "unperturbed" part, $\mathbf{H}_{bb}$ is the "minor" part, and $\mathbf{H}_{ab}$ and $\mathbf{H}_{ba}$ are "interaction" components.

The equations corresponding to the equation

$$\mathbf{HC} = \mathbf{EC} \tag{9-184}$$

are given by

$$\mathbf{H}_{aa}\mathbf{c}_a + \mathbf{H}_{ab}\mathbf{c}_b = E\mathbf{c}_a \tag{9-185}$$

$$\mathbf{H}_{ba}\mathbf{c}_a + \mathbf{H}_{bb}\mathbf{c}_b = E\mathbf{c}_b, \tag{9-186}$$

where

$$\mathbf{c} = \mathbf{c}_a + \mathbf{c}_b. \tag{9-187}$$

If Eq. (9-186) is solved for $\mathbf{c}_b$, we obtain:

$$\mathbf{c}_b = (E\mathbf{I} - \mathbf{H}_{bb})^{-1}\mathbf{H}_{ba}\mathbf{c}_a, \tag{9-188}$$

which gives, when substituted[33] into Eq. (9-185),

$$[\mathbf{H}_{aa} + \mathbf{H}_{ab}(E\mathbf{I} - \mathbf{H}_{bb})^{-1}\mathbf{H}_{ba}]\mathbf{c}_a = E\mathbf{c}_a. \tag{9-189}$$

---

[33] As implied in Eq. (9-188), the existence of the inverse $(E\mathbf{I} - \mathbf{H}_{bb})^{-1}$ is assumed in this analysis. Similar assumptions will be made for other inverses that occur.

As in the previous case, the above equation can be solved iteratively by, e.g., diagonalizing $\mathbf{H}_{aa}$ to obtain initial estimates for E, substituting into the left-hand side of Eq. (9-189), solving for new values of E and $\mathbf{c}_a$ via Eq. (9-189), and iterating until convergence.

However, it is also possible to develop a perturbation theory approach further by defining an "effective" potential $[\mathbf{V}(E)]$ as

$$\mathbf{V}(E)=\mathbf{H}_{ab}(E\mathbf{I}-\mathbf{H}_{bb})^{-1}\mathbf{H}_{ba}, \tag{9-190}$$

so that Eq. (9-189) can be written as

$$[\mathbf{H}_{aa}+\mathbf{V}(E)]\mathbf{c}_a=E\mathbf{c}_a. \tag{9-191}$$

In order to generate a perturbation expansion, we consider the term in greater detail. In particular, if we define $(M \times M)$ matrices $\mathbf{D}$ and $\mathbf{F}$ as

$$\mathbf{D}=\begin{pmatrix} E_b-H_{N+1,N+1} & 0 & \cdots & 0 \\ 0 & E_b-H_{N+2,N+2} & \cdots & 0 \\ \vdots & \vdots & \ddots & \vdots \\ 0 & 0 & & E_b-H_{N+M,N+M} \end{pmatrix} \tag{9-192}$$

and

$$\mathbf{F}=\begin{pmatrix} 0 & H_{N+1,N+2} & \cdots & H_{N+1,N+M} \\ H_{N+2,N+1} & 0 & & \\ \vdots & & \ddots & \\ H_{N+M,N+1} & & & 0 \end{pmatrix}, \tag{9-193}$$

we have

$$(E\mathbf{I}-\mathbf{H}_{bb})=(\mathbf{D}-\mathbf{F}). \tag{9-194}$$

For aid in the analysis to follow, we note the following:

$$(\mathbf{D}-\mathbf{F})^{-1}(\mathbf{D}-\mathbf{F})=\mathbf{I}$$

$$(\mathbf{D}-\mathbf{F})^{-1}(\mathbf{D}-\mathbf{F})\mathbf{D}^{-1}=\mathbf{D}^{-1}$$

or

$$(\mathbf{D}-\mathbf{F})^{-1}(\mathbf{I}-\mathbf{F}\mathbf{D}^{-1})=\mathbf{D}^{-1},$$

which can be written as

$$(\mathbf{D}-\mathbf{F})^{-1}(\mathbf{I}-\mathbf{F}\mathbf{D}^{-1}(\mathbf{I}-\mathbf{F}\mathbf{D}^{-1})^{-1}=\mathbf{D}^{-1}(\mathbf{I}-\mathbf{F}\mathbf{D}^{-1})^{-1},$$

or

$$(\mathbf{D}-\mathbf{F})^{-1}=\mathbf{D}^{-1}(\mathbf{I}-\mathbf{F}\mathbf{D}^{-1})^{-1}. \tag{9-195}$$

By using an analogy to the binomial expansion:

$$(1-x)^{-1}=1+x+x^2+x^3+\cdots, \tag{9-196}$$

we can write Eq. (9-195) as

$$(E\mathbf{I}-\mathbf{H}_{bb})^{-1}=(\mathbf{D}-\mathbf{F})^{-1}$$
$$=\mathbf{D}^{-1}[\mathbf{I}+\mathbf{F}\mathbf{D}^{-1}+\mathbf{F}\mathbf{D}^{-1}\mathbf{F}\mathbf{D}^{-1}+\mathbf{F}\mathbf{D}^{-1}\mathbf{F}\mathbf{D}^{-1}\mathbf{F}\mathbf{D}^{-1}+\cdots]. \tag{9-197}$$

This allows the expression for $\mathbf{V}(E)$ in Eq. (9-190) to be written as

$$\mathbf{V}(E)=\mathbf{H}_{ab}\left[\mathbf{D}^{-1}\sum_{k=0}^{\infty}(\mathbf{F}\mathbf{D}^{-1})^{k}\right]\mathbf{H}_{ba} \tag{9-198}$$

$$=\mathbf{H}_{ab}\mathbf{D}^{-1}\mathbf{H}_{ba}+\mathbf{H}_{ab}\mathbf{D}^{-1}\mathbf{F}\mathbf{D}^{-1}\mathbf{H}_{ba}$$
$$+\mathbf{H}_{ab}\mathbf{D}^{-1}\mathbf{F}\mathbf{D}^{-1}\mathbf{F}\mathbf{D}^{-1}\mathbf{H}_{ba}+\cdots. \tag{9-199}$$

To illustrate a connection of BWPT to the variational approach, let us consider the case of a three-term variational expansion and the special case in which the secular determinant arising from it has the following form:

$$\begin{vmatrix} H_{11}-E & H_{12} & H_{13} \\ H_{21} & H_{22}-E & 0 \\ H_{31} & 0 & H_{33}-E \end{vmatrix}=0. \tag{9-200}$$

Expansion of the above determinant gives

$$(H_{11}-E)(H_{22}-E)(H_{33}-E)-H_{12}H_{21}(H_{33}-E)-H_{13}H_{31}(H_{22}-E)=0, \tag{9-201}$$

which can be seen to be exactly the same result as truncation of the perturbation expansion in Eq. (9-182) after the second order terms [i.e., after the second terms on the right-hand side of Eq. (9-182)]. Thus, there is a direct connection between BWPT and the variational approach, at least at low orders of approximation.[34]

Problems

1. Show that the first order perturbation correction to the ground state of the hydrogen atom due to the presence of an external electric field is zero.
2. Consider a trial wavefunction of the form

$$\varphi=Ne^{-\rho r^2}$$

for the hydrogen atom, where N is a normalization constant.

[34] As an illustration of the application of BWPT-type techniques see, for example, G. A. Segal and R. W. Wetmore, *Chem. Phys. Letters*, **32**, 556–560 (1975); J. J. Diamond, G. A. Segal and R. W. Wetmore, *J. Phys. Chem.*, **88**, 3532–3538 (1984); for alternative approaches, see R. J. Bartlett, J. C. Bellum, and E. J. Brändas, *Int. J. Quantum Chem.*, **57**, 449–462 (1973).

a. Find the optimum value of ρ and associated energy by direct calculation of the energy for several ρ values.
b. Find the optimum ρ by solving $\partial E/\partial\rho = 0$ directly.

3. Derive equations for the second- and third-order corrections to the energy ($E_n^{(2)}$ and $E_n^{(3)}$) in terms of unperturbed wavefunctions.
4. Show that the third-order correction to the energy [Eq. (9-112)] can be written using Dirac notation as

$$E_n^{(3)} = \sum_{\substack{m,p\\(\neq n)}} \frac{\langle n|\mathcal{H}'|m\rangle\langle m|\mathcal{H}'|p\rangle\langle p|\mathcal{H}'|n\rangle}{(E_n^{(0)}-E_m^{(0)})(E_n^{(0)}-E_p^{(0)})} - E_n^{(1)} \sum_{\substack{m\\(\neq n)}} \frac{|\langle n|\mathcal{H}'|m\rangle|^2}{(E_n^{(0)}-E_m^{(0)})^2} .$$

5. Consider a single particle of mass M in a one-dimensional box of length a, as discussed in Chapter 5, having a potential:

$$V=\infty \qquad -\infty<x<0$$
$$V=0 \qquad 0\leq x\leq a$$
$$V=\infty \qquad a<x<\infty.$$

a. For the approximate function

$$\Psi(x)=Nx(a-x)$$

1. Determine the normalization constant, N.
2. Show that $\Psi(x)$ is an acceptable function for an approximation to the lowest state.
3. Use the Variational Theorem to evaluate the energy, and compare it with the exact energy of the lowest state,

b. For the approximate function

$$\Psi(x)=Nx(a-x)\left(\frac{a}{2}-x\right)$$

1. Determine N and show that $\Psi(x)$ also satisfies the boundary conditions of the problem.
2. Assuming the function in part "a" is used to approximate the lowest state, show that it is justified to use $\Psi(x)$ as an approximation to an excited state, and use the Variational Theorem to determine the approximate energy.

6. For a particle in a one-dimensional box considered in Problem 5, consider two basis functions

$$\varphi_1=N_1x(a-x)$$
$$\varphi_2=N_2x^2(a-x)^2$$

a. Show that the function

$$\Psi=c_1\varphi_1+c_2\varphi_2$$

satisfies the boundary conditions of the problem (c_1 and c_2 are any expansion coefficients).

b. Use the Linear Variation Method to determine approximate energies and wavefunctions for the two states. Compare the results with the approximations in Problem 5. (Hint: The basis functions are not orthonormal.)

7. Devise, if possible, another function for an approximation to another excited state of the particle in the one-dimensional box, assuming that the two functions in Problem 5 have been used to approximate the ground and one excited state. If not possible, indicate the reasons in detail.
8. The wavefunction for the ground state of a hydrogen-like atom of nuclear charge Z is (in atomic units)

$$\Psi = 2Z^{3/2}e^{-Zr},$$

with

$$\mathcal{H} = -\frac{1}{2}\nabla^2 - \frac{Z}{r},$$

and

$$E = -Z/2.$$

Apply first-order Rayleigh–Schrödinger perturbation theory to a hydrogen-like atom with nuclear charge $(Z + 1)$ to determine an approximate energy, assuming the system with nuclear charge Z is a zeroth-order approximation. Under what conditions is this result a good approximation? How does the approximate energy expression obtained compare with the exact equation?

9. Verify Eq. (9-76) for the case $N = 2$ and $N = 3$.
10. Derive explicit expressions for $f(\epsilon)$ from Eq. (9-75) for the cases $N = 2$ [when $f(\epsilon)$ is a parabola] and $N = 3$, and sketch the results as in Fig. 9.1.
11. Apply the variational theorem to the ground state of helium atom (see Section 9-4) using the trial wavefunction

$$\Psi(1, 2) = \chi(1)\chi(2),$$

with

$$\chi(1) = Ne^{-\alpha r_1^2}$$
$$\chi(2) = Ne^{-\alpha r_2^2}$$

and α is a variable parameter. Minimize the energy with respect to α. [Hint: Use Eqs. (9-85) and (9-87) to evaluate integrals.] Also, use the formula

$$\int\int e^{-ar_1^2}e^{-ar_2^2}\left(\frac{1}{r_{12}}\right)e^{-ar_1^2}e^{-ar_2^2}\,d\tau_1\,d\tau_2$$

$$= 2\pi^{-1/2}\langle e^{-ar_1^2}|e^{-ar_1^2}\rangle\langle e^{-ar_2^2}|e^{-ar_2^2}\rangle\left(\frac{1}{a}\right)^{-1/2}$$

12. Let $\mathbf{H}$ be a 2×2 matrix of the form

$$\mathbf{H} = \begin{pmatrix} 1+\epsilon & \epsilon \\ \epsilon & 1+\epsilon \end{pmatrix},$$

where ϵ is a small real number. Use the principles of first-order Rayleigh–Schrödinger perturbation theory (RSPT) to find approximate eigenvalues and eigenvectors of $\mathbf{H}$.

Hints: let $\mathbf{H} = \mathbf{H}_0 + \mathbf{H}'$ where

$$\mathbf{H}_0 = \begin{pmatrix} 1 & \epsilon \\ \epsilon & 1 \end{pmatrix}$$

and

$$\mathbf{H}' = \begin{pmatrix} \epsilon & 0 \\ 0 & \epsilon \end{pmatrix}.$$

Determine the eigenvalues and eigenvectors of $\mathbf{H}_0$. Then, develop first-order RSPT in terms of zero- and first-order eigenvalues and eigenvectors (assume zero-order and first-order eigenvectors are orthogonal). Use these results to determine first-order corrections to the eigenvalue and eigenvector of $\mathbf{H}_0$.

13. For the perturbation expansion of the wavefunction in Eq. (9-99), i.e.,

$$\Psi_n = \Psi_n^{(0)} + \lambda \Psi_n^{(1)} + \lambda^2 \Psi_n^{(2)} + \cdots$$

with

$$\langle \Psi_n^{(0)} | \Psi_n^{(0)} \rangle = 1$$

show that, when Ψ_n is taken to be "intermediately normalized," i.e.,

$$\langle \Psi_n | \Psi_n^{(0)} \rangle = 1,$$

all kth order corrections to $\Psi_n^{(0)}$ are orthogonal to $\Psi_n^{(0)}$, i.e.,

$$\langle \Psi_n^{(k)} | \Psi_n^{(0)} \rangle = 0, \qquad k = 1, 2, \cdots.$$

Use this fact instead of the condition in Eq. (9-101) to develop energy expressions for RSPT through third order, and show that they are identical to Eqs. (9-107) through (9-109).

14. Verify Eqs. (9-180) and (9-181).
15. Consider a 4×4 Hamiltonian matrix of the form

$$\mathbf{H} = \begin{pmatrix} H & 1 & 0 & 0 \\ 1 & H' & 1 & 0 \\ 0 & 1 & H' & 1 \\ 0 & 0 & 1 & H' \end{pmatrix}, \qquad \text{where } |H| > |H'|$$

a. Using a partition of $\mathbf{H}$ into

$$\mathbf{H}_{aa} = H \qquad \mathbf{H}_{ab} = (1 \quad 0 \quad 0)$$

$$\mathbf{H}_{ba} = \begin{pmatrix} 1 \\ 0 \\ 0 \end{pmatrix} \qquad \mathbf{H}_{bb} = \begin{pmatrix} H' & 1 & 0 \\ 1 & H' & 1 \\ 0 & 1 & H' \end{pmatrix},$$

use partitioning perturbation theory to develop an expression for the energy (E) through fourth order in Brillouin–Wigner correction terms.

b. With H and H' equal to -100 and -99, respectively, obtain approximate energies to two decimal places by iteratively solving the expression obtained in part (a), to both second and fourth order (a simple computer program will be useful not but necessary). Note that the exact lowest root of H, to six decimal places, is -100.879385.

c. Determine an approximate eigenvector corresponding to the fourth order energy obtained in (b). Note: the exact, normalized, eigenvector is

$$\begin{pmatrix} 0.656539 \\ -0.577350 \\ 0.428525 \\ -0.228013 \end{pmatrix}.$$

Chapter 10

General Considerations for Many Electron Systems

Thus far our considerations have in general been limited to systems containing only a single electron. While we have seen that many important principles and techniques can be developed using those cases, we shall now see that a major new concept is needed for systems containing more than one electron. That concept (the Pauli Exclusion Principle) will be developed in the sections to follow, along with a number of analyses and techniques that are of substantial importance in contemporary applications of quantum mechanics to chemistry. Before doing that, however, it is useful to introduce several conceptual approaches to many electron systems that were developed early, as well as the basic concepts of group theory. These will help us to understand and motivate the discussions to follow, as well as to provide useful tools for incorporating and understanding symmetry properties of molecules and wavefunctions.

10-1. Early Computational Concepts and Procedures

When moving from the hydrogen atom to multielectron molecules, not only it is necessary to utilize approximation methods to solve the Schrödinger equation, it is necessary to reexamine the conceptual model itself. In particular, while atomic orbitals are the natural choice for describing an atomic system, it is not immediately obvious that those same functions will be appropriate for describing multielectron molecular systems where the electron distribution may be located in bonding regions between nuclei as well as at nuclear locations. Also, the interactions among electrons must be considered. This has led to several conceptual approaches to the description of molecular systems, and we shall introduce two of them here that have both historical significance and current use.

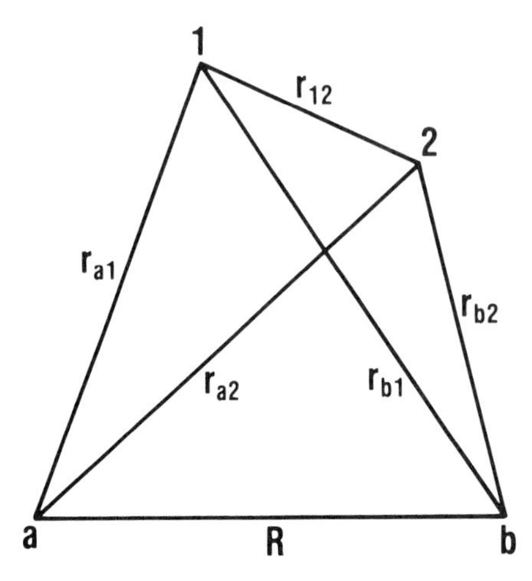

Figure 10.1. Coordinates for H_2 molecule.

A. Valence Bond Theory

To provide a frame of reference for the discussions, we begin by considering perhaps the most studied molecule in all of theoretical chemistry, H_2. Within the Born–Oppenheimer approximation, and assuming no external fields, the quantum mechanical Hamiltonian for H_2 is given (in atomic units) by

$$\mathcal{H} = -\frac{1}{2}\nabla_1^2 - \frac{1}{2}\nabla_2^2 - \sum_{i=1}^{2}\left(\frac{1}{r_{ai}} + \frac{1}{r_{bi}}\right) + \frac{1}{r_{12}} + \frac{1}{R}, \tag{10-1}$$

where the various distances are identified in Fig. 10.1, and R is assumed to be constant. It is the electron repulsion term $(1/r_{12})$ in the Hamiltonian that is the source of the computational difficulties in multielectron systems. The reason for these difficulties is that, contrary to the one-electron case, the electron variables cannot be separated. In particular,

$$r_{12} = [r_1^2 + r_2^2 - 2(\mathbf{r}_1 \cdot \mathbf{r}_2)]^{1/2}, \tag{10-2}$$

and the square root results in the nonseparability. In addition, the electron repulsion term suggests that the wavefunction for the system should be sensitive to the *relative position of the electrons* as well as the position of the electrons relative to the nuclei. Since more than one nucleus will also be present in molecules, questions of how to account for relative motion of electrons that may be delocalized over multiple nuclei as well as their interaction with each other were thus major new issues that needed examination when molecules were considered.

Two separate approaches were designed to deal with these problems, which appeared to be quite different conceptually. We shall introduce each of them below, and will see that they have many more similarities than are expected at first glance.

The *valence bond method* that was introduced by Heitler and London[1] assumes that the description of a molecule ought to start by considering the separated atoms. For example, in the case of H_2 the separated atoms are two hydrogen atoms, each containing one electron. If *no* interaction occurred between the two atoms, then the *exact* solution for H_2 would simply be the *product* of the solutions for each individual atom. To see why this is the case, let us assume that the two atoms are infinitely separated and noninteracting, in which case the Hamiltonian operator is given by

$$\mathcal{H} = -\frac{1}{2}\nabla_1^2 - \frac{1}{r_{a1}} - \frac{1}{2}\nabla_2^2 - \frac{1}{r_{b2}}. \tag{10-3}$$

If the exact solution to this Hamiltonian (for the ground state) is a product of the normalized solutions for individual atoms, we have

$$\Psi(1, 2) = 1s_a(1) \cdot 1s_b(2), \tag{10-4}$$

where we have labeled the two hydrogen atoms by a and b. Using Eqs. (10-3) and (10-4), we have

$$\begin{aligned}
E &= \int\int \Psi^*(1, 2)\mathcal{H}\Psi(1, 2)\, d\tau_1\, d\tau_2 \qquad (10\text{-}5)\\
&= \int\int 1s_a^*(1)1s_b^*(2)\left[-\frac{1}{2}\nabla_1^2 - \frac{1}{r_{a1}} - \frac{1}{2}\nabla_2^2 - \frac{1}{r_{b2}}\right] 1s_a(1)1s_b(2)\, d\tau_1\, d\tau_2\\
&= \int 1s_a^*(1)\left[-\frac{1}{2}\nabla_1^2 - \frac{1}{r_{a1}}\right] 1s_a(1)\, d\tau_1 \int 1s_b^*(2)1s_b(2)\, d\tau_2\\
&\quad + \int 1s_a^*(1)1s_a(1)\, d\tau_1 \int 1s_b^*(2)\left[-\frac{1}{2}\nabla_2^2 - \frac{1}{r_{b2}}\right] 1s_b(2)\, d\tau_2\\
&= E_a + E_b\\
&= -0.5 \quad + (-0.5)\\
&= -1.0 \quad \text{Hartree}, \qquad (10\text{-}6)
\end{aligned}$$

where E_a and E_b refer to the energies of isolated hydrogen atoms a and b. Thus, when the atoms are noninteracting and infinitely separated, the total energy of the system is indeed simply the sum of the energies of the individual atoms.

Heitler and London's valence bond method used this idea as the basis for forming an approximate wavefunction for the H_2 molecule, even though the atoms are no longer infinitely apart and do interact. In this case, the Hamiltonian is given by Eq. (10-1), and an approximate wavefunction is taken as[2] a product

[1] W. Heitler and F. London, *Z. Phys.*, **44**, 455 (1927).

[2] This formulation appears to have ignored spin. In Section 10-3 it will be seen that, for two-electron problems such as H_2, the spin dependence can be factored out and solved for explicitly, resulting in a wavefunction that is a *separable product* of space and spin coordinates. Thus, treating only the spatial portion here does not restrict the generality of the arguments.

of "separated atom" wavefunctions, i.e.,

$$\Psi(1, 2) = N[1s_a(1)1s_b(2) + 1s_b(1)1s_a(2)], \tag{10-7}$$

where N is a normalization constant. The second term on the right hand side of Eq. (10-7) is added to account for the fact that the electrons are indistinguishable,[3] e.g., electron 1 can reside in the vicinity of nucleus b while electron 2 resides near nucleus a, and the wavefunction must reflect this indistinguishability.

When this approximation to the exact wavefunction was used with $R = 1.67$ Bohrs, a total energy of -1.116 Hartree was obtained. To place this result in context, it is useful to remember as noted above that the exact energy of each hydrogen atom is -0.5 Hartree, and so the predicted energy of H_2 is -0.116 Hartree more stable than that of two separated hydrogen atoms. This increased stability over that of the separated atoms is called the *binding energy,* and the above result implies that H_2 has a binding energy of 66.5 kcal/mol. For comparison, the experimentally observed binding energy of H_2 is 109.5 kcal/mol (from a total energy of -1.17447498301730 Hartrees at $R = 1.401080$ Bohrs[4]), indicating that the remarkably simple valence bond wavefunction describes 61% of the binding energy of H_2.

Given the simplicity of the initial wavefunction approximation, it is not surprising that improvements were made easily. For example, Wang[5] noted that the orbital exponents of the $1s$ orbitals would not be expected to have the same value as for the isolated atoms, since each electron now is affected by both nuclei. Hence, by optimizing the value of the orbital exponents, he found that, instead of $\zeta = 1.0$, an optimum value of $\zeta = 1.6875$ was found. This gave an energy of

$$E = -1.139 \text{ Hartrees},$$

and a minimum distance of $R = 1.404$ Bohrs. This energy corresponds to a binding energy of 79.7 kcal/mol. Hence, both the calculated energy and the predicted internuclear distance were improved substantially by a rather simple (conceptually) improvement in the trial wavefunction.

Another rather simple conceptual improvement was also made a few years later. In particular, it was noted by Weinbaum[6] that the Wang wavefunction for H_2 did not allow for the possibility of both electrons being in the vicinity of the same nucleus. These so-called *ionic terms* were easily added to the *covalent terms* of the Wang wavefunction [Eq. (10-7)] as follows:

$$\Psi(1, 2) = N\{1s_a(1)1s_b(2) + 1s_b(1)1s_a(2) + \lambda[1s_a(1)1s_a(2) + 1s_b(2)1s_b(2)]\}, \tag{10-8}$$

[3] Additional discussion of the indistinguishability of electrons is given in Section 10-3.

[4] W. Kolos and L. Wolniewicz, *J. Chem. Phys.*, **49**, 404–410 (1968).

[5] S. Wang, *Phys. Rev.*, **31**, 579 (1927).

[6] S. Weinbaum, *J. Chem. Phys.*, **1**, 593 (1933).

where λ is a parameter to be determined. The optimized energy of this wavefunction was found to be

$$E = -1.14796 \text{ Hartrees},$$

corresponding to a binding energy of 84.8 kcal/mol. Hence, even a relatively simple approximate wavefunction as given in Eq. (10-8) was seen to be quite effective in describing the H_2 molecule, having a total energy that is 97.7% of the true energy. However, the remaining error (0.0265 Hartree or 16.6 kcal/mol) is still large on a chemical scale, and further improvements in the wavefunction will be needed if absolute accuracy to 1–2 kcal/mol is to be obtained.

B. Molecular Orbital Theory

During the same period, another approach was developed by Mulliken, Hund, and Lennard-Jones,[7] which appeared conceptually to be quite opposite from that of valence bond theory. This second approach started by assuming that the nuclei were at their *equilibrium molecular positions* (as opposed to infinitely apart, as in valence bond theory), and constructed orbitals for motion of the electrons that encompassed the entire molecule, and not simply a single atom. These orbitals were called *molecular orbitals*, in order to reflect the fact that they encompassed the entire molecule (at least in principle).

For the case of H_2, there are only two nuclei, and construction of a molecular orbital that allows an electron to move about both nuclei is easily accomplished via

$$\sigma_{1s}(1) = (1s_a + 1s_b)(1). \tag{10-9}$$

Thus, σ_{1s} is a *molecular* orbital, composed of the sum of $1s$ atomic orbitals on each nucleus.

The total wavefunction is thus given by

$$\Psi(1, 2) = N\sigma_{1s}(1) \cdot \sigma_{1s}(2). \tag{10-10}$$

Using $1s$ atomic orbitals with $\zeta = 1$, the energy obtained from this wavefunction was

$$E = -1.0985 \text{ Hartrees},$$

corresponding to a binding energy of 56.6 kcal/mol.

Since the molecular orbital (MO) and valence bond (VB) approaches both appeared during the same period, comparisons as to the relative merits of the two approaches were inevitable. For example, since the MO wavefunction

---

[7] See for example, R. S. Mulliken, *Phys. Rev.*, **32**, 186, 388, 761(1928); E. Hund, *Z. Phys.*, **51**, 759 (1928) and *Z. Electrochem.*, **34**, 437 (1928); and J. E. Lennard-Jones, *Trans. Faraday Soc.*, **25**, 688 (1928).

energy (– 1.0985 Hartrees) was higher than the corresponding VB wavefunction energy (– 1.116 Hartrees), it was argued that the VB approach was "better."

However, as we now understand, the two approaches have many similarities, and each can be used as a basis from which accurate wavefunctions can be determined. To illustrate the similarities, let us consider the MO wavefunction of Eq. (10-10) in greater detail. Expanding Eq. (10-10) and rearranging terms gives

$$\begin{aligned}\Psi(1, 2) &= N[(1s_a + 1s_b)(1) \cdot (1s_a + 1s_b)(2)] \\ &= N[1s_a(1)1s_b(2) + 1s_b(1)1s_a(2) + 1s_a(1)1s_a(2) + 1s_b(1)1s_b(2)].\end{aligned} \tag{10-11}$$

Thus, the MO wavefunction is seen to be nothing more than the Weinbaum VB wavefunction for the special case in which $\lambda = 1$. Additional aspects of MO and VB theory will be discussed in later chapters,[8] but suffice it to say that both MO and VB theory have made significant contributions to the conceptual and computational development of quantum chemistry, and continue to occupy a prominent role in current research.

10-2. Symmetry Considerations and Group Theory

As even the earliest attempts to describe multielectron, multinuclear systems illustrate, symmetry properties (e.g., indistinguishability of electrons) play an important role in constructing approximate wavefunctions. In this section and the next we shall discuss several kinds of symmetry of importance to molecular systems. The first of these, shown through the use of *group theory,* will illustrate how to utilize molecular and wavefunction symmetry both to simplify calculations and to gain insight not otherwise easily possible. Another kind, the *Pauli Exclusion Principle*, deals with the indistinguishability of electrons and the symmetry constraints placed on wavefunctions as a result. As we shall see, consideration of both kinds of symmetry is important. More specifically, however, we shall see that group theory is important to our efforts in at least three ways: (1) to deduce the symmetry of the molecule, (2) to construct symmetry eigenfunctions, and (3) to construct matrix element theorems that simplify calculations significantly.

In order to introduce the topic of *group theory*,[9] which is the primary mathematical framework for discussing symmetry properties, we note first that the discussion will be limited to the *symmetry properties of individual*

---

[8] See Section 12-3.

[9] For additional discussion see, for example, M. Tinkham, *Group Theory and Quantum Mechanics*, McGraw-Hill, New York, 1964; L. H. Hall, *Group Theory and Symmetry in Chemistry*, McGraw-Hill, New York, 1969; and G. G. Hall, *Applied Group Theory*, Longmans, Green and Co., London, 1967.

molecules (called *point groups*), and will not be extended to symmetry properties of infinite systems (e.g., crystals, where *space groups* are used to describe the symmetry relationships between molecules). When discussing point groups, we shall be interested in identifying mathematical operations that, after they have been applied, leave the system (e.g., molecule, molecular orbital wavefunction, etc.) indistinguishable from its status before the application of the mathematical operation. Such operations are called *symmetry operations*, and there are five kinds of particular interest.

The first kind is called the *identity operation* (E), which simply leaves the object alone. However, it has more than simply formal interest, as we shall soon see. The second kind of symmetry operation is *rotation about an axis of symmetry.*[10] For example, if rotation about a particular axis in a molecule by ($2\pi/n$) degrees results in a molecule that is indistinguishable from the molecule prior to the rotation, then we say that the molecule has an n-fold axis of symmetry, and we speak of a C_n symmetry operation. Thus, rotation about an axis bisecting the H–O–H angle in H_2O by 180° represents a 2-fold axis of symmetry. Incidentally, if there is more than one axis of symmetry, then the axis having the largest value of n is known as the *principal axis of symmetry.*[11]

The third type of symmetry operation is known as a *reflection*, and it can occur in a vertical (σ_v), horizontal (σ_h) or dihedral (σ_d) plane. A "vertical" plane is defined as one passing through the principal axis of highest symmetry, while a "horizontal" plane is the one through the origin and perpendicular to the principal axis. A "dihedral" plane (sometimes called a "diagonal" plane) is one containing the symmetry axis and bisecting the dihedral angles between the planes defining two σ_v s. For example, reflection through the plane bisecting the HOH angle is an allowed reflection operation (σ_v) in H_2O, as is reflection through the HOH plane (σ_v'). In benzene, the plane bisecting planes passing through the center and two nuclei separated by an intervening carbon atom (e.g., C_1 and C_3), is an example of a σ_d plane.

The fourth type of symmetry operation is known as *inversion through a center of symmetry*, and is given the symbol[12] i. In this case the operation consists of moving a point from an arbitrary position through a center of symmetry to a point along the same line and at the same distance, but on the opposite side of the center of symmetry.[13] The center of the benzene molecule in a regular hexagonal geometry illustrates a center of symmetry through which

---

[10] As is illustrated here, not only symmetry operations per se are of interest, but there also is a *symmetry element* (which is the point, line, or plane with respect to which the symmetry operation is carried out) associated with each symmetry operation.

[11] If there is more than one axis of symmetry, but each has the same value of n, then *any* of the axes may in principle be selected as the "principal" axis.

[12] Care must be taken to distinguish (via the context in which the symbol is used) between the inversion symmetry operator and $i = \sqrt{-1}$.

[13] Another way to view the inversion operation is that it takes a point (x, y, z) and moves it to ($-x$, $-y$, $-z$), where the origin is the inversion center.

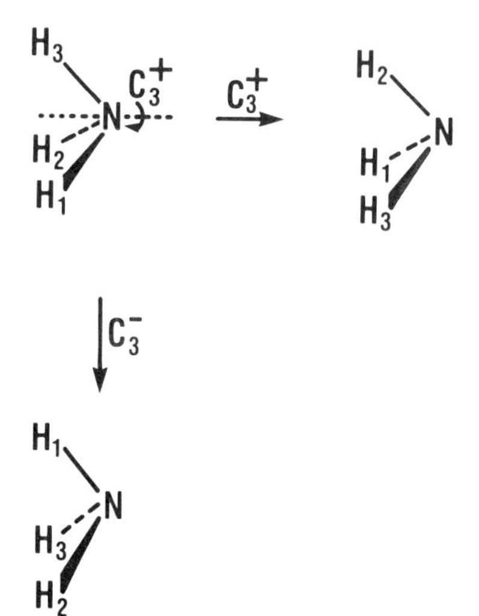

Figure 10.2. Illustration of C_3^+ and C_3^- symmetry operations using ammonia as an example.

inversion leaves the molecule unchanged. The last type of symmetry operation is known as an *"improper rotation" about an axis of symmetry.* This operation consists of a rotation by $(2\pi/n)$ about the axis of symmetry, followed by reflection through a plane of symmetry perpendicular to the rotation axis, and is given the symbol S_n. An example in which three such symmetry operations are present is the methane molecule (CH_4) in a tetrahedral geometry, where S_4 improper rotations each involving an axis bisecting an HCH angle is present.

One of the reasons for introducing the concept of symmetry operations is that it allows us to classify molecules[14] according to their symmetry. "Groups" form the mathematical language convenient for such classifications, and it turns out that there are a relatively small number of collections of symmetry operations that satisfy the requirements of a mathematical *group* (hence the term *group theory*), which are as follows:

1. The "product" of any two elements (i.e., symmetry operations) is another element (symmetry operation). Because of this property we say that the set of elements is *closed under group multiplication.*[15]

   To illustrate this characteristic, consider the rotation symmetry operations (C_3^+) and (C_3^-) for the vertices represented by the hydrogen nuclei of the ammonia molecule (see Fig. 10.2). In that case we see that the "product" of

[14] Throughout the discussion here we shall assume the Born–Oppenheimer approximation, i.e., the nuclei will be fixed in relation to each other. The symmetry characteristics that result, however, are applicable both to the nuclear symmetry properties and to wavefunctions describing electron motion in the molecule, as we shall see shortly.

[15] We shall utilize the terminology "multiplication" in this context as the law of combination between any two elements of the group. This "law of combination" between two elements $\mathcal{A}$ and $\mathcal{B}$ in the form $\mathcal{A}\mathcal{B}$ means the successive application of symmetry operation $\mathcal{B}$, followed by $\mathcal{A}$.

C_3^+ with itself is equivalent to another symmetry operation, i.e.,

$$C_3^+ C_3^+ = C_3^-, \tag{10-12}$$

where C_3^+ corresponds to a positive rotation by 120° and C_3^- is a negative rotation by 120° (see Fig. 10.2). Other symmetry operations that are present are reflections (σ_v, σ_v', σ_v'') in three vertical planes that bisect the vertices of the triangle. For these operations the "product rule" also applies, e.g.,

$$\sigma_v C_3^+ = \sigma_v'. \tag{10-13}$$

2. The associative law must hold for any pair of symmetry operations,[16] i.e., if $\mathcal{A}$, $\mathcal{B}$, and $\mathcal{C}$ are symmetry operations,

$$\mathcal{A}(\mathcal{B}\mathcal{C}) = (\mathcal{A}\mathcal{B})\mathcal{C}. \tag{10-14}$$

3. There must be an identity symmetry (E) element (operation) such that, for any symmetry operation $\mathcal{A}$,

$$E\mathcal{A} = \mathcal{A} = \mathcal{A}E. \tag{10-15}$$

4. There is, for each element $\mathcal{A}$ (symmetry operation), an *inverse* element ($\mathcal{A}^{-1}$) such that

$$\mathcal{A}\mathcal{A}^{-1} = \mathcal{A}^{-1}\mathcal{A} = E. \tag{10-16}$$

It can easily be verified that the above four properties are true for each of the six symmetry operations of the ammonia example introduced above.

Given these properties, we can construct what is known as a *group multiplication table*, which is a convenient way of summarizing the results of all possible applications of the "product rule" described above. Considering the collection of 3-fold rotations and reflections of the ammonia molecule once again, we can easily construct the following "multiplication table:"[17]

	E	C_3	C_3^2	σ_v	σ_v'	σ_v''
E	E	C_3	C_3^2	σ_v	σ_v'	σ_v''
C_3	C_3	C_3^2	E	σ_v''	σ_v	σ_v'
C_3^2	C_3^2	E	C_3	σ_v'	σ_v''	σ_v
σ_v	σ_v	σ_v'	σ_v''	E	C_3	C_3^2
σ_v'	σ_v'	σ_v''	σ_v	C_3^2	E	C_3
σ_v''	σ_v''	σ_v	σ_v'	C_3	C_3^2	E

There are several points of interest here. First, the table illustrates each of the

[16] Sometimes all products of symmetry operations also commute, i.e, $\mathcal{A}\mathcal{B} = \mathcal{B}\mathcal{A}$ for all $\mathcal{A}$ and $\mathcal{B}$ in the group, in which case the group is called *Abelian*. In general, however, the order of operation is important, e.g., in Eq. (10-13) the C_3^+ operation is carried out first, followed by the σ_v operation.

[17] In this example we use C_3^2 instead of C_3^- for convenience. In general, the operation listed along the top row is performed first, followed by the operation listed in the left hand column.

four requirements cited above, and therefore the collection of symmetry elements forms a group. Next, the results that are summarized in this table are not restricted only to, e.g., the ammonia molecule in its equilibrium configuration. Specifically, any collection of elements having these symmetry operations (and thus is isomorphous to the ammonia case) will satisfy the group multiplication table. The group that is represented by this multiplication table is known as the C_{3v} point group.

This "multiplication table" also allows another concept of interest to be illustrated. The concept is that of a *class*, which helps to relate various elements of a group. In particular, two elements ($\mathcal{R}$ and $\mathcal{R}'$) of a group are considered to be in the same class if they are *conjugate to each other*, i.e., if

$$\mathcal{R}' = \mathcal{S}\mathcal{R}\mathcal{S}^{-1} \tag{10-17}$$

for some element ($\mathcal{S}$) of the group.

For the C_{3v} point group just discussed, it is easily verified that

$$E^{-1} = E, \qquad (C_3)^{-1} = C_3^2, \qquad (C_3^2)^{-1} = C_3,$$

$$(\sigma_v)^{-1} = \sigma_v, \qquad (\sigma_v')^{-1} = \sigma_v', \qquad (\sigma_v'')^{-1} = \sigma_v''.$$

From these relations and the C_3 "multiplication table," it can be seen that[18] E forms a class by itself, C_3 and C_3^2 form another class, while σ_v, σ_v', and σ_v'' form a third class. This ability to distinguish among various symmetry elements by means of the "class" to which they belong will prove to be quite useful, as we shall see shortly.

Using the definitions and concepts introduced above, we are now in a position to classify molecules[19] according to their symmetry. The way in which this is done is to identify all of the symmetry operations that are applicable to the molecule, followed by assignment of the name of the group that contains those symmetry elements.[20] As seen in Table 10.1, there are 11 types of groups with no fewer than 43 specific examples that are important in chemistry.

To illustrate how the classification of molecules takes place, we shall consider the equilateral triangle in Fig. 10.3. The H_3^+ molecule in its equilibrium geometry is known to have this shape,[21] for example. Such a molecule contains the E, C_3, C_3^2, σ_v, σ_v', and σ_v'' symmetry operations discussed earlier, but also

---

[18] See Problem 10-1.

[19] Although the discussion here will use the nuclei in molecules to illustrate the concepts, far wider applicability is present that will be illustrated in later discussions (e.g., to molecular orbital symmetries, state symmetries, etc.).

[20] The symbols used here to designate group names are known as the *Schoenflies notation*. Other systems of nomenclature are also available, e.g., the Hermann–Mauguin nomenclature used for crystallographic point groups. For a detailed procedure to determine the point group for any molecule see, for example, L. H. Hall, *Group Theory and Symmetry in Chemistry*, McGraw Hill, New York, 1969, Chapter 4; see also F. A. Cotton, *Chemical Applications of Group Theory*, John Wiley, New York, 1963, Chapter 3, Sect. 10.

[21] R. E. Christoffersen, S. Hagstrom, and F. P. Prosser, *J. Chem. Phys.*, **40**, 236 (1964).

Table 10.1. Symmetry Groups of Importance in Chemistry.

Type of group	Group name	Symmetry operations contained in group
1. Non-Axial	C_1	E
	C_s	E, σ_h
	C_i	E, i
2. C_n	C_2	E, C_2
	C_3	E, C_3, C_3^2
	C_4	E, C_4, C_2, C_4^3
	C_5	$E, C_5, C_5^2, C_5^3, C_5^4$
	C_6	$E, C_6, C_3, C_2, C_3^2, C_6^5$
	C_7	$E, C_7, C_7^2, C_7^3, C_7^4, C_7^5, C_7^6$
	C_8	$E, C_8, C_4, C_2, C_4^3, C_8^3, C_8^5, C_8^7$
3. D_n	D_2	$E, C_2(z), C_2(y), C_2(x)$
	D_3	$E, 2C_3, 3C_2$
	D_4	$E, 2C_4, C_2, 2C_2', 2C_2''$
	D_5	$E, 2C_5, 2C_5^2, 5C_2$
	D_6	$E, 2C_6, 2C_3, C_2, 3C_2', 3C_2''$
4. C_{nv}	C_{2v}	$E, C_2, \sigma_v(xz), \sigma_v'(yz)$
	C_{3v}	$E, 2C_3, 3\sigma_v$
	C_{4v}	$E, 2C_4, C_2, 2\sigma_v, 2\sigma_d$
	C_{5v}	$E, 2C_5, 2C_5^2, 5\sigma_v$
	C_{6v}	$E, 2C_6, 2C_3, C_2, 3\sigma_v, 3\sigma_d$
5. C_{nh}	C_{2h}	E, C_2, i, σ_h
	C_{3h}	$E, C_3, C_3^2, \sigma_h, S_3, S_3^5$
	C_{4h}	$E, C_4, C_2, C_4^3, i, S_4^3, \sigma_h, S_4$
	C_{5h}	$E, C_5, C_5^2, C_5^3, C_5^4, \sigma_h, S_5, S_5^7, S_5^3, S_5^9$
	C_{6h}	$E, C_6, C_3, C_2, C_3^2, C_6^5, i, S_3^5, S_6^5, \sigma_h, S_6, S_3$
6. D_{nh}	D_{2h}	$E, C_2(z), C_2(y), C_2(x), i, \sigma(xy), \sigma(xz), \sigma(yz)$
	D_{3h}	$E, 2C_3, 3C_2, \sigma_h, 2S_3, 3\sigma_v$
	D_{4h}	$E, 2C_4, C_2, 2C_2', 2C_2'', i, 2S_4, \sigma_h, 2\sigma_v, 2\sigma_d$
	D_{5h}	$E, 2C_5, 2C_5^2, 5C_2, \sigma_h, 2S_5, 2S_5^3, 5\sigma_v$
	D_{6h}	$E, 2C_6, 2C_3, C_2, 3C_2', 3C_2'', i, 2S_3, 2S_6, \sigma_h, 3\sigma_d, 3\sigma_v$
7. D_{nd}	D_{2d}	$E, 2S_4, C_2, 2C_2', 2\sigma_d$
	D_{3d}	$E, 2C_3, 3C_2, i, 2S_6, 3\sigma_d$
	D_{4d}	$E, 2S_8, 2C_4, 2S_8^3, C_2, 4C_2', 4\sigma_d$
	D_{5d}	$E, 2C_5, 2C_5^2, 5C_2, i, 2S_{10}^3, 2S_{10}, 5\sigma_d$
	D_{6d}	$E, 2S_{12}, 2C_6, 2S_4, 2C_3, 2S_{12}^5, C_2, 6C_2, 6\sigma_d$
8. S_n	S_4	E, S_4, C_2, S_4^3
	S_6	$E, C_3, C_3^2, i, S_6^5, S_6$
	S_8	$E, S_8, C_4, S_8, C_2, S_8^5, C_4^3, S_8^7$
9. Cubic	T_d	$E, 8C_3, 3C_2, 6S_4, 6\sigma_d$
	O_h	$E, 8C_3, 6C_2, 6C_4, 3C_2, i, 6S_4, 8S_6, 3\sigma_h, 6\sigma_d$
10. Linear	$C_{\infty v}$	$E, 2C_\infty^\Phi \cdots \infty\sigma_v$
	$D_{\infty h}$	$E, 2C_\infty^\Phi \cdots \infty\sigma_{v1}, 2S_\infty^\Phi \cdots \infty C_2$
11. Icosahedral	I_h	$E, 12C_5, 12C_5^2, 20C_3, 15C_2, i, 12S_{10}, 12S_{10}^3, 20S_6, 15\sigma$

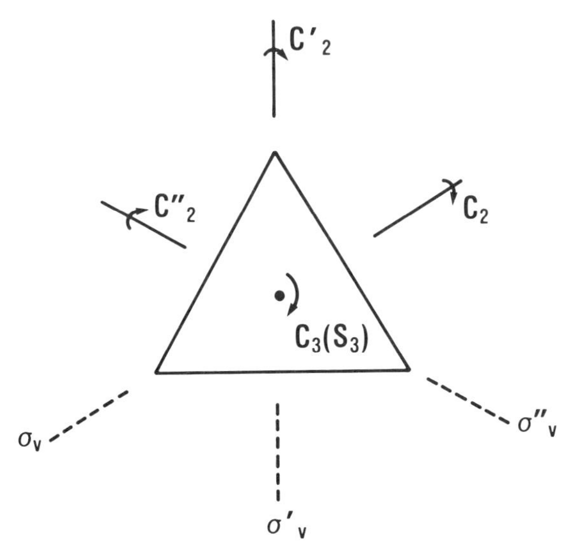

Figure 10.3. Symmetry operations of an equilateral triangle.

several more. In particular, we see in Fig. 10.3 that the full set of symmetry operations for a molecule such as H_3^+ also includes three C_2 axes of rotation, a horizontal reflection plane (σ_h), and two S_3 improper rotations. Inspection of Table 10.1 indicates that the group designation of a molecule such as H_3^+ which includes the full set of symmetry operations, is D_{3h}. The total number of symmetry operations in the group (12 in this case) is known as the *order* of the group. Since other molecules (e.g., BF_3) also contain the same symmetry operations, it is clear that we can derive symmetry information about BF_3 or any other molecule having D_{3h} symmetry from a study of H_3^+ (or vice versa).

However, classification of moleculcs according to their nuclear symmetry is not the only utility of group theory to quantum chemistry. In particular, the concepts of group theory also apply to the Schrödinger equation itself, and thus to wavefunctions resulting from the solution (whether exact or approximate) of the Schrödinger equation. To understand this application, we consider a general molecular Hamiltonian operator ($\mathcal{H}$) for a molecule containing N electrons and M nuclei in the Born–Oppenheimer approximation, i.e.,

$$\mathcal{H} = -\frac{1}{2}\sum_i^N \nabla_i^2 - \sum_i^N \sum_\alpha^M \frac{Z_\alpha}{r_{i\alpha}} + \sum_{i<j}^N \frac{1}{r_{ij}} + \sum_{\alpha<\beta}^M \frac{Z_\alpha Z_\beta}{R_{\alpha\beta}} .$$

Considering the effects of a nuclear symmetry operator ($\mathcal{R}$) on $\mathcal{H}$ we note that, from the definition of $\mathcal{R}$, it will have no effect on $\mathcal{H}$ since $\mathcal{R}$ represents operations that leave the atom positions in the molecule invariant. In mathematical terms, the lack of effect of $\mathcal{R}$ on $\mathcal{H}$ can be represented as

$$\mathcal{R}\mathcal{H} = \mathcal{H}\mathcal{R}, \tag{10-18}$$

i.e., $\mathcal{H}$ and $\mathcal{R}$ commute. Extending this idea to include all symmetry operators that comprise a group for the molecule of interest means that the entire set of symmetry operators ($\mathcal{R}_i$) representing a group ($\mathcal{G}$) commute with $\mathcal{H}$, and we can (without loss of generality) seek solutions to the Schrödinger equation that are simultaneous eigenfunctions of $\mathcal{H}$ and the operations of the group $\mathcal{G}$. This statement applies both to exact or approximate solutions to the Schrödinger equation,[22] and will be of significant assistance to us in constructing approximate wavefunctions in the sections and chapters to follow.

Thus far we have used the symmetry operations themselves to extract symmetry information. However, just as we can use a matrix representation of vectors using a basis as an alternative way of obtaining information about vector properties,[23] we shall usually find it convenient to utilize a matrix representation of symmetry operations to represent symmetry groups.[24] Of course, this implies that a basis must be chosen for the matrix representation. We shall now discuss several important properties of groups using matrix representations, and show how appropriate bases for such matrix representations can be constructed. To begin, let us construct a matrix representation of several of the fundamental symmetry operations (E, C_2, i and σ_h) on an arbitrary vector ($\mathbf{r}$) in three dimensions. This will allow us to obtain a matrix representation of the C_{2h} group. We choose unit vectors in the x, y, and z direction as a basis for any vector in this space, and we can write our arbitrary vector ($\mathbf{r}$) in this basis as

$$\mathbf{r} = \begin{pmatrix} x \\ y \\ z \end{pmatrix}. \tag{10-19}$$

It is easily verified that a matrix representation of E and i in this basis is given by

$$\mathbf{D}(E) = \begin{pmatrix} 1 & 0 & 0 \\ 0 & 1 & 0 \\ 0 & 0 & 1 \end{pmatrix} \tag{10-20}$$

$$\mathbf{D}(i) = \begin{pmatrix} -1 & 0 & 0 \\ 0 & -1 & 0 \\ 0 & 0 & -1 \end{pmatrix}. \tag{10-21}$$

If we wish to construct a matrix representation of the C_2 symmetry operation

---

[22] The analysis as just presented applies directly to cases in which the eigenvalues of $\mathcal{H}$ are nondegenerate. For cases in which degeneracies are present, modification in the analysis is needed, but the result is the same.

[23] See Chapter 3, Section 3-6.

[24] In this context, the use of matrices to represent group operations means that the matrices obey the relationships given in the group multiplication table, using matrix multiplication as the rule of combination of any two matrices of the representation.

consisting of rotation by 180° in the xy plane (i.e., rotation around the z axis), it is easily verified that the following matrix representation will suffice.

$$\mathbf{D}(C_2)=\begin{pmatrix} -1 & 0 & 0 \\ 0 & -1 & 0 \\ 0 & 0 & 1 \end{pmatrix}. \tag{10-22}$$

In a similar fashion it can be seen that reflection through the plane perpendicular to C_2 is accomplished via

$$\mathbf{D}(\sigma_\text{h})=\begin{pmatrix} 1 & 0 & 0 \\ 0 & 1 & 0 \\ 0 & 0 & -1 \end{pmatrix}. \tag{10-23}$$

It can also been seen that products of these matrices give rise to matrices representing other operations, e.g.,

$$\mathbf{D}(\sigma_\text{h})=\mathbf{D}(i)\mathbf{D}(C_2). \tag{10-24}$$

By forming each of the products of matrices in Eqs. (10-20)–(10-23), it can be seen that the entire multiplication table of the $C_{2\text{h}}$ group is created.

What we have uncovered via the preceding example is that the matrices form a *representation of the group,* since they have the same effect and multiply together in exactly the same way as the symmetry operations themselves. However, the size of the matrices (known as the *dimension of the representation*) depends on the size of the basis used, and a large number of matrix representations of a given group is thus possible. To aid in the classification of various matrix representations, we define the *dimension of the representation* as equal to the dimensionality of the basis that is used (e.g., the number of functions in the basis). The question that then arises is: How many different representations are there, and which is best to use? In answering these questions we shall introduce the ideas of *reducible* and *irreducible representations* and other concepts.

To begin with we shall find it useful to speak of the *character of a matrix* representing a symmetry operation, which is simply the trace of the matrix. For example, the characters[25] of the matrices forming the representing the $C_{2\text{h}}$ group described above are given by

$$\begin{aligned} \chi(E)&=\text{tr }\mathbf{D}(E)=+3 \\ \chi(i)&=\text{tr }\mathbf{D}(\mathcal{I})=-3 \\ \chi(C_2)&=\text{tr }\mathbf{D}(C_2)=-1 \\ \chi(\sigma_\text{h})&=\text{tr }\mathbf{D}(\sigma_\text{h})=+1. \end{aligned} \tag{10-25}$$

[25] See Chapter 3, Section 3-3 for a discussion of properties of the trace of a matrix.

Next, since we have the possibility of many matrix representations corresponding to the use of different basis sets, it is useful to recall several theorems and properties associated with changes of bases that were discussed earlier.[26] In particular, we note that if **C** is a matrix that transforms one basis into another, then the matrices representing symmetry operators in the two bases (e.g., **D** and **D**′) are related via the following similarity transformation:

$$\mathbf{D}(R)=\mathbf{C}\mathbf{D}'(R)\mathbf{C}^{-1} \tag{10-26}$$

where R is any symmetry operation in the group. It is also of interest to note that, if **C** is a matrix representation of a symmetry operation in the group, then the matrices **D** and **D**′ are said to be *conjugate* to each other.[27] This case is of particular interest to us, as we shall now see.

In particular, the complete set of matrix representations of symmetry operators that are conjugate to each other is said to form a *class*. This result is the matrix analog of the result found earlier using symmetry operations directly [see Eq. (10-17)]. It is of interest to note (although we shall not prove it) that the number of elements in a class is an integral divisor of the total number of elements (g) in the group, i.e.,

$$\left(\frac{g}{c}\right)=m, \tag{10-27}$$

where c is the number of elements in a class, and m is an integer.

Finally, it is also useful to recall[28] that the character (i.e., trace) of a matrix representation will be unaffected by a similarity transformation as in Eq. (10-26), i.e.,

$$\operatorname{tr}\mathbf{D}(R)=\operatorname{tr}\mathbf{D}'(R). \tag{10-28}$$

To introduce the notion of reducible and irreducible representations, we note that each of the matrices in the representation of the C_{2h} group that was introduced earlier [Eqs. (10-20)–(10-23)] is a diagonal matrix. However, such a situation does not automatically arise in every case, and it is frequently found that the matrix representation contains many off-diagonal components. In this latter case it is frequently possible to "reduce" the original representation into several smaller "irreducible" representations, and it is through the use of similarity transformations such as illustrated in Eq. (10-26) that this task will be

---

[26] See Chapter 3, Section 3-7.

[27] Since matrix algebra is also preserved under similarity transformations (see Problems 3-16 and 3-25), the matrices $\mathbf{D}(R)$ and $\mathbf{D}'(R)$ obey the same group multiplication table as the symmetry operations they represent, and are therefore equivalent representations of the group.

[28] See Theorem 3-12.

accomplished. Pictorially, we wish to accomplish the following:

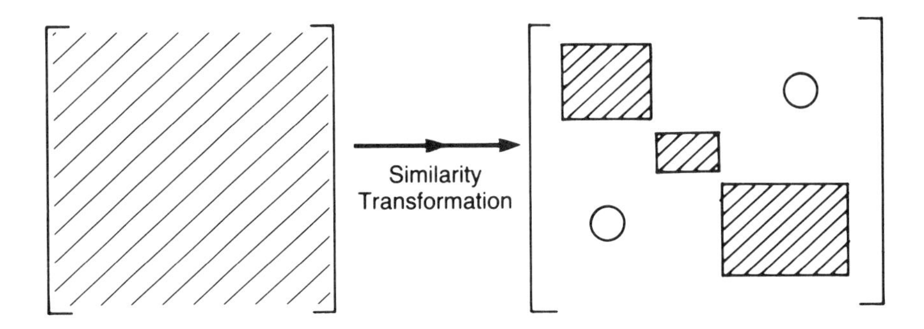

where each of the smaller blocks on the right-hand side cannot be reduced further in dimension via similarity transformations. In the latter case, each of the smaller blocks produced by the similarity transformations is said to be an *irreducible representation* of the group. More formally, reducible and irreducible representations are defined as follows. A *reducible representation* is one in which each of the matrices of the representation can be brought simultaneously to the same block diagonal form by a similarity transformation (or series of them). If a representation is not reducible, then it is said to be an irreducible representation.

There are two theorems that provide important tools in the process of constructing irreducible representations, which we shall state[29] and discuss below. The first of these is known as the *Great Orthogonality Theorem*, and can be stated as follows:

$$\sum_{\mathcal{R}=1}^{g} (\mathbf{D}^*(\mathcal{R}))_{ij}^{(\alpha)}(\mathbf{D}(\mathcal{R}))_{kl}^{(\beta)} = \left(\frac{g}{d_\alpha}\right) \delta_{ik}\delta_{jl}\delta_{\alpha\beta}, \tag{10-29}$$

where g is the number of symmetry operations in the group, d_α is the dimension of the αth irreducible representation, $\mathbf{D}^*(\mathcal{R})_{ij}^{(\alpha)}$ is the complex conjugate of the i, j element of the unitary matrix[30] representation (α) of the symmetry operation $\mathcal{R}$, and $D(\mathcal{R})_{kl}^{(\beta)}$ is the k, l element of a different unitary matrix representation (β) of the symmetry operation $\mathcal{R}$. While ominous in appearance, Eq. (10-29) is easy to understand when viewed pictorially as in Fig. 10.4. To carry out the sum in Eq. (10-29), we select the i, j element of the first matrix in the α representation, multiply it by the k, l element of the β matrix representation

---

[29] These theorems will not be proved here, but the proofs are available in many group theory texts. See, for example, E. P. Wigner, *Group Theory*, Academic Press, New York, 1959, Chapter 9.

[30] It should be emphasized that Eq. (10-29) applies only when a unitary matrix representation is employed. However, it can also be shown (e.g., see M. Tinkham, *Group Theory and Quantum Mechanics*, McGraw Hill, New York, 1964, p. 20) that any nonsingular matrix representation by means of a similarity transformation. Hence, Eq. (10-29) has quite broad applicability.

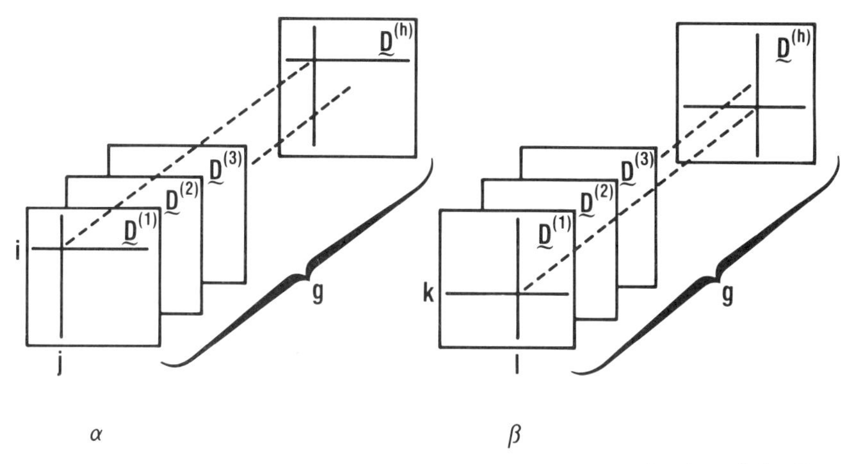

Figure 10.4. Pictorial representation of elements involved in application of the Great Orthogonality Theorem.

corresponding to the same symmetry operation, and sum each of those products over all matrices in the group [i.e., all symmetry operations (g) in the group].

The next bit of mathematical machinery that we need (but will not prove) is derivable from the Great Orthogonality Theorem, and states that the number of irreducible representations that are possible is given by

$$\sum_{i}^{\substack{\text{all} \\ \text{irred.} \\ \text{reps.}}} d_i^2 = g, \tag{10-30}$$

where d_i is the dimension of the ith irreducible representation.

The final theorem of importance (frequently called the *Little Orthogonality Theorem*) is obtained from Eq. (10-29) by summing over diagonal elements. In particular, we set $i = k$ and $j = l$, and obtain

$$\sum_{\mathcal{R}=1}^{g} \chi^{(\alpha)}(\mathcal{R})^* \chi^{(\beta)}(\mathcal{R}) = g\delta_{\alpha\beta}, \tag{10-31}$$

where we have utilized the definition of the character (i.e., trace) of a matrix. Since the character for each member of a class is the same, we can rewrite Eq. (10-31) in terms of classes as

$$\sum_{\mathcal{C}=1}^{r} g_{\mathcal{C}} \chi^{(\alpha)}(\mathcal{C})^* \chi^{(\beta)}(\mathcal{C}) = g\delta_{\alpha\beta}, \tag{10-32}$$

where g_c is the number of symmetry elements in the class C.

Working with the characters in Eq. (10-32) is much easier in general than working with the matrices in Eq. (10-29). One of the reasons for this is related to the invariance of the character to similarity transformations.[31] In particular, if two matrix representations of a group are related via similarity transformations [see Eq. (10-26)], we saw [see Eq. (10-28)] that their characters are identical. This is very helpful, for in general it allows us to seek (and work with) only the characters associated with different classes, and not have to worry about which basis is chosen, as we shall see below.

Now let us show how the mathematical apparatus just described can be utilized in a practical manner. To do that, we shall work with the characters almost exclusively, and do so within the framework of a *character table*. Such a table exists for each group, and has the form:

	$C_1 = E$	C_2	C_3	$\cdots$	C_r
Γ_1	$\chi^1(E)$	$\chi^1(C_2)$	$\chi^1(C_3)$	$\cdots$	$\chi^1(C_r)$
Γ_2	$\chi^2(E)$	$\chi^2(C_2)$	$\chi^2(C_3)$	$\cdots$	$\chi^2(C_r)$
Γ_3	$\chi^3(E)$	$\chi^3(C_2)$	$\chi^3(C_3)$	$\cdots$	$\chi^3(C_r)$
$\vdots$	$\vdots$	$\vdots$	$\vdots$		$\vdots$
Γ_r	$\chi^r(E)$	$\chi^r(C_2)$	$\chi^r(C_3)$	$\cdots$	$\chi^r(C_r)$

where $\Gamma_1 \ldots \Gamma_r$ represent the r irreducible representations of the group and C_1, C_2, ..., C_r represent the r different classes in the group. Such a table also indicates the utility of the "class" concept, since we deal only with the number of different classes in such a table, and not all symmetry elements.[32]

To form such a table, we shall utilize Eqs. (10-31) or (10-32) (which can be seen to be an orthogonality relationship for the rows of the above table) and Eq. (10-30), plus the facts that the elements of the first column are always easy to construct, since

$$\chi^{(i)}(E) = d_i, \tag{10-33}$$

along with the fact that there is always a one-dimensional irreducible representation (i.e., with all ones as characters, and which we shall always take as Γ_1), i.e.,

$$\chi^1(C_k) = 1, \qquad k = 1, 2, \cdots, r. \tag{10-34}$$

Thus, Γ_1 is always taken as the fully symmetric representation.

To illustrate the construction of a character table, let us consider the C_{2h} point group whose matrix representations of the four symmetry operations (E, i, C_2,

[31] See Theorem 3-12.

[32] The above table also illustrates the general fact (although not proven here) that the number of irreducible representations equals the number of classes.

and σ_h) were given in Eqs. (10-20)–(10-23). While this is a simplified example because each class contains only one symmetry operation, it will be sufficient to illustrate how character tables are formed. In particular, we wish to find entries for the following table:

C_{2h}	E	i	C_2	σ_h
Γ_1	1	1	1	1
Γ_2				
Γ_3				
Γ_4				

where the first row is given by Eq. (10-34). The remaining elements in the first column are also seen to be equal to $+1$, since we have [from Eq. (10-30)]

$$\sum_{i=1}^{4} d_i^2 = 4,$$

and the only way to satisfy this equation when the first irreducible representation is one-dimensional is for the other three to be one-dimensional as well. Thus, at this point we have

C_{2h}	E	i	C_2	σ_h
Γ_1	1	1	1	1
Γ_2	1			
Γ_3	1			
Γ_4	1			

We now construct the remainder of the second row by use of the orthogonality relation of Eq. (10-31) as applied to Γ_1 and Γ_2, i.e.,

$$\sum_{\mathcal{R}=1}^{4} \chi^{(1)}(\mathcal{R})\chi^{(2)}(\mathcal{R}) = 0,$$

or

$$(1)(1) + (1)(\ \ ) + (1)(\ \ ) + (1)(\ \ ) = 0.$$

The simplest choice of elements that will satisfy the above equation is $\Gamma_2 = (1, 1, -1, -1)$. Similar application of Eq. (10-31) for Γ_3 and Γ_4 results in the following table:

C_{2h}	E	C_2	i	σ_h
A_g	1	1	1	1
A_u	1	1	-1	-1
B_g	1	-1	1	-1
B_u	1	-1	-1	1

where we have introduced the *Mulliken notation for irreducible representations* in which A and B are used to designate one-dimensional irreducible representations. Also, A is used to refer to irreducible representations that are symmetric with respect to C_n rotations, while B is used for irreducible representations that are antisymmetric with respect to C_n rotations. The "g" and "u" subscripts refer to symmetric and antisymmetric behavior, respectively, with respect to the inversion operation. Finally, although there are no cases in the above example, the symbols E and T are used[33] to represent two- and three-dimensional irreducible representations, respectively.

To illustrate multidimensional irreducible representations, let us consider the C_3 point group as a second example. The ammonia molecule (NH_3) is a case in which such a point group is applicable, and the symmetry elements include E, two C_3 axes, and three vertical reflection planes (σ_v). Using methods entirely analogous to those used above for the C_{2h} group, the following character table can be easily constructed:

C_{3v}	E	$2C_3$	$3\sigma_v$
$\Gamma_1 = A_1$	1	1	1
$\Gamma_2 = A_2$	1	1	-1
$\Gamma_3 = E$	2	-1	0

In this example, however, we find that the third irreducible representation ($\Gamma_3 = E$) is two dimensional, since $\chi^{(3)}(E) = 2$. To illustrate this situation, let us return to the use of the unit vector [Eq. (10-19)] to illustrate a representation for the C_{3v} group. Application of the C_3^+ operation to $\mathbf{r}$ is seen to produce a new

[33] Unfortunately, the symbol E is typically used both to describe the identity symmetry operation and a doubly degenerate irreducible representation. While the context usually is sufficient to distinguish the two uses, the student should be altered to the potential ambiguity.

vector $\mathbf{r}'$ with components

$$\begin{aligned} x' &= (\cos \omega)x - (\sin \omega)y \\ y' &= (\sin \omega)x + (\cos \omega)y \\ z' &= z \end{aligned} \tag{10-35}$$

where $\omega = 2\pi/3$, which means that the matrix representation for the C_3^+ operation is given by

$$\mathbf{D}(C_3^+) = \begin{pmatrix} \cos \omega & -\sin \omega & 0 \\ \sin \omega & \cos \omega & 0 \\ 0 & 0 & 1 \end{pmatrix}.$$

Thus, we see that, for the case of the C_3^+ operation, the x component is transformed into a linear combination of x and y components (with a similar result for the y component), i.e., *both* x and y components are needed to form a representation of C_3^+. This result means that the 2×2 component of $\mathbf{D}(C_3^+)$ is not reducible to two one-dimensional components, and the irreducible representation must be two-dimensional. Irreducible representations having dimensions greater than one are found not only in the C_3 group, for every group in which a greater than two-fold C or S symmetry axis is found.

The existence of irreducible representations that are greater than one-dimensional also has implications regarding the degeneracies to be expected in quantum mechanical studies. In particular, it can be shown[34] that the degree of degeneracy of a set of functions is equal to the dimension of the irreducible representation that they span.

While a discussion as above indicates how character tables can be constructed, such procedures are not required in practice since character tables have already been constructed for each of the groups of usual interest in chemistry.[35] Of greater interest is the *use* of these character tables to assure that approximate wavefunctions that are constructed have symmetry properties that are appropriate. We shall work through an example shortly in detail that will illustrate how appropriate symmetry properties can be assured, but prior to that it is necessary to develop several additional mathematical aspects of group theory.

First, to aid in the process of "reducing" a reducible representation [whose character we shall take as $\chi(R)$], we note that $\chi(R)$ can be written in terms of its irreducible representation components [$\chi^{(\alpha)}(R)$] as

$$\chi(R) = \sum_{\beta} a_{\beta} \chi^{(\beta)}(R), \tag{10-36}$$

[34] See, for example, M. Tinkham, *Group Theory and Quantum Mechanics*, McGraw-Hill, New York, 1964, Chapter 3.

[35] See, for example, F. A. Cotton, *Chemical Applications of Group Theory*, John Wiley, New York, 1963, Appendix 2.

where a_β is the number of times the βth irreducible representation appears in the reducible representation $\chi(R)$. To find the value of a_β, we multiply Eq. (10-36) by $\chi^{(\alpha)}(R)^*$ and sum over all symmetry elements (R) to give

$$\sum_{R=1}^{g} \chi^{(\alpha)}(R)^*\chi(R) = \sum_{\beta}\sum_{R} a_\beta \chi^{(\alpha)}(R)^*\chi^{(\beta)}(R)$$

$$= g a_\alpha, \tag{10-37}$$

or

$$a_\alpha = \left(\frac{1}{g}\right) \sum_{R=1}^{g} \chi^{(\alpha)}(R)^*\chi(R), \tag{10-38}$$

where we have used the Little Orthogonality Theorem [Eq. (10-31)] to obtain Eq. (10-37). The relationship given above will be quite useful in decomposing reducible representations, as we shall see below.

The next mathematical idea of interest is that of a *direct product.* To illustrate this idea, let us return to the example of the matrix representation of the C_{2h} group given by Eqs. (10-20)–(10-23), using unit vectors in x, y, and z directions as the basis set. In general, what we saw in that case is that applying a symmetry operation to the basis can be described by the following relationship:

$$\mathcal{R} r_k = \sum_{l=1}^{m} a_{lk} r_l, \tag{10-39}$$

where $r = (r_1, r_2, r_3) = (x, y, z)$ and $m = 3$ for the C_{2h} example. Also, the a_{lk} in Eq. (10-39) for the C_{2h} example are particularly simple, i.e., a_{lk} has values of $0, \pm 1$. However, if we had chosen a different basis $\mathbf{r} = (r_1' r_2' r_3')$ that resulted in a different reducible representation, then the a_{lk} would have been more complicated, but would still satisfy an equation of the form

$$\mathcal{R} r_k' = \sum_{l'=1}^{m'} b_{l'k'} r_{l'}. \tag{10-40}$$

To introduce the notion of direct products, we consider the effect of applying the symmetry operation $(\mathcal{R})$ successively to each of the members of the *product* of the basis vector components, i.e.,

$$\mathcal{R} r_k r_k' = \sum_{l=1}^{m} a_{lk} r_l \mathcal{R} r_{k'}$$

$$= \sum_{l=1}^{m} a_{lk} \sum_{l'=1}^{m'} b_{l'k'} r_l r_{l'}$$

$$= \sum_{l=1}^{m}\sum_{l'=1}^{m'} a_{lk} b_{l'k'} r_l r_{l'}. \tag{10-41}$$

In other words, operating on the product $(r_k r_{k'})$ by $\mathcal{R}$ produces a linear combination of the same products $(r_l r_{l'})$. In fact, it can be shown that such products form a representation for the group, although in practice it may be a reducible representation. The matrix form of this representation is seen from Eq. (10-41) to be given by matrices whose elements are

$$(\mathbf{C})_{ll',kk'} = a_{lk} b_{l'k'}, \tag{10-42}$$

and $\mathbf{C}$ is said to provide a "direct product representation" of the group.

To illustrate the direct product concept, consider the following (2×2) matrices:

$$\mathbf{A} = \begin{pmatrix} a_{11} & a_{12} \\ a_{21} & a_{22} \end{pmatrix}, \qquad \mathbf{B} = \begin{pmatrix} b_{11} & b_{12} \\ b_{21} & b_{22} \end{pmatrix}.$$

The direct product of these two matrices is a 4×4 matrix, and is given by

$$\mathbf{C} = \mathbf{A} \otimes \mathbf{B} = \begin{pmatrix} a_{11}b_{11} & a_{11}b_{12} & a_{12}b_{11} & a_{12}b_{12} \\ a_{11}b_{21} & a_{11}b_{22} & a_{12}b_{21} & a_{12}b_{22} \\ a_{21}b_{11} & a_{21}b_{12} & a_{22}b_{11} & a_{22}b_{12} \\ a_{21}b_{21} & a_{21}b_{22} & a_{22}b_{21} & a_{22}b_{22} \end{pmatrix},$$

where the symbol $\otimes$ has been used to designate the direct product operation.

One of the reasons this is a convenient mathematical construct can be seen by examination of the character of $\mathbf{C}$, i.e.,

$$\begin{aligned} \chi(\mathbf{C}) &= \sum_{l,k} (\mathbf{C})_{lk,lk} = \sum_{l} \sum_{k} a_{ll} b_{kk} \\ &= \sum_{l} a_{ll} \sum_{k} b_{kk} \\ &= \chi(\mathbf{A}) \chi(\mathbf{B}). \end{aligned} \tag{10-43}$$

In other words, the character of a direct product representation of a group is equal to the product of the characters of the individual representations. For the C_{2h} case with a matrix representation given by Eqs. (10-20)–(10-23) and whose characters are given in Eq. (10-25), characters for the direct product representation are found by straightforward application of Eq. (10-43). For example,

$$\chi(\mathbf{I})\chi(\mathbf{C}_2) = (-3)(-1) = \chi(\mathbf{E}). \tag{10-44}$$

If it had been the case that the irreducible representations were two- or three-dimensional, then the character direct products would not be a single term [as in Eq. (10-44)], but would be a reducible representation containing a linear combination of characters from several irreducible representations.[36]

Finally, we shall be interested in starting with a given basis, and using it to

[36] This type of example is illustrated in the discussion of H_3^+, beginning with Eq. (10-52).

construct approximate eigenfunctions having symmetry properties appropriate to the molecule of interest. We shall now show how to construct projection operators[37] that can accomplish that task.

Let us begin with a function $\varphi_k^{(j)}$ that is one of the functions that forms the basis for an l_j-dimensional irreducible representation (j) of the group of interest. If we operate on that function with any of the symmetry operations of the group, we shall generate (at most) a linear combination of the other members of the basis, i.e.,

$$\mathcal{R}\varphi_k^{(j)} = \sum_{\lambda=1}^{l_j} \gamma_{\lambda k}^{(j)}(\mathcal{R})\varphi_\lambda^{(j)}, \tag{10-45}$$

where $\gamma_{\lambda k}^{(j)}$ is the (λ, k) element of the matrix representation $(\boldsymbol{\gamma})$ of the operator $\mathcal{R}$, and l_j is the dimensionality of the jth irreducible representation. To convert this equation into a more useful form, we multiply on the left of both sides of Eq. (10-45) by $\gamma_{\lambda' k'}^{(j)}(\mathcal{R})^*$ and sum over all group elements (g) to give

$$\begin{aligned}\sum_{\mathcal{R}=1}^{g} \gamma_{\lambda' k'}^{(j)}(\mathcal{R})^*\mathcal{R}\varphi_k^{(j)} &= \sum_{\mathcal{R}=1}^{g}\sum_{\lambda=1}^{l_j} \gamma_{\lambda' k'}^{(j)}(\mathcal{R})^*\gamma_{jk}^{(j)}(\mathcal{R})\varphi_k^{(j)} \\ &= \sum_{\lambda=1}^{l_j} \varphi_k^{(j)}\left[\sum_{\mathcal{R}=1}^{g} \gamma_{\lambda' k'}^{(j)}(\mathcal{R})\gamma_{\lambda k}^{(j)}(\mathcal{R})\right].\end{aligned} \tag{10-46}$$

Using the Great Orthogonality Theorem [Eq. (10-29)], we can write the above equation as

$$\varphi_\lambda^{(i)} = \left[\left(\frac{l_i}{g}\right)\sum_{\mathcal{R}=1}^{g} \gamma_{\lambda k}^{(i)}(\mathcal{R})^*\mathcal{R}\right]\varphi_k^{(i)}. \tag{10-47}$$

By defining a *projection operator* $(\mathcal{P}_{\lambda k}^{(i)})$ as

$$\mathcal{P}_{\lambda k}^{(i)} = \left(\frac{l_i}{g}\right)\sum_{\mathcal{R}=1}^{g} \gamma_{\lambda k}^{(i)}(\mathcal{R})^*\mathcal{R}, \tag{10-48}$$

Eq. (10-47) can be written as

$$\varphi_\lambda^{(i)} = \mathcal{P}_{\lambda k}^{(i)}\varphi_k^{(i)}. \tag{10-49}$$

Since it is easily seen that $\mathcal{P}_{\lambda k}^{(i)}$ is a linear operator and is idempotent (i.e., $[\mathcal{P}_{\lambda k}^{(i)}]^2 = \mathcal{P}_{\lambda k}^{(i)}$), it is appropriate to refer to $\mathcal{P}_{\lambda k}^{(i)}$ as a projection operator.[38] Also, since it is seen from Eq. (10-49) that application of $\mathcal{P}_{\lambda k}^{(i)}$ onto $\varphi_k^{(i)}$ projects out *another* member $[\varphi_\lambda^{(i)}]$ of the basis representing the ith irreducible representation, the rationale for the terminology is illustrated.

---

[37] Such operators are discussed in Section 2-13.

[38] See Chapter 2, Section 2-13.

The other point to emphasize regarding projection operators as defined in Eq. (10-48) has to do with the utility of such operators to *generate* other basis functions belonging to the irreducible representation of interest. In particular, λ in Eq. (10-49) is valid for all basis functions of the ith irreducible representation, and $\mathcal{P}^{(i)}_{\lambda k}$ can therefore be utilized to generate all of the basis functions of the irreducible representation, starting from any one of the functions $(\varphi^{(i)}_k)$.

To convert Eq. (10-48) into a more convenient form, we set $\lambda = k$ and sum over k to obtain

$$\mathcal{P}^{(i)} = \sum_{k=1}^{l_i} \left(\frac{l_i}{g}\right) \sum_{\mathcal{R}=1}^{g} \gamma^{(i)}_{kk}(\mathcal{R})^* \mathcal{R}$$

$$= \left(\frac{l_i}{g}\right) \sum_{\mathcal{R}=1}^{g} \left[\sum_{k=1}^{l_i} \gamma^{(i)}_{kk}(\mathcal{R})^*\right] \mathcal{R}, \tag{10-50}$$

or

$$\mathcal{P}^{(i)} = \left(\frac{l_i}{g}\right) \sum_{\mathcal{R}=1}^{g} \chi^{(i)}(R)^* R. \tag{10-51}$$

Thus, an alternative projection operator can be expressed in terms of the character of the irreducible representation, thus making manipulations using the latter projection operator considerably simpler. Such operators are very useful for, as we shall see, it means that we can begin with any arbitrary function in the space, and use Eq. (10-51) to generate all of the members of the basis that are needed to form a representation of the irreducible representation of interest.

To indicate how these ideas can be utilized in practical and important ways, let us consider a specific example and carry it through in detail from beginning to end in order to illustrate just how powerful these techniques are and how easy they are to use. Specifically, let us consider the H_3^+ molecule again, which we shall assume has an equilateral triangle geometry for the nuclei. This will be an especially useful example, for it will illustrate not only the application of group theory concepts, but will show how to extend the ideas of MO and VB theory of the previous section to a more complicated 2-electron problem.

Examination of the symmetry properties of the three nuclei can be shown to include the following symmetry operations: E, $2C_3$, $3C_2$, σ_h, $2S_3$, and $3\sigma_v$ (see Fig. 10.3). This means that the molecule belongs to the D_{3h} point group, whose character table is given in Table 10.2 for use in our analysis. Of the various irreducible representations of the D_{3h} group, we shall be interested in the H_3^+ electronic states whose wavefunctions have A_1' symmetry. (It turns out that the ground state has A_1' symmetry.)

Let us now consider how we can construct a MO description of the A_1' state of H_3^+. We expect that the wavefunction will have the form

$$\Phi(1, 2) = \varphi(1)\varphi(2), \tag{10-52}$$

Table 10.2. Character Table for the D_{3h} Point Group

D_{3h}	E	$2C_3$	$3C_2$	σ_h	$2S_3$	$3\sigma_v$
A_1'	1	1	1	1	1	1
A_2'	1	1	−1	1	1	−1
E'	2	−1	0	2	−1	0
A_1''	1	1	1	−1	−1	−1
A_2''	1	1	−1	−1	−1	1
E''	2	−1	0	−2	1	0

where φ is a molecular orbital suitably devised for H_3^+. Since $\Phi(1, 2)$ must have A_1' symmetry overall, we see from Eq. (10-52) that one of the requirements on is that the product $\varphi(1)\varphi(2)$ must have A_1' symmetry. In group theoretic language, this means that the *direct product* of the MO with itself must have A_1' symmetry.

To see which MO symmetries satisfy this requirement, we look at all the possibilities,[39] by examining the direct product of the various characters, i.e.,

$$a_1' \otimes a_1' = (1\ \ 1\ \ 1\ \ 1\ \ 1\ \ 1) \otimes (1\ \ 1\ \ 1\ \ 1\ \ 1\ \ 1)$$
$$= (1\ \ 1\ \ 1\ \ 1\ \ 1\ \ 1). \tag{10-53}$$

Thus, the direct product of a_1' with itself also has A_1' symmetry. For the other irreducible representations, we have

$$a_2' \otimes a_2' = (1\ \ 1\ \ -1\ \ 1\ \ 1\ \ -1) \otimes (1\ \ 1\ \ -1\ \ 1\ \ 1\ \ -1)$$
$$= (1\ \ 1\ \ 1\ \ 1\ \ 1\ \ 1), \tag{10-54}$$

$$e' \otimes e' = (2\ \ -1\ \ 0\ \ 2\ \ -1\ \ 0) \otimes (2\ \ -1\ \ 0\ \ 2\ \ -1\ \ 0)$$
$$= (4\ \ 1\ \ 0\ \ 4\ \ 1\ \ 0), \tag{10-55}$$

$$a_1'' \otimes a_1'' = (1\ \ 1\ \ 1\ \ 1\ \ 1\ \ 1), \tag{10-56}$$

$$a_2'' \otimes a_2'' = (1\ \ 1\ \ 1\ \ 1\ \ 1\ \ 1), \tag{10-57}$$

$$e'' \otimes e'' = (4\ \ 1\ \ 0\ \ 4\ \ 1\ \ 0). \tag{10-58}$$

From the above equations, we see that the direct product of a_1', a_2', a_1'' and a_2'' with themselves each have A_1' symmetry. This means that use of MOs having any of these symmetries in Eq. (10-52) would be acceptable from a group theory point of view. It will turn out that these four MO symmetries are not equivalent

[39] In the following discussion we shall use capital letters to describe symmetries of states (e.g., A_1') and small letters to describe symmetries of individual MOs (e.g., a_1'). Also the symbol "$\otimes$" will be used to indicate the direct product operation.

from an energetic point of view, but that cannot in general be determined from symmetry considerations alone.

The direct product of e' and e'' with themselves has produced reducible representations, and we shall use Eq. (10-38) to accomplish the "decomposition." In particular, we can find out whether an A_1' component is present by forming

$$\left(\frac{1}{12}\right)\sum_{\mathcal{R}=1}^{12}\chi^{(a_1')}(\mathcal{R})\chi(\mathcal{R})=\left(\frac{1}{12}\right)(1\ \ 2\ \ 3\ \ 1\ \ 2\ \ 3)\begin{pmatrix}4\\1\\0\\4\\1\\0\end{pmatrix}$$

$$=1, \tag{10-59}$$

where the vector multiplication has been simplified by noting the number of members in each class (e.g., there are $2C_3$ elements, $3C_2$ elements, etc.) Thus, MOs having e' and e'' symmetry also have a component of their direct product with A_1' symmetry, although we shall have to project out only the A_1' in such a case if we are to assure that Φ has *only* A_1' symmetry.

We now know that, if suitably projected, MOs having the symmetries of *any* of the six irreducible representations of H_3^+ can be used to construct a wavefunction (Φ) having A_1' symmetry. The next step is to find out how to form MOs that have the symmetries of the six irreducible representations. To do this, we must choose a basis.

To begin with, let us examine the simplest possible basis, i.e., a normalized $1s$ orbital on each nucleus: $1s_a$, $1s_b$, and $1s_c$. To see if this set will be suitable, let us apply the symmetry operations to each orbital, to be sure that no new orbital is needed, i.e., the set is closed under D_{3h} symmetry operations. Using the $1s_a$ orbital as an example, we have

$$E\,1s_a=1s_a$$

$$C_3^+\,1s_a=1s_b$$

$$(C_3^+)^2 1s_a=1s_c,\ \text{etc.}$$

If continued for all D_{3h} symmetry operations on each orbital, it is easily seen that the set of three $1s$ orbitals *is* closed under D_{3h} symmetry operations. In fact, the character vector that results from application of D_{3h} operations on such a basis is easily seen to be[40]

$$\Gamma_{1s}=(3\ \ 0\ \ 1\ \ 3\ \ 0\ \ 1). \tag{10-60}$$

[40] The character vector of Eq. (10-60) is based on the use of classes, not individual symmetry elements. If the latter had been used, the character vector would have contained 18 elements.

Since this character vector does not appear in the character table of D_{3h} irreducible representations (Table 10-2), the representation must be reducible.

To decompose Γ_{1s} into its components, we use Eq. (10-38) again to give

$$\left(\frac{1}{12}\right)\sum_{\mathcal{R}=1}^{12}\Gamma_{1s}(\mathcal{R})\chi^{(a_1')}(\mathcal{R})=\left(\frac{1}{12}\right)301301\begin{pmatrix}1\\2\\3\\1\\2\\3\end{pmatrix}=1 \tag{10-61}$$

$$\left(\frac{1}{12}\right)\sum_{\mathcal{R}}\Gamma_{1s}(\mathcal{R})\chi^{(e')}(\mathcal{R})=1, \tag{10-62}$$

and

$$\left(\frac{1}{12}\right)\sum_{\mathcal{R}}\Gamma_{1s}\chi^{(a_2')}=\left(\frac{1}{12}\right)\sum_{\mathcal{R}}\Gamma_{1s}\chi^{(a_1'')}=\left(\frac{1}{12}\right)\sum_{\mathcal{R}}\Gamma_{1s}(\mathcal{R})\chi^{(a_2'')}$$

$$=\left(\frac{1}{12}\right)\sum_{\mathcal{R}}\Gamma_{1s}\chi^{(a'')}=0. \tag{10-63}$$

Thus, the group of three $1s$ orbitals can be used to form MOs that have symmetry properties of two irreducible representations (a_1' and e'), but no others.

The next step is to determine the particular linear combination of $1s_a$, $1s_b$, and $1s_c$ orbitals that gives the MO of a_1' symmetry, and which linear combination gives e' symmetry. To do this, we shall use the projection operator approach, as defined in Eq. (10-51). In particular, we wish to utilize a projection operator that will determine the coefficients a, b, and c in

$$\varphi=a\cdot 1s_a+b\cdot 1s_b+c\cdot 1s_c, \tag{10-64}$$

to give φs of a_1' and e' symmetry.

For a_1' symmetry, we have

$$\mathcal{P}^{(a_1')}\varphi=\left(\frac{1}{12}\right)\sum_{\mathcal{R}=1}^{12}\chi^{(a_1')}(\mathcal{R})^*\mathcal{R}\varphi$$

$$=\left(\frac{1}{12}\right)\{1\cdot E\varphi+1\cdot C_3^+\varphi+1\cdot C_3^-\varphi+1\cdot C_2\varphi$$

$$+1\cdot C_2'\varphi+1\cdot C_2''\varphi+1\cdot\sigma_\lambda\varphi+1\cdot S_3^+\varphi$$

$$+1\cdot S_3^-\varphi+1\cdot\sigma_v\varphi+1\cdot\sigma_v'\varphi+1\cdot\sigma_v''\varphi\}$$

$$= \left(\frac{1}{12}\right) \{(a\,1s_a + b\,1s_b + c\,1s_c) + (a\,1s_b + b\,1s_c + c\,1s_a)$$

$$+ (a\,1s_c + b\,1s_a + c\,1s_b) + (a\,1s_a + b\,1s_c + c\,1s_b) + (a\,1s_c + b\,1s_b + c\,1s_a)$$

$$+ (a\,1s_b + b\,1s_a + c\,1s_c) + (a\,1s_a + b\,1s_b + c\,1s_c) + (a\,1s_b + b\,1s_c + c\,1s_a)$$

$$+ (a\,1s_c + b\,1s_a + c\,1s_b) + (a\,1s_a + b\,1s_c + c\,1s_b) + (a\,1s_c + b\,1s_b + c\,1s_a)$$

$$+ (a\,1s_b + b\,1s_a + c\,1s_c)\}$$

$$= \left(\frac{1}{12}\right) \{(4a + 4b + 4c)\,1s_a + (4a + 4b + 4c)\,1s_b$$

$$+ (4a + 4b + 4c)\,1s_c\}$$

or

$$\varphi^{(a_1')} = \mathcal{P}^{(a_1')}\varphi = N(1s_a + 1s_b + 1s_c). \tag{10-66}$$

Thus, aside from determination of a normalization constant,[41] we see that a MO of a_1' is obtained simply by picking $a = b = c$ in Eq. (10-64). The overall wavefunction having A_1' symmetry is then constructed simply as

$$\Phi(1, 2) = N\varphi^{(a_1')}(1)\varphi^{(a_1')}(2). \tag{10-67}$$

Now let us turn to the construction of an MO having e' symmetry. Actually, since the e' irreducible representation is two-dimensional, we shall need to find two (linearly independent) functions having e' symmetry. To do this, we shall employ the projection operator approach of Eq. (10-51) again, but in a slightly different manner. In the a_1' case, we employed a linear combination of all three basis orbitals. However, Eq. (10-51) can be applied equally well to any one of the basis orbitals [which can be considered as "arbitrary functions" relative to the discussion surrounding Eqs. (10-49)–(10-51)], as we shall now see. Using Eq. (10-51), we have (choosing the $1s_a$ orbital for convenience)

$$\varphi_1(e') = \left(\frac{2}{12}\right) \sum_{\mathcal{R}=1}^{12} \chi^{(e')}(\mathcal{R})\mathcal{R}\,1s_a \tag{10-68}$$

$$= N\{2 \cdot 1s_a - 1s_b - 1s_c\}. \tag{10-69}$$

To obtain the second MO, we simply start with a different basis function, e.g., $1s_b$, which results in

$$\varphi_2(e') = N\{2 \cdot 1s_b - 1s_c - 1s_a\}. \tag{10-70}$$

Two additional steps are needed, however, to complete the process. First, the process we have used to generate φ_1 and φ_2 does not assure orthogonality. It is easily verified that the following combination of φ_1 and φ_2 is orthogonal[42]:

$$\varphi_1'(e') = N(\varphi_1 + \varphi_2),$$

[41] See Problem 10-3.

[42] More generally, the Schmidt orthogonalization procedure or another general orthogonalization procedure would be used to create orthogonal orbitals.

$$\varphi_2'(e') = N(\varphi_1 - \varphi_2).$$

When Eqs. (10-69) and (10-70) are substituted into Eqs. (10-71) and (10-72) and the resulting orbitals are normalized, the following result is obtained:

$$\varphi_1^{(e')} = \frac{1}{[6(1-S)]^{1/2}} \cdot [2 \cdot 1s_c - 1s_a - 1s_b] \tag{10-73}$$

$$\varphi_2^{(e')} = \frac{1}{[2(1-S)]^{1/2}} \cdot [1s_a - 1s_b], \tag{10-74}$$

where S is the overlap integral between any pair of orbitals from the ($1s_a$, $1s_b$, $1s_c$) set.

Finally, while products φ_1^2 and φ_2^2 are assured to contain an a_1' component by the process we used; the process does not guarantee that a ''pure'' A_1' state will result when an overall wavefunction is formed. In particular, it is easily verified that the following combination must be used if a configuration having *only* an A_1' component is to be obtained[43]:

$$\Phi(1, 2) = N[3\varphi_2^{(e')}(1)\varphi_2^{(e')}(2) + \varphi_1^{(e')}(1)\varphi_1^{(e')}(2)]. \tag{10-75}$$

Thus, we have seen that, with a basis set of three $1s$ orbitals, a wavefunction having A_1' symmetry can be constructed using either a_1' or e' MOs. However, we saw in Eqs. (10-54) and (10-56)–(10-58) that A_1' states are also possible by the use of a_2', a_1'', a_2'', and e'' MOs, yet no MOs of those symmetries have been found in the procedure just completed. The reason for this is that $1s$ orbitals lack the symmetry properties necessary to construct MOs with a_2', a_1'', a_2'', and e'' symmetries, and orbitals such as $2p_x$, $2p_y$, and $2p_z$ would be needed in the basis if MOs having these other symmetries are to be constructed. In any case, we see that the techniques of group theory provide powerful tools for assuring that wavefunctions contain appropriate symmetry characteristics. We shall see additional applications in the sections and chapters to follow.

Before leaving this section, let us examine how construction of orbitals that have the symmetry properties of irreducible representations of the group of interest can be helpful in evaluation of integrals. In particular, let us return to Eq. (10-45) and consider two functions, $\varphi_k^{(\kappa)}$ and $\varphi_l^{(\lambda)}$, which respectively are members of sets of functions which form the bases for l_κ- and l_λ-dimensional irreducible representations κ and λ of a group of interest. If we form the scalar product of these functions, such a scalar product should be unaffected by any symmetry operation ($\mathcal{R}$) of the group, i.e.,

$$\begin{aligned} \langle \varphi_k^{(\kappa)} | \varphi_l^{(\lambda)} \rangle &= \mathcal{R} \langle \varphi_k^{(\kappa)} | \varphi_l^{(\lambda)} \rangle \\ &= \langle \mathcal{R}\varphi_k^{(\kappa)} | \mathcal{R}\varphi_l^{(\lambda)} \rangle. \end{aligned} \tag{10-76}$$

[43] In general, projection operator techniques can be used to obtain these results, as we showed earlier in this section when dealing with individual MO symmetries.

Using Eq. (10-45), the above equation becomes

$$\langle \varphi_k^{(\kappa)} | \varphi_l^{(\lambda)} \rangle = \sum_{s=1}^{l_\kappa} \sum_{t=1}^{l_\lambda} \langle \gamma_{sk}^{(\kappa)}(\mathfrak{R}) \varphi_s^{(\kappa)} | \gamma_{tl}^{(\lambda)}(\mathfrak{R}) \varphi_t^{(\lambda)} \rangle.$$

Since the above result is valid for any symmetry operation $(\mathfrak{R})$, we can rewrite the above equation equivalently as

$$\begin{aligned} \langle \varphi_k^{(\kappa)} | \varphi_l^{(\lambda)} \rangle &= \left(\frac{1}{g}\right) \sum_{\mathfrak{R}=1}^{g} \sum_{s=1}^{l_\kappa} \sum_{t=1}^{l_\lambda} \gamma_{sk}^{(\kappa)}(\mathfrak{R})^* \gamma_{tl}^{(\lambda)}(\mathfrak{R}) \langle \varphi_s^{(\kappa)} | \varphi_t^{(\lambda)} \rangle \\ &= \left(\frac{1}{g}\right) \sum_{s=1}^{l_\kappa} \sum_{t=1}^{l_\lambda} \langle \varphi_s^{(\kappa)} | \varphi_t^{(\lambda)} \rangle \left[\sum_{\mathfrak{R}=1}^{g} \gamma_{sk}^{(\kappa)}(\mathfrak{R})^* \gamma_{tl}^{(\lambda)}(\mathfrak{R}) \right], \end{aligned} \tag{10-77}$$

where g is the number of symmetry elements in the group. Use of the Great Orthogonality Theorem [Eq. (10-29)] allows the above equation to be written as

$$\begin{aligned} \langle \varphi_k^{(\kappa)} | \varphi_l^{(\lambda)} \rangle &= \left(\frac{1}{g}\right) \sum_{s=1}^{l_\kappa} \sum_{t=1}^{l_\lambda} \langle \varphi_s^{(\kappa)} | \varphi_t^{(\lambda)} \rangle \delta_{st} \delta_{kl} \delta_{\kappa\lambda} \left(\frac{g}{l_\kappa}\right) \\ &= \left(\frac{1}{l_\kappa}\right) \sum_{s=1}^{l_\kappa} \langle \varphi_s^{(\kappa)} | \varphi_s^{(\lambda)} \rangle \delta_{\kappa\lambda} \delta_{kl} \\ &= (1/l_\kappa) \cdot (C) \cdot \delta_{\kappa\lambda} \cdot \delta_{kl}, \end{aligned} \tag{10-78}$$

where C is a constant. This equation provides two important results. First if $\kappa \neq \lambda$ and $k = l$, Eq. (10-78) shows that functions belonging to different irreducible representations are orthogonal. The second case of interest, i.e., when $\kappa = \lambda$ and $k \neq l$, shows that different members of an l_κ-dimensional set of basis functions of the same irreducible representation are also orthogonal.

To illustrate this result, let us return to the H_3^+ equilateral triangle example, and consider the following scalar product:

$$\langle \varphi^{(a_1')} | \varphi'^{(e')} \rangle \cong \langle (1s_a + 1s_b + 1s_c) | (1s_a - 1s_b) \rangle,$$

where we have utilized Eqs. (10-67) and (10-74) and ignored normalization constants for convenience. Expansion of the above equation gives

$$\begin{aligned} \langle \varphi^{(a_1')} | \varphi'^{(e')} \rangle &\cong \langle 1s_a | 1s_a \rangle + \langle 1s_b | 1s_a \rangle + \langle 1s_c | 1s_a \rangle \\ &\quad - \langle 1s_a | 1s_b \rangle - \langle 1s_b | 1s_b \rangle - \langle 1s_c | 1s_b \rangle \\ &= 0. \end{aligned}$$

In other words, use of Eq. (10-78) would have allowed us to conclude that the above scalar product was zero without having to carry out the analysis given in the above equation. Thus, for MOs representing the other irreducible representations in H_3^+, we can conclude

$$\begin{aligned}\langle\varphi^{(a_1')}|\varphi^{(a_2')}\rangle &= \langle\varphi^{(a_1')}|\varphi^{(a_1'')}\rangle = \langle\varphi^{(a_1')}|\varphi^{(a_2'')}\rangle \\ &= \langle\varphi^{(a_1')}|\varphi^{(e')}\rangle = \langle\varphi^{(a_1')}|\varphi^{(e')}\rangle \\ &= 0,\end{aligned}$$

with similar results for scalar products over all other pairs of irreducible representations.

We can summarize the general result of Eq. (10-78) using group theory language as follows: unless the direct product of the irreducible representations for which two functions transform contains the totally symmetric representation, the scalar product of the two functions will be zero, i.e.,

$$\langle\varphi^{(i)}|\varphi^{(j)}\rangle = 0 \tag{10-79}$$

unless[44]

$$\Gamma_i \otimes \Gamma_j \propto \Gamma_1. \tag{10-80}$$

This result can now be generalized easily to integrals involving operators. Specifically, we can see by analogy to Eqs. (10-79) and (10-80) that

$$\langle\varphi^{(i)}|f|\varphi^{(j)}\rangle = 0, \tag{10-81}$$

for any operator f, unless

$$\Gamma_i \otimes \Gamma_f \otimes \Gamma_j = \Gamma_{A_1}. \tag{10-82}$$

The relationships in Eqs. (10-79)–(10-82) will be quite useful in the evaluation of matrix elements needed in later discussions.

In particular, if $f = \mathcal{H}$, where $\mathcal{H}$ is a Hamiltonian operator that is invariant under all operations of a group, we see that $\mathcal{H}$ transforms as the totally symmetric irreducible representation (A_1) of the group. This means that evaluation of matrix elements over $\mathcal{H}$ is determined by examination of the direct product of the irreducible representations of the functions involved, i.e.,

$$\langle\varphi^{(i)}|\mathcal{H}|\varphi^{(j)}\rangle = 0 \tag{10-83}$$

unless

$$\Gamma_i \otimes \Gamma_{\mathcal{H}} \otimes \Gamma_j = \Gamma_i \otimes \Gamma_{A_1} \otimes \Gamma_j = \Gamma_i \otimes \Gamma_j \propto \Gamma_{A_1}. \tag{10-84}$$

Such a result is very useful in the evaluation of Hamiltonian matrix elements, as we shall see in subsequent discussions. Furthermore, if the functions ($\varphi^{(i)}$ and $\varphi^{(j)}$) are chosen to be symmetry functions, i.e., contain only components of a *single* irreducible representation, Eq. (10-83) can be applied directly without concern about contributions from multiple irreducible representations. Imposing such a constraint is frequently done, and simplifies matrix element calculation

[44] The symbol "$\propto$" in Eqs. (10-80) and elsewhere in this section means "contains."

considerably. In particular, they make the resulting Hamiltonian matrix block diagonal, with nonzero blocks arising only where functions transform to the same irreducible representation.

10-3. Antisymmetry and the Pauli Exclusion Principle

Our earlier considerations showed that a single electron has many quantum mechanical properties that have classical analogs (e.g., kinetic energy, orbital angular momentum, etc.), and at least one property (spin) for which there is no classical analog. We have also seen that, at least in the Dirac formulation, spin as well as the other components, are explicitly present in the one-electron wavefunction.

However, we have also seen that the nonrelativistic, time-independent Hamiltonian operator in the absence of external fields (which is the form of the Hamiltonian that is most often used) does not contain any spin terms. In other words, wavefunctions determined for this form of Hamiltonian will be insensitive to the spin state of the various electrons in the system as well as the system as a whole unless other constraints are added. Our next task, therefore, is to see if constraints other than those due to spatial symmetry need to be placed on approximate wavefunctions so that an appropriate description of electron spin is incorporated.

That additional constraints should be needed for multielectron wavefunctions should not come as a complete surprise. In particular, the discussions in Chapter 5 concerning the Uncertainty Principle and the measurement process for microscopic systems provide strong evidence that identical microscopic particles (e.g., electrons) will be much less distinguishable than their classical mechanical analogs. Specifically, the detailed path of motion can be described for any classical particle, hence allowing us to know precisely where the various classical particles are at a given time, thus allowing us to distinguish them from one another.

However, we have seen that, for a microscopic particle, the position and momentum cannot be measured simultaneously without incurring an uncertainty of at least $(\hbar/2)$ in the momentum-position product. This means that we cannot expect to determine the path of motion of a microscopic particle to arbitrary accuracy as was the case for classical particles. Furthermore, if more than one identical microscopic particle is present (e.g., two electrons), there will be an uncertainty in locating each of them. When extrapolated, this leads to the conclusion that, at the microscopic level, we cannot distinguish one identical particle from another. Thus, in contrast to classical particles, the Uncertainty Principle applied to microscopic systems requires us to create a description that does not distinguish among identical particles. In other words, a wavefunction that is to describe a collection of identical particles (e.g., an N-electron wavefunction) must not distinguish among them.

To consider how this can be accomplished, let us consider a wavefunction for

N identical particles, i.e.,

$$\Psi(1\ 2\ \cdots\ N) = \Psi(\tau_1, \tau_2, \cdots \tau_N), \tag{10-85}$$

where $\tau_1, \tau_2, \ldots \tau_N$ represent both the space and spin coordinate dependence of ψ on each particle. If each particle is to be indistinguishable from the others, the exchange of any pair of particles should not affect any observable that is to be measured. This means that if $\mathcal{P}_{ij}$ is an operator that exchanges particles i and j,

$$\mathcal{P}_{ij}\Psi(\tau_1\ \tau_2\ \cdots\ \tau_i\ \cdots\ \tau_j\ \cdots\ \tau_N) = e^{ik}\Psi(\tau_1\ \tau_2\ \cdots\ \tau_j\ \cdots\ \tau_i\ \cdots\ \tau_N), \tag{10-86}$$

where k is a real number. The reason that only multiplication of ψ by e^{ik} will result from applying $\mathcal{P}_{ij}$ is that the normalization of ψ as well as expectation values of ψ need to be invariant to the application of $\mathcal{P}_{ij}$. In other words, if $\mathcal{A}$ is an operator corresponding to the observable A, the expectation value of $\mathcal{A}$ should be invariant to $\mathcal{P}_{ij}$, i.e.,

$$\begin{aligned}
\langle \mathcal{A} \rangle &= \int \cdots \int (\mathcal{P}_{ij}\Psi)^* \mathcal{A} (\mathcal{P}_{ij}\Psi)\, d\tau_1 \cdots d\tau_N \\
&= e^{-ik}e^{ik} \int \cdots \int \Psi^* \mathcal{A} \Psi\, d\tau_1 \cdots d\tau_N \\
&= \int \cdots \int \Psi^* \mathcal{A} \Psi\, d\tau_1 \cdots d\tau_N,
\end{aligned}$$

and the necessary behavior of $\langle \mathcal{A} \rangle$ is assured by inclusion of the factor e^{ik}.

If we take this analysis further, we can learn even more about the factor e^{ik} and its allowable form. Let us apply the operator $\mathcal{P}_{ij}$ twice, i.e., restore the original wavefunction,

$$\mathcal{P}_{ij}\mathcal{P}_{ij}\Psi = \Psi.$$

Using Eq. (10-86), the above equation becomes

$$\begin{aligned}
\mathcal{P}_{ij}\mathcal{P}_{ij}\Psi &= \mathcal{P}_{ij}[e^{ik}\Psi(\tau_1\ \tau_2\ \cdots\ \tau_j\ \cdots\ \tau_i\ \cdots\ \tau_N)] \\
&= e^{2ik}\Psi(\tau_1\ \tau_2\ \cdots\ \tau_i\ \cdots\ \tau_j\ \cdots\ \tau_N) \qquad (10\text{-}87) \\
&= \Psi(\tau_1\ \tau_2\ \cdots\ \tau_i\ \cdots\ \tau_j\ \cdots\ \tau_N). \qquad (10\text{-}88)
\end{aligned}$$

From Eqs. (10-87) and (10-88), we see that

$$e^{2ik} = +1$$

which implies that

$$e^{ik} = \pm 1.$$

Thus, we have discovered that

$$\mathcal{P}_{ij}\Psi = \pm\Psi, \tag{10-89}$$

i.e., interchanging any pair of identical particles must produce the original wavefunction back again, multiplied by ± 1. In other words, the wavefunction must be either antisymmetric or symmetric with respect to exchange of identical particles.

Since either result will give an acceptable wavefunction from the point of view of the Uncertainty Principle, it should not be surprising to find that both kinds of systems are found at the microscopic level. Specifically, it is found experimentally that the antisymmetric property is associated with particles called *fermions* (which include electrons), and which obey what are known as *Fermi–Dirac statistics*. Particles whose wavefunctions are symmetric with respect to exchange are known as *bosons* (e.g., α particles), and which obey *Bose-Einstein statistics*.

This antisymmetric wavefunction behavior for electrons was formulated first by Wolfgang Pauli in 1925,[45] when he postulated that, "There never exist two or more equivalent electrons in an atom which, in strong fields, agree in all quantum numbers." This postulate is known as the *Pauli Exclusion Principle.* That this postulate is equivalent to the antisymmetric wavefunction requirement developed above will be shown in the following section.

10-4. Multielectron Systems and Slater Determinants

We are now in a position to show in general how to construct wavefunctions for multielectron multinuclear systems that satisfy both spatial and spin symmetry requirements. However, it is important to note that, as illustrated in the H_2 example of Section 10-1, the wavefunctions we shall be constructing from this point on will in general be *approximate*, not exact, *wavefunctions*. This is due to the inability thus far to be able to find exact solutions to the time-independent Schrödinger equation for systems containing more than one electron. Hence, each wavefunction will reflect approximations due to the investigator's choice of theoretical model (i.e., overall wavefunction form) as well as basis set, even though satisfying space and symmetry requirements rigorously.

When we search for a way to incorporate the antisymmetry requirement of the previous section into a wavefunction, we find that our previous mathematical discussions have provided a convenient mechanism. In particular, we saw earlier (Theorem 3-1) that exchanging two rows or two columns of a determinant changes only the sign of the determinant. Thus, the use of determinants (or sums of them) provides as easy way to incorporate the antisymmetry requirement.

However, we must also decide *how* to represent electrons in a multielectron system. From the discussion of the historical MO and VB approaches to the H_2 molecule, it should not be surprising to find that the choice that has enjoyed overwhelming popularity is that of assigning individual electrons to individual functions (orbitals). For example, a three-electron atom might be described by a wavefunction containing three atomic orbitals, one for each electron. More

[45] W. Pauli, *Z. Phys.*, **31**, 765–785 (1925).

generally, the approach is to choose a set of orbitals, and utilize the first N of those to describe an N-electron system.

This idea can be generalized in a fashion that incorporates space and spin symmetry constraints by the use of *determinantal wavefunctions*. The use of determinantal wavefunctions was formulated by Heisenberg[46] and Dirac[47] and, since the appearance of a paper by Slater,[48] has been known as the *Slater determinant* approach. In this approach, individual electrons are assigned to individual orbitals, with the total determinant containing as many orbitals as there are electrons. For example, an N-electron system would be represented in Slater determinant form as

$$\psi(1,2,\cdots,N) = \frac{1}{[N!]^{1/2}} \begin{bmatrix} \psi_1(1) & \psi_2(1) & \psi_3(1) & \cdots & \psi_N(1) \\ \psi_1(2) & \psi_2(2) & \psi_3(2) & \cdots & \psi_N(2) \\ \psi_1(3) & \psi_2(3) & \psi_3(3) & \cdots & \psi_N(3) \\ \vdots & \vdots & \vdots & \ddots & \vdots \\ \psi_1(N) & \psi_2(N) & \psi_3(N) & \cdots & \psi_N(N) \end{bmatrix}, \tag{10-90}$$

where $\psi_j(k)$ is the jth orbital describing the kth electron, and $(1/N!)^{1/2}$ is required for normalization purposes.[49] It is important to note that the orbitals $[\psi_j(k)]$ are functions of both space and spin,[50] and hence are known as *spin orbitals*. However, it should be noted that a particular form for the $\psi_j(k)$ has not been chosen at this point, e.g., $\psi_j(k)$ might be chosen to be an atomic orbital or it might be chosen to be an MO. In any case, we note that each row corresponds to a different electron, and each column to a different orbital.

If the spin orbitals are taken as products of spatial and spin components (which is nearly universally done), then each spatial orbital may appear twice in a Slater determinant, once with α-spin and once with β-spin. If both appear, we refer to the spatial orbital as being *doubly* occupied. If only one of the spin components appears with a given spatial orbital, we say that the spatial orbital is *singly occupied*.

---

[46] W. Heisenberg, *Z. Phys.*, **38**, 411–426 (1926).

[47] P. A. M. Dirac, *Proc. R. Soc. (London)*, **A112**, 661–677 (1926).

[48] J. C. Slater, *Phys. Rev.*, **34**, 1293–1322 (1929).

[49] The normalization constant given in Eq. (10-90) applies only when the orbitals are orthonormal, which is the case we shall consider in virtually all applications. While the overall wavefunction (Ψ) in Eq. (10-90) is unaffected by the choice of orthogonal or non-orthogonal orbitals, the use of nonorthogonal orbitals produces energy expressions which are terribly unwieldy in general.

[50] It should also be noted that the choice of orbitals must also be such that the resulting wavefunction must in general possess the other necessary properties of wavefunctions that were identified in earlier discussions, e.g., continuity, single-valuedness, square integrability, etc.

For example, if the hydrogen-like orbitals of Eq. (7-47) were used to describe an N-electron atom, each ψ_i in Eq. (10-90) would have the form

$$\psi_i(1) = \phi_{nlm}(r_1, \vartheta_1, \varphi_1) \cdot \chi(s_1) \tag{10-91}$$

where ψ_i is characterized by n, l, m, and the spin function (χ), and the spatial portion (ϕ_{nlm}) can be either singly or doubly occupied in the Slater determinant. Clearly, such a description cannot be expected to provide an exact description of the N-electron system, since it assumes that the behavior of each electron is described by an orbital that is independent of all the others. However, while not exact, such an *independent particle model* is found to provide considerable insight into chemical systems.

Using the analysis presented earlier,[51] an alternate way of expressing the Slater determinant of Eq. (10-90) is seen to be

$$\Psi(1\ 2\ \cdots\ N) = \mathcal{A}\{\psi_1(1)\ \psi_2(2)\ \cdots\ \psi_N(N)\}, \tag{10-92}$$

where $\mathcal{A}$ is known as the *antisymmetrization operator*, and is given by

$$\mathcal{A} = \frac{1}{(N!)^{1/2}} \sum_{\mathcal{P}} (-1)^p \mathcal{P}, \tag{10-93}$$

where $\mathcal{P}$ is a permutation operator that permutes the electrons in the orbitals[52], p is the parity of the permutation, and the factor $(N!)^{-1/2}$ is introduced to maintain normalization of Ψ. Incidentally, it is easily seen[53] that the antisymmetrization operator ($\mathcal{A}$) is an example of a projection operator, since

$$\mathcal{A}^2 = \mathcal{A}. \tag{10-94}$$

To illustrate Eq. (10-92) and the use of permutation operators, let us consider the three-electron case, i.e.,

$$\begin{aligned}\Psi(1, 2, 3) &= \left(\frac{1}{3!}\right)^{1/2} \sum_{\mathcal{P}} (-1)^p \mathcal{P}\{\psi_1(1)\psi_2(2)\psi_3(3)\} \\ &= \left(\frac{1}{3!}\right)^{1/2} [\mathcal{I} - \mathcal{P}_{12} - \mathcal{P}_{13} - \mathcal{P}_{23} + \mathcal{P}_{12}\mathcal{P}_{23} + \mathcal{P}_{23}\mathcal{P}_{12}] \\ &\quad \cdot \{\psi_1(1)\psi_2(2)\psi_3(3)\} \\ &= \left(\frac{1}{3!}\right)^{1/2} \{\psi_1(1)\psi_2(2)\psi_3(3) - \psi_1(2)\psi_2(1)\psi_3(3) \\ &\quad - \psi_1(3)\psi_2(2)\psi_3(1) - \psi_1(1)\psi_2(3)\psi_3(2) \\ &\quad + \psi_1(2)\psi_2(3)\psi_3(1) + \psi_1(3)\psi_2(1)\psi_3(2)\},\end{aligned} \tag{10-95}$$

---

[51] See Section 3-2, Eq. (3-50).

[52] As an alternative to permuting electrons, it can be shown that the same result represented by Eqs. (10-92) and (10-93) can be obtained by permuting the indices of the basis orbitals. See Problem 10-8.

[53] See Problem 10-9.

where we have used the convention:[54]

$$\begin{aligned}\mathcal{P}_{12}\mathcal{P}_{23}\{\psi_1(1)\psi_2(2)\psi_3(3)\}\\ &= -\mathcal{P}_{12}\{\psi_1(1)\psi_2(3)\psi_3(2)\}\\ &= +\{\psi_1(2)\psi_2(3)\psi_3(1)\}.\end{aligned}$$

The representation of a Slater determinant used in Eq. (10-92) will be of substantial convenience in later analyses.

To see how the Slater determinant formulation is consistent with Pauli's formulation of the Exclusion Principle, consider the possibility that two spin orbitals have all quantum numbers (n, l, m, s) alike, i.e.,

$$\begin{aligned}\psi_i^{(1)} = \psi_j^{(2)} &= R_{nl}(r_1)\,Y_{lm}(\vartheta_1,\ \varphi_1)\chi(s_1)\\ &= R_{nl}(r_2)\,Y_{lm}(\vartheta_2,\ \varphi_2)\chi(s_2).\end{aligned} \tag{10-96}$$

When these are substituted into Eq. (10-90), it is seen that two rows are identical, which means that the determinant is identically zero. In other words, use of a Slater determinant automatically guarantees that the Pauli Exclusion Principle will be satisfied.

Let us now investigate the effect that inclusion of spin has on calculated properties, using the example of a two-electron atomic system to illustrate the point. For the ground state of such a system, the wavefunction can be represented as

$$\Psi(1,\ 2) = \frac{1}{\sqrt{2}}\begin{vmatrix} 1s(1)\alpha(1) & 1s(1)\beta(1) \\ 1s(2)\alpha(2) & 1s(2)\beta(2) \end{vmatrix}, \tag{10-97}$$

where

$$1s(i)\alpha(i) = R_{10}(r_i)\,Y_{00}(\vartheta_i,\ \varphi_i)\alpha(i), \tag{10-98}$$

and the two spin orbitals are the same except for the choice of spin function. In other words, each electron is described by a $1s$ spatial orbital, coupled with an α-spin for one electron and β-spin for the other electron.

Expanding the determinant gives

$$\Psi(1,\ 2) = \frac{1}{\sqrt{2}}\,[1s(1)\alpha(1)1s(2)\beta(2) - 1s(1)\beta(1)1s(2)\alpha(2)] \tag{10-99}$$

$$= \frac{1}{\sqrt{2}}\cdot\ 1s(1)1s(2)[\alpha(1)\beta(2) - \beta(1)\alpha(2)]. \tag{10-100}$$

Thus, the wavefunction has been *factored* into a product of spatial and spin functions. To see what effect such a factorization has upon the energy, let us

---

[54] Sometimes the operator form $\mathcal{P}_{ijk}$ is used to represent all permutations of three electrons. This is identical to the result obtained from the product of two-electron permutations ($\mathcal{P}_{12}\mathcal{P}_{23}$ and $\mathcal{P}_{23}\mathcal{P}_{12}$) in Eq. (10-95).

consider the expectation value of the nonrelativistic, time-independent Hamiltonian operator ($\mathcal{H}$) with a wavefunction of the form

$$\Psi(1, 2)=\varphi(\mathbf{r}_1, \mathbf{r}_2) \cdot \chi(1, 2), \tag{10-101}$$

which is the form of wavefunction represented by Eq. (10-100). The energy expectation value is given by

$$\begin{aligned} E=\langle\mathcal{H}\rangle &= \int\int \Psi^*(1, 2)\mathcal{H}\Psi(1, 2)\, d\tau_1\, d\tau_2 \\ &= \int\int\int\int \varphi^*(\mathbf{r}_1, \mathbf{r}_2)\chi^*(1, 2)\mathcal{H}\varphi(\mathbf{r}_1, \mathbf{r}_2)\chi(1, 2)\, dV_1\, dV_2\, ds_1\, ds_2. \end{aligned} \tag{10-102}$$

Since $\mathcal{H}$ is independent of spin, we have

$$E=\int\int \chi^*(1, 2)\chi(1, 2)\, ds_1\, ds_2 \cdot \int\int \varphi^*(\mathbf{r}_1, \mathbf{r}_2)\mathcal{H}\varphi(\mathbf{r}_1, \mathbf{r}_2)\, dV_1\, dV_2$$

or, since χ is normalized,

$$E=\int\int \varphi^*(\mathbf{r}_1, \mathbf{r}_2)\mathcal{H}\varphi(\mathbf{r}_1, \mathbf{r}_2)\, dV_1\, dV_2. \tag{10-103}$$

Thus, we see that the expectation value of the energy is unaffected by the choice of spin function, even though an antisymmetric overall wavefunction is required by the Pauli Exclusion Principle.[55] Even though the idea of factoring the wavefunction into a product of space and functions has been illustrated for only the two-electron case, it can be generalized[56] to the N-electron case for the choice of Hamiltonian used here. In other words, even though we have chosen a formulation using Slater determinants that includes the spin dependence explicitly in the orbitals, it is always possible for the case of the nonrelativistic, time-independent Hamiltonian to formulate an alternative approach that separates the wavefunction into a product of space and spin functions.

10-5. Expansion Theorem and Slater Determinant Expansions

We have now seen that the use of a Slater determinant assures that an approximate wavefunction contains the appropriate antisymmetric behavior of electrons but cannot, by itself, be an exact solution to the Schrödinger equation because of its independent particle model formulation. While developing the characteristics and techniques of the independent particle model and corrections to it will comprise much of the remainder of this text, it is of importance first to

[55] The fact that the spin can be factored from the spatial portion of the wavefunction is the reason why the considerations at the beginning of this chapter concerning valence bond and molecular orbital theory for two-electron systems are valid as presented.

[56] See, for examples, F. A. Matsen, *Adv. Quantum Chem.*, **1**, 60–114 (1964).

show that, at least *in principle*, the appropriate use of Slater determinants can provide an *exact* solution to the nonrelativistic Schrödinger equation.

To do this, we recall from earlier discussions[57] that, if a complete set of functions (vectors) is available that spans the space, the set forms a basis for the space, and any arbitrary function in the space can be expressed in terms of the basis functions. In other words, the orbitals discussed thus far will now be used as a basis set with which to expand the wavefunction. Utilizing this concept for the case of a *single* electron wavefunction,[58] we have

$$\Psi(\tau)=\sum_{k} c_k \psi_k(\tau), \tag{10-104}$$

where c_k are the expansion coefficients to be determined, and τ includes both the spatial ($\mathbf{r}$) and spin (s) coordinates

$$\tau=f(\mathbf{r}, s), \tag{10-105}$$

and the set of basis functions $\{\psi_k(\tau)\}$ is complete. This latter point implies that the summation in Eq. (10-104) may contain an infinite number of terms (e.g., if the basis is composed entirely of discrete functions), or may contain continuum functions, or both.

If we now consider the case of *two* electrons, we require a wavefunction $\Psi(\tau_1, \tau_2)$ of two coordinates (τ_1, τ_2). If τ_2 is considered as a *fixed* parameter for the moment, we can expand $\Psi(\tau_1, \tau_2)$ as before, i.e.,

$$\Psi(\tau_1, \tau_2)=\sum_{k} c_k(\tau_2)\psi_k(\tau_1). \tag{10-106}$$

The difference between Eqs. (10-106) and (10-104) is that the coefficients $[c_k(\tau_2)]$ in Eq. (10-106) are functions of τ_2, and must be determined for all values of τ_2. However, as a function of τ_2, $c_k(\tau_2)$ can also be expanded in terms of the basis functions to give:

$$\begin{aligned}\Psi(\tau_1, \tau_2)&=\sum_{k}\left[\sum_{l} d_{kl}\psi_l(\tau_2)\right]\psi_k(\tau_1)\\ &=\sum_{k,l} d_{kl}\psi_k(\tau_1)\psi_l(\tau_2),\end{aligned} \tag{10-107}$$

where d_{kl} are coefficients to be determined that are *independent* of τ_1 and τ_2.

For the case of N electrons, Eq. (10-107) is easily seen to be generalized to

$$\Psi(\tau_1 \cdots \tau_N)=\sum_{k_1,k_2,\cdots,k_N} d(k_1 \cdots k_N)\psi_{k_1}(\tau_1)\psi_{k_2}(\tau_2) \cdots \psi_{k_N}(\tau_N). \tag{10-108}$$

---

[57] See Section 2-2 through 2-5.

[58] The presentation in this section is based upon the analysis of P. O. Löwdin, *Adv. Chem. Phys.*, **2**, 207–322 (1959).

However, we have just seen that not every product, but only the antisymmetrized product of spin orbitals forms an acceptable wavefunction. To assure antisymmetry we apply the antisymmetrization operator ($\mathcal{A}$) of Eq. (10-93) to both sides of Eq. (10-108) to give

$$\Psi(\tau_1 \cdots \tau_N) = \sum_{k_1 \cdots k_N} d(k_1 \cdots k_N)\mathcal{A}\{\psi_{k_1}(\tau_1)\psi_{k_2}(\tau_2) \cdots \psi_{k_N}(\tau_N)\}, \tag{10-109}$$

where $\Psi(\tau_1 \tau_2 \ldots \tau_N)$ now refers to an appropriately antisymmetrized wavefunction. Using the explicit form of $\mathcal{A}$ given in Eq. (10-93) plus the definition of a determinant [Eq. (3-50)] gives

$$\Psi(\tau_1, \tau_2 \cdots \tau_N) = \frac{1}{(N!)^{1/2}} \sum_{k_1 \cdots k_N} d(k_1 \cdots k_N) \cdot \det\{\psi_{k_1}(\tau_1)\psi_{k_2}(\tau_2) \cdots \psi_{k_N}(\tau_N)\}. \tag{10-110}$$

This indicates that the *exact wavefunction* for an N-electron system can be written as a sum of Slater determinants formed from a complete basis set of one-electron functions. This principle is frequently referred to as the *Expansion Theorem,* and forms the basis for much of computational quantum chemistry as we know it today.

Several additional points should be noted, however. First, there is no guarantee as to how fast the expansion in Eq. (10-110) will converge. Indeed, much of the "art" of computational quantum chemistry is associated with formulating approaches to deal with the usual slow convergence of the expansion.

The other major point of interest is that, if either a noncomplete basis set or a truncated basis set (e.g., a finite basis set, whether or not the total basis is complete) is used, there is no guarantee that the expansion will converge. Thus, even though we are assured that the use of Slater determinants and complete set of one-electron orbitals can give exact answers in principle, it is important to estimate the errors involved if the expansion is used in truncated form.

10-6. Matrix Elements between Slater Determinants

In order to prepare for the use of Slater determinants in the calculation of approximate wavefunctions and properties, it is useful to carry out a general analysis of the various kinds of matrix elements that will arise, as well as to establish notation that will be useful later.

Let us begin with the operators that are involved. For the case of the nonrelativistic Hamiltonian in the Born–Oppenheimer approximation, we have seen that it can be written as follows for an N-electron molecule containing M nuclei:

$$\mathcal{H} = \mathcal{H}_0 + \mathcal{H}_1 + \mathcal{H}_2 \tag{10-111}$$

where $\mathcal{H}_0$ is a constant term, $\mathcal{H}_1$ contains only one-electron operators, and $\mathcal{H}_2$ contains only two-electron operators, i.e.,

$$\mathcal{H}_0 = \sum_{\alpha<\beta}^{M} \frac{Z_\alpha Z_\beta}{R_{\alpha\beta}}, \tag{10-112}$$

where Z_k is the charge on the αth nucleus, and $R_{\alpha\beta}$ is the (fixed) distance between nucleus α and β.

$$\mathcal{H}_1 = \sum_{i=1}^{N} \left[-\frac{1}{2}\nabla_1^2 - \sum_{\alpha=1}^{M} \frac{Z_\alpha}{r_{\alpha i}} \right] = \sum_{i=1}^{N} h(i), \tag{10-113}$$

where the first term contains the kinetic energy operator for each electron, and the second term represents the attraction of each electron to each nucleus. Note that $\mathcal{H}_1$ is symmetric in the electron coordinates.

The final term is

$$\mathcal{H}_2 = \sum_{i<j}^{N} \frac{1}{r_{ij}} \tag{10-114}$$

and represents the electron–electron repulsion terms. As in the previous case, $\mathcal{H}_2$ is symmetric in the electron coordinates.

Even for operators other than the nonrelativistic Hamiltonian,[59] it was seen that operators involving more than two electrons never occur. Thus, treatment of matrix elements using the terms represented in Eq. (10-111) represents the general case of interest.

Therefore, we shall be interested in evaluating three kinds of matrix elements:[60]

$$\langle D|\mathcal{H}_0|D'\rangle = \int \cdots \int D^* \mathcal{H}_0 D' \, d\tau_1 \cdots d\tau_N, \tag{10-115}$$

$$\langle D|\mathcal{H}_1|D'\rangle = \int \cdots \int D^* \mathcal{H}_1 D' \, d\tau_1 \cdots d\tau_N, \tag{10-116}$$

and

$$\langle D|\mathcal{H}_2|D'\rangle = \int \cdots \int D^* \mathcal{H}_2 D' \, d\tau_1 \cdots d\tau_N, \tag{10-117}$$

where the integration is taken over both spatial and spin coordinates, and the

---

[59] See Chapter 8, Section 8-3B.

[60] The method of analysis used here is essentially that used in P. O. Löwdin, *Phys. Rev.*, **97**, 1474–1490 (1955). See also J. C. Slater, *Quantum Theory of Atomic Structure*, Vol. I, McGraw-Hill, New York, 1960, Chapter 12.

Slater determinants (D and D') are defined as

$$D = \frac{1}{(N!)^{1/2}} \begin{vmatrix} \psi_1(1) & \psi_2(1) & \cdots & \psi_N(1) \\ \psi_1(2) & \psi_2(2) & \cdots & \psi_N(2) \\ \vdots & \vdots & \ddots & \vdots \\ \psi_1(N) & \psi_2(N) & \cdots & \psi_N(N) \end{vmatrix} \tag{10-118}$$

$$= \frac{1}{(N!)^{1/2}} \sum_{\mathcal{P}} (-1)^p \mathcal{P}\{\psi_1(1)\psi_2(2) \cdots \psi_N(N)\}, \tag{10-119}$$

and

$$D' = \frac{1}{(N!)^{1/2}} \sum_{\mathcal{P}} (-1)^p \mathcal{P}\{\psi_{1'}(1)\psi_{2'}(2) \cdots \psi_{N'}(N)\}, \tag{10-120}$$

where the ψ_i and $\psi_{i'}$ are chosen from a single complete set of orthonormal orbitals, i.e., their overlap integrals are given by

$$\int \psi_i^*(1)\psi_j(1)\, d\tau_1 = \delta_{ij}. \tag{10-121}$$

Thus, the same set of orbitals is used in constructing both D and D', although the particular choice of orbitals used in D may be different from those used in D'.

Before dealing with the matrix elements in Eqs. (10-115) through (10-117) individually, it is useful first to show how each of them can be simplified. Since the permutations in Eqs. (10-119) and (10-120) involve "zero" permutations (i.e., the identity permutation), single permutations, double permutations, etc., let us begin by noting the form of the integral over the determinant (D) with the determinant (D') and a general operator ($\mathfrak{F}$) that is symmetric in the electron coordinates, i.e.,

$$\langle D|\mathfrak{F}|D'\rangle = \left(\frac{1}{N!}\right) \left[\int \cdots \int \psi_1^*(1)\psi_2^*(2) \cdots \psi_N^*(N)\mathfrak{F} \cdot \begin{vmatrix} \psi_{1'}(1) & \psi_{2'}(1) & \cdots & \psi_{N'}(1) \\ \psi_{1'}(2) & \psi_{2'}(2) & \cdots & \psi_{N'}(2) \\ \vdots & \vdots & \ddots & \vdots \\ \psi_{1'}(N) & \psi_{2'}(N) & \cdots & \psi_{N'}(N) \end{vmatrix} d\tau_1 \cdots d\tau_N + \cdots \right], \tag{10-122}$$

where only the "zero permutation" term of the $N!$ terms arising from expansion of D has been written explicitly. For each single permutation term of D, a pair of electrons (e.g., i and j) is interchanged in the product $\psi_1^* \dots \psi_N^*$, and a factor of (-1) is attached. However, an interchange of rows i and j in D' also results in a factor of -1, and the resulting term has exactly the same form as the first term in Eq. (10-122), including sign, except for the label on electrons. Since $\mathcal{H}$ is symmetric and since the electron labels are simply integration variables, it is clear that each of the single permutation terms is exactly the same as the first term in Eq. (10-122). A similar analysis can be applied to the other permutations, with the result that there are $N!$ identical terms that are given by the first term in Eq. (10-122), that will exactly cancel the $(1/N!)$ that is present in Eq. (10-122). Thus, we can reduce the $(N!)^2$ terms with which we started, without loss of generality, to

$$\langle D|\mathfrak{F}|D'\rangle = \int \cdots \int \psi_1^*(1)\psi_2^*(2) \cdots \psi_N^*(N)\mathfrak{F} \cdot \begin{vmatrix} \psi_{1'}(1) & \psi_{2'}(1) & \cdots & \psi_{N'}(1) \\ \psi_{1'}(2) & \psi_{2'}(2) & \cdots & \psi_{N'}(2) \\ \vdots & \vdots & \ddots & \vdots \\ \psi_{1'}(N) & \psi_{2'}(N) & \cdots & \psi_{N'}(N) \end{vmatrix} d\tau_1 \cdots d\tau_N.$$

Thus, we have just proven the general result that can be stated alternatively using Eq. (10-94) as

$$\begin{aligned} \langle D|\mathfrak{F}|D'\rangle &= \langle \mathcal{A}\{\psi_1(1) \cdots \psi_N(N)\}|\mathfrak{F}|\mathcal{A}\{\psi_1'(1) \cdots \psi_N'(N)\}\rangle \\ &= \langle \psi_1(1) \cdots \psi_N(N)|\mathfrak{F}|\mathcal{A}^2\{\psi_1'(1) \cdots \psi_N'(N)\}\rangle \\ &= \langle \psi_1(1) \cdots \psi_N(N)|\mathfrak{F}|\mathcal{A}\{\psi_1'(1) \cdots \psi_N'(N)\}\rangle \\ &= \langle \psi_1(1) \cdots \psi_N(N)|\mathfrak{F}|D'\rangle. \end{aligned} \qquad (10\text{-}123)$$

A. Zero-Electron Operators

Turning now to the matrix elements for the various operators in Eqs. (10-115) through (10-117), the "zero-electron" operator matrix element is given by

$$\langle D|\mathcal{H}_0|D'\rangle = \mathcal{H}_0 \int \cdots \int \psi_1^*(1)\psi_2^*(2) \cdots \psi_N^*(N) \cdot \begin{vmatrix} \psi_{1'}(1) & \psi_{2'}(1) & \cdots & \psi_{N'}(1) \\ \psi_{1'}(2) & \psi_{2'}(2) & \cdots & \psi_{N'}(2) \\ \vdots & \vdots & \ddots & \vdots \\ \psi_{1'}(N) & \psi_{2'}(N) & \cdots & \psi_{N'}(N) \end{vmatrix} d\tau_1 \cdots d\tau_N.$$

Using Theorem 3-2, we can rewrite the above equation as

$$\langle D|\mathcal{H}_0|D'\rangle$$

$$= \mathcal{H}_0 \int \cdots \int \psi_2^*(2) \cdots \psi_N^*(N)$$

$$\cdot \begin{bmatrix} \int \psi_1^*(1)\psi_{1'}(1)\, d\boldsymbol{\tau}_1 & \int \psi_1^*(1)\psi_{2'}(1)\, d\boldsymbol{\tau}_1 & \cdots & \int \psi_1^*(1)\psi_{N'}(1)\, d\boldsymbol{\tau}_1 \\ \psi_{1'}(2) & \psi_{2'}(2) & \cdots & \psi_{N'}(2) \\ \vdots & \vdots & \ddots & \vdots \\ \psi_{1'}(N) & \psi_{2'}(N) & \cdots & \psi_{N'}(N) \end{bmatrix} \cdot p\boldsymbol{\tau}_2 \cdots d\boldsymbol{\tau}_N$$

where each element of the first row has been multiplied by $\psi_1^*(1)$ and integrated over $d\boldsymbol{\tau}_1$. Continuing the process for $\psi_2^* \ldots \psi_N^*$ gives

$$\langle D|\mathcal{H}_0|D'\rangle$$

$$= \mathcal{H}_0$$

$$\cdot \begin{bmatrix} \int \psi_1^*(N)\psi_{1'}(1)\, d\boldsymbol{\tau}_1 & \int \psi_1^*(1)\psi_{2'}(1)\, d\boldsymbol{\tau}_1 & \cdots & \int \psi_1^*(1)\psi_{N'}(1)\, d\boldsymbol{\tau}_1 \\ \int \psi_2^*(2)\psi_{1'}(2)\, d\boldsymbol{\tau}_2 & \int \psi_2^*(2)\psi_{2'}(2)\, d\boldsymbol{\tau}_2 & \cdots & \int \psi_2^*(2)\psi_{N'}(2)\, d\boldsymbol{\tau}_2 \\ \vdots & \vdots & \ddots & \vdots \\ \int \psi_N^*(N)\psi_{1'}(1)\, d\boldsymbol{\tau}_N & \int \psi_N^*(N)\psi_{2'}(N)\, d\boldsymbol{\tau}_N & \cdots & \int \psi_N^*(N)\psi_{N'}(N)\, d\boldsymbol{\tau}_N \end{bmatrix}$$

But, the basis set is orthonormal, which means that, unless *each* primed orbital is equal to its unprimed counterpart, the entire determinant will be zero. To see this, assume that all primed orbitals are the same as their unprimed orbital counterparts except for the Nth orbital. That implies that the elements of the last column are given by

$$\int \psi_i^* \psi_{N'}\; d\boldsymbol{\tau}_i = 0, \qquad i = 1, 2, \cdots, N,$$

which means that the entire determinant vanishes. For the special case where $D' \equiv D$, we have

$$\langle D|\mathcal{H}_0|D\rangle = \mathcal{H}_0 \begin{bmatrix} 1 & 0 & 0 & \cdots & 0 \\ 0 & 1 & 0 & \cdots & 0 \\ 0 & 0 & 1 & & \\ \vdots & \vdots & & \ddots & \vdots \\ 0 & 0 & \cdots & \cdots & 1 \end{bmatrix} = \mathcal{H}_0, \tag{10-124}$$

indicating simply that D is normalized.

B. One-Electron Operators

Next we consider matrix elements over one-electron operators [Eq. (10-116)], i.e.,

$$\langle D|\mathcal{H}_1|D'\rangle = \int \cdots \int \psi_1^*(1)\psi_2^*(2)\cdots\psi_N^*(N)\left(\sum_{i=1}^{N} h(i)\right) \cdot \begin{bmatrix} \psi_{1'}(1) & \psi_{2'}(1) & \cdots & \psi_{N'}(1) \\ \psi_{1'}(2) & \psi_{2'}(2) & \cdots & \psi_{N'}(2) \\ \vdots & \vdots & \ddots & \vdots \\ \psi_{1'}(N) & \psi_{2'}(N) & \cdots & \psi_{N'}(N) \end{bmatrix} d\boldsymbol{\tau}_1\cdots d\boldsymbol{\tau}_N. \tag{10-125}$$

Taking each of the N one-electron operators separately, the analysis then proceeds in a manner similar to that of the "zero-electron" operator. For the ith one-electron operator, we have

$$\langle D|h(i)|D'\rangle = \begin{bmatrix} \int \psi_1^*(1)\psi_{1'}(1)\,d\boldsymbol{\tau}_1 & \int \psi_1^*(1)\psi_{2'}(1)\,d\boldsymbol{\tau}_1 & \cdots & \int \psi_1^*(1)\psi_{i'}(1)\,d\boldsymbol{\tau}_1 & \cdots & \int \psi_1^*(1)\psi_{N'}(1)\,d\boldsymbol{\tau}_1 \\ \int \psi_2^*(2)\psi_{1'}(2)\,d\boldsymbol{\tau}_2 & \int \psi_2^*(2)\psi_{2'}(2)\,d\boldsymbol{\tau}_2 & \cdots & \int \psi_2^*(2)\psi_{i'}(2)\,d\boldsymbol{\tau}_2 & \cdots & \int \psi_2^*(2)\psi_{N'}(2)\,d\boldsymbol{\tau}_2 \\ \vdots & \vdots & \ddots & \vdots & & \vdots \\ \int \psi_i^*(i)h(i)\psi_{1'}(i)\,d\boldsymbol{\tau}_i & \int \psi_i^*(i)h(i)\psi_{2'}(i)\,d\boldsymbol{\tau}_i & \cdots & \int \psi_i^*(i)h(i)\psi_{i'}(i)\,d\boldsymbol{\tau}_i & \cdots & \int \psi_i^*(i)h(i)\psi_{N'}(i)\,d\boldsymbol{\tau}_i \\ \vdots & \vdots & & \vdots & \ddots & \vdots \\ \int \psi_N^*(N)\psi_{1'}(N)\,d\boldsymbol{\tau}_N & \int \psi_N^*(N)\psi_{2'}(N)\,d\boldsymbol{\tau}_N & \cdots & \int \psi_N^*(N)\psi_{i'}(N)\,d\boldsymbol{\tau}_N & \cdots & \int \psi_N^*(N)\psi_{N'}(N)\,d\boldsymbol{\tau}_N \end{bmatrix}.$$

As in the previous case, except for the ith row, all of the integrals in the determinants (and hence the determinant itself) will be zero unless each of the orbitals in D' is identical to their counterpart in D. For the case of the ith row, the integrals will in general be nonzero even if $\psi_{i'} \neq \psi_i$, since the orbital orthogonality property cannot be applied in the presence of the operator $h(i)$. Hence, we have two cases to consider. First, for $D' \equiv D$, we have

$$\langle D|h(i)|D\rangle = \begin{bmatrix} 1 & 0 & \cdots & 0 & \cdots & 0 \\ 0 & 1 & & & & \vdots \\ & 0 & & & & \vdots \\ \vdots & \vdots & \ddots & & & \vdots \\ 0 & 0 & & & & 0 \\ \langle\psi_i(i)|h(i)|\psi_1(i)\rangle & \langle\psi_i(i)|h(i)|\psi_2(i)\rangle & \cdots & \langle\psi_i(i)|h(i)|\psi_i(i)\rangle & \cdots & \langle\psi_i(i)|h(i)|\psi_N(i)\rangle \\ 0 & 0 & & & & 0 \\ \vdots & \vdots & & & \ddots & \vdots \\ 0 & 0 & \cdots & 0 & \cdots & 1 \end{bmatrix}$$

$$= \langle\psi_i(i)|h(i)|\psi_i(i)\rangle.$$

Summing over i, we obtain

$$\langle D|\mathcal{H}_1|D\rangle = \sum_{i=1}^{N} \langle \psi_i(i)|h(i)|\psi_i(i)\rangle = \sum_{i=1}^{N} \langle \psi_i(1)|h(1)|\psi_i(1)\rangle, \quad (10\text{-}126)$$

where the last term in the above equation reflects the fact that the integration variable label is arbitrary.

For the case when D' is identical to D except for one orbital, e.g., the ith orbital, we find

$$\langle D|h(i)|D'(i' \neq i)\rangle$$

$$= \begin{bmatrix} 1 & 0 & \cdots & 0 & \cdots & 0 \\ 0 & & 1 & & & \\ & 0 & & & & \vdots \\ \vdots & & & & & \\ & \vdots & \ddots & & & \\ 0 & 0 & & & & \\ \langle \psi_i(i)|h(i)|\psi_1(i)\rangle & \langle \psi_i(i)|h(i)|\psi_2(i)\rangle & \cdots & \langle \psi_i(i)|h(i)|\psi_{i'}(i)\rangle & \cdots & \langle \psi_i(i)|h(i)|\psi_N(i)\rangle \\ 0 & 0 & & & & 0 \\ \vdots & \vdots & & & & \vdots \\ 0 & 0 & \cdots & 0 & \cdots & 1 \end{bmatrix} \quad (10\text{-}127)$$

or

$$\langle D|h(i)|D'(i' \neq i)\rangle = \langle \psi_i(i)|h(i)|\psi_{i'}(i)\rangle = \langle \psi_i(1)|h(1)|\psi_{i'}(1)\rangle. \quad (10\text{-}128)$$

However, for the other $h(j)$ with $j \neq i$, it can be easily seen[61] that

$$\langle D|h(j \neq i)|D'(i' \neq i)\rangle = 0, \quad (10\text{-}129)$$

and we have the general result:

$$\langle D|\mathcal{H}_1|D'(i' \neq i)\rangle = \langle \psi_i(i)|h(i)|\psi_{i'}(i)\rangle = \langle \psi_i(1)|h(1)|\psi_{i'}(1)\rangle, \quad (10\text{-}130)$$

i.e., only a single term results in this case, instead of N terms for the case $D = D'$ that is represented in Eq. (10-126). Clearly, if D' differs by *more* than one orbital from D, the matrix element in Eq. (10-124) will be zero.

C. Two-Electron Operators

Finally, let us examine the two-electron terms of Eq. (10-117). The first case of interest is when $D' \equiv D$, i.e., when each orbital in D' is the same as its corresponding orbital in D. In that case, if $f(i, j)$ is a general two-electron

[61] See Problem 10-11.

operator that is symmetric in the electron coordinates, we have

$$\langle D|f(i,j)|D\rangle$$

$$= \int \cdots \int \psi_1^*(1)\psi_2^*(2)\psi_N^*(N)f(i,j)$$

$$\cdot \begin{bmatrix} \psi_1(1) & \psi_2(1) & \cdots & \psi_i(1) & \cdots & \psi_j(1) & \cdots & \psi_N(1) \\ \psi_1(2) & \psi_2(2) & \cdots & \psi_i(2) & \cdots & \psi_j(2) & \cdots & \psi_N(2) \\ \vdots & \vdots & \ddots & \vdots & & \vdots & & \vdots \\ \psi_1(i) & \psi_2(i) & \cdots & \psi_i(i) & \cdots & \psi_j(i) & \cdots & \psi_N(i) \\ \vdots & \vdots & & \vdots & \ddots & \vdots & \ddots & \vdots \\ \psi_1(j) & \psi_2(j) & \cdots & \psi_i(j) & \cdots & \psi_j(j) & \cdots & \psi_N(j) \\ \vdots & \vdots & & \vdots & & \vdots & \ddots & \vdots \\ \psi_1(N) & \psi_2(N) & \cdots & \psi_i(N) & \cdots & \psi_j(N) & \cdots & \psi_N(N) \end{bmatrix} \cdot d\boldsymbol{\tau}_1 \cdots d\boldsymbol{\tau}_N$$

(10-131)

Multiplying row 1 by $\psi_1^*(1)$ and integrating over $d\tau_1$, multiplying row 2 by $\psi_2^*(2)$ and integrating over $d\tau_2$, etc., except for rows i and j, we have

$$\langle D|f(i,j)|D\rangle$$

$$= \iint \psi_i^*(i)\psi_j^*(j)f(i,j)$$

$$\cdot \begin{bmatrix} \langle\psi_1|\psi_1\rangle & \langle\psi_1|\psi_2\rangle & \cdots & \langle\psi_1|\psi_i\rangle & \cdots & \langle\psi_1|\psi_j\rangle & \cdots & \langle\psi_1|\psi_N\rangle \\ \langle\psi_2|\psi_1\rangle & \langle\psi_2|\psi_2\rangle & \cdots & \langle\psi_2|\psi_i\rangle & \cdots & \langle\psi_2|\psi_j\rangle & \cdots & \langle\psi_2|\psi_N\rangle \\ \vdots & \vdots & \ddots & \vdots & & \vdots & & \vdots \\ \psi_1(i) & \psi_2(i) & \cdots & \psi_i(i) & \cdots & \psi_j(i) & \cdots & \psi_N(i) \\ \vdots & \vdots & & \vdots & \ddots & \vdots & & \vdots \\ \psi_1(j) & \psi_2(j) & \cdots & \psi_i(j) & \cdots & \psi_j(j) & \cdots & \psi_N(j) \\ \vdots & \vdots & & \vdots & & \vdots & \ddots & \vdots \\ \langle\psi_N|\psi_1\rangle & \langle\psi_N|\psi_2\rangle & \cdots & \langle\psi_N|\psi_i\rangle & \cdots & \langle\psi_N|\psi_j\rangle & \cdots & \langle\psi_N|\psi_N\rangle \end{bmatrix} \cdot d\boldsymbol{\tau}_i d\boldsymbol{\tau}_j$$

(10-132)

or

$$\langle D|f(i,j)|D\rangle$$

$$= \iint \psi_i^*(i)\psi_j^*(j) f(i,j) \cdot \begin{bmatrix} 1 & 0 & & 0 & & 0 & \cdots & 0 \\ 0 & & & & & & & \\ \vdots & \vdots & \ddots & \vdots & & \vdots & & \vdots \\ \psi_1(i) & \psi_2(i) & \cdots & \psi_i(i) & \cdots & \psi_j(i) & \cdots & \psi_N(i) \\ \vdots & \vdots & & \vdots & & \vdots & & \vdots \\ \psi_1(j) & \psi_2(j) & \cdots & \psi_i(j) & \cdots & \psi_j(j) & \cdots & \psi_N(j) \\ \vdots & \vdots & & \vdots & & \vdots & \ddots & \vdots \\ 0 & 0 & \cdots & 0 & \cdots & 0 & \cdots & 1 \end{bmatrix} \cdot d\tau_i d\tau_j \, . \tag{10-133}$$

Expanding the determinant along the first row, we obtain

$$\langle D|f(i,j)|D\rangle = \int\int \psi_i^*(i)\psi_j^*(j) f(i,j)[\psi_i(i)\psi_j(j) - \psi_j(i)\psi_i(j)]\, d\tau_i\, d\tau_j$$

$$= \langle ij|f|ij\rangle - \langle ij|f|ji\rangle, \tag{10-134}$$

where

$$\langle ij|f|kl\rangle = \int\int \psi_i^*(1)\psi_j^*(2) f(1,\, 2)\psi_k(1)\psi_l(2)\; d\tau_1\; d\tau_2, \tag{10-135}$$

and where we have called the integration variables "1" and "2." In general, we have[62]

$$\langle D|\mathcal{H}_2|D\rangle = \sum_{i<j}^{N} [\langle ij|f|ij\rangle - \langle ij|f|ji\rangle]. \tag{10-136}$$

The next case of interest is when one orbital of D' differs from its corresponding orbital in D (i.e., $i' \neq i$), and all others are alike. Proceeding as above, we obtain

---

[62] Equation (10-136) as written is valid for an operator of the type $f = \Sigma_{i<j}^N f_{ij}$, as used in Eq. (10-114). If f has the form $f = \frac{1}{2}\Sigma_{i,j}^N f_{ij}$, then the summation in Eq. (10-136) is written as $\frac{1}{2}\Sigma_{i,j}^N$ instead of $\Sigma_{i<j}^N$.

$$\langle D|f|D'(i' \neq i)\rangle$$

$$= \iint \psi_i^*(i)\psi_j^*(j)f(i,j)$$

$$\cdot \begin{bmatrix} 1 & 0 & & 0 & & 0 & 0 & 0 \\ 0 & 1 & & & & & & \\ \vdots & & & \vdots & & \vdots & & \vdots \\ \psi_1(i) & \psi_2(i) & \cdots & \psi_{i'}(i) & \cdots & \psi_j(i) & \cdots & \psi_N(i) \\ \vdots & \vdots & & & \ddots & \vdots & & \\ \psi_1(j) & \psi_2(j) & \cdots & \psi_{i'}(j) & \cdots & \psi_j(j) & \cdots & \psi_N(j) \\ \vdots & \vdots & & \vdots & & \vdots & \ddots & \vdots \\ 0 & 0 & & 0 & \cdots & 0 & \cdots & 1 \end{bmatrix} \cdot d\boldsymbol{\tau}_i d\boldsymbol{\tau}_j . \tag{10-137}$$

It should be noted that, if the differing orbital does not correspond to one of the electrons in $f(i, j)$, the matrix element will be zero. Expanding the determinant as before, we obtain

$$\langle D|f|D'(i' \neq i)\rangle = \int\int \psi_i^*(i)\psi_j^*(j)f(i,j)[\psi_{i'}(i)\psi_j(j) - \psi_j^{(i)}\psi_{i'}(j)\}]\, d\tau_i\, d\tau_j$$

$$= \langle ij|f|i'j\rangle - \langle ij|f|ji'\rangle. \tag{10-138}$$

In general, the result is

$$\langle D|\mathcal{H}_2|D'(i' \neq i)\rangle = \sum_{j=1}^{N} \{\langle ij|f|i'j\rangle - \langle ij|f|ji'\rangle\}. \tag{10-139}$$

For the case when *two* orbitals in D' differ from their counterparts in D (and all other orbitals are identical), an analysis similar to that carried out above[63] gives the result

$$\langle D|\mathcal{H}_2|D'(i' \neq i, j' \neq j)\rangle = \langle ij|f|i'j'\rangle - \langle ij|f|j'i'\rangle. \tag{10-140}$$

If more than two orbitals in D' differ from their counterparts in D, the matrix element is zero.[64]

---

[63] See problem 10-12.

[64] See Problem 10-13.

D. Energy Expression for a Single Slater Determinant

In addition to the general results that have been obtained above for matrix elements between Slater determinants, a special case of those results is also of interest. In particular, for the Hamiltonian operator of Eq. (10-111), the expectation value of the energy associated with a trial wavefunction consisting of a single Slater determinant is seen from Eqs. (10-124), (10-126), and (10-136) to be given by

$$E = \langle D|\mathcal{H}|D\rangle$$
$$= \langle D|\mathcal{H}_0|D\rangle + \langle D|\mathcal{H}_1|D\rangle + \langle D|\mathcal{H}_2|D\rangle \tag{10-141}$$
$$= \mathcal{H}_0 + \sum_{i=1}^{N} \langle \psi_i^{(i)}|h(1)|\psi_i^{(1)}\rangle + \sum_{i<j}^{N} \left[\left\langle ij\left|\frac{1}{r_{12}}\right| ij\right\rangle - \left\langle ij\left|\frac{1}{r_{12}}\right| ji\right\rangle\right]. \tag{10-142}$$

This result[65] will be of direct utility in discussions of Hartree–Fock theory in the next chapter. For the moment, however, it is sufficient to note that, in order to evaluate the energy associated with wavefunctions built from Slater determinants, it is necessary only to evaluate integrals over the coordinates of one electron [in the case of the $h(1)$ operator] or two electrons [in the case of the $f(1, 2)$ operator]. Thus, even though we are treating an N-electron problem, the particular form of the Hamiltonian operator (in which, at most, two-electron operators appear) assures that the most complicated integral to arise is a two-electron integral.

10-7. Virial Theorem, Hypervirial Theorem, and Hellmann–Feynman Theorem

At this point we shall take up a number of theorems related to forces in atoms and molecules. Some of these concepts have direct classical mechanical analogs (e.g., the virial theorem), and each of them is useful in understanding the forces acting in atoms and molecules (and therefore the nature of chemical bonds).

Since the quantum mechanical virial theorem can be obtained as a special case of a *"hypervirial" theorem*,[66] we consider the latter first. The hypervirial theorem deals with expectation values of the Hamiltonian and other operators, and can be seen as follows. If $\mathcal{H}$ is the nonrelativistic, time-independent Hamiltonian of Eq. (10-111), and $\mathcal{A}$ is a linear operator that is time-independent, we shall be interested in the commutator of $\mathcal{H}$ and $\mathcal{A}$, i.e.,

$$[\mathcal{H}, \mathcal{A}] = \mathcal{H}\mathcal{A} - \mathcal{A}\mathcal{H}, \tag{10-143}$$

[65] It should be emphasized again that the relatively simple energy expression in Eq. (10-142) is obtained only when orthogonal orbitals are used.

[66] See J. O. Hirschfelder, *J. Chem. Phys.*, **33**, 1462 (1960) for a discussion and applications of the hypervirial theorem.

and expectation values of the commutator. In particular, let us consider expectation values of $[\mathcal{H}, \mathcal{A}]$ with one of the stationary state eigenfunctions (Ψ) of $\mathcal{H}$, i.e.,

$$\langle[\mathcal{H}, \mathcal{A}]\rangle = \int \Psi^*[\mathcal{H}, \mathcal{A}]\Psi \, d\tau, \tag{10-144}$$

where

$$\mathcal{H}\Psi = E\Psi. \tag{10-145}$$

Writing out Eq. (10-144) in detail, we have

$$\begin{aligned}\langle[\mathcal{H}, \mathcal{A}]\rangle &= \int \Psi^*\mathcal{H}\mathcal{A}\Psi \, d\tau - \int \Psi^*\mathcal{A}\mathcal{H}\Psi \, d\tau \\ &= \int (\mathcal{H}\Psi)^*(\mathcal{A}\Psi) \, d\tau - E\int \Psi^*\mathcal{A}\Psi \, d\tau,\end{aligned} \tag{10-146}$$

or

$$\langle[\mathcal{H}, \mathcal{A}]\rangle = 0, \tag{10-147}$$

where the Hermitian property of $\mathcal{H}$ has been used along with Eq. (10-145). Equation (10-147) is known as the *hypervirial theorem,* and there are several points to be observed about it. First, it is a theorem that is valid for *exact* eigenfunctions of $\mathcal{H}$. If approximate eigenfunctions of $\mathcal{H}$ are used, Eq. (10-147) cannot be guaranteed to hold and, even if $\langle[\mathcal{H}, \mathcal{A}]\rangle$ is very small, it does not necessarily imply that Ψ is approaching the exact wavefunction of $\mathcal{H}$. In other words, the hypervirial theorem is a necessary, but not sufficient, condition in the determination of the exact eigenfunction (Ψ).

The other point to be emphasized relates to the restrictions used in deriving Eq. (10-147). In particular, although not shown, the hypervirial theorem as derived above implies only to the Hamiltonian of Eq. (10-105), and to stationary states associated with it.

To see how the *virial theorem* is a special case of the hypervirial theorem, let us consider first a one-dimensional case, and choose

$$\mathcal{A} = x\frac{d}{dx}, \qquad \begin{aligned}\mathcal{H} &= -\frac{\hbar^2}{2m}\frac{d^2}{dx^2} + \mathcal{V}(x) \\ &= \mathcal{T}_x + \mathcal{V}_x.\end{aligned} \tag{10-148}$$

Forming the commutator of $\mathcal{H}$ and $\mathcal{A}$ gives

$$\begin{aligned}[\mathcal{H}, \mathcal{A}] = &\left\{\left(\frac{-\hbar^2}{2m}\right)\frac{d^2}{dx^2} + \mathcal{V}(x)\right\}\left\{x\left(\frac{d}{dx}\right)\right\} \\ &- \left\{x\frac{d}{dx}\right\}\left\{\frac{-\hbar^2}{2m}\frac{d^2}{dx^2} + \mathcal{V}(x)\right\}.\end{aligned} \tag{10-149]}$$

Carrying out the indicated operations yields

$$[\mathcal{H}, \mathcal{A}] = -2\left(\frac{\hbar^2}{2m}\right)\frac{d^2}{dx^2} - x\left(\frac{d\mathcal{V}}{dx}\right) + \mathcal{V}x\frac{d}{dx}, \tag{10-150}$$

and substitution of this result into Eq. (10-147) gives

$$2\langle \mathfrak{T}_x \rangle = \left\langle x \frac{\partial \mathcal{V}}{\partial x} \right\rangle . \tag{10-151}$$

This result is known as the *quantum mechanical virial theorem* (in one dimension).

When generalized to three dimensions and N particles, $\mathcal{Q}$ is chosen to be

$$\mathcal{Q} = \sum_{j}^{3N} q_j \frac{\partial}{\partial q_j} , \tag{10-152}$$

where q_j are the $3N$ Cartesian coordinates of the N particles, and the Hamiltonian is

$$\mathcal{H} = \mathfrak{T} + \mathcal{V} = -\frac{\hbar^2}{2} \sum_{j}^{N} \left(\frac{1}{m_j} \right) \nabla_j^2 + \frac{1}{2} \sum_{i \neq j}^{N} \mathcal{V}_{ij}, \tag{10-153}$$

which results[67] in the three-dimensional form of the quantum mechanical virial theorem[68]:

$$2\langle \mathfrak{T} \rangle = \left\langle \sum_{j=1}^{3N} q_j \frac{\partial \mathcal{V}}{\partial q_j} \right\rangle . \tag{10-154}$$

This is a general result, and applies equally well to atoms or molecules.

However, under certain circumstances simplifications occur, which allow Eq. (10-154) to be rewritten in a simpler form. These simplifications arise primarily because of the form that the potential energy operator takes when dealing with atomic and molecular problems. Specifically, there are two facets of $\mathcal{V}$ that are of interest here. First, for the time-independent, nonrelativistic case under consideration, the potential energy operator contains only multiplicative operators, and therefore behaves as an ordinary function. Second, we shall see below that $\mathcal{V}$ not only behaves like an ordinary function, but it has a form that allows the differentiations in Eq. (10-154) to be replaced by a simpler expression.

In preparation for those discussions, it is useful to note a theorem known as *Euler's theorem*. This theorem states that, for functions that have the characteristic

$$f(kx_1, kx_2, \cdots, kx_n) = k^m f(x_1, x_2, \cdots, x_n), \tag{10-155}$$

where m is an integer, k is a constant, the following relationship holds:

$$\sum_{j=1}^{n} x_j \frac{\partial f}{\partial x_j} = mf, \tag{10-156}$$

[67] See problem 10-14.

[68] The original derivations were given by V. Fock, *Z. Phys.*, **63**, 855 (1930), and for molecules within the Born–Oppenheimer approximation, by J. C. Slater, *J. Chem. Phys.*, **1**, 687 (1933).

which is Euler's Theorem.[69] Functions that have the property illustrated in Eq. (10-155) are also known as *homogeneous functions of degree m*.

Now let us see how this theorem is useful in simplifying the virial theorem [Eq. (10-154), beginning with atomic systems. For nonrelativistic, time-independent atomic systems containing N electrons in the absence of external fields, the potential energy operator of the Hamiltonian can be written in in atomic units and Cartesian coordinates as

$$\mathcal{V} = -\sum_{i=1}^{N} \frac{Z}{[x_i^2 + y_i^2 + z_i^2]^{1/2}} + \frac{1}{2} \sum_{i \neq j}^{N} \frac{1}{[(x_i - x_j)^2 + (y_i - y_j)^2 + (z_i - z_j)^2]^{1/2}}, \tag{10-157}$$

where Z is the charge on the nucleus. If each coordinate (x, y, z) is replaced by (kx, ky, kz), it is easily seen that

$$\mathcal{V}(kx, ky, kz) = \left(\frac{1}{k}\right) \mathcal{V}(x, y, z), \tag{10-158}$$

which means that the potential energy operator is a homogeneous function of the Cartesian electron coordinates of degree -1. Applying Euler's theorem to this result allows the virial theorem for atoms to be written as

$$2\langle \mathfrak{T} \rangle = -\langle \mathcal{V} \rangle, \tag{10-159}$$

or, since

$$E = \langle \mathfrak{T} \rangle + \langle \mathcal{V} \rangle, \tag{10-160}$$

we can write

$$E = -\langle \mathfrak{T} \rangle, \tag{10-161}$$

or

$$E = \frac{1}{2} \langle \mathcal{V} \rangle. \tag{10-162}$$

Now let us consider molecular systems. Two cases arise, due to the effect of the Born–Oppenheimer approximation on the applicability of Euler's theorem. Let us consider first the case when the Born–Oppenheimer approximation is *not* invoked. In this case the general nonrelativistic, time-independent molecular Hamiltonian for a system of N electrons and M nuclei in the absence of external fields is given by

$$\mathcal{H} = \mathfrak{T} + \mathcal{V}, \tag{10-163}$$

---

[69] A proof of this theorem can be found in a number of mathematics texts, e.g., A. E. Taylor, *Advanced Calculus* Ginn and Co., New York, 1955, p. 184.

where

$$\mathfrak{J} = -\frac{1}{2}\sum_{i=1}^{N+M}\left(\frac{1}{m_i}\right)\nabla_i^2, \tag{10-164}$$

$$\begin{aligned}\mathcal{V} = &-\sum_{i=1}^{N}\sum_{\alpha=1}^{M}\frac{Z_\alpha}{[(x_i-x_\alpha)^2+(y_i-y_\alpha)^2+(z_i-z_\alpha)^2]^{1/2}}\\ &+\frac{1}{2}\sum_{i\neq j}^{N}\frac{1}{[(x_i-x_j)^2+(y_i-y_j)^2+(z_i-z_j)^2]^{1/2}}\\ &+\frac{1}{2}\sum_{\alpha\neq\beta}\frac{Z_\alpha Z_\beta}{[(x_\alpha-x_\beta)^2+(y_\alpha-y_\beta)^2+(z_\alpha-z_\beta)^2]^{1/2}},\end{aligned} \tag{10-165}$$

where Z_α is the charge on the αth nucleus. Since both nuclear and electronic coordinates are considered as variables (i.e., neither is fixed), it is easily seen that $\mathcal{V}$ is a homogeneous function of the coordinates $(x_i, y_i, z_i, x_\alpha, y_\alpha, z_\alpha)$ of degree -1, analogous to the atomic case. Thus, when the Born–Oppenheimer approximation is not invoked, the virial theorem for molecules takes on the simplified form of Eq. (10-159).

The other case to be considered is when the Born–Oppenheimer approximation *is* invoked, i.e., the nuclear coordinates are fixed and are treated only as parameters in Eq. (10-165). For this case complications occur, because the first term in Eq. (10-165) is no longer homogeneous in the coordinates, and Euler's theorem cannot be used to simplify the virial theorem in Eq. (10-154) as we have done in previous cases. However, a modification of the previous analysis will be seen to be appropriate.

When the Born–Oppenheimer approximation is used along with the Hamiltonian of Eqs. (10-163) through (10-165), the wavefunction takes the form

$$\Psi(\mathbf{r}_i, \mathbf{r}_\alpha) = \Psi_e(\mathbf{r}_i, \mathbf{r}_\alpha)\cdot\Psi_N(\mathbf{r}_\alpha) \tag{10-166}$$

and the Schrödinger equation is solved for Ψ_e, i.e.,

$$\mathcal{H}\Psi_e = E\Psi_e \tag{10-167}$$

is solved for fixed values of the nuclear coordinates $(\mathbf{r}_\alpha)$. Even though $\mathcal{V}$ is not a homogeneous function in this case, the general form of the virial theorem (without simplifications from the use of Euler's theorem) still applies i.e.,

$$2\langle\Psi_e|\mathfrak{J}|\Psi_e\rangle = \left\langle\Psi_e\left|\sum_i q_i\frac{\partial\mathcal{V}}{\partial q_i}\right|\Psi_e\right\rangle, \tag{10-168}$$

where the integration as well as the summation is over electronic coordinates only.

This expression can be simplified somewhat by recalling that, in the absence of the Born–Oppenheimer approximation, $\mathcal{V}$ *is* homogeneous of degree -1, and we can write

$$\sum_i q_i \frac{\partial \mathcal{V}}{\partial q_i} + \sum_\alpha q_\alpha \frac{\partial \mathcal{V}}{\partial q_\alpha} = -\mathcal{V} \tag{10-169}$$

where the term over all coordinates has been split, in which the first term includes a summation over all electronic Cartesian coordinates, and the second term includes all nuclear Cartesian coordinates (which are variables and not constants). Multiplication of Eq. (10-169) by ψ_e on the left and right and integration only over electronic coordinates gives

$$\left\langle \psi_e \left| \sum_i q_i \frac{\partial \mathcal{V}}{\partial q_i} \right| \psi_e \right\rangle + \left\langle \psi_e \left| \sum_\alpha q_\alpha \frac{\partial \mathcal{V}}{\partial q_\alpha} \right| \psi_e \right\rangle = -\langle \psi_e | \mathcal{V} | \psi_e \rangle. \tag{10-170}$$

Comparison of Eq. (10-170) with Eq. (10-168) allows the latter to be rewritten as

$$2\langle \psi_e | \mathcal{T} | \psi_e \rangle = -\langle \psi_e | \mathcal{V} | \psi_e \rangle - \left\langle \psi_e \left| \sum_\alpha q_\alpha \frac{\partial \mathcal{V}}{\partial q_\alpha} \right| \psi_e \right\rangle. \tag{10-171}$$

Thus, we see that use of the Born–Oppenheimer approximation has the effect of adding a term to the previous virial theorem result. We shall return to the last term in Eq. (10-171) shortly after discussing the Hellmann–Feynman theorem and scaling concept, which will be useful in simplifying the expression further.

The *generalized Hellmann–Feynman theorem*[70] is one that shows how to relate changes in the parameters in the Hamiltonian to changes that can be expected in the energy. To prove this theorem, we assume that the Hamiltonian (and wavefunction) depends upon a parameter λ. We shall be interested in the effect of changes in λ on the energy, i.e., we wish to examine

$$\frac{\partial E}{\partial \lambda} = \frac{\partial}{\partial \lambda} \langle \Psi | \mathcal{H} | \Psi \rangle, \tag{10-172}$$

where Ψ is the exact solution of the Schrödinger equation, i.e., it satisfies

$$\mathcal{H}\Psi = E\Psi. \tag{10-173}$$

Carrying out the indicated operations in Eq. (10-172) gives

$$\frac{\partial E}{\partial \lambda} = \left\langle \frac{\partial \Psi}{\partial \lambda} \left| \mathcal{H} \right| \Psi \right\rangle + \left\langle \Psi \left| \frac{\partial \mathcal{H}}{\partial \lambda} \right| \Psi \right\rangle + \left\langle \Psi \left| \mathcal{H} \right| \frac{\partial \Psi}{\partial \lambda} \right\rangle \tag{10-174}$$

$$= E \left\{ \frac{\partial}{\partial \lambda} \langle \Psi | \Psi \rangle \right\} + \left\langle \Psi \left| \frac{\partial \mathcal{H}}{\partial \lambda} \right| \Psi \right\rangle, \tag{10-175}$$

where Eq. (10-173) and the Hermitian property of $\mathcal{H}$ have been used. Since Ψ is normalized for all λ, the normalization integral is invariant to changes in λ and

[70] See J. I. Musher, *Am. J. Phys.*, **34**, 267 (1966).

the first term on the right-hand side of Eq. (10-175) is zero. We thus have the result

$$\frac{\partial E}{\partial \lambda} = \left\langle \Psi \left| \frac{\partial \mathcal{H}}{\partial \lambda} \right| \Psi \right\rangle, \tag{10-176}$$

which is known as the *generalized Hellmann-Feynman theorem.*[71]

Let us now return to our discussion o the virial theorem within the context of the Born–Oppenheimer approximation to see that simplifications are possible. If we take λ to be one of the nuclear coordinate components, i.e.,

$$\lambda = q_\alpha$$

then the generalized Hellmann–Feynman theorem can be written within the Born–Oppenheimer approximation as

$$\frac{\partial E_e}{\partial q_\alpha} = \left\langle \psi_e \left| \frac{\partial \mathcal{H}}{\partial q_\alpha} \right| \psi_e \right\rangle, \tag{10-177}$$

where the nuclei are fixed and the integration is taken over electronic coordinates only. However, from Eqs. (10-633)–(10-165), we have

$$\frac{\partial \mathcal{H}}{\partial q_\alpha} = \frac{\partial}{\partial q_\alpha} [\mathfrak{T} + \mathcal{V}] = \frac{\partial \mathcal{V}}{\partial q_\alpha}, \tag{10-178}$$

since the kinetic energy operator for the electrons is independent of the nuclear coordinates. This allows Eq. (10-177) to be rewritten as

$$\frac{\partial E_e}{\partial q_\alpha} = \left\langle \psi_e \left| \frac{\partial \mathcal{V}}{\partial q_\alpha} \right| \psi_e \right\rangle \tag{10-179}$$

and Eq. (10-171) to be rewritten as

$$2\langle \psi_e | \mathfrak{T} | \psi_e \rangle = -\langle \psi_e | \mathcal{V} | \psi_e \rangle - \sum_\alpha q_\alpha \frac{\partial E_e}{\partial q_\alpha}. \tag{10-180}$$

We shall return shortly to these theorems and how they can be applied when approximate wavefunctions are used. First, however, the concept of wavefunction scaling needs to be discussed.

10-8. Scaling

Since in general the trial wavefunctions that are chosen for molecular systems will not be exact, we cannot expect these trial wavefunctions to satisfy the theorems that were discussed in the previous section. Fortunately, there exists a

---

[71] Although this theorem has been proved for the exact wavefunction, it applies to some approximate wavefunctions as well. For example, see R. E. Stanton, *J. Chem. Phys.*, **36**, 1298 (1962).

straightforward procedure for "scaling" a trial wavefunction that will assure that the virial theorem is satisfied.

To see how this is done,[72] let us consider a system of N particles (electrons and nuclei), whose nonrelativistic, time-independent, field-free Hamiltonian operator in atomic units is given by

$$\mathcal{H} = \mathfrak{T} + \mathcal{V}, \tag{10-181}$$

with

$$\mathfrak{T} = -\frac{1}{2}\sum_{k=1}^{N}\left(\frac{1}{m_k}\right)\nabla_k^2 \tag{10-182}$$

and

$$\mathcal{V} = \frac{1}{2}\sum_{k\neq l}^{N}\frac{Z_k Z_l}{r_{kl}} \tag{10-183}$$

where m_k is the mass of the kth particle and Z_k is the charge. Note that the Born-Oppenheimer approximation has not been involved.

If $\varphi(\mathbf{r}_1 \cdots \mathbf{r}_N)$ is an arbitrary, normalized trial wavefunction for the system, let us consider the following "*scaled*" wavefunction φ_η, given by

$$\varphi_\eta = \eta^{3N/2} \cdot \varphi(\eta\mathbf{r}_1, \eta\mathbf{r}_2, \cdots, \eta\mathbf{r}_N), \tag{10-184}$$

where η is referred to as the *scale factor*. The normalization of φ_η is easily verified, i.e.,

$$\begin{aligned}
&\int \cdots \int \varphi_\eta^*(\eta\mathbf{r}_1, \eta\mathbf{r}_2, \cdots, \eta\mathbf{r}_N)\varphi_\eta(\eta\mathbf{r}_1, \eta\mathbf{r}_2, \cdots, \eta\mathbf{r}_N)\, d\mathbf{r}_1\, d\mathbf{r}_2 \cdots d\mathbf{r}_N \\
&\quad = \eta^{3N}\int \cdots \int \varphi^*(\eta\mathbf{r}_1, \eta\mathbf{r}_2, \cdots, \eta\mathbf{r}_N)\varphi(\eta\mathbf{r}_1, \eta\mathbf{r}_2, \cdots, \eta\mathbf{r}_N) \\
&\qquad \cdot (\eta^{-3N})(\eta^3 d\mathbf{r}_1)(\eta^3 d\mathbf{r}_2) \cdots (\eta^3 d\mathbf{r}_N) \\
&\quad = \int \cdots \int \varphi^*(\eta\mathbf{r}_1, \eta\mathbf{r}_2, \cdots, \eta\mathbf{r}_N)\varphi(\eta\mathbf{r}_1, \eta\mathbf{r}_2, \cdots, \eta\mathbf{r}_N) \\
&\qquad \cdot \eta\, dx_1 \eta\, dy_1 \cdots \eta\, dz_N.
\end{aligned} \tag{10-185}$$

In Eq. (10-184) it is seen that, if a change of variables $x_1' = \eta x_1$, etc., is carried out, the form of the integral is exactly the same as the normalization integral over the original function, except for the arbitrary labels on the integration variables. Hence, since the original function φ was normalized, we are assured that φ_η is also normalized for all choices of η.

This scaling process can be thought of as a uniform "stretching" or

---

[72] The analysis used in this section was given first by E. Hylleraas, *Z. Phys.*, **54**, 347 (1929) and V. Fock, *Z. Phys.*, **63**, 855 (1930), and the form used here was outlined by P. O. Löwdin, *Adv. Chem. Phys.*, **2**, 207 (1959).

"shrinking" of the vectors in the space by the same amount (η). It is easily applied, and we shall now examine the relationship between "scaled" and "unscaled" expectation values and with the virial theorem. We begin by examining expectation values of the total kinetic energy operator [Eq. (10-182)] using a scaled trial wavefunction of the form given in Eq. (10-184), i.e.,

$$\langle \varphi_\eta | \mathfrak{T} | \varphi_\eta \rangle = \eta^{3N} \int \cdots \int \varphi^*(\eta \mathbf{r}_1, \cdots, \eta \mathbf{r}_N) \left[-\frac{1}{2} \sum_{k=1}^{N} \left(\frac{1}{m_k} \right) \nabla_k^2 \right]$$
$$\cdot \varphi(\eta \mathbf{r}_1, \cdots, \eta \mathbf{r}_N)\, d\mathbf{r}_1 \cdots d\mathbf{r}_N$$
$$= \int \cdots \int \varphi^*(\eta \mathbf{r}_1, \cdots, \eta \mathbf{r}_N) \left[-\frac{1}{2} \sum_{k=1}^{N} \left(\frac{1}{m_k} \right) \nabla_k^2 \right]$$
$$\cdot \varphi(\eta \mathbf{r}_1, \cdots, \eta \mathbf{r}_N)(\eta^3 d\mathbf{r}_1) \cdots (\eta^3 d\mathbf{r}_N). \tag{10-186}$$

Thusfar, all the variables in the integral have been expressed in terms of "scaled" coordinates except the operator itself. We can write the ∇_k^2 operator as

$$\nabla_k^2 = \frac{d^2}{dx_k^2} + \frac{d^2}{dy_k^2} + \frac{d^2}{dz_k^2} = \eta^2 \left[\frac{d^2}{d(\eta x_k)^2} + \frac{d^2}{d(\eta y_k)^2} + \frac{d^2}{d(\eta z_k)^2} \right], \tag{10-187}$$

which means that Eq. (10-186) can be rewritten as

$$\langle \varphi_\eta | \mathfrak{T}_\eta | \varphi_\eta \rangle = \eta^2 \langle \varphi | \mathfrak{T}(\eta = 1) | \varphi \rangle. \tag{10-188}$$

Thus, simple multiplication of the unscaled kinetic energy integral by η^2 will give the scaled kinetic energy integral.

In a similar manner it can be seen[73] that

$$\langle \varphi_\eta | \mathcal{V}_\eta | \varphi_\eta \rangle = \eta \langle \varphi | \mathcal{V}(\eta = 1) | \varphi \rangle, \tag{10-189}$$

which means that the total energy using scaled expectation values is given by

$$E = \langle \varphi_\eta | \mathcal{H}_\eta | \varphi_\eta \rangle$$
$$= \eta^2 \langle \varphi | \mathfrak{T}(\eta = 1) | \varphi \rangle + \eta \langle \varphi | \mathcal{V}(\eta = 1) | \varphi \rangle. \tag{10-190}$$

If we treat the above energy expression as a function of η, we can minimize it with respect to the choice of η, which gives

$$\frac{\partial E}{\partial \eta} = 0 = 2\eta \langle \varphi | \mathfrak{T}(\eta = 1) | \varphi \rangle + \langle \varphi | \mathcal{V}(\eta = 1) | \varphi \rangle. \tag{10-191}$$

There are two cases of importance relative to Eq. (10-191). The first corresponds to the case when φ is the exact solution (Ψ). In that case the function is obviously scaled properly, and $\eta = 1$. This means that Eq. (10-191) can be rewritten as

$$2\langle \Psi | \mathfrak{T}(\eta = 1) | \Psi \rangle = -\langle \Psi | \mathcal{V}(\eta = 1) | \Psi \rangle, \tag{10-192}$$

---

[73] See Problem 10-16.

and we see that the virial theorem is automatically satisfied for the exact solution.

The other case of interest is when φ is not exact eigenfunction, but is an arbitrary trial function. In that case, the optimum choice for η is seen from Eq. (10-191) to be given by

$$\eta = \frac{-\langle \varphi | \mathcal{V}(\eta=1) | \varphi \rangle}{2\langle \varphi | \mathfrak{T}(\eta=1) | \varphi \rangle} .$$

Using this value for η, the total (scaled) energy can be seen from Eq. (10-190) to be given by

$$E = \langle \varphi_\eta | \mathfrak{T}_\eta | \varphi_\eta \rangle + \langle \varphi_\eta | \mathcal{V}_\eta | \varphi_\eta \rangle \tag{10-194}$$

$$= \frac{[\langle \varphi | \mathcal{V}(\eta=1) | \varphi \rangle]^2}{4\langle \varphi | \mathfrak{T}(\eta=1) | \varphi \rangle} - \frac{[\langle \varphi | \mathcal{V}(\eta=1) | \varphi \rangle]^2}{2\langle \varphi | \mathfrak{T}(\eta=1) | \varphi \rangle} = -\frac{[\langle \varphi | \mathcal{V}(\eta=1) | \varphi \rangle]^2}{4\langle \varphi | \mathfrak{T}(\eta=1) | \varphi \rangle} . \tag{10-195}$$

Several points of interest are exhibited in Eq. (10-195). First, calculation of the energy of a properly scaled trial wavefunction is easily accomplished from the integrals calculated using the unscaled trail wavefunction, as shown in Eq. (10-195). Second, it is easily seen from Eq. (10-195) that the scaled wavefunction satisfies the virial theorem, even though it is not the exact solution. This is an important practical consideration, for it shows how to assure that trial wavefunctions are both properly scaled and satisfy the virial theorem.

To illustrate this concept, let us return to several examples that were considered earlier. In Chapter 9, Section 9-3, we considered a trial wavefunction for the hydrogen atom of the form

$$\varphi(r) = Ne^{-\zeta r} \tag{10-196}$$

and found the energy expression

$$E = N^2 \left(\frac{-\zeta^2}{2} - \zeta \right) . \tag{10-197}$$

Minimizing E with respect to ζ gave $\zeta = 1$, which was also the exact answer in that case. However, minimization of E with respect to the choice of ζ has also accomplished the appropriate scaling of the wavefunction, since it is the same process that was indicated in Eq. (10-191).

To illustrate the concept when the trail wavefunction is not exact, even after scaling, let us consider the following trial wavefunction for the hydrogen atom:

$$\varphi_\eta = Ne^{-\zeta r^2}, \qquad N = \left[\frac{2\eta}{\pi} \right]^{3/4} . \tag{10-198}$$

Evaluation of the energy expectation value is easily accomplished using the formulas of Eqs. (9-73) through (9-75), and gives:

$$\langle \varphi_\eta | \mathfrak{T}_\eta | \varphi_\eta \rangle = \left(\frac{3\eta}{2} \right) \tag{10-199}$$

and

$$\langle \varphi_\eta | \mathcal{V}_\eta | \varphi_\eta \rangle = -2 \left[\frac{2\eta}{\pi} \right]^{1/2}. \tag{10-200}$$

If we had not attempted to choose the appropriate scale factor and had, e.g., chosen $\eta = 1$, the calculated energy would be

$$E(\eta = 1) = \frac{3}{2} - 2 \sqrt{\frac{2}{\pi}} = -0.09577 \text{ Hartrees.} \tag{10-201}$$

The value of η that optimizes E is seen from Eq. (10-193) to be

$$\eta_{\text{opt}} = 0.2829. \tag{10-202}$$

From Eq. (10-195) we see that the energy calculated for the optimum scale factor is

$$E(\eta_{\text{opt}}) = \frac{-(8/\pi)}{4(3/2)} = -0.4244 \text{ Hartrees.} \tag{10-203}$$

Thus, an energy improvement of 0.3287 Hartrees (206 kcal/mol!) has been obtained simply through optimization of the scale factor for the system.[74] Furthermore, the scale factor is easily calculated, once the unscaled integrals have been evaluated.

When a trail wavefunction contains more than one term, the process to be used is quite similar. For example, a multiterm Gaussian trail wavefunction for the hydrogen atom can be written as

$$\varphi = N \sum_{i=1}^{I} c_i e^{-a_i r^2} \tag{10-204}$$

where I represents the number of basis functions used in the trial wavefunction. If arbitrary values of the a_i are chosen, φ will not be properly scaled. However, if we take φ to have the form

$$\varphi_\eta = N \sum_{i=1}^{I} c_i e^{-a_i \eta r^2}, \tag{10-205}$$

where η is a scale factor applied to each basis orbital, then the appropriately scaled energy and wavefunction can be obtained through the use of Eqs. (10-193) and (10-195).

Before leaving this discussion it is appropriate to consider how these theorems and concepts can be applied to molecular systems. To begin with, let

[74] Of course, it should be remembered that, if η had been optimized in advance (e.g., through study of other prototype systems), the energy improvement due to further scaling will be reduced substantially.

us consider an N-electron diatomic molecule,[75] with fixed internuclear distance R, and a scaled trial wavefunction φ_η, i.e.,

$$\varphi_{\eta,R} = \eta^{3N/2}\varphi(\eta\mathbf{r}_1, \eta\mathbf{r}_2, \cdots, \eta\mathbf{r}_N; \rho), \tag{10-206}$$

where

$$\rho = \eta R. \tag{10-207}$$

In Eq. (10-206), the subscripts (η, R) refer to the quantities that are parameters (not variables) to be chosen. Since the internuclear distance is fixed in this example, it is treated as a parameter. However, it is scaled along with the electronic coordinates, and by the same amount, as indicated in Eqs. (10-206) and (10-207).

As in the previous discussion of scaling, we need to determine the effect that introduction of the scale factor in Eq. (10-206) has on the kinetic and potential energy expectation values. An analysis similar to that in Eqs. (10-186) through (10-188) yields

$$\langle \varphi_{\eta,R} | \mathfrak{T}_{\eta,R} | \varphi_{\eta,R} \rangle = \eta^2 \langle \varphi_{1,\rho} | \mathfrak{T}_{1,\rho} | \varphi_{1,\rho} \rangle = \eta^2 T(1, \rho). \tag{10-208}$$

The notation on the right-hand side of Eq. (10-208) indicates that the effect of introducing the scale factor can be included by evaluating the kinetic energy integral as if the scale factor were absent (or equal to unity), except for using an internuclear distance of ηR instead of R, and multiplying the result by η^2.

Similarly, it can be seen that the scaled potential energy expectation value is given by

$$\langle \varphi_{\eta,R} | \mathfrak{V}_{\eta,R} | \varphi_{\eta,R} \rangle = \eta \langle \varphi_{1,\rho} | \mathfrak{V}_{1,\rho} | \varphi_{1,\rho} \rangle = \eta V(1, \rho). \tag{10-209}$$

The total energy for the N-electron diatomic molecule is thus given by

$$E(\eta, R) = \eta^2 T(1, \eta R) + \eta V(1, \eta R). \tag{10-210}$$

If we optimize E with respect to the choice of scale factor, we obtain

$$\frac{\partial E(\eta, R)}{2\eta} = 0 = 2\eta T(1, \eta R) + V(1, \eta R) + \eta^2 \frac{\partial}{\partial \eta}[T(1, \eta R)] + \eta \frac{\partial}{\partial \eta}[V(1, \eta R)]. \tag{10-211}$$

Since

$$\frac{\partial \rho}{\partial \eta} = R, \tag{10-212}$$

[75] The analysis here is based on the work of J. O. Hirschfelder and J. F. Kincaid, *Phys. Rev.*, **52**, 658 (1937), and the modifications suggested by P. O. Löwdin, *Adv. Chem. Phys.*, **2**, 207 (1959).

we can rewrite Eq. (10-211) as

$$0 = 2\eta T(1, \eta R) + V(1, \eta R) + \eta^2 R \frac{\partial}{\partial \rho} [T(1, \eta R)] + \eta R \frac{\partial}{\partial \rho} [V(1, \eta R)]. \tag{10-213}$$

As in the previous analyses, there are two cases to consider in Eq. (10-213). The first case is when φ is exact solution, i.e., $\eta = 1$, in which case Eq. (10-213) becomes

$$0 = 2T + V + R \frac{\partial}{\partial \rho} [T + V],$$

or

$$0 = 2T + V + R \frac{\partial E}{\partial R}. \tag{10-214}$$

This is the result obtained by Slater[76] for the case of fixed nuclear positions. We see that it differs from the atomic case by the presence of a new term that results from the use of the Born–Oppenheimer approximation. Of course, there is a way in which Eq. (10-214) will reduce to the simpler form that was found earlier [Eq. (10-193)], and that is when

$$\frac{\partial E}{\partial R} = 0. \tag{10-215}$$

This result occurs either when the two nuclei are infinitely apart or when they are at their equilibrium internuclear distance. In all other cases the more complicated expression in Eq. (10-214) applies.

For the other case, in which φ is not the exact solution but is only an approximate wavefunction, we do not have $\eta = 1$, but wish to obtain the optimum value of η from Eq. (10-213). Using Eq. (10-207), we can solve for η using Eq. (10-213) as follows:

$$\eta = -\frac{\left[V(1, \rho) + \rho \frac{\partial}{\partial \rho} V(1, \rho)\right]}{\left[2T(1, \rho) + \rho \frac{\partial}{\partial \rho} T(1, \rho)\right]}. \tag{10-216}$$

It is easily verified[77] that this choice of η, when substituted into Eq. (10-213), satisfies the virial theorem as expressed in Eq. (10-214).

However, before leaving this discussion it should also be noted that scaling of wavefunctions is not in general a serious consideration in current molecular

[76] J. C. Slater, *J. Chem. Phys.*, **1**, 687 (1933).

[77] See Problem 10-17.

studies. In part this is due to the possibility of direct optimization of geometries and the use of large, flexible approximate wavefunctions, which will result frequently in satisfaction of the virial theorem without further calculation.

Nevertheless, both for exact and approximate solutions to the Schrödinger equation, the concepts embodied in the virial theorem, the Hellmann–Feynman theorem, and scaling provide useful tools for improving and/or interpreting wavefunctions.[78] In fact, interesting applications of these concepts to the understanding of chemical bonding have occurred,[79] which have provided considerable insight into the nature of chemical bonds even when only approximate wavefunctions are available.

10-9. Coupling of Angular Momenta

As we have seen in a number of ways in this chapter, moving to a multiparticle system from a single particle system is not always a simple matter of larger scale. Indeed, the Pauli Exclusion Principle and other ideas introduced in earlier sections have shown that new concepts and/or constraints may be needed as a multiparticle system is considered. It is therefore appropriate to reconsider the concept of angular momentum that was discussed from the point of view of a single particle in Chapter 6 to see what modifications, if any, are appropriate when dealing with multiparticle systems. We shall, however, restrict our considerations to electrons only at this point, and will not consider effects such as those from nuclear spin.

As a first step, we define the *total orbital angular momentum operator* ($\mathcal{L}$) for a system of N electrons as a sum of the orbital angular momentum operators for each electron [$\mathcal{L}(i)$], i.e.,

$$\mathcal{L} = \sum_{j=1}^{N} \mathcal{L}(j), \tag{10-217}$$

where the components of $\mathcal{L}$ are given by

$$\mathcal{L}_z = \sum_{j=1}^{N} \mathcal{L}_z(j), \tag{10-218}$$

with similar expressions for $\mathcal{L}_x$ and $\mathcal{L}_y$. The commutation relations among the components of $\mathcal{L}$ are given by

$$[\mathcal{L}_p, \mathcal{L}_q] = i\hbar\mathcal{L}_r, \tag{10-219}$$

---

[78] For application of the virial theorem to polyatomic molecules see, for example, R. G. Parr and J. E. Brown, *J. Chem. Phys.*, **49**, 4849 (1968).

[79] See, for example, R. F. W. Bader and P. M. Beddall, *J. Chem. Phys.*, **56**, 3320 (1972); R. F. W. Bader, P. M. Beddall, and J. J. Peslak, *J. Chem. Phys.*, **58**, 557 (1973); R. F. W. Bader, A. J. Duke, and R. R. Messer, *J. Amer. Chem. Soc.*, **95**, 7715 (1973); R. F. W. Bader, *J. Chem. Phys.*, **73**, 2871 (1980); and *Acct. Chem. Res.*, **8**, 34 (1975).

where (p, q, r) represent (x, y, z) or any cyclic permutation of (x, y, z). The square of $\mathcal{L}$ is defined as

$$\mathcal{L}^2 = \mathcal{L}_x^2 + \mathcal{L}_y^2 + \mathcal{L}_z^2. \tag{10-220}$$

Thus, even though $\mathcal{L}$ is an orbital angular momentum operator for N electrons, it still has the form (and hence the properties) of the general angular momentum operator ($\mathfrak{M}$) defined in Chapter 6 [Eqs. (6-9) and (6-10)].

In a similar manner, the *total spin angular momentum operator* ($\mathcal{S}$) for a system of N electrons, and its square ($\mathcal{S}^2$), are defined as

$$\mathcal{S} = \sum_{j=1}^{N} \mathcal{S}(j), \tag{10-221}$$

$$\mathcal{S}^2 = \mathcal{S}_x^2 + \mathcal{S}_y^2 + \mathcal{S}_z^2, \tag{10-222}$$

and

$$[\mathcal{S}_p, \mathcal{S}_q] = i\hbar \mathcal{S}_r, \tag{10-223}$$

with (p, q, r) representing (x, y, z) or any cyclic permutation of (x, y, z).

The ways in which these different forms of angular momentum can couple and what measurements can be made will be of particular interest. In order to explore these possibilities, we define a *total angular momentum operator* ($\mathcal{J}$) as

$$\mathcal{J} = \mathcal{L} + \mathcal{S}, \tag{10-224}$$

with

$$[\mathcal{J}_p, \mathcal{J}_q] = i\hbar \mathcal{J}_r, \tag{10-225}$$

with (p, q, r) representing (x, y, z) or any cyclic permutation of (x, y, z). The square of $\mathcal{J}$ is defined as

$$\mathcal{J}^2 = (\mathcal{L} + \mathcal{S}) \cdot (\mathcal{L} + \mathcal{S}) = \mathcal{L}^2 + \mathcal{S}^2 + 2(\mathcal{L} \cdot \mathcal{S}), \tag{10-226}$$

where have have used the fact that

$$[\mathcal{L}, \mathcal{S}] = 0. \tag{10-227}$$

In order to determine the opportunities for simultaneous measurement of these quantities, we note several commutation relations that can be easily proven:[80]

$$\begin{aligned}[][\mathcal{J}^2, \mathcal{L}^2] = [\mathcal{J}^2, \mathcal{S}^2] = [\mathcal{J}^2, \mathcal{J}_z] = [\mathcal{S}^2, \mathcal{J}_z] \\ = [\mathcal{L}^2, \mathcal{S}^2] = [\mathcal{L}^2, \mathcal{J}_z] = 0\end{aligned} \tag{10-228}$$

Thus, we see that the four operators $\mathcal{J}^2$, $\mathcal{L}^2$, $\mathcal{S}^2$, and $\mathcal{J}_z$ mutually commute, which means that it is possible to find simultaneous eigenfunctions of all four

[80] See Problem 10-18.

operators, and, hence, to measure each observable corresponding to the operators simultaneously.

However, not only the operators listed above commute. In fact, it can also be shown[81] that the four operators $\mathcal{L}^2$, $\mathcal{L}_z$, $\mathcal{S}^2$, and $\mathcal{S}_z$ mutually commute. Thus, we have an alternative set of commuting operators whose observables are simultaneously measurable. We therefore need to decide which of the two sets is more convenient for use. However, our decision in this regard will depend on a number of factors, including the Hamiltonian to be used, the wavefunction form that is chosen, and the overall symmetry of the system. In fact, for molecular systems we shall see that, in general, $[\mathcal{H}, \mathcal{L}^2] \neq 0$, and only spin eigenfunction construction will be important to consider.

However, for field-free atomic problems described by a nonrelativistic, time-independent Hamiltonian, each of the operators just mentioned commutes with the Hamiltonian, and either set of operators is an equally good choice. It should be noted that if a Hamiltonian with relativistic effects and external fields is considered, only $\mathcal{J}^2$, $\mathcal{J}_z$, and $\mathcal{H}$ will commute, and the choice of operators is reduced substantially. For the more common former case, the form of trial wavefunction that is usually chosen will help determine the most convenient set of four operators to use.

Since Slater determinants are utilized in trial wavefunctions as a convenient means of satisfying the Pauli Exclusion Principle, let us examine the effect of $\mathcal{L}_z$ on a Slater determinant. Specifically, let us consider an N-electron atom with $\mathcal{L}_z$ given by Eq. (10-218) and a Slater determinant given by

$$\Psi(1, 2, \cdots, N) = \frac{1}{(N!)^{1/2}} \sum_{\mathcal{P}} (-1)^p \mathcal{P}\{\psi_1(1)\psi_2(2) \cdots \psi_N(N)\} \qquad (10\text{-}229)$$

where the atomic orbitals in the determinant are chosen as products of radial functions, spherical harmonics and spin functions, i.e.,

$$\psi_i(j) = \psi_i(r_j, \vartheta_j, \phi_j, \sigma_j) = R_{n_i}(r_j) Y_{l_i m_i}(\vartheta_j, \phi_j) S_{s_i}(\sigma_j), \qquad (10\text{-}230)$$

where the subscript (i) on ψ_i refers to a set of four quantum numbers (n_i, l_i, m_i, s_i), and

$$\mathcal{L}_z(j)\psi_i(r_j, \vartheta_j, \phi_j, \sigma_j) = m_i \psi_i(r_j, \vartheta_j, \phi_j, \sigma_j). \qquad (10\text{-}231)$$

Let us begin by considering the effect of $\mathcal{L}_z$ on the diagonal term of the Slater determinant, i.e.,

$$\begin{aligned}
&\mathcal{L}_z\{\psi_1(1)\psi_2(2) \cdots \psi_N(N)\} \\
&= \{\mathcal{L}_z(1) + \mathcal{L}_z(2) + \cdots + \mathcal{L}_z(N)\}\{\psi_1(1)\psi_2(2) \cdots \psi_N(N)\} \\
&= (m_1 + m_2 + \cdots + m_N)\{\psi_1(1)\psi_2(2) \cdots \psi_N(N)\} \\
&= \left(\sum_{j=1}^{N} m_j\right) \{\psi_1(1)\psi_2(2) \cdots \psi_N(N)\}.
\end{aligned} \qquad (10\text{-}232)$$

[81] See Problem 10-19.

The next term of interest is that of a single permutation of Eq. (10-229), e.g.,

$$-\mathcal{L}_z\{\psi_1(1)\psi_2(2)\cdots\psi_i(j)\cdots\psi_j(i)\cdots\psi_N(N)\}$$
$$=-\{\mathcal{L}_z(1)+\mathcal{L}_z(2)+\cdots+\mathcal{L}_z(i)+\cdots+\mathcal{L}_z(j)+\cdots+\mathcal{L}_z(N)\}$$
$$\cdot\{\psi_1(1)\psi_2(2)\cdots\psi_i(j)\cdots\psi_j(i)\cdots\psi_N(N)\}$$
$$=-\{m_1+m_2+\cdots+m_j+\cdots+m_i+\cdots+m_N\}$$
$$\cdot\{\psi_1(1)\psi_2(2)\cdots\psi_i(j)\cdots\psi_j(i)\cdots\psi_N(N)\}$$
$$=-\left(\sum_{j=1}^{N} m_j\right)\{\psi_1(1)\psi_2(2)\cdots\psi_i(j)\cdots\psi_j(i)\cdots\psi_N(N)\}. \tag{10-233}$$

Thus, the single permutation terms will give the same result [i.e., multiplication by $\Sigma_j\, m_j$] as the diagonal term, along with the sign appropriate for that term in the expansion of the determinant.

Following similar reasoning for the other permutations in Eq. (10-229), it is seen that

$$\mathcal{L}_z\Psi(1,2,\cdots,N)=\left(\sum_{j=1}^{N} m_j\right)\Psi(1,2,\cdots,N)=M_L\Psi(1,2,\cdots,N), \tag{10-234}$$

i.e., the Slater determinant with basis orbitals chosen by Eq. (10-230) is an eigenfunction of $\mathcal{L}_z$, with eigenvalue

$$M_L=\sum_{j=1}^{N} m_j. \tag{10-235}$$

In an entirely similar manner it can be shown[82] that the Slater determinant of Eq. (10-229) is an eigenfunction of $\mathcal{S}_z$ with eigenvalue $\mathcal{S}_z$, where

$$\mathcal{S}_z=\sum_{j=1}^{N}\mathcal{S}_z(j), \tag{10-236}$$

$$\mathcal{S}_z(j)\psi_i(r_j,\vartheta_j,\phi_j,\sigma_j)=s_i\psi_i(r_j,\vartheta_j,\phi_j,\sigma_j), \tag{10-237}$$

and

$$S_Z=\sum_{j=1}^{N} s_j. \tag{10-238}$$

Thus, since Slater determinants have found widespread use and are eigenfunctions of $\mathcal{L}_z$ and $\mathcal{S}_z$ without modification, it is clear that the set of four operators

[82] See Problem 10-20.

that is most convenient to use is ($\mathcal{L}^2$, $\mathcal{L}_z$, $\mathcal{S}^2$, $\mathcal{S}_z$). For that choice, we must now examine the effect of $\mathcal{L}^2$ and $\mathcal{S}^2$ on a Slater determinant to determine what steps, if any, need to be taken to make eigenfunctions of $\mathcal{L}^2$ and $\mathcal{S}^2$.

We begin that analysis by considering Eq. (10-220), which can be written as

$$\begin{aligned}\mathcal{L}^2 &= \mathcal{L}_x^2 + \mathcal{L}_y^2 + \mathcal{L}_z^2 \\ &= [\mathcal{L}_x(1) + \mathcal{L}_x(2) + \cdots + \mathcal{L}_x(N)]^2 \\ &\quad + [\mathcal{L}_y(1) + \mathcal{L}_y(2) + \cdots + \mathcal{L}_y(N)]^2 \\ &\quad + [\mathcal{L}_z(1) + \mathcal{L}_z(2) + \cdots + \mathcal{L}_z(N)]^2 \\ &= \mathcal{L}^2(1) + \mathcal{L}^2(2) + \cdots + \mathcal{L}^2(N) \\ &\quad + 2\mathcal{L}_z(1)\mathcal{L}_z(2) + 2\mathcal{L}_z(1)\mathcal{L}_z(3) + \cdots + 2\mathcal{L}_z(N-1)\mathcal{L}_z(N) \\ &\quad + \mathcal{L}_+(1)\mathcal{L}_-(2) + \mathcal{L}_-(1)\mathcal{L}_+(2) + \cdots + \mathcal{L}_+(N-1)\mathcal{L}_-(N) \\ &\quad + \mathcal{L}_-(N-1)\mathcal{L}_+(N) \\ &= \sum_{j=1}^{N} \mathcal{L}^2(j) + \sum_{j<k}^{N} [2\mathcal{L}_z(j)\mathcal{L}_z(k) + \mathcal{L}_+(j)\mathcal{L}_-(k) + \mathcal{L}_-(j)\mathcal{L}_+(k)],\end{aligned} \tag{10-239}$$

where

$$\mathcal{L}^2(j) = \mathcal{L}_x^2(j) + \mathcal{L}_y^2(j) + \mathcal{L}_z^2(j), \tag{10-240}$$

$$\mathcal{L}_+(j) = \mathcal{L}_x(j) + i\mathcal{L}_y(j) \tag{10-241}$$

and

$$\mathcal{L}_-(j) = \mathcal{L}_x(j) - i\mathcal{L}_y(j). \tag{10-242}$$

In a similar manner, $\mathcal{S}^2$ can be written as

$$\mathcal{S}^2 = \sum_{j=1}^{N} \mathcal{S}^2(j) + \sum_{j<k}^{N} [2\mathcal{S}_z(j)\mathcal{S}_z(k) + \mathcal{S}_+(j)\mathcal{S}_-(k) + S_-(j)\mathcal{S}_+(k)], \tag{10-243}$$

with

$$\mathcal{S}^2(j) = \mathcal{S}_x^2(j) + \mathcal{S}_y^2(j) + \mathcal{S}_z^2(j), \tag{10-244}$$

$$\mathcal{S}_+(j) = \mathcal{S}_x(j) + i\mathcal{S}_y(j), \tag{10-245}$$

and

$$\mathcal{S}_-(j) = \mathcal{S}_x(j) - i\mathcal{S}_y(j). \tag{10-246}$$

Alternatively, since Slater determinants have been shown to be eigenfunctions of S_z, it is typically more convenient from a computational point of view to use the following equivalent expression for $\mathcal{S}^2$:

$$\mathcal{S}^2 = \sum_{i=1}^{N} \mathcal{S}_z(i) + \sum_{i=1}^{N}\sum_{j=1}^{N} [\mathcal{S}_-(i)\mathcal{S}_+(j) + \mathcal{S}_z(i)\mathcal{S}_z(j)]. \quad (10\text{-}247)$$

Returning to our examination of the effect of $\mathcal{L}^2$ on a Slater determinant, we use Eqs. (6-41), (6-51), (6-52), (6-58), and (6-60) along with (10-234) to give the following result for the effect of $\mathcal{L}^2$ on the diagonal term of a Slater determinant:

$$\mathcal{L}^2\{\psi_1(1)\psi_2(2) \cdots \psi_N(N)\}$$
$$= \left\{ \sum_{j=1}^{N} \mathcal{L}^2(j) + \sum_{j<k}^{N} [2\mathcal{L}_z(j)\mathcal{L}_z(k) + \mathcal{L}_+(j)\mathcal{L}_-(k) + \mathcal{L}_-(j)\mathcal{L}_+(k)] \right\}$$
$$\cdot \{\psi_1(1)\psi_2(2) \cdots \psi_N(N)\},$$

$$\mathcal{L}^2\{\psi_1(1)\psi_2(2) \cdots \psi_N(N)\}$$
$$= [l_1(l_1+1) + l_2(l_2+1) + \cdots + l_N(l_N+1)]\{\psi_1(1)\psi_2(2) \cdots \psi_N(N)\}$$
$$+ 2[m_1m_2 + m_1m_3 \cdots + m_{N-1}m_N]\{\psi_1(1)\psi_2(2) \cdots \psi_N(N)\}$$
$$+ [l_1(l_1+1) - m_1(m_1+1)]^{1/2}[l_2(l_2+1) - m_2(m_2-1)]^{1/2}$$
$$\cdot \{(n_1l_1m_1+1,\, s_1)(1)(n_2l_2m_2-1,\, s_2)(2)\psi_3(3) \cdots \psi_N(N)\}$$
$$+ \cdots + [l_{N-1}(l_{N-1}+1) - m_{N-1}(m_{N-1}+1)]^{1/2}[l_N(l_N+1) - m_N(m_N-1)]^{1/2}$$
$$\cdot \{\psi_1(1)\psi_2(2) \cdots\cdots (n_{N-1}l_{N-1}m_{N-1}-1,\, s_{N-1})(N-1)$$
$$\cdot (n_Nl_Nm_N+1,\, s_N)(N)\}, \quad (10\text{-}248)$$

where the notation

$$\psi_i(i) \equiv (n_il_im_is_i)(i) \quad (10\text{-}249)$$

has been used.

If the other terms in the expansion of the Slater determinant are examined, it is found that, analogous to the considerations with $\mathcal{L}_z$, the same terms generated in Eq. (10-248) are generated for the other terms in the expansion of the Slater determinant (along with the appropriate sign). Thus, the general result can be written as

$$\mathcal{L}^2\Psi(1, 2, \cdots, N)$$
$$= \left\{ \sum_{j=1}^{N} \left[l_j(l_j+1) + 2\sum_{k>j}^{N} m_jm_k \right] \right\} \Psi(1, 2, \cdots, N)$$
$$+ \frac{1}{2}\sum_{j\neq k}^{N} \{[l_j(l_j+1) - m_j(m_j+1)][l_k(l_k+1) - m_k(m_k-1)]\}^{1/2}$$
$$\cdot \det\{\psi_1(1) \cdots (n_jl_jm_j-1,\, s_j)(j) \cdots (n_kl_km_k+1,\, s_k)(k) \cdots \psi_N(N)\}. \quad (10\text{-}250)$$

Of course, since $\mathcal{S}^2$ is an angular momentum operator analogous to $\mathcal{L}^2$ (except for the values of its eigenvalues), we can without further analysis write the result of operating on a Slater determinant by $\mathcal{S}^2$ as

$$\begin{aligned}
&\mathcal{S}^2\Psi(1, 2, \cdots, N)\\
&\quad=\left\{\frac{3N}{4}+2\sum_{k>j}^{N} s_j s_k\right\}\Psi(1, 2, \cdots, N)\\
&\quad+\frac{1}{2}\sum_{j\neq k}^{N}\left\{\left[\frac{1}{2}\left(\frac{1}{2}+1\right)-s_j(s_j+1)\right]\left[\frac{1}{2}\left(\frac{1}{2}+1\right)-s_k(s_k-1)\right]\right\}^{1/2}\\
&\quad\cdot\det\{\psi_1(1)\cdots(n_j l_j m_j s_j-1)(j)\cdots(n_k l_k m_k s_k+1)(k)\cdots\psi_N(N)\},
\end{aligned}\tag{10-251}$$

where the eigenvalues of $\mathcal{S}^2(i)$ are the same for all i, i.e.,

$$S(S+1)=\frac{1}{2}\left(\frac{1}{2}+1\right)=\frac{3}{4},\tag{10-252}$$

and has been inserted in Eq. (10-251). We can rewrite Eq. (10-251) in a form more convenient for computation as

$$\begin{aligned}
&\mathcal{S}^2\Psi(1, 2, \cdots, N)\\
&\quad=\left\{\frac{3N}{4}+2\sum_{k<j}^{N} s_j s_k\right\}\Psi(1, 2, \cdots, N)\\
&\quad+\frac{1}{2}\sum_{j\neq k}^{N}\left\{\left(\frac{3}{2}+s_j\right)\left(\frac{1}{2}-s_j\right)\left(\frac{3}{2}-s_k\right)\left(\frac{1}{2}+s_k\right)\right\}^{1/2}\\
&\quad\cdot\det\{\psi_1(1)\cdots(n_j l_j m_j s_j-1)(j)\cdots(n_k l_k m_k s_k+1)(k)\cdots\psi_N(N)\}.
\end{aligned}\tag{10-253}$$

Thus, we see that, while the choice of a single Slater determinant using basis orbitals as described in Eq. (10-230) automatically assures that the trial wavefunction will be an eigenfunction of $\mathcal{L}_z$ and $\mathcal{S}_z$, this trail wavefunction will not in general be an eigenfunction of $\mathcal{L}^2$ and $\mathcal{S}^2$. This means that, even though the operators $\mathcal{L}^2$, $\mathcal{L}_z$, $\mathcal{S}^2$, and $\mathcal{S}_z$ commute with each other and with an atomic Hamiltonian, a *single* Slater determinant trail wavefunction will not in general be suitable for the simultaneous description of the observables associated with $\mathcal{L}^2$ or $\mathcal{S}^2$.

Let us now turn our attention to the application of these results to systems of interest. Since we shall in general be interested in *molecular*, as opposed to *atomic* systems, the question of whether or not a trial wavefunction is an eigenfunction of $\mathcal{L}^2$ is not nearly as important as the analogous question for $\mathcal{S}^2$.

There are several reasons for such an observation. First, in an atomic system, the spherically symmetric environment allows the use of $\mathcal{L}^2$ and $\mathcal{S}^2$ eigenfunctions to classify atomic states in a manner that is very convenient for comparison to and classification of experimental atomic spectra.[83] Second, constructing trial wavefunctions that are eigenfunctions of both $\mathcal{L}^2$ and $\mathcal{S}^2$ has the effect of simplifying and reducing the number of elements to be calculated in the Hamiltonian matrix, which was of important practical significance prior to large-scale digital computer capabilities.

However, the reduced spatial symmetry of molecular systems (which means that $[\mathcal{L}^2, \mathcal{H}] \neq 0$ in general), coupled with availability of computer programs that diagonalize large-size matrices efficiently, has in general meant that construction of $\mathcal{L}^2$ eigenfunctions is not of importance for molecular wavefunctions. Hence, we shall consider only the construction of $\mathcal{S}^2$ eigenfunctions in detail, although the techniques apply equally well[84] to $\mathcal{L}^2$.

It is also useful at this point, and for future use, to distinguish between a single Slater determinant (that may or may not be an eigenfunction of $\mathcal{S}^2$) and combinations of Slater determinants that *are* eigenfunctions of $\mathcal{S}^2$. In particular, we define a *single configuration wavefunction* as one that is an eigenfunction of both $\mathcal{S}^2$ and $\mathcal{S}_z$. It is, in general, a *linear combination* of single Slater determinants in which each determinant contains the *same* set of doubly occupied MOs, and a set of singly occupied MOs constructed from a common set of spatial orbitals but with differing spins in each determinant. Thus, the case when a single Slater determinant is also an eigenfunction of $\mathcal{S}^2$ is a special case of a single configuration wavefunction.

Let us now consider several examples, beginning with a two-electron system. First, let us consider the simplest case, i.e.,

$$\Phi_1(1, 2) = N \det\{\varphi_1(1)\alpha(1) \;\; \varphi_1(2)\beta(2)\}, \tag{10-254}$$

where N is a normalization constant and φ_1 is a spatial molecular orbital. Such a case, in which each spatial orbital contains two electrons, is known as a *closed shell* determinant. Such a case[85] is also usually referred to as a *restricted closed shell* determinant, because of the restriction that φ_1 be doubly occupied.

Applying $\mathcal{S}^2$ using Eq. (10-253) to the wavefunction of Eq. (10-254) gives

$$\begin{aligned}\mathcal{S}^2\Phi_1(1, 2) &= \Phi_1(1, 2) + N \det\{\varphi_1(1)\beta(1) \;\; \varphi_1(2)\alpha(2)\} \\ &= \Phi_1(1, 2) - \Phi_1(1, 2) \\ &= 0.\end{aligned} \tag{10-255}$$

---

[83] Such an approach is frequently referred to as the "vector model" of the atom. For a detailed discussion see, for example, A. R. Edmonds, *Angular Momentum in Quantum Mechanics*, Princeton University Press, Princeton, 1960.

[84] For a comprehensive discussion and listing of eigenfunctions of $\mathcal{L}^2$ and $\mathcal{S}^2$ using Slater determinants, see J. C. Slater, *Quantum Theory of Atomic Structure*, Volume 2, McGraw-Hill, New York, 1960, Chapter 21.

[85] For $2N$ electrons, there would be N spatial orbitals each forming two spinorbitals, one with α spin and the other with β spin.

Table 10.3. Spin State Terminology

Spin quantum number (S)	Eigenvalue of $\mathcal{S}^2$ $[S(S+1)]$	Multiplicity $(2S+1)$	State Terminology
0	0	1	Singlet
1/2	3/4	2	Doublet
1	2	3	Triplet
3/2	15/4	4	Quartet
2	6	5	Quintet

This means that the wavefunction in Eq. (10-254) is an eigenfunction of $\mathcal{S}^2$, with eigenvalue:

$$S(S+1)=0,$$

or

$$S=0. \tag{10-256}$$

The *multiplicity,* which is defined as $2S + 1$, equals 1 for this case, and $\Phi_1(1, 2)$ is said to be a *singlet* trial wavefunction. Other values of S and the associated terminology are summarized in Table 10.3.

The next example of interest is obtained by relaxing the requirement that both α and β spin functions must have the same spatial function. In that case, the trial wavefunction has the form

$$\Phi_2(1, 2)=N \det\{\varphi_1(1)\alpha(1) \;\; \varphi_2(2)\beta(2)\}. \tag{10-257}$$

Such a case, in which $\varphi_1 \neq \varphi_2$, is referred to as an *unrestricted* determinant.[86] Applying $\mathcal{S}^2$ from Eq. (10-253) to Φ_2 gives

$$\begin{aligned}\mathcal{S}^2\Phi_2(1, 2)=&\left[2\left(\frac{3}{4}\right)+2\left(\frac{1}{2}\right)\left(-\frac{1}{2}\right)\right]\Phi_2+0\\ &+N\left[\left(\frac{3}{2}-\frac{1}{2}\right)\left(\frac{1}{2}+\frac{1}{2}\right)\left(\frac{3}{2}-\frac{1}{2}\right)\left(\frac{1}{2}+\frac{1}{2}\right)\right]^{1/2}\\ &\cdot\det\{\varphi_1(1)\beta(1) \;\; \varphi_2(2)\alpha(2)\}.\end{aligned} \tag{10-258}$$

If we define

$$\Phi_3(1, 2)=N \det\{\varphi_1(1)\beta(1) \;\; \varphi_2(2)\alpha(2)\}, \tag{10-259}$$

we can rewrite Eq. (10-258) as

$$\mathcal{S}^2\Phi_2=\Phi_2+\Phi_3. \tag{10-260}$$

[86] This formulation is also known as the "different orbitals for different spins" approach. See, for example, D. R. Hartree, *Proc. R. Soc. (London)*, **A154**, 588 (1936).

Thus, we see that Φ_2 is *not* an eigenfunction of $\mathcal{S}^2$. On the other hand, if we use the linear combination of $(\Phi_2 + \Phi_3)$ as a trial wavefunction, it is easily seen[87] that

$$\mathcal{S}^2(\Phi_2+\Phi_3)=2(\Phi_2+\Phi_3). \qquad (10\text{-}261)$$

Thus, $(\Phi_2 + \Phi_3)$ *is* an eigenfunction of $\mathcal{S}^2$ (i.e., a *single configuration* wavefunction) with eigenvalue

$$S(S+1)=2,$$

or

$$S=1, \qquad (10\text{-}262)$$

which corresponds to a *triplet state* $(2S + 1 = 3)$.

If can also be seen[88] that Φ_4 and Φ_5, defined as

$$\Phi_4(1,2)=N\det\{\varphi_1(1)\alpha(1)\ \ \varphi_2(2)\alpha(2)\} \qquad (10\text{-}263)$$

$$\Phi_5(1,2)=N\det\{\varphi_1(1)\beta(1)\ \ \varphi_2(2)\beta(2)\} \qquad (10\text{-}264)$$

are eigenfunctions of $\mathcal{S}^2$, with $S = 1$, i.e., Φ_4 and Φ_5 also correspond to trial wavefunctions that are triplet state eigenfunctions. The difference between $(\Phi_2 + \Phi_3)$, Φ_4, and Φ_5 is that each corresponds to different eigenvalues[88] of $\mathcal{S}_z$, i.e.,

$$\mathcal{S}_z(\Phi_2+\Phi_3)=0, \qquad (10\text{-}265)$$

$$\mathcal{S}_z\Phi_4=(+1)\Phi_4, \qquad (10\text{-}266)$$

and

$$\mathcal{S}_z\Phi_5=(-1)\Phi_5. \qquad (10\text{-}267)$$

Furthermore, $\Phi_2 - \Phi_5$ are examples of *open shell wavefunctions* in which one or more spatial orbital contains only *one* electron (i.e., all spatial orbitals are not doubly occupied).

The results obtained thus far are summarized in Table 10.4, including the further observation that $(\Phi_2 - \Phi_3)$ is an eigenfunction of $\mathcal{S}^2$ with eigenvalue

$$S(S+1)=0,$$

or

$$S=0. \qquad (10\text{-}268)$$

The important points that have been illustrated by these examples are as follows:

1. All Slater determinants are automatically eigenfunctions of $\mathcal{S}_z$, but are not necessarily eigenfunctions of $\mathcal{S}^2$.
2. Eigenfunctions of $\mathcal{S}^2$ that correspond to a given total spin (S) and to each of

---

[87] See Problem 10-21.

[88] See Problem 10-22.

Table 10.4. Two-Electron Trial Wavefunctions That Are Spin Eigenfunctions

Wavefunction	Eigenvalue of $\mathcal{S}_z$	Total spin (S)	Multiplicity
$\Phi_1 = N \det\{\varphi_1(1)\alpha(1)\ \varphi_1(2)\beta(2)\}$	0	0	Singlet
$\Phi_2 + \Phi_3 = N \det\{\varphi_1(1)\alpha(1)\ \varphi_2(2)\beta(2)\} + N \det\{\varphi_1(1)\beta(1)\ \varphi_2(2)\alpha(2)\}$	0	1	Triplet
$\Phi_4 = N \det\{\varphi_1(1)\alpha(1)\ \varphi_2(2)\alpha(2)\}$	1	1	Triplet
$\Phi_5 = N \det\{\varphi_1(1)\beta(1)\ \varphi_2(2)\beta(2)\}$	-1	1	Triplet
$\Phi_2 - \Phi_3 = N \det\{\varphi_1(1)\alpha(1)\ \varphi_2(2)\beta(2)\} - N \det\{\varphi_1(1)\beta(1)\ \varphi_2(2)\alpha(2)\}$	0	0	Singlet

the allowed $\mathcal{S}_z$ eigenvalues can be formed by taking linear combinations of Slater determinants to form single configuration wavefunctions.

The specific result that $S = 0$ for a two-electron closed shell wavefunction [Eqs. (10-255) and (10-256)] is also generalizable to the N-electron closed shell case. Specifically, we consider the N-electron (N even) closed shell wavefunction

$$\begin{aligned}\Psi(1, 2, \cdots, N) &= K \det\{\varphi_1(1)\alpha(1)\ \varphi_1(2)\beta(2) \cdots \varphi_{i/2}(i-1)\alpha(i-1)\ \varphi_{i/2}(i)\beta(i) \\ &\quad \cdots \varphi_{N/2}(N-1)\alpha(N-1)\ \varphi_{N/2}(N)\beta(N)\},\end{aligned} \tag{10-269}$$

with i even and where K is a normalization constant. Thus, each spatial orbital (φ_k) is doubly occupied, i.e., appears with both an α and a β spin function in the Slater determinant. Either Eq. (10-247) or (10-253) can be used to evaluate the effect of $\mathcal{S}^2$ on the wavefunction of Eq. (10-269), and we shall use Eq. (10-247) for convenience. For the first and third terms of the $\mathcal{S}^2$ operator, we can easily see that

$$\sum_{i=1}^{N} \mathcal{S}_z(i)\Psi(1, 2, \cdots, N) = 0, \tag{10-270}$$

and

$$\sum_{i=1}^{N}\sum_{j=1}^{N} \mathcal{S}_z(i)\mathcal{S}_z(j)\Psi(1, 2, \cdots, N) = 0, \tag{10-271}$$

since every $+\frac{1}{2}$ eigenvalue for $\varphi_{i/2}(i-1)\alpha(i-1)$ will exactly cancel the $-\frac{1}{2}$ eigenvalue associated with $\varphi_{i/2}(i)\beta(i)$ in Eq. (10-270), and similarly for the products $(s_i s_j)$ in Eq. (10-271).

For the $\mathcal{S}_-(i)\mathcal{S}_+(j)$ terms in Eq. (10-247), cancellations also occur, as well as some terms making zero contributions. In particular, it is easily seen that the only nonzero contributions from the $\Sigma_{i=1}^{N} \Sigma_{j=1}^{N} \mathcal{S}_-(i)\mathcal{S}_+(j)$ terms acting on $\Psi(1, 2, \ldots N)$ in Eq. (10-269) arise from two kinds of terms; $\mathcal{S}_-(i-1)\mathcal{S}_+(i)\Psi$ and $\mathcal{S}_-(i)\mathcal{S}_+(i)\Psi$, with i even. However, each of these terms cancels, and the net result is

$$\mathcal{S}^2\Psi(1, 2, \cdots, N)=0$$

whenever $\Psi(1, 2, \cdots, N)$ represents a closed shell single Slater determinant.

Of course, as illustrated in the two-electron case, single Slater determinants are not in general eigenfunctions of S^2, and linear combinations are required. While the determination of which coefficients are necessary in order to obtain the correct linear combination of Slater determinant that is an eigenfunction of S^2 may be complicated, a variety of procedures are now available[89] for accomplishing this task.

10-10. Orbital Transformations[90]

In Section 10-3 it was noted[91] that the form of the energy expression for a single Slater determinant or single configuration wavefunction is greatly simplified by the use of orthonormal orbitals. However, before leaving this discussion it is appropriate to inquire as to what effect the orthogonalization process has on both the energy and wavefunction. For example, if the process of orthogonalization is viewed as a constraint on the orbitals, does that mean that the use of nonorthogonal orbitals in energy optimization processes will lead to a lower energy? Also, is it always possible to carry out the orthogonalization process without affecting the spin state or other properties of the wavefunction? We shall see that answers to questions such as these cannot be given in general without consideration of the overall spin state. Since such a statement is surprising at first glance, we shall consider several important cases that illustrate the nature of the problem.

Let us consider first an "unrestricted" single Slater determinant wavefunction for an N-electron system having the form

$$\Psi(1, 2, \cdots, N)=K \det\{\chi_1(1)\chi_2(2) \cdots \chi_N(N)\}, \tag{10-272}$$

where K is a normalization constant and

$$\chi_i(i)=\begin{cases} \varphi_i(i)\alpha(i) \\ \quad \text{or} \\ \varphi_i(i)\beta(i) \end{cases}, \tag{10-273}$$

and each φ_i is, in general, different.

[89] See, for example, R. Paunz, *Spin Eigenfunctions*, Plenum Press, New York, 1979, for a discussion of a number of the methods that are available.

[90] The author is indebted to J. Petke for the approach used in this section.

[91] See the discussion associated with Eq. (10-90).

To address the questions raised above, let us consider the following linear (nonsingular) transformation of the spatial orbitals $(\varphi_1\varphi_2 \cdots \varphi_N)$:

$$\varphi_i' = \sum_{k=1}^{N} \varphi_k t_{ki}, \qquad i=1, 2, \cdots, N, \tag{10-274}$$

which can be expressed in matrix form as

$$(\varphi_1' \ \varphi_2' \ \cdots \ \varphi_N') = (\varphi_1 \ \varphi_2 \ \cdots \ \varphi_N) \begin{pmatrix} t_{11} & t_{12} & \cdots & t_{1N} \\ t_{21} & t_{22} & \cdots & t_{2N} \\ \vdots & \vdots & \ddots & \vdots \\ t_{N1} & t_{N2} & \cdots & t_{NN} \end{pmatrix},$$

or

$$\boldsymbol{\varphi}' = \boldsymbol{\varphi}\mathbf{T}. \tag{10-275}$$

A similar expression applies to the spin orbitals (χ), i.e.,

$$\boldsymbol{\chi}' = \boldsymbol{\chi}\mathbf{T}. \tag{10-276}$$

If we form a new unrestricted single Slater determinant using the primed orbitals, we have

$$\Psi'(1, 2, \cdots, N) = K' \det\{\chi_1'(1)\chi_2'(2) \cdots \chi_N'(N)\}$$

$$= K' \begin{vmatrix} \chi_1'(1) & \chi_2'(1) & \cdots & \chi_N'(1) \\ \chi_1'(2) & \chi_2'(2) & \cdots & \chi_N'(2) \\ \vdots & \vdots & \ddots & \vdots \\ \chi_1'(N) & \chi_2'(N) & \cdots & \chi_N'(N) \end{vmatrix}. \tag{10-277}$$

Using Eq. (10-276) we can rewrite Eq. (10-277) as

$$\Psi'(1, 2, \cdots, N) = K' \begin{vmatrix} \chi_1(1) & \chi_2(1) & \cdots & \chi_N(1) \\ \chi_1(2) & \chi_2(2) & \cdots & \chi_N(2) \\ \vdots & \vdots & \ddots & \vdots \\ \chi_1(N) & \chi_2(N) & \cdots & \chi_N(N) \end{vmatrix} \cdot |\mathbf{T}| \tag{10-278}$$

$$= (\text{const.}) \cdot \Psi(1, 2, \cdots, N), \tag{10-279}$$

where Theorem 3-5 has also been used. In other words, the process of orthogonalization (or any other nonsingular transformation $\mathbf{T}$) produces a new wavefunction (Ψ') that differs from the original wavefunction (Ψ) by at most a constant. Hence, such a transformation is not an additional constraint in this case, and orthogonalization or any other nonsingular transformation having the form of Eq. (10-276) of the orbitals can be carried out without affecting either the wavefunction or energy.

The next case of interest is that of a closed-shell, restricted single Slater determinant. In particular, let us consider the following wavefunction[92] for N electrons and $M = N/2$ doubly occupied spatial orbitals $\{\varphi_i\}$:

$$\Psi(1, 2, \cdots, N) = K \det\{\chi_1\chi_2 \cdots \chi_m\bar{\chi}_1\bar{\chi}_2 \cdots \bar{\chi}_m\}, \quad (10\text{-}280)$$

where

$$\begin{aligned} \chi_i &= \varphi_i\alpha \\ \bar{\chi}_i &= \varphi_i\beta. \end{aligned} \quad (10\text{-}281)$$

For this case we consider the following linear (nonsingular) transformation (**T**):

$$\boldsymbol{\chi}' = \boldsymbol{\chi}\mathbf{T} \quad (10\text{-}282)$$

and

$$\bar{\boldsymbol{\chi}}' = \bar{\boldsymbol{\chi}}\mathbf{T}, \quad (10\text{-}283)$$

where the same transformation (**T**) is used for both α and β orbitals. The wavefunction (Ψ') formed from the transformed orbitals (χ', χ'), can be written as

$$\begin{aligned} \Psi'(1, 2, \cdots, N) &= K' \det\{\chi_1'\chi_2' \cdots \chi_M'\bar{\chi}_1'\bar{\chi}_2' \cdots \bar{\chi}_M'\}, \\ &= K' \det\left\{\boldsymbol{\chi}\ \bar{\boldsymbol{\chi}} \begin{pmatrix} \mathbf{T} & \mathbf{0} \\ \mathbf{0} & \mathbf{T} \end{pmatrix}\right\}, \\ &= K' \cdot \det^2(\mathbf{T}) \cdot \det\{\boldsymbol{\chi}\ \bar{\boldsymbol{\chi}}\} \\ &= (\text{const.}) \cdot \Psi(1, 2, \cdots, N). \end{aligned} \quad (10\text{-}284)$$

Hence, in this case, as in the previous one, transformation (e.g, orthogonalization) of the orbitals leaves the wavefunction (and hence energy) invariant. Furthermore, both Ψ' and Ψ represent singlet states, so the transformation has not affected the spin properties of the wavefunction. This case corresponds to the important case of the closed shell Hartree–Fock description of an N electron (N even) system that will be discussed in the next chapter.

As our next case, let us consider an open shell example, i.e., the case of a "high spin" open shell, single Slater determinant description of an $(2M + m)$ electron system:

$$\Psi(1, 2, \cdots, 2M+m) = K \det\{\bar{\chi}_1 \cdots \bar{\chi}_M\chi_1 \cdots \chi_M\chi_{M+1}\chi_{M+2} \cdots \chi_{M+m}\}, \quad (10\text{-}285)$$

where

$$\chi_i = \varphi_i\alpha \qquad i = 1, 2, \cdots, M+m \quad (10\text{-}286)$$

$$\bar{\chi}_i = \varphi_i\beta \qquad i = 1, 2, \cdots, M. \quad (10\text{-}287)$$

---

[92] This wavefunction is the same as the closed shell wavefunction of Eq. (10-269), except that all of the orbitals with α-spin have been collected together for convenience in the current analysis.

Thus, in this case the first M orbitals are doubly occupied, and the remaining m orbitals are each singly occupied with α spin.

Proceeding as in previous cases, we form a new wavefunction using transformed orbitals as follows:

$$\Psi'(1, 2, \cdots, N) = K' \det\{\bar{\chi}_1' \cdots \bar{\chi}_M' \chi_1' \cdots \chi_M' \chi_{M+1}' \chi_{M+2}' \cdots \chi_{M+m}'\}, \tag{10-288}$$

with $N = 2M + m$, and where the transformation applied to the orbitals is given by

$$\bar{\chi}_i' = \sum_{k=1}^{M+m} \bar{\chi}_k \bar{t}_{ki} \qquad i = 1, 2, \cdots, M, \tag{10-289}$$

for the orbitals with β spin, and

$$\chi_i' = \sum_{k=1}^{M+m} \chi_k t_{ki} \qquad i = 1, 2, \cdots, M+m. \tag{10-290}$$

for the orbitals with α spin. Inserting these expressions into the wavefunction of Eq. (10-288) gives

$$\Psi'(1, 2, \cdots, N) = K' \det\left\{\left(\sum_{k_1=1}^{M+m} \bar{\chi}_{k_1} \bar{t}_{k_1,1}\right)\left(\sum_{k_2=1}^{M+m} \bar{\chi}_{k_2} \bar{t}_{k_2,2}\right) \cdots \left(\sum_{k_{M+m}=1}^{M+m} \chi_{k_{M+m}} t_{k_{M+m},M+m}\right)\right\}, \tag{10-291}$$

which can be rewritten as:

$$\Psi'(1, 2, \cdots, N) = K' \sum_{k_1=1}^{M+m} \sum_{k_2=1}^{M+m} \cdots \sum_{k_{M+m}=1}^{M+m} \bar{t}_{k_1,1} \bar{t}_{k_2,2} \cdots \bar{t}_{k_{M+m},M+m} \cdot \det\{\bar{\chi}_{k_1} \bar{\chi}_{k_2} \cdots \bar{\chi}_{k_{M+m}}\}. \tag{10-292}$$

Since the transformation of the orbitals [Eqs. (10-289) and (10-290)] involves *all* ($M = m$) orbitals, there will be some nonzero terms in Eq. (10-292) in which one or more of the original orbitals with β spin [$\bar{\chi}_k$, $k = 1, 2, \cdots, M$] has been replaced by one of the orbitals from the singly occupied original set ($\bar{\chi}_L$, $L = M + 1, M + 2, \cdots, M + m$). Such terms are not present in the original wavefunction, and we see that, for the "high spin" open shell case, the wavefunction is *not* invariant to transformations (e.g., orthogonalization) such as given in Eqs. (10-289) and (10-290).

There are, however, two types of more limited transformations that do leave the wavefunction invariant. First, a transformation only among the closed shell orbitals ($\bar{\chi}_i$ and χ_i, $i = 1, 2, \ldots M$) will leave the wavefunction invariant, i.e., the closed shell orbitals can be orthogonalized using such a process. Alternatively, the open shell orbitals (χ_i, $i = M + 1, \ldots, M + m$) can be transformed (e.g., orthogonalized) among themselves only, but still leave the wavefunction invariant.

As a final example, let us consider another case in which orthogonal orbitals cannot be constructed in a way that leaves the wavefunction invariant. Specifically, let us consider the case of an open shell singlet wavefunction for a system of $N = 2M + 2$ electrons, which is written as a linear combination of two Slater determinants as follows:

$$\Psi(1, 2, \cdots, 2M+2) = K\,[\det\{\chi_1\bar{\chi}_1 \cdots \chi_M\bar{\chi}_M\chi_{M+1}\bar{\chi}_{M+2}\} - \det\{\chi_1\bar{\chi}_1 \cdots \chi_M\bar{\chi}_M\bar{\chi}_{M+1}\chi_{M+2}\}]. \quad (10\text{-}293)$$

Thus, the first M orbitals are doubly occupied in each determinant, and the singly occupied orbitals (χ_{M+1} and χ_{M+2}) are spin coupled via a singlet spin function of the form ($\alpha\beta - \beta\alpha$).

As described in the previous example, we may assume, without loss of generality or wavefunction restriction, that the closed shell orbitals are orthonormal. Next, let us attempt to orthogonalize the singly occupied orbitals to the doubly occupied set as done in the previous example. Using the Schmidt orthogonalization procedure,[93] for example, we can construct an orbital χ'_{M+1} that is orthogonal to the $2M$ closed shell orbitals via

$$\chi'_{M+1} = \chi_{M+1} - \sum_{i=1}^{M} c_i\chi_i, \quad (10\text{-}294)$$

with the c_is are chosen so that $\langle\chi'_{M+1}|\chi_i\rangle = 0$, $i = 1, 2, \ldots, M$. If we solve the above equation for χ_{M+1} (or an analogous equation for $\bar{\chi}_{M+1}$) and substitute it into Eq. (10-293), we obtain

$$\Psi(1, 2, \cdots, 2M+2) = K\left[\det\left\{\chi_1\bar{\chi}_1 \cdots \chi_M\bar{\chi}_M\left(\chi'_{M+1} + \sum_{i=1}^{M} c_i\chi_i\right)\bar{\chi}_{M+2}\right\} - \det\left\{\chi_1\bar{\chi}_1 \cdots \chi_M\bar{\chi}_M\left(\bar{\chi}'_{M+1} + \sum_{i=1}^{M} c_i\bar{\chi}_i\right)\chi_{M+2}\right\}\right]. \quad (10\text{-}295)$$

Using Theorem 3-4, the above equation can be rewritten as:

$$\Psi(1, 2, \cdots, 2M+2) = K\,[\det\{\chi_1\bar{\chi}_1 \cdots \chi_M\bar{\chi}_M\chi'_{M+1}\bar{\chi}_{M+2}\} - \det\{\chi_1\bar{\chi}_1 \cdots \chi_M\bar{\chi}_M\bar{\chi}'_{M+1}\chi_{M+2}\}]. \quad (10\text{-}296)$$

[93] See Chapter 2, Section 2-4.

Thus, it has been possible to create an orbital χ'_{M+1} that is orthogonal to each of the doubly occupied orbitals ($\chi_1 \ldots \chi_M$) yet leaving the wavefunction invariant.

However, the difficulty arises when we attempt to construct the next orthogonal orbital. In particular, we use the Schmidt orthogonalization process to construct an orbital χ'_{M+2} which is orthogonal to each of the other orbitals in the wavefunction via

$$\chi'_{M+2} = \chi_{M+2} - \sum_{i=1}^{M} a_i \chi_i - a_{M+1} \chi'_{M+1}. \qquad (10\text{-}297)$$

Solving the above equation for χ_{M+2} and substituting into Eq. (10-269) gives

$$\begin{aligned}
&\Psi(1, 2, \cdots, 2M+2) \\
&= K \left[\det\left\{ \chi_1 \bar{\chi}_1 \cdots \chi_M \bar{\chi}_M \chi'_{M+1} \left(\bar{\chi}'_{M+2} + \sum_{i=1}^{M} a_i \bar{\chi}_i + a_{M+1} \bar{\chi}'_{M+1} \right) \right\} \right. \\
&\quad \left. - \det\left\{ \chi_1 \bar{\chi} \cdots \chi_M \bar{\chi}_M \bar{\chi}'_{M+1} \left(\chi'_{M+2} + \sum_{i=1}^{M} a_i \chi_i + a_{M+1} \chi'_{M+1} \right) \right\} \right] \qquad (10\text{-}298)
\end{aligned}$$

Using Theorem 3-4, we can rewrite Eq. (10-298) as:

$$\begin{aligned}
\Psi(1, 2, \cdots, 2M+2) = K\, [&\det\{\chi_1 \bar{\chi}_1 \cdots \chi_M \bar{\chi}_M \chi'_{M+1} \bar{\chi}'_{M+2}\} \\
&- \det\{\chi_1 \bar{\chi}_1 \cdots \chi_M \bar{\chi}_M \bar{\chi}'_{M+1} \chi'_{M+2}\}] \\
&+ a_{M+1} \det\{\chi_1 \bar{\chi}_1 \cdots \chi_M \bar{\chi}_M \chi'_{M+1} \bar{\chi}'_{M+1}\} \\
&- a_{M+1} \det\{\chi_1 \bar{\chi}_1 \cdots \chi_M \bar{\chi}_M \bar{\chi}'_{M+1} \chi'_{M+1}\}]. \qquad (10\text{-}299)
\end{aligned}$$

Thus, while the first two terms on the right-hand side of Eq. (10-299) are the desired terms, the last two terms with coefficient (a_{M+1}) are new terms, i.e., are *not* present in the original wavefunction of Eqn. (10-296). We therefore see that, for this open shell example as well as the previous one, it has not been possible to construct a wavefunction that contains orthonormal orbitals that is invariant to linear transformations among the orbitals. In other words, while it is possible to construct a wavefunction that contains orthonormal orbitals, the process of orthogonalization has placed an arbitrary constraint on the wavefunction. This means that the energy associated with such a wavefunction will be higher in general than for a wavefunction in which such orthogonality constraints have not been applied. However, the complications that typically arise in the energy expression when nonorthogonal orbitals are used are so significant that the orthogonal orbital ‘‘constraint’’ must in general be accepted.

The other general conclusion of importance is that open shell problems need to be approached carefully and on a case-by-case basis, taking care to assure that spin, spatial symmetry, or other wavefunction properties are not affected by transformations such as orthogonalization of the orbitals.

Problems

1. Verify that the three classes of the C_{3v} point group contain E, C_3 and C_3^2, and σ_v, σ_v', and σ_v'', respectively.
2. Determine the point group name (from Table 10-1) that is appropriate for each of the following molecules:
   a. $H_2C{=}C{=}CH_3$ (allene)
   b. Benzene
   c. CH_4
   d. SF_6
   e. NH_3
3. a. Show that the normalization constant in Eq. (10-66) is given by

$$N = \frac{1}{[3(1+2S)]^{1/2}},$$

   where

$$S = \int 1s_a 1s_b \, d\tau.$$

   b. Verify Eq. (10-69) by carrying out the operations indicated in Eq. (10-68).
   c. Verify the normalization constant in Eq. (10-73) where S is given in part "*a*" of this problem.
4. Construct matrix representations of the symmetry operations for the E irreducible representation of the C_{3v} group, as begun for the C_3^+ operator [Eq. (10-35)]. Also, verify that the functions given in Eqs. (10-73) and (10-74) are (apart from normalization) basis functions for the x and y rows of the same matrix representation of E.
5. As an extension of Eq. (10-78), prove the following more general matrix element theorem. Let $\mathcal{H}$ be a symmetric operator, with $\varphi_k^{(\kappa)}$ and $\varphi_l^{(\lambda)}$ defined as in Eqs. (10-76)–(10-78). Prove that

$$\langle \varphi_k^{(\kappa)} | \mathcal{H} | \varphi_l^{(\lambda)} \rangle = \left(\frac{1}{l_\kappa} \right) \sum_{S=1}^{l_\kappa} \langle \varphi_S^{(\kappa)} | \mathcal{H} | \varphi_S^{(\kappa)} \rangle \delta_{\kappa\lambda} \delta_{kl},$$

   where l_κ is the dimension of the κth irreducible representation.
6. For the group C_{3v}, let Ψ_{A_1}, Ψ_{A_2} and Ψ_E represent functions that transform as A_1, A_2 and E irreducible representations of the C_{3v} group. Determine which of the integrals $\langle \Psi_I | \mathcal{H} | \Psi_J \rangle$ are zero, where $\mathcal{H}$ is a quantum-mechanical Hamiltonian for a molecule having C_{3v} nuclear symmetry, and, $I, J = A_1, A_2$, or E.
7. Using Eq. (10-78), prove that an l_j-dimensional set of functions $\{\varphi_m^{(j)}\}$ belonging to different rows of a unitary group representation j are degenerate.
8. Show that an appropriately antisymmetrized wavefunction can be obtained

by permitting the indices on basis orbitals [cf. Eq. (10-92)] instead of electrons.

9. Prove that the antisymmetrization operator ($\mathcal{A}$) is a projection operator, i.e., show that

$$\mathcal{A}^2 = \mathcal{A}.$$

10. Show that the wavefunction in Eq. (10-90) is normalized to unity, assuming that individual basis orbitals (Ψ_i) are each orthonormal.
11. Prove Eq. (10-129).
12. Prove Eq. (10-140).
13. Show that if more than two orbitals in a Slater determinant (D') differ from their counterparts in another Slater determinant (D),

$$\langle D | \mathcal{H} | D' \rangle = 0.$$

14. Prove the three-dimensional quantum mechanical, N-particle virial theorem, i.e.,

$$2\langle \mathcal{T} \rangle = \left\langle \sum_{j=1}^{3N} q_j \frac{\partial \mathcal{V}}{\partial q_j} \right\rangle,$$

where $\mathcal{T}$ and $\mathcal{V}$ are defined by Eq. (10-153), and the q_j are the $3N$ Cartesian coordinates of the N particles.

15. Show that the virial theorem for atoms, i.e.,

$$2\langle \mathcal{T} \rangle = -\langle \mathcal{V} \rangle$$

applies to the ground and first excited states of the hydrogen atom.

16. Prove Eq. (10-189).
17. Using the optimized η as given by Eq. (10-216), show that the virial theorem as given by Eq. (10-214) is satisfied by the scaled trial wavefunction of Eq. (10-206).
18. Using the commutation relations and definitions in Eqs. (10-217) through (10-225), show that

$$[\mathcal{J}^2, \mathcal{L}^2] = [\mathcal{J}^2, \mathcal{S}^2] = [\mathcal{J}^2, \mathcal{J}_z]$$
$$= [\mathcal{S}^2, \mathcal{J}_z] = [\mathcal{L}^2, \mathcal{S}^2] = [\mathcal{L}^2, \mathcal{J}_z] = 0.$$

19. Using the commutation relations and definitions in Eqs. (10-217) through (10-225), show that the operators, $\mathcal{L}^2, \mathcal{L}_z$, $\mathcal{S}^2$, and $\mathcal{S}_z$ mutually commute.
20. Show that the N-electron Slater determinant given by

$$\Psi(1, 2, \cdots, N) = \frac{1}{(N!)^{1/2}} \sum_{\mathcal{P}} (-1)^p \mathcal{P} \{ \psi_1(1) \psi_2(2) \cdots \psi_N(N) \}$$

is an eigenfunction of $\mathcal{S}_z$ with eigenvalue

$$S_Z = \sum_{j=1}^{N} s_j$$

21. Prove Eq. (10-261).
22. For the Slater determinants ($\phi_2 - \phi_5$) defined in Eqs. (10-257), (10-259), (10-263), and (10-264),
    a. Show that ϕ_2 and ϕ_5 are eigenfunctions of $\mathcal{S}^2$ corresponding to triplet states (i.e., $2S + 1 = 3$).
    b. Show that $(\phi_2 + \phi_3)$, ϕ_4 and ϕ_5 are eigenfunctions of $\mathcal{S}_z$, with eigenvalues 0, +1, and −1, respectively.
23. For a three-electron problem having trial wavefunctions composed of three Slater determinants defined as

$$\Phi_1(1, 2, 3) = N \det\{\varphi_1(1)\alpha(1) \;\; \varphi_2(2)\alpha(2) \;\; \varphi_3(3)\beta(3)\}$$

$$\Phi_2(1, 2, 3) = N \det\{\varphi_1(1)\beta(1) \;\; \varphi_2(2)\alpha(2) \;\; \varphi_3(3)\alpha(3)\}$$

$$\Phi_3(1, 2, 3) = N \det\{\varphi_1(1)\alpha(1) \;\; \varphi_2(2)\beta(2) \;\; \varphi_3(3)\alpha(3)\}.$$

    a. Show that

$$\mathcal{S}^2\Phi_1 = \frac{7}{4}\,\Phi_1 + \Phi_2 + \Phi_3$$

$$\mathcal{S}^2\Phi_2 = \frac{7}{4}\,\Phi_2 + \Phi_1 + \Phi_3$$

$$\mathcal{S}^2\Phi_3 = \frac{7}{4}\,\Phi_3 + \Phi_1 + \Phi_2.$$

    b. Show that the following linear combinations of Slater determinants are eigenfunctions of S^2, i.e.,

$$\mathcal{S}^2(\Phi_1 + \Phi_2 + \Phi_3) = \frac{15}{4}\,(\Phi_1 + \Phi_2 + \Phi_3) \qquad \text{(quartet state)}$$

$$\mathcal{S}^2(-\Phi_1 + \Phi_2) = \frac{3}{4}\,(-\Phi_1 + \Phi_2) \qquad \text{(doublet state)}$$

$$\mathcal{S}^2(\Phi_1 + \Phi_2 - 2\Phi_3) = \frac{3}{4}\,(\Phi_1 + \Phi_2 - 2\Phi_3) \qquad \text{(doublet state).}$$

24. Consider the following "high-spin," single Slater determinant, open shell wavefunction:

$$\Psi(1, 2, \cdots, N+m) = K \det\{\varphi_1\bar{\varphi}_1 \cdots \varphi_M(N-1)\bar{\varphi}_M(N)\varphi_{M+1}(N+1) \cdot \varphi_{M+2}(N+2) \cdots \varphi_{M+m}(M+m)\},$$

    where $M = N/2$ with N even, while φ_i refers to a spatial orbital with α spin and $\bar{\varphi}_i$ refers to the same spatial orbital with β spin. Hence, in the above wavefunction the first $(N/2)$ orbitals are doubly occupied, and the remaining m orbitals $\varphi_{M+1} \cdots \varphi_{M+m}$ are singly occupied with α spin.

a. Show that the closed shell portion of the wavefunction (i.e., the first M doubly occupied orbitals) contributes nothing to $\mathcal{S}_z$ or $\mathcal{S}^2$, and that

$$\mathcal{S}_z\Psi = \left(\frac{m}{2}\right)\Psi$$

$$\mathcal{S}^2\Psi = \left(\frac{m}{2}\right)\left(\frac{m}{2}+1\right)\Psi.$$

25. Let

$$D = (N!)^{-1/2}\det\{(\text{core})\ \varphi_1\bar{\varphi}_2\varphi_3\bar{\varphi}_4\},$$

where "core" represents any set of doubly occupied orbitals. Also consider a determinant D', derived from D by exchanging spin functions for the noncore orbitals, i.e.,

$$D' = (N!)^{-1/2}\det\{(\text{core})\ \bar{\varphi}_1\varphi_2\bar{\varphi}_3\varphi_4\}.$$

Show that

$$\Psi_+ = (D + D')/\sqrt{2}$$

is not an eigenfunction of S^2, but that

$$\Psi_- = (D - D')/\sqrt{2}$$

is a triplet state eigenfunction.

26. Consider two closed shell single determinant wavefunctions

$$\Psi_1 = \det\{\varphi_1\bar{\varphi}_1\varphi_2\bar{\varphi}_2\varphi_3\bar{\varphi}_3\}$$

$$\Psi_2 = \det\{\varphi_1\bar{\varphi}_1\varphi_2\bar{\varphi}_2\varphi_4\bar{\varphi}_4\}$$

constructed from the same set of orbitals $\{\varphi_1\varphi_2\varphi_3\varphi_4\}$.

a. Using the Schmidt orthogonalization process, show that

$$\Psi = c_1\Psi_1 + c_2\Psi_2,$$

where c_1 and c_2 are constants, is *not* invariant to orthonormalization of the set $\{\varphi_1\varphi_2\varphi_3\varphi_4\}$.

b. Construct a multideterminantal wavefunction (Ψ) using the orbitals $\{\varphi_1\varphi_2\varphi_3\varphi_4\}$ plus Ψ_1, and Ψ_2 that *is* invariant to Schmidt orthogonalization of the set $\{\varphi_1\varphi_2\varphi_3\varphi_4\}$ and conserves singlet properties for Ψ.

Chapter 11

Computational Techniques for Many-Electron Systems Using Single Configuration Wavefunctions

In the last chapter we found that one-electron orbitals can be used as a basis for describing multielectron systems. Furthermore, the use of Slater determinants of these one-electron orbitals forms a convenient means of satisfying the Pauli Exclusion principle as well as creating eigenfunctions of operators such as $\mathcal{S}^2$, $\mathcal{L}^2$, etc. In this chapter and the next we shall develop these ideas further in ways that are important both from a conceptual and practical point of view.

In particular, in this chapter we shall use the concepts and results of previous chapters in the development of Hartree–Fock theory. These techniques form the basis for a very large fraction of contemporary applied molecular quantum mechanics, and have allowed a great deal of chemical insight to be gained. Put another way, a thorough study of Hartree–Fock theory is a necessary prerequisite to understanding the last several decades of quantum chemistry literature. It is to this task that this chapter is addressed.

11-1. Hartree-Fock Theory for Closed Shell Systems

Hartree-Fock theory[1] is really quite simple conceptually. It seeks to find *the best single configuration orbital description of multielectron systems*, whether atomic or molecular. To see how this is accomplished, we begin with a few general considerations.

We shall consider an N-electron M-nucleus system with a nonrelativistic, field-free, time-independent Hamiltonian operator ($\mathcal{H}$) within the Born–

D. R. Hartree, *Proc. Cambridge Philos. Soc.*, **24**, 89 (1928); V. Fock, *Z. Physik*, **61**, 126 (1930). Also, see S. M. Blinder, *Am. J. Phys.*, **33**, 431 (1965) for further discussion.

Oppenheimer approximation given by:

$$\mathcal{H} = -\frac{1}{2}\sum_{j=1}^{N} \nabla_j^2 - \sum_{\alpha=1}^{M}\sum_{j=1}^{N} \frac{Z_\alpha}{r_{\alpha j}} + \sum_{i<j}^{N} \frac{1}{r_{ij}}, \tag{11-1}$$

where Z_α is the charge on the α nucleus, and the constant nuclear–nuclear repulsion term $[\Sigma_{\alpha<\beta}^{M} ((Z_\alpha Z_\beta)/R_{\alpha\beta})$ has been omitted for convenience.[2] This Hamiltonian, frequently called the "electronic Hamiltonian" because of the omission of the nuclear repulsion terms, is sometimes rewritten in the form

$$\mathcal{H} = \sum_{j=1}^{N} h(j) + \sum_{i<j}^{N} \frac{1}{r_{ij}}, \tag{11-2}$$

where $h(j)$ is a one-electron operator given by

$$h(j) = -\frac{1}{2}\nabla_j^2 - \sum_{\alpha=1}^{M} \frac{Z_\alpha}{r_{\alpha j}}, \tag{11-3}$$

and represents the operators for the kinetic energy of the jth electron plus the attraction of the jth electron to each of the nuclei.

Since the Schrödinger equation associated with the Hamiltonian of Eq. (11-1) cannot be solved exactly, the question of importance is how to construct an approximate wavefunction that is both an adequate representation of the chemical system and is easy to interpret in chemical terms. Hartree–Fock theory[3] provides a major step toward answering these questions, as we shall see.

The form of approximate wavefunction that is chosen for many cases in Hartree–Fock theory is that of a single Slater determinant,[4] made up of orthonormal one-electron orbitals, i.e.,

$$\Psi(1, 2, \cdots, N) = \frac{1}{(N!)^{1/2}} \begin{vmatrix} \chi_1(1) & \chi_2(1) & \cdots & \chi_N(1) \\ \chi_1(2) & \chi_2(2) & \cdots & \chi_N(2) \\ \vdots & \vdots & \ddots & \vdots \\ \chi_1(N) & \chi_2(N) & \cdots & \chi_N(N) \end{vmatrix} \tag{11-4}$$

$$= \mathcal{A}\{\chi_1(1)\ \chi_2(2)\ \cdots\ \chi_N(N)\}, \tag{11-5}$$

[2] The nuclear–nuclear repulsion term can simply be added at the end of the considerations, since it is a constant. It will therefore be omitted in general from the discussions for convenience.

[3] For additional discussion of this topic, see, e.g., A. Szabo and N. Ostlund, *Modern Quantum Chemistry*, Macmillan, New York, 1982.

[4] We have already seen that a single Slater determinant is not always sufficient to obtain an appropriate Hartree–Fock wavefunction that is also an eigenfunction of $\mathcal{S}^2$, and sums of Slater determinants are sometimes required. In such cases we refer to Hartree–Fock theory as a *single configuration* theory, with a configuration being defined as the linear combination of Slater determinants needed to form an eigenfunction of S^2. To illustrate the concept, however, a single Slater determinant is sufficient, and is used here.

where[5]

$$\langle \chi_i | \chi_j \rangle = \delta_{ij}. \tag{11-6}$$

Thus, the conceptual basis of Hartree–Fock theory is quite simple. It asserts that the behavior of each electron can be described by its own orbital, and the effect of other electrons on the orbital is included only through the terms of Hamiltonian.

These orbitals (χ_i), which are functions of both the space and spin of the electron, are referred to as *spin orbitals*. The energy expectation value for a wavefunction having the form of Eq. (11-4) has already been derived in Section 10-6, and is given by:[6]

$$E_{\text{el}} = \sum_{i=1}^{N} \langle \chi_i | h | \chi_i \rangle + \frac{1}{2} \sum_{i,j}^{N} \left\{ \left\langle \chi_i \chi_j \left| \frac{1}{r_{12}} \right| \chi_i \chi_j \right\rangle - \left\langle \chi_i \chi_j \left| \frac{1}{r_{12}} \right| \chi_j \chi_i \right\rangle \right\}, \tag{11-7}$$

where the subscript on E_{el} is used to indicate that only the electronic energy is being calculated (omitting the nuclear repulsion terms) and where h is given by Eq. (11-3).

The last two terms in Eq. (11-7) are known as *electron repulsion integrals* and are given by:

$$\left\langle \chi_i \chi_j \left| \frac{1}{r_{12}} \right| \chi_i \chi_j \right\rangle = \int \int \chi_i^*(1) \chi_j^*(2) \frac{1}{r_{12}} \chi_i(1) \chi_j(2) \, d\tau_1 \, d\tau_2, \tag{11-8}$$

and

$$\left\langle \chi_i \chi_j \left| \frac{1}{r_{12}} \right| \chi_j \chi_i \right\rangle = \int \int \chi_i^*(1) \chi_j^*(2) \frac{1}{r_{12}} \chi_j(1) \chi_i(2) \, d\tau_1 \, d\tau_2, \tag{11-9}$$

where the integration is taken over both space and spin coordinates. The integral in Eq. (11-8) is known as a *Coulomb integral*, since it describes the (integrated) Coulombic interaction between two charge distributions, $\chi_i^*(1)\chi_i(1)$ and $\chi_j^*(2)\chi_j(2)$, separated by a distance (r_{12}). The integral in Eq. (11-9) is known as an *exchange* integral and has no direct classical analog.

The basic problem that we must now address is to devise a way to determine the spin orbitals (χ) so that they represent the *optimum* orbital description of the system. To do this we shall utilize the Method of Undetermined Multipliers that was described in Chapter 2.[7] This method is particularly appropriate for use here, because we want to retain orthonormality among orbitals even after

[5] It should be recalled that the orthonormality condition of Eq. (11-6) is not a restriction on the generality of the wavefunction, since the transformation to an orthonormal basis leaves the wavefunction (and hence energy) invariant. See Chapter 10, Section 10-10.

[6] The second summation in Eq. (11-7), while appearing different than the expression used in Eq. (10-142), i.e., $\Sigma_{i>j}^{N}$, can easily be seen to be identical.

[7] See Chapter 2, Section 2-10.

optimization and thus retain a tractable energy expression. Also, we shall need the generality of the method, since the form of the orbitals as well as the parameters within the orbitals will be optimized. Thus, we shall be interested in minimizing the energy expression of Eq. (11-7), subject to the constraints represented by Eq. (11-6). Written in a form suitable for carrying out the desired analysis, we wish to minimize the following expression:

$$G(\chi)=E_{\text{el}}(\chi)-\sum_{i,j}^{N}\lambda_{ij}(\langle\chi_i|\chi_j\rangle-\delta_{ij}), \tag{11-10}$$

where $E_{\text{el}}(\chi)$ is given by Eq. (11-7) and the λ_{ij} are Lagrange multipliers to be determined in a way that guarantees orthonormality of the optimized orbitals.

To carry out the minimization of Eq. (11-10), we change each of the orbitals (χ) by an infinitessimal amount $(\delta\chi)$, i.e., replace each χ_i by:

$$\chi_i\rightarrow\chi_i+\delta\chi_i. \tag{11-11}$$

Then, the choice of χ_i which will minimize Eq. (11-10) will be the one in which such a change in orbitals gives no change to G, i.e., G is at a minimum.[8] This requires that

$$\delta G=0=\delta[E_{\text{el}}(\chi)]-\delta\left[\sum_{i,j}^{N}\lambda_{ij}(\langle\chi_i|\chi_j\rangle-\delta_{ij})\right]. \tag{11-12}$$

Carrying out the indicated operations in Eq. (11-12), we obtain

$$\begin{aligned}0=&\sum_{i=1}^{N}[\langle\delta\chi_i|h|\chi_i\rangle+\langle\chi_i|h|\delta\chi_i\rangle]\\&+\frac{1}{2}\sum_{i,j}^{N}\left[\left\langle\delta\chi_i\chi_j\left|\frac{1}{r_{12}}\right|\chi_i\chi_j\right\rangle+\left\langle\chi_i\delta\chi_j\left|\frac{1}{r_{12}}\right|\chi_i\chi_j\right\rangle\right.\\&\left.+\left\langle\chi_i\chi_j\left|\frac{1}{r_{12}}\right|\delta\chi_i\chi_j\right\rangle+\left\langle\chi_i\chi_j\left|\frac{1}{r_{12}}\right|\chi_i\delta\chi_j\right\rangle\right]\\&-\frac{1}{2}\sum_{i,j}^{N}\left[\left\langle\delta\chi_i\chi_j\left|\frac{1}{r_{12}}\right|\chi_j\chi_i\right\rangle+\left\langle\chi_i\delta\chi_j\left|\frac{1}{r_{12}}\right|\chi_j\chi_i\right\rangle\right.\\&\left.+\left\langle\chi_i\chi_j\left|\frac{1}{r_{12}}\right|\delta\chi_j\chi_i\right\rangle+\left\langle\chi_i\chi_j\left|\frac{1}{r_{12}}\right|\chi_j\delta\chi_i\right\rangle\right]\\&-\sum_{i,j}^{N}\lambda_{ij}[\langle\delta\chi_i|\chi_j\rangle+\langle\chi_i|\delta\chi_j\rangle].\end{aligned} \tag{11-13}$$

[8] The variation in χ expressed in Eq. (11-11) can be guaranteed only to result in G being *stationary*, i.e., it could be a maximum as well as a minimum. For cases of interest to us, however, a minimum of G is the likely result.

After some manipulation,[9] Eq. (11-13) can be written as

$$0=\sum_{i=1}^{N}[\langle\delta\chi_i|h|\chi_i\rangle$$

$$+\sum_{i,j}^{N}\left[\left\langle\delta\chi_i\chi_j\left|\frac{1}{r_{12}}\right|\chi_i\chi_j\right\rangle-\left\langle\delta\chi_i\chi_j\left|\frac{1}{r_{12}}\right|\chi_j\chi_i\right\rangle\right]$$

$$-\sum_{i,j}^{N}\lambda_{ij}\langle\delta\chi_i|\chi_j\rangle+\text{(complex conjugate)}. \qquad (11\text{-}14)$$

At this point it is convenient to simplify the notation by introducing several operators. The *Coulomb operator* is defined by:

$$\mathcal{J}_j(1)\chi_i(1)=\left[\int d\tau_2\chi_j^*(2)\chi_j(2)\frac{1}{r_{12}}\right]\chi_i(1). \qquad (11\text{-}15)$$

This operator, $\mathcal{J}_j(1)$, which is an integral operator, takes the product of orbitals $[\chi_j^*(2)\chi_j(2)]$, multiplies by $1/r_{12}$ and integrates over the coordinates of electron two. The result operates on electron one, and is dependent only on the choice of orbital (χ_j). From the discussions in Chapter 5, we also recognize that $\chi_j^*(2)\chi_j(2)d\tau_2$ is a charge distribution,[10] and represents the contribution of orbital j to the probability of finding electron two in the space of orbital χ_j. Forming the Coulomb operator, $\mathcal{J}_j(1)$, is thus to integrate the charge distribution over all possible space and spin coordinates of electron two, weighted by the factor $1/r_{12}$. The result is an *average local potential* at point r_1, which means that the Hartree–Fock orbitals are determined in the *averaged* Coulomb field of the other electrons.

In a similar manner, an *exchange operator,* $\mathcal{K}_j$, is defined as

$$\mathcal{K}_j(1)\chi_i(1)=\left[\int d\tau_2\chi_j^*(2)\chi_i(2)\frac{1}{r_{12}}\right]\chi_j(1). \qquad (11\text{-}16)$$

In this case we also have an integral operator whose result is a function only of electron one. However, the potential thus generated is now dependent on the value of $\chi_i(1)$ at *all* points in space, not simply at a single point (as was the case of the Coulomb operator). The exchange operator is therefore referred to as a *nonlocal operator.* Furthermore, the operator is defined by its action on *both* χ_i and χ_j, and thus does not have a convenient classical analog. In any case, both the Coulomb and exchange operators will be helpful in carrying out the analysis of Eq. (11-14). Insertion of Eqs. (11-15) and (11-16) into Eq. (11-14) and

---

[9] See Problem 11-1.

[10] See the discussion following Eq. (11-73) for additional details.

writing out the integrals in detail gives

$$0=\sum_{i=1}^{N}\int d\tau_1 \delta\chi_i^*(1)\Bigg[h(1)\chi_i(1) + \sum_{j=1}^{N}(\mathcal{J}_j(1)-\mathcal{K}_j(1))\chi_i(1)-\sum_{j=1}^{N}\lambda_{ij}\chi_j(1)\Bigg]+(\text{complex conjugate}). \quad (11\text{-}17)$$

If Eq. (11-7) is to hold for arbitrary variations in χ_i^* (and χ_i in the complex conjugate term), we must require that the term that multiplies $\delta\chi_i^*$ (and $\delta\chi_i$ from the complex conjugate term) be equal to zero, i.e.,

$$\left[h(1)+\sum_{j=1}^{N}(\mathcal{J}_j(1)-\mathcal{K}_j(1))\right]\chi_i(1)=\sum_{j=1}^{N}\lambda_{ij}\chi_j(1), \qquad i=1, 2, \cdots N. \quad (11\text{-}18)$$

An identical equation (except for complex conjugation) arises from the complex conjugate term in Eq. (11-17), so satisfying Eq. (11-18) will automatically assure that Eq. (11-17) is satisfied.

The term in brackets in Eq. (11-18) is usually referred to as the *Fock operator*, defined as

$$f(1)=h(1)+\sum_{j=1}^{N}(\mathcal{J}_j(1)-\mathcal{K}_j(1)). \quad (11\text{-}19)$$

Note that the Fock operator depends on the orbitals to be determined, through the Coulomb and exchange operators. Using the Fock operator definition allows Eq. (11-18) to be rewritten as

$$f\chi_i=\sum_{j=1}^{N}\lambda_{ij}\chi_j, \qquad i=1, 2, \cdots, N. \quad (11\text{-}20)$$

These equations are known as the *Hartree–Fock equations*, and their solution will give the optimum single determinant, orbital description of an atomic or molecular system. The orbitals (χ) that result from the solution of Eq. (11-20) are known as *Hartree–Fock orbitals*. Before proceeding, however, there are several simplifications that can be introduced into Eq. (11-20) that will convert it to a more common eigenvalue-like equation.

To facilitate this discussion, we note that the N integrodifferential equations in Eq. (11-20) can be written in matrix form as

$$f\boldsymbol{\chi}=\boldsymbol{\lambda}\boldsymbol{\chi}, \quad (11\text{-}21)$$

where $\boldsymbol{\chi}$ is a $(1 \times N)$ column vector of the orbitals to be determined, and $\boldsymbol{\lambda}$ is an $(N \times N)$ Hermitian matrix. Since it can be shown that $\boldsymbol{\lambda}$ is Hermitian,[11] we are

[11] See Problem 11-2.

assured[12] that it is always possible to find a unitary transformation that diagonalizes $\boldsymbol{\lambda}$, i.e.,

$$\mathbf{U}\boldsymbol{\lambda}\mathbf{U}^{\dagger}=\boldsymbol{\epsilon}=\begin{pmatrix} \epsilon_1 & & & \\ & \epsilon_2 & & 0 \\ & 0 & \ddots & \\ & & & \epsilon_N \end{pmatrix}, \qquad (11\text{-}22)$$

where

$$\mathbf{U}^{\dagger}\mathbf{U}=\mathbf{U}\mathbf{U}^{\dagger}=\mathbb{1}. \qquad (11\text{-}23)$$

If we introduce such a transformation into Eq. (11-21), we obtain

$$\begin{aligned}\mathbf{U}f\mathbf{U}^{\dagger}\mathbf{U}\boldsymbol{\chi}&=\mathbf{U}\boldsymbol{\lambda}\mathbf{U}^{\dagger}\mathbf{U}\boldsymbol{\chi}\\ &=\boldsymbol{\epsilon}\mathbf{U}\boldsymbol{\chi}.\end{aligned} \qquad (11\text{-}24)$$

By defining

$$\boldsymbol{\chi}'=\mathbf{U}\boldsymbol{\chi}. \qquad (11\text{-}25)$$

and

$$\mathbf{f}'=\mathbf{U}f\mathbf{U}^{\dagger} \qquad (11\text{-}26)$$

we obtain

$$\mathbf{f}'\boldsymbol{\chi}'=\boldsymbol{\epsilon}\boldsymbol{\chi}', \qquad (11\text{-}27)$$

where we have apparently simplified the equation to be solved by eliminating the off-diagonal Lagrangian multipliers (λ_{ij}) by means of the unitary transformation. However, before we can be certain that we have not changed the nature of the problem by this transformation, we must see what effect the transformation has had on the Fock operator (f) and Hartree–Fock orbitals (χ).

First of all, we saw earlier* that a unitary transformation of a set of orthonormal functions does not change the orthonormality characteristics. In other words, if the original set of orbitals (χ) is orthonormal, the use of the unitary transformation in Eq. (11-25) will not affect the orthonormality characteristics, and we are assured that:

$$\langle\chi_i'\,|\,\chi_j'\rangle=\delta_{ij}. \qquad (11\text{-}28)$$

As for the effect of the unitary transformation on the Fock operator, let us examine Eq. (11-26) in greater detail. Since $h(1)$ does not depend upon the orbitals, it is obviously independent of the unitary transformation, and we need only to examine the Coulomb and exchange terms. For the Coulomb term, the

[12] See Chapter 3, Section 3-8.

* See Chapter 2, Section 12.

transformed operator can be written as:

$$\sum_{j=1}^{N} \mathcal{J}_j' = \sum_{j=1}^{N} \int d\tau_2 \chi_j'^*(2)\chi_j'(2)\frac{1}{r_{12}}$$

$$= \sum_{j=1}^{N} \sum_{k=1}^{N} \sum_{l=1}^{N} \int d\tau_2 [(\mathbf{U}^\dagger)_{jk}\chi_k^*(2)][\mathbf{U}_{lj}\chi_l(2)]\frac{1}{r_{12}}$$

$$= \sum_{k,l}^{N}\left[\sum_{j}^{N} \mathbf{U}_{lj}(\mathbf{U}^\dagger)_{jk}\right]\int d\tau_2 \chi_k^*(2)\chi_l(2)\frac{1}{r_{12}} \tag{11-29}$$

But

$$\sum_{j=1}^{N} \mathbf{U}_{lj}(\mathbf{U}^\dagger)_{jk} = \delta_{lk}, \tag{11-30}$$

which gives:

$$\sum_{j=1}^{N} \mathcal{J}_j' = \sum_{k=1}^{N} \mathcal{J}_k, \tag{11-31}$$

and the Coulomb operator is seen to be invariant to the unitary transformation. In a similar manner,[13] it can be shown that the exchange operator also is invariant to the unitary transformation. Thus, the entire Fock operator is invariant to the unitary transformation (as well as the resulting Hartree–Fock wavefunction), and we have

$$f' = f. \tag{11-32}$$

These results mean that we are free to choose to solve for that set of Hartree–Fock orbitals that diagonalize the Lagrangian multiplier matrix, without loss of generality. This means that we can write the Hartree–Fock equations of Eq. (11-20) in the form

$$f\chi_i = \epsilon_i\chi_i, \qquad i = 1, 2, \cdots, N \tag{11-33}$$

where we have dropped the primes for convenience, and ϵ_i can now be interpreted as the *orbital energy* of the Hartree–Fock orbital (χ_i). This equation looks much more like an eigenvalue equation than Eq. (11-20), but it still is not a "standard" eigenvalue equation. In particular, the Fock operator (f) depends on the Hartree–Fock orbitals that we are attempting to determine, and Eq. (11-33) is therefore not a "standard" eigenvalue equation of the kind studied in Chapter 2.[14] In practice, the problem of the Fock operator dependence on the orbitals can be treated iteratively, i.e., assume a set of orbitals, form the Fock operator,

---

[13] See Problem 11-3.

[14] See Chapter 2, Section 2-13.

solve Eq. (11-33) for a new set of orbitals, form a new Fock operator, etc. If the process converges (which cannot be guaranteed), the orbitals used to form the Fock operator will be the same as those resulting from the solution of Eq. (11-33), and the desired solution has been obtained. We shall discuss computational aspects of Hartree–Fock theory in Section 11-2, but for the moment, let us assume that satisfactory computational procedures are available to solve the Hartree–Fock equations in Eq. (11-33).

The orbitals that result from the solution of Eq. (11-33) are usually referred to as the *canonical Hartree–Fock* orbitals, to distinguish them from the other possible solutions to the Hartree–Fock equations [Eq. (11-20)]. Indeed, there are an infinite number of acceptable solutions to the Hartree–Fock equations, that differ from each other only by a unitary transformation of $\boldsymbol{\lambda}$ or, equivalently, the orbitals χ. Thus, while the canonical Hartree–Fock orbitals are perhaps the easiest to determine, it is possible to transform these orbitals (via a unitary transformation) to other sets of orbitals that have the same properties as the canonical Hartree–Fock orbitals, but whose shape and/or localization sometimes correspond more closely to chemical intuition.[15]

A. Koopmans' Theorem

Assuming that the Hartree–Fock equations can be solved, there are a number of interesting concepts that can be developed and properties whose values can be estimated. Let us therefore now turn our attention to interpretation of the results.

If the Hartree–Fock equations of Eq. (11-33) are multiplied by χ_i^* and integrated, we obtain

$$\epsilon_i = \langle \chi_i | h | \chi_i \rangle + \sum_{j=1}^{N} \left\{ \left\langle \chi_i \chi_j \left| \frac{1}{r_{12}} \right| \chi_i \chi_j \right\rangle - \left\langle \chi_i \chi_j \left| \frac{1}{r_{12}} \right| \chi_j \chi_i \right\rangle \right\} . \qquad (11\text{-}34)$$

If we sum over all electrons, we obtain

$$\sum_{i=1}^{N} \epsilon_i = \sum_{i=1}^{N} \langle \chi_i | h | \chi_i \rangle + \sum_{i,j}^{N} \left\{ \left\langle \chi_i \chi_j \left| \frac{1}{r_{12}} \right| \chi_i \chi_j \right\rangle - \left\langle \chi_i \chi_j \left| \frac{1}{r_{12}} \right| \chi_j \chi_i \right\rangle \right\} . \qquad (11\text{-}35)$$

It should be noted that the sum of orbital energies in the above equation is *not* equal to the total energy. Indeed, comparison with Eq. (11-7) shows that

$$E_{\text{el}} = \sum_{i=1}^{N} \epsilon_i - \frac{1}{2} \sum_{i,j}^{N} \left\{ \left\langle \chi_i \chi_j \left| \frac{1}{r_{12}} \right| \chi_i \chi_j \right\rangle - \left\langle \chi_i \chi_j \left| \frac{1}{r_{12}} \right| \chi_j \chi_i \right\rangle \right\} . \qquad (11\text{-}36)$$

Therefore, it is not appropriate to attempt to relate orbital energies[16] to the total

---

[15] See Section 11-2.D.

[16] Unless otherwise specified, we shall assume that electrons will be placed in the lowest energy orbitals available. Also, orbitals into which electrons have been placed are referred to as *occupied orbitals*.

energy,[17] unless the correction terms in Eq. (11-36) are considered. However, if orbital energies cannot be related to the total energy, the next question is whether they can be related to *any* physical property of the system under consideration.

To answer this question, let us now examine the energy expression for an $N-1$ electron system, i.e.,

$$E_{N-1}=\sum_{i=1}^{N-1}\langle\chi_i|h|\chi_i\rangle+\frac{1}{2}\sum_{j,j}^{N-1}\left\{\left\langle\chi_i\chi_j\left|\frac{1}{r_{12}}\right|\chi_i\chi_j\right\rangle-\left\langle\chi_i\chi_j\left|\frac{1}{r_{12}}\right|\chi_j\chi_i\right\rangle\right\}. \quad (11\text{-}37)$$

If we assume that the Hartree–Fock orbitals for the N-electron system from Eq. (11-33) are also appropriate for the description of the $(N-1)$ electron system whose energy is described Eq. (11-37), then we can calculate the first ionization potential for the N-electron system by forming the energy difference:[18]

$$\mathrm{IP}=E_{N-1}-E_N \quad (11\text{-}38)$$

$$=-\langle\chi_N|h|\chi_N\rangle-\sum_{j=1}^{N}\left\{\left\langle\chi_N\chi_j\left|\frac{1}{r_{12}}\right|\chi_N\chi_j\right\rangle-\left\langle\chi_N\chi_j\left|\frac{1}{r_{12}}\right|\chi_j\chi_N\right\rangle\right\}. \quad (11\text{-}39)$$

$$=-\epsilon_N. \quad (11\text{-}40)$$

Thus, the ionization potential can be estimated directly from knowledge of the orbital energy, provided a single determinant is an appropriate choice of wavefunction for both the neutral and ionized systems. Of course, it should be recognized that the Hartree–Fock orbitals that are solutions to the N-electron problem will not, in general, be exact solutions to the $(N-1)$ electron problem, and estimation of ionization potentials from orbital energies can be expected to result in some error, due to the "frozen orbital" approximation. Estimation of ionization potentials in this way is known as *Koopmans' theorem.*[19]

Let us now turn to a discussion of the errors that can be expected from the use of Koopmans' theorem. To do this requires consideration of the errors inherent in Hartree–Fock theory itself. We shall at this point discuss these errors qualitatively, and save more quantitative discussions for later sections. Since Hartree–Fock theory is a single particle (orbital) theory, and determines these orbitals only in the averaged field of the other electrons, it is to be expected that even the exact Hartree–Fock solution to a problem will not give the exact total energy for the system. In fact, the difference between Hartree–Fock results

[17] In developing methods to obtain approximate solutions to the Hartree–Fock equations, this caveat is sometimes ignored, which leads to substantial difficulties in establishing a sound theoretical basis for the method.

[18] The analysis given here is based on removal of the electron from the *highest* occupied orbital (i.e., to estimate the first ionization potential), but is is applicable in general to removal of an electron from any occupied orbital.

[19] T. A. Koopmans, *Physica*, **1**, 104 (1933).

(whether approximate or exact) and the exact answer, and devising ways to estimate or calculate this difference, has been a major topic of interest for many years. In order to quantify this concept further before using it, we define[20] the difference between the Hartree–Fock energy and the exact eigenvalue of the Hamiltonian for the state under consideration as the *correlation energy*, i.e.,

$$\Delta E_{\text{correl}} = E_{\text{exact}} - E_{\text{HF}}. \tag{11-41}$$

At least in a qualitative sense, we expect the correlation energy error to increase as the number of electrons increases.[21]

With such a definition, we see that use of orbital energies to estimate ionization potentials has two major sources of error, in addition to the error associated with the use of a single configuration as the wavefunction. The first is due to the "frozen orbital" approximation, where it is expected that E_{N-1} will be improved (i.e., lowered) by relaxation of the "frozen orbital" approximation. However, correction of the correlation energy error would be expected to lower E_N more than E_{N-1}. Such a correction would tend to cancel the "frozen orbital" error in the case of ionization potentials. These corrections and their effect are depicted qualitatively in Fig. 11.1.[22] The net result is that, as a general rule, the first several ionization potentials estimated from Koopmans' Theorem are frequently reliable. However, when orbital energies are close, care must be taken even in ionization potential estimates, since interchanges of ordering can occur.[23]

It should also be noted that Koopmans' Theorem, as illustrated for ionization potentials in Eq. (11-40), applies only to single determinant wavefunctions. This means that, in open shell situations where there is one or more unpaired electron and even the Hartree–Fock wavefunction is a sum of Slater determinants,[24] the analysis becomes substantially more complicated and may break down altogether. More generally, application of Koopmans' Theorem with reliable results can be expected only for closed shell ground states, and when the total wavefunction for each state is dominated by a single configuration.

---

[20] This definition was suggested by P. O. Löwdin, and has found widespread acceptance and use. See P. O. Löwdin, *Adv. Chem. Phys.*, **2**, 207, (1959).

[21] It should also be noted that the correlation energy defined in this manner represents the difference between two theoretically calculated numbers. It does not compare Hartree-Fock results to experimental results, since the latter include relativistic corrections and zero point nuclear vibration contributions.

[22] Also indicated in Fig. 11-1 are electron affinities, whose estimate via Koopmans' Theorem is not reliable, as can be inferred from the figure.

[23] For additional discussion of ionization potential studies on molecular systems, see L. Radom, *Modern Theoretical Chemistry*, Volume 4, Plenum Press, New York, 1977, pp. 333–356, and M. E. Schwartz, Modern Theoretical Chemistry, Volume 4, Plenum Press, New York, 1977, pp. 357–380; see also K. Wittel and S. P. McGlynn, *Chem. Rev.*, **77**, 745-771 (1977).

[24] These situations will be discussed in Section 11-3.

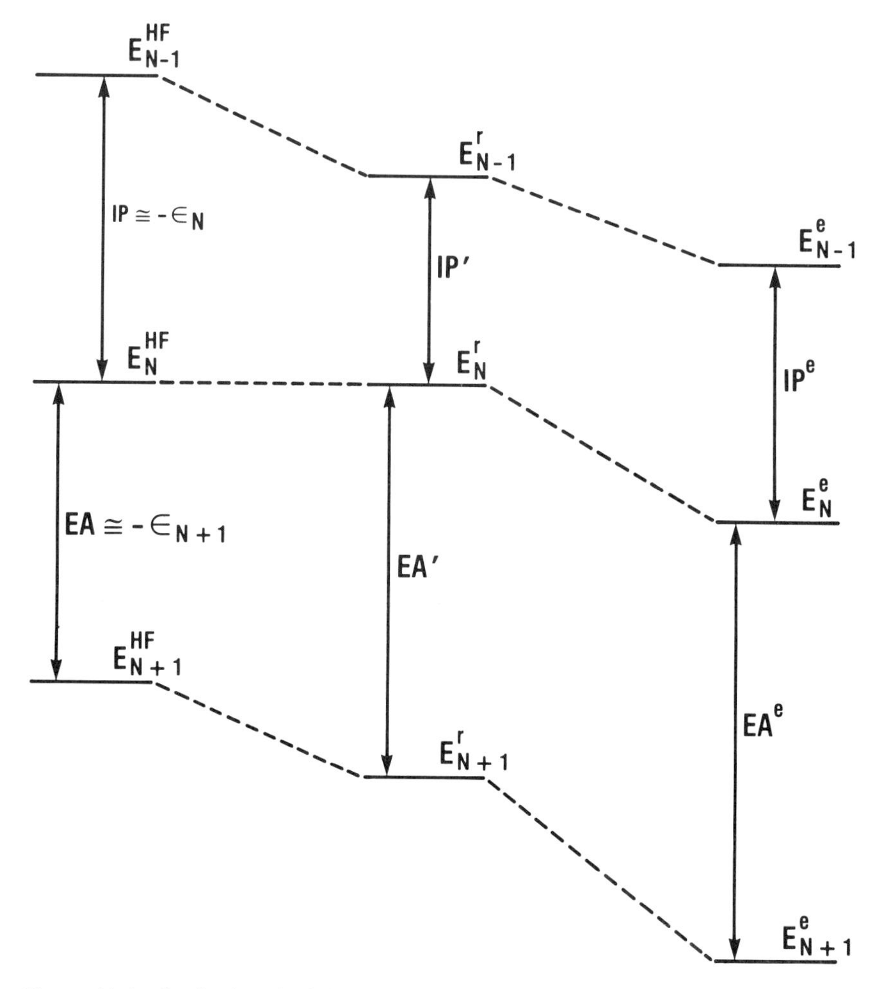

Figure 11.1. Qualitative depiction of errors in estimating ionization potentials and electron affinities using Hartree–Fock theory. The columns refer to calculations using Hartree–Fock theory (“frozen orbital” approximation), relaxation of the “frozen orbital” approximation, and the exact energies, respectively.

B. Correlation Diagrams and the Molecular Aufbau Principle

In addition to Koopmans' Theorem, other qualitative conclusions regarding electronic structural features of molecules can be extracted from the theory, even without detailed computational studies. Indeed, prior to the advent of significant computational capabilities, many of the insights obtained were possible only through qualitative analyses. In this section we shall introduce two examples of

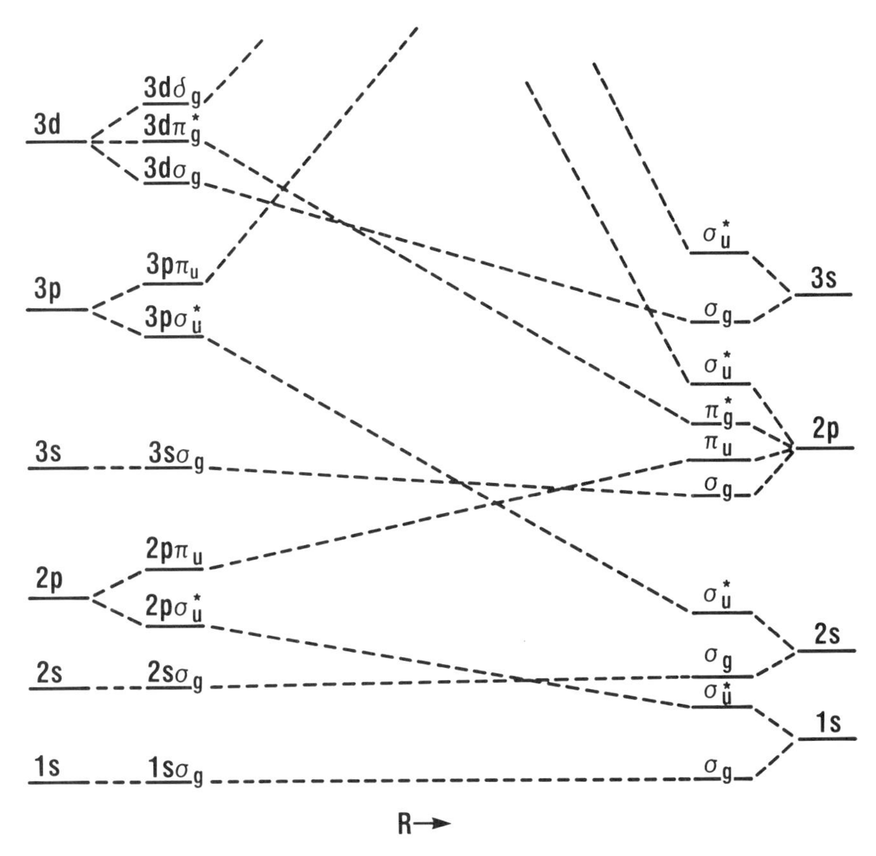

Figure 11.2. Correlation diagram for hypothetical homonuclear diatomic molecule. Energy level spacings are arbitrary. The "united atom" limit is depicted on the left ($R = 0$), and the "separated atom" limit ($R = \infty$) is depicted on the right.

qualitative analyses that have significant importance both historically and currently, due to Mulliken[25] and Walsh.[26]

To illustrate these ideas, we begin by recalling that, for H_2, the quantum number m and the "g" or "u" symmetry designation are constant for all internuclear distances (R). Then, Mulliken constructed a *correlation diagram* by listing the states of the united and separated atoms in order of increasing energy at opposite sides of the diagram. States having the same (g, u) character and m quantum number are connected, but using the "noncrossing" rule.

Such a diagram is given in Fig. 11.2 for a hypothetical homonuclear diatomic

---

[25] R. S. Mulliken, *Rev. Mod. Phys.*, **14**, 204 (1942).

[26] A. D. Walsh, *J. Chem. Soc.*, 2260 (1953) and following articles; see also R. J. Buenker and S. D. Peyerimhoff, *Chem. Rev.*, **74**, 127–188 (1974).

molecule. It not only indicates how "united atom" orbitals correlate with "separated atom" orbitals, but also gives a rough idea about the possible energy level ordering for orbitals in the diatomic molecule. Of course, since energy level spacings cannot be drawn to scale for a hypothetical molecule, and since only the two end points along the R axis ($R = 0$ and ∞) can actually be specified, the figure cannot be used in any quantitative sense. However, as we shall now see, a correlation diagram of this type is quite useful in constructing an "aufbau principle" for diatomic molecules analogous to that used for atoms. Specifically, we shall construct MO electronic structural descriptions of diatomic molecules by filling lowest energy orbitals first, allowing two electrons in each σ-orbital, four electrons in each π-orbital, etc., until all electrons have been placed in MOs.

With this aufbau concept and correlation diagram, let us now consider several homonuclear diatomic molecules made from first row atoms, to see if their electronic structural features can be qualitatively understood. For H_2, we expect a ground state configuration of $(\sigma_g 1s)^2$, using the separated atom notation. Also, since the $\sigma_g 1s$ orbital is a bonding orbital,[27] we expect a $(\sigma_g 1s)^2$ configuration to represent a stable molecule, which is indeed what is observed[28] for the H_2 ground state (dissociation energy = 2.6 eV).

Proceeding next to He_2, we expect a ground state configuration of $(\sigma_g 1s)^2$ $(\sigma_u^* 1s)^2$, which contains as many filled antibonding orbitals as bonding orbitals. Since no excess of filled bonding orbitals over antibonding orbitals is found, we do not expect to find significant binding forces. Hence, a stable molecule is not expected nor is found experimentally.

In the case of Li_2, the ground state configuration is expected to be $(\sigma_g 1s)^2$ $(\sigma_u^* 1s)^2$ $(\sigma_g 2s)^2$, in which a stable ground state containing a pair of bonding electrons in a Li–Li single bonding MO $[(\sigma 2s)^2]$ is expected, and is found experimentally (dissociation energy = 1.0 eV). On the other hand, the ground state of the Be_2 molecule would be expected to have the configuration $(\sigma_g 1s)^2$ $(\sigma_u^* 1s)^2(\sigma_g 2s)^2$ $(\sigma_u^* 2s)^2$, which is unstable (and confirmed experimentally).

The remaining first row homonuclear diatomics include B_2, C_2, N_2, O_2, F_2, and Ne_2, and involve energy levels that cross on the correlation diagram of Fig. 11.2. Therefore, in considering these molecules, we shall need to make use of additional symmetry properties to distinguish among the various possible electronic configurations. One such distinction can be made on the basis of spin state, using the techniques described earlier to calculate the multiplicity of the various possible states. In addition, we shall use other spatial symmetry properties, the notation for which we shall now summarize. First, the total orbital angular momentum (L) is simply the sum of individual orbital angular momenta. The resulting states are labeled Σ (for $L = 0$), Π (for $L = 1$), etc. In

---

[27] See Section 7-5 for definitions of bonding and antibonding orbitals.

[28] In the following discussion, experimental dissociation energies will in general be taken from G. Herzberg, *Molecular Spectra and Molecular Structure. I. Spectra of Diatomic Molecules*, D. Van Nostrand, New York, 1950.

addition, the state will be labeled "g" or "u," depending on whether the total electronic configuration is symmetric or antisymmetric, respectively, on inversion of electron coordinates through the midpoint of the bond.

Using this classification, we now consider the B_2 molecule, where a dissociation energy of approximately 3.0 eV is observed, and a ground state electronic configuration of $(\sigma_g 1s)^2$ $(\sigma_u^* 1s)^2(\sigma_g 2s)^2$ $(\sigma_u^* 2s)^2$ $(\sigma_g 2p)^2$ or $(\sigma_g 1s)^2$ $(\sigma_u^* 1s)^2(\sigma_g 2s)^2$ $(\sigma_u^* 2s)^2$ $(\pi_u 2p)^2$ or $(\sigma_g 1s)^2$ $(\sigma_u^* 1s)^2(\sigma_g 2s)^2$ $(\sigma_u^* 2s)^2$ $(\sigma_g 2p)^1$ $(\pi_u 2p)^1$ is possible. In each of these electronic configurations a net of two filled bonding orbitals is found, corresponding to an expected stable ground state with a single B–B bond, in general agreement with experiment. However, the ground state of B_2 is also known to be a triplet state, with two unpaired electrons. This means that either the $(\pi_u 2p)^2$ or $(\sigma_g 2p)^1$ $(\pi_u 2p)^1$ configuration must be correct. In addition, the lowest state is found experimentally to have Σ symmetry, meaning that the $(\sigma_g 1s)^2$ $(\sigma_u^* 1s)^2(\sigma_g 2s)^2$ $(\sigma_u^* 2s)^2$ $(\pi_u 2p)^2$ configuration actually represents the ground state.

For the C_2 molecule, the found state electronic configuration is observed experimentally to be $^3\Pi_u$, which means that its electronic configuration must be $(\sigma_g 1s)^2$ $(\sigma_u^* 1s)^2(\sigma_g 2s)^2$ $(\sigma_u^* 2s)^2$ $(\pi_u 2p)^3$ $(\sigma_g 2p)^1$. In a similar manner it is found that the ground state electronic configuration of N_2, O_2, F_2, and Ne_2, when compared to experiment, is given by

$$N_2: \quad (\sigma_g 1s)^2(\sigma_u^* 1s)^2(\sigma_g 2s)^2(\sigma_u^* 2s)^2(\pi_u 2p)^4(\sigma_g 2p)^2 \qquad (^1\Sigma_g) \tag{11-42}$$

$$O_2: \quad (\sigma_g 1s)^2(\sigma_u^* 1s)^2(\sigma_g 2s)^2(\sigma_u^* 2s)^2(\pi_u 2p)^4(\sigma_g 2p)^2(\pi_g^* 2p)^2 \qquad (^3\Sigma_g)$$

$$F_2: \quad (\sigma_g 1s)^2(\sigma_u^* 1s)^2(\sigma_g 2s)^2(\sigma_u^* 2s)^2(\pi_u 2p)^4(\sigma_g 2p)^2(\pi_g^* 2p)^4 \qquad (^1\Sigma_g) \tag{11-43}$$

$$Ne_2: \quad (\sigma_g 1s)^2(\sigma_u^* 1s)^2(\sigma_g 2s)^2(\sigma_u^* 2s)^2(\pi_u 2p)^4(\sigma_g 2p)^2(\pi_g^* 2p)^4(\sigma_u^* 2p)^2.$$

For N_2, we see a net of three pairs of bonding electrons, corresponding to a triple bond, while O_2 has a double bond and F_2 a single bond. Since Ne_2 has no net bonding electrons, no stable ground state is expected or found. Thus, correlation diagrams are seen to be quite useful in interpreting electronic structural features of diatomic molecules, requiring only quite simple theoretical descriptions.

Before leaving this discussion, it is useful to note that early models were not only capable of rationalizing electronic structural features, but could make qualitative geometric predictions as well. Among the most successful of these was the work of Walsh, known usually as *Walsh's rules.* The basic idea in this model is that molecular geometry is determined to a large extent simply by the number of valence electrons in the molecule. The theoretical tool that is used to create an "aufbau" procedure for geometric structural predictions is one again a correlation diagram, but in this case the diagram correlates orbital energies with *angular* variables.

Typical examples of such correlation diagrams are given in Fig. 11.3 for two general molecular classes, AH_2 and BAB molecules, where A and B are nonhydrogen "heavy" atoms such as C, N, O, etc. In these diagrams, the key

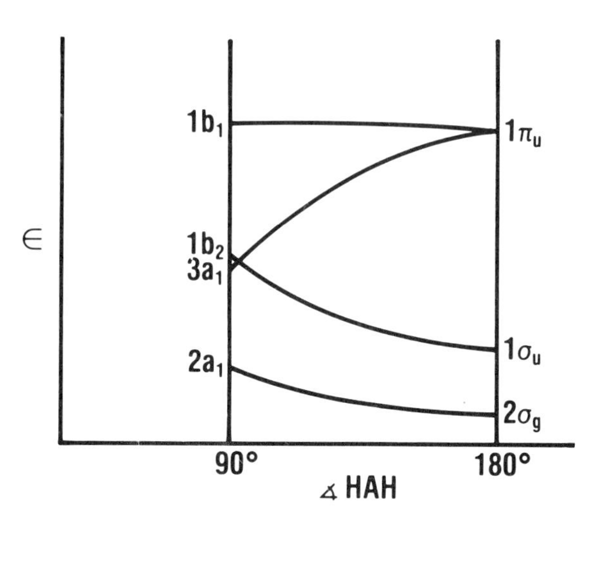

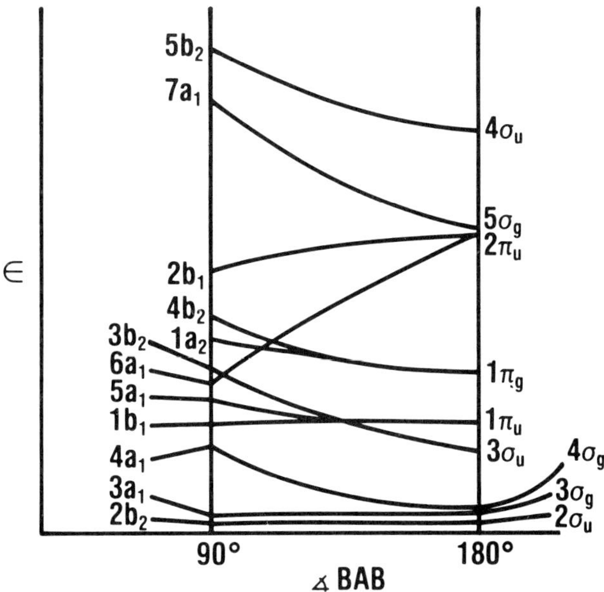

Figure 11.3. Empirical angular correlation diagrams for general AH_2 and BAB systems, after the work of Mulliken and Walsh [Reprinted with permission from R. J. Buenker and S. D. Peyerimhoff, *Chem. Rev.*, **74**, 127–188. Copyright (1974) American Chemical Society].

point of interest is whether the sum of occupied orbital energies[29] increases, decreases, or stays constant as the angle between the nuclei changes. In Walsh's early studies, simple concepts were used since detailed computational studies

[29] For additional rationale behind consideration of this quantity, see footnote 31 of this chapter.

were not possible, that related the expected effects of atomic orbitals on the stability of molecular orbitals in question.

For example, in the HAH case, the $1\pi_u$ orbital is a pair of doubly degenerate orbitals that can be represented pictorially as

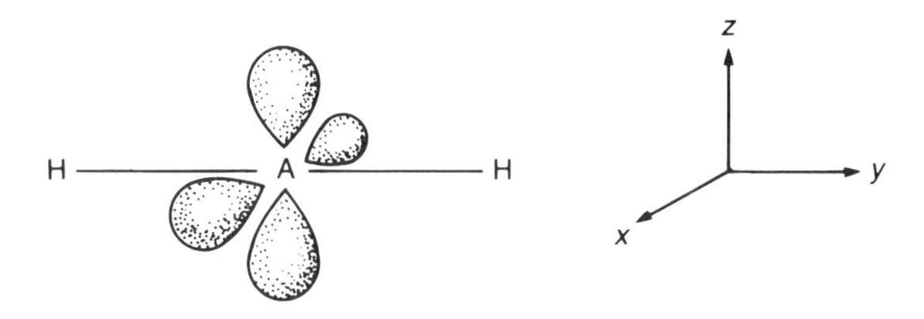

As the HAH angle is decreased from 180° to 90° in the x–y plane, the p_z orbital is expected to be relatively unaffected, and simply becomes a lone pair orbital[30] ($1b_1$) at ∡ HAH = 90°. On the other hand, the p_x orbital will interact with and overlap the $1s$ orbitals on the hydrogens increasingly as ∡ HAH is decreased, thus stabilizing (i.e., decreasing) the orbital energy as the HAH angle is decreased from 180° ($1\pi_u$) to 90° ($3a_1$). In a similar manner, the $1\sigma_u$ orbital, which represents a bonding orbital formed from a $2p_y$–$1s$ interaction, is expected to increase in energy as the HAH angle is decreased from 180° to 90°. This is due to the decreased ability of the $2p_y$ orbital to overlap and interact with the $1s$ orbitals as the hydrogen nuclei move away from a linear relationship.

Such largely empirical arguments, based on elementary atomic orbital interaction concepts, were used by Walsh to construct a variety of angular correlation diagrams such as depicted in Fig. 11.3. Predictions of the geometry of molecules based on these diagrams can then be easily accomplished simply be noting which of the molecular orbitals are expected to be occupied in a particular electronic state.[31] In those considerations, occupation at a level whose orbital energy is decreasing with diminishing angle is expected to produce a bent molecule, while a linear molecule is expected when a molecular orbital is populated whose energy increases with decreasing nuclear angle.

To illustrate the use of such diagrams, let us consider AH_2 molecules first. If there are four valence electrons,[32] both the $2\sigma_g$ and $1\sigma_u$ orbitals will be doubly occupied (i.e., a $2\sigma_g^2 1\sigma_u^2$ configuration). Given the energy increase of each of these orbitals as HAH decreases, a linear arrangement would be expected for the

[30] In general, the MO notation used will correspond to the point group of the molecule. For the case of a bent AH_2 molecule, the C_{2v} notation is used.

[31] Since we have seen that the total energy is not equal simply to a sum of orbital energies [Eq. (11-36)], predicting geometries only on the basis of MO energy considerations is not expected to be satisfactory. However, cancellation of errors and other effects occur [e.g., see R. J. Buenker and S. D. Peyerimhoff, *Chem. Rev.*, **74**, 127–188 (1974)] that allow rather elementary approaches as described here to be useful in a number of cases.

[32] Inner shell electrons of "heavy atoms" are ignored in these discussions, since they are not expected to affect molecular geometry.

ground state of all HAH molecules with four valence electrons. This result is found experimentally, with BeH_2, BH_2^+, LiH_2^- and MgH_2 each being observed as linear molecules.

As an example with a different result, consider the case where there are eight valence electrons. In this case the decrease in $3a_1$ orbital energy as ∡HAH decreases is more rapid than the increase in $1b_2$ as ∡HAH decreases, and a configuration $2a_1^2 3a_1^2 1b_2^2 1b_1^2$ is expected for all molecules with eight valence electrons. Such is the case, as illustrated by NH_2^-, H_20, H_2F^+, H_2S, H_2Se, and H_2Te, all of which are strongly bent.

Considering the BAB case, where all components are nonhydrogen atoms, the MOs expected to play a major role in determining geometry are the $3\sigma_u$, $1\pi_g$ and $2\pi_u$ MOs, since their orbital energies are strong functions of the BAB angle. If we consider the case of 16 valence electrons, all MOs up to (but not including) the $2\pi_u$ MO will be filled. Since the filled orbitals are either unchanged or increase in energy as ∡BAB decreases, we expect a linear structure for the ground state of BAB molecules containing 16 valence electrons. Such is indeed the case, as seen in CO_2, N^3, NO_2^+, NNO, OCS, CS_2, NCCl, BeF_2, and $CaCl_2$.

On the other hand, molecules with 20 valence electrons will fully occupy the $2\pi_u$ MO with four electrons, and the strong MO energy dependence of this MO as a function of the BAB angle is expected to counteract the effects of BAB angle bending seen for the lower lying MOs. Hence, bent ground state structures would be expected for molecules with 20 valence electrons, as is observed for F_2O, Cl_2O and Cl_2S.

Hence, even though the model used is quite simple and qualitative, it is quite remarkable in its ability to rationalize geometric features of molecules in easily understood concepts. Further analyses, calculations, and refinements have taken place by a number of authors,[33] which have helped place the basic ideas on a sounder theoretical footing and help to explain the apparent exceptions. Nevertheless, the ideas of Mulliken and Walsh have been very helpful in providing a qualitative framework for understanding both electronic and geometric structural features of a wide range of small molecules.

However, the advent of large-scale, high-speed digital computers plus the formulation of suitable computational algorithms have allowed detailed, quantitative studies (e.g., on potential energy surfaces) to be carried out. It is to a discussion of those algorithms and computational studies to which we now turn our attention.

11-2. Hall–Roothaan LCAO–MO–SCF Theory for Closed Shell Systems

Thus far we have seen that the best single determinant orbital wavefunction can be obtained by solving the Hartree–Fock equations of Eq. (11-33). Even though these equations have been made more tractable by the analysis in Section 11-1,

[33] See R. J. Buenker and S. D. Peyerimhoff, *Chem. Rev.*, **74**, 127–188 (1974) for a review of many such studies.

the fact remains that Eq. (11-33) represents a set of integrodifferential equations that are difficult to solve directly (e.g., to seek a numerical solution). In fact, direct solutions have been obtained in general only for atomic systems. Since we are interested primarily in molecular systems, we shall therefore seek alternative ways to approach the solution of the Hartree–Fock equations.

Fortunately, G. G. Hall and C. C. J. Roothaan each proposed[34] (independently) an approach that is quite useful, and has received widespread acceptance. The basic idea that Hall and Roothaan introduced was that, if a basis set which is complete is used to expand the molecular orbitals, the orbitals to be determined (or any other vector in the space) can be expressed in terms of the complete basis set.[35] This allows conversion of the problem of determining the Hartree–Fock orbitals into a much simpler problem, i.e., determining the expansion coefficients of the basis set.

Such an approach is essentially a generalization of the MO ideas introduced by Mulliken, Hund, and Leonard-Jones, and illustrated earlier[36] for the case of H_2. In that example, only two orbitals were included in the basis ($1s_a$ and $1s_b$, a clearly incomplete basis), and the expansion coefficients are determined by symmetry. Hall and Roothaan's approach allows extension of that concept to many basis orbitals using procedures for determining expansion coefficients via energetic considerations, as we shall now see.

The systems that we shall analyze first are somewhat simplified from the general case discussed in Section 11-1. In particular, we shall be interested in those molecules whose electrons are in a *closed shell configuration*. As discussed in the previous chapter,[37] we know that a closed shell (singlet state) case is one in which, for every spinorbital with α-spin, there is a corresponding spin orbital whose spatial description is identical, but has a β-spin. In other words an N-electron molecule (with N even) would be described in this case by a Slater determinant of N spin orbitals that, in turn, are described by $(N/2)$ spatial orbitals, each having an α- and β-spin component. Expressed mathematically, we have

$$\Psi(1, 2, \cdots, N) = A\{\chi_1(1)\,\chi_2(2)\,\cdots\,\chi_N(N)\}, \tag{11-44}$$

where the spin orbitals (χ) are expressed as

$$\chi_i(1) = \varphi_k(\mathbf{r}_1)\begin{cases}\alpha(\sigma_1)\\ \beta(\sigma_1),\end{cases} \tag{11-45}$$

where $i = 2k$ when i is even (and a β-spin is required), and $i + 1 = 2k$ when i is

---

[34] See G. G. Hall, *Proc. R. Soc. (London)*, **A205**, 541 (1951), and C. C. J. Roothaan, *Rev. Mod. Phys.*, **23**, 69 (1951).

[35] See Chapter 2, Sections 2-1 through 2-4. This is a particularly convenient procedure when the set is also finite.

[36] See Chapter 10, Section 10-1.B.

[37] See Chapter 10, Section 10-9.

Table 11.1. Relationships between Spin Orbitals and Space Orbitals and Their Corresponding Orbital Energies for a Six-Electron System

Spin orbitals (χ_i) and orbital energies (ϵ_i)		Spatial orbital (φ_i)	Spatial orbital energies (ϵ_i')
χ_1	ϵ_1	$\varphi_1\alpha \rightarrow \varphi_1$	$\epsilon_1' = \epsilon_1$
χ_2	ϵ_2	$\varphi_1\beta \rightarrow \bar{\varphi}_1$	$\epsilon_1' = \epsilon_2$
χ_3	ϵ_3	$\varphi_2\alpha \rightarrow \varphi_2$	$\epsilon_2' = \epsilon_3$
χ_4	ϵ_4	$\varphi_2\beta \rightarrow \bar{\varphi}_2$	$\epsilon_2' = \epsilon_4$
χ_5	ϵ_5	$\varphi_3\alpha \rightarrow \varphi_3$	$\epsilon_3' = \epsilon_5$
χ_6	ϵ_6	$\varphi_3\beta \rightarrow \bar{\varphi}_3$	$\epsilon_3' = \epsilon_6$

odd (and an α-spin is needed). Such a case is usually referred to as the *restricted closed shell* formulation of Hartree–Fock theory. In general, such a case is found in a vast majority of ground states of molecules, so the analysis, even though it represents a special case, is of rather broad utility in chemisty.

For this case, the Hartree–Fock wavefunction can be written as

$$\Psi(1, 2, \cdots N) = \mathcal{A}\{\varphi_1(1)\bar{\varphi}_1(2)\varphi_2(3)\bar{\varphi}_2(4) \cdots \bar{\varphi}_{N/2}(N)\}, \qquad (11\text{-}46)$$

where φ_i refers to a spatial orbital with α-spin and $\bar{\varphi}_i$ refers to the same spatial orbital but with β-spin.

The particular choice of a closed shell chemical system allows further simplification of the Hartree–Fock equations, which we shall carry out before proceeding. Starting with Eqs. (11-33) and (11-45), let us select the case[38] when $i = 2k$ (i.e., i even, which means each χ_i has β-spin), multiply Eq. (11-33) by $\beta^*(\sigma_1)$, and integrate over $d\sigma_1$, which gives

$$\begin{aligned}\int d\sigma_1 \beta^*(\sigma_1) f(1)\chi_i(1) &= \epsilon_i \int d\sigma_1 \beta^*(\sigma_1)\chi_i(1) \\ &= \epsilon_{i/2}' \int d\sigma_1 \beta^*(\sigma_1)\varphi_{i/2}(\mathbf{r}_1)\beta(\sigma_1) \\ &= \epsilon_{i/2}'\varphi_{i/2}(\mathbf{r}_1) \qquad i = 2, 4, \cdots, N. \end{aligned} \qquad (11\text{-}47)$$

In other words, the right-hand side of Eq. (11-47) has the same form as for the spin orbital case, except that only half as many orbitals (and orbital energies) need to be determined. Table 11.1 illustrates the relationships between space and spin orbitals and the orbital energies described above for a six-electron system.

Examining the left-hand side of Eq. (11-47) in more detail, we have

$$\int d\sigma_1 \beta^*(\sigma_1) f(1)\chi(1) = \int d\sigma_1 \beta^*(\sigma_1) \left[h(1) + \sum_{j=1}^{N} (\mathcal{J}_j(1) - \mathcal{K}_j(1)\right] \chi_i(1). \qquad (11\text{-}48)$$

[38] The case $i + 1 = 2k$ (i odd) gives identical results.

The first term on the right-hand side of Eq. (11-48) gives

$$\int d\sigma_1 \beta^*(\sigma_1) h(1) \varphi_{i/2}(\mathbf{r}_1) \beta(\sigma_1) = h(1) \varphi_{i/2}(\mathbf{r}_1), \tag{11-49}$$

since $h(1)$ does not contain spin-dependent terms.

The Coulomb and exchange terms are more complicated, however, since they depend on spin orbitals (and hence, spin). In order to analyze those terms, it is convenient to write the sum in Eq. (11-48) as two sums, i.e.,

$$\sum_{j=1}^{N} \Rightarrow \sum_{\substack{j'=1,3,5,\cdots \\ (j \text{ odd})}}^{N-1} + \sum_{\substack{j'=2,4,\cdots \\ (j \text{ even})}}^{N},$$

in which the spin orbitals in the "j' odd" terms each will have α-spin functions, and those in the "j' even" terms each will have β-spin functions. Examining the Coulomb terms, we obtain

$$\sum_{j' \text{ odd}}^{N-1} \int d\sigma_1 \beta^*(\sigma_1) \left[\int d\sigma_2 \int d\mathbf{r}_2 \chi_{j'}^*(2) \chi_{j'}(2) \frac{1}{r_{12}} \right] \varphi_{i/2}(\mathbf{r}_1) \beta(\sigma_1)$$

$$= \sum_{j' \text{ odd}}^{N-1} \int d\sigma_1 \beta^*(\sigma_1) \beta(\sigma_1) \int d\sigma_2 \beta^*(\sigma_2) \beta(\sigma_2)$$

$$\cdot \left[\int d\mathbf{r}_2 \varphi_{(j'+1)/2}^*(\mathbf{r}_2) \varphi_{(j'+1)/2}(\mathbf{r}_2) \frac{1}{r_{12}} \right] \varphi_{i/2}(\mathbf{r}_1)$$

$$= \sum_{k=1}^{N/2} \left[\int d\mathbf{r}_2 \varphi_k^*(\mathbf{r}_2) \varphi_k(\mathbf{r}_2) \frac{1}{r_{12}} \right] \varphi_{i/2}(\mathbf{r}_1). \tag{11-50}$$

In a similar manner, it can be shown that[39]

$$\sum_{j' \text{ even}}^{N} \int d\sigma_1 \beta^*(\sigma_1) \left[\int d\sigma_2 \int d\mathbf{r}_2 \chi_{j'}^*(2) \chi_{j'}(2) \frac{1}{r_{12}} \right] \varphi_{i/2}(\mathbf{r}_1) \beta(\sigma_1)$$

$$= \sum_{k=1}^{N/2} \left[\int d\mathbf{r}_2 \varphi_k^*(\mathbf{r}_2) \varphi_k(\mathbf{r}_2) \frac{1}{r_{12}} \right] \varphi_{i/2}(\mathbf{r}_1). \tag{11-51}$$

Thus, both sums give the same result [Eqs. (11-50) and (11-51)] for the case of the Coulomb operator.

[39] See Problem 11-4.

Now let us examine the exchange terms:

$$\sum_{j' \text{ odd}}^{N-1} \int d\sigma_1 \beta^*(\sigma_1) \left[\int d\sigma_2 \int d\mathbf{r}_2 \chi_{j'}^*(2) \chi_{i/2}(2) \frac{1}{r_{12}} \right] \chi_{j'}(1)$$

$$= \sum_{j' \text{ odd}}^{N-1} \int d\sigma_1 \beta^*(\sigma_1) \alpha(\sigma_1) \int d\sigma_2 \alpha^*(\sigma_2) \beta(\sigma_2)$$

$$\cdot \left[\int d\mathbf{r}_2 \varphi_{j'/2}^*(\mathbf{r}_2) \varphi_{i/2}(\mathbf{r}_2) \frac{1}{r_{12}} \right] \varphi_{j'/2}(\mathbf{r}_1)$$

$$= 0, \tag{11-52}$$

where the orthogonality of the spin functions causes the terms to vanish. For the other portion of the summation, we have:

$$\sum_{j' \text{ even}}^{N} \int d\sigma_1 \beta^*(\sigma_1) \left[\int d\sigma_2 \int d\mathbf{r}_2 \chi_{j'}^*(2) \chi_{i/2}(2) \frac{1}{r_{12}} \right] \chi_{j'}(1)$$

$$= \sum_{j' \text{ even}}^{N} \int d\sigma_1 \beta^*(\sigma_1) \beta(\sigma_1) \int d\sigma_2 \beta^*(\sigma_2) \beta(\sigma_2)$$

$$\cdot \left[\int d\mathbf{r}_2 \varphi_{(j'+1)/2}^*(\mathbf{r}_2) \varphi_{i/2}(\mathbf{r}_2) \frac{1}{r_{12}} \right] \varphi_{(j'+1)/2}(\mathbf{r}_1)$$

$$= \sum_{k=1}^{N} \left[\int d\mathbf{r}_2 \varphi_k^*(\mathbf{r}_2) \varphi_{i/2}(\mathbf{r}_2) \frac{1}{r_{12}} \right] \varphi_k(\mathbf{r}_1). \tag{11-53}$$

Summarizing the results, we see that the Hartree–Fock equations for spin orbitals [(Eq. (11-33)] can be simplified somewhat in the special case of a closed shell system, in which integration over the spin coordinates reduces the number of equations to $N/2$, and half of the exchange terms vanish. This gives the following set of equations to be solved for the spatial orbitals:

$$\left[\mathscr{h}(\mathbf{r}_1) + \sum_{k=1}^{N/2} (2\mathcal{J}_k'(\mathbf{r}_1) - \mathcal{K}_k'(\mathbf{r}_1) \right] \varphi_l(\mathbf{r}_1) = \epsilon_l' \varphi_l(\mathbf{r}_1),$$

$$l = 1, 2, \cdots, N/2 \tag{11-54}$$

where the primes indicate spatial orbital energies and operators, i.e.,

$$\mathcal{J}_k'(\mathbf{r}_1)\varphi_l(\mathbf{r}_1) = \left[\int d\mathbf{r}_2 \varphi_k^*(\mathbf{r}_2) \varphi_k(\mathbf{r}_2) \frac{1}{r_{12}} \right] \varphi_l(\mathbf{r}_1), \tag{11-55}$$

and

$$\mathcal{K}_k'(\mathbf{r}_1)\varphi_l(\mathbf{r}_1) = \left[\int d\mathbf{r}_2 \varphi_k^*(\mathbf{r}_2) \varphi_l(\mathbf{r}_2) \frac{1}{r_{12}} \right] \varphi_k(\mathbf{r}_1). \tag{11-56}$$

If we define the closed shell Fock operator over spatial orbitals as

$$f'(\mathbf{r}_1)=h(\mathbf{r}_1)+\sum_{k=1}^{N/2}(2\mathcal{J}_k'(\mathbf{r}_1)-\mathcal{K}_k(\mathbf{r}_1)), \qquad (11\text{-}57)$$

then Eq. (11-54) can be written as

$$f'(\mathbf{r}_1)\varphi_l(\mathbf{r}_1)=\epsilon_l'\varphi_l(\mathbf{r}_1), \qquad l=1, 2, \cdots, N/2. \qquad (11\text{-}58)$$

Thus, the form of the Hartree–Fock equations for the closed shell case looks the same as in the general spin orbital case, except that only spatial orbitals are involved, and a factor of two times the Coulomb term is needed.

A. SCF Equations

Given the modified Hartree–Fock equations for the spatial orbitals in Eq. (11-58), let us now turn our attention to the method of solving those equations that was introduced by Hall and Roothaan. As indicated earlier, the basic idea in their approach is that each orbital in Eq. (11-58) is expressed in terms of a basis set, i.e.,

$$\varphi_l=\sum_t^M c_{tl}\phi_t, \qquad l=1, 2, \cdots, M \qquad (11\text{-}59)$$

where the basis set (ϕ_t) is completely determined beforehand, and is assumed to be taken from a complete or suitable incomplete set.[40] In that case, determination of the Hartree–Fock orbitals (φ_l) is accomplished by finding the optimum linear coefficients (c_{tl}). In practice, determination of these linear coefficients is much easier than attempting to solve the integrodifferential equations of Eq. (11-58) numerically, as we shall see below.

Since we shall be interested primarily in molecular systems, it is of importance to note that the φ_l resulting from the solution of the Hartree–Fock equations will in general extend throughout the space of the entire molecule, and not simply be localized in the region surrounding one nucleus. This is also expected from a chemical perspective, where the formation of molecules from atoms is expected to be accompanied at least in some cases by a delocalization of the electronic charge distribution. Because of this, the φ_l are usually referred to as *Hartree–Fock molecular orbitals* (MO).[41] Furthermore, since the basis sets

[40] Questions regarding choice of basis sets will be explored in greater detail in Section 11-2C of this chapter.

[41] The idea that electrons should be considered as belonging to the molecule as a whole, i.e., the concept of "molecular orbitals," was developed early in the history of quantum mechanics by Robert Mulliken and others. See R. W. Mulliken, *Phys. Rev.*, **32**, 186, 388, 761 (1928), E. Hund, *Z. Physik*, **51**, 759 (1928) and *Z. Elektrochem.*, **34**, 437 (1928), and J. E. Lennard-Jones, *Trans. Faraday Soc.*, **25**, 688 (1929). For an excellent review and tribute to Mulliken's contributions, see P. O. Löwdin and B. Pullman, eds., *Molecular Orbitals in Chemistry, Physics, and Biology*, Academic Press, New York, 1964. For initial discussions of MO theory applied to H_2, see Section 10-1.

$\{\phi_t\}$ that are used to form these MOs are typically atomic orbitals[42] or functions similar to atomic orbitals, Eq. (11-59) is seen to represent the formation of molecular orbitals using a *linear combination of atomic orbitals* (LCAO). The net result is that, when speaking of the solution to the Hartree–Fock equations via a basis set expansion as indicated in Eq. (11-59), the process is usually known as the HF–LCAO–MO approach.

Before proceeding, it is of interest to note that the summation in Eq. (11-59) is written with a finite limit (M). If the complete basis set that is chosen is infinite, use of only a finite number from that set will mean that an exact solution to Eq. (11-58) will not be obtained. Indeed, one of the areas where significant research efforts have taken place is in choosing truncated basis sets that are relatively small, but provide sufficient accuracy in the solution of Eq. (11-58) to be useful for chemical interpretation and prediction. We shall return to this point in Section 11-2C, but for the moment we shall assume that a suitable finite basis set is available,[43] with $M > (N/2)$.

If we introduce Eq. (11-59) into the Hartree–Fock equations [Eq. (11-58)], we obtain

$$\sum_t^M c_{tl} f'(\mathbf{r}_1)\phi_t(\mathbf{r}_1) = \epsilon_l' \sum_t^M c_{tl}\phi_t, \qquad l = 1, 2, \cdots, M \tag{11-60}$$

where we have not made any assumptions about the orthogonality of the basis set, i.e., the basis set has been assumed to be normalized, but nonorthogonal. Also, we see that M molecular orbitals will be determined by the solution of Eq. (11-60). Since $M > N/2$, the solutions to Eq. (11-60) will in general provide some Hartree–Fock MOs that contain electrons (*occupied orbitals*—the first $N/2$ molecular orbitals), and others that are empty (*virtual orbitals*—the remaining $M - N/2$ molecular orbitals). The number of unfilled molecular orbitals will increase, therefore, as the size of the basis set is increased.

If we multiply Eq. (11-60) by $\phi_s^*(\mathbf{r}_1)$ and integrate over $d\mathbf{r}_1$, we can convert the integrodifferential equations into matrix equations, i.e.,

$$\sum_t^M c_{tl} \int d\mathbf{r}_1 \phi_s'(\mathbf{r}_1) f'(\mathbf{r}_1)\phi_t(\mathbf{r}_1)$$

$$= \epsilon_l' \sum_t^M c_{tl} \int \phi_s^*(\mathbf{r}_1)\phi_t(\mathbf{r}_1)\, d\mathbf{r}_1 \qquad l = 1, 2, \cdots, M \tag{11-61}$$

or

$$\sum_t^M F_{st} c_{tl} = \sum_t^M S_{st} c_{tl} \epsilon_l', \qquad l = 1, 2, \cdots, M \tag{11-62}$$

---

[42] A simple example of the use of atomic orbitals to form MOs was given for the H_2 molecule in Chapter 10, Section 1B.

[43] A basis set having $M = N/2$ functions is also possible, but will not result in an energy-minimized wavefunction since the coefficients will be fully determined by orthonormality constraints.

where we have defined the elements of the *Fock matrix* (**F**) as

$$F_{st}=\int d\mathbf{r}_1\phi_s^*(\mathbf{r}_1)f'(\mathbf{r}_1)\phi_t(\mathbf{r}_1), \tag{11-63}$$

and the elements of the *overlap matrix* (**S**) as

$$S_{st}=\int \phi_s^*(\mathbf{r}_1)\phi_t(\mathbf{r}_1)\,d\mathbf{r}_1. \tag{11-64}$$

Equation (11-62) is seen to be the component form of the following matrix equation:

$$\mathbf{FC}=\mathbf{SC}\boldsymbol{\epsilon} \tag{11-65}$$

where $\boldsymbol{\epsilon}$ is a diagonal matrix of orbital energies, i.e.,

$$\boldsymbol{\epsilon}=\begin{pmatrix} \epsilon_1' & & & \\ & \epsilon_2' & & 0 \\ & 0 & \ddots & \\ & & & \epsilon_M' \end{pmatrix}. \tag{11-66}$$

We shall return shortly to the procedures used to solve Eq. (11-65), but let us first examine the elements of the Fock matrix more closely. Writing out Eq. (11-63) in detail gives

$$\begin{aligned} F_{st}&=\int d\mathbf{r}_1\phi_s^*(\mathbf{r}_1)\left[h(r_1)+\sum_k^{N/2}(2\mathcal{J}_k'(\mathbf{r}_1)-\mathcal{K}_k'(\mathbf{r}_1))\right]\phi_t(\mathbf{r}_1)\\ &=\int d\mathbf{r}_1\phi_s^*(\mathbf{r}_1h(\mathbf{r}_1)\phi_t(\mathbf{r}_1)+\sum_k^{N/2}\sum_u^{M}\sum_v^{M}c_{uk}^*c_{vk}\\ &\quad\cdot\left\{2\int d\mathbf{r}_1\phi_s^*(\mathbf{r}_1)\left[\int d\mathbf{r}_2\phi_u^*(\mathbf{r}_2)\phi_v(\mathbf{r}_2)\frac{1}{r_{12}}\right]\phi_t(\mathbf{r}_1)\right.\\ &\quad\left.-\int d\mathbf{r}_1\phi_s^*(\mathbf{r}_1)\left[\int d\mathbf{r}_2\phi_u^*(\mathbf{r}_2)\phi_t(\mathbf{r}_1)\frac{1}{r_{12}}\right]\phi_v(\mathbf{r}_1)\right\}. \end{aligned} \tag{11-67}$$

By defining

$$H_{st}^{\text{core}}=\int d\mathbf{r}_1\phi_s^*(\mathbf{r}_1)h(\mathbf{r}_1)\phi_t(\mathbf{r}_1), \tag{11-68}$$

and the general electron repulsion integral[44] as

$$\langle\phi_i\phi_j|\phi_k\phi_l\rangle=\int d\mathbf{r}_1\int d\mathbf{r}_2\phi_i^*(\mathbf{r}_1)\phi_j^*(\mathbf{r}_2)\frac{1}{r_{12}}\phi_k(\mathbf{r}_1)\phi_l(\mathbf{r}_2), \tag{11-69}$$

---

[44] An alternative notation for electron repulsion integrals is often used, i.e.,

$$(ij|kl)=\int d\mathbf{r}_1\int d\mathbf{r}_2\phi_i^*(\mathbf{r}_1)\phi_j(\mathbf{r}_1)\frac{1}{r_{12}}\phi_k^*(\mathbf{r}_2)\phi_l(\mathbf{r}_2),$$

and care should be taken to determine which notation is being used in a given context.

we can rewrite Eq. (11-67) as

$$F_{st} = H_{st}^{\text{core}} + \sum_{u}^{M} \sum_{v}^{M} \sum_{k}^{N/2} c_{uk}^{*} c_{vk} \{2\langle \phi_s \phi_u | \phi_t \phi_v \rangle - \langle \phi_s \phi_u | \phi_v \phi_t \rangle \}. \quad (11\text{-}70)$$

It is useful to make one additional definition, i.e.,

$$P_{uv} = 2 \sum_{k}^{N/2} c_{uk}^{*} c_{vk}, \quad (11\text{-}71)$$

where the matrix **P** is known as the *charge and bond order matrix* (or *density* matrix). Using this definition, the Fock matrix elements can be written finally as

$$F_{st} = H_{st}^{\text{core}} + \sum_{u}^{M} \sum_{v}^{M} P_{uv} \{ \langle \phi_s \phi_u | \phi_t \phi_v \rangle - \frac{1}{2} \langle \phi_s \phi_u | \phi_v \phi_t \rangle \}. \quad (11\text{-}72)$$

The rationale for choosing the "charge and bond order matrix" title for **P** can be seen via the following slight digression. In particular, let us consider the expression:

$$\rho(1) = N \int \cdots \int \Psi^*(1, 2, \cdots, N) \Psi(1, 2, \cdots N)\, d\tau_2 \cdots d\tau_N, \quad (11\text{-}73)$$

where the Hartree–Fock wavefunction of Eq. (11-4) is the wavefunction to be employed. First, we note that

$$\int \rho(1)\, d\tau_1 = N,$$

implying that ρ is a measure of the total electron density of the N electrons in the system.

If we consider the Hartree–Fock wavefunction in greater detail, we can see that

$$\begin{aligned} \Psi(1, 2, \cdots, N) = \left(\frac{1}{\sqrt{N!}}\right) [&\chi_1(1) \det\{\chi_2(2)\chi_3(3) \cdots \chi_N(N)\} \\ &- \chi_2(1) \det\{\chi_1(2)\chi_3(3) \cdots \chi_N(N)\} \\ &+ \cdots + \chi_N(1) \det\{\chi_1(2)\chi_2(3) \cdots \chi_{N-1}(N)\}], \end{aligned}$$

with an analogous expression for Ψ^*. When the expressions for Ψ and Ψ^* are multiplied together, it can be seen that the only nonzero terms from the integration over $d\tau_2 \cdots d\tau_N$ are those in which the term from Ψ precisely matches the term from Ψ^* (i.e., in assignment of electrons to orbitals). This gives

$$\rho(1) = \left(\frac{N}{N!}\right) \sum_{j=1}^{N} (N-1)! \chi_i^*(1) \chi_i(1),$$

or

$$\rho(1)=\sum_{i=1}^{N}\chi_i^*(1)\chi_i(1). \tag{11-74}$$

Since this analysis holds for any electron (not just electron "1"), we see that for Hartree–Fock theory, ρ has the form of a probability density[45] for an electron.

If the closed shell Hartree–Fock case [i.e., Eq. (11-46)] is considered, the above equation becomes

$$\rho(1)=2\sum_{i=1}^{N/2}\varphi_i^*(1)\varphi_i(1). \tag{11-75}$$

We note in the above equation that ρ is comprised of a sum of terms, each of which can be thought of an "orbital probability" contribution to ρ. If the expansion of φ_i in a basis set [Eq. (11-59)] is introduced, we obtain

$$\rho_k(\mathbf{r})=2\sum_{u}^{M}\sum_{v}^{M}c_{uk}^*c_{vk}\phi_u^*(\mathbf{r})\phi_v(\mathbf{r}),$$

or

$$\begin{aligned}\rho(\mathbf{r})&=\sum_{k}^{N/2}\rho_k(\mathbf{r})\\&=\sum_{u}^{M}\sum_{v}^{M}P_{uv}\phi_u^*(\mathbf{r})\phi_v(\mathbf{r}).\end{aligned} \tag{11-76}$$

Thus, $\mathbf{P}$ is a matrix representation of the density matrix in the $\{\phi\}$ basis, and the diagonal elements of $\mathbf{P}$ can be interpreted as the charge residing in an orbital, thus rationalizing at least a portion of the name given to $\mathbf{P}$. Interpretation of the off-diagonal elements of $\mathbf{P}$ as "bond orders" has a rationale that can be similarly developed, and will be taken up in Section 11-2E.

Turning to the practical problem of solving the matrix equations represented in Eq. (11-65), we note first that, by orthogonalization of the basis set using any of a variety of methods,[46] Eq. (11-65) can be transformed into a "pseudostandard" eigenvalue equation, i.e.,

$$\mathbf{FC}'=\mathbf{C}'\epsilon, \tag{11-77}$$

where the orbital energies and the form of the Fock matrix are invariant to the transformations that transform the basis functions to an orthonormal basis. Thus,

---

[45] See Chapter 4, Section 4-1 for comparison.

[46] See Chapter 2, Section (2-4) and Chapter 9, Section 9-5 for examples of several procedures that could be used. Also, the original coefficients ($\mathbf{C}$) can be obtained from the expression $\mathbf{C} = \mathbf{TC}'$, where $\mathbf{T}$ is the matrix transformation that converts $\{\phi\}$ into an orthonormal set.

the overlap and Fock matrices can be constructed using Eqs. (11-64) and (11-72), followed by transformation to an orthogonalized basis and use of Eq. (11-77) to solve the Hartree–Fock equations.

The next point to note is that Eq. (11-77), like Eq. (11-33), is not a "standard" matrix eigenvalue equation, since the Fock matrix depends on the Hartree–Fock orbitals that are being sought. This problem is dealt with in practice by solving Eq. (11-77) iteratively. In other words, the necessary integrals are calculated, a first guess for values of the coefficients (c_{ij}) is made, and orthogonalization is carried out. These values plus the calculated integral values are used to compute the matrix elements in Eqs. (11-64) and (11-72), followed by solution of the matrix equations in Eq. (11-77). The result is a new set of coefficients (obtained by back-transformation of the new orthonormal basis $\mathbf{C}'$), which are used to go through the process again. Assuming that this process converges,[47] there will be a point at which the coefficients resulting from the solution of Eq. (11-77) are the same as those used to construct the Fock matrix. The resulting orbitals are known as *approximate Hartree–Fock orbitals*, which will be the same as the *exact* Hartree–Fock orbitals only when the basis set used is complete.

This iterative process is known generally as the *self-consistent-field* (SCF) approach. It gets its name from the fact that the initial guess for coefficients is used to calculate a Hartree–Fock "field," i.e., the noncore part of Eq. (11-72), in which the new orbitals are obtained. When this field has become self-consistent, i.e., the field calculated from the orbitals resulting from the solution of Eq. (11-77) is the same as the field in which the orbitals were determined, the process has converged. When combined with the other terminology introduced earlier, the overall process is known as the *HF–SCF–LCAO approach*.

In practice, this convergence is measured by monitoring the convergence of the total energy, the charge and bond order matrix, or the MO coefficients. In the case of energy, a difference of 10^{-6}–10^{-8} Hartrees between successive iterations is typically considered to have reached convergence. For the charge and bond order matrix, convergence is usually assumed when the largest change in individual matrix elements in successive iterations[48] is less than 10^{-4}. This

---

[47] Depending on the system under consideration as well as the initial choice of coefficients, it is possible that the iterative procedure may not converge, but oscillate or even diverge. Even in the latter cases, damping or extrapolation techniques are available, which usually result in convergence for closed shell problems. See, for example, B. O'Leary, B. J. Duke, and J. E. Eilers, *Adv. Quantum Chem.*, **9**, 1–68 (1975), and V. R. Saunders and I. H. Hillier, *Int. J. Quantum. Chem.*, **7**, 699–705 (1973). For examples of methods useful for initial charge and bond order matrix choice, see J. H. Letcher, I. Assar, and J. R. Van Wazer, *Int. J. Quantum Chem.*, **56**, 451 (1972), J. Letcher, *J. Chem. Phys.*, **54**, 3215 (1971), L. L. Shipman and R. E. Christoffersen, *Chem. Phys. Letters*, **15**, 469 (1972).

[48] Alternatively, the mean square deviation of all charge and bond order matrix elements from the ith to $(i+1)$ iteration, $[\Sigma_{u,v}(P^{i+1}_{uv} - P^{i}_{uv})^2]^{1/2}$, is sometimes used to monitor convergence. Also, it may be necessary to converge more fully than 10^{-4} if the wavefunction is to be used in processes such as geometry optimization, since derivatives of the energy will be needed in such cases.

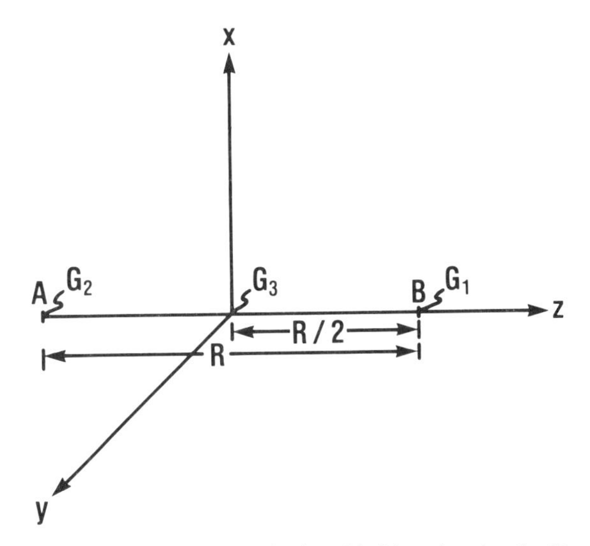

Figure 11.4. Coordinate system and basis orbital location for the H_2 molecule.

latter method is usually the method of choice, since the wavefunction typically converges slower than the energy, and monitoring of only the energy may give a false impression of the actual convergence of the process.

B. Application to H_2

In order to illustrate the concepts and techniques described in the previous section, a sample calculation on the H_2 molecule will be undertaken at this point. While the basis set chosen will be very small in order to keep the discussion and calculations relatively uncomplicated, it will be sufficient to illustrate the mechanics and various steps typically encountered in a Hartree–Fock calculation on an actual chemical system.

We shall be interested in describing the H_2 molecule placed on the z axis (see Fig. 11.4), with the nuclei fixed at the equilibrium internuclear distance (R = 1.40 Bohrs), and will use a basis set consisting of 3 Gaussian 1s-type orbitals,[49] i.e.,

$$\begin{aligned} G_1 &= N_1 \exp\{-\zeta_1[x^2+y^2+(z+0.7)^2]\} \\ G_2 &= N_2 \exp\{-\zeta_2[x^2+y^2+(z-0.7)^2]\} \\ G_3 &= N_3 \exp\{-\zeta_3[x^2+y^2+z^2]\}. \end{aligned} \tag{11-78}$$

It is thus seen that G_1 is located on nucleus B, G_2 is located on nucleus A, and G_3

[49] Such orbitals are known as *Floating Spherical Gaussian Orbitals* (FSGO), and will be discussed further in the next section.

is located at the mid-point of the bond. Comparing to the original MO description[50] of H_2, we see that G_1 and G_2 are approximations to atomic orbitals, while G_3 is a nonatomic orbital added to aid in the description of electron density in the bonding region. Also, the normalization constants are given by

$$N_i = \left(\frac{2\zeta_i}{\pi}\right)^{3/4} \tag{11-79}$$

and the orbital exponents are chosen as

$$\zeta_1 = \zeta_2 = 0.2710 \tag{11-80}$$

and

$$\zeta_3 = 0.3184, \tag{11-81}$$

where the values chosen for ζ_1 and ζ_2 give the maximum overlap of a single Gaussian with an STO (with $\zeta = 1.0$), and ζ_3 is taken as the value found by Frost[51] for a Gaussian orbital placed in the middle of the bond in H_2.

With these basis orbitals, the various integrals were calculated using the formulas given by Boys,[52] and are tabulated in Table 11.2. Since there are three basis orbitals, there will be three MOs produced by the calculation (of which one will be occupied by a pair of electrons). However, since only the occupied MOs are affected by the energy minimization, and unoccupied MOs are determined through orthonormality constraints, an initial guess is needed only for φ_1 in order to begin the SCF interactions.[53] For this example, any of a large number of initial guesses will result in convergence of the SCF process, and the particular choice used for φ_1 is given below. Also given below for information are the orthonormal orbitals for φ_2 and φ_3 that were formed via symmetric and antisymmetric combinations of the basis orbitals, i.e.,

$$\begin{aligned} \varphi_1^{\text{init}} &= N_1(\phi_1 + \phi_2) \\ \varphi_2^{\text{init}} &= N_2(2\phi_3 - \phi_1 - \phi_2) \\ \zeta_3^{\text{init}} &= N_3(\phi_1 - \phi_2). \end{aligned} \tag{11-82}$$

When normalization constants are included, the initial eigenvector matrix that results is[54]

$$\mathbf{C} = \begin{pmatrix} 0.531981 & -3.09034 & 1.46415 \\ 0.531981 & -3.09034 & -1.46415 \\ 0 & 5.89457 & 0 \end{pmatrix}. \tag{11-83}$$

[50] See Chapter 10, Section 10-1B.

[51] A. D. Frost, *J. Chem. Phys.*, **47**, 3714–3716 (1967).

[52] S. F. Boys, *Proc. Roy. Soc. (London)*, **A200**, 542–554 (1950).

[53] For an example of a procedure that can be used to obtain initial estimates for charge and bond order matrices for orbitals of the type used here, see L. L. Shipman and R. E. Christoffersen, *Chem Phys. Letters*, **15**, 469–474 (1972).

[54] See Problem 11-6.

Table 11.2. Overlap (**S**), Kinetic Energy (**T**), Nuclear Attraction (**V**), and Electron Repulsion Integrals for Sample H_2 Calculations.*

$$\mathbf{S} = \begin{pmatrix} 1.000000 & & \\ 0.766761 & 1.000000 & \\ 0.926259 & 0.926529 & 1.000000 \end{pmatrix}$$

$$\mathbf{T} = \begin{pmatrix} 0.406500 & & \\ 0.256503 & 0.406500 & \\ 0.387350 & 0.387350 & 0.477600 \end{pmatrix}$$

$$\mathbf{V} = \begin{pmatrix} -1.441472 & & \\ -1.169596 & -1.441472 & \\ -1.444635 & -1.444635 & -1.629882 \end{pmatrix}$$

(11/11) = 0.587408	(31/21) = 0.421722	(32/32) = 0.525546
(21/11) = 0.431235	(31/22) = 0.505300	(33/11) = 0.582587
(21/21) = 0.345350	(31/31) = 0.525546	(33/21) = 0.468162
(22/11) = 0.498081	(31/11) = 0.505300	(33/22) = 0.582587
(22/21) = 0.431235	(32/21) = 0.421722	(33/31) = 0.572193
(22/22) = 0.587408	(32/22) = 0.547988	(33/32) = 0.572193
(31/11) = 0.547988	(32/31) = 0.504915	(33/33) = 0.636710

* Since **S**, **T**, and **V** are symmetric, only the lower half is listed for convenience. The notation used for electron repulsion integrals is

$$(ij|kl) = \iint d\mathbf{r}_1\, d\mathbf{r}_2 \phi_i^*(1)\phi_j(1)\frac{1}{r_{12}}\phi_k^*(2)\phi_l(2).$$

The Fock matrix that results from combining the integrals of Table 11-2 with the initial eigenvector guess is[55]

$$\mathbf{F} = \begin{pmatrix} -0.440470 & -0.516764 & -0.520502 \\ -0.516764 & -0.440470 & -0.520502 \\ -0.520502 & -0.520502 & -0.546068 \end{pmatrix} \tag{11-84}$$

and the total energy that results from diagonalization of **F** is

$$E^{(1)} = -0.930135 \text{ Hartrees.}$$

The resulting lowest eigenvector is then used to form a new Fock matrix, and the process is continued to convergence. In Table 11.3 the value of the total energy for each iteration is given, along with the value of the quantity used to measure convergence, i.e., the mean square deviation of eigenvector coefficients from one iteration to the next:

$$I = \left[\sum_{j,k} |c_{j,k}^{i+1} - c_{j,k}^{i}|^2\right]^{1/2}, \tag{11-85}$$

[55] See Problem 11-7.

Table 11.3. Convergence Data for SCF Calculation on H_2

Iteration	Total energy[a] (Hartrees)	Energy difference	I[b]
1	−0.930135	—	—
2	−0.957088	−0.026952	0.0003889
3	−0.958249	−0.001161	0.0000181
4	−0.958303	−0.000054	0.0000008
5	−0.958306	−0.000003	4.07×10^{-8}
6	−0.958306	0	1.93×10^{-9}

[a] Includes the (constant) nuclear repulsion term.

[b] Defined in Eq. (11-85).

where $c^i_{j,k}$ is the kth coefficient of the jth MO in the ith iteration. As the data in the table indicate, the process has converged nicely after six iterations.

Of particular interest is the converged wavefunction and its properties. The converged eigenvector matrix is given by

$$\mathbf{C} = \begin{pmatrix} -0.156521 & -3.131889 & 1.464146 \\ -0.156521 & -3.131889 & -1.464146 \\ -0.708795 & 5.851797 & 0 \end{pmatrix}, \tag{11-86}$$

which shows clearly (e.g., in MO #1) how the Gaussian in the bond has mixed strongly with those on the nuclei in the final MO description. Indeed, the large coefficient of the mid-point basis function is indicative of σ-bond formation, as well as to imply that the basis set needs functions capable of providing considerable amplitude between the two nuclei.

The orbital energy of the filled MO is -0.529058 Hartrees, which corresponds to an ionization potential estimate (via Koopmans' Theorem) of approximately 14.4 eV. This compares remarkably well with the experimental value of 15.9 eV, especially when it is recognized that the total energy (-0.930135 Hartrees) is substantially above the true Hartree–Fock energy of -1.134. Thus, even with very approximate wavefunctions, it is possible to extract information of chemical interest, as long as care is taken to identify the errors.

C. Basis Orbitals and Complete Sets

Now that computational procedures for solving the Hartree–Fock equations have been outlined and illustrated, only one step remains before a large number of applications can be carried out. That step consists of choosing a basis set, which in practice is not a trivial task, and is a step on which the accuracy and usefulness of the calculation may strongly depend. Indeed, as we shall see in the following discussion, devising and/or choosing basis sets for molecular studies is an area of quantum chemistry that has received considerable attention in the last several decades, and is likely to continue to be of interest in the years ahead.

Of course, this problem would be easily solved if there existed a complete set of functions that is both computationally convenient to use and is also the exact solution to one or more prototype chemical problems of interest. A set having each of these properties has not been found, although progress has been made. In the following discussion we shall discuss a number of basis sets that have been found to be useful, as well as criteria that can be used to quantify the accuracy of a basis set.

The number of basis sets described in the quantum chemistry literature[56] is nearly 100, and they possess a wide range of accuracy, flexibility, and ease of use. Perhaps the greatest driving force behind the development of such a large number of basis sets is that computations are strongly dependent on the number of basis functions used. For example, electron repulsion integrals have the form

$$\left\langle \phi_i \phi_j \left| \frac{1}{r_{12}} \right| \phi_k \phi_l \right\rangle,$$

and evaluation of them with a basis set $\{\phi_i\}$ having N orbitals requires computation of the order of N^4 integrals. Hence, balancing computational cost with desired accuracy of the results must be considered carefully.

1. Choice of Primitive Functions

There are a number of different kinds of functions used to construct basis sets for use in LCAO-MO-SCF and other studies, and it is useful first to define them. To distinguish between the *kinds* of functions used and the particular *combinations of them* that have been found to be useful, we shall refer below to the former as *primitives*, and the latter as *contractions*. More specifically, individual functions of a particular type are called "primitive" functions, and an MO may be expressed as a linear combination of primitives. Alternatively, an MO may be expanded as a linear combination of "contracted" basis functions,

[56] For an excellent review of available basis sets, see E. R. Davidson and D. Feller, *Chem. Rev.*, **86**, 681–696 (1986). The discussion in this section draws heavily from that analysis. See also S. Huzinaga, *Comp. Phys. Rep.*, **2**, 279 (1985) and S. Wilson, *Methods of Computational Physics*, G. H. Dierksen and S. Wilson (eds.), Reidel, Dordrecht, The Netherlands, 1983; T. H. Dunning, Jr., *J. Chem. Phys.*, **53**, 2382 (1970); T. H. Dunning, Jr. and P. J. Hay, *Methods of Electronic Structure Theory*, Volume 3, Plenum Press, New York, 1977, pp. 1–28; A. A. Frost, *Methods of Electronic Structure Theory*, Volume 3, Plenum Press, New York, 1977, pp. 29–50; S. Wilson, *Adv. Chem. Phys.*, **67**, 429 (1987). In addition, a number of detailed tabulations of basis orbitals are also available, including R. Poirer, R. Kari, and I. G. Csizmadia, *Handbook of Gaussian Basis Sets*, Elsevier Science, 1985; S. Huzinaga, H. Andzelm, M. Klobukowski, E. Radzio-Andzelm, Y. Sakai, and H. Tatewaki, *Gaussian Basis Sets of Molecular Calculations*, Elsevier, Amsterdam, 1984; and W. J. Hehre, L. Radom, P. v R. Schleyer, and J. A. Pople, *Ab Initio Molecular Orbital Theory*, Wiley-Interscience, New York, 1986.

where some or all of the basis functions are constructed as linear combinations of primitive functions with fixed coefficients. Within that context, we shall begin by discussing the types of primitive functions that are available.

Historically, the functions arising from the solution of the Schrödinger equation for the H-atom and hydrogen-like atoms were obvious choices for primitives, since they were exact solutions to several prototype problems of interest. As we saw earlier,[57] the *hydrogen-like functions* are defined as

$$\phi_{nlm}(r, \vartheta, \varphi) = N_{nl}(\eta)\left(\frac{2\eta r}{n}\right) e^{-(\eta r/n)} L_{n+l}^{2l+1}\left(\frac{2\eta r}{n}\right) \cdot Y_{lm}(\vartheta, \varphi), \qquad n = 1, 2, \cdots, \tag{11-87}$$

where n, l and m are quantum numbers, L_{n+l}^{2l+1} are the Laguerre functions of order $(2l + 1)$, η is a scale factor, Y_{lm} is a spherical harmonic, and N_{nl} is a normalization constant.

However, while the discrete basis given above is infinite, it is *not* complete. For example, when applied to the helium atom, Shull and Löwdin[58] found that, regardless of the number of functions utilized in the set described in Eq. (11-87), no more than 56.5% of the exact energy could be obtained, regardless of the level of complexity of the wavefunction. The reason for this is that the continuum contributions have not been included. In other words, the complete description of the hydrogen atom requires *two* parts, i.e., description of the bound states (an infinite number) using the discrete set of Eq. (11-87), plus a description of the continuum corresponding to a free electron scattered by a proton. However, use of a basis set having both a discrete and continuous part is certainly not computationally convenient, and other sets were devised.

An alternative set of primitives that is similar in form to the original hydrogen-like set but does not have the same exponential dependence is the set[59]

$$\phi_{nlm}(r, \vartheta, \varphi) = N_{nl}(\eta)(2\eta r)^l e^{-\eta r} L_{n+l+1}^{2l+2}(2\eta r) Y_{lm}(\vartheta, \varphi). \tag{11-88}$$

This set is both orthogonal *and* complete as an atomic basis set,[60] as well as discrete, and is obviously better suited as a basis set for larger atomic problems than the hydrogen-like functions of Eq. (11-87).

Since we have seen that nonorthogonality of a basis is not necessarily a serious computational problem as long as linear dependence can be avoided, other exponential basis sets have also been devised. For example, a primitive

---

[57] See Chapter 7, Section 7-2.

[58] H. Shull and P. O. Löwdin, *J. Chem. Phys.*, **23**, 1362 (1955). These studies were not restricted to Hartree–Fock theory and, except for basis set limitations, could have achieved the exact result at least in principle.

[59] See P. O. Löwdin, *Adv. Chem. Phys.*, **2**, 207 (1959) for a review of basis sets such as these.

[60] It should be noted that, while this basis set is orthonormal when used in atomic problems, it will not be orthonormal in other circumstances, e.g., if used on each nucleus within a molecule.

basis set composed of functions known as *Slater-type orbitals* (STO) (or *exponential-type functions*) is given by

$$\phi_{nlm}(r, \vartheta, \varphi) = Nr^{n-1}e^{-\zeta r}Y_{lm}(\vartheta, \varphi). \tag{11-89}$$

This basis, which uses powers of r instead of Laguerre functions, and frequently uses the real (instead of complex) form of Y_{lm}, has been found to be quite useful in the description of atomic and small molecular systems.[61]

To illustrate the use of an STO basis set, as well as to introduce some notation that will be useful later, let us consider how an STO basis for the Hartree–Fock (HF) molecule would be constructed. This molecule contains a total of 10 electrons, so at least 5 basis orbitals (each doubly occupied) are needed. In practice, the usual choice is to take an atomic basis consisting of a $1s$ STO on the hydrogen nucleus plus a set of 5 STOs on the F nucleus consisting of $1s$, $2s$, $2p_x$, $2p_y$, and $2p_z$ STOs, where the *orbital exponents* (ζ) for each of the $2p$ orbitals are constrained to be equal, and the "x, y, and z" notation refers to $2p$ orbitals formed using the real form of Y_{lm} and placed on x, y, z axes of a local coordinate system on the F nucleus. This gives a total basis set of 6 functions, and is usually referred to as a *minimal basis set* (even though it actually contains one function more than the minimum number needed), since it contains the minimum number of functions needed to describe 10 electrons, consistent with maintaining local spherical symmetry of $2p$ functions on the F atom. It also is referred to as a *single zeta exponential basis*, since only one nonlinear parameter (ζ) to be optimized is present for *each* of the basis function types ($1s$, $2s$, $2p$).

If one desires (or needs) a larger basis set, a *double zeta exponential basis* can be chosen, which consists of *two* functions of each type. For HF this would consist of $1s$, $1s'$, STOs on the hydrogen nucleus, and $1s$, $1s'$, $2s$, $2s'$, $2p_x$, $2p_x'$, $2p_y$, $2p_y'$, $2p_z$, $2p_z'$ STOs on the flourine nucleus, giving a total of 12 basis functions. In addition, another kind of basis that is referred to as a *split-valence basis* is frequently used. In this case, a single function is used to describe inner shells, and two functions are used to described the valence shell. For the HF example, such a basis would consist of $1s$, $1s'$ STOs on hydrogen, plus $1s$, $2s$, $2s'$, $2p_x$, $2p_x'$, $2p_y$, $2p_y'$, $2p_z$, $2p_z'$ STOs on fluorine.

However, in spite of major efforts, exponential basis sets such as described in Eqs. (11-88) and (11-89) have not been used extensively in polyatomic molecules. The main reason that their use has been restricted primarily to atoms or diatomic molecules is that electron repulsion integrals of the form

$$\iint d\mathbf{r}_1\, d\mathbf{r}_2 \phi_a^*(\mathbf{r}_1)\phi_b^*(\mathbf{r}_2)\,\frac{1}{r_{12}}\,\phi_c(\mathbf{r}_1)\phi_d(\mathbf{r}_2),$$

[61] For an example of application of such a basis to a small molecule system (H_3^+) see R. E. Christoffersen, *J. Chem. Phys.*, **41**, 960 (1964). For application to diatomics and other molecules, see W. G. Richards, T. E. H. Walker and R. K. Hinkley, *A Bibliography of Ab Initio Molecular Wavefunctions*, Oxford University Press, Oxford, 1971; see also the Supplements for 1970–1973 and for 1974–1977, Oxford University Press, and E. Clementi, *IBM J. Res. Dev.*, **9**, 2 (1965), plus E. Clementi and C. Roetti, *Atom. Data Nuclear Data Tables*, **14**, 177 (1974).

where ϕ_a, ϕ_b, ϕ_c, and ϕ_d are STO basis functions located at centers a, b, c, and d, respectively, arise in polyatomic molecules containing multiple nuclei. Evaluation of these integrals is both difficult and time-consuming and, even though the functional form of STO basis functions has several advantages, the difficulty of multicenter integral evaluation[62] has nearly eliminated their use on contemporary quantum chemistry research on molecular systems.

An alternative to the use of exponential-based basis sets was suggested by Boys[63] in 1950, which consists of functions of the form

$$\phi_{tuv} = Nx^t y^u z^v e^{-\zeta r^2} \qquad t, u, v, = 0, 1, 2, \cdots. \tag{11-90}$$

These functions are known as *Cartesian Gaussians* [or *Gaussian-type orbitals* (GTO)], and various choices of t, u, v give rise to functions having similarities to STO basis functions. For example, $t = 1$, $u = v = 0$ is referred to as a $2p_x$-Gaussian type orbital (*GTO*), due to the similarity of its angular dependence to a $2p_x$ STO.

This set of both complete and discrete, but it also has several significant disadvantages. In particular, we have seen earlier[64] that both the cusp behavior near the origin and the long range behavior of Gaussians are not correct even for the hydrogen atoms. In addition, appropriate radial nodes (e.g., that are present in a $2s$ hydrogen-like function) are not present in most cases. This requires the use of multiple Gaussians in order to mimic the correct behavior in these regions. However, the redeeming feature that Boys showed in 1950 was that *all* of the integrals that arise in the calculation of the energy or other molecular properties can be evaluated analytically in closed form. It has also turned out that highly efficient algorithms for integral evaluation have been developed using these functions. Hence, both high integral accuracy and computational convenience are assured,[65] even though larger numbers of basis functions (compared to exponential basis sets) are expected in order to obtain high accuracy for the energy and molecular properties.

The basic idea that dominates approaches to use GTO (or other) primitives is that molecules can be viewed as a collection of distorted atoms. Hence, most primitives are chosen by requiring that they provide an accurate description of atoms of interest. In practice, most Gaussian primitive sets are constructed by optimization of the Hartree–Fock energy of the atom, or by least-square fitting

---

[62] For examples of techniques devised for STO integral evaluation see, for example, F. E. Harris, *Adv. Chem. Phys.*, **13**, 205 (1967); R. E. Christoffersen and K. Ruedenberg, *J. Chem. Phys.*, **49**, 4825 (1968).

[63] S. F. Boys, *Proc. R. Soc. (London)*, **A200**, 542–554 (1968).

[64] See Chapter 9, Section 9-6.

[65] For reviews of techniques that can be used for Gaussian integral evaluation see, for example, I. Shavitt, *Methods Comput. Phys.*, **3**, 1–45, (1963); V. R. Saunders, in *Methods in Computational Molecular Physics*, G. H. F. Diercksen and S. Wilson (eds.), D. Reidel, Boston, 1983, pp. 1–36, D. Hegarty and G. Van der Velde, *Int. J. Quantum Chem.*, **23**, 1135–1153 (1983).

to exponential functions, or both. Fortunately, such basis functions are also useful in higher accuracy studies,[66] and thus have wide applicability.

The first optimized primitive GTO basis set for atomic SCF studies was produced by Huzinaga,[67] who constructed s- and p-type GTO basis sets of varying size and accuracy for atoms from H to Ne by minimizing the energy of each atom. The orbital exponents and total energy for the largest of the basis sets (the $10s$, $6p$ basis) are given in Table 11.4. The $10s$ and $6p$ basis is quite accurate, e.g., the error is only 0.006 Hartree in the calculated Hartree–Fock energy for Ne. Since these studies optimized the orbital exponents (ζ) for each GTO of Eq. (11-90), determination of new terms to be included is computationally expensive.

However, it was noted by Ruedenberg and colleagues[68] that the ratio of successive orbital exponents for functions outside the inner shell (usually referred to as *valence orbitals*) was very nearly constant. This has given rise to the idea of *"even-tempered" orbitals* (primitives) that provides a good estimate for optimized orbital exponents without having to carry out nonlinear parameter optimizations. In particular, orbital exponents are chosen by

$$\zeta_i = \alpha\beta^i, \qquad i = 1, 2, \cdots, \tag{11-91}$$

where different values of α and β are chosen for $s, p, d, f, \ldots$ functions. Thus, orbital exponent determination is simplified greatly, e.g., all exponents in an (s, p) basis for the carbon atom are specified by only four parameters (α_s, β_s, α_p, β_p), regardless of the number of primitive functions used. Other similar approaches have also been devised,[69] and it is clear that such approaches comprise quite effective ways to construct primitive GTOs for atomic and molecular calculations.

Another alternative to STOs that avoids utilizing higher angular momentum GTO basis functions directly was devised by Whitten.[70] In particular, Whitten noted that integrals over $1s$-type Gaussians are particularly easy and convenient to evaluate, and that higher order basis functions can be approximated using $1s$-type Gaussians, as long as the $1s$-type Gaussians are not required to be located on a nucleus. For example, a $2p_z$-type Gaussian basis function in this

---

[66] See the discussion of configuration interaction techniques in Chapter 12, for example.

[67] S. Huzinaga, *J. Chem. Phys.*, **42**, 1293–1302 (1965); see also K. O. O-Ohata, H. Taketa, and S. Huzinaga, *J. Phys. Soc., Japan*, **21**, 2306 (1966).

[68] K. Ruedenberg, R. C. Raffenetti and R. D. Bardo, *Energy Structure and Reactivity, Proceedings of the 1972 Boulder Conference*, John Wiley, New York, 1973, p. 164; see also R. D. Bardo and K. Ruedenberg, *J. Chem. Phys.*, **60**, 918 (1974).

[69] See, for example, S. Huzinaga, M. Klobukowski, and H. Tatewaki, *Can. J. Chem.*, **63**, 1812 (1985); D. M. Silver and W. C. Nieuwpoort, *Chem. Phys. Lett.*, **57**, 421 (1978); D. M. Silver, S. Wilson, and W. C. Nieuwpoort, *Int. J. Quantum Chem.*, **14**, 635 (1978); E. Clementi and G. Corongiu, *Chem. Phys. Lett.*, **90**, 359 (1982).

[70] J. L. Whitten, *J. Chem. Phys.*, **39**, 349 (1963); **44**, 359 (1966).

Table 11.4. Orbital Exponents and Total Energies of First Row Atoms in the (10*s*, 6*p*) Basis Sets of Huzinaga

Orbital type	Li (^2S)	Be (^1S)	B (^2P)	C (^3P)	N (^4S)	O (^3P)	F (^2P)	Ne (^1S)
1*s*	1.90603	3.66826	6.25286	9.40900	13.4578	17.8966	23.3705	29.1672
1s	16.7798	32.6562	55.8340	84.5419	120.899	160.920	209.192	261.476
1s	60.0718	117.799	202.205	307.539	439.998	585.663	757.667	946.799
1s	267.096	532.280	916.065	1397.56	1998.96	2660.12	3431.25	4262.61
1s	0.71791	1.35431	2.31177	3.50002	4.99299	6.63901	8.62372	10.7593
1s	0.26344	0.38905	0.68236	1.06803	1.56866	2.07658	2.69163	3.34255
1s	0.077157	0.15023	0.26035	0.40017	0.580017	0.77360	1.00875	1.24068
1s	0.540327	10.4801	17.8587	26.9117	38.4711	51.1637	66.7261	83.3433
1s	1782.90	3630.38	6249.59	9470.52	13515.3	18045.3	23342.2	28660.2
1s	0.028536	0.052406	0.089400	0.13512	0.19230	0.25576	0.33115	0.40626
2p			0.15033	0.24805	0.37267	0.48209	0.62064	0.78526
2p			0.39278	0.65771	0.99207	1.32052	1.73193	2.21058
2p			1.06577	1.78730	2.70563	3.60924	4.78819	6.21877
2p			3.48347	5.77636	8.48042	11.4887	15.2187	19.7075
2p			15.4594	25.3655	35.9115	49.8279	65.6593	84.8396
2p			0.057221	0.091064	0.13460	0.16509	0.20699	0.25665
E (Hartree)	−7.4325033	−14.572579	−24.528282	−37.687324	−54.398909	−74.806295	−99.404870	−128.54094

formulation is written as

$$\phi_{2p_z} = N(\phi^+_{1s} - \phi^-_{1s}),$$

where ϕ^+_{1s} and ϕ^-_{1s} are $1s$-type GTOs located on opposite sides of the nucleus along the z-axis. These basis functions are usually referred to as *lobe functions.* In practice, however, the number of terms needed to describe higher orbital angular momentum orbitals and the loss of accuracy due to differencing in computing the integrals has led to alternative approaches.

Before leaving the discussion of primitive functions, it is important to note that high accuracy molecular calculations require the use of more than just primitives that are occupied in atoms (e.g., s and p GTOs in first-row atoms). For example, description of the distortion of atomic orbitals that occurs when they are placed in a molecular environment is needed, and can be accomplished through a number of approaches. The primitives thus produced are usually referred to as *polarization functions.*

One obvious way to devise such primitives is to utilize GTOs corresponding to higher orbital momentum, e.g., d, f, ... GTOs. Optimization of the orbital exponents or such primitives is typically carried out through studies on prototype molecules,[71] and comprehensive lists of polarization exponents for first- and second-row atoms have been compiled by various authors.[72]

Another way of introducing polarization effects into primitive basis sets is to continue to utilize *low* angular momentum GTOs (e.g., s-type GTOs), but to locate them off the nuclei (e.g., at the mid-point of bonds). Such functions are sometimes referred to as *bond functions*, although their location need not be restricted to bonding regions.

To see how the use of, e.g., s-type GTOs in off-nuclei locations introduces polarization components, consider a normalized s-type GTO located at some point ($\mathbf{R}$) from an arbitrary origin, i.e.,

$$G(\mathbf{r}-\mathbf{R}) = \left(\frac{2}{\pi\rho^2}\right)^{3/4} \exp\left\{-\frac{(\mathbf{r}-\mathbf{R})^2}{\rho^2}\right\}, \tag{11-92}$$

where ρ is the *orbital radius.*[73] Such orbitals are known as *floating spherical*

[71] Molecules instead of atoms are typically used in these studies, since atomic exponents determined this way are typically very diffuse and not useful as polarization functions.

[72] See, for example, B. Roos and P. Siegbahn, *Theoret. Chim. Acta*, **17**, 208 (1970); T. H. Dunning, Jr., *J. Chem. Phys.*, **55**, 3958–3966 (1971); I. Abscar and J. R. vanWazer, *Chem. Phys. Letters*, **11**, 310 (1971); P. C. Hariharan and J. A. Pople, *Theoret. Chim. Acta*, **28**, 213 (1973); R. Alrichs, F. Keil, H. Lischka, B. Zurawski, and W. A. Kutzelnigg, *J. Chem. Phys.*, **63**, 4685 (1975); T. H. Dunning and P. J. Hay, *Modern Theoretical Chemistry*, Volume 3, H. F. Schaefer III (ed.), Plenum Press, New York, 1977, p. 1; M. Keeton and D. P. Santry, *Chem Phys. Lett.*, **1**, 105 (1970).

[73] See Problem 11-8. See Fig. 11.5 for a depiction of this FSGO. These are also the orbitals used in the H_2 example that was discussed in Section 11-2B.

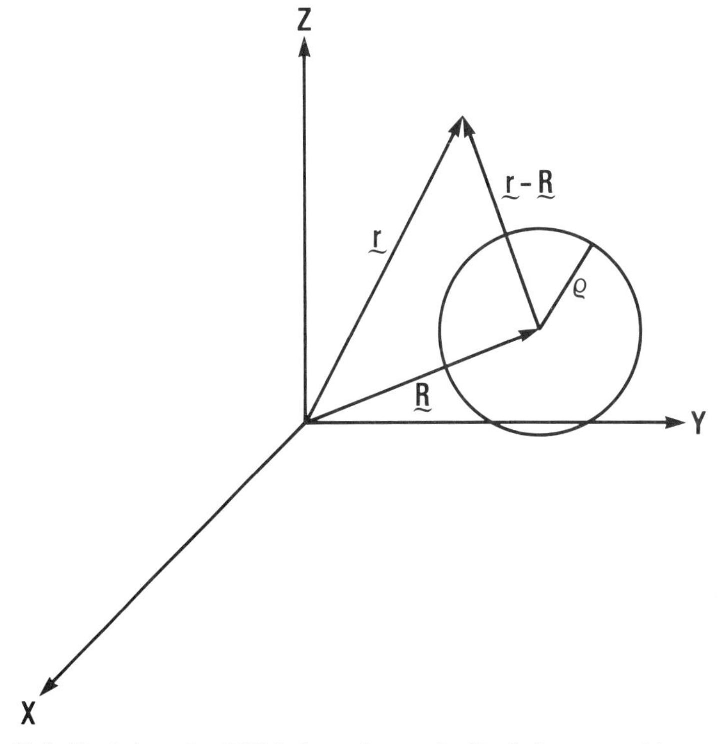

Figure 11.5. Depiction of an FSGO, located at a point **R** relative to an arbitrary origin.

Gaussian orbitals[74] *(FSGO)*, and are simply *s*-type GTOs that are allowed to be located away from nuclei. While such an orbital is spherically symmetric about the origin (**R**), we shall see shortly that this spherical symmetry does not translate to other origins. In particular, we rewrite Eq. (11-92) as follows:

$$G(\mathbf{r}-\mathbf{R})=\left(\frac{2\pi}{\rho^2}\right)^{3/4}\cdot\exp\{-[(x-X)^2+(y-Y)^2+(z-Z)^2]/\rho^2\} \quad (11\text{-}93)$$

$$=G^0(r)\exp\left\{-\left(\frac{R^2}{\rho^2}\right)\right\}\exp\{(\epsilon_X x+\epsilon_Y y+\epsilon_z z)/\rho\}, \quad (11\text{-}94)$$

where $G^0(r)$ is an FSGO whose location is at the origin of the coordinate system, and the ϵ's are unitless parameters defined by

$$\epsilon_X=\frac{2X}{\rho}, \qquad \epsilon_Y=\frac{2Y}{\rho}, \qquad \epsilon_Z=\frac{2z}{\rho}. \quad (11\text{-}95)$$

---

[74] For initial development and application of such orbitals see, for example. H. Preuss, *Z. Naturforsch*, **11**, 283 (1956); A. A. Frost, *J. Chem. Phys.*, **47**, 3707–3714 (1964); *J. Am. Chem. Soc.*, **89**, 3064 (1967); R. E. Christoffersen and G. M. Maggiora, *Chem. Phys. Letters*, **3**, 419 (1969).

While the first two terms on the right-hand side of Eq. (11-94) are spherically symmetric about the origin, the last term is not. To see this, we expand the exponential of the last term as follows:

$$
\begin{aligned}
&\exp\{(\epsilon_X x+\epsilon_Y y+\epsilon_Z z)/\rho\} \\
&\quad = 1+[\epsilon_X x+\epsilon_Y y+\epsilon_Z z]/\rho \\
&\quad + \left(\frac{1}{2}\right)[\epsilon_X^2 x^2+\epsilon_Y^2 y^2+\epsilon_Z^2 z^2+2\epsilon_X\epsilon_Y xy+2\epsilon_X\epsilon_Z xz+2\epsilon_Y\epsilon_Z yz]/\rho^2 \\
&\quad + \left(\frac{1}{6}\right)[\epsilon_X^3 x^3+\epsilon_Y^3 y^3+\epsilon_Z^3 z^3+3\epsilon_X\epsilon_Y^2 xy^2+3\epsilon_X\epsilon_Z^2 xz^2+3\epsilon_X^2\epsilon_Y x^2 y \\
&\quad + 3\epsilon_X^2\epsilon_Z x^2 z+3\epsilon_Y^2\epsilon_Z y^2 z+3\epsilon_Y\epsilon_Z^2 yz^2+6\epsilon_X\epsilon_Y\epsilon_Z xyz]/\rho^3 \\
&\quad + \vartheta(\epsilon^4).
\end{aligned}
\qquad (11\text{-}96)
$$

When Eq. (11-96) is substituted into Eq. (11-94), we obtain

$$
\begin{aligned}
G(\mathbf{r}-\mathbf{R}) &= e^{-(\epsilon_X^2+\epsilon_Y^2+\epsilon_Z^2)/4} \\
&\cdot \left\{\phi_{000}(\mathbf{r})+\left(\frac{1}{2}\right)[\epsilon_X\phi_{100}(\mathbf{r})+\epsilon_Y\phi_{010}(\mathbf{r})+\epsilon_Z\phi_{001}(\mathbf{r})]\right. \\
&+ \left(\frac{1}{4}\right)\left[\left(\frac{\sqrt{3}}{2}\right)\epsilon_X^2\phi_{200}(\mathbf{r})+\left(\frac{\sqrt{3}}{2}\right)\epsilon_Y^2\phi_{020}(\mathbf{r})+\left(\frac{\sqrt{3}}{2}\right)\epsilon_Z^2\phi_{002}(\mathbf{r})\right. \\
&\left.+\epsilon_X\epsilon_Y\phi_{110}(\mathbf{r})+\epsilon_X\epsilon_Z\phi_{101}(\mathbf{r})+\epsilon_Y\epsilon_Z\phi_{011}(\mathbf{r})\right]+\cdots,
\end{aligned}
\qquad (11\text{-}97)
$$

where the ϕ_{ijk} are GTOs located at the origin of the coordinate system, and defined in Eq. (11-90).

We now can see from Eq. (11-97) that, while $G(\mathbf{r} - \mathbf{R})$ is spherically symmetric with respect to the local origin $\mathbf{R}$, expansion of the orbital onto another origin gives rise to many components. In the case illustrated above we see that, with respect to the origin of the coordinate system, $G(\mathbf{r} - \mathbf{R})$ is not spherically symmetric, but contains components of *all* possible GTOs (s, p, d, ...). In principle this is a real advantage, for it allows polarization components corresponding to all GTOs to be added to a basis by simply including an FSGO not located at the origin of the GTOs. On the other hand, we see from Eq. (11-97) that the coefficients of the GTO components are *fixed*, which means that the relative contribution of any GTO type (e.g., d-type GTOs) cannot be altered from that given by the coefficient of that GTO in Eq. (11-97). In addition, the importance of the s, p, d, ... components is also determined by the orbital radius (ρ), which is typically fixed in advance. Hence, the context of the particular problem must be used to determine whether such off-center functions are appropriate for inclusion in the basis set.

The ease of integral evaluation over FSGOs, when combined with determina-

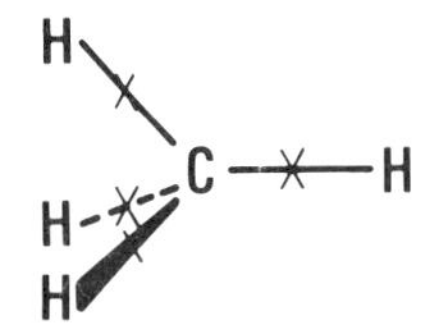

Figure 11.6. Depiction of an sp^3 environment for determination of FSGO basis orbitals.

tion of optimum nonlinear parameters in molecular (instead of atomic) environments, has led to yet another method for determination of primitives for molecular calculations. This method, known as the *molecular fragment approach*,[75] determines primitives by examination of small molecular fragments or molecules that are expected to be found in larger molecular systems. In the simplest basis, N FSGO are used to describe a $2N$ electron molecular fragment. For example, the description of a carbon atom in an sp^3-type hybridization environment is given by four FSGO,[76] as illustrated in Figure 11.6. The position and size of each FSGO is determined by energy minimization of the CH_4 molecule, and the FSGO thus determined are used directly in the study of large molecules. By studying a variety of molecular fragments that reflect anticipated molecular environments, we see that primitive orbitals can be determined that already reflect the polarized environment expected for a large molecule. While small basis sets such as these are not expected to be used for high accuracy, small molecule studies, they have been very useful in studying large chemical systems (e.g., with >200 electrons) where many of the important questions are relatively qualitative and can be answered through small basis set studies. We shall see examples of such applications later in this section as well as in the following chapter.

Before leaving this section, it should be noted that, for some excited states of atomic and molecular systems, the charge distribution is quite dissimilar to the ground state. For such cases, different kinds of basis functions need to bc added if accuracy comparable to that obtained for closed shell ground state studies is to be achieved. For example, when the charge distribution is different than the ground state, but still valence-like, addition of polarization functions such as described earlier in this section is an effective way to include the needed flexibility.

For other cases, e.g., when the charge distribution for the state of interest is diffuse compared to the ground state, new types of primitive orbitals are needed. Such situations arise in several ways, including anionic states (e.g., where an electron has been added to a closed shell ground state) and *Rydberg states*

[75] R. E. Christoffersen and G. M. Maggiora, *Chem. Phys. Lett.*, **3**, 419–423 (1969); R. E. Christoffersen, *Adv. Quantum Chem.*, **6**, 333 (1972); G. M. Maggiora and R. E. Christoffersen, *J. Am. Chem. Soc.*, **98**, 8325 (1976); D. Spangler and R. E. Christoffersen, *Int. J. Quantum Chem., Quantum Biol. Symp.*, **5**, 127 (1978); *Int. J. Quantum Chem.*, **17**, 1075 (1980).

[76] A fifth FSGO is also placed on the C nucleus to describe the inner schell electrons, but has been omitted from the figure for simplicity.

(where the overall charge distribution is quite diffuse). In such cases, functions with large orbital radii are needed, and can be chosen in various ways, e.g., by performing Hartree–Fock excited state calculations on atoms.[77] While the results are frequently found not to be particularly sensitive to the value of the diffuse orbital exponent, inclusion of such types of functions in the basis can be quite important.

2. Contraction Procedures

While the use of primitive Gaussian basis functions has been shown to be an effective alternative to STOs and can be used to obtain accurate energies and other properties for small systems if a large enough basis set is utilized, such basis sets are not usable for large systems in general. The reason is simply that the number of basis functions rises too fast to be manageable. In the case of benzene, for example, a ($9s$, $5p$) basis for each carbon and a ($4s$) basis for each hydrogen gives rise to a total basis set of 168 GTOs, and nearly 100 million electron repulsion integrals. Handling of such a large number of integrals, e.g., in Fock matrix formation, is a significant computational liability, leading to the search for alternative approaches.

Such difficulties in applying GTO basis functions to large systems have led to the construction of smaller basis sets from the original GTO bases, the former of which are known as *contracted basis sets*. As the name implies, such basis sets "contract" the original GTO basis, and do so by constructing *fixed* linear combinations of GTOs that give rise to a smaller group of basis functions. To illustrate how this can be done,[78] let us consider how a split valence basis can be constructed for oxygen, starting from a ($9s$, $5p$) GTO basis of Huzinaga.[79] The orbital exponents for the ($9s$, $5p$) basis are given in Table 11.5, along with a contraction suggested by Dunning.[80]

The notation used here and throughout is that the original GTO basis is listed in parentheses, while the contracted basis is given in brackets. Also, the nonhydrogen functions are listed first, with hydrogen functions listed (when present) following a "slash." Thus, an uncontracted basis of nine s-type GTOs, five p-type GTOs (with an x, y and z-component for each p-type GTO) for each oxygen plus four s-type GTOs for each hydrogen would be written as ($9s5p/4s$). If that basis is contracted to three s-type functions and two p-type functions (each

[77] See, for example, T. H. Dunning, Jr. and P. J. Hay, *Methods of Electronic Structure*, Volume 3, H. F. Schaefer III, (ed.), Plenum Press, New York, 1977, pp. 1–27; see also W. L. Hunt and W. A. Goddard III, *Chem. Phys. Lett.*, **3**, 414 (1969).

[78] Several authors have suggested such procedures, i.e., E. Clementi and D. R. Davis, *J. Comput. Phys.*, **1**, 223 (1966); J. L. Whitten, *J. Chem. Phys.*, **44**, 359 (1966); H. Taketa, S. Huzinaga and K. O-ohata, *J. Phys. Soc. Japan*, **21**, 2313 (1966).

[79] S. Huzinaga, *J. Chem. Phys.*, **42**, 1293–1302 (1965).

[80] T. H. Dunning, Jr., *J. Chem. Phys.*, **53**, 2823–2833 (1970); *J. Chem. Phys.*, **55**, 716–723 (1971).

Table 11.5. Exponents and Expansion Coefficients for Contraction of a (9*s*, 5*p*) Basis for Oxygen to a [3*s*, 2*p*] Basis[a]

ζ_s	$[s_1]$	$[s_2]$	$[s_3]$
7817.	0.001176	—	—
1176.	0.008968	—	—
273.2	0.042868	—	—
81.17	0.143930	—	—
27.18	0.355630	—	—
9.532	0.461248	−0.154153	—
3.414	0.140206	—	—
0.9398	—	1.056914	—
0.2846	—	—	1.000000
ζ_p	$[p_1]$	$[p_2]$	
35.18	0.019580	—	
7.904	0.124200	—	
2.305	0.394714	—	
0.7171	0.627376	—	
0.2137	—	1.000000	

[a] See T. H. Dunning, Jr. and P. J. Hay, *Methods of Electronic Structure Theory*, Volume 3, H. F. Schaefer, III (ed.), Plenum Press, New York, 1977, p. 1.

with x, y, z components) for each oxygen plus two s-type functions for each hydrogen using fixed linear combinations of the $(9s5p/4s)$ basis as indicated in Table 11.5, the resulting contracted basis would be represented by $[3s2p/2s]$. In general, a contracted basis function (χ_i) can be represented by

$$\chi_i = N_i \sum_j c_{ji} g_j, \tag{11-97a}$$

where N_i is a normalization constant, g_j are the primitive GTOs, and c_{ji} are the contraction coefficients.

It should be noted that, while the notation $[3s2p/2s]$ is an accurate representation of the number of contracted functions that are used, it is not fully indicative of the quality of the basis, since the number of primitives used in the contraction is not indicated. For example, the $[3s, 2p]$ portion of the above contraction might have been formed from a $(6s, 3p)$ contraction instead of a $(9s, 5p)$ contraction, but the set of primitives that has been used is not revealed by the $[3s, 2p]$ notation. Hence, consideration of both the number of primitives as well as contracted functions should take place when assessing the quality of a basis set.

A number of other points of interest are revealed in Table 11.5. First, the seven GTOs included in $[s_1]$ have large orbital exponents, meaning that the functions have small orbital radii. Thus, this contracted orbital is clearly designed to describe an inner shell s-function, which is not expected to change

Table 11.6. Comparison of Atomic Excitation Energies, Lowest Ionization Potential, and Electron Affinity for Oxygen Using a [3*s*2*p*] Basis and a Near-Hartree–Fock Basis

State	(9*s*5*p*)/[3*s*2*p*]	ΔE (eV)	Near Hartree–Fock[a]	ΔE (eV)
^{3}P	−74.79884	—	−74.80938	—
^{1}D	−74.71845	2.19	−74.72921	2.18
^{1}S	−74.59938	5.43	−74.61096	5.40
Cation (^{4}S)	−74.36055	11.93	−74.37256	11.89
Anion (^{2}P)	−74.77788	−0.57	−74.78948	−0.54

[a] H. F. Schaefer III, R. A. Klemm, and F. E. Harris, *J. Chem. Phys.*, **51**, 4643–4650 (1969).

much in molecular formation. Next, the GTOs with smaller orbital exponents are least contracted (or uncontracted), reflecting the need to maintain greatest flexibility (i.e., have more functions) for functions used to describe electrons in bonding regions where distortion of atomic density occurs in molecule formation. Also, we see that one of the *s*-type GTOs is used in both [s_1] and [s_2], indicating that the GTO is important to both inner and outer shell descriptions, but in different ways. Finally, as the "split-valence" designation implies, we note that the [3*s*2*p*/2*s*] contracted basis corresponds to a double zeta basis for non-inner shell and hydrogen functions, and a single zeta basis for non-hydrogen inner shells. Such a contraction process is clearly of computational help, since only three *s*-functions and two *p*-functions[81] are used in the SCF calculations, instead of nine *s*-functions and five *p*-functions.

However, we must be sure that the contraction process has not significantly affected the calculated properties of oxygen. To illustrate this point, Table 11.6 compares atomic excitation energies, lowest ionization potential, and electron affinity calculated using the [3*s*2*p*] basis with near-Hartree–Fock values. Since each of the calculated properties is within 0.1 eV of the value obtained using a near-Hartree–Fock basis, we have some confidence that the contraction procedure has not reduced the flexibility of the basis inappropriately.

Of course, it should also be recognized that a contracted basis can be constructed in other ways, e.g., the (9*s*) basis for oxygen could be contracted into a [4*s*] or [5*s*] basis, etc. Alternatively, criteria other than atomic Hartree–Fock studies could be used to determine contraction coefficients. However, the Dunning contractions have been found to provide a reasonable balance between retaining basis set flexibility and computational efficiency, and have found rather widespread use for describing bonding situations in at least semiquantitative ways.[82]

Another approach to basis set contraction is that employed by Pople and co-

[81] It should be remembered that each "*p*-function" refers to a group of p_x, p_y and p_z functions.

[82] For a compilation of a wide variety of Gaussian orbital exponents and contraction coefficients, see R. Poirier, R. Kari and I. G. Csizmadia, *Handbook of Gaussian Basis Sets*, Elsevier Science, New York, 1985.

workers,[83] in which STOs are used to construct contracted GTO basis sets. The specific form chosen for an *STO-nG expansion* is that each STO (ϕ_μ) with orbital exponent (ζ) is replaced by an atomic orbital ϕ_μ that is represented by a sum of K GTOs. In particular,

$$\phi_\mu'(\zeta, \mathbf{r}) = \zeta^{3/2}\phi_\mu(1, \zeta\mathbf{r}), \tag{11-98}$$

where, for example,

$$\phi_{1s}'(1, \mathbf{r}) = \sum_{k=1}^{K} d_{1s,k} g_{1s}(\alpha_{1k}, \mathbf{r}), \tag{11-99}$$

$$\phi_{2s}'(1, \mathbf{r}) = \sum_{k=1}^{K} d_{2s,k} g_{1s}(\alpha_{2k}, \mathbf{r}), \tag{11-100}$$

and

$$\phi_{2p}'(1, \mathbf{r}) = \sum_{k=1}^{K} d_{2p,k} g_{2p}(\alpha_{2k}, \mathbf{r}). \tag{11-101}$$

In the above equations, g_{1s} and g_{2p} are normalized, atomic $1s$-type and $2p$-type GTOs with orbital exponents (α), e.g.,

$$g_{1s}(\alpha, \mathbf{r}) = \left(\frac{2\alpha}{\pi}\right)^{3/4} \exp[-\alpha r^2], \tag{11-102}$$

$$g_{2p_z}(\alpha, \mathbf{r}) = \left(\frac{128\alpha^5}{\pi^3}\right)^{1/4} r \exp[-\alpha r^2] \cos\vartheta, \tag{11-103}$$

with similar expressions to Eq. (11-103) for the $2p_x$ and $2p_y$ components. In addition, it should be noted that, as with GTO basis sets in general, the $2s$ STO is represented by $1s$-type GTOs in Eq. (11-100). However, each of the exponents (α_{2k}) in the $2s$ and $2p$ expansions of Eqs. (11-100) and (11-101) is constrained to be *equal* (for ease of computation).

The orbital exponents (α) and coefficients (d) are then determined by least squares fitting[84] of the ϕ_μ's to STOs. Optimized values of these parameters are

---

[83] W. J. Hehre, R. F. Stewart, and J. A. Pople, *Symp. Faraday Soc.*, **2**, 15 (1968); *J. Chem. Phys.*, **51**, 2657 (1969); W. J. Hehre, R. Ditchfield, R. F. Stewart, and J. A. Pople, *J. Chem. Phys.*, **52**, 2769 (1970); W. J. Pietro, B. A. Levi, W. J. Hehre, and R. F. Stewart, *Inorg. Chem.*, **19**, 2225 (1980); W. J. Pietro and W. J. Hehre, *J. Comput. Chem.*, **4**, 241 (1981); J. A. Pople, in *Modern Theoretical Chemistry*, Volume 4, H. F. Schaefer III (ed.), Plenum, New York, 1976. For earlier reports using this approach see K. O-ohata, H. Taketa, and S. Huzinaga, *J. Phys. Soc. Japan*, **21**, 2306 (1966) and C. M. Reeves and R. Fletcher, *J. Chem. Phys.*, **42**, 4073 (1965).

[84] It is possible to use methods other than least-squares fitting to obtain contracted GTO expansions of STOs. For example, energetic criteria were applied in a variational approach by S. Huzinaga, *J. Chem. Phys.*, **42**, 1293 (1965); see also, S. Huzinaga, J. Andzelm, M. Klobukowski, E. Radzio-Andzelm, Y. Sakar and H. Tatewaki, *Gaussian Basis Sets for Molecular Calculations*, Elsevier, New York, 1984. For comparison of this approach with the least squares approach, see M. Klessinger, *Theoret. Chim. Acta*, **15**, 353 (1969).

Table 11.7. STO-nG Expansion Coefficients and Orbital Exponents for K = 2–6

K	α_{1s}	d_{1s}	α_2	d_{2s}	d_{2p}
2	0.151623	0.678914	0.0974545	0.963782	0.612820
	0.851819	0.430129	0.384244	0.0494718	0.511541
3	0.109818	0.444635	0.0751386	0.700115	0.391957
	0.405771	0.535328	0.231031	0.399513	0.607684
	2.22766	0.154329	0.994203	−0.0999672	0.155916
4	0.0880187	0.291626	0.0628104	0.497767	0.246313
	0.265204	0.532846	0.163541	0.558855	0.583575
	0.954620	0.260141	0.502989	0.297680[a]	0.286379
	5.21686	0.0567523	2.32350	−0.0622071	0.0436843
5	0.0744527	0.193572	0.0544949	0.346121	0.156828
	0.197572	0.482570	0.127920	0.612290	0.510240
	0.578648	0.331816	0.329060	0.128997	0.373598
	2.07173	0.113541	1.03250	−0.0653275	0.107558
	11.3056	0.0221406	5.03629	−0.0294086	0.0125561
6	0.00651095	0.130334	0.0485690	0.240706	0.101708
	0.158088	0.416492	0.105960	0.595117	0.425860
	0.407099	0.370563	0.243977	0.250242	0.418036
	1.18506	0.168538	0.634142	−0.0337854	0.173897
	4.23592	0.0493615	2.04036	−0.0469917	0.0376794
	12.1030	0.00916360	10.3087	−0.0132528	0.00375970

[a] This entry is listed as 2.97680×10^{-5} by Hehre, Stewart, and Pople, but is believed to be a misprint.

given in Table 11.7 for the cases K = 2–6. In addition, optimized ζ-values for L-shell orbitals based on atom and small molecule studies are given in Table 11.8, along with corresponding total energies. Also listed in this table are "standard" ζ-values that are thought to be best suited for an "average molecular environment." While we shall discuss the adequacy of basis sets in the context of energies and molecular properties of several important prototype molecules in the next section, it is useful here to make a number of initial basis set comparisons.

One of the most popular basis sets is the STO-3G minimal basis set (i.e., "single-zeta" type) of Pople and co-workers whose parameters are listed in Table 11.7. Its main utility is in the prediction of geometries. For example, in a study of many different kinds of molecules containing, H, C, N, O and F, the predicted SCF bond lengths had a mean absolute deviation from experiment of 0.030 Å, and angles were predicted within ~4°. Similar results have been reported by others. Of course, it must be recognized that the excellent results for geometric predictions that have been found using the STO-3G basis are really due to a cancellation of errors.[85] Nevertheless, the net balance of the basis is quite reasonable for geometry studies.

[85] See, for example, E. R. Davison and D. Feller, *Chem. Rev.*, **86**, 681–696 (1986).

Table 11.8. Optimized ζ-Values for STO-nG Expansions.

	STO-3G			STO-4G		STO-5G		STO-6G		"Standard ζ-values"	
Atom	$\zeta(K)$[a]	ζ(L)	E (a.u.)	ζ(L)	E (a.u.)	ζ(L)	E (a.u.)	ζ(L)	E (a.u.)	ζ(K)	$\zeta(L)$
$H(^2S)$		1.00	−0.49491	1.00	−0.49848	1.00	−0.49951	1.00	−0.49983	1.24	—
$Li(^2S)$		0.65	−7.32823	0.64	−7.39185	0.64	−7.40971	0.64	−7.41536	2.69	0.75
$Be(^1S)$		0.97	−14.39180	0.96	−14.50884	0.96	−14.54080	0.96	−14.55098	3.68	1.10
$B(^2P)$		1.28	−24.23160	1.27	−24.42216	1.27	−24.47226	1.27	−24.48829	4.68	1.45
$C(^3P)$	5.67	1.60	−37.22866	1.59	−37.51069	1.59	−37.58578	1.59	−37.60906	5.67	1.72
$N(^4S)$	6.67	1.93	−53.72010	1.92	−54.11585	1.92	−54.21972	1.92	−54.25155	6.67	1.95
$O(^3P)$	7.66	2.24	−73.80425	2.24	−74.33740	2.24	−74.47555	2.23	−74.51749	7.66	2.25
$F(^2P)$	8.65	2.56	−97.98709	2.56	−98.68185	2.56	−98.85976	2.56	−98.91327	8.65	2.55

[a] Taken from "free atom" values of E. Clementi and D. Raimondi, *J. Chem. Phys.*, **38**, 2686 (1963), and used in all STO-*n*G expansions.

The first way in which STO-nG basis sets can be extended is to construct "split-valence shell" basis sets as described earlier. In other words, in SCF calculations two basis orbitals are used instead of a single basis orbital for each valence atomic orbital in a given shell (K, L, M ...). Using the example introduced earlier,[86] the HF molecule is expected to have 12 basis orbitals in a double zeta basis. What Pople and co-workers have created in the split-valence category is well illustrated using what is called a "4-31G" basis as an example. In this basis, the inner shell orbital is described by a contraction of four GTOs, while there are *two* sets of $2s$ and $2p$ orbitals to describe valence shell orbitals. The first set of $2s$ and $2p$ orbitals is each described by a contraction of three GTOs, while the second set of $2s$ and $2p$ orbitals is described by a single GTO. Thus, a true "double zeta" basis has not been created, since the inner shell is still described by a single basis function (a contraction of four GTOs). The valence shell *is* of "double zeta" nature, however, since two basis functions of each kind are used. Using our HF example, the 4-31G basis would contain the following orbitals.

H: $1s$, $1s'$

F: $1s$, $2s$, $2s'$, $2p_x$, $2p_y$, $2p_z$, $2p_x$, $2p_y$, $2p_z$.

A series of basis functions of the "M-N1G" category have been created by Pople and co-workers, e.g., 3-21G, 4-21G, 5-21G, and 6-21G, where increasing numbers of GTOs are used for the inner shell description. Also, the series 4-31G, 5-31G and 6-31G have been created, in which an additional GTO has been used to describe the first valence shell orbital. In general, the M-N1G bases contain two basis functions for H and He, nine basis functions for Li through Ne, etc., but have differing numbers of GTOs in the contractions for each basis orbital depending on the compromise between speed and accuracy that is desired.

In obtaining optimized parameters for the M-N1G basis sets, atomic SCF calculations were carried out[87] (instead of simply fitting the GTOs to STOs as in the STO-nG case). However the exponents of $2s$ and $2p$ GTOs were still constrained to be equal for computational convenience, as in the "single zeta" STO-nG case. Geometric predictions using such basis sets are generally improved further from results using STO-nG basis sets. For example, the 3-21G basis gives calculated bond lengths for molecules containing first-row atoms that have mean absolute deviations of 0.016 Å. In general, addition of more functions to the 3-21G basis within the "split-valence shell" framework improves energetic properties somewhat, at the expense of increased computer time.

---

[86] See the discussion following Eq. (11-89).

[87] P. Pulay, G. Forarasi, F. Pang, and J. E. Boggs, *J. Am. Chem. Soc.*, **101**, 2550 (1979); J. S. Binkley, J. A. Pople, and W. J. Hehre, *J. Am. Chem. Soc.*, **102**, 939 (1980); M. S. Gordon, J. S. Binkley, J. A. Pople, J. Pietro, and W. J. Hehre, *J. Am. Chem. Soc.*, **104**, 2797 (1982).

Addition of polarization functions to the basis comprises the next step taken by Pople and co-workers in expanding the basis. These sets consist of adding only a set of six *d*-functions to Li through Ne basis sets,[88] or adding the set of 6 *d*-functions plus a set of 3*p*-functions to H or He basis sets. These additions of polarization functions are usually used in conjunction with the 6-31G basis, and are referred to as the "6-31G*" and "6-31G**" bases, respectively. Other modifications are also possible, and a variety of extended basis sets of this type have been constructed and used.[89]

D. Energies and Molecular Property Calculations

After illustrating the actual mechanics of a Hartree–Fock calculation on H_2 using a simple basis set and describing the kinds of basis sets that are available, it is now useful to review more generally some of the research results that have been obtained using the HF–SCF–LCAO–MO approach. Unless otherwise stated, the calculations to be discussed have been carried out using the Born–Oppenheimer approximation on the ground state of the molecule. It should be recognized, however, that there have been a large number of calculations using this approach that have been reported in the literature, and the discussion below will provide only a small sample of the work that has been done in this area.[90]

Perhaps the most appropriate place to begin this discussion is with calculated total energies. Since the total energy emerges directly from the solution of the Hartree–Fock equations, it has been used extensively as a measure of the "goodness" of the study. This has been done even though it is known that the correlation error that is inherent in the theory will prevent even the exact Hartree–Fock energy from approaching the true energy.

In Table 11.9, a summary of near-Hartree–Fock (i.e., large basis set) results for a number of relatively small molecules is presented. This summary is of interest in part because it contains various kinds of molecules (diatomics, polyatomics, linear molecules, nonlinear molecules, etc.), and in part because a number of different basis sets (e.g., both STOs and GTOs were used) were employed in essentially reaching the estimated Hartree–Fock limit. It also means that we shall be able to assess the characteristics of Hartree–Fock theory itself, and not have to be concerned in general with basis set completeness.

---

[88] Only five *d*-functions are actually needed to form a full atomic set, but d_{xx}, d_{yy}, d_{zz}, d_{xy}, d_{xz}, d_{yz} functions are used for computational conveience.

[89] See, for example, D. J. DeFrees, B. A. Levi, J. S. Binkley, and J. A. Pople, *J. Amer. Chem. Soc.*, **101**, 4085 (1979).

[90] Extensive bibliographys of *ab initio* calculations such as these have been given in several places. See, for example, M. Krauss, *Compendium of Ab Initio Calculations of Molecular Energies and Properties*, NBS Tech. Note 438, U.S. Dept. of Commerce, Washington, D.C., 1967. See also W. G. Richards, T. E. H. Walker, and R. K. Hinkley, *A Bibliography of Ab Initio Molecular Wave Functions*, Oxford University Press, Oxford, 1971. Supplements for 1970–73 and 1974–77 have also been published. See also K. Ohno and K. Morokuma, *Quantum Chemistry Literature Data Base*, Elsevier, New York, 1982.

Table 11.9. Near Hartree-Fock Calculations of Total Energy, Dissociation Energies, and Correlation Energy Error Estimates for Selected Molecules

Molecule	Total energy (Hartrees)	Dissociation energy (eV)[a]	Correlation energy error (Hartrees)
H_2	−1.134[b]	3.64(4.75)	0.0409[c]
CH_4	−40.225[d]	14.33(18.25)[e]	0.295[e]
NH_3	−56.225[d]	8.74(12.77)[f]	0.357[f]
H_2O	−76.065[d]	6.94(10.08)[g]	0.370[g]
HF	−100.071[d]	4.38(6.12)	0.459[h]
N_2	−108.997[d]	5.27(9.90)[i]	0.589[i]
CO	−112.791[d]	7.89(11.24)[f]	0.588[j]
F_2	−198.7683[k]	−1.37(1.68)	0.884[k]

[a] Experimental values are given in parentheses.

[b] J. M. Schulman and D. N. Kaufman, *J. Chem. Phys.*, **53**, 477 (1970).

[c] W. Kolos and L. Wolniewicz, *J. Chem. Phys.*, **49**, 404 (1968).

[d] P. C. Hariharan and J. A. Pople, *Theoret. Chim. Acta*, **28**, 213 (1973).

[e] W. Meyer, *J. Chem. Phys.*, **58**, 1017 (1973).

[f] A. Rauk, L. C. Allen, and E. Clementi, *J. Chem. Phys.*, **52**, 4133 (1970).

[g] B. J. Rosenberg and I. Shavitt, *J. Chem. Phys.*, **63**, 2162 (1975).

[h] P. E. Cade and W. Huo, *J. Chem. Phys.*, **47**, 614 (1967).

[i] P. E. Cade, K. D. Sales, and A. C. Wahl, *J. Chem. Phys.*, **44**, 1973 (1966).

[j] D. B. Neumann and J. W. Moskowitz, *J. Chem. Phys.*, **50**, 2216 (1969).

[k] G. Das and A. C. Wahl, *J. Chem. Phys.*, **44**, 87 (1966); see also A. C. Wahl, *J. Chem. Phys.*, 41, 2600 (1966).

Among the points of interest in Table 11.9 is that the correlation energy error per electron tends to increase as the number of electrons increases, going from 0.041 Hartree per electron pair in H_2 to 0.0982 Hartree per pair in F_2. However, for isoelectronic systems the correlation error is much closer to being constant. In any case, this observation makes it clear that, while Hartree–Fock theory may form a useful starting point for small molecules, it will become less and less reliable as the number of electrons in the molecule increases. Furthermore, this observation is relatively independent of choice of basis set, since the basis sets used produced near-Hartree–Fock results. Of course, these results are not totally unexpected since *inter*pair and other correlations can be expected to increase as the number of electrons increases. In other words, we should not be surprised that an independent particle model has limitations that increase in severity as the number of electrons increases.

The situation with dissociation energies is not much different, with the Hartree–Fock limit producing results of substantial variability (from 53% of the experimental value in HF or 78.5% in CH_4). Furthermore, in at least one case

Table 11.10. Geometric Prediction from Hartree–Fock Calculations[a]

Molecule	Predicted bond length (a.u.)/angle	Experimental bond length (a.u.)/angle	Absolute error (%)
H_2	$1.39(R_{HH})$	$1.40(R_{HH})$	0.7
CH_4	$2.048(R_{CH})$	$2.050(R_{CH})$	0.1
NH_3	$1.890(R_{NH})$	$1.912(R_{NH})$	1.2
	$107.2°(\measuredangle HNH)$	$106.7°(\measuredangle HNH)$	0.5
H_2O	$1.776(R_{OH})$	$1.809(R_{OH})$	1.8
	$106.1°(\measuredangle HOH)$	$104.5°(\measuredangle HOH)$	1.5
HF	$1.696(R_{HF})$	$1.733(R_{HF})$	2.1
N_2	$2.013(R_{NN})$	$2.074(R_{NN})$	2.9
CO	$2.081(R_{CO})$	$2.132(R_{CO})$	2.4
F_2	$2.50(R_{FF})$	$2.68(R_{FF})$	6.7

[a] In general, these data are reported in the footnotes of Table 11.9.

(F_2), even the sign of the dissociation energy is incorrectly predicted. The reasons for these results will be discussed later, but the main point of interest here is that, in spite of having produced energies that are approximately 99% of the exact answers, the Hartree–Fock energy of the molecule compared to that of the separated atoms cannot be expected in general to provide reliable dissociation energy predictions. Indeed, the errors are typically comparable to the binding energies.

However, experience has shown that Hartree–Fock calculations are quite useful in predicting and understanding other properties,[91] as we shall now see. Consider, for example, the prediction of molecular geometries. In Table 11.10 we see that, for the same group of molecules discussed previously, the results are quite good. Both distances and angles are well predicted, with bond distance errors typically in the 0.1–0.2 a.u. range and bond angle errors in the 1–5° range. Such errors make Hartree–Fock calculations quite useful for understanding and predicting molecular geometries, and a number of quite successful applications have occurred in this area as we saw in the previous section.

Not only molecular geometries, but several features of the electronic structure of molecules are well predicted using approximate Hartree–Fock calculations. For example, it was indicated earlier that, due to the approximate

[91] The following discussion provides examples of the general principle that Hartree-Fock wavefunctions will give one-electron properties that are correct to first order. See, for example, M. Cohen and A. Dalgarno, *Proc. Phys. Soc. London*, **77**, 748–750 (1961); K. F. Freed, *Chem. Phys. Lett.*, **2**, 255–256 (1968); J. Goodisman and W. Klemperer, *J. Chem. Phys.*, **38**, 721–725 (1963).

Table 11.11. Predicted and Observed First Ionization Potentials from Hartree–Fock Calculations

Molecule	Calculated ionization potential[a] (eV)	Observed ionization potential (eV)
H_2	16.2	15.9
CH_4	14.9	14.4
NH_3	11.6	10.9
H_2O	13.8	12.6
HF	17.7	15.8
N_2	17.3	15.6
CO	15.0	13.9
F_2	18.0	16.3

[a] These values were obtained using Koopmans' Theorem and the near-Hartree–Fock wavefunctions reported in the footnotes of Table 11.9.

cancellation of errors, at least the first several ionization potentials could be expected to be well predicted using Hartree–Fock theory. The first ionization potential for the set of molecules considered earlier, calculated by the use of Koopmans' Theorem is given in Table 11.11. In general, the agreement is good, with Hartree–Fock values up to 10% higher than experimentally observed.

As an example of assessing results of Hartree–Fock calculations using a different approach, let us consider a single molecule and the results obtained using a variety of basis sets. In particular, let us consider the ground state ($^1A_1$) of the formaldehyde molecule (H_2CO) at its experimentally observed geometry (R_{CO} = 1.20785 Å, R_{CH} = 1.1160 Å, $\measuredangle$HCH = 116.52°). Results for energy-related properties using nine different basis sets are given in Table 11.12, while other properties calculated for the same basis sets are listed in Table 11.13. For properties other than energetic, the following expectation values were used:

$$\langle \mu z \rangle = \left\langle \Psi \left| \sum_i q_i z_i \right| \Psi \right\rangle , \qquad (11\text{-}104)$$

$$\vartheta_{xx} = \left\langle \Psi \left| \left(\frac{1}{2}\right) \sum_i q_i(3x_i^2 - r_i^2) \right| \Psi \right\rangle , \qquad (11\text{-}105)$$

$$\vartheta_{yy} = \left\langle \Psi \left| \left(\frac{1}{2}\right) \sum_i q_i(3y_i^2 - r_i^2) \right| \Psi \right\rangle , \qquad (11\text{-}106)$$

$$Q(0) = \left\langle \Psi \left| \sum_i{}' \, q_i(3x_i^2 - r_i^2)/r_i^5 \right| \Psi \right\rangle , \qquad (11\text{-}107)$$

where the sums run over all electrons and nuclei [except Eq. (11-107), in which

Table 11.12. Selected Energy-Related Properties of Formaldehyde ($^1A_1$) Using Different Basis Sets (Energies in Hartrees)[a]

Basis set[a]	Number of Basis functions	Total energy	Atomization energy	Orbital energies $\epsilon\ (1b_1)$	Orbital energies $\epsilon\ (2b_2)$
Dunning/Hay (split valence)[b]	22	−113.8292	0.349	−0.538	−0.443
Dunning/Hay (split valence plus polarization)[c]	40	−113.8939	0.413	−0.534	−0.440
Dunning (double zeta plus polarization)[d]	42	−113.8940	0.413	−0.534	−0.440
STO-3G	12	−112.3537	0.418	−0.447	−0.352
3-21G	22	−113.2204	0.400	−0.527	−0.430
6-31G	22	−113.8069	0.355	−0.534	−0.440
6-31G**	40	−113.8680	0.415	−0.531	−0.436
Hartree–Fock atomic orbital (single-zeta STO basis)	12	−113.6604	0.163	−0.596	−0.498
(s, p, d, f)-Extended[e]	120	−113.9202	0.420	−0.537	−0.443
Experimental	—	−114.562[f]	0.600	0.529[g]	0.400[g]

[a] See E. R. Davidson and D. Feller, *Chem. Rev.*, **86**, 681 (1986) for further description of the basis sets and discussion of the results.

[b] Formed using a ($9s$, $5p/4s$) → [$3s$, $2p/2s$] contraction.

[c] Formed using a ($9s$, $5p$, $1d/4a$, $1p$) → [$3s$, $2p$, $1d/2s$, $1p$] contraction.

[d] Formed using a ($9s$, $5p$, $1d/4s$, $1p$) → [$4s$, $2p$, $1d/2s$, $1p$] contraction.

[e] Formed using a ($19s$, $10p$, $2d$, $1f/10s$, $2p$, $1d$) → [$10s$, $5p$, $2d$, $1f/4s$, $2p$, $1d$] contraction.

[f] Includes relativistic effects.

[g] Experimental ionization potentials. See R. Colle, R. Montagnani, P. Riani, and O. Salvetti, *Theoret. Chim. Acta*, **49**, 37 (1978). See also M. Robin, N. Kuebler and C. R. Brundle, *Electron Spectroscopy*, D. A. Shirely, (ed.), North Holland, Amsterdam, 1972, pp. 351–378.

Table 11.13. Selected Origin-Centered and Atom-Centered Properties of Formaldehyde ($^1A_1$) Using Different Basis Sets[a]

Basis set[a]	$\langle \mu \rangle$	ϑ_{xx}	ϑ_{yy}	$Q(O)$	$\eta(O)$	$q(H)$	$\eta(H)$
Dunning/Hay (split valence)	−1.224	−0.015	0.433	−2.608	0.475	0.287	0.007
Dunning/Hay (split valence plus polarization)	−1.124	−0.181	0.219	−2.343	0.515	0.256	0.013
Dunning (double zeta plus polarization)	−1.124	−0.182	0.218	−2.342	0.515	0.256	0.013
STO-3G	−0.600	+0.287	0.220	−2.443	0.252	0.277	0.000
3-21G	−1.038	0.042	0.258	−2.223	0.641	0.279	0.009
6-31G	−1.184	0.057	0.287	−2.428	0.546	0.280	0.000
6-31G**	−1.085	−0.061	0.093	−2.219	0.562	0.247	0.012
Hartree–Fock atomic orbital (minimal STO basis)	−0.785	−0.107	0.405	−3.441	0.557	0.381	0.004
(s, p, d, f)-Extended	−1.124	−0.155	0.239	−2.272	0.642	0.245	0.007
Experimental	−0.917 ± 0.008	1.35 ± 1.0	−0.45 ± 0.5	−2.194	0.695	0.261	0.018

[a] See E. R. Davidson and D. Feller, *Chem. Rev.*, **86**, 681 (1986) for further discussion of the basis sets and discussion of the results. Values listed are in atomic units, and are measured relative to the center of mass.

the sum omits terms for oxygen]. The properties thus computed are dipole moment (μz), quadrupole moment (ϑ), field gradient at the nucleus (Q), and associated asymmetry parameter (η).

Examination of these results illustrates a number of interesting points regarding basis sets and their applicability. First, from an energetic point of view, the Dunning basis sets provide better total energies than the Pople basis sets of comparable size. At the same time, the basis set using Hartree–Fock atomic orbitals (formed from a single zeta STO basis) is substantially better energetically than the STO-3G basis. On the other hand, any of the Dunning bases or 6-31G bases are within ~ 3 kcal/mol of the extended basis set results, and so it is difficult to distinguish among these basis sets on the basis of total energy alone.

The atomization energy provides an example of the calculation of an energy difference where it is usually assumed that better results will be obtained through cancellation of errors, as opposed to calculation of a property via an absolute energy. As is seen from Table 11.12, none of the basis sets used at the Hartree–Fock level provides a satisfactory description, with even the extended basis having an error of $\sim 30\%$ compared to the experimental value. In large part, this error is a deficiency of the theory itself. On the other hand, the minimum STO, Hartree–Fock atomic orbital basis appears to provide a substantially poorer description of this property than any of the Dunning or Pople basis sets, even when compared only to Hartree–Fock results. This is attributed usually to cancellation of errors in the STO-3G and other basis sets. It also indicates that only with extensive basis sets (e.g., with split-valence plus polarization functions included) can one expect to obtain atomization energies that are accurate for the "right reasons."

Considering orbital energies (and ionization potentials), we see from Table 11.12 that, except for the STO-3G and minimum STO basis sets, each of the other basis sets provides a good estimate of the first two ionization potentials. The data also indicate the frequently observed finding that small basis sets (e.g., STO-3G) underestimate ionization potentials, while larger basis sets tend to overestimate ionization potentials.

Considering other properties, we see from the data in Table 11.13 that, in general, the STO-3G and minimum STO basis sets provide poor results. In addition, for the dipole moment, it is seen that the corresponding Dunning and Pople basis provide essentially comparable results (higher than experimentally observed), while the STO-3G basis is correspondingly below the experimentally observed dipole moment. The quadrupole moments are poorly predicted by all basis sets, although the experimental error estimates are quite high, making comparisons to calculated values difficult.

Since dipole and quadrupole moment calculations weight regions away from nuclei [e.g., see Eqs. (11-104)–(11-106)], it is expected that more extensive basis sets (e.g., including long-range functions) will be needed to assure reliable results.

Field gradient (Q) calculations produce somewhat better results, except for the minimum STO basis, with essentially all results seen to be within $\sim 10\%$ of the experimentally observed values. It is also seen that asymmetry parameters are less reliably calculated, with errors of 20–30%.

Perhaps the first general point of the above analysis is that substantial differences exist among basis sets. Next, the nature and magnitude of the expected error will depend upon the property of interest. Also, the examples cited thus far include only closed shell systems, and the errors found for open shell cases can be larger, as we shall see in Section 11-3. Hence, broad conclusions and generalizations concerning the "merits" of one basis set compared to another need to be viewed as having many caveats. However, even with such qualifications, some generalizations are appropriate. For example, the geometric structure of most closed shell molecules is predicted quite well with any of the split valence or split valence plus polarization basis sets. Also, ionization potential estimates, at least the first several, can usually be expected to be predicted adequately.

Before leaving this section it is useful to consider what options exist when the molecules of interest are so large that essentially none of the highly contracted or more extensive basis sets described above can be used. As we shall now see, the situation is far from hopeless, although care must be taken to focus on properties that are relatively insensitive to basis set size.

To illustrate this concept, let us use the FSGO basis of Eq. (11-92) as primitives, plus the "molecular fragment" approach described earlier.[92] In particular, FSGOs (G_i) determined through studies on molecular fragments are used to form MOs (φ_i) for a closed shell molecule containing N electrons via

$$\varphi_j = \sum_{i=1} c_i G_i, \qquad j = 1, 2, \cdots, N/2 \tag{11-108}$$

where the sum is taken over all FSGO from each molecular fragment.

As an example of the utility of such a basis, consider the calculation of ionization potentials for large molecular systems via orbital energies and Koopmans' theorem. Since such a property was found to be relatively invariant to basis set size, use of FSGO basis sets for calculation of such properties on very large molecular systems is an interesting possibility. To prepare for such studies, a large number of smaller molecules were studied,[93] in order to compare with other basis sets. Interestingly, such studies illustrated a basis set balance that was quite remarkable. In particular, comparison to large basis set studies on more than 20 molecules[93] indicated that the MO energies obtained using the molecular fragment basis are *linearly* related to those of larger basis set studies. In particular, a relation of the form

$$\epsilon_i^{\text{ref}} = a\epsilon_i^{\text{MF}} + b, \tag{11-109}$$

[92] See the discussion following Eq. (11-96).

[93] See, for example, R. E. Christoffersen, D. Spangler, G. G. Hall, and G. M. Maggiora, *J. Am. Chem. Soc.*, **95**, 8526–8536 (1973).

was found, where ϵ_i^{MF} are valence MO energies obtained using the molecular fragment basis, and ϵ_i^{ref} are corresponding quantities taken from more extensive basis set calculations. The values of the slope range from 0.7 to 0.9, and the intercepts range from -0.14 to -0.40. The existence of such a relationship is of considerable importance, for it indicates that both the energetic ordering of MOs (except when near-degeneracies occur) and their relative spacing can be expected to be correctly predicted for large molecular systems, even with a small basis set. This can be of direct assistance in the prediction and interpretation of properties such as ionization potentials and photoelectron spectra.

To illustrate this type of application to a large system, let us consider the studies of LeBreton *et al.* on biological purines and pyrimidines.[94] In one of these studies, UV photoelectron spectroscopy was used to examine guanine, hypoxanthine, and several methyl-substituted derivatives in the gas phase. The observed spectra are shown in Fig. 11.7, and are seen to provide a rich description of the first 5–10 vertical ionization potentials in these systems.

For comparison, the calculated MO energies [scaled via Eq. (11-109)] are also listed along the horizontal axis for each molecule. It is seen that qualitative agreement between calculated MO energies as extrapolated and observed ionization potentials exists, since the ordering of the first seven or eight calculated MO energies is found to be identical to the assignment of experimental ionization potentials.

It should be emphasized, however, that results on systems as complex as these must be considered as only suggestive and not definitive, even though they can be quite helpful when experimental interpretations are difficult. Furthermore, there are limitations on the theory itself, since charge redistribution and other effects are expected to make use of Koopmans' Theorem less reliable for higher IPs. Nevertheless, this procedure is seen to be of particular utility in the assignment of relative positions of π-orbitals and lone pair orbitals in these cases, and indicates the general usefulness and synergism that are possible even for relatively large molecular systems between experimental results and theoretical calculations. Other examples of the utility of these small basis sets will be seen in Section 12-3C.

E. Interpretation of Wavefunctions

In addition to dealing with energies and other properties calculated directly from the wavefunction, it is also highly desirable to be able to translate results of calculations into traditional concepts and ideas in chemistry. While we shall discuss this topic further in the next chapter, it is useful here to introduce two types of procedures that illustrate, even within Hartree–Fock calculations, how

[94] J. Lin, C. Yu, S. Peng, I. Akiyama, K. Li, L. K. Lee, and P. R. LeBreton, *J. Phys. Chem.*, **84**, 1006–1012 (1980) and references contained therein. For a recent example, see J. P. Boutique, J. Riga, J. J. Verbist, J. G. Fripiat, R. C. Haddon, and M. L. Kaplan, *J. Am. Chem. Soc.*, **106**, 312–318 (1984).

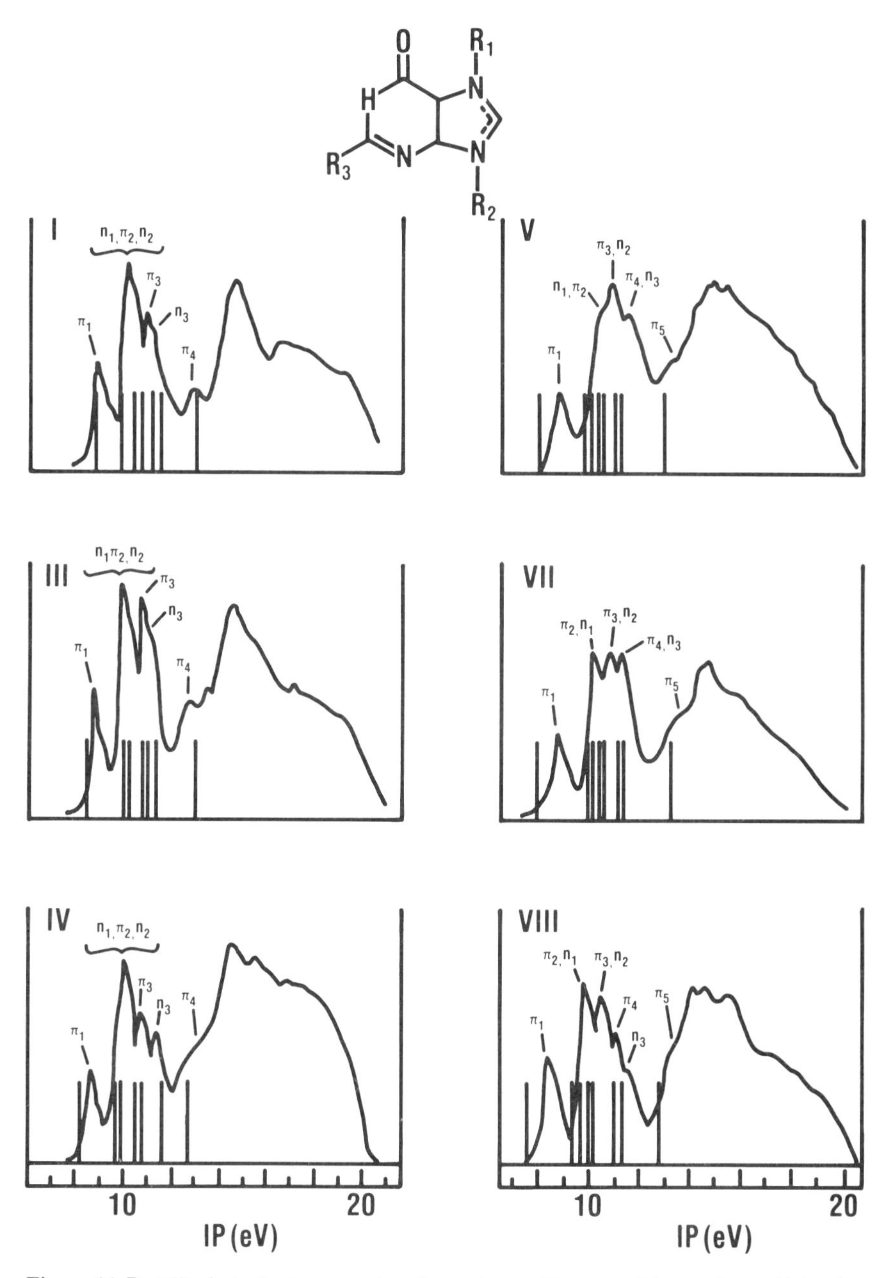

Figure 11.7. UV photoelectron spectra of guanine and hypoxanthine analogs. (I) has $R_1 = H$, $R_3 = H$; (III) has $R_1 = CH_3$, $R_3 = H$; (IV) has $R_2 = CH_3$, $R_3 = H$; V has $R_1 = H$, $R_3 = HN_2$; VII has $R_1 = CH_3$, $R_3 = NH_2$; (VIII) has $R_2 = CH_3$, $R_3 = NH_2$. Extrapolated values of calculated MO energies are also given for each molecule (with arbitrary intensity). [Reprinted with permission from Lin *et al.*, *J. Phys. Chem.*, **84**, 1006–1013. Copyright 1980, Americal Chemical Society.]

wavefunction information can be interpreted in terms of charge distributions in molecules and their relation to chemical properties.

1. Charge Distributions, Contour Maps and Point Charge Models

Probably the most direct and unambiguous way of examining the charge distribution in a molecule is by means of *contour maps*. To illustrate the idea, consider the kth filled MO:

$$\varphi_k(\mathbf{r}) = \sum_{\mu=1}^{M} c_{\mu k}\phi_\mu(\mathbf{r}), \tag{11-110}$$

whose coefficients ($c_{\mu k}$) have been determined via a Hartree–Fock calculation. The probability of finding an electron at $\mathbf{r}$ within the volume element $d\mathbf{r}$ was seen earlier[95] to be given by

$$P_k(\mathbf{r})\, d\mathbf{r} = |\varphi_k(\mathbf{r})|^2\, d\mathbf{r} = \rho_k(\mathbf{r})\, d\mathbf{r}, \tag{11-111}$$

where $\rho_k(\mathbf{r})$ is the probability density function for the kth orbital. Also, as we saw earlier,[95] $\rho_k(\mathbf{r})$ can be written as

$$\rho_k(\mathbf{r}) = N_k \sum_{\mu}^{M}\sum_{\nu}^{M} c_{\mu k}c_{\nu k}\phi_\mu^*(\mathbf{r})\phi_\nu(\mathbf{r}), \tag{11-112}$$

where N_k is the number of electrons in the kth MO. In order to understand the shape of this probability density function, there are several steps that are possible. For example, a plot of $\rho_k(\mathbf{r})$ along a given axis can be made. Alternatively, the points at which $\rho_k(\mathbf{r})$ has the same value can be computed and connected. This latter process, referred to as creating *contour maps*, has been found to be quite informative.[96] As an example, Fig. 11.8 shows a total charge density contour map for the LiF molecule, in which the expected asymmetry of charge is evident.

Alternatively, a mathematical analysis of the charge distribution is frequently used. There are a variety of ways to undertake such an analysis, and some of these use the Hartree–Fock results directly. Others use the fact that the Hartree–Fock equations are invariant to a unitary transformation, and construct such a transformation to create orbitals whose shape corresponds more closely to chemical intuition (e.g., *localized* orbitals). We shall now examine each of these approaches.

The mathematical analyses that has found the most widespread use is known

---

[95] See Section 11-2.

[96] See, for example, A. C. Wahl, *Science*, **151**, 961 (1966) for a series of charge density contour maps for homonuclear diatomic molecules, both for individual MOs and the total charge density.

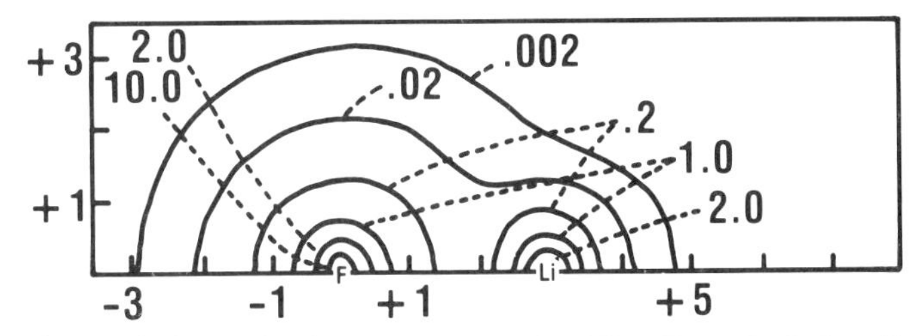

Figure 11.8. Total charge density contour map for LiF. [From B. J. Ransil and J. J. Sinai, *J. Chem. Phys.*, **46**, 4050–4074 (1967)]. Electron density is expressed in electrons per $(\text{a.u.})^3$ and spatial coordinates and expressed in Bohrs.

as the *Mulliken population analysis*.[97] To illustrate the approach, let us consider first the simple example of a diatomic molecule MO having one atomic orbital on each nucleus, i.e.,

$$\varphi = c_A \phi_A + c_B \phi_B. \tag{11-113}$$

If we square φ, integrate and multiply by the number of electrons ($N\varphi$) in the MO, we obtain

$$N_\varphi = N_\varphi c_A^2 + 2N_\varphi c_A c_B S_{AB} + N_\varphi c_B^2, \tag{11-114}$$

where

$$S_{AB} = \int \phi_A^* \phi_B \, d\tau. \tag{11-115}$$

In the Mulliken approach, the terms (Nc_A^2) and (Nc_B^2) are referred to as the *net atomic population* [$N(\phi_A)$ and $N(\phi_B)$] on atom A and B, respectively, and the cross-term [$2Nc_A c_B S_{AB}$] is referred to as the *overlap population*.

Also, to describe the bond between atoms A and B, the *bond order* (P_{AB}) between atoms A and B is defined as

$$p_{AB} = N_\varphi c_A c_B S_{AB}.$$

Furthermore, if one assumes that the cross-term an be divided *equally* between atoms A and B, then a *gross atomic population* on each atom can be defined as

$$N(A) = N(c_A^2 + c_A c_B S_{AB})$$
$$N(B) = N(c_B^2 + c_A c_B S_{AB}).$$

Next, *gross charges* are defined as

$$\begin{aligned} Q(\phi_A) &= N_0(\phi_A) - N(\phi_A) \\ Q(A) &= N_0(A) - N(A) \end{aligned} \tag{11-117}$$

[97] R. S. Mulliken, *J. Chem. Phys.*, **3**, 573 (1935); **23**, 1833, 1841, 2338, 2343 (1955); **36**, 3428 (1962); *J. Chim. Phys.*, **46**, 497, 675 (1949). See also P. O. Löwdin, *J. Chem. Phys.*, **18**, 365 (1950) for an alternative formulation.

with analogous definitions for nucleus B. In the above equations $Q(\phi_A)$ is referred to as the *gross charge in the ϕ_A atomic orbital,* $Q(A)$ is the *gross charge on atom A*, $N_0(\phi_A)$ is the number of electrons in ϕ_A when A is a separate atom, and $N(A)$ is the number of electrons in the ground state of the neutral atom. Generalization of these concepts to extended basis sets and polyatomic molecules is straightforward. As is obvious, however, such definitions contain several arbitrary assumptions, e.g., the manner in which the overlap density is allocated to atoms, as well as to depend on the basis set that has been chosen.[98] Nevertheless, this approach has provided useful insights into the charge distributions of molecules.

As a simple example used by Mulliken to illustrate the concepts, we shall consider the H_2O molecule, using an early approximate wavefunction determined by Ellison and Shull.[99] A minimum STO basis set of $1s$, $2s$, $2p_x$, $2p_y$, and $2p_z$ atomic orbitals on oxygen plus one $1s$ orbital on each hydrogen was used. The SCF calculation resulted in five filled MOs ($1a_1$, $2a_1$, $3a_1$, $1b_2$, and $1b_1$), and the distribution of charge resulting from the Mulliken analysis is given in Table 11.14. As the data indicate, the effective electron population in each orbital can be represented as

$$1s_0^{2.00}2s_0^{1.85}2p_0^{4.50}; \qquad 1s_{H1}^{0.82};\ 1s_{H2}^{0.82},$$

indicating that there is a "deficit" of approximately $0.15e$ in the $2s_0$ orbital in H_2O, compared to the oxygen atom. This is observed in spite of the considerable electron transfer from the hydrogens to the oxygen nucleus. In particular, the gross atomic population of oxygen is seen to be 8.349, for an overall charge (Q) of -0.35. While such analyses are admittedly approximate, they are seen to provide useful aids in understanding the charge distributions in molecular systems.

An alternative to population analyses for prediction and/or understanding of chemical reactivity is the use of molecular *electrostatic potential* (MEP) *maps*.[100] These maps are based on the concept that the electrostatic potential at any point ($\mathbf{R}$) in space that is generated by an electronic charge distribution (ρ) and nuclear charges (Z_α) located at positions ($\mathbf{R}_\alpha$) is given by

$$V(\mathbf{R}) = \sum_\alpha \frac{Z_\alpha}{|\mathbf{R}-\mathbf{R}_\alpha|} - \int \frac{\rho(\mathbf{r})}{|\mathbf{R}-\mathbf{r}|}\, d\mathbf{r}. \tag{11-118}$$

---

[98] For examples of studies designed to remove some of the difficulties of the Mulliken analysis, see R. E. Christoffersen and K. A. Baker, *Chem. Phys. Lett.*, **8**, 4–9 (1971); see also B. T. Thole and P. T. vanDuijnen, *Theoret. Chim. Acta*, **63**, 209–222 (1983); and F. Fliazar, *Charge Distribution and Chemical Effects*, Springer-Verlag, Berlin, 1983.

[99] F. O. Ellison and H. Shull, *J. Chem. Phys.*, **21**, 1420 (1953); **23**, 2348 (1955).

[100] See, for example, H. Weinstein, R. Osman, J. P. Green, and S. Topiol, in *Chemical Applications of Atomic and Molecular Electrostatic Potentials*, P. Politzer and D. G. Truhlar (eds.), Plenum Press, New York, 1981, pp. 309–334. See also R. Lavery and A. Pullman, in *New Horizons of Quantum Chemistry*, P. O. Löwdin and B. Pullman, (eds.), D. Reidel, Dordrecht, Holland, 1983, pp. 439–451.

Table 11.14. Gross Atomic Populations and Charges in H_2O Using a Mulliken Population Analysis

	AO								
MO	$1s_0$	$2s_0$	$2p_{z0}$	$2p_{y0}$	$2p_{x0}$	$1s_{H_1}$	$1s_{H_2}$	$N(\varphi)_O$	$N(\varphi)_H$
$1a_1$	2.0002	0.0005	0.0000	—	—	−0.0005	—	2.0007	−0.0005
$2a_1$	0.0008	1.638	0.049	—	—	0.309	—	1.688	0.309
$1b_2$	—	—	—	0.918	—	—	1.080	0.918	1.080
$3a_1$	−0.0001	0.209	1.534	—	—	0.257	—	1.743	0.257
$1b_1$	—	—	—	—	2.000	—	—	2.000	—
$N(\phi)$	2.0009	1.847	1.583	0.918	2.000	0.565	1.080	$N(O) =$ 8.349	$N(2H) =$ 1.645
$Q(\phi)$	0	+0.15	−0.58	+0.08	0	+0.43	−0.08	$Q(0) =$ −0.35	$Q(2H) =$ +0.35

Thus, $V(\mathbf{R})$ represents the potential associated with the interaction of a point charge at $\mathbf{R}$ with the charge distribution of the molecule.

Since $\rho(\mathbf{r})$ can be either the total electronic charge distribution or the distribution of, e.g., individual MOs, it is possible to use $V(\mathbf{R})$ to examine the molecular electrostatic potential in several ways. Furthermore, if points where $V =$ constant are connected, electrostatic potential contour maps can be created which indicate regions of positive and/or negative electrostatic potentials. Since a region of large negative electrostatic potential indicates, at least to a first approximation, a region that would be likely for attack by a positive species, it is possible to use electrostatic potential maps to find regions in molecules that are expected to be chemically reactive, at least for early stages of chemical reactions. Among the significant advantages of the use of quantities such as MEPs is that an arbitrary subdivision or assignment of charges onto specific nuclei or regions as is required in population analyses is not needed. Instead, the MEP is calculated directly from the wavefunction and nuclear positions without further approximation. Of course, the MEP will still reflect inadequacies in the wavefunction itself and ignores polarization, charge transfer, etc., that may accompany the interaction of reactants, but at least one significant level of approximation has been avoided.

Applications of this approach have been applied to a variety of systems, including systems of biological interest,[101] and appear to provide good insight into chemical reactivity and, in some cases, biological action. As an example, electrostatic potential maps for the adenine molecule[102] are given in Fig. 11.9, both in the plane of the molecule and in a plane perpendicular to the plane of the molecule. The basis set used to construct these maps was a $(4s, 2p/3s)$ basis contracted to a split-shell $[2s, 1p/2s]$ basis.[103] As is seen from the figure, three minima are seen in the plane and two additional minima reflect the greater basicity of pyridine-type nitrogens over that of amine-type nitrogens. Also, N_1 and N_3 are indicated to be preferred positions for electrophilic attack, which is confirmed experimentally.[104]

The Mulliken population and modifications thereof have one important characteristic in common: they use mathematical analyses to allow assignment of electron density and net charge to each of the atoms in a molecule. Thus, these approaches can be thought of as "atoms-in-molecules" approaches. However, point charges other than on nuclei can also be defined from calculated wavefunctions, and we shall now explore several approaches of this type that

---

[101] See, for example, H. Weinstein, R. Osman, S. Topiol and J. P. Green, *Ann. N.Y. Acad. Sci.*, **367**, 434–451 (1981).

[102] E. Scrocco and J. Tomasi, *Topics Curr. Chem.*, **42**, 97–120 (1973); R. Bonaccorsi, A. Pullman, E. Scrocco and J. Tomasi, *Theoretic. Chim. Acta.*, **24**, 52 (1972).

[103] B. Mely and A. Pullman, *Theoret. Chim. Acta*, **13**, 278 (1969).

[104] J. J. Christianson, J. T. Rytting, and R. M. Izatt, *Biochemistry*, **9**, 4907 (1970); B. C. Pal, *Biochemistry*, **1**, 558 (1962).

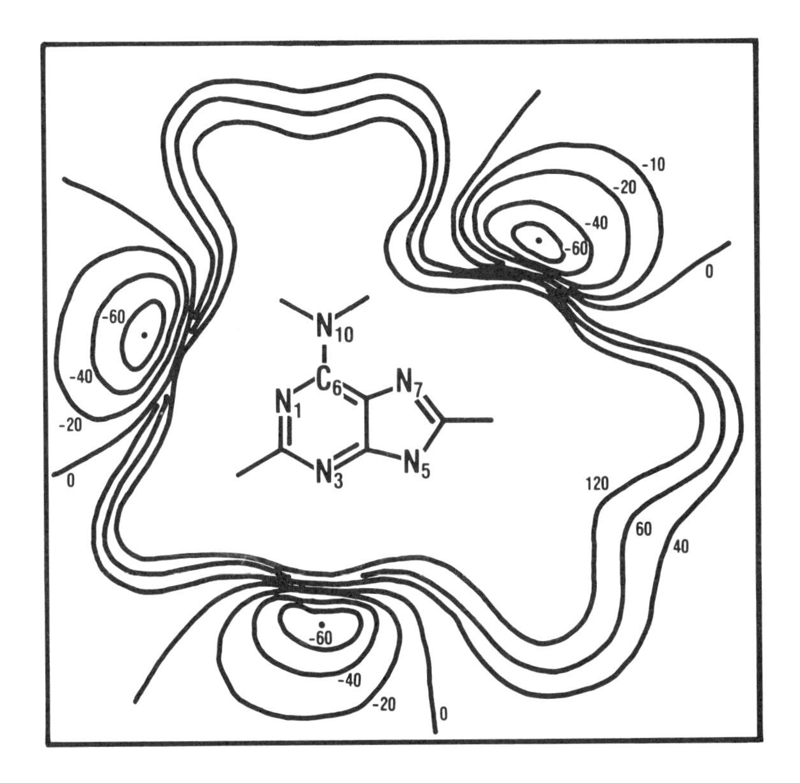

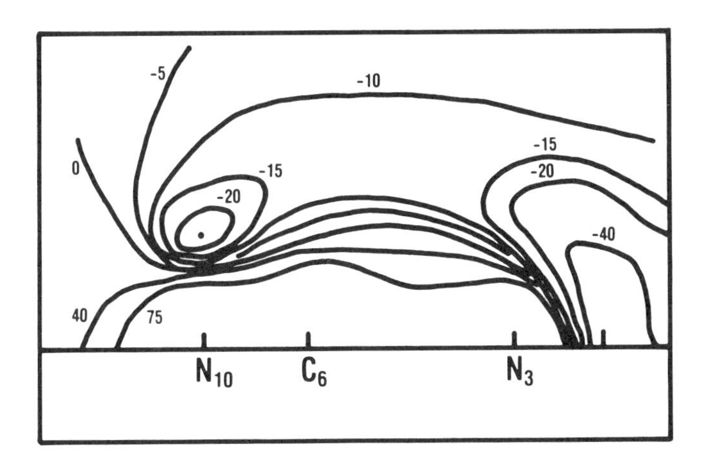

Figure 11.9. Molecular electrostatic maps for adenine in the molecular plane and in a plane perpendicular to the molecular plane. [From E. Scrocco and J. Tomasi (1973)].

have been found to be useful. Also, we shall see how these point charges can be used to create convenient representations of MEPs and other properties.

To illustrate the approaches, let us assume that for an SCF calculation, the basis set consisted of N FSGOs.[105] In general, the electron density **(r)** at a point **r** resulting from closed shell SCF calculation in which there are M doubly occupied MOs can be expressed as

$$\rho(\mathbf{r}) = 2 \sum_{t}^{M} \varphi_t^*(\mathbf{r}) \varphi_t(\mathbf{r}). \tag{11-119}$$

If each MO is a linear combination of N FSGOs (G_i), i.e.,

$$\varphi_t = \sum_{i=1}^{N} d_{ti} G_i, \tag{11-120}$$

where the d_{ti} are determined via the SCF study, and normalized FSGO are defined in general as

$$G_i(\mathbf{r}_A) = \left(\frac{2\alpha_i}{\pi}\right)^{3/4} \exp\{-\alpha_i(\mathbf{r} - \mathbf{R}_A)^2\}. \tag{11-121}$$

Using Eq. (11-120), we can rewrite Eq. (11-119) as

$$\rho(\mathbf{r}) = \sum_{i}^{N} \sum_{j}^{N} c_{ij} G_i(\mathbf{r}) G_j(\mathbf{r}), \tag{11-122}$$

where

$$c_{ij} = 2 \sum_{t=1}^{M} d_{ti}^* d_{tj}.$$

The key to finding procedures less arbitrary than the Mulliken population analysis lies in the use of properties of the FSGO in Eq. (11-122) to rewrite the equation in a different form. In particular, it is easily shown[106] that the product of two FSGOs centered at points A and B is a third FSGO, i.e.,

$$e^{-\alpha_i(\mathbf{r}-\mathbf{R}_A)^2} e^{-\alpha_j(\mathbf{r}-\mathbf{R}_B)^2} = e^{-(\alpha_i\alpha_j/(\alpha_i+\alpha_j))R_{AB}^2} \cdot e^{-\alpha_k(\mathbf{r}-\mathbf{R}_c)^2}, \tag{11-123}$$

where the first term on the right-hand side of Eq. (11-123) is a constant, and

$$\alpha_k = \alpha_i + \alpha_j, \tag{11-124}$$

$$R_{AB}^2 = (X_A - X_B)^2 + (Y_A - Y_B)^2 + (Z_A - Z_B)^2, \tag{11-125}$$

$$\mathbf{R}_c = (X_c, Y_c, Z_c), \tag{11-126}$$

---

[105] The ideas presented here apply only to FSGOs. To see how higher angular momentum basis orbitals can be utilized in this approach see, for example, D. Martin and G. G. Hall, *Theoret. Chim. Acta*, **59**, 281 (1981), and G. G. Hall, *Adv. Atom. Mol. Phys.*, **20**, 41 (1985).

[106] See Problem 11-9.

where

$$X_c = \frac{\alpha_i X_A + \alpha_j X_B}{\alpha_i + \alpha_j}, \tag{11-127}$$

with analogous expressions for Y_c and Z_c.

Insertion of Eq. (11-123) into Eq. (11-122) gives:[107]

$$\rho(\mathbf{r}) = \sum_i^N \sum_j^N c_{ij} N_i N_j \exp\left[-\left(\frac{\alpha_i \alpha_j}{\alpha_i + \alpha_j}\right) R_{AB}^2\right] e^{-(\alpha_k) r_c^2}$$

$$= \sum_i^N \sum_j^N c_{ij} S_{ij} G_{ij}, \tag{11-128}$$

where we have used

$$S_{ij} = \int G_i^* G_j \, d\tau = N_i N_j \int e^{-\alpha_i r_A^2} e^{-\alpha_j r_B^2} \, d\tau$$

$$= N_i N_j \left(\frac{\pi}{\alpha_i + \alpha_j}\right)^{3/2} \exp\left[-\left(\frac{\alpha_i \alpha_j}{\alpha_i + \alpha_j}\right) R_{AB}^2\right], \tag{11-129}$$

plus

$$N_i = \left(\frac{2\alpha_i}{\pi}\right)^{3/4}, \tag{11-130}$$

and

$$G_{ij} = \left(\frac{\alpha_i + \alpha_j}{\pi}\right)^{3/4} \exp[-\alpha_k r_c^2]. \tag{11-131}$$

The representation of electron density using Eq. (11-128) allows an interesting new type of charge distribution analysis to be carried out. In particular, we note first that the total number of electrons is given by

$$N = \int \rho(r) \, d\tau = \sum_i^N \sum_j^N c_{ij} S_{ij}. \tag{11-132}$$

The similarity of the above equation to Eq. (11-128) is evident, and suggests that an approximation to the electron density that may be useful is

$$\rho(\mathbf{r}) \cong \sum_i^N \sum_j^N c_{ij} S_{ij} \delta(\mathbf{r} - \mathbf{R}_{ij}), \tag{11-133}$$

where δ is a Dirac delta function. In other words, by "shrinking" G_{ij} in Eq. (11-128) to a delta function we have created a representation of the total charge

---

[107] The analysis here follows the work of G. G. Hall, *Chem. Phys. Lett.*, **20**, 501, (1973).

density that is made up of $N(N + 1)/2$ point charges located at $\mathbf{R}_{ij}$ and whose charge is given by

$$q_{ij}=\begin{cases}(c_{ij}+c_{ji})S_{ij} & i\neq j\\ c_{ii} & i=j.\end{cases} \tag{11-134}$$

Thus, the arbitrariness of the Mulliken analysis is entirely avoided, but still results in a point charge description of the molecule. Of course, there are now $N(N + 1)/2$ point charges, which is substantially greater than the number of nuclei in general, which complicates both the computations and the interpretation using a Mulliken-like analysis.

An alternative to the Hall model that obviates both of the above difficulties (at the expense of accuracy, as we shall see) was given by Shipman,[108] who began by considering the expression for the electric dipole moment[109] in an FSGO basis, i.e.,

$$\boldsymbol{\mu}_{\text{el}}=\int \mathbf{r}\rho(\mathbf{r})\,d\tau \tag{11-135}$$

$$=\sum_{i}^{N}\sum_{j}^{N} c_{ij}S_{ij}\mathbf{R}_{ij}, \tag{11-136}$$

where

$$\mathbf{R}_{ij}=\frac{\alpha_i\mathbf{R}_i+\alpha_j\mathbf{R}_j}{\alpha_i+\alpha_j}. \tag{11-137}$$

Manipulating Eq. (11-136) gives

$$\boldsymbol{\mu}_{el}=\sum_{i}^{N}\sum_{j}^{N}\left(\frac{c_{ij}S_{ij}\alpha_i\mathbf{R}_i}{\alpha_i+\alpha_j}\right)+\sum_{i}^{N}\sum_{j}^{N}\left(\frac{c_{ij}S_{ij}\alpha_j\mathbf{R}_j}{\alpha_i+\alpha_j}\right)$$

$$=\sum_{i}^{N}\mathbf{R}_i\left[\sum_{j}^{N}\frac{2c_{ij}S_{ij}\alpha_i}{\alpha_i+\alpha_j}\right]$$

$$=\sum_{i}^{N}\mathbf{R}_i q_i, \tag{11-138}$$

where we have defined

$$q_i=\sum_{j=1}^{N}\left(\frac{2c_{ij}S_{ij}\alpha_i}{\alpha_i+\alpha_j}\right). \tag{11-139}$$

---

[108] L. L. Shipman, *Chem. Phys. Lett.*, **31**, 361 (1975).

[109] The full dipole moment also includes a nuclear component, i.e., $\boldsymbol{\mu} = \boldsymbol{\mu}_N - \boldsymbol{\mu}_{\text{el}}$.

Table 11.15. MEP Values for Porphyrin.

MEP	r^a	Hall	Shipman	Wavefunction
Above ring N	4	−0.029146	−0.019769	−0.029144
	8	−0.014080	−0.006129	−0.014080
In plane (along axis containing two Ns)	4	+0.015523	−0.010874	+0.015543
	8	+0.007911	+0.003506	+0.007911

[a] For points above the ring, r refers to the perpendicular distance (in a.u.) above the ring N at which the MEP was calculated. For points in the porphyrin plane, r refers to the distance from the outer edge of the pyrrole ring containing the axis to the point of interest.

Thus, this analysis has produced a point charge description of the dipole moment that contains only N point charges, defined by Eq. (11-139).

To illustrate the ability of these point charge models to describe electronic structural features, it is important to examine a property other than total charge or dipole moment, since the models were designed to reproduce those properties exactly. One such property of importance is the molecular electrostatic potential (MEP) discussed earlier. Using the point charge definitions of Hall [Eq. (11-133)] and Shipman [Eq.(11-139)], calculated MEP values were compared to the values calculated directly from the wavefunction at several points in and above a plane containing the porphyrin molecule, and the results[110] are illustrated in Table 11.15. From these data we see that, both qualitatively and quantitatively, the Hall model provides an excellent representation of the MEP for porphyrin, relative to the quantum mechanical density of the FSGO basis. On the other hand, the simpler model of Shipman appears unsuitable for calculation of the MEP, with even sign reversals occurring. However, the use of the Shipman model as a qualitative guide in interpreting wavefunctions is an important contribution, since it provides a very informative point charge model using a small number of points and avoids the need for arbitrary division of charge among atoms as in the Mulliken model. Other approaches that avoid the difficulties of the Mulliken model have also been proposed,[111] and the subject remains of active research interest.

2. Localized Orbitals

Another technique that has been found to be useful in the interpretation of wavefunctions is based on the concept of *localized orbitals*. In particular, the various computational techniques for obtaining approximate wavefunctions by optimizing the energy do not necessarily result in wavefunctions that are easy to

[110] These calculations were carried out by J. D. Petke, *Chem. Phys. Lett.*, **126**, 26 (1986).

[111] See, for example, C. M. Smith and G. G. Hall, *Int. J. Quantum Chem.*, **31**, 685 (1987).

interpret. Yet, much of contemporary chemical structure and its interpretation is based on a localized bond, ball and stick-type visualization of chemical bonds. This has led to development of a number of approaches that are designed to take a given set of orbitals and rearrange them mathematically to localize them, in the hope that their interpretation will be easier and more in line with traditional chemical thinking. In this section we shall outline two such approaches,[112] and provide illustrations of their use in practical situations. In general, the approaches to be considered are based on the use of a unitary transformation of the molecular orbital basis, which is a transformation known to leave a closed shell Slater determinant invariant.[113] Hence, the processes for producing localized orbitals will not in general affect the energetic optimization of most approximate wavefunction of interest, but will improve its interpretability.

Two of the approaches have seen reasonably widespread use. The first of these, suggested by Foster and Boys, produces what were referred to initially as *oscillator orbitals* and in a later form as *exclusive orbitals*. The criterion used to construct these orbitals is that the sum of the quadratic repulsions between the orbitals should be minimized. This can be represented mathematically as the minimization of

$$I=\sum_{a}^{N} \langle \varphi_a\varphi_a | r_{12}^2 | \varphi_a\varphi_a \rangle \tag{11-140}$$

for a set of N occupied orthonormal spatial orbitals $\{\varphi_a\}$ in a molecule. This definition is equivalent to the maximization of the sum of squares of the distances of the orbitals from each other, as we shall now see. Since the centroid $(\mathbf{R}_a)$ of an orbital φ_a is given by

$$\mathbf{R}_a=\langle \varphi_a | \mathbf{r}_1 | \varphi_a \rangle, \tag{11-141}$$

where

$$\mathbf{r}_{1a}=\mathbf{r}_1-\mathbf{R}_a, \tag{11-142}$$

---

[112] A number of approaches to obtaining localized orbitals according to various criteria have been proposed, including F. Hund, *Z. Phys.*, **73**, 1 (1931); *Z. Phys.*, 565 (1932); C. A. Coulson, *Trans. Faraday Soc.*, **38**, 433 (1942); J. E. Lennard-Jones, *Proc. R. Soc. (London)*, **A198**, 1, 14 (1949); G. G. Hall and J. E. Lennard-Jones, *Proc. R. Soc. (London)*, **A202**, 155 (1950); *Proc. R. Soc. (London)*, **A205**, 357 (1951); J. E. Lennard-Jones and J. A. Pople, *Proc. R. Soc. (London)*, **A202**, 166 (1950); *Proc. R. Soc. (London)*, **A210**, 190 (1951); J. M. Foster and S. F. Boys, *Rev. Mod. Phys.*, **32**, 300 (1960); C. Edmiston and K. Ruedenberg, *Rev. Mod. Phys.*, **32**, 457 (1963); *J. Chem. Phys.*, **43**, 597 (1965); in *Quantum Theory of Atoms, Molecules and the Solid State*, P. O. Löwdin (ed.), Academic Press, New York, 1966, p. 263; S. F. Boys, in *Quantum Theory of Atoms, Molecules and the Solid State*, P. O. Löwdin (ed.), Academic Press, New York, 1966, p. 253.

[113] See Chapter 10 and the discussion involving Eqs. (10-280)–(10-284).

we can rewrite the first criterion [Eq. (11-140)] as

$$I=\sum_a^N \langle\varphi_a\varphi_a|(\mathbf{r}_{1a}-\mathbf{r}_{2a})^2|\varphi_a\varphi_a\rangle$$

$$=\sum_a^N \langle\varphi_a\varphi_a|(r_{1a}^2+r_{2a}^2-2\mathbf{r}_{1a}\cdot\mathbf{r}_{2a})|\varphi_a\varphi_a\rangle$$

$$=2\sum_a^N \langle\varphi_a|r_{1a}^2|\varphi_a\rangle\langle\varphi_a|\varphi_a\rangle$$

$$-2\sum_a^N \langle\varphi_a|\mathbf{r}_1-\mathbf{R}_a|\varphi_a\rangle\langle\varphi_a|\mathbf{r}_1-\mathbf{R}_a|\varphi_a\rangle \tag{1-143}$$

$$=2\sum_a^N \langle\varphi_a|r_{1a}^2|\varphi_a\rangle, \tag{1-144}$$

where Eq. (11-141) has been used to eliminate the second term in Eq. (11-143). This last expression is just the sum of the spherical quadratic moments of each exclusive orbital about its centroid. In other words, maximization of the squares of the distances of the orbitals from each other is equivalent to maximizing the distance between the centroids of each orbital. Thus, this process is equivalent to making each exclusive orbital contract as much as possible around its own centroid, consistent with maintenance of the original orthonormality condition. Since the terms in Eq. (11-144) are one-electron integrals, their evaluation is straightforward in general, and application of this procedure requires only the evaluation of $\sim N^2$ matrix elements.

The other procedure in widespread use was developed by Edmiston and Ruedenberg, and is one in which the overall interorbital electron repulsion interactions are minimized. The particular manner in which this is formulated is to maximize the sum of individual orbital "self-energies," i.e., to maximize

$$D=\sum_a^N \left\langle\varphi_a\varphi_a\left|\frac{1}{r_{12}}\right|\varphi_a\varphi_a\right\rangle. \tag{11-145}$$

This process is computationally more complex and time-consuming than the Boys procedure, since the evaluation of approximately N^4 two-electron integrals is required. On the other hand, it is applicable to all types of orbitals while, for example, the Boys procedure is not easily applied to two s-type orbitals on the same atom or two orbitals of like symmetry in a homonuclear diatomic molecule.

In any case, each of the two methods has been found to be quite useful in improving the interpretability of approximate wavefunctions. To illustrate this,

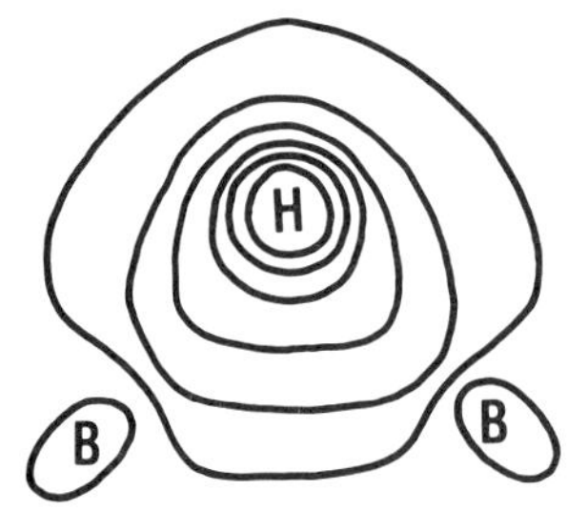

Figure 11.10. Electron density contours for the localized molecular orbital of the three-center bond in B_2H_6. Density is given in units of *e*/Bohr^3, and the values (from B to H) are 0.005, 0.005, 0.025, 0.045, 0.065, 0.085, and 0.105.

let us consider a case in which traditional discussions of bonding have had difficulty, i.e., in "electron-deficient" molecules such as boranes. For these cases, Lipscomb and colleagues[114] have used the Edmiston–Ruedenberg approach to obtain significant insight into the bonding in boron hydrides. For the simplest of this class of molecules, diborane (B_2H_6), it was possible to show that, instead of resembling the bonding in ethane, no direct B–B bonding occurred, and rationalization of the bridged diborane structure through the use of multicenter bonds was possible. In particular, each boron atom was found to have a localized inner shell orbital plus two "normal" B–H bonds (similar to C–H bonds). However, the remaining two orbitals were found to be three-center bonds, as suggested by Longuet-Higgins.[115] An example of these three-center bonds is given in Fig. 11.10, from the work of Switkes *et al.* Thus, both traditional and "nontraditional" chemical concepts can be included in the depiction of wavefunctions using techniques such as these.

F. Deficiencies of the Theory

Although some of the limitations inherent in closed shell Hartree–Fock theory have been pointed out in previous discussions by means of specific examples, it is useful to summarize and analyze these limitations further, so that directions appropriate for improvement of the theory are more evident. The limitations to be discussed here are those inherent in the theory itself, and are separate from practical limitations, such as basis set size and flexibility that have already been discussed.

Most of the limitations of Hartree–Fock theory are discussed within the concept of the correlation energy, which was defined earlier[116] as

$$\Delta E_{\text{correl}} = E_{\text{exact}} - E_{\text{HF}}, \tag{11-146}$$

where E_{HF} is the Hartree–Fock energy calculated using a complete basis, and E_{exact} is the exact eigenvalue of the nonrelativistic Hamiltonian of the system.

---

[114] See, for example, E. Switkes, R. M. Stevens, and W. N. Lipscomb, *J. Chem. Phys.*, **51**, 2085 (1969), and Fig. 11.10.

[115] H. C. Longuet-Higgins, *J. Chim. Phys.*, **46**, 175 (1949); *J. R. Inst. Chem.*, **77**, 197 (1953).

[116] See Section 11-1A.

Experience has shown that the magnitude of the correlation error as calculated in Eq. (11-146) is typically quite small when compared to the total energy. For example, the near Hartree–Fock calculations reported in Table 11.9 indicate correlation errors that, except for H_2, are in the range of 0.4–0.7%. However, we have also seen[117] that chemical properties of interest (e.g., dissociation energies) frequently are of similar magnitude to the correlation error, leading to the inability of Hartree–Fock theory to describe a number of properties of interest.

One of the principal sources of this correlation error is associated with the independent particle model that forms the basis of Hartree–Fock theory. In particular, the wavefunction form that has been chosen in Hartree–Fock theory is an antisymmetrized product of one-electron orbitals. Thus, the wavefunction does not explicitly consider the detailed motion of electrons with each other. Instead, the orbital describing the motion of a given electron is determined in an *averaged* field of the other electrons. In particular, the Hartree–Fock Hamiltonian includes the Coulomb and exchange operators [Eqs. [11-15) and [11-16)] for an electron that *integrate* over all spatial and spin coordinates of the electrons in other orbitals to produce an *averaged* potential for determination of the Hartree–Fock orbitals. Thus, the optimized filled orbitals for an N-electron system are determined by a Fock operator that contains a kinetic energy operator, operators representing the attraction of each electron to each of the nuclei, plus the *averaged* Coulomb interactions of each electron with $(N - 1)$ other electrons, and an *averaged* exchange interaction between electrons *only having the same spin.*[118] Therefore, in addition to having an averaged interelectronic interaction, we also see that the *motion of electrons having opposite spins is uncorrelated.*

These conceptual limitations of closed shell Hartree–Fock theory having to do with the lack of a detailed description of correlated electron–electron motions (usually called "dynamic" correlation errors[119]) means that properties sensitive to these motions and changes in them will not be predicted well within closed shell Hartree–Fock theory regardless of basis set choice. As we shall see below, this includes the calculation of reaction surfaces, spectral transition energies, and even equilibrium geometries in some cases.[120]

As an illustration of the problems that can arise in the calculation of reaction surfaces, let us reconsider the H_2 example that was discussed earlier in Section 11-2B. In particular, if we utilize the 3-FSGO basis set again and perform Hartree–Fock calculations at several values of R, we obtain the potential curve

[117] See Section 11-2D.

[118] This Coulomb and exchange interaction does guarantee, however, that the probability of finding two electrons having the *same* spin at the same point in space is zero as desired conceptually, and is known as the *Fermi Hole*.

[119] See, for example, O. Sinanouglu, *Proc. Natl. Sci. Acad. U.S.A.* **47**, 1217 (1961).

[120] See for example, D. J. DeFrees, K. Raghavachari, H. B. Schlegel, and J. A. Pople, *J. Am. Chem. Soc.*, **104**, 5576–5580 (1982).

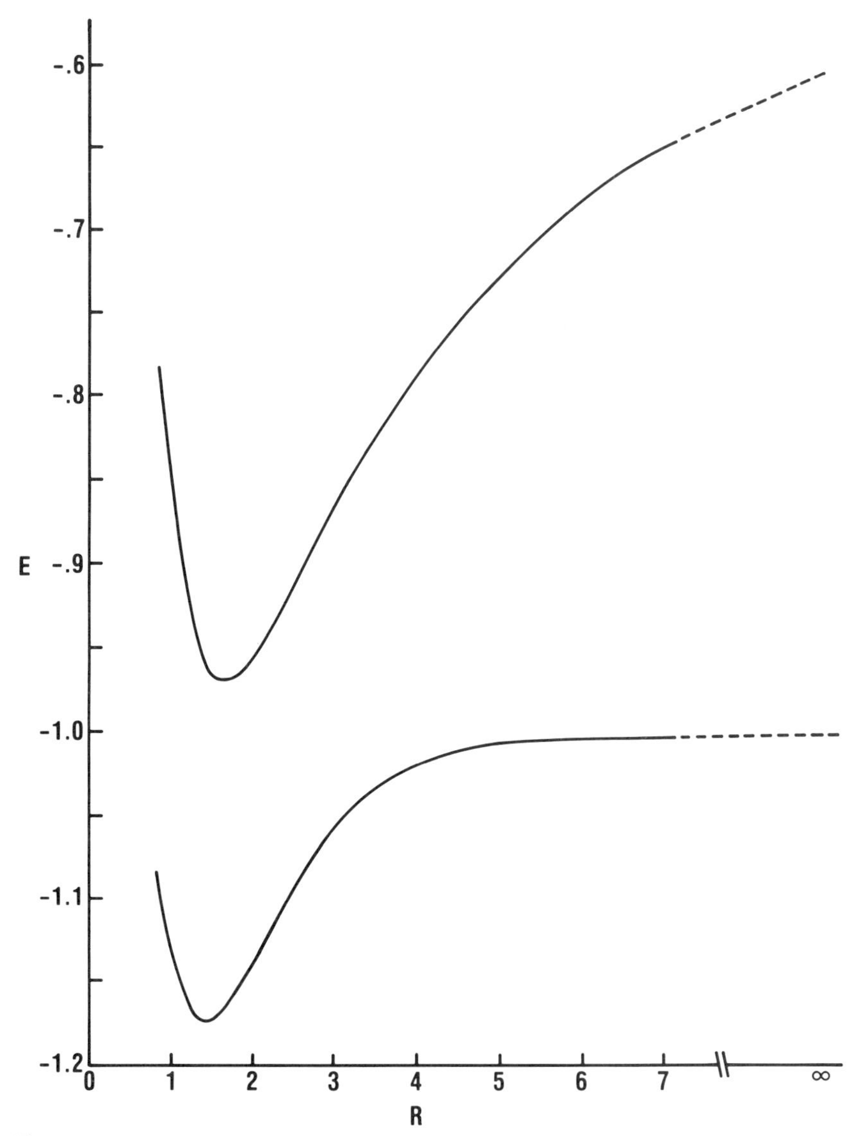

Figure 11.11. H_2 molecule potential curves calcuated with an accurate wavefunction (lower curve) compared to a small basis set Hartree–Fock wavefunction (upper curve).

shown in Fig. 11.11. Also shown in this figure is the very accurate potential curve that was calculated by Kolos and Wolniewicz[121] using a very flexible wavefunction that was not constrained by Hartee–Fock theory.

Several points of interest are illustrated in this example. First, we see that the

[121] W. Kolos and L. Wolniewicz, *J. Chem. Phys.*, **49**, 404–410 (1968). This wavefunction will be discussed in greater detail in Section 12-2B.

predicted equilibrium distance is shortened from $R = 1.6$ Bohrs to $R = 1.4$ Bohrs in going from the use of the approximate Hartee-Fock wavefunction to the accurate wavefunction. This lengthened bond distance prediction from Hartree-Fock wavefunctions is observed frequently and with a variety of basis sets, and is generally attributed to the lack of appropriate electron correlation description in Hartree-Fock theory. More important, however, is the difference in shape of the curves as R increases. Instead of producing potential curves that are parallel, it is seen that the limited FSGO basis set potential curve continues to rise as R increases, instead of flattening out as the accurate calculation indicates. This difference in shape is not merely a reflection of basis set inadequacies, although that is a factor.

The main reason for the difference in shape, however, is due to the inadequacy of the closed shell Hartree-Fock wavefunction to describe the process of bond dissociation even in principle. To see this, we note first that the correct description of two hydrogen atoms that are infinitely separated requires a wavefunction of the form

$$\Psi(1, 2) = N[\chi_A(1)\chi_B(2) + \chi_B(1)\chi_A(2)][\alpha(1)\beta(2) - \beta(1)\alpha(2)], \quad (11\text{-}147)$$

where χ_A and χ_B are the exact H-atom wavefunctions located on nuclei A and B, respectively. The symmetric combination of the χ_A, χ_B product merely reflects the requirement that the electrons must be indistinguishable, and the spin function is that needed for a singlet state.

On the other hand, the MO wavefunction found at the equilibrium distance is a doubly occupied MO, i.e.,

$$\begin{aligned}\Psi_{MO}(1, 2) &= N'\varphi(1)\varphi(2)[\alpha(1)\beta(2) - \beta(1)\alpha(2)] \\ &= N'(\chi_A + \chi_B)(1)(\chi_A + \chi_B)(2)[\alpha(1)\beta(2) - \beta(1)\alpha(2)], \quad (11\text{-}148)\end{aligned}$$

where χ_A and χ_B in this case represent atomic orbitals[122] on nuclei A and B, respectively. Expanding Eq. (11-148) gives

$$\begin{aligned}\Psi_{MO}(1, 2) = N'[&\chi_A(1)\chi_B(2) + \chi_B(1)\chi_A(2) + \chi_A(1)\chi_A(2) + \chi_B(1)\chi_B(2)] \\ &\cdot [\alpha(1)\beta(2) - \beta(1)\alpha(2)]. \quad (11\text{-}149)\end{aligned}$$

We now see that the functional form required by the MO approach includes a pair of "ionic" terms [$\chi_A(1)\chi_A(2)$ and $\chi_B(1)\chi_B(2)$], corresponding to both electrons on a single nucleus], which are present at all distances with *equal weighting* as the terms [$\chi_A(1)\chi_B(2) + \chi_B(1)\chi_A(2)$]. While such "ionic" terms are indeed suitable for distances close to the equilibrium distance, they are clearly inappropriate when the two hydrogen atoms are infinitely apart. Thus, the closed shell Hartree-Fock wavefunction cannot, even in principle, be expected to give an appropriate description of the bond dissociation process in cases such as the one considered above.

[122] These orbitals in general could be linear combinations of Gaussians, or single Gaussians as in the example. The third Gaussian located in the mid-point of the "bond" in the example will have a zero coefficient when $R = \infty$.

Another type of difficulty in the application of Hartree–Fock theory in general (whether for closed shell or open shell systems) is computational, and is associated with situations where descriptions of more than one state of the same symmetry are desired. For example, if an excited state having the same spatial symmetry and spin state as the ground state exists, determination of the Hartree–Fock description of the excited state will be difficult in practice, since straightforward minimization of the energy via the techniques described in this chapter will converge only to the lowest state (i.e., the ground state in this example). For such cases, special techniques such as maintaining strict orthogonality between the states are needed in order to carry out calculations successfully.

Other examples that illustrate the inability of closed shell Hartree–Fock theory to describe some aspects of chemical structure and behavior can be cited, and which require inclusion of correlation effects in the wavefunction if they are to be described properly. These include electron affinities, electronic spectra, reaction surfaces, and a number of other properties. These properties, and the kinds of wavefunctions that are needed to describe them, will be discussed in the sections to follow and in the next chapter.

11-3. Hartree–Fock Theory for Open Shell Systems

Thus far we have focused our attention on Hartree–Fock studies of closed shell systems, i.e., where all orbitals are doubly occupied. However, there are numerous examples of *open shell systems* of importance, which are systems *in which one or more orbital is singly occupied*. Such systems include radical anions, doublets, triplets, along with a number of important chemical situations involving these species. It is these systems to which we shall now focus our attention, with the intent of obtaining a Hartree–Fock description comparable to that obtained in the previous section for closed shell systems. As we shall discover below, however, there are many approaches that can be used for open shell systems.

A. Unrestricted Hartree–Fock Theory[123]

The approach described below was developed by Slater[124] plus Pople and Nesbet,[125] and can be considered as a generalization of the closed shell formulation of Hall and Roothaan that was described in the previous section.

---

[123] The approach described here is not the only approach that is applicable to open shell systems. For a review of various approaches to open shell SCF methods, see G. Berthier, in *Molecular Orbitals in Chemistry, Physics and Biology*, P. O. Löwdin and B. Pullman (eds.), Academic Press, New York, 1964, pp. 57–82. Also, see Section 11-3C.

[124] J. C. Slater, *Phys. Rev.*, **35**, 210 (1930).

[125] J. A. Pople and R. K. Nesbet, *J. Chem. Phys.*, **22**, 571 (1954). See also G. Berthier, *Compt. Rend. Acad. Sci.*, **238**, 91 (1954); *J. Chim. Phys.*, **51**, 363 (1954).

While it will be seen to have some drawbacks, it retains Koopmans' Theorem[126] and is therefore of interest to us here.

Specifically, instead of utilizing a single Slater determinant of doubly occupied orbitals [Eqs. (11-45) and (11-46)] as the approximate wavefunction, two important modifications are introduced in *unrestricted Hartree-Fock* (UHF) theory. First, the number of orbitals with α-spin is *not* restricted to be equal to the number of orbitals with β-spin. Also, the form of spatial orbitals describing electrons with α-spin is *not* constrained to be the same as for β-electrons. In other words, each orbital is now independent of all others. Incorporating these generalizations gives rise to a single Slater determinant of the form

$$\begin{aligned}\Psi(1, 2, \cdots, N) = \mathcal{A}\{&\varphi_1^\alpha(\mathbf{r}_1)\alpha(\sigma_1)\varphi_1^\beta(\mathbf{r}_2)\beta(\sigma_2)\varphi_2^\alpha(\mathbf{r}_3)\alpha(\sigma_3)\varphi_2^\beta(\mathbf{r}_4)\beta(\sigma_4)\\ &\cdots \varphi_q^\alpha(\mathbf{r}_{2q-1})\alpha(\sigma_{2q-1})\varphi_q^\beta(\mathbf{r}_{2q})\beta(\sigma_{2q})\varphi_{q+1}^\alpha(\mathbf{r}_{2q+1})\alpha(\sigma_{2q+1})\\ &\cdots \varphi_p^\alpha(\mathbf{r}_{q+p})\alpha(\sigma_{q+p})\}, \qquad (11\text{-}150)\end{aligned}$$

$$\begin{aligned}= \mathcal{A}\{&\chi_1^\alpha(1)\chi_1^\beta(2) \cdots \chi_q^\alpha(2q-1)\chi_q^\beta(2q)\chi_{q+1}^\alpha(2q+1)\\ &\cdots \chi_p^\alpha(q+p)\}, \qquad (11\text{-}151)\end{aligned}$$

where the number of electrons (N) is

$$N = N_\alpha + N_\beta$$

and

$$N_\alpha = p$$
$$N_\beta = q.$$

The wavefunction in Eq. (11-150) or Eq. (11-151) is known as the *unrestricted Hartree–Fock (UHF) wavefunction*. Also, each spatial orbital in Eq. (11-150) is independent of the others. Thus, we have in general that

$$\varphi_i^\alpha \neq \varphi_i^\beta. \qquad (11\text{-}152)$$

To illustrate the UHF wavefunction, a seven-electron molecular system having three open shells[127] could be represented as

$$\begin{aligned}\Psi(1, 2, \cdots, 7) = \mathcal{A}\{&\varphi_1^\alpha(\mathbf{r}_1)\alpha(\sigma_1)\varphi_1^\beta(\mathbf{r}_2)\beta(\sigma_2)\varphi_2^\alpha(\mathbf{r}_3)\alpha(\sigma_3)\varphi_2^\beta(\mathbf{r}_4)\beta(\sigma_4)\\ &\varphi_3^\alpha(\mathbf{r}_5)\alpha(\sigma_5)\varphi_4^*(\mathbf{r}_6)\alpha(\sigma_6)\varphi_5^\alpha(\mathbf{r}_7)\alpha(r_7)\}. \qquad (11\text{-}153)\end{aligned}$$

with

$$\begin{aligned}N &= 7\\ N_\alpha &= 5 = p \qquad (11\text{-}154)\\ N_\beta &= 2 = q,\end{aligned}$$

[126] Brillouin's Theorem is also retained in this approach, but will not be discussed here.

[127] This example is known as the "high spin" UHF case for three open shells.

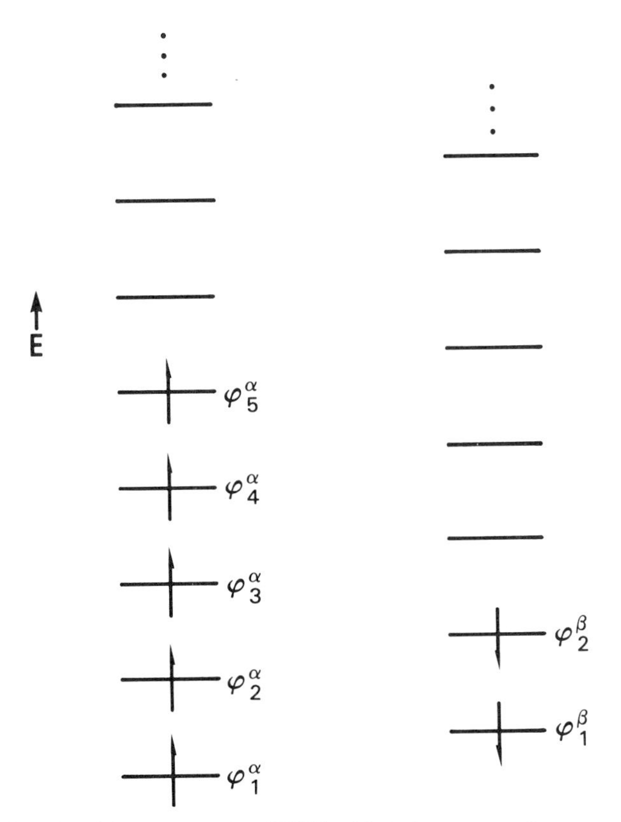

Figure 11.12. Pictorial representatin of UHF orbitals for a seven-electron system having three open shells.

and, in general

$$\begin{aligned} \varphi_1^\alpha &\neq \varphi_1^\beta \\ \varphi_2^\alpha &\neq \varphi_1^\beta. \end{aligned} \tag{11-155}$$

The relative energies of the orbitals of Eq. (11-152) are depicted pictorially in Fig. 11.12.

Analogous to the case of closed shell Hartree–Fock theory, we desire to optimize the UHF wavefunction of Eq. (11-151), subject to the constraint[128] that the resulting set of spin orbitals

$$\chi_1^\alpha \chi_2^\alpha \cdots \chi_p^\alpha \chi_1^\beta \chi_2^\beta \cdots \chi_q^\beta$$

are mutually orthonormal. Of course, members of the α-spin orbital set will be automatically orthogonal to those of the β-spin orbital set, so the orthonormality constraint needs to be included only for spin orbitals *within* the α (or β) spin orbital set.

[128] The invariance of the wavefunction to this constraint was shown in Chapter 10, Section 10-10.

If the energy expectation value of the UHF wavefunction of Eq. (11-155) is evaluated and optimized, it is found[129] that the closed shell Fock operator [Eq. (11-57)] is generalized to *two* Fock operators, i.e.,

$$f^{\alpha}(1)=h(1)+\sum_{a}^{N_\alpha}[\mathcal{J}_a^{\alpha}(1)-\mathcal{K}_a^{\alpha}(1)]+\sum_{b}^{N_\beta}\mathcal{J}_b^{\beta}(1), \qquad (11\text{-}156)$$

and

$$f^{\beta}(1)=h(1)+\sum_{b}^{N_\beta}[\mathcal{J}_b^{\beta}(1)-\mathcal{K}_b^{\beta}(1)]+\sum_{a}^{N_\alpha}\mathcal{J}_a^{\alpha}(1), \qquad (11\text{-}157)$$

where $h(1)$ is defined as before [Eq. (11-13)], and the UHF Coulomb and exchange operators are defined as:

$$\mathcal{J}_a^{\alpha}(1)\varphi_a^{\alpha}(\mathbf{r}_1)=\left[\int d\mathbf{r}_2\varphi_a^{\alpha *}(\mathbf{r}_2)\varphi_a^{\alpha}(\mathbf{r}_2)\frac{1}{r_{12}}\right]\varphi_a^{\alpha}(\mathbf{r}_1) \qquad (11\text{-}158)$$

and

$$\mathcal{K}_a^{\alpha}(1)\varphi_{a'}^{\alpha}(1)=\left[\int d\mathbf{r}_2\varphi_a^{\alpha *}(\mathbf{r}_2)\varphi_{a'}^{\alpha}(\mathbf{r}_2)\frac{1}{r_{12}}\right]\varphi_a^{\alpha}(\mathbf{r}_1) \qquad (11\text{-}159)$$

with analogous definitions for $\mathcal{J}_b^{\beta}$ and $\mathcal{K}_b^{\beta}$.

The Hartree–Fock equations that must be solved in order to obtain the optimized UHF orbitals now consist of two sets of equations, i.e.,:

$$f^{\alpha}(1)\varphi_{a'}(1)=\epsilon_{a'}^{\alpha}\varphi_{a'}^{\alpha}(1) \qquad a'=1,2,\cdots,p \qquad (11\text{-}160)$$

$$f^{\beta}(1)\varphi_{b'}^{\beta}(1)=\epsilon_{b'}^{\beta}\varphi_{b'}^{\beta}(1) \qquad b'=1,2,\cdots,q, \qquad (11\text{-}161)$$

where $\epsilon_{a'}^{\alpha}$ and $\epsilon_{b'}^{\beta}$ are the orbital energies of $\varphi_{a'}^{\alpha}$ and $\varphi_{b'}^{\beta}$, respectively.

Aside from the obvious complication that is introduced as a result of having to deal with two sets of integrodifferential equations instead of one, there is another important difference compared to restricted Hartree–Fock theory. In particular, the Fock operator $f^{\alpha}(1)$ not only contains contributions from all α spin orbitals, but also includes a coulomb contribution from each β-spin orbital. From a conceptual point of view, Eqs. (11-156) and (11-157) indicate that the Hartree–Fock potential that determines a given orbital consists of the usual kinetic and nuclear attraction terms, plus a coulombic interaction with *all* electrons (α plus β) and a (nonclassical) exchange interaction with all orbitals of the same spin.

From a practical point of view, the additional Coulomb operator terms have complicated the solution to the Hartree–Fock equations. In particular, the equations for α-spin orbitals are now *coupled* with the equations for the β-spin orbitals. Thus, the iterative solution that is required to determine, e.g., the α-spin orbitals, must also be coupled with the iterative solution for the β-spin orbitals, until all equations are self-consistent.

---

[129] See Problem 11-10.

Using the definitions of coulomb and exchange operators just introduced, the total electronic energy associated with the UHF wavefunction of Eq. (11-150) can be written[130] as

$$E_{\text{el}}=\sum_{a}^{N_\alpha}\langle\varphi_a^\alpha|h|\varphi_a^\alpha\rangle+\sum_{b}^{N_\beta}\langle\varphi_b^\beta|h|\varphi_b^\beta\rangle$$

$$+\frac{1}{2}\sum_{a}^{N_\alpha}\sum_{a'}^{N_\alpha}[\langle\varphi_a^\alpha\varphi_{a'}^\alpha|\varphi_a^\alpha\varphi_{a'}^\alpha\rangle-\langle\varphi_a^\alpha\varphi_{a'}^\alpha|\varphi_{a'}^\alpha\varphi_a^\alpha\rangle]$$

$$+\frac{1}{2}\sum_{b}^{N_\beta}\sum_{b}^{N_\beta}[\langle\varphi_b^\beta\varphi_{b'}^\beta|\varphi_b^\beta\varphi_{b'}^\beta\rangle-\langle\varphi_b^\beta\varphi_{b'}^\beta|\varphi_{b'}^\beta\varphi_b^\beta\rangle]$$

$$+\sum_{a}^{N_\alpha}\sum_{b}^{N_\beta}\langle\varphi_a^\alpha\varphi_b^\beta|\varphi_a^\alpha\varphi_b^\beta\rangle, \qquad (11\text{-}162)$$

where the electron repulsion integral notation of Eq. (11-69) hs been used.

In order to determine the UHF orbitals via the solution fo Eqs. (11-160) and (11-161), Pople and Nesbet proposed expansion in a basis, analogous to the Hall–Roothaan procedure used in closed shell Hartree–Fock theory. In this case, we have

$$\varphi_a^\alpha=\sum_{t}^{T}c_{ta}^\alpha\phi_t \qquad a=1,2,\cdots,T \qquad (11\text{-}163)$$

$$\varphi_b^\beta=\sum_{t}^{T}c_{tb}^\beta\phi_t \qquad b=1,2,\cdots,T. \qquad (11\text{-}164)$$

Note that the same basis $\{\phi_t\}$ is used for both sets of MOs, and that the solution of the UHF equations will result in T α-orbitals and T β-orbitals. Of that total set of MOs, p of the α-orbitals will be occupied and q of the β-orbitals will be occupied, with the balance [$(T-p)$ α orbitals and $(T-p)$ β orbitals] of the MOs unoccupied.

If the expansions in Eqs. (11-163) and (11-164) are substituted into the UHF equations [Eqs. (11-160) and (11-161)], we obtain

$$\sum_{t}^{T}c_{ta'}^\alpha f^\alpha(\mathbf{r}_1)\phi_t(\mathbf{r}_1)=\epsilon_{a'}^\alpha\sum_{t}^{T}c_{ta}^\alpha\phi_t(\mathbf{r}_1) \qquad a'=1,2,\cdots,T \qquad (11\text{-}165)$$

$$\sum_{t}^{T}c_{tb'}^\beta f^\beta(\mathbf{r}_1)\phi_t(\mathbf{r}_1)=\epsilon_{b'}^\beta\sum_{t}^{T}c_{tb'}^\beta\phi_t(\mathbf{r}_1) \qquad b'=1,2,\cdots,T \qquad (11\text{-}166)$$

---

[130] See Problem 11-11.

Multiplication of Eqs. (11-165) and (11-166) by $\phi_s^*(r_1)$ and integrating yields

$$\sum_t^T c_{ta'}^{\alpha} \int d\mathbf{r}_1 \phi_s^*(\mathbf{r}_1) f^{\alpha}(\mathbf{r}_1)\phi_t(\mathbf{r}_1)$$

$$= \epsilon_{a'}^{\alpha} \sum_t^T c_{ta'}^{\alpha} \int \phi_s^*(\mathbf{r}_1)\phi_t(\mathbf{r}_1)\, d\mathbf{r}_1, \qquad a' = 1, 2, \cdots, T \qquad (11\text{-}167)$$

$$\sum_t^T c_{tb'}^{\beta} \int d\mathbf{r}_1 \phi_s^*(\mathbf{r}_1) f^{\beta}(\mathbf{r}_1)\phi_t(\mathbf{r}_1)$$

$$= \epsilon_{b'}^{\beta} \sum_t^T c_{tb'}^{\beta} \int \phi_s^*(\mathbf{r}_1)\phi_t(\mathbf{r}_1)\, d\mathbf{r}_1, \qquad b' = 1, 2, \cdots, T \qquad (11\text{-}168)$$

or

$$\sum_t^T F_{st}^{\alpha} c_{ta'}^{\alpha} = \epsilon_{a'}^{\alpha} \sum_t^T S_{st} c_{ta'}^{\alpha}, \qquad a' = 1, 2, \cdots, T \qquad (11\text{-}169)$$

$$\sum_t^T F_{st}^{\beta} c_{tb'}^{\beta} = \epsilon_{b'}^{\beta} \sum_t^T S_{st} c_{tb'}^{\beta}, \qquad b' = 1, 2, \cdots, T \qquad (11\text{-}170)$$

in which we have defined:

$$F_{st}^{j} = \int d\mathbf{r}_1 \phi_s^*(\mathbf{r}_1) f^{j}(\mathbf{r}_1)\phi_t(\mathbf{r}_1), \qquad j = \alpha, \beta, \qquad (11\text{-}171)$$

and S_{st} is the overlap integral matrix element that was introduced earlier [Eq. (11-64)]. In matrix form, Eqs. (11-169) and (11-170) can be written as

$$\mathbf{F}^{\alpha}\mathbf{C}^{\alpha} = \mathbf{S}\mathbf{c}^{\alpha}\boldsymbol{\epsilon}^{\alpha}, \qquad (11\text{-}172)$$

$$\mathbf{F}^{\beta}\mathbf{C}^{\beta} = \mathbf{S}\mathbf{c}^{\beta}\boldsymbol{\epsilon}^{\beta}, \qquad (11\text{-}173)$$

where $\boldsymbol{\epsilon}^{\alpha}$ and $\boldsymbol{\epsilon}^{\beta}$ are diagonal matrices of orbital eigenvalues, and the colums of $\mathbf{C}^{\alpha}$ and $\mathbf{C}^{\beta}$ are the MO coefficients for α and β MOs, respectively.

As in the closed shell case, the solutions to Eqs. (11-172) and (11-173) are obtained iteratively, as we shall see shortly. In preparation, we need first to modify the definition of the charge and bond order matrix [Eq. (11-71)] to allow us to distinguish between α and β components. Specifically, we define the elements of the *charge and bond order matrices* ($\mathbf{P}^{\alpha}$ and $\mathbf{P}^{\beta}$) for α and β electrons as follows:

$$P_{uv}^{\alpha} = \sum_a^{N_{\alpha}} (c_{ua}^{\alpha})^*(c_{va}^{\alpha}), \qquad (11\text{-}174)$$

$$P_{uv}^{\beta} = \sum_b^{N_{\beta}} (c_{ub}^{\beta})^*(c_{vb}^{\beta}). \qquad (11\text{-}175)$$

From these definitions, it is seen that the *total charge and bond order matrix* (**P**) is given by

$$\mathbf{P} = \mathbf{P}^{\alpha} + \mathbf{P}^{\beta}. \tag{11-176}$$

Since the number of α electrons will not, in general, be the same as the number of β electrons, the opportunity to discern asymmetries in associated properties now becomes a possibility. In particular, the spin density created by α-spins may not be the same as that generated by β-spins, thus providing a mechanism for predicting spin distribution asymmetry for use in interpreting NMR spectroscopic results. It is therefore of interest to define a *spin density matrix* ($\mathbf{P}^s$) as

$$\mathbf{P}^s = \mathbf{P}^{\alpha} - \mathbf{P}^{\beta}. \tag{11-177}$$

Using the charge and bond order matrix definitions of Eqs. (11-174) and (11-175), we can obtain an explicit expression for the Fock matrix elements in Eq. (11-171) in terms of integrals over basis orbitals. In particular, using methods entirely analogous to those used in the closed shell case, it can be seen[131] that

$$F_{st}^{\alpha} = H_{st}^{\text{core}} + \sum_{u,v}^{T} P_{u,v}^{\alpha}[\langle \phi_s \phi_u | \phi_t \phi_v \rangle - \langle \phi_s \phi_u | \phi_v \phi_t \rangle] + \sum_{u,v}^{T} P_{u,v}^{\beta} \langle \phi_s \phi_u | \phi_t \phi_v \rangle, \tag{11-178}$$

with an analogous equation for F_{st}^{β}.

Solving the coupled UHF equations [Eqs. (11-172) and (11-173) is achieved by calculating the necessary integrals, making an initial guess of the charge and bond order matrices for α and β spin electrons ($\mathbf{P}^{\alpha}$ and $\mathbf{P}^{\beta}$), constructing the Fock matrices [Eq. (11-178) and its analog for F_{st}^{β}], and solving the UHF equations. The result is used to obtain a new approximation to $\mathbf{P}^{\alpha}$ and $\mathbf{P}^{\beta}$, and the process is continued (alternating the iterations between α and β orbital equations), until self-consistency is achieved. As expected, the coupling of the α and β equations adds not only to the computational complexity, but to the possibility that convergence difficulties may be encountered. Nevertheless, this approach has been found to provide techniques for open shell systems that are quite comparable to those for closed shell systems.

B. Restricted Hartree–Fock Theory

As noted in the previous sections, use of a single determinantal approach to open shell problems has the potentially significant disadvantage that, in general, the resulting UHF wavefunction is not an eigenfunction of $\mathcal{S}^2$. One obvious way in

[131] See Problem 11-12.

which this difficulty can be overcome is to construct a linear combination of Slater determinants that is an eigenfunction of $\mathcal{S}^2$ and $\mathcal{S}_z$, and optimize the orbitals subsequently. Such an approach is usually known as *restricted Hartree–Fock theory*. Of course, working with linear combinations of determinants complicates both the analysis and computations, as we shall see shortly. However, in this formulation of Hartree–Fock theory we see the importance of viewing Hartree–Fock theory as a *single configuration theory*, and not necessarily a single determinant theory.

There are a number of alternative approaches that have been suggested. In the analysis to follow, we shall describe one of the general approaches that has been given by several groups.[132] However, a variety of alternative approaches are also possible, and are recommended for further information.[133]

To begin with, it should be noted that, contrary to the closed shell case, where only a single energy expression arises, many possible different cases arise when open shell situations are to be described, depending on the number of open shells and their degeneracy. However, it is still possible to write an energy expression that is general enough to encompass nearly all of the cases of interest, i.e.,

$$E=\sum_{i=1}^{N} \alpha_i h_i+\sum_{i=1}^{N}\sum_{j=1}^{N} (a_{ij}J_{ij}-b_{ij}K_{ij}), \tag{11-179}$$

where N is the number of MOs that are occupied (either fully or partially) and α_i, a_{ij} and b_{ij} are real constants whose values are assigned based on the electronic state and wavefunction under consideration. In addition, h_i, J_{ij} and K_{ij} are one-electron, Coulomb and exchange integrals, respectively, over orthonormal spatial MOs (φ), defined as

$$h_i=\int \varphi_i^*(1)h(1)\varphi_i(1)\, dV_1, \tag{11-180}$$

$$J_{ij}=\int\int \varphi_i^*(1)\varphi_j^*(2)\,\frac{1}{r_{12}}\,\varphi_i(1)\varphi_j(2)\, dV_1\, dV_2, \tag{11-181}$$

$$K_{ij}=\int\int \varphi_i^*(1)\varphi_i^*(2)\,\frac{1}{r_{12}}\,\varphi_j(1)\varphi_j(2)\, dV_1\, dV_2, \tag{11-182}$$

---

[132] K. Hirao and H. Nakatsuji, *J. Chem. Phys.*, **59**, 1457 (1973); R. Carbo, R. Gallifa, and J. M. Riera, *Chem. Phys. Lett.*, **30**, 43 (1975); R. Carbo and J. M. Riera, *A General SCF Theory*, Lecture Notes in Chemistry, Volume 5, Springer-Verlag, Berlin, 1978.

[133] The first paper in this area was that of C. C. J. Roothaan, *Rev. Mod. Phys.*, **32**, 179 (1960). See also S. Huzinaga, *Phys. Rev.*, **120**, 866 (1960); **122**, 131 (1961); *J. Chem. Phys.*, **51**, 3971 (1969); F. W. Birss and S. Fraga, *J. Chem. Phys.*, **38**, 2562 (1963); **40**, 3203, 3207, 3212, (1964); W. Laidlaw and R. W. Birss, *Theoret. Chim. Acta*, **2**, 181 (1964); W. A. Goddard III, T. H. Dunning, and W. J. Hunt, *Chem. Phys. Lett.*, **4**, 231 (1969); D. Peters, *J. Chem. Phys.*, **57**, 4351 (1972); R. Caballol, R. Carbo, R. Gallifa, and J. M. Riera, *Int. J. Quantum Chem.*, **8**, 373 (1974); K. Hirao, *J. Chem. Phys.*, **60**, 3215 (1974); M. F. Guest and V. R. Saunders, *Mol. Phys.*, **28**, 819 (1974); R. Albat and N. Gruen, *Chem. Phys. Lett.*, **18**, 572 (1973); H. J. Silverstone, *J. Chem. Phys.*, **67**, 4172 (1977); E. R. Davidson, *Chem. Phys. Lett.*, **21**, 565 (1973); H. L. Hsu, E. R. Davidson, and R. M. Pitzer, *J. Chem. Phys.*, **65**, 609, (1976).

with

$$\langle \varphi_i | \varphi_j \rangle = \delta_{ij}. \tag{11-183}$$

Specific values of α_i, a_{ij} and b_{ij} that correspond to several of the common states of interest are given in Table 11.16. In general, it should be noted that

$$\begin{aligned} a_{ij} &= a_{ji} \\ b_{ij} &= b_{ji} \\ J_{ij} &= I_{ji} \\ K_{ij} &= K_{ji} \\ J_{ii} &= K_{ii}. \end{aligned} \tag{11-184}$$

Also, the number of occupied orbitals (N) includes all members of degenerate states, and the energy expression of Eq. (11-179) is sometimes an average expression of degenerate states.

We now wish to optimize E with respect to the choice of MOs, i.e., assure a stationary energy, while maintaining the orthonormality of the MOs. To do this we use Lagrange's method of undetermined multipliers, and define

$$E' = E - \sum_{i=1}^{N} \sum_{j=1}^{N} \epsilon_{ji} [\langle \varphi_i | \varphi_j \rangle - \delta_{ij}], \tag{11-185}$$

where E is given by Eq. (11-179), and the ϵ_{ji} are Lagrange multipliers. From this expression, we know that E will be stationary when E' is stable relative to a first-order variation in the MOs, i.e.,

$$\delta E' = 0. \tag{11-186}$$

To accomplish this, let us first examine the variation in E. From Eq. (11-179), we have

$$\begin{aligned} \delta E &= \sum_{i=1}^{N} \alpha_i \delta(h_i) + \sum_{i=1}^{N} \sum_{j=1}^{N} \delta(a_{ij} J_{ij} - b_{ij} K_{ij}) \qquad (11\text{-}187) \\ &= \sum_{i=1}^{N} \alpha_i \left[\int \delta\varphi_i^*(1) h \varphi_i(1)\, dV_1 + \int \varphi_i^*(1) h \delta\varphi_i(1)\, dV_1 \right] \\ &\quad + \sum_{i=1}^{N} \sum_{j=1}^{N} a_{ij} \left[\iint \delta\varphi_i^*(1) \varphi_j^*(2) \frac{1}{r_{12}} \varphi_i(1) \varphi_j(2)\, dV_1\, dV_2 \right. \\ &\quad + \iint \varphi_i^*(1) \delta\varphi_j^*(2) \frac{1}{r_{12}} \varphi_i(1) \varphi_j(2)\, dV_1\, dV_2 \\ &\quad + \iint \varphi_i^*(1) \varphi_j^*(2) \frac{1}{r_{12}} \delta\varphi_i(1) \varphi_j(2)\, dV_1\, dV_2 \end{aligned}$$

Table 11.16. Values of α_i, a_{ij} and b_{ij} corresponding to various states of interest[a]

State description	α_i	a_{ij}	b_{ij}
1. Closed shell; singlet state; nondegenerate orbitals	$\alpha_1 = 2$	$a_{11} = 2$	$b_{11} = 1$
2. One open shell, singly occupied; doublet state; non-degenerate open shell orbital	$\alpha_1 = 2$ $\alpha_2 = 1$	$a_{11} = 2, a_{12} = 1$ $a_{21} = 1, a_{22} = 0$	$b_{11} = 1, b_{12} = 1/2$ $b_{21} = 1/2, b_{22} = 0$
3. Two open shells, each singly occupied; singlet state; nondegenerate open shell orbitals	$\alpha_1 = 2$ $\alpha_2 = 1$ $\alpha_3 = 1$	$a_{11} = 2, a_{12} = 1, a_{13} = 1$ $a_{21} = 1, a_{22} = 0, a_{23} = 1/2$ $a_{31} = 1, a_{32} = 1/2, a_{33} = 0$	$b_{11} = 1, b_{12} = 1/2, b_{13} = 1/2$ $b_{21} = 1/2, b_{22} = 0, b_{23} = 1/2$ $b_{31} = 1/2, b_{32} = 1/2, b_{33} = 0$
4. One open shell, triplet state; nondegenerate open shell orbitals	$\alpha_1 = 2$ $\alpha_2 = 1$	$a_{12} = 2, a_{21} = 1$ $a_{21} = 1, a_{22} = 1/2$	$b_{12} = 1, b_{22} = 1/2$ $b_{21} = 1/2, b_{22} = 1/2$
5. Two 2-fold degenerate orbitals; one electron; doublet state	$\alpha_1 = 2$ $\alpha_2 = 1/2$	$a_{11} = 2, a_{12} = 1/2$ $a_{21} = 1/2, a_{22} = 0$	$b_{11} = 1, b_{12} = 1/4$ $b_{21} = 1/4, b_{22} = 0$
6. Two 2-fold degenerate orbitals; three electrons; *doublet state*	$\alpha_1 = 2$ $\alpha_2 = 3/2$	$a_{11} = 2, a_{12} = 3/2$ $a_{21} = 3/2, a_{22} = 1$	$b_{11} = 1, b_{12} = 3/4$ $b_{21} = 3/4, b_{22} = 1/2$

[a] See Eq. (11-163).

$$+\iint \varphi_i^*(1)\varphi_j^*(2)\,\frac{1}{r_{12}}\,\varphi_i(1)\delta\varphi_j(2)\,dV_1\,dV_2\Bigg]$$

$$-\sum_{i=1}^{N}\sum_{j=1}^{N} b_{ij}\Bigg[\iint \delta\varphi_i^*(1)\varphi_j^*(2)\,\frac{1}{r_{12}}\,\varphi_i(2)\varphi_j(1)\,dV_1\,dV_2$$

$$+\iint \varphi_i^*(1)\delta\varphi_j^*(2)\,\frac{1}{r_{12}}\,\varphi_i(2)\varphi_j(1)\,dV_1\,dV_2$$

$$+\iint \varphi_i^*(1)\varphi_j^*(2)\,\frac{1}{r_{12}}\,\delta\varphi_i(2)\varphi_j(1)\,dV_1\,dV_2$$

$$+\iint \varphi_i^*(1)\varphi_j^*(2)\,\frac{1}{r_{12}}\,\varphi_i(2)\delta\varphi_j(1)\,dV_1\,dV_2\Bigg]\,. \qquad (11\text{-}188)$$

Rearranging the above expression, using the definition of Coulomb and exchange operators [Eqs. (11-15) and (11-16)] and their Hermitian properties, and utilizing Eq. (11-184) allows Eq. (11-188) to be rewritten as:

$$\delta E=\sum_{i=1}^{N}\alpha_i\int \delta\varphi_i h^*\varphi_i^*\,dV_1$$

$$+2\sum_{i=1}^{N}\sum_{j=1}^{N}\Bigg[a_{ij}\int \delta\varphi_i \mathcal{J}_j^*\varphi_i^*\,dV_1-b_{ij}\int \delta\varphi_i\mathcal{K}_j^*\varphi_i^*\,dV_1\Bigg]$$

$$+\sum_{i=1}^{N}\alpha_i\int \delta\varphi_i^* h\varphi_i\,dV_1+2\sum_{i=1}^{N}\sum_{j=1}^{N}\Bigg[a_{ij}\int \delta\varphi_i^*\mathcal{J}_j\varphi_i\,dV_1$$

$$-b_{ij}\int \delta\varphi_i^*\mathcal{K}_j\varphi_i\,dV_1\Bigg]\,. \qquad (11\text{-}189)$$

Performing a similar analysis on the second set of terms in Eq. (11-185) and combining the results with Eq. (11-189) allows Eq. (11-186) to be written as

$$\delta E'=0=\sum_{i=1}^{N}\int \delta\varphi_i^*\Bigg[\alpha_i h+2\sum_{j=1}^{N}a_{ij}\mathcal{J}_j-2\sum_{j=1}^{N}b_{ij}\mathcal{K}_j\Bigg]\varphi_i\,dV_1$$

$$-\sum_{i=1}^{N}\delta\varphi_i^*\sum_{j=1}^{N}\epsilon_{ji}\varphi_j\,dV_1$$

$$+\sum_{i=1}^{N}\int \delta\varphi_i\Bigg[\alpha_i h^*+2\sum_{j=1}^{N}a_{ij}\mathcal{J}_j^*-2\sum_{j=1}^{N}b_{ij}\mathcal{K}_j^*\Bigg]\varphi_i^*\,dV_1$$

$$-\sum_{i=1}^{N}\delta\varphi_i\sum_{j=1}^{N}\epsilon_{ij}\varphi_j^*\,dV_1. \qquad (11\text{-}190)$$

For the above equation to be valid for arbitrary variations in $\delta\varphi_i^*$ and $\delta\varphi_i$, the terms multiplying $\delta\varphi_i^*$ and $\delta\varphi_i$ must each be equal to zero independently, i.e.,

$$\alpha_i h + 2\sum_{j=1}^{N}(a_{ij}\mathcal{J}_j - b_{ij}\mathcal{K}_j)\varphi_i = \sum_{j=1}^{N}\epsilon_{ji}\varphi_i, \qquad i=1, 2, \cdots, N, \quad (11\text{-}191)$$

$$\alpha_i h^* + 2\sum_{j=1}^{N}(a_{ij}\mathcal{J}_j^* - b_{ij}\mathcal{K}_j^*)\varphi_i^* = \sum_{j=1}^{N}\epsilon_{ij}\varphi_i^*, \qquad i=1, 2, \cdots, N. \quad (11\text{-}192)$$

If we define the Fock operator for orbital i as

$$\mathfrak{F}_i \equiv \frac{1}{2}\alpha_i h + \sum_{j=1}^{N}(a_{ij}\mathcal{J}_j - b_{ij}\mathcal{K}_j), \quad (11\text{-}193)$$

then Eqs. (11-191) and (11-192) can be rewritten as

$$\mathfrak{F}_i\varphi_i = \sum_{j=1}^{N}\epsilon_{ji}\varphi_j, \qquad i=1, 2, \cdots, N, \quad (11\text{-}194)$$

$$\mathfrak{F}_i^*\varphi_i^* = \sum_{j=1}^{N}\epsilon_{ij}\varphi_j^*, \qquad i=1, 2, \cdots, N. \quad (11\text{-}195)$$

By subtracting the complex conjugate of Eq. (11-195) from Eq. (11-194) and using the linear independence of the set of MOs $\{\varphi_i\}$, we obtain

$$(\epsilon_{ji} - \epsilon_{ij}^*) = 0 \qquad i=1, 2, \cdots, N,$$

or

$$\epsilon_{ji} = \epsilon_{ij}^*. \quad (11\text{-}196)$$

This result means that, if the above equation is satisfied, Eqs. (11-194) and (11-195) are equivalent. In other words, the correct variational conditions for determination of optimized MOs $\{\varphi_i\}$ are guaranteed only if *both*

$$\mathfrak{F}_i\varphi_i = \sum_{i=1}^{N}\epsilon_{ji}\varphi_j, \qquad i=1, 2, \cdots N \quad (11\text{-}197)$$

and

$$\epsilon_{ji} = \epsilon_{ij}^*, \qquad i=1, 2, \cdots N \quad (11\text{-}198)$$

are satisfied.

While the above two sets of equations could be used as the basis of an LCAO-type expansion method, it is useful to take the analysis further in order to achieve at least two additional goals: (1) to combine Eqs. (11-197) and (11-198) into a single set of variational equations to be solved, and (2) to construct a single pseudoeigenvalue equation to be solved for all orbitals. The latter point will be particularly helpful since $\mathfrak{F}_i$ in Eq. (11-197) is in general different for each occupied "shell"[134] ($i = 1, 2, \ldots, N$).

To accomplish these goals, we note first from Eq. (11-197) that

$$\epsilon_{ki} = \langle \varphi_k | \mathfrak{F}_i | \varphi_i \rangle \tag{11-199}$$

and

$$\begin{aligned} \epsilon_{ik}^* &= \langle \varphi_i | \mathfrak{F}_k | \varphi_k \rangle^* \\ &= \langle \varphi_k | \mathfrak{F}_k | \varphi_i \rangle. \end{aligned} \tag{11-200}$$

Substituting these results into Eqs. (11-197) and (11-198) gives the following equivalent pair of equations (using Dirac notation):

$$\mathfrak{F}_i | \varphi_i \rangle = \sum_{j=1}^{N} | \varphi_j \rangle \langle \varphi_j | \mathfrak{F}_i | \varphi_i \rangle, \qquad i = 1, 2, \cdots N \tag{11-201}$$

and

$$\langle \varphi_j | \mathfrak{F}_i - \mathfrak{F}_j | \varphi_i \rangle = 0, \qquad \begin{matrix} i = 1, 2, \cdots N \\ j = 1, 2, \cdots N \end{matrix}, \tag{11-202}$$

in which Lagrange multipliers no longer explicitly appear.

By noting the following equivalent representations of Eq. (11-202):

$$\begin{aligned} 0 &= \langle \varphi_j | \mathfrak{F}_i - \mathfrak{F}_j | \varphi_i \rangle, && \begin{matrix} i = 1, 2, \cdots N \\ j = 1, 2, \cdots N \end{matrix} \\ &= \sum_{j=1}^{N} | \varphi_j \rangle \langle \varphi_j | \mathfrak{F}_i - \mathfrak{F}_j | \varphi_i \rangle, && i = 1, 2, \cdots N \\ &= \sum_{j=1}^{N} \lambda_{ji} [| \varphi_j \rangle \langle \varphi_j | \mathfrak{F}_i - \mathfrak{F}_j | \varphi_i \rangle] && i = 1, 2, \cdots N \end{aligned} \tag{11-203}$$

with $\lambda_{ji} \neq 0$ and λ_{ji} real and arbitrary, we can add Eq. (11-203) to Eq. (11-201), which will incorporate Eq. (11-202) into Eq. (11-198) in a general way to give

$$\left[\mathfrak{F}_i - \sum_{j=1}^{N} | \varphi_j \rangle \langle \varphi_j | \mathcal{G}_{ji} \right] | \varphi_i \rangle = 0, \qquad i = 1, 2, \cdots N, \tag{11-204}$$

where we have defined the operator $\mathcal{G}_{ji}$ as:

$$\mathcal{G}_{ji} = \lambda_{ji} \mathfrak{F}_j + (1 - \lambda_{ji}) \mathfrak{F}_i, \qquad \begin{matrix} i = 1, 2, \cdots N \\ j = 1, 2, \cdots, N \end{matrix} \tag{11-205}$$

---

[134] A "shell" in this context is the group of orbitals for which $\mathfrak{F}_i$ is the same, e.g., s-shells, p-shells, etc. For example, in closed shell Hartree-Fock theory, $\mathfrak{F}_i$ is the same for all occupied electrons, and there is only *one* shell. The reader must take care to distinguish this use of the term "shell" and use of it to describe, e.g., "s-shells," "p-shells," etc., of electrons.

with $\lambda_{ij} \neq \lambda_{ji}$ and $\lambda_{ij} \neq 0$. Thus, we have accomplished our first goal, since Eq. (11-204) incorporates both Eqs. (11-197) and (11-198) into a single equation. However, we still have a different operator ($\mathfrak{F}_i$) for each shell, and it is to the creation of a *single* operator for all occupied orbitals to which we now turn.

By adding $|\varphi_i\rangle\langle\varphi_i|F_i|\varphi_i\rangle$ to both sides of Eq. (11-204) and noting that $\mathcal{G}_{ii} = \mathfrak{F}_i$, we obtain

$$\left[\mathfrak{F}_i - \sum_{j\neq 1}^{N} |\varphi_j\rangle\langle\varphi_j|\mathcal{G}_{ji}\right]|\varphi_i\rangle = |\varphi_i\rangle\langle\varphi_i|\mathfrak{F}_i|\varphi_i\rangle$$

$$= |\varphi_i\rangle\epsilon_i, \tag{11-206}$$

which can be written equivalently as

$$\left[\mathfrak{F}_i|\varphi_i\rangle\langle\varphi_i| - \sum_{j\neq 1}^{N} |\varphi_j\rangle\langle\varphi_j|\mathcal{G}_{ji}|\varphi_i\rangle\langle\varphi_i|\right]|\varphi_i\rangle = |\varphi_i\rangle\epsilon_i. \tag{11-207}$$

We now "symmetrize" the operator in brackets in the above equation that guarantees Hermicity but whose action on $|\varphi_i\rangle$ is the same as the operator in Eq. (11-207). In particular, we define

$$r_i \equiv [\mathfrak{F}_i|\varphi_i\rangle\langle\varphi_i| + |\varphi_i\rangle\langle\varphi_i|\mathfrak{F}_i] - |\varphi_i\rangle\langle\varphi_i|\mathfrak{F}_i|\varphi_i\rangle\langle\varphi_i|$$

$$-\sum_{j\neq 1}^{N} [|\varphi_j\rangle\langle\varphi_j|\mathcal{G}_{ji}|\varphi_i\rangle\langle\varphi_i| + |\varphi_i\rangle\langle\varphi_i|\mathcal{G}_{ji}|\varphi_j\rangle\langle\varphi_j|], \tag{11-208}$$

along with

$$\mathcal{R} = \sum_{i=1}^{N} r_i. \tag{11-209}$$

Hence, $\mathcal{R}$ is a single operator over all occupied orbitals, while r_i and $\mathcal{R}$ are each Hermitian.[135] Inserting Eq. (11-208) into Eq. (11-209) and combining the double sums allows $\mathcal{R}$ to be written as

$$\mathcal{R} = \sum_{i=1}^{N} [\mathfrak{F}_i|\varphi_i\rangle\langle\varphi_i| + |\varphi_i\rangle\langle\varphi_i|\mathfrak{F}_i] - \sum_{i=1}^{N} |\varphi_i\rangle\langle\varphi_i|\mathfrak{F}_i|\varphi_i\rangle\langle\varphi_i|$$

$$-\sum_{j=1}^{N}\sum_{j\neq i}^{N} |\varphi_i\rangle\langle\varphi_i|\mathcal{G}_{ij} + \mathcal{G}_{ji}|\varphi_j\rangle\langle\varphi_j|. \tag{11-210}$$

If the effect of $\mathcal{R}$ operating on one of the occupied orbitals $|\varphi_k\rangle$ is considered, we obtain

$$\mathcal{R}|\varphi_k\rangle = \sum_{i=1}^{N} [\mathfrak{F}_i|\varphi_i\rangle\langle\varphi_i|\varphi_k\rangle + |\varphi_i\rangle\langle\varphi_i|\mathfrak{F}_i|\varphi_k\rangle - |\varphi_i\rangle\langle\varphi_i|\mathfrak{F}_i|\varphi_i\rangle\langle\varphi_i|\varphi_k$$

$$-\sum_{j=1}^{N}\sum_{j\neq i}^{N} |\varphi_i\rangle\langle\varphi_i|\mathcal{G}_{ij} + \mathcal{G}_{ji}|\varphi_j\rangle\langle\varphi_j|\varphi_k\rangle, \qquad k = 1, 2, \cdots N \tag{11-211}$$

---

[135] See Problem 11-14.

which, after insertion of the definition of $\mathcal{G}_{ij}$ and $\mathcal{G}_{ji}$ and some manipulation, can be written as

$$\mathcal{R}|\varphi_k\rangle = |\varphi_k\rangle\langle\varphi_k|\mathcal{F}_k|\varphi_k\rangle, \quad k=1, 2, \cdots N$$

or

$$\mathcal{R}|\varphi_k\rangle = |\varphi_k\rangle\epsilon_k, \qquad k=1, 2, \cdots N. \tag{11-212}$$

Thus, we have achieved our second goal as well, since Eq. (11-212) represents a single pseudoeigenvalue equation with a single Fock operator ($\mathcal{R}$) that is valid for all occupied orbitals, and the equations also satisfy all of the conditions for a stationary energy. Furthermore, the form of Eq. (11-212) is the same as for the closed shell case (although the operator is substantially more complex), and the Hall–Roothaan "LCAO" approach used for the closed shell case can be applied directly to Eq. (11-212) to obtain equations suitable for determination of open shell MOs using basis sets such as those already described.[136]

C. Examples

To illustrate the use of open shell Hartree–Fock theory, let us consider first the case of the methoxide radical (CH_3O). Although this radical was known to be produced photochemically by reactions such as

$$CH_3COOCH_3 \xrightarrow{h\nu} CH_3CO + CH_3O$$

neither geometric nor electronic structural data were available in 1974 when the studies of Yarkony, Schaefer, and Rothenberg[137] were reported. In these studies a double zeta basis of contracted Gaussians was used in a restricted Hartree–Fock study, and both geometric and electronic structural features of the ground and low-lying excited states were explored.

The optimized geometric data that resulted is as follows:

	R_{CH}	R_{CO}	$\measuredangle$ OCH
^2E state	1.08 Å	1.44 Å	109°
$^2A_1$ state	1.08 Å	1.65 Å	112°

[136] It should be noted that, while not discussed here, excited state SCF studies are complicated further when the excited state symmetry is the same as the ground state. For examples of approaches that are applicable in such cases, see H. Hsu, E. R. Davidson, and R. M. Pitzer, *J. Chem. Phys.*, **65**, 609 (1976); T. P. Hamilton and P. Pular, *J. Chem. Phys.*, **84**, 5728 (1986); R. Colle, A. Fortunelli, and O. Salvetti, *Theoret. Chim. Acta*, **71**, 467 (1987); E. R. Davidson and L. Z. Stenkamp, *Int. J. Quantum Chem. Symp.*, **10**, 21 (1976).

[137] D. R. Yargony, H. F. Schaeffer III, and S. Rothenberg, *J. Am. Chem. Soc.*, **96**, 656–659 (1974).

Table 11.17. Mulliken Atomic and Overlap Populations for the ^{2}E State (Ground State) of the CH_3O Radical[a]

Atomic Populations

AO		MO				
		$3a_1$	$4a_1$	$1e$	$5a_1$	$2e$
C	s	0.29	1.08	0.00	0.01	0.00
	p	0.07	0.10	2.01	0.52	0.31
Total		0.36	1.18	2.01	0.53	0.31
H	s	0.01	0.45	1.38	0.14	0.55
O	s	1.56	0.26	0.00	0.10	0.00
	p	0.06	0.10	0.61	1.22	3.14
Total		1.62	0.36	0.61	1.32	3.14

Overlap Populations

	C-H	C-O	O-H
$3a_1$	0.01	0.37	0.00
$4a_1$	0.19	−0.19	−0.01
$1e$	0.32	0.21	0.04
$5a_1$	0.04	0.22	−0.03
$2e$	0.20	−0.28	−0.10

[a] Data were obtained from the calculations of D. Yarkony, H. F. Schaefer III, and S. Ruthenberg, *J. Am. Chem. Soc.*, **96**, 656–659 (1974). The $1a_1$ and $2a_1$ orbitals have been omitted for convenience, since they are essentially oxygen $1s$ and carbon $1s$ orbitals.

From these data it is seen that a major lengthening of the C–O bond distance occurs from the ground state to the first excited state, while the other geometric parameters remain approximately constant.

In Table 11.17, a Mulliken population analysis of the ground (^2E) state is given, corresponding to the electron configuration:

$$1a_1^2\, 2a_1^2\, 3a_1^2\, 4a_1^2\, 1e^4\, 5a_1^2\, 2e^3.$$

From these data it is seen that the $3a_1$ orbital is composed of approximately 75% oxygen $2s$ and 15% carbon $2s$, and is a C–O bonding MO (as indicated in the positive C–O overlap population). The $4a_1$ MO is the antibonding analog of $3a_1$, while the $1e$ orbital is primarily a pair of C–H bonding MOs. The $5a_1$ MO is composed of $2p^{0.55}2p^{1.2}$, and is a C–O bonding MO, while the $2e$ MO is predominantly a pair of nonbonding $2p_0$ orbitals. Thus, the large C–O lengthening in the transition from the ^2E (ground) state to the $^2A_1$ (first excited) state can be understood qualitatively as being due to the excitation of an electron from the $5a_1$ C–O bonding MO to the $2e$ nonbonding MO.

As a different kind of example, let us consider spin properties. In particular, the spin density matrix ($\mathbf{P}^s$) introduced in Eq. (11-177) provides a means of

Table 11.18. Experimental Splittings and Calculated Spin Densities for the Pentadienyl Radical[a]

Atom	ρ_{sd}	a_{sd}	ρ_{aa}	a_{aa}	$a_{expt.}$
1	0.545	−13.30	0.383	−9.34	−8.99
2	−0.307	+ 8.30	−0.094	+2.55	+2.65
3	0.524	−14.15	0.422	−11.39	−13.40
$\langle \mathcal{S}^2 \rangle$	0.95625		0.76762		

[a] Data taken from L. C. Snyder and A. T. Amos, *J. Chem. Phys.*, **42**, 3670–3683 (1965). For C_1, $Q = 24$, while for C_2 and C_3, $Q = 27$ Gauss. The quantities ρ_{sd} and a_{sd} are calculated using a single determinantal UHF wavefunction, while ρ_{aa} and a_{aa} are the same quantities calculated from the UHF wavefunction after annihilation of higher spin components using Eq. (11-183).

calculating net spin densities at various points in a molecule, which can be compared to experimentally observed proton isotropic hyperfine splittings. In particular, it has been noted by McConnell, Weissman, and Bersohn[138] that the isotropic hyperfine splittings by protons (a_{Hi}) in doublet π-electron radicals should be approximately proportional to the π-electron spin density (ρ_i) on the carbon atom to which the proton is bonded, i.e.,

$$a_{Hi} = -Q\rho_i, \tag{11-213}$$

where Q is a constant.

Snyder and Amos[139] used UHF theory within the framework of semiempirical calculations to calculate the spin densities (ρ_i) for π-electrons in a large number of organic radicals. As an example of their study, the pentadienyl radical had calculated spin densities (ρ_{sd}) and splittings (a_{sd}) as indicated in Table 11.18. It is seen that the signs and relative magnitude of the calculated splittings agree with the experimentally observed values.

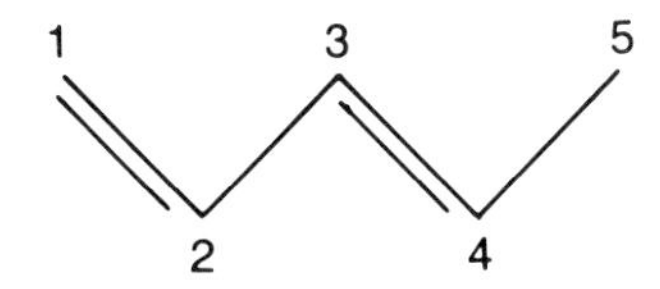

However, the calculated expectation value of $\langle \mathcal{S}^2 \rangle$ for the UHF single determinant (0.95625) is also given in Table 11.18, and is seen to be substantially different than the value of 0.75 (which is the value to be expected for a doublet eigenfunction of $\mathcal{S}^2$). Thus, the single determinant of UHF theory is not an eigenfunction of $\mathcal{S}^2$, which was to be expected based on the analysis given earlier.[140]

[138] H. M. McConnell, *J. Chem. Phys.*, **24**, 764 (1956); S. J. Weissman, *J. Chem. Phys.*, **25**, 890 (1956); R. Bersohn, *J. Chem. Phys.*, **24**, 1066 (1956).

[139] L. C. Synder and A. T. Amos, *J. Chem. Phys.*, **42**, 3670–3683 (1965).

[140] See Chapter 10, Section 10-7.

To examine the nature of this "spin contamination" further, we note that the UHF single determinant can be constructed to be an eigenfunction of $\mathcal{S}_z$, and the range of eigenvalues that is possible is $\frac{1}{2}(p + q)$ to $\frac{1}{2}(p - q)$. This therefore is also the range of possible $\mathcal{S}^2$ values, so we see that, for the case of pentadienyl radical with 5 π-electrons, the UHF wavefunctions can be written as

$$\Psi_{\text{UHF}} = c_1 \Psi_{\text{(doublet)}} + c_2 \Psi_{\text{(quintet)}} + c_3 \Psi_{\text{(sextet)}}. \tag{11-214}$$

In general, it has been shown[141] that the expectation value of $\mathcal{S}^2$ for a single determinant is given by

$$\langle S^2 \rangle = \frac{1}{4}(p-q)^2 + \frac{1}{2}(p+q) - \text{tr}(\mathbf{P}^\alpha \mathbf{P}^\beta). \tag{11-215}$$

Thus, the value of 0.95625 found by Snyder and Amos is simply a reflection of the quartet and sextet contaminants in Ψ_{UHF}.

To alleviate this problem, the most general approach[142] is to apply the pure spin state projection operator ($\mathcal{O}_{2S+1}$) to Ψ_{UHF}.

$$\vartheta_{2S+1} = \prod_{k \neq s} \left[\frac{\mathcal{S}^2 - k(k+1)}{s(s+1) - k(k+1)} \right], \tag{11-216}$$

which will project out the state of multiplicity $2S + 1$ from Ψ_{UHF}. Since the contaminations are not found to be large in general, it has been found that simply eliminating the largest component of contamination is usually sufficient to obtain a reasonably pure state. For example, the annihilator

$$\mathcal{Q}_{3/2} = \left(\mathcal{S}^2 - \frac{15}{4} \right) \tag{11-217}$$

was applied to the UHF wavefunction for the pentadienyl radical, and the expectation value of $\langle \mathcal{S}^2 \rangle$ is seen in Table 11.18 to be reduced to 0.76762, with an improvement in the calculated splittings compared to experiment.

Problems

1. Derive Eq. (11-14) from (11-13).
2. Show that λ in Eq. (11-21) is a Hermitian matrix.
3. Show that the exchange operator, $\mathcal{K}_j$, defined as

$$\mathcal{K}_j(1)\chi_j(1) = \left[\int d\tau_2 \chi_j^*(2)\chi_i(2) \frac{1}{r_{12}} \right] \chi_j(1)$$

---

[141] See, for example, P. O. Löwdin, *Adv. Chem. Phys.*, **2**, 207–322 (1959).

[142] P. O. Löwdin, *Phys. Rev.*, **97**, 1509 (1955); A. T. Amos and G. G. Hall, *Proc. R. Soc. (London)*, **A263**, 483 (1961).

is invariant to a unitary transformation of the basis:

$$\boldsymbol{\chi}' = \mathbf{U}\boldsymbol{\chi}, \text{ where } \mathbf{U}^\dagger\mathbf{U} = \mathbf{U}\mathbf{U}^\dagger = \mathbb{1}.$$

4. Show that integration over the spin coordinates in the Coulomb operator in Eq. (11-51) gives the result

$$\sum_{j' \text{ even}}^{N} \int d\mathbf{r}_1 \beta^*(\sigma_1) \left[\int d\sigma_2 \int d\mathbf{r}_2 \chi_j^*(\mathbf{r}_2)\chi_{j'}(\mathbf{r}_2) \frac{1}{r_{12}} \right] \varphi_{i/2}(\mathbf{r}_1)\beta(r_1)$$

$$= \sum_{k=1}^{N/2} \left[\int d\mathbf{r}_2 \varphi_k^*(\mathbf{r}_2)\varphi_k(\mathbf{r}_2) \frac{1}{r_{12}} \right] \varphi_{i/2}(\mathbf{r}_1).$$

5. By direct calculation show that, for the charge and bond order matrix that results from a solution to the Hartree–Fock equations,

$$\sum_u^M \sum_v^M P_{uv} \int \phi_u^*(\mathbf{r})\phi_v(\mathbf{r}) = N$$

where N is the number of electrons and there are M basis orbitals $\{\phi\}, j = 1, 2, \ldots, M$.

6. Verify Eq. (11-83) using Eqs. (11-78)–(11-82).
7. Verify Eq. (11-84) using Eq. (11-83) and the integrals from Table 11.2.
8. For an s-type GTO as defined in Eq. (11-92), show that the rationale for calling ρ an "orbital radius" is appropriate. Accomplish this by proving that ρ is the radius of a sphere that contains approximately 74% of the orbital density of the GTO, i.e., show that

$$4\pi \int_0^\rho G^2(r)r^2\, dr \cong 0.74.$$

9. Prove Eq. (11-123).
10. Using methods analogous to the closed shell case, derive Eqs. (11-156) and (11-157) for the UHF wavefunction.
11. For the UHF wavefunction of Eq. (11-150), show that the total electronic energy can be written as

$$E = \sum_a^{N_\alpha} \langle \varphi_a^\alpha | h | \varphi_a^\alpha \rangle + \sum_b^{N_\beta} \langle \varphi_b^\beta | h | \varphi_b^\beta \rangle$$

$$+ \frac{1}{2} \sum_a^{N_\alpha} \sum_{a'}^{N_{\alpha'}} [\langle \varphi_a^\alpha \varphi_{a'}^\alpha | \varphi_a^\alpha \varphi_{a'}^\alpha \rangle - \langle \varphi_a^\alpha \varphi_{a'}^\alpha | \varphi_{a'}^\alpha, \varphi_a^\alpha \rangle]$$

$$+ \frac{1}{2} \sum_b^{N_\beta} \sum_b^{N_\beta} [\langle \varphi_b^\beta \varphi_{b'}^\beta | \varphi_b^\beta \varphi_{b'}^\beta \rangle - \langle \varphi_b^\beta \varphi_{b'}^\beta | \varphi_{b'}^\beta, \varphi_b \rangle]$$

$$+ \sum_a^{N_\alpha} \sum_b^{N_\beta} \langle \varphi_a^\alpha \varphi_b^\beta | \varphi_a^\alpha \varphi_b^\beta \rangle.$$

12. Show that, for the UHF wavefunction and the matrix elements of Eq. (11-171), the elements of the Fock matrix can be written as

$$F_{st}^{\alpha}=H_{st}^{\text{core}}+\sum_{u,v}^{T} P_{uv}^{\alpha}[\langle\phi_s\phi_u|\phi_t\phi_v\rangle-\langle\phi_s\phi_u|\phi_v\phi_t\rangle]$$

$$+\sum_{u,v}^{T} P_{u,v}^{\beta}\langle\phi_s\phi_u|\phi_t\phi_v\rangle,$$

and an analogous equation for F_{st}^{β}, where the elements of the charge and bond order matrices are given in Eqs. (11-174) and (11-175).

13. Show that, if one-electron energies for α-orbitals are given by

$$\epsilon_i^{\alpha}=h_{ii}^{\alpha}+\sum_{j}^{N_\alpha}[\langle\varphi_i^{\alpha}\varphi_j^{\alpha}|\varphi_i^{\alpha}\varphi_j^{\alpha}\rangle-\langle\varphi_i^{\alpha}\varphi_j^{\alpha}|\varphi_j^{\alpha}\varphi_i^{\alpha}\rangle]$$

$$+\sum_{j}^{N_\beta}\langle\varphi_i^{\alpha}\varphi_j^{\beta}|\varphi_i^{\alpha}\varphi_j^{\beta}\rangle$$

with a similar expression for ϵ_i^{β}, and with

$$h_{ii}^{p}=\langle\varphi_i^{p}|h|\varphi_i^{p}\rangle \qquad p=\alpha,\ \beta,$$

the total electronic energy of a UHF wavefunction can be written as

$$E=\frac{1}{2}\sum_{a}^{N_\alpha}(\epsilon_a^{\alpha}+h_{aa}^{\alpha})+\frac{1}{2}\sum_{b}^{N_\beta}(\epsilon_b^{\beta}+h_{bb}^{\beta}).$$

14. Show that $\mathfrak{F}_j$, $\mathcal{G}_{ji}$, r_i and $\mathcal{R}$, defined by Eqs. (11-198), (11-210), (11-213), and (11-214), respectively, are Hermitian. Also show that, if $\lambda_{ji} = 0$, Eq. (11-211) reduces to Eq. (11-206) and Eq. (11-201) cannot be guaranteed to be satisfied.

Chapter 12

Beyond Hartree–Fock Theory

12-1. Electron Correlation: General Comments

In the previous chapter we saw that Hartree–Fock theory provides a remarkably good description of molecular systems, especially when the conceptual simplicity of the model is considered. Many properties are accurately predicted, and total energies can be obtained that are accurate to within >99% of the experimental value. However, we have also seen that the 1% error (i.e., the "correlation energy error") results in incorrect conclusions in a number of important cases such as dissociation energies, electronic spectra, and potential surfaces.

Since the formulation of quantum mechanics as a theory is intended to provide calculated values for observables of *any* desired accuracy (at least in principle), it is of interest to see what steps can be taken to go beyond the limitations of Hartree–Fock theory. As we shall see, it is not difficult to formulate a number of approaches that are satisfactory in principle. However, translation of these approaches into computationally convenient and viable processes has turned out to be a formidable task, and one that still occupies a prominent place in quantum chemistry research laboratories today.

An introduction to several of the available techniques is given in the sections to follow, along with a description of several concepts that are useful in general, regardless of the particular technique utilized. However, it should be noted that these techniques have not matured to the point where straightforward use of any one of them will always give unambiguous and correct results. Thus, there is still considerable "art" involved when designing studies that go beyond Hartree–Fock theory. Indeed, we shall see examples where large basis sets and extensive studies give excellent results, as well as cases in which similar efforts provide poor results. We shall also encounter examples where small basis sets and modest studies can provide important insights, as well as cases in which such studies produce erroneous conclusions. In other words, the size and flexibility of the overall wavefunction and basis set that are required to answer questions of chemical interest are not always the same. Also, while absolute accuracy of a few kcal/mol is always desirable, it is seldom achievable and frequently not essential. It is to a discussion of the available approaches and their accuracy to which much of the following discussion will focus, so as to identify the options available for a given molecule and property of interest.

From a historical perspective, there are two main kinds of approaches that provide a theoretical framework sufficient to allow results of arbitrary high accuracy to be obtained, at least in principle. These approaches are known as *configuration interaction theory* and *perturbation theory* (i.e., extensions of the discussions in Chapters 9 and 10), and each will be described in the sections to follow.

12-2. Configuration Interaction

A. General Formulation

To see the general formulation we recall that,[1] at least in principle, the exact wavefunction for an N-electron system (in the Born–Oppenheimer approximation) can be written as an expansion in antisymmetrized products of one-electron functions, which we shall write as

$$\Psi(1, 2, \cdots, N) = \sum_k d_k \Phi_k(1, 2, \cdots, N) \qquad (12\text{-}1)$$

where each Φ_k is referred to as a *configuration* or (*configuration state function*), with

$$\Phi_k(1, 2, \cdots, N) = N_k \det\{\chi_{1k}(1)\chi_{2k}(2) \cdots \chi_{Nk}(N)\}, \qquad (12\text{-}2)$$

where N_k is a normalization constant and the χ_{jk} are orbitals taken from a complete basis set.

It is important to reemphasize that the concept of a configuration may require use of more than the single Slater determinant used in Eq. (12-2). In particular, constructing wavefunctions that represent pure spin states (i.e., where the wavefunction is an eigenfunction of $\mathcal{S}^2$ and $\mathcal{S}_z$ with specified eigenvalues) is usually accomplished by assuring that *each* configuration of the wavefunction represents the pure spin state of interest. As illustrated earlier,[2] a linear combination of Slater determinants is frequently needed to assure that specific spin state eigenfunctions have been constructed. For those cases, a configuration Φ_k will consist of a *sum* of Slater determinants with coefficients chosen to assure that the configuration is an eigenfunction of $\mathcal{S}^2$. In cases in which spatial

[1] Chapter 10, Section 10-4.

[2] See Section 10-9.

symmetry is also present, linear combinations of Slater determinants may also be needed to form configurations having suitable spatial symmetry. Hence, while we shall continue to utilize a single Slater determinant to represent a configuration, it should be remembered that linear combinations of them are required in general to construct a configuration that has appropriate space and spin symmetry.

Within the caveats just stated, the process of determining the Φ_k and d_k is referred to as the *configuration interaction* (CI) approach. It is of sufficient generality to provide results of arbitrarily high accuracy if applied in sufficient detail,[3] and will occupy a major component of the remainder of this chapter.

Given a CI wavefunction of the form given in Eq. (12-1), minimization of the energy expression can be carried out as in Section 9-5 (assuming the Φ_k have already been chosen), giving rise to the following secular determinant that can be solved to obtain energy levels:

$$\det\{\mathbf{H}-E\mathbf{S}\}=0 \tag{12-3}$$

where

$$H_{st}=\int \Phi_s^* \mathcal{H} \Phi_t \, d\tau, \tag{12-4}$$

and

$$S_{st}=\int \Phi_s^* \Phi_t \, d\tau. \tag{12-5}$$

We shall, in general, construct the orbitals and configurations to be orthonormal,[4] in which case the overlap matrix (**S**) is diagonal, i.e.,

$$S_{st}=\delta_{st}, \tag{12-6}$$

and the results of Section 10-5 allow us to rewrite the elements of **H** as

$$H_{st}=\sum_{i,j} a_{ij}^{st} h_{ij} + \sum_{i,j,k,l} b_{ijkl}^{st} g_{ijkl}, \tag{12-7}$$

where

$$h_{ij}=\int \chi_i^* h \chi_j \, d\tau, \tag{12-8}$$

and

$$g_{ijkl}=\left\langle ij \left| \frac{1}{r_{12}} \right| kl \right\rangle \tag{12-9}$$

$$=\int\int \chi_i^*(1)\chi_j^*(2) \frac{1}{r_{12}} \chi_k(1)\chi_l(2) \, d\tau_1 \, d\tau_2. \tag{12-10}$$

---

[3] For a review of configuration interaction techniques, see, for example, I. Shavitt, *Methods of Electronic Structure Theory*, H. F. Schaefer III, ed., Plenum Press, New York, 1977, pp. 189–276; see also P. O. Löwdin, *Adv. Chem. Phys.*, **2**, 207 (1959).

[4] As shown earlier, the configurations will be orthonormal if the MOs used to form them are orthonormal.

The coefficients a_{ii}^{st} and b_{ijkl}^{st} differ depending upon the orbitals of configurations s and t, but are fixed constants once s and t are chosen.[5] Thus, the CI approach in principle consists of nothing more than determining a suitable set of orthogonal orbitals, selecting configurations, evaluating integrals, and solving the secular equation to determine energy levels and associated wavefunctions. The result will provide, since McDonald's Theorem[6] applies to the CI expansion, not only a ground state wavefunction and an upper bound estimate to the ground state energy but also upper bound energy estimates and associated wavefunctions for *all* excited states described by the CI expansion. However, as noted earlier, carrying out that process in practice is not a trivial undertaking.

Before describing the CI approach in greater detail, it is appropriate first to introduce some notation that will be useful in subsequent discussions. We begin by noting that, in addition to the filled MOs that result from Hartree–Fock studies, a set of unfilled *"virtual" orbitals* is also determined. For example, if a basis set of M spin orbitals is used to describe the MOs of an N-electron system ($M > N$), the HF-SCF-MO process will determine M MOs of which N are occupied. The remaining (M-N) MOs are unoccupied, and are called virtual MOs. Such a situation is depicted in Fig. 12.1.

While the Hartree–Fock determinant is obviously the most important configuration for the ground state in the CI expansion of Eq. (12-1), there are many others than can be formed, e.g., if we utilize the unfilled (virtual) MOs resulting from the Hartree–Fock calculation.[7] For example, if the electron that was in χ_N (the highest occupied MO) in the Hartree–Fock configuration is placed instead in χ_{N+1}, we can construct a new configuration. Specifically, if Φ_0 is the Hartree–Fock configuration:

$$\Phi_0(1, 2, \cdots, N) = N_0 \det\{\chi_1(1)\chi_2(2) \cdots \chi_N(N)\}, \qquad (12\text{-}11)$$

then Φ_1, is given in this example by

$$\Phi_1(1, 2, \cdots, N) = N_1 \det\{\chi_1(1)\chi_2(2) \cdots \chi_{N-1}(N-1)\chi_{N+1}(N)\}, \qquad (12\text{-}12)$$

and is also depicted in Fig. 12.1.

Clearly, Φ_1 is only one of a large number of configurations that can be formed in this manner. In particular, if a basis set of M spin orbitals is used ($M > N$), then a total (T) of

$$T = \binom{M}{N} = \frac{M!}{N!(M-N)!} \qquad (12\text{-}13)$$

configurations can be formed.[8]

---

[5] See Problem 12-1.

[6] See Chapter 9, Section 9-5, Eq. (9-79).

[7] It should be noted that there are many ways to create unfilled orbitals for use in CI studies, and the use of virtual MOs from Hartree–Fock calculations in the discussion here is done for convenience only.

[8] See Problem 12-2.

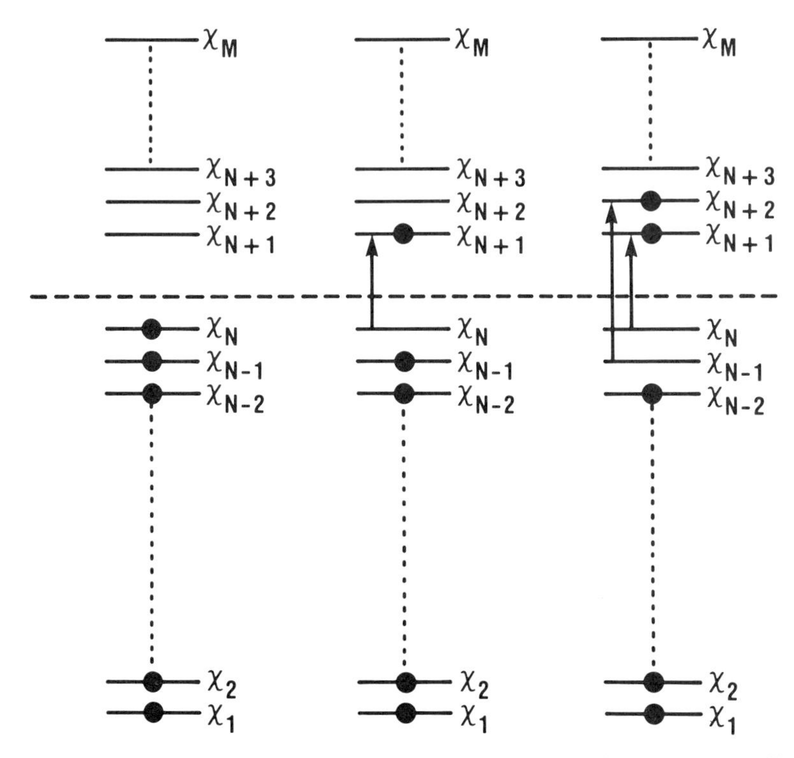

Figure 12.1. Depiction of spin orbitals (filled and unfilled) resulting from Hartree–Fock calculations on an N-electron system using an M spin orbital basis. Also depicted is Φ_1 and Φ_2 illustrating single and double excitations.

For the case above, Φ_1 is usually referred to as a *singly excited configuration*,[9] since only one electron has been promoted. If two electrons are promoted simultaneously, e.g., into χ_{N+1} and χ_{N+2} (see Fig. 12.1), we have

$$\Phi_2(1, 2, \cdots, N) = N_2 \det\{\chi_1(1)\chi_2(2) \cdots \chi_{N-2}(N-2)\chi_{N+1}(N-1)\chi_{N+2}(N)\}, \qquad (12\text{-}14)$$

which is referred to as a *doubly excited configuration*. Extension of the terminology to triple, quadruple, and higher excitations is obvious.

Using this approach, the exact wavefunction of Eq. (12-1) can be written

[9] It should be emphasized again that, for simplicity of presentation, it is assumed in this discussion that each configuration is composed of a single Slater determinant, and the terms "configuration" and "determinant" are used interchangeably. It should be remembered, as pointed out earlier, that a linear combination of Slater determinants is required in general to form a "configuration" that has specified spatial and/or spin symmetry.

using a slightly modified notation as

$$\Psi(1, 2, \cdots, N) = c_0\Phi_0 + \sum_{a,\alpha} c_a^\alpha \Phi_a^\alpha + \sum_{\substack{a<b\\ \alpha<\beta}} c_{ab}^{\alpha\beta}\Phi_{ab}^{\alpha\beta} + \sum_{\substack{a<b<c\\ \alpha<\beta<\gamma}} c_{abc}^{\alpha\beta\gamma}\Phi_{abc}^{\alpha\beta\gamma} + \cdots, \tag{12-15}$$

where Φ_a^α is a configuration in which a single excitation of an electron from the filled spin orbital (χ_a) to the unfilled spin orbital (χ_α) has taken place. Similarly, $\Phi_{ab}^{\alpha\beta}$ is a configuration in which a double excitation from χ_a to χ_α and from χ_b to χ_β has taken place, $\Phi_{abc}^{\alpha\beta\gamma}$ is a configuration in which a triple excitation from Φ_0 has taken place, etc. If all possible configurations that can be formed from a given basis set are used in Eq. (12-15), it is referred to as a *full CI* calculation.[10]

It is important to note that the size of the CI calculation that is implied by "full CI" depends on the number of orbitals that is used. For example, a *full CI* calculation on an eight-electron system with a basis of 10 spin orbitals will involve 45 configurations, while a *full CI* calculation for the same system using a basis of 20 spin orbitals will involve, 1,511,640 configurations.

B. General Concepts

1. Brillouin's Theorem

One of the theorems that is useful in simplifying some of the computations that arise in CI calculations is known as *Brillouin's Theorem*. To see this theorem and its applicability, let us consider an N-electron system and a Hamiltonian matrix element between the Hartree–Fock configuration (Φ_0) and a configuration (Φ_a^α) containing a single excitation to an unfilled Hartree–Fock orbital, that arises in the determination of the coefficients (c_a^α) in the CI expansion of Eq. (12-1):

$$\langle\Phi_0|\mathcal{H}|\Phi_a^\alpha\rangle = N_0 N_a^\alpha \int \cdots \int \det\{\chi_1^*(1) \cdots \chi_a^*(A) \cdots \chi_N^*(N)\} \cdot \mathcal{H} \det\{\chi_1(1) \cdots \chi_\alpha(A) \cdots \chi_N(N)\}\, d\tau_1 \cdots d\tau_N. \tag{12-16}$$

Using the techniques for matrix element evaluation that were developed in Chapter 11 (Section 11-4), it is easily seen[11] that

$$\langle\Phi_0|\mathcal{H}|\Phi_a^\alpha\rangle = \langle\chi_a|h|\chi_\alpha\rangle + \sum_j^N \left[\left\langle\chi_a\chi_j\left|\frac{1}{r_{12}}\right|\chi_\alpha\chi_j\right\rangle - \left\langle\chi_a\chi_j\left|\frac{1}{r_{12}}\right|\chi_j\chi_\alpha\right\rangle\right], \tag{12-17}$$

[10] See Problem 12-4.

[11] See Problem 12-5.

or

$$\langle \Phi_0 | \mathcal{H} | \Phi_a^\alpha \rangle = \langle \chi_a | f | \chi_\alpha \rangle, \tag{12-18}$$

where f is the Fock operator.

However, since χ_a is an eigenfunction of the Fock operator, and since the filled and virtual Hartree–Fock orbitals are mutually orthonormal, we have

$$\begin{aligned}\langle \chi_a | f | \chi_\alpha \rangle &= \langle \chi_\alpha | f | \chi_a \rangle^* = \epsilon_a^* \langle \chi_\alpha | \chi_a \rangle^* \\ &= 0, \quad \alpha \neq a.\end{aligned} \tag{12-19}$$

This means that we have proven the general result:

$$\langle \Phi_0 | \mathcal{H} | \Phi_a^\alpha \rangle = 0. \tag{12-20}$$

This result is known as *Brillouin's Theorem.*[12] It is of obvious theoretical importance, although its impact on actual computations is limited due to its applicability only to Hartree–Fock orbitals and the fact (that will be demonstrated later) that single excitation configurations from a Hartree–Fock reference configuration (Φ_0) for a state generally are small contributors to a CI wavefunction (provided Φ_0 dominates the expansion).

For example, if we start from the ground state Hartree–Fock wavefunction and wish to make corrections to it via a CI expansion, we expect from Brillouin's Theorem that the most important correction terms to be considered will be *double excitation* terms, since single excitations give no contribution by direct interaction with the reference configuration (Φ_0). This is the case in general, and some CI approaches consider only double excitations.

It should also be noted that Brillouin's Theorem applies only to interactions of singly excited configurations with the Hartree–Fock reference configuration. Therefore, Hamiltonian matrix elements involving single excitation configurations with higher excitation configurations (e.g., double, triple, ... excitation configurations) will *not* be zero in general, since Brillouin's Theorem does not apply to these cases. However, contributions from matrix elements such as these are generally small compared to, e.g., those from double excitations from the Hartree–Fock configuration. Hence, single excitation configurations play a very secondary role in ground state CI calculations when the Hartree–Fock wavefunction is used as a starting point. Of course, in the case of excited states, where the main configuration is not the Hartree–Fock (ground state) configuration, and the configurations that dominate the state wavefunction are themselves single excitations from Φ_0, these comments do not apply.

---

[12] L. Brillouin, *Actualites sci. et ind.*, No. 71 (1933); No. 159 (1934), Hermann et Cie, Paris; see also C. Moeller and M. S. Plesset, *Phys. Rev.*, **46**, 618 (1934).

2. Density Matrices and Natural Orbitals

Another concept that is helpful in CI studies both from a computational and conceptual point of view is that of *reduced density matrices*.[13] To see why this is the case, we begin by defining the pth *order reduced density matrix* $[\Gamma^{(p)}(\boldsymbol{\tau}|\boldsymbol{\tau}')]$ for an N-electron system that is described by a normalized, antisymmetric wavefunction (Ψ) as

$$\Gamma^{(p)}(\boldsymbol{\tau}|\boldsymbol{\tau}') \equiv \binom{N}{p} \int \cdots \int \Psi(\tau_1, \tau_2 \cdots \tau_p; \tau_{p+1} \cdots \tau_N) \cdot \Psi^*(\tau_1', \tau_2', \cdots \tau_p'; \tau_{p+1} \cdots \tau_N)\, d\tau_{p+1} \cdots d\tau_N, \quad (12\text{-}21)$$

where $\boldsymbol{\tau}$ or $\boldsymbol{\tau}'$ (without a subscript) represent both space ($\mathbf{r}$) and spin (σ) components of the first p electron coordinates. Thus, $\Gamma^{(p)}$ is a function of the space and spin coordinates of p electrons, where the space and spin coordinates of the remaining $(N - p)$ electrons are averaged over all space. Also, $\Gamma^{(p)}$ can be thought of as a multidimensional matrix, where the indices $(\boldsymbol{\tau}, \boldsymbol{\tau}') \equiv (\tau_1, \tau_2 \cdots \tau_p; \tau_1' \tau_2' \cdots \tau_p')$ are continuous (not discrete) indices.

Another frequently used form of density matrices is known as the *spinless* pth *order reduced density matrix* $[P^{(p)}(r|r')]$, which is obtained from $\Gamma^{(p)}(\boldsymbol{\tau}|\boldsymbol{\tau}')$ by integrating over the spin coordinates of the first p electrons, $\sigma_1, \sigma_2, \ldots, \sigma_p$, i.e.,

$$P^{(p)}(\mathbf{r}|\mathbf{r}') = \int \cdots \int \Gamma^{(p)}(\boldsymbol{\tau}|\boldsymbol{\tau}')\, d\sigma_1 \cdots d\sigma_p. \quad (12\text{-}22)$$

Thus, $P^{(p)}$ is a function of only the *spatial* coordinates of p electrons.

To see part of the rationale for introduction of such concepts and quantities, it is of interest to note that some of the reduced density matrices have direct physical interpretations as well as being of computational assistance. For example, the *first-order reduced density matrix* (sometimes called the "1-matrix") is usually denoted by the special symbol γ, i.e.,

$$\Gamma^{(1)}(\tau_1|\tau_1') \equiv \gamma(\tau_1|\tau_1') = N \int \cdots \int \Psi(\tau_1 \tau_2 \cdots \tau_N) \Psi^*(\tau_1' \tau_2 \cdots \tau_N)\, d\tau_2 \cdots d\tau_N. \quad (12\text{-}23)$$

This matrix has only two (continuous) indices, and the diagonal element (i.e., $\tau_1 = \tau_1'$) of γ represents the probable number of electrons per Bohr^3 (in atomic units) in the vicinity of the point ($\mathbf{r}$) with spin (σ). The spin-less version of γ is

[13] For a more extensive discussion of properties of quantum mechanical density matrices, see P. O. Löwdin, *Phys. Rev.*, **97**, 1474 (1955); see also *Reduced Density Matrices with Applications to Physical and Chemical Systems*, A. J. Coleman and R. M. Erdahl (eds.), Queen's Papers on Pure and Applied Mathematics, No. 11, Queens University, Kingston, Ontario, 1968. See also E. R. Davidson, *Reduced Density Matrices in Quantum Chemistry*, Academic Press, New York 1976.

usually denoted by $\rho(\mathbf{r}_1|\mathbf{r}_1')$, and is given by

$$\rho(\mathbf{r}_1|\mathbf{r}_1') = \int \gamma(\tau_1|\tau_1')\, d\sigma_1. \qquad (12\text{-}24)$$

Thus, we see that an alternative to charge and bond order matrices[14] for chemical interpretation that has a rigorous and basis-independent definition is provided by the first-order reduced density matrix. Furthermore, these quantities can be calculated for any wavefunction, whether exact or approximate.

Before proceeding further, it is of interest to make a point about notation. In particular, each of the matrices defined in this section has continuous indices, and formation of the trace of such matrices implies the need for integration and not simply summation over discrete indices. For example,

$$\mathrm{Tr}[\rho(\mathbf{r}_1|\mathbf{r}_1')] = \int \rho(\mathbf{r}_1|\mathbf{r}_1)\, d\mathbf{r}_1 = N, \qquad (12\text{-}25)$$

where the primes have been removed (i.e., the diagonal elements of ρ have been formed) and integration carried out subsequently over unprimed coordinates. Thus, the reduced density matrix can be thought of as the kernel of an integral operator, and the general result illustrating this point is given by

$$\mathrm{Tr}\ \Gamma^{(p)}(\boldsymbol{\tau}|\boldsymbol{\tau}') = \int \Gamma^{(p)}(\boldsymbol{\tau}|\boldsymbol{\tau}')\, d\boldsymbol{\tau} = \binom{N}{p}. \qquad (12\text{-}26)$$

Thus far our discussion has not introduced basis sets, but has dealt with the overall wavefunction (whether approximate or exact). In practice, however, calculations involve limited basis set expansions, and different basis sets will give different representations of the various reduced density matrices. This difference can be used to considerable advantage, however, as we shall see shortly. Specifically, we shall see how the use of reduced density matrices can provide a way to analyze various approximate wavefunctions regardless of basis set choice or configuration interaction size.

Since much of the analysis is based on the reduced first-order density matrix, let us consider it in greater detail. We consider an arbitrary configuration interaction wavefunction for an N-electron system, written as follows:

$$\Psi(1, 2, \cdots N) = \sum_K C_K \Phi_K(1, 2, \cdots N), \qquad (12\text{-}27)$$

where

$$\Phi_K(1, 2, \cdots N) = \frac{1}{(N!)^{1/2}} \det\{\varphi_1^K(1)\varphi_2^K(2) \cdots \varphi_N^K(N)\}. \qquad (12\text{-}28)$$

The $\{\varphi_k^K\}$ may be either individual basis orbitals or molecular orbitals. We can

[14] See Section 11-2.E, where the relationship of the first-order density matrix to probability density is discussed further.

use Eq. (12-23) to write the reduced first order density matrix as

$$\gamma(\boldsymbol{\tau}_1|\boldsymbol{\tau}_1')=N\sum_{K,L} C_K^* C_L \int \cdots \int \Phi_K^*(\boldsymbol{\tau}_1'\boldsymbol{\tau}_2 \cdots \boldsymbol{\tau}_N)$$

$$\cdot \Phi_L(\boldsymbol{\tau}_1\boldsymbol{\tau}_2 \cdots \boldsymbol{\tau}_N)\, d\boldsymbol{\tau}_2 \cdots d\boldsymbol{\tau}_N. \tag{12-29}$$

It should be noted that the orbitals used in the construction of Φ_K will not necessarily be the same as those used in Φ_L, although they all come from the same basis. By expanding each of the Slater determinants in the above equation using a Laplace expansion along the first row, we can rewrite Eq. (12-29) as

$$\gamma(\boldsymbol{\tau}_1|\boldsymbol{\tau}_1')=\sum_{K,L} C_K^* C_L \sum_{k,l} \varphi_k^K(\boldsymbol{\tau}_1')^* \varphi_l^L(\boldsymbol{\tau}_1) D_{KL}(k|l)$$

$$=\sum_k^{(K)}\sum_l^{(L)} \varphi_k^K(\boldsymbol{\tau}_1')^*\varphi_l^L(\boldsymbol{\tau}_1)\left(\sum_K^{(k)}\sum_L^{(l)} C_K^* D_{KL}(k|l) C_L\right), \tag{12-30}$$

where the summations over K and k include only orbitals from the set $\{\varphi_k^K\}$, the summations over L and l include only orbitals in the set $\{\varphi_l^L\}$, and where $D_{KL}(k|l)$ is the minor of the determinant

$$D=\begin{vmatrix} S_{11} & S_{12} & \cdots & S_{1N} \\ S_{21} & S_{22} & \cdots & S_{2N} \\ \vdots & \vdots & \ddots & \vdots \\ S_{N1} & S_{N2} & \cdots & S_{NN} \end{vmatrix}, \tag{12-31}$$

formed by removing the kth row and lth column from D, and where

$$S_{pq}=\int \varphi_p^K(\boldsymbol{\tau})^*\varphi_q^L(\boldsymbol{\tau})\, d\boldsymbol{\tau}. \tag{12-32}$$

The quantity in parentheses (which is also known as the *first-order reduced density matrix*) in Eq. (12-30) is seen to be completely determined by the choice of orbitals and configurations, and is given the symbol γ_R, i.e.,

$$\gamma_R(k|l)=\sum_K^{(k)}\sum_L^{(l)} C_K^* D_{KL}(k|l) C_L. \tag{12-33}$$

Inserting this definition into Eq. (12-30) gives

$$\gamma(\boldsymbol{\tau}_1|\boldsymbol{\tau}_1')=\sum_{k,l}^{(K)(L)} \varphi_k^K(\boldsymbol{\tau}_1')^*\gamma_R(k|l)\varphi_l^L(\boldsymbol{\tau}_1). \tag{12-34}$$

Since the elements of $\gamma_R(k|l)$ form a Hermitian matrix,[15] we may use a unitary transformation to bring $\boldsymbol{\gamma}_R$ to diagonal form,[16], i.e.,

$$\mathbf{U}^\dagger\boldsymbol{\gamma}_R\mathbf{U}=n, \tag{12-35}$$

[15] See Problem 12-7.

[16] See Theorem 3-18.

where $\mathbf{U}$ is unitary and n is diagonal and contains the eigenvalues of $\boldsymbol{\gamma}_R$.

If we now define a new set of orbitals via

$$\chi = \mathbf{U}\varphi, \tag{12-36}$$

we can rewrite Eq. (12-34) as

$$\begin{aligned}\gamma(\tau_1|\tau_1') &= \varphi^{\dagger}\gamma_R\varphi \\ &= \varphi^{\dagger}\mathbf{U}^{\dagger}\mathbf{U}\gamma_R\mathbf{U}^{\dagger}\mathbf{U}\varphi \\ &= \chi^{\dagger}n\chi,\end{aligned} \tag{12-37}$$

or

$$\gamma(\tau_1|\tau_1') = \sum_k n_k\chi_k^*(\tau_1')\chi_k(\tau_1). \tag{12-38}$$

The functions $\{\chi_k\}$ are known as *natural orbitals* and the $\{n_k\}$ are referred to as the *occupation numbers* of the natural orbitals.[17] If the natural orbitals are used to form a new set of configurations, the *natural expansion of the wavefunction* is accomplished, i.e.,

$$\Psi(1, 2, \cdots N) = \sum_L d_L X_L(1, 2, \cdots N), \tag{12-39}$$

where

$$d_L = \sum_K c_K U_{KL}, \tag{12-40}$$

and the $\{X_L\}$ are configurations using natural orbitals as the basis.

There are a number of advantages (and some disadvantages as well) to finding the natural orbitals and natural expansion of a wavefunction. First, it allows direct comparisons to be made for any set of approximate wavefunctions, regardless of the basis sets or configurations used. In other words, the natural orbitals and natural expansion are basis independent, and thus provide an excellent framework for assessing both similar and disparate approaches to approximate wavefunction construction.

In addition, there is substantial evidence that *convergence of a configuration interaction expansion*, at least for small molecules, is enhanced by the use of natural orbitals. To see why such behavior is plausible and expected, consider

---

[17] The concept of natural orbitals and natural orbital expansions was introduced by Löwdin and Shull, i.e., P. O. Löwdin, *Phys. Rev.*, **97**, 1474, 1509 (1955); P. O. Löwdn and H. Shull, *Phys. Rev.*, **101**, 1730 (1956). For additional analysis and reviews of this approach, see B. C. Carlson and J. M. Keller, *Phys. Rev.*, **121**, 659 (1961); E. R. Davidson, *J. Chem. Phys.*, **37**, 577 (1962); *Adv. Chem. Phys.*, **6**, 235–266 (1972); *Rev. Mod. Phys.*, **44**, 451 (1972); *Reduced Density Matrices in Quantum Chemistry*, Academic Press, New York, 1966; and W. A. Bingel and W. Kutzelnigg, *Adv. Quantum Chem.*, **5**, 201 (1970).

the following measure of convergence of an N-electron CI expansion:

$$\vartheta = \left(\frac{1}{N}\right) \sum_k [1 - \gamma_R(k|k)]\gamma_R(k|k), \tag{12-41}$$

where $\gamma_R(k|k)$ is the kth element along the diagonal of the first order reduced density matrix. Since we know that[18]

$$0 \leq n_k \leq 1, \qquad \sum_k n_k = N, \tag{12-42}$$

let us consider the limiting case where

$$n_k = 1, \qquad k = 1, 2, \cdots N, \tag{12-43}$$

which implies that all other $n_k = 0$ $(k > N)$. Thus, for this case only N terms are needed to describe the system, and we have

$$\vartheta = \left(\frac{1}{N}\right) \sum_{k=1}^{N} [1 - \gamma_R(k|k)]\gamma_R(k|k) \tag{12-44}$$

$$= \left(\frac{1}{N}\right) \left\{ \sum_{k=1}^{N} \gamma_R(k|k) - \sum_{k=1}^{N} [\gamma_R(k|k)]^2 \right\}$$

$$= 0. \tag{12-45}$$

The other case of interest is when each of the n_k is vanishingly small, in which a very large number of orbitals are needed to describe the system. In that case, we have

$$\vartheta = \left(\frac{1}{N}\right) \sum_{k=1}^{(\infty)} [1 - \gamma_R(k|k)]\gamma_R(k|k)$$

$$= \lim_{n_k \to 0} \left[1 - \left(\frac{1}{N}\right) \sum_{k=1}^{(\infty)} n_k^2\right] \to 1. \tag{12-46}$$

Thus, we have

$$0 \leq \vartheta < 1, \tag{12-47}$$

which implies that ϑ can be used as a measure of the convergence of the approximate wavefunction to the exact wavefunction. For example, if one compares several trial wavefunctions, the one having the smallest value of ϑ is considered to be the one having the most rapid convergence.

To see why natural orbitals are considered to be important in improving convergence, we note that, when natural orbitals are employed, the last term in

[18] See Problem 12-8.

Eq. (12-44) is given by

$$\sum_{k=1}^{N} [\gamma_R(k|k)]^2 = \mathrm{Tr}(\gamma_R^2)$$
$$= \sum_k n_k^2. \tag{12-48}$$

Now let us calculate the above expression again, using orbitals other than natural orbitals, i.e.,

$$\sum_k [\gamma_R(k|k)]^2 = \sum_k (\gamma_R^\dagger \gamma_R)_{kk} = \sum_k \left(\sum_l (\gamma_R^\dagger)_{kl} (\gamma_R)_{lk} \right)$$
$$= \sum_{k,l} |(\gamma)_{lk}|^2$$
$$= \sum_k |\gamma_{kk}|^2 + \sum_{k \neq l} |\gamma_{lk}|^2, \tag{12-49}$$

where the simplified notation $\gamma_{lk} \equiv (\gamma_R)_{lk}$ has been used. Comparing Eqs. (12-48) and (12-49) yields

$$\sum_k n_k^2 = \sum_k |\gamma_{kk}|^2 + \sum_{k \neq l} |\gamma_{lk}|^2, \tag{12-50}$$

or

$$\sum_k n_k^2 \geq \sum_k |\gamma_{kk}|^2. \tag{12-51}$$

Thus, we see that, for a given number of terms, use of natural orbitals will maximize the contribution of the second term on the right-hand side of Eq. (12-44), and therefore provide the best (i.e., most efficient) description.

As a practical example, a 10-term CI wavefunction for the ground state of the He atom was considered,[19] which gave $\vartheta = 0.0732$. When converted to a natural orbital expansion, the original $M(M + 1)/2$ terms ($=10$) are reduced to M terms ($=4$), and the 4 term expansion produced $\vartheta = 0.0087$.

Not only is an efficient expansion assured, there are also conceptual advantages associated with the use of natural orbitals. For example, in the He atom case, use of widely different types of basis sets and/or CI expansions give a remarkably similar description in terms of natural orbitals, as shown in Table 12.1. These data, taken from the review by Davidson,[20] represent the coefficients (d_{il}) of the natural expansion written in the form:

$$\Psi(1, 2) = \sum_{i,l} d_{il} \sum_m g_{ilm}(\mathbf{r}_1) g_{ilm}^*(\mathbf{r}_2). \tag{12-52}$$

[19] For examples including both the He atom and H_2 molecule, see H. Shull and P. O. Löwdin, *J. Chem. Phys.*, **23**, 1565 (1955); **30**, 617 (1959); and H. Shull, *J. Chem. Phys.*, **30**, 1405 (1959).

[20] E. R. Davidson, *Adv. Quantum Chem.*, **6**, 235–266 (1972).

Table 12.1. Coefficients in the Natural Expansion of the He Atom Ground State

	Davidson[a]	Ahlrichs *et al.*[b]	Shull and Löwdin[c]	Banyard and Baker[d]
1*s*	0.99599	0.99622	0.99592	0.99598
1*p*	−0.03563	−0.03467	−0.03582	−0.03574
2*s*	−0.06148	−0.06003	−0.06204	−0.06163
1*d*	−0.00566	−0.00545	−0.00599	−0.00566
2*p*	−0.00638	−0.00552	−0.00673	−0.00643
3*s*	−0.00786	−0.00681	−0.00735	−0.00790
1*f*	−0.00169	−0.00161	−0.00112	−0.00169
2*d*	−0.00178	−0.00148	−0.00099	−0.00174
3*p*	−0.00180	−0.00134	−0.00124	−0.00189
4*s*	−0.00197	−0.00144	−0.00221	−0.00192

[a] E. R. Davidson, *J. Chem. Phys.*, **39**, 875 (1963).

[b] R. Ahlrichs, W. Kutzelnigg, and W. A. Bingel, *Theoret. Chim. Acta*, **5**, 289 (1966).

[c] H. Shull and P. O. Löwdin, *J. Chem. Phys.*, **30**, 617 (1959).

[d] K. E. Banyard and C. C. Baker, *J. Chem. Phys.*, **51**, 2680 (1969).

Such invariance of occupation numbers to basis set or CI choice is not only convenient for comparison of different approximate wavefunctions, but also suggests a kind of "universality" of natural orbitals for describing atomic and molecular systems.

Furthermore, examination of the natural orbitals themselves leads to understanding of the role that each natural orbital plays. In the He example, the first natural orbital is essentially the Hartree–Fock orbital, while the first correlation orbital introduces "angular correlation" between the electrons and the second correlation orbital introduces "in–out" correlation. Other two-electron systems have also been analyzed using natural orbitals with analogous results.[21]

The substantial computational efficiency and chemical insight obtained in studying two-electron systems led to considerable optimism concerning application of these ideas to many-electron systems. Experience has indeed shown that faster convergence can be obtained by use of procedures based on natural orbitals. To illustrate this, it is of interest to consider a comparison that was made[22] of the correlation energy recovery for two different CI expansions for

[21] For H_2, see for example, S. Hagstrom and H. Shull, *Rev. Mod. Phys.*, **35**, 624 (1963), E. R. Davidson and L. C. Jones, *J. Chem. Phys.*, **37**, 2966 (1962), and S. Rothenberg and E. R. Davidson, *J. Chem. Phys.*, **45**, 2560 (1966). For HeH^+, see for example, B. G. Anex and H. Shull, in *Molecular Orbitals in Chemistry, Physics, and Biology*, P. O. Löwdin and B. Pullman (eds.), Academic Press, New York, 1964, p. 227. For H_3^+, see for example, R. E. Christoffersen and H. Shull, *J. Chem. Phys.*, **48**, 1790 (1968) and W. Kutzelnigg, R. Ahlrichs, I. Labib-Iskander, and W. A. Bingel, *Chem. Phys. Lett.*, **1**, 447 (1967).

[22] I. Shavitt, B. J. Rosenberg, and S. Palalikit, *Int. J. Quantum Chem.*, **S10**, 33 (1976).

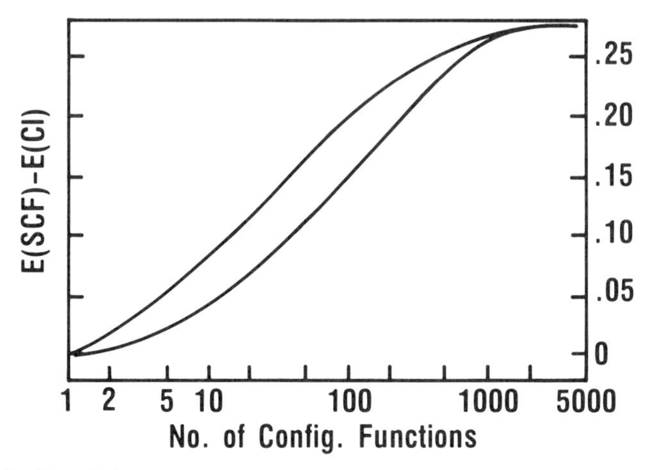

Figure 12.2. Correlation energy recovery in H_2O from CI wavefunctions based on natural orbitals (NO-upper curve) and canonical SCF orbitals (MO-lower curve). The configurations were chosen on the basis of decreasing energy contributions. Data are taken from I. Shavitt, *Methods of Electronic Structure Theory*, H. F. Schaefer (ed.), Plenum Press, New York, 1977, pp. 189–275.

the ground state of the H_2O molecule. One expansion used natural orbitals, while the other used canonical SCF orbitals. In this study, a set of 39 STO basis orbitals was used to form a CI wavefunction consisting of all single and double excitations from the SCF configuration (4120 configurations total). Then a series of truncated CI calculations were carried out, using natural orbitals in one case and canonical SCF orbitals in the other. The results of this study are given in Fig. 12.2. As the data indicate, the correlation energy recovery is greater for the natural orbital-based CI wavefunction at each length of CI expansion, thus indicating the more rapid convergence of the natural orbital approach.

a. Nature of the Chemical Bond

In addition to the computational properties of natural orbitals already mentioned, they have also proved to be powerful tools for understanding the nature of chemical bonding in molecules. To illustrate this, we shall consider a natural spin orbital analysis of various H_2 ground state wavefunctions that was carried out by Shull.[23]

As we have seen earlier,[24] the space and spin portions of a two-electron wavefunction can be factored, i.e., any two-electron wavefunction can be written in the form

$$\Psi(1, 2) = \psi(\mathbf{r}_1, \mathbf{r}_2) \cdot \left[\frac{\alpha(1)\beta(2) - \beta(1)\alpha(2)}{\sqrt{2}}\right]. \tag{12-53}$$

[23] H. Shull, *J. Chem. Phys.*, **30**, 1405–1413 (1959); see also S. Hagstrom and H. Shull, *Rev. Mod. Phys.*, **35**, 624 (1963).

[24] See Chapter 10, Section 10-3.

for the singlet case. If an M-term CI wavefunction of orthonormal orbitals is also used, then the spatial part of the above equation can be written as

$$\psi(\mathbf{r}_1, \mathbf{r}_2) = \sum_{k,l}^{M} C_{kl}[\psi_k(\mathbf{r}_1)\psi_l(\mathbf{r}_2) + \psi_l(\mathbf{r}_1)\psi_k(\mathbf{r}_2)]/2, \tag{12-54}$$

where the requirement that $\psi(\mathbf{r}_1, \mathbf{r}_2)$ be totally symmetric (for the H_2 ground state) has been explicitly assured. Since the spin part can be factored, we shall concentrate only on the space part of the wavefunction, and rewrite Eq. (12-54) as:

$$\Psi(\mathbf{r}_1, \mathbf{r}_2) = \sum_{k}^{M} C_{kk}\psi_k(1)\psi_k(2) + \sum_{k<l}^{M} C_k C_l[\psi_k(1)\psi_l(2) + \psi_l(1)\psi_k(2)], \tag{12-55}$$

where it is easily seen that there are $\frac{1}{2}M(M+1)$ terms in Eq. (12-55).

The first order density matrix for this wavefunction is given by

$$\gamma(\mathbf{r}_1'|\mathbf{r}_1) = \sum_{k,l}^{M} \psi_k^*(\mathbf{r}_1')\psi_l(\mathbf{r}_1)\gamma_R(l|k), \tag{12-56}$$

where the reduced first order density matrix (γ_R) is given by

$$\gamma_R(l|k) = \sum_{m}^{M} C_{lm}C_{km}^*, \tag{12-57}$$

or

$$\boldsymbol{\gamma}_R = \mathbf{C}\mathbf{C}^\dagger. \tag{12-58}$$

If $\mathbf{V}$ is the matrix that diagonalizes $\boldsymbol{\gamma}_R$, then the natural orbital occupation numbers are given by:[25]

$$\mathbf{V}\gamma_R\mathbf{V}^\dagger = \mathbf{V}\mathbf{C}^2\mathbf{V}^\dagger = \text{diagonal matrix} = \mathbf{c}^2. \tag{12-59}$$

Thus, for this special case,

$$n_k = c_k^2 \qquad \text{or} \qquad c_k = \pm n_k^{1/2}, \tag{12-60}$$

and

$$\chi = \Psi\mathbf{V}. \tag{12-61}$$

Inserting Eqs. (12-60) and (12-61) into Eq. (12-55) gives

$$\Psi(\mathbf{r}_1, \mathbf{r}_2) = \sum_{k}^{M} c_k\chi_k(\mathbf{r}_1)\chi_k(\mathbf{r}_2), \tag{12-62}$$

[25] See Problem 12-11.

where the c_k are given by Eq. (12-60), and Eq. (12-62) is the explicit expression for the natural expansion of the two-electron wavefunction.

In Table 12.2 are given the energies and occupation numbers that result from an analysis of a variety of approximate wavefunctions for H_2, using the procedure outlined above. In spite of great differences in complexity and completeness of the approximate wavefunctions, there are a number of interesting and revealing observations that can be made from the data. First, only the first two natural orbitals have significant occupation numbers, and essentially the same occupation numbers result regardless of basis set choice and extent of CI. This reinforces the idea that natural orbital occupation numbers are essentially independent of the basis orbitals that are chosen, even when the basis set is truncated and therefore not complete.

Next, the first natural orbital is an approximation to the SCF orbital, and the fact that the occupation number remains near unity even as other natural orbitals are added illustrates the importance of the SCF configuration, and makes apparent why the MO picture has been so successful in describing many aspects of chemical structure.

The second occupation number $[n_2(\sigma_u)^2]$ has an associated natural orbital coefficient that is opposite in sign to the first term. Also, analysis of the shape of this natural orbital indicates that the type of correlation introduced by this term can be classified as "in–out," in which "in" and "out" refer to radial distances from the internuclear axis. The remaining occupation numbers are seen to be relatively small compared to the first two, and suggest that a two-term natural orbital expansion may provide a satisfactory description of at least the homopolar two-electron chemical bond illustrated by the H_2 molecule.

This idea was developed further by Shull and colleagues,[26] in which a two-term truncated natural expansion was postulated to contain essentially all of the information associated with a chemical bond. For the two-electron case, the truncated expansion was taken as

$$\begin{aligned}\Psi(1, 2) &= n_1^{1/2}\chi_1(1)\chi_1(2) - n_2^{1/2}\chi_2(1)\chi_2(2)\\ &= c_1\chi_1(1)\chi_1(2) + c_2\chi_2(1)\chi_2(2),\end{aligned} \tag{12-63}$$

where the spin dependence has been omitted and $|c_i| = +n_i^{1/2}$. We also note that, since χ_1 and χ_2 are orthogonal,

$$n_1 + n_2 = 1. \tag{12-64}$$

In addition to the independence from basis set choice and other data cited above for H_2 that supports such a choice, there are additional reasons, intuitive

[26] H. Shull, *J. Am. Chem. Soc.*, **82**, 1287–1295 (1960); *J. Phys. Chem.*, **66**, 2320 (1962); S. Hagstrom and H. Shull, *Rev. Mod. Phys.*, **35**, 624 (1963); H. Shull and F. Prosser, *J. Chem. Phys.*, **40**, 233 (1964); H. Shull, *J. Am. Chem. Soc.*, **86**, 1469 (1964); B. Anex and H. Shull, in *Molecular Orbitals in Chemistry, Physics and Biology*, P. O. Löwdin and B. Pullman (eds.), Academic Press, New York, 1964, p. 227; R. E. Christoffersen and H. Shull, *J. Chem. Phys.*,**48**, 1790–1797 (1968).

Table 12.2. Energies and Occupation Numbers for Approximate H_2 Wavefunctions[a]

Type of wavefunction	Energy (Hartree)	$n_1(\sigma_g)^2$	$n_2(\sigma_u)^2$	$n_3(\pi)$	$n_4(\sigma_g)^2$	$n_5(\sigma)^2$
Unscaled Heitler–London–Sigiura[b]	−1.1160	0.9663 (+0.9830)	0.0337 (−0.1836)	—	—	—
Scaled Heitler–London–Wang[c]	−1.13910	0.9667 (+0.9832)	0.0333 (−0.1824)	—	—	—
Wang + ionic (Weinbaum)[d]	−1.14796	0.9867 (+0.9933)	0.0133 (−0.1153)	—	—	—
Wang + polar (Rosen)[e]	−1.1485	0.9845 (+0.9922)	0.0155 (−0.1245)	—	—	—
Floating Weinbaum[f]	−1.1501	0.9811 (+0.9905)	0.0189 (−0.1374)	—	—	—
Inui + ionic[g]	−1.154	0.9908 (+0.9954)	0.0092 (−0.0959)	—	—	—
Wang + ionic + polar[d]	−1.1515	0.9908 (+0.9954)	0.0092 (−0.0959)	—	0.0000	0.0000
Hirschfelder–Linnett[h] ($R = 1.4731$)	−1.1560	0.9772 (+0.9885)	0.0184 (−0.1356)	0.0041	0.0002	0.0001
Hagstrom–Shull[i]	−1.161409	0.9815 (+0.9907)	0.0099 (−0.0995)	0.0045	0.0037	0.0002
Newell[j]	−1.16781	0.9813 (+0.9906)	0.0104 (−0.1020)	0.0050	0.0033	0.0000

[a] $R = 1.4$ Bohrs unless otherwise specified, and coefficients (c_k) of natural spin orbitals are given in parentheses.

[b] Y. Sigiura, *Z. Phys.*, **45**, 484 (1927).

[c] S. C. Wang, *Phys. Rev.*, **31**, 579 (1928).

[d] S. Weinbaum, *J. Chem. Phys.*, **1**, 593 (1933).

[e] N. Rosen, *Phys. Rev.*, **38**, 2099 (1931).

[f] H. Shull and D. D. Ebbing, *J. Chem. Phys.*, **28**, 866 (1958).

[g] C. R. Mueller and H. Eyring, *J. Chem. Phys.*, **19**, 1495 (1951).

[h] J. O. Hirschfelder and J. W. Linnett, *J. Chem. Phys.*, **18**, 130 (1950).

[i] S. Hagstrom and H. Shull, *J. Chem. Phys.*, **30**, 1314 (1959).

[j] G. F. Newell, *Phys. Rev.*, **78**, 711 (1950).

and otherwise, for such a choice. For example, we saw for H_2 that the first natural configuration (σ_g^2) served the purpose of assuring that the overall electron distribution was correct, while the second natural configuration introduced an "in–out" (or "left–right")-type correlation effect. Qualitatively, the second term can be thought of as allowing the situation where, when one electron is in the vicinity of one nucleus, the second electron will be in the vicinity of the other nucleus. Also, since the third natural configuration had "π-symmetry," and since the bonding in H_2 and other "normal" single bonds is of "σ-type," we expect π-type contributions to be of lesser importance. We shall now see how these numerical observations and intuitive justifications can be used to provide a theoretical basis for the concepts of "ionic" and "covalent" character of two-electron chemical bonds that was introduced earlier on intuitive grounds by Pauling.[27]

To aid in the analysis, we define several types of wavefunctions, e.g., *"atomic"* (Ψ_A):

$$\Psi_A = N_+ \{u(1)v(2) + v(1)u(2)\}, \tag{12-65}$$

and *"ionic"* (Ψ_{I+} and Ψ_{I-}):

$$\Psi_{I+} = N_+ \{u(1)u(2) + v(1)v(2)\}, \tag{12-66}$$

$$\Psi_{I-} = N_- \{u(1)u(2) - v(1)v(2)\}, \tag{12-67}$$

where u and v are *arbitrary*, normalized space orbitals, and

$$N_\pm = [2(1 \pm S_{uv})]^{-1/2}, \tag{12-68}$$

with

$$S_{uv} = \int uv \, d\tau. \tag{12-69}$$

We shall now see how such a choice of names can have significance, irrespective of the choice of functions u and v. To do this we shall utilize natural orbitals, which we know to be independent of basis set choice, and write u and v as

$$u = (\cos \alpha)\chi_1 + (\sin \alpha)\chi_2$$

$$v = (\sin \beta)\chi_1 - (\cos \beta)\chi_2, \tag{12-70}$$

where α and β are parameters to be chosen shortly. It is easily seen that

$$S_{uv} = \sin(\beta - \alpha) \tag{12-71}$$

and that u and v can be made orthogonal by the choice

$$\beta - \alpha = n\pi \qquad (n = 0, 1, 2, \cdots). \tag{12-72}$$

We shall deal only with the principal case of $n = 0$, which gives

$$\alpha = \beta. \tag{12-73}$$

[27] L. Pauling, *Nature of the Chemical Bond*, Cornell University Press, Ithaca, NY, 1939.

By this choice not only u and v, but also Ψ_A, Ψ_{I+} and Ψ_{I-} can be seen to be mutually orthogonal.[28]

At this point we see that at least two of the major deficiencies in early ideas of ionic and covalent character have been removed. First, the functions to be used do not refer to any particular basis set, i.e., they are independent of basis set choice. Second, the functions to be used to develop concepts analogous to ionic and covalent character are orthogonal, thus assuring independence of new concepts that may arise.

By finding χ_1 and χ_2 in terms of u, v, α and β, plus substitution of expressions for χ_1 and χ_2 into Eq. (12-63), it can be shown[29] that

$$\Psi(1, 2) = \lambda_A \Psi_A + \lambda_{I+} \Psi_{I+} + \lambda_{I-} \Psi_{I-}, \tag{12-74}$$

in which

$$\lambda_A = \left[\frac{N_+}{2\cos^2(\alpha-\beta)}\right] \{(n_1^{1/2}+n_2^{1/2})\sin(\alpha+\beta) + (n_1^{1/2}-n_2^{1/2})\sin(\alpha-\beta)\}, \tag{12-75}$$

$$\lambda_{I+} = \left[\frac{N_+}{2\cos^2(\alpha-\beta)}\right] \{(n_1^{1/2}-n_2^{1/2})\cos(\alpha-\beta), \tag{12-76}$$

and

$$\lambda_{I-} = \left[\frac{N_-}{2\cos^2(\alpha-\beta)}\right] \{(n_1^{1/2}+n_2^{1/2})\cos(\alpha+\beta)\}. \tag{12-77}$$

Shull has also shown that the particular choice

$$\alpha = \beta = \pi/4. \tag{12-78}$$

will give $\lambda_{I-} = 0$ for the symmetric (homopolar) case, and that this choice also gives Ψ_A and Ψ_{I+} optimal "atomic" and "ionic" characteristics. This includes the property

$$\lambda_A^2 + \lambda_{I+}^2 = 1, \tag{12-79}$$

which allows interpretation of λ_A^2 and λ_{I+}^2 as the *fractional atomic* and *ionic character*, respectively. The choice of α and β in Eq. (12-78) also leads to

$$u = \left(\frac{1}{\sqrt{2}}\right)(\chi_1 + \chi_2)$$
$$v = \left(\frac{1}{\sqrt{2}}\right)(\chi_1 - \chi_2), \tag{12-80}$$

[28] See Problem 12-12.

[29] See Problem 12-13.

Table 12.3. Atomic and Ionic Character of Wang Function for H_2 as a Function of Internuclear Distance

R	λ_A^2	λ_{I+}^2	$2\lambda_A\lambda_{I+}$
0.0000	0.5361	0.4639	0.9974
0.7339	0.5758	0.4242	0.9885
1.404 [a]	0.6795	0.3205	0.9334
1.7976	0.7441	0.2559	0.8727
2.9916	0.8917	0.1083	0.6215
4.0274	0.9654	0.0346	0.3654
5.050	0.99076	0.00924	0.1914
10.02	0.999996	0.000004	0.004

[a] Equilibrium distance.

and

$$\Psi_A = \left(\frac{1}{\sqrt{2}}\right)[\chi_1(1)\chi_1(2) - \chi_2(1)\chi_2(2)]$$
$$\Psi_{I+} = \left(\frac{1}{\sqrt{2}}\right)[\chi_1(1)\chi_1(2) + \chi_2(1)\chi_2(2)]. \qquad (12\text{-}81)$$

To illustrate the applicability of these concepts, the Wang function

$$\Psi_{WA} = N[1s_a(1)1s_b(2) + 1s_b(1)1s_a(2)]$$

and Weinbaum function

$$\Psi_{WI} = N[\Psi_{WA} + \lambda(1s_a(1)1s_a(2) + 1s_b(1)1s_b(2))]$$

for H_2 were examined by Shull, and relevant data are summarized in Table 12.3. Among the points of interest in these data are the gradual decrease of ionic character and corresponding increase in atomic character as R increases, as would be expected in satisfactory definitions of atomic and ionic character. This variation is largely ignored in the Pauling analysis, and adds to the appropriateness of using concepts such as "atomic" and "ionic" in this context.

Next, we note that, at the equilibrium separation ($R = 1.4$ Bohrs), approximately 68% atomic character and 32% ionic character is indicated for H_2. This is a striking difference from the Pauling approach, where a 5% ionic character is found for H_2. However, to place these observations in context, it is important to include energetic considerations as well. In particular, the following energy components are found for the Wang function at the equilibrium internuclear distance:

$$\lambda_A^2 H_{AA} = -0.38030$$
$$2\lambda_A\lambda_{I+} H_{AI+} = -0.68738$$
$$\lambda_{I+}^2 H_{II} = -0.07138$$
$$E\text{ (total)} = -1.13905$$

Thus, we see that the "cross-term," $2\lambda_A \lambda_I H_{AI}$, is the largest contributor to the energy. It is this term that is argued to be most closely connected with the "covalent" concept. The H_{AA} and H_{II} terms have negative energies, but far less than -1.0 Hartree (the energy of the separated atoms), and Ψ_A or Ψ_{I+} are thus neither strongly bonding nor strongly antibonding.

In any case, it is seen that the use of natural orbitals greatly facilitates the development of concepts that can be applied without reference to any particular basis set, and allows extraction of principles of chemical bonding using independent orthogonal parts of wavefunctions that can be closely identified with intuitive chemical principles.

In spite of these advantages, there are still several difficulties that are encountered in the use of natural orbitals for CI studies. Among them is that the convergence for systems beyond two electrons has not been found to be as fast as for two-electron systems, thus reducing the expected computational advantages of natural orbitals.

Another difficulty has to do with the way in which natural orbitals are defined. In particular, we have defined them in terms of the final CI wavefunction, implying in effect that the natural orbitals are not needed after all, except for interpretive purposes. This has led to a number of attempts to obtain natural orbitals in different ways. One such method is the "iterative natural orbital" method,[30] in which a series of successively larger CI calculations are carried out using the natural orbitals of the preceding iteration. Others methods for obtaining natural orbitals or approximations thereto have also been devised,[31] and a variety of relatively practical approaches are available for obtaining approximate natural orbitals without having the ultimate CI wavefunction. However, regardless of the utility of natural orbitals as a means of constructing CI wavefunctions, their role in obtaining insight and understanding of wavefunctions and the chemistry they represent is significant indeed.

C. Conceptual and Computational Considerations

When considering how to carry out CI studies using the general techniques described in Section 12-2A while maintaining suitable accuracy, there are a number of conceptual and computational questions that arise,[32] in addition to those regarding basis set choice that were discussed in Section 11-2C and the

[30] C. F. Bender and E. R. Davidson, *J. Chem. Phys.*, **47**, 4792 (1967); *J. Phys. Chem.*, **70**, 2675 (1966); *Phys. Rev.*, **183**, 23 (1969); H. F. Schaefer III, *J. Chem. Phys.*, **54**, 2207 (1971).

[31] See, for example, P. J. Hay, *J. Chem. Phys.*, **59**, 2468 (1973); A. K. Siu and E. F. Hayes, *J. Chem. Phys.*, **61**, 37 (1974); C. Edmiston and M. Krauss, *J. Chem. Phys.*, **45**, 1833 (1966); W. Meyer, *J. Chem. Phys.*, **58**, 1017 (1973); S. A. Houlden and I. G. Csizmadia, *Theoret. Chim. Acta*, **30**, 209 (1973).

[32] For an excellent discussion of these aspects of CI calculations, see I. Shavitt, in *Methods of Electronic Structure Theory*, H. F. Schaefer III (ed.), Plenum Press, New York, 1977, pp. 189–275.

general considerations of the previous sections. These questions can be grouped conveniently into several categories[33]: (1) What kinds/number of orbitals (MOs or otherwise) should be used for CI studies? (2) What should be the nature and extent of the CI expansion itself [e.g., should Eq. (12-1) be used directly and straightforwardly]? (3) What algorithms can be used to reduce the steps needed to carry out the large number of manipulations needed for CI studies? (4) How can appropriate "size consistency" and "size extensiveness" be assured, i.e., how can we assure that the techniques scale up appropriately with the size of molecule being treated? We shall discuss each of these issues below, and will see that a number of different options are available, both conceptually and computationally.

1. Orbitals for CI Studies

Since questions relating to which kinds of basis orbitals are available and should be used for SCF studies has been addressed earlier, we shall focus here on the question of which orbitals are best for use in CI studies. In principle, the answer to this question is straightforward, i.e., the orbitals that are optimum for correlation purposes are those that are located in the same regions of space as the electrons whose motion they are to correlate. In practice, however, the situation is substantially more complex than the previous statement would lead the reader to believe. Part of the reason for this is that the region of space of interest varies with the kind of problem under investigation, e.g., correlating orbitals that are appropriate for potential surface studies may be localized in quite different regions than those required for excited state studies.

To examine some of the common situations in which appropriate correlation orbitals are sought, let us begin by considering ground states of molecules at (or near) their equilibrium nuclear configuration. In this case the canonical SCF filled and virtual orbitals frequently form quite a reasonable starting point. The orbitals are automatically orthogonal, as we saw in the discussion of SCF theory in Chapter 11, which is of significant computational importance. It also generally implies that the virtual orbitals are located in the same regions of space as the filled orbitals, and differ primarily from their filled counterparts in the number of nodes that are present. However, it should be noted that the filled Hartree–Fock orbitals are each determined in an effective field of $N - 1$ electrons, while the corresponding virtual orbitals are determined in an effective field of N electrons. The general effect of this, especially for large basis sets, is that low-energy canonical SCF virtual orbitals are more diffuse than their filled counterparts, and care must be exercised both in the choice of virtual orbitals and the nature of the CI expansion used. Nevertheless, the canonical virtual orbitals have been found generally to be quite useful and convenient as ground state correlation orbitals, especially when used in large-scale CI studies. For

[33] For additional discussion on criteria that may be used to measure the applicability and suitability of theoretical approaches see, e.g., J. A. Pople, J. S. Binkley, and R. Seeger, *Int. J. Quantum Chem.*, **10**, 1 (1976).

example, as noted by Davidson,[34] over 50% of the correlation energy can usually be obtained for small molecules using the virtual orbitals from a double zeta plus polarization function basis set.

For other types of studies, such as spectra or potential surfaces, the choice of orbitals appropriate for CI studies is complicated by several factors. For example, let us consider the case in which spectral studies involving transitions from a localized ground state to a large number of excited states having varied character (e.g., varied degrees of charge distribution reorganization, localized states, Rydberg states, etc.) are desired. In such a case, construction of orbitals for correlation purposes that satisfy *all* situations is in general not possible. Instead, a variety of approaches have been developed that generally apply well in some circumstances and not so well in others. In addition, the orbitals and their applicability are also tied frequently to a particular CI formulation, making their general evaluation more complicated.

Perhaps the most straightforward approach is the one just described for ground state correlation studies, i.e., to use the canonical Hartree–Fock virtual orbitals directly in the CI study of each excited state. Whitten and Hackmeyer,[35] for example, have used such as approach successfully within a specific CI formulation to describe singlet and triplet $\pi \rightarrow \pi^*$ and $n \rightarrow \pi^*$ spectra of organic molecules. While simple and straightforward conceptually and computationally, it must be remembered that states whose charge distribution are dissimilar to the ground state will not generally be well described by the use of ground state SCF virtual orbitals, regardless of CI approach used, unless the basis is augmented by orbitals that reflect the excited state charge redistribution.

Many other approaches have also been devised for obtaining appropriate correlation orbitals. For example, several approaches begin from the observation that it does not matter *how* the correlation orbitals are determined, as long as they are located in appropriate regions of space and are effective in CI expansions. As a result, some studies have used Hamiltonians *different* than the SCF Hamiltonian[36] to determine correlation orbitals. Of course, maintenance of orthogonality to SCF-determined orbitals is an important component of such approaches.

Alternatively, other investigators have created *localized orbitals* for use in CI studies, e.g., using the techniques introduced earlier (Section 10-9). This approach is conceptually quite different than using modified Hamiltonians for determining correlation orbitals, in that the localization process is separate from

---

[34] E. R. Davidson, *The World of Quantum Chemistry*, R. Daudel and B. Pullman (eds.), D. Reidel, Dordrecht, Holland, 1974, p. 17.

[35] J. L. Whitten and M. Hackmeyer, *J. Chem. Phys.*, **51**, 5584–5596 (1969); M. Hackmeyer and J. L. Whitten, *J. Chem. Phys.*, **54**, 3739–3750 (1971).

[36] See, for example, S. Huzinaga and C. Arnan, *Phys. Rev.*, **A1**, 1285 (1970): *J. Chem. Phys.*, **54**, 1948 (1971); E. R. Davidson, *J. Chem. Phys.*, **57**, 1999 (1972); H. P. Kelly, *Phys. Rev.*, **136**, B896 (1964); W. J. Hunt and W. A. Goddard III, *Chem. Phys. Lett.*, **3**, 414 (1969); N. Björna, *J. Phys.*, **56**, 1412 (1973); L. R. Kahn, P. J. Hay, and I. Shavitt, *J. Chem. Phys.*, **61**, 3530 (1974).

any Hartree–Fock SCF studies that may be done. The results are similar, in the sense that correlation orbitals are obtained that are localized in appropriate regions of space for CI studies. However, one of the practical difficulties that typically arises when localized orbitals are used is that the computational conveniences associated with automatically orthogonal orbitals or orbitals that have well-defined symmetry properties relative to the filled SCF orbitals are lost. Such complications can significantly restrict the range of applicability of these approaches.

Of course, it is to be expected from the discussion in Section 12-2B2 that natural orbitals or some approximation thereto will also be effective as correlation orbitals. However, since natural orbitals are determined generally by retrospective analysis of the density matrix formed from an already completed CI wavefunction, construction of such orbitals in advance of a CI study is problematical. It is possible in practice to carry out the process iteratively, and a number of approaches have been suggested. For example, use of SCF virtuals in an initial CI study followed by diagonalization of the resulting first-order density matrix will give approximate natural orbitals that can be used directly or augmented by other orbitals and used in subsequent CI study from which new natural orbitals can be obtained, etc. Such *iterative natural orbital* approaches,[37] or *pseudonatural orbital* approaches,[38] or *average natural orbital* approaches[39] each provides ways to utilize the conceptual advantages of natural orbitals to construct correlation orbitals for CI studies.

Yet another approach is to carry out separate SCF studies on excited states, and combine the orbitals thus generated with ground state SCF orbitals to create correlation orbitals for CI studies.[40] However, problems of lack of orthogonality generally arise, and the presence of many different excited state symmetries and disparate charge distributions may make such approaches unwieldy.[41]

Finally, a general CI wavefunction can be formed in which both the form of the orbitals *and* configuration coefficients are varied. While this approach will be discussed in Section 12-3A in greater detail, it is useful to note here that the orbitals for each configuration may be different for each configuration in principle. This allows for the possibility of determining better suited correlation orbitals that are also automatically orthogonal, although the computational complexity is increased substantially.

For studies of reaction surfaces, the question of how to choose orbitals for

---

[37] See, for example, E. R. Davidson, *The World of Quantum Chemistry*, R. Daudel and B. Pullman (eds.), Dordrecht, Holland, 1974, p. 17.

[38] C. Edmiston and M. Krauss, *J. Chem. Phys.*, **45**, 1833 (1966); W. Meyer, *J. Chem. Phys.*, **58**, 1017 (1973).

[39] S. A. Houlden and I. G. Csizmadia, *Theoret. Chim. Acta*, **36**, 275 (1973).

[40] See, for example, R. J. Buenker, and S. D. Peyerimhoff, *J. Chem. Phys.*, **53**, 1368 (1970); S. Shih, R. J. Buenker and S. D. Peyerimhoff, *Chem. Phys. Lett.*, **16**, 244 (1972).

[41] For approaches to excited state SCF calculations on states having the same symmetry as the ground state, see the references following Eq. (11-217).

general use is not answerable without also considering the form of the CI expansion that is to be used. In particular, the orbitals in general will be required to change along the reaction surface,[42] and cannot be determined only once in advance.

Thus, when attempting to determine orbitals appropriate for CI studies, universal solutions are not generally available, and each case should be considered separately.

Another issue that arises particularly when reaction surfaces, dissociation energies, or intermolecular interactions in general are to be calculated is known as the *basis set superposition error*. This type of error arises, e.g., when the interaction of two atoms is investigated, and when the basis set for the first atom is incomplete. In that case, the basis set for the second atom will, especially for close interatomic distances, be used by the SCF or CI procedure to attempt to make up for deficiencies in the first basis (and vice versa). For a diatomic molecule, for example, an artificial lowering of the two atomic energies in the vicinity of the equilibrium distance will appear as an artificial lowering of the true binding energy, and skew the calculated binding energy as a result.

Unfortunately, the elimination or even estimation of the magnitude of basis set superposition errors is a difficult problem. The ''counterpoise method'' of Boys and Bernardi[43] represents one approach to estimation of this error, in which atomic energies are corrected by computing the atoms in the full molecular basis set. While such errors are expected in all calculations involving multiple atomic basis sets, they can be particularly troubling when interaction energies are small, e.g., in studies of van der Waals complexes.

2. CI Expansions, "Size Consistency," and "Size Extensivity"

If we assume that orbitals appropriate for the problem of interest can be found using techniques as described in the previous section, questions such as the length of the CI expansion needed and the nature of configurations to be included still remain. To address such questions, we begin by identifying in greater detail the various numbers of configurations that can arise. We have already seen [Eq. (12-13)] that the *total* number (T) of determinants/configurations[44] that can be

[42] See, for example, K. Morokuma, S. Kato, K. Kitaura, S. Obara, K. Ohta, and M. Hanamura, in *New Horizons of Quantum Chemistry*, P. O. Löwdin and B. Pullman (eds.), D. Reidel, Dordrecht, Holland, 1983, pp. 221–241.

[43] S. F. Boys and F. Bernardi, *Mol. Phys.*, **19**, 553 (1970); for application of this approach see, for example, D. W. Schwenke and G. D. Truhlar, *J. Chem. Phys.*, **82**, 2418 (1985); **84**, 4113 (1985), and M. Gutowski, J. H. vanLenthe, J. Veerbeck, and G. Chalasinki, *Chem. Phys. Lett.*, **124**, 370 (1985).

[44] Throughout this discussion we shall, for convenience, assume that each configuration can be represented by a single determinant. In practice, linear combinations of determinants may be required to guarantee particular spin and spatial symmetry properties, but we shall use the terms ''configuration'' and ''determinant'' interchangeably to aid in simplifying the concepts involved.

formed from a basis of M spin orbitals ($M > N$) is

$$T = \binom{M}{N}, \tag{12-82}$$

without regard to constraints due to spin and spatial symmetry. In order to understand the implications of this relationship more fully, as well as several of the specialized approaches to CI calculations to be described later, let us consider the above relationship in greater detail using the H_2 molecule in its ground state as an example.

As a simple example of HF–SCF calculations, we earlier considered the ground state ($S = S_z = 0$) of H_2 using three FSGO as basis orbitals (see Chapter 11, Section 11-2B). This gives rise to three MOs and six spin orbitals, into which two electrons must be accommodated. Use of Eq. (12-82) results in a total of

$$T = \binom{6}{2} = 15 \tag{12-83}$$

determinants that are possible, prior to consideration of spin or spatial constraints. Using the notation $k = (m, n)$ to represent the kth configuration comprised of spin orbitals m and n, we can write the 15 possible determinants explicitly as

$$\begin{array}{lll} 1=(1,\ 2) & 2=(1,\ 3) & 3=(1,\ 4) \\ 4=(1,\ 5) & 5=(1,\ 6) & 6=(2,\ 3) \\ 7=(2,\ 4) & 8=(2,\ 5) & 9=(2,\ 6) \\ 10=(3,\ 4) & 11=(3,\ 5) & 12=(3,\ 6) \\ 13=(4,\ 5) & 14=(4,\ 6) & 15=(5,\ 6). \end{array} \tag{12-84}$$

If we now identify the six spin orbitals with their spatial MO counterparts (φ) as follows:

$$\begin{array}{lll} \chi_1 = \varphi_1\alpha, & \chi_2 = \varphi_1\beta, & \chi_3 = \varphi_2\alpha \\ \chi_4 = \varphi_2\beta, & \chi_5 = \varphi_3\alpha, & \chi_6 = \varphi_3\beta, \end{array} \tag{12-85}$$

with configuration 1 taken as the SCF configuration, the 15 determinants of Eq. (12-84) can be grouped into single and double excitation determinants as follows:

Single Excitations: Configurations 2, 3, 4, 5, 6, 7, 8, 9
Double Excitations: Configurations 10, 11, 12, 13, 14, 15

Such an example illustrates several other relationships of interest between the total number of possible determinants and the number of single and double excitation determinants. Specifically, if we choose p orbitals from the total of N occupied orbitals, it can be done in $\binom{N}{p}$ different ways. Similarly, we can choose p orbitals from among the unoccupied orbitals in $\binom{M-N}{p}$ different ways. This means that the total number of determinants (T_p) that can be formed from replacement of p occupied orbitals by p unoccupied orbitals is given by

$$T_p = \binom{N}{p}\binom{M-N}{p} . \qquad (12\text{-}86)$$

In the example cited above, we verify that

$$\begin{aligned} T_0 &= 1 \\ T_1 &= 8 \\ T_2 &= 6. \end{aligned} \qquad (12\text{-}87)$$

Thus far we have not considered either spin or spatial constraints and the effect that they have on the number of determinants of each type that can be formed. For example, in the current H_2 example, consideration of the ground state requires the overall wavefunction spin state to have $S = S_z = 0$ eigenvalues. The usual way in which this is accomplished is to require each configuration to satisfy the $S = S_z = 0$ condition. In this example, satisfying the spin conditions eliminates determinants numbered 2, 4, 7, 9, 11, and 14, leaving a CI wavefunction which contains the SCF determinant (#1), four single excitation determinants (3, 5, 6 and 8), and four double excitation determinants (#10, 12, 13, and 15). This group of nine configurations represents a *full CI* calculation (assuming no spatial symmetry constraints) for the chosen basis, since we have considered all possible excitations from the SCF determinant.

While the previous example illustrates how to determine the number of various types of configurations that can arise, it is useful to extend this example to a larger basis to illustrate just how fast the computational complexity increases. In particular, let us consider a "minimum STO" basis for H_2, i.e., a $1s$, $2s$, $2p_x$, $2p_y$, and $2p_z$ orbital on each atom. This gives rise to 10 spatial orbitals, and thus 10 MOs or 20 spin orbitals for potential use in CI studies. Use of Eq. (12-86) shows that the following configurations are possible (assuming no additional spatial symmetry constraints):

$$\begin{aligned} T_0 &= 1 \\ T_1 &= \binom{2}{1}\binom{18}{1} = 36 \\ T_2 &= \binom{2}{2}\binom{18}{2} = 153, \end{aligned} \qquad (12\text{-}88)$$

for a total of 190 configurations compared to 15 for the 3 FSGO basis. A rapid increase in the relative number of double excitation configurations is seen, and for large basis sets the number of double excitation configurations typically exceeds the number of single excitation configurations by a wide margin.

If spin constraints to assure a singlet state are imposed as in the previous example, i.e., we require each configuration to be a spin eigenfunction having $S = S_z = 0$, the "full CI" wavefunction for the basis (assuming no spatial symmetry constraints) is found to be comprised of the SCF configuration, 18 single excitation configurations, and 45 double excitation configurations, for a total of 64 configurations.

It should also be noted that, since the example includes only two electrons, configurations having greater than double excitations have not been required. For systems with >2 electrons, triple, quadruple, etc., excitations need also to be considered, which complicates the CI expansion even further.[45]

In addition to spin and spatial symmetry constraints that in general reduce the size of the CI expansion, there are also a number of computational considerations that help in this regard. For example, we can draw directly from first order Rayleigh–Schrödinger perturbation theory[46] to obtain an estimate of the CI coefficient of a given configuration (t) by the use of

$$c_{\mathrm{t}} = \frac{H_{\mathrm{t}0}}{H_{00} - H_{\mathrm{tt}}}, \tag{12-89}$$

where the "0" subscript refers to a reference configuration (e.g., the SCF configuration). Alternatively, the energetic contribution of configuration "t" is given by

$$\Delta E = \frac{|H_{\mathrm{t}0}|^2}{H_{00} - H_{\mathrm{tt}}}. \tag{12-90}$$

Since expressions such as Eq. (12-89) or (12-90) can be used in advance of construction and diagonalization of the CI matrix itself, they can be used to eliminate configurations from the CI expansion that are likely to provide only minor energetic contributions. Typically, threshold values of $\Delta E \geq 10^{-4}$ or 10^{-5} Hartrees are used to choose configurations for inclusion in the overall CI wavefunction.

Also, all matrix elements between single excitation configurations and a reference Hartree–Fock configuration (Φ_0) will be zero by Brillouin's Theorem. While this simplifies computations involving the reference configuration, it does not imply that single excitation configurations will be unimportant to an overall CI study of, e.g., the electronic absorption of a molecule. For such a case, as noted in the earlier discussion of Brillouin's Theorem, it frequently occurs that a single excitation configuration is the dominant configuration in the description of one or more excited states, and whose inclusion is therefore of great importance.

Even for the ground state wavefunction it is possible for single excitation configurations to contribute, through matrix elements such as H_{tp} where t and p refer, for example, to single and double excitation configurations, respectively.

---

[45] See Problem 12-14.

[46] See Section 9-8. See also C. Møller and M. S. Plesset, *Phys. Rev.*, **46**, 618 (1934); R. K. Nesbet, *Proc. R. Soc. London*, **A230**, 312 (1955); P. Claverie, S. Diner, and J. P. Malrieu, *Int. J. Quantum Chem.*, **1**, 751 (1967); J. L. Whitten and M. Hackmeyer, *J. Chem. Phys.*, **51**, 5584 (1969).

While such contributions are frequently small from an energetic point of view, they should not be assumed to be zero automatically, and may contribute substantially to properties other than the energy.

In addition, it is sometimes necessary to utilize more than one reference configuration, e.g., in examining potential surfaces in regions where bonds are being formed or broken. In such cases, care must be taken not to assume inadvertently or inappropriately that CI matrix elements will "automatically" be small.

Finally, it should be noted that triple, quadruple, and higher excitation configurations will not interact with the reference configuration. This is due to the fact that the Hamiltonian operator couples two electrons at most, and in a formulation employing orthogonal orbitals, any configuration with greater than two orbitals replaced by unfilled orbitals will have zero interaction with the reference configuration due to orthogonality of the orbitals.

Even with simplifying procedures and concepts such as described above, construction of a suitably flexible, accurate, and appropriate CI wavefunction is still difficult and tedious. For example, even with the above procedures, it is difficult to tell when "enough" configurations have been added. In order to help in that regard, other procedures have been developed.

For example, Davidson[47] has suggested that the total correlation energy can be estimated for small systems by the use of

$$\Delta E_{\text{correl}} \cong (1 - c_0^2)\Delta E_d, \tag{12-91}$$

where ΔE_d is correlation energy recovery achieved through single and double excitations only, and c_0 is the coefficient of the SCF configuration in the normalized CI expansion. In addition, it has been noted in a number of studies of two-electron systems[48] that the correlation energy per electron pair is remarkably constant at ~ 0.042 Hartrees/pair, which is also of help in estimating total correlation energy.

A final aspect of the CI approach of interest here has to do with the notion of *size consistency* and *size extensivity*. To illustrate these ideas, it is clearly intuitively expected that a CI description of a system containing two fragments (e.g., A and B) will be satisfactory only if the energy of subsystems A and B when calculated as a single system far apart is equal to the sum of the energies of A and B computed separately using the same method. Put more generally in the language used by Primas,[49] "If a system is composed of two subsystems without any interaction, then this fact should at any stage of the theory be in full evidence." Expressed another way, a method is said to be *size consistent* if the

---

[47] E. R. Davidson, in *The World of Quantum Chemistry*, R. Daudel and B. Pullman (eds.), Reidel, Dordrecht, Holland, 1974, p. 17; S. R. Langhoff and E. R. Davidson, *Int. J. Quantum Chem.*, **8**, 61 (1974).

[48] See, for example, T. L. Allen and H. Shull, *J. Chem. Phys.*, **35**, 1644 (1961).

[49] H. Primas, in *Modern Quantum Chemistry*, Volume 2, O. Sinanoglu, (ed.), Academic Press, New York, 1965, p. 45.

energy of a system being dissociated into subsystems reaches the sum of energies of the isolated subsystems in the limit of infinite separation of the subsystems.[50]

In the context of CI studies, Primas' or the more general statement should be applicable regardless of whether "full CI" or some subset of "full CI" is used (e.g., CI with all single and double excitations). Unfortunately, this is not always the case, as a simple example will illustrate.

Let us consider a system consisting of a pair of H_2 molecules. If a CI approach consisting of inclusion of all single and double excitations from a single parent, closed shell (Hartree–Fock) configuration is used, the result can be seen to be *not* "size consistent." To see this conceptually, we note that a product of double excitation configurations that can arise for each of the two separated hydrogen molecules will correspond to a quadruple excitation in the two H_2 "supermolecule" system. Since quadruple excitations are excluded in the example we are considering, comparison of CI studies of this type on two H_2 separated molecules versus a two-H_2 "supermolecule" system at large distances will not give the same results even if *all* single and double excitations are included.

The other notion of importance here is that of *size extensivity*, which refers to appropriate scaling of a method with the number of electrons and/or size of molecule. This concept arose first in nuclear and solid-state physics, where, e.g., studies of a single atom are not necessarily useful in the description of a solid containing an infinite number of atoms unless the results on the isolated atom can be appropriately scaled to the infinite system.

For example, even with a good CI wavefunction, it has been shown for N H_2 molecules that

$$E_{\text{correl}}[N(\text{H}_2)] \neq N[E_{\text{correl}}(\text{H}_2)],$$

unless a size-extensive method is used. In fact, for this particular problem, it has been shown[51] that the correlation energy error is proportional to $N^{1/2}$ as $N \to \infty$. For example, by the time $N = 10$ (20-electrons), the error amounts to 31 kcal/mol.

For larger molecular problems, similar results are seen for truncated CI studies.[52] As an illustration, it has been found that the "CI truncation error" in the reaction

$$2\text{BH}_3 \to \text{B}_2\text{H}_6$$

---

[50] For additional discussion of this concept see, for example, J. A. Pople, J. S. Binkley, and R. Seeger, *Int. J. Quantum Chem. Symp.*, **10**, 1 (1976); E. R. Davidson and D. W. Silver, *Chem. Phys. Lett.*, **52**, 403 (1977); W. Kutzelnigg, A. Meunier, B. Levy, and G. Berthier, *Int. J. Quantum Chem.*, **12**, 77 (1977); R. J. Bartlett, *Annu. Rev. Phys. Chem.*, **32**, 359–401 (1981).

[51] R. J. Bartlett and G. D. Purvis, *Ann. N.Y. Acad. Sci.*, **367**, 62 (1981).

[52] See R. Alrichs *Theoret. Chim. Acta*, **35**, 59 (1974) and F. Keil and R. Alrichs, *J. Chem. Soc.*, **98**, 4787 (1976).

is approximately 9 kcal/mol, and for the reaction

$$CH_3F + F^- \rightarrow CH_3F_2^-,$$

the error is approximately 15 kcal/mol.

Thus, we see via several examples that limited types of CI expansions, e.g., use of only single and double excitation CI or perturbation theory studies from a single reference configuration (e.g., a restricted Hartree–Fock wavefunction), will *not* in general produce size consistent or size extensive results. To alleviate such difficulties, a number of options are available. Of course, if *full CI* is carried out, all possible excitations will have been included and the results will be size extensive. Alternatively, many-body perturbation theory approaches[53] typically are size consistent. Finally, use of multireference configurations even with only single and double excitations can achieve essentially size-extensive (although not exact) results, as can multiconfiguration SCF approaches.[54] The main point, however, is that size extensivity as well as size consistency should not be ignored when devising or applying limited CI approaches.

3. Computational Considerations

Even assuming that a suitable correlation orbital basis and CI formulation is available, a number of substantial computational issues still remain. For example, formation of integrals over correlation orbitals from integrals calculated using the original basis set is a significant task, as we shall see below.

As discussed in the previous chapter,[55] it is necessary to form electron repulsion integrals over MOs as part of the Hartree–Fock iteration process, starting from integrals over basis orbitals. Unfortunately, such a process gives rise to significant computational problems, as we shall see below. Expressed mathematically, we write the molecular orbitals (χ_i) as a linear combination of M basis orbitals (ϕ), i.e.,

$$\chi_i = \sum_p^M c_{pi}\phi_p. \tag{12-92}$$

If this expansion is inserted for each of the orbitals in Eq. (12-10), we obtain

$$g_{ijkl} = \sum_p^M \sum_q^M \sum_r^M \sum_s^M c_{pi}^* c_{qj}^* c_{rk} c_{sl} \cdot \int\int \phi_p^*(1)\phi_q^*(2) \frac{1}{r_{12}} \phi_r(1)\phi_s(2)\, d\tau_1\, d\tau_2. \tag{12-93}$$

[53] See Section 12-4.

[54] See Section 12-3.A.

[55] See Section 11-2.B.

This means that, for *each* g_{ijkl}, there are M^4 terms that must be computed if no additional analysis is done. In addition, there are M^4g's to be calculated in a full transformation of all basis function integrals to MO integrals, so it appears that the time to calculate the electron repulsion integrals in a form useful for forming Hamiltonian matrix elements in CI studies is proportional to M^8. For large basis sets such a large number of terms becomes prohibitively expensive to calculate, and alternatives are necessary.

A number of preferable alternatives to the M^8 problem have been developed. For example, due to the index symmetry properties of one- and two-electron integrals, it is necessary to calculate (and store) only slightly more than half of the one-electron integrals and approximately $(1/8)M^4$ of the two-electron integrals.[56] If that is done, convenient storage of the integrals can be achieved by using the "lower (or upper) triangle" form. For the one-electron integral case, for example, the lower triangle form can be represented as a linear array having the following elements:

$$h_{11}h_{21}h_{22}h_{31}h_{32}h_{33}h_{41} \cdots h_{MM}.$$

For the case of the two-electron integrals, the corresponding linear array is given by the following elements:

$$g_{1111}g_{2111}g_{2121}g_{2211}g_{2221}g_{2222} \cdots g_{MMMM}.$$

In this case, the order of storing g_{ijkl} elements is determined by

$$i \geq j, \qquad k \geq l, \qquad [ij] \geq [kl], \tag{12-94}$$

where

$$[ij] = \frac{1}{2}\, i(i-1) + j, \qquad [kl] = \frac{1}{2}\, k(k-1) + l. \tag{12-95}$$

Of course, if there is molecular symmetry in addition, the number of unique integrals that need to be calculated is reduced even further.

In addition, while the number of basis function integrals that can be ignored because they are numerically insignificant is relatively few for a small molecule and large basis set (and $\sim N^4/8$ integrals need to be calculated), the situation may change substantially as the molecule gets larger and basis sets are typically smaller on a relative scale (compared to small molecules). In particular, a substantial fraction (as high as 80–90% in some large molecule calculations) of basis function electron repulsion integrals may become essentially zero from a numerical point of view for large molecules. Thus, by making relatively easy checks of integral magnitude in advance, the calculation, storage, and transformation of a very large number of basis function electron repulsion integrals can frequently be avoided.

Also, a word regarding numerical accuracy is in order. Since a relatively large number of mathematical operations are involved in Eq. (12-93), it is

[56] See Problem 12-15.

possible (and usually is the case) that loss of accuracy in g_{ijkl} will occur because of the addition of many (positive and negative) terms. In practice this can be avoided by the use of high accuracy in the calculation of integrals over basis functions.[57]

The process of transforming basis function electron repulsion integrals to MO integrals has also been improved substantially. In particular, if the calculation of the g_{ijkl} in Eq. (12-93) is carried out in the following manner:

$$\left\langle \phi_p \phi_q \left| \frac{1}{r_{12}} \right| \phi_r \chi_l \right\rangle = a_{pqrl} = \sum_{s=1}^{M} c_{sl} \left\langle \phi_p \phi_q \left| \frac{1}{r_{12}} \right| \phi_r \phi_s \right\rangle \tag{12-96}$$

$$\left\langle \phi_p \phi_q \left| \frac{1}{r_{12}} \right| \chi_k \chi_l \right\rangle = b_{pqkl} = \sum_{r=1}^{M} c_{rk} a_{pqrl}, \tag{12-97}$$

$$\left\langle \phi_p \chi_j \left| \frac{1}{r_{12}} \right| \chi_k \chi_l \right\rangle = d_{pjkl} = \sum_{q=1}^{M} c_{qj} b_{pqkl}, \tag{12-98}$$

$$\left\langle \chi_i \chi_j \left| \frac{1}{r_{12}} \right| \chi_k \chi_l \right\rangle = g_{ijkl} = \sum_{p=1}^{M} c_{pi} d_{pjkl}, \tag{12-99}$$

it can be seen that substantial savings in computational time results.[58] The reason for this savings is that each of the partial sums in Eqs. (12-96)-(12-99) is proportional to M^5 (and not M^8).

In addition, a full transformation of all basis function integrals to integrals over all possible MOs is not always required. For example, if some MOs are not used in the CI study, electron repulsion integrals over those MOs are not needed. In such a case, if K represents the number of MOs used in the CI study, the overall time for integral transformation is proportional to M^4K^4. Thus, when this is used along with the algorithm expressed in Eqs. (12-96)–(12-99), significant improvements in the "M^8" problem occur, with the overall process being proportional to MK^4.

Finally, in addition to simplifications in the computational algorithms such as described above, there are other techniques that have been devised that allow simplification of Hamiltonian matrix element [Eq, (12-4)] computation. In particular, for molecular systems where substantial symmetry is present, it is possible to utilize this symmetry to show that many of the matrix elements will be zero, as well as to simplify the calculation of nonzero matrix elements. This

---

[57] For example, basis orbital integrals having ≥ 12 decimal place accuracy are frequently used in Gaussian basis set calculations to assure satisfactory accuracy in g_{ijkl} elements.

[58] See, for example, K. C. Tong and C. Edmiston, *J. Chem. Phys.*, **52**, 997 (1970); C. F. Bender, *J. Comput. Phys.*, **9**, 547 (1972); P. S. Bagus, B. Liu, A. D. McLean, and M. Yoshimime, in *Energy, Structure and Reactivity*, D. W. Smith and W. B. McRae (eds.), John Wiley, New York, 1973, p. 130; G. H. F. Diercksen, *Theoret. Chim. Acta*, **33**, 1 (1974); P. Pendergast and W. H. Fink, *J. Comput. Phys.*, **14**, 286 (1974).

includes the use of both spatial and spin symmetry, and a large body of literature is available that deals with various aspects of these techniques.[59] Furthermore, the use of unitary group techniques[60] has allowed very large ground state CI studies to take place, including more than a million configurations.[61]

Once the CI matrix has been formed, finding the eigenvalues and eigenvectors is also a formidable task, especially when the CI Hamiltonian matrix is large (e.g., 10,000 × 10,000). However, a number of general techniques have been developed that facilitate diagonalization of the CI Hamiltonian matrix.[62] In addition, as the CI Hamiltonian matrix becomes large, it also typically becomes sparse (i.e., many elements are zero), and in practice only a few low-lying eigenvalues and eigenvectors are typically desired. For such cases, additional techniques have been devised,[63] which aid significantly in reducing computa-

---

[59] For reviews that discuss the use of spin symmetry to construct configurations and Hamiltonian matrix elements see, for example, I. Shavitt, in *Methods of Electronic Structure Theory*, H. F. Schaefer III, (ed.), Plenum Press, New York, 1977, pp. 189–275; P. O. Löwdin, *Calcul des Fonctions d'Onde Moléculaire*, CNRS, Paris, 1958, p. 23; M. Kotani, A. Amemiya, E. Ishiguro and T. Kimura, *Tables of Molecular Integrals*, 2nd ed., Maruzen, Tokyo, 1963; F. A. Matsen, *Adv. Quantum Chem.*, **1**, 59 (1964); R. Pauncz, *Alternant Molecular Orbital Method*, Saunders, Philadelphia, 1967; F. E. Harris, in *Energy, Structure and Reactivity*, D. W. Smith and W. B. McRae (eds.), John Wiley, New York, 1973, p. 112; J. I. Musher, *J. Phys. (Paris)*, **31**, Suppl. C4, 51 (1970); K. Ruedenberg and R. D. Poshusta, *Adv. Quantum Chem.*, **6**, 267 (1972); W. I. Salmon, *Adv. Quantum Chem.*, **8**, 37 (1974); R. K. Nesbet, *J. Math. Phys.*, **2**, 701 (1961); E. R. Davidson, *Int. J. Quantum Chem.*, **8**, 83 (1974).

For examples of the use of spatial symmetry in the construction of configurations and simplification of Hamiltonian matrix elements see M. A. Melvin, *Rev. Mod. Phys.*, **28**, 18 (1956); P. O. Löwdin, *Rev. Mod. Phys.*, **39**, 259 (1967); J. Killingbeck, *J. Math. Phys.*, **11**, 2268 (1970); A. Golebiewski, *Mol. Phys.*, **20**, 481 (1971); G. A. Gallup, *Int. J. Quantum Chem.*, **8**, 267 (1974); R. McWeeny, *Symmetry—An Introduction to Group Theory and Its Applications*, Pergamon, London, 1963; R. Moccio, *Theoret. Chim. Acta*, **7**, 85 (1967). For a thorough review of matrix diagonalization techniques, see I. Shavitt, *Methods of Electronic Structure Theory*, Volume 3, H. F. Schaefer III (ed.), Plenum Press, New York, 1977, p. 189.

[60] J. Paldus, in *Theoretical Chemistry: Advances and Perspectives*, Volume 2, H. Eyring and D. G. Henderson (eds.), Academic Press, New York, 1976, p. 131; I. Shavitt, in *New Horizons of Quantum Chemistry*, P. W. Löwdin and B. Pullman (eds.), D. Reidel, Holland, 1983, pp. 279–293.

[61] D. Fox, N. Handy, P. Saze, and H. F. Schaefer III, 183st American Chemical Society Meeting, Las Vegas, Nevada, 1982.

[62] See, for example, J. H. Wilkinson, *The Algebraic Eigenvalue Problem*, Oxford University Press, London, 1965; J. H. Wilkinson and C. Reinsch, *Linear Algebra*, Springer-Verlag, New York, 1971; H. R. Schwarz, H. Rutishauser, and E. Steifel, *Numerical Analysis of Symmetric Matrices*, Prentice-Hall, Englewood Cliffs, NJ, 1973; and G. W. Stewart, *Information Processing 74*, North-Holland, Amsterdam, 1974. For a thorough review of matrix diagonalization techniques, see I. Shavitt, *Methods of Electronic Structure Theory*, Volume 3, H. F. Schaefer III (ed.), Plenum Press, New York, 1977, p. 189.

[63] For a review, see G. W. Stewart, *Proc. IFIP Congress 74, Stockholm*, p. 666, North-Holland, Amsterdam, 1974; see also I. Shavitt, C. F. Bender, A. Pipano, and R. P. Hosteny, *J. Comput. Phys.*, **11**, 90 (1973); S. Falk, *Z. Angew. Math. Mech.*, **53**, 73 (1973); E. R. Davidson, *J. Comput. Phys.*, **17**, 87 (1975); R. K. Nesbet, *J. Chem. Phys.*, **43**, 311 (1965).

tional requirements. However, such developments are beyond the scope of the current discussions, and the reader is referred to references cited below for additional information.

Finally, before leaving this section it is appropriate to note that techniques have also been devised that avoid the construction of the CI Hamiltonian matrix entirely[64] for some cases. Instead, procedures have been developed in which CI expansion coefficients are calculated *directly* from the list of one- and two-electron integrals for a number of cases of interest. As we shall see, this approach not only avoids constructing and diagonalizing large CI matrices, it is especially useful when very large CI expansions (e.g., full CI with large numbers of basis orbitals) are used.

While the method involves a straightforward use of the CI expansion [Eq. (12-1)], the only integral evaluations needed are the one- and two-electron integrals over all basis functions. The way in which formation and diagonalization of the Hamiltonian matrix are avoided is to note that many of the procedures for diagonalization of the Hamiltonian matrix utilize an iterative process that takes an initial trial eigenvector $[\mathbf{d}^{(k-1)}]$ to generate a new trial eigenvector $[\sigma^{(k)}]$ using the Hamiltonian matrix (H) via the following relation:

$$\sigma^{(k)} = \mathbf{H}\mathbf{d}^{(k-1)}. \tag{12-100}$$

However, Roos and colleagues noted that the Hamiltonian matrix elements are simply linear combinations of one- and two-electrons and, at least for several special cases, it is possible to formulate explicit expressions for the elements of $\sigma^{(k)}$ via Eq. (12-100) without having to calculate the Hamiltonian matrix beforehand. Specifically, they showed that the changes in σ due to a particular integral (ab/cd) can be represented by an expression of the form

$$\Delta\sigma_i^{(k)} = A(ab|cd)(\mathbf{d})_i^{(k-1)}, \tag{12-101}$$

where A is a coupling coefficient that depends on the spin symmetry of the configurations. It is therefore the identification and explicit derivation of the coupling coefficients for various cases of interest that allows this method to be successfully developed.

The procedure for identification of coupling coefficients begins by dividing the basis orbitals into several types: (1) orbitals that describe the *internal* space, i.e., the occupied HF orbitals plus all orbitals needed to describe degeneracy and near-degeneracy effects; and (2) orbitals describing the *external* space, i.e., those that describe correlation effects. In the CI study, the internal space is subdivided one step further, in which the first group contains internal orbitals that are fully occupied in all states. Only excitations *from* these orbitals is

---

[64] For a more detailed discussion of this approach see, for example, B. O. Roos and P. E. M. Siegbahn, *Methods of Electronic Structure Theory*, Volume 3, H. F. Schaefer, III (ed.), Plenum Press, New York, 1977, p. 277; see also B. O. Roos, *Chem. Phys. Letters*, **15**, 153 (1972); R. F. Hausman, Jr., S. D. Bloom and C. F. Bender, *Chem. Phys. Lett.*, **32**, 483 (1975); C. F. Bender, *J. Comp. Phys.*, **30**, 324 (1979).

Table 12.4. The Fourteen Types of Two-Electron Integrals That Arise When All Orbitals Belong to the Same Classification

Type number	Index relations among orbitals in (ab/cd) integral
1	$a = b = c = d$
2	$a = b = c$
3	$b = c = d$
4	$a = b, c = d$
5	$a = c, b = d$
6	$a = b$
7	$c = d, b > d$
8	$c = d, b < d$
9	$a = c$
10	$b = d$
11	$b = c$
12	$b > c$
13	$b < c, b > d$
14	$b < d$

possible. The remainder of the internal orbitals comprise the second group. Excitations *to* and *from* these orbitals are possible. Of course, for external orbitals, only excitations *to* these orbitals are possible. In any case, a total of three types of orbitals is considered.

These orbital classifications are then used to classify the types of two-electron integrals that arise. To illustrate the procedure, the data in Table 12.4 show that, if each of the orbitals in a two-electron integral (ab/cd) is in the *same* orbital classification, only 14 different types of two-electron integrals arise. An important example of such a case is when "full CI" is carried out, i.e., when all orbitals are treated equally and all possible configurations are formed. For such a case only orbitals of the second kind (i.e., those for which excitations either *to* or *from* are possible) arise, and only the 14 types of two-electron integrals indicated in Table 12-4 arise.

By contrast, if a CI study is to be carried out using a closed shell configuration as a *reference "state"* from which other selected configurations are formed by excitations from the closed shell configuration, all three types of orbitals can arise, and Roos and Siegbahn showed that 53 different integral types are found. For the case when the reference state has one open shell (e.g., a low-lying excited state), 59 new integral types are added to the 53 from the previous case. It is therefore clear that each kind of CI study must be analyzed in advance using this approach, to determine the various kinds of two-electron integrals that arise. It is also clear that the types of integrals are minimized in a full CI study, making such an approach perhaps most helpful in high accuracy studies.

Once the reference states and CI procedure to be used are selected and the types of two-electron integrals that arise have been determined, it is still necessary to determine the coupling coefficients (A) in Eq. (12-101) that reflect the interactions of the various configurations in the wavefunction. Since such a determination must be carried out for each type of CI study, we shall utilize the case of single and double excitations from a closed shell single-determinant reference state to illustrate how the method works.

The trial wavefunction for such a case is written using the notation of Eq. (12-15) as

$$\Psi = \Phi_0 + \sum_{a,\alpha} c_a^\alpha \Phi_a^\alpha + \sum_{\substack{a<b \\ \alpha<\beta}} c_{ab}^{\alpha\beta} \Phi_{ab}^{\alpha\beta}, \tag{12-102}$$

in which orbitals a or b are considered as "internal" orbitals, while orbitals α and β are considered to be "external" orbitals. In such a wavefunction, there are six specific types of configurations that arise, i.e.,

$$\Phi_a^\alpha = \mathcal{S} \det\{\cdots \varphi_a \bar{\varphi}_\alpha \cdots\}, \tag{12-103}$$

$$\Phi_{aa}^{\alpha\alpha} = \mathcal{S} \det\{\cdots \varphi_\alpha \bar{\varphi}_\alpha \cdots\}, \tag{12-104}$$

$$\Phi_{ab}^{\alpha\alpha} = \mathcal{S} \det\{\cdots \varphi_a \bar{\varphi}_\alpha \cdots \varphi_\alpha \bar{\varphi}_b \cdots\}, \tag{12-105}$$

$$\Phi_{aa}^{\alpha\beta} = \mathcal{S} \det\{\cdots \varphi_\alpha \bar{\varphi}_\beta \cdots\}, \tag{12-106}$$

$${}^1\Phi_{ab}^{\alpha\beta} = \mathcal{S} \det\{\cdots \varphi_a \varphi_\alpha \cdots \bar{\varphi}_\beta \bar{\varphi}_b \cdots\}, \tag{12-107}$$

$$\begin{aligned} {}^2\Phi_{ab}^{\alpha\beta} = \mathcal{S}\ [&\det\{\cdots \varphi_a \bar{\varphi}_\alpha \cdots \varphi_b \bar{\varphi}_\beta \cdots\}, \\ &+ \det\{\cdots \varphi_\alpha \bar{\varphi}_a \cdots \varphi_b \bar{\varphi}_\beta \cdots\}], \end{aligned} \tag{12-108}$$

where normalization constants have been omitted, $\mathcal{S}$ is a singlet state spin-projection operator, and only the orbitals in which replacements have occurred are shown explicitly.

Roos and Siegbahn showed[65] that 21 different types of matrix elements arise with the above configurations, all of which can be written as linear combinations of one- and two-electron integrals. In particular, the contributions to $\Delta\sigma$ in Eq. (12-101) can be written as one of three different types, i.e., for a double replacement state, the contributions from interactions with other double

[65] B. O. Roos and P. E. M. Siegbahn, *Methods of Electronic Structure Theory*, H. F. Schaefer, III (ed.), Plenum Press, New York, 1977, pp. 277–183.

replacements is given by

$$\Delta\sigma_{ij\to ab} = \sum_c j_2(\mu, \nu)[f_{bc}C_{ij\to ac} + f_{ac}C_{ij\to cb}]$$

$$-\sum_k j_1(\mu, \nu)[f_{kj}C_{ik\to ab} + f_{ik}C_{kj\to ab}]$$

$$+\sum_{k\geq l} [j_1(\mu, \nu)(ik|jl) + k_1(\mu, \nu)(il|kj)]C_{kl\to ab}$$

$$+\sum_{c\geq d} [j_2(\mu, \nu)(ac|bd) + k_2(\mu, \nu)(ad|bc)]C_{ij\to cd}$$

$$+\sum_{k,c} [j_3(\mu, \nu)(ac|ik) + k_3(\mu, \nu)(ai|ck)]C_{kj\to cb}$$

$$+\sum_{k,c} [j_4(\mu, \nu)(bc|jk) + k_4(\mu, \nu)(bj|ck)]C_{ik\to ac}$$

$$+\sum_{k,c} [j_5(\mu, \nu)(bc|ik) + k_5(\mu, \nu)(bi|ck)]C_{kj\to ac}$$

$$+\sum_{k,c} [j_6(\mu, \nu)(ac|jk) + k_6(\mu, \nu)(aj|ck)]C_{jk\to cb}, \qquad (12\text{-}109)$$

where the coupling constants $j_i(\mu, \nu)$ and $k_i(\mu, \nu)$, $(\mu, \nu = 1, \ldots, 5)$, are tabulated explicitly, and:

$$f_{pq} = \int \varphi_p^* \hat{f} \varphi_q \, d\tau, \qquad (12\text{-}110)$$

and

$$(pq|rs) = \int\int \varphi_p^*(1)\varphi_q(1) \frac{1}{r_{12}} \varphi_r^*(2)\varphi_s(2) \, d\tau_1 \, d\tau_2, \qquad (12\text{-}111)$$

where $\hat{f}$ is the Fock operator. Expressions analogous to Eq. (12-109) are found for the interaction between double replacements and single replacements, and between single replacements and other single replacements, i.e.,

$$\Delta\sigma_{ij\to ab} = i_1(\mu)f_{bj}C_{i\to a} + i_2(\mu)f_{aj}C_{i\to b} + i_1(\mu)f_{bi}C_{j\to a} + i_2(\mu)f_{ai}C_{j\to b}$$

$$+\sum_c [i_1(\mu)(ac|bj) + i_2(\mu)(aj|bc)]C_{i\to c}$$

$$+\sum_c [i_1(\mu)(ai|bc) + i_2(\mu)(ac|bi)]C_{j\to c}$$

$$-\sum_k [i_1(\mu)(ai|jk) + i_2(\mu)(aj|ik)]C_{k\to b}$$

$$-\sum_k [i_1(\mu)(bj|ik) + i_2(\mu)(bi|jk)]C_{k\to a}, \qquad (12\text{-}112)$$

Table 12.5. Correlation Energy for NH_3 Obtained with Different Basis Sets

Basis set[a]	Number of external orbitals	Number of configurations	Correlation energy (Hartrees)[b]
(4,2/2)	11	875	−0.1328
(5,3/3)	17	2,271	−0.1424
(4,2,1/2,1)	25	2,690	−0.1909[c]
(5,3,1/3,1)	32	4,335	−0.2033[c]
(7,6,1/2,1)	40	10,800	−0.2428

[a] Contracted orbitals are listed, which are obtained from a primitive basis of (N: 10,6,1/H: 5,1).

[b] Estimated exact value $\simeq -0.328$ Hartrees.

[c] 1s core electrons "frozen."

and

$$\Delta\sigma_{i\to a} = \sum_b f_{ab} C_{i\to b} - \sum_j f_{ij} C_{j\to a} + \sum_{j,b} [2(ai\,|\,bj) - (ab\,|\,ij)] C_{j\to b}. \quad (12\text{-}113)$$

Thus, construction of $\Delta\sigma$ has indeed been reduced to contributions from one- and two-electron integrals, coupling constants, and estimates of CI coefficients from the previous iteration.

In practice, six to eight iterations have been found necessary in order to converge to approximately 10^{-6} a.u. in the correlation energy for the closed shell case with one dominant configuration. The actual correlation energy found for various basis sets is illustrated for the NH_3 molecule in Table 12.5, where ~74% of the correlation energy is obtained from a "double zeta plus polarization" basis set and a 10,800 configuration wavefunction.

Thus, we see that this method accommodates long CI expansions, and thus is useful in high accuracy studies. On the other hand, it is not easily generalizable, e.g., to studies on various open shell cases. However, for those cases where explicit expressions such as Eqs. (12-111)-(12-113) have been worked out, considerable computational simplification can be seen to have resulted compared to "conventional" CI approaches.

D. Examples

As illustrated in the previous discussion, there are a very large number of configurations that arise if full CI is to be carried out using a large basis. Not surprisingly, there have been a substantial number of approaches devised to reformulate the CI approach using restricted forms of the CI expansion, the basis set, or both. Before describing some of those approaches, it is useful to illustrate the kinds of results that have been obtained with a straightforward use of the CI expansion in Eq. (12-15). In such illustrations, it will be instructive to calculate properties not well described by Hartree–Fock theory, i.e., electronic spectra and potential surface calculations.

Table 12.6. Number of Configurations Used in CI Study of Low-Lying Singlet States in H_2O

	$^1A_1$	$^1A_2$	$^1B_1$	$^1B_2$
Ground State (Φ_0)	4	1	3	1
Φ_0 + single excitations	306	116	304	98
Φ_0 + single and double excitations	6371	2844	6332	2388

Let us start by considering electronic spectral calculations[66] via a study of low-lying singlet states of the H_2O molecule as performed by Hausman and Bender.[67] For this study, several basis sets were employed, the largest consisting of a contracted double zeta basis[68] discussed earlier, augmented by four diffuse orbitals on oxygen. This gives rise to the following set of 18 basis orbitals:

Oxygen $1s$, $1s'$, $2s$, $2s'$, $2p_x$, $2p_x'$, $2p_y$, $2p_y'$, $2p_z$, $2p_z'$, $2s^d$, $2p_x^d$, $2p_y^d$, $2p_z^d$

Each hydrogen $1s$, $1s'$

Diffuse orbitals were added to the basis because it was known that several of the excited states exhibited a very diffuse charge distribution (*Rydberg states*), and the original double-zeta basis would not provide an adequate description of these states.

Given this basis, an SCF calculation was carried out on the ground electronic state using the equilibrium ground state geometry ($\vartheta_{HOH} = 104.45°$, $R_{OH} =$ 1.8111 Bohr), followed by CI calculations using single and double excitations from the ground state SCF MOs to unfilled SCF MOs (excluding excitations from the two lowest-lying MOs, since they were not likely to be of importance to the states of interest). In all, nine different, low-lying singlet states were studied (four $^1A_1$ states, including the ground state, one $^1A_2$ state, three $^1B_1$ states, and one $^1B_2$ state), and the number of configurations used in the various calculations is given in Table 12.6.

It should be noted that the excited state calculations were carried out using the *ground state* nuclear geometry. Such a choice represents a restriction on the accuracy of the excited state descriptions, especially if the excited state charge distribution is changed substantially from the ground state. In the latter case, it

[66] For examples of early application of CI techniques to atomic and molecular problems see, for example, G. R. Taylor and R. G. Parr, *Proc. U.S. Nat. Acad. Sci. U.S.A.*, **38**, 154 (1952); M. J. M. Bernal and S. F. Boys, *Philos. Trans. R. Soc., (London)*, **A245**, 139 (1952); S. F. Boys, *Proc. Roy. Soc. (London)*, **A217**, 136, 235 (1953); D. Kastler, *J. Chim. Phys.*, **50**, 556 (1953); A. Meckler, *J. Chem. Phys.*, **21**, 1750 (1953).

[67] R. F. Hausman, Jr. and C. F. Bender, *Methods of Electronic Structure Theory*, H. F. Schaefer, III (ed.), Plenum Press, New York, 1977, Chapter 8, 319–338.

[68] T. H. Dunning, *J. Chem. Phys.*, **53**, 2823 (1970).

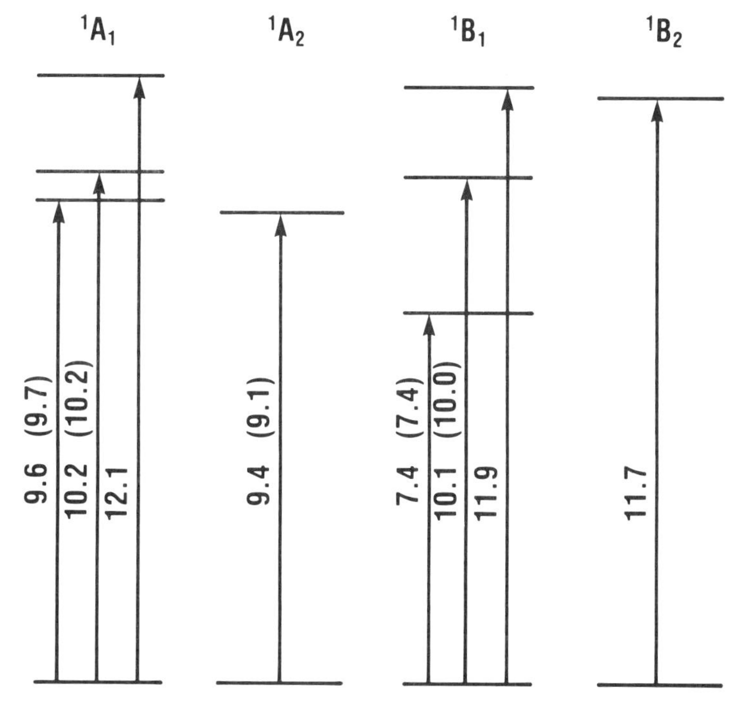

Figure 12.3. Calculated and observed vertical transition energies (in eV) from the ground state to low-lying singlet states in H_2O. (Experimental values are given in parentheses).

would be expected that a change in geometry would accompany the charge distribution change when equilibrium has been reached.

However, it is also expected that nuclear relaxation will take place on a slower time scale than electronic charge distribution rearrangement. Such an observation is usually expressed in a more formal manner as the *Franck-Condon Principle*,[69] which, for our purposes, states that the highest intensity electronic transitions (e.g., absorption) occur from the minimum of a potential curve (i.e., the equilibrium configuration) vertically upward (i.e., at the geometry of the lower state). Hence, for spectral purposes, use of the ground state equilibrium geometry for excited state descriptions is appropriate for comparisons to *vertical transitions* (i.e., Franck-Condon transitions).

For the case of interest here, the calculated vertical transition energies for the nine states of interest are shown in Fig. 12.3, along with comparison to experimental transition energies when available. As the data indicate, excellent agreement with experimental values is obtained, indicating that spectral features might be expected to be well predicted by a straightforward extension of the Hartree-Fock wavefunction using single and double excitation CI. While such

[69] J. Franck, *Trans. Faraday. Soc.*, **21**, 536 (1925); E. U. Condon, *Phys. Rev.*, **32**, 858 (1928).

Table 12.7. Energetic Contributions from Various Configurations in a Full CI Study of H_2O

Wavefunction description	Number of configurations	Total energy (Hartrees)	Correlation energy (Hartrees)
SCF	1	−76.009838	0.0
SCF + all double excitations (D)	342	−76.149178	−0.139340
SCF + all single (S) and double (D) excitations	361	−76.150015	−0.140177
SCF + S + D + all triple (T) excitations	3,203	−76.151156	−0.141318
SCF + S + D + T + all quadruple (Q) excitations	17,678	−76.157603	−0.147765
Full CI	256,473	−76.157866	−0.148048

an assertion is true in some cases, it is of interest to examine the nature of the difficulties encountered if higher accuracy is sought.

For the H_2O example just considered, a study has been carried out that addresses the importance of configurations beyond single and double excitations (i.e., explores "many-body" effects). In particular, Saxe *et al.*[70] also used the double zeta basis set of Dunning discussed in the previous example, but without the four diffuse orbitals on oxygen.

With this basis, a *full CI* study on the ground state of H_2O was performed. This means that not only single and double excitations were included, but configurations containing triple, quadruple, ..., up to 10-fold excitations were included in the wavefunction. Such a full CI study gives rise to 256,473 configurations, and the energetic contributions from the various types of configurations are given in Table 12.7.

A number of points of interest are illustrated in Table 12.7. First, the number of configurations needed for a full CI study on H_2O is very large (256,473), even with a double zeta basis. Furthermore, most of the configurations arise beyond the single and double excitation level, with 2842 triple excitations, 14,475 quadruple excitations, 41,952 quintuple excitations, 72,365 sextuple excitations, 71,434 7-fold excitations, 40,046 8-fold excitations, 11,492 9-fold excitations, and 1506 10-fold excitations.

These data also illustrate why CI studies using only single and double excitations are frequently used, since 94.7% of the correlation energy has already been recovered (−0.140177 out of −0.148028) by the addition of

[70] P. Saxe, H. F. Schaefer, III, and N. C. Handy, *Chem. Phys. Lett.*, **79**, 202 (1981).

single plus double excitation configurations (361 configurations). We also see that triple and quadruple excitations provide 5.13% of the correlation energy in this basis set, while all higher excitations (238,795 out of 256,473) provide only 0.18% of the correlation energy.

Such results illustrate not only that high accuracy studies are possible, at least for small molecular systems, but also that higher order correlation effects are relatively unimportant, at least for closed shell ground states of small systems. Results such as these have led to significant efforts to develop CI techniques using only single and double excitation CI, as we shall see.

Before turning to those efforts, let us consider a second example, i.e., calculation of a potential surface. While calculation of spectra using CI or other techniques may be difficult and time-consuming, the calculation of an accurate potential surface is even more difficult, due to the need to calculate the energy for many different geometries (and not simply the ground state geometry). However, at least for simple cases, it is possible to obtain very high accuracy, as we shall now see.

The particular example that we shall consider is the ground state dissociation of the H_2 molecule, as carried out by Kolos and Wolniewicz.[71] Using the Born–Oppenheimer approximation, they employed a basis set that is very well suited for diatomic molecules. Specifically, a basis of *elliptical functions* introduced earlier by James and Coolidge[72] was used, which has the form

$$\varphi_{rs\bar{r}\bar{s}\mu}=\left(\frac{1}{2\pi}\right)\xi_1^r\eta_1^s\xi_2^{\bar{r}}\eta_2^{\bar{s}}\rho^{\mu}\exp\{-\alpha(\xi_1+\xi_2)\}, \tag{12-114}$$

where

$$\rho=\frac{2r_{12}}{R}, \tag{12-115}$$

$$\xi_i=\frac{r_{ai}+r_{bi}}{R}, \qquad i=1,\,2 \tag{12-116}$$

$$\eta_i=\frac{r_{ai}-r_{bi}}{R}, \qquad i=1,\,2. \tag{12-117}$$

R is the internuclear distance, r_{12} is the interelection distance, r_{ai} is the distance of electron one from nucleus a, r, s, $\bar{r}$, $\bar{s}$ and μ are integers, and α is a variational parameter to be determined. As is evident in Eq. (12-114), this basis (when $\mu \neq 0$) is *not* an orbital basis, since eash $\varphi_{rs\bar{r}\bar{s}\mu}$ is in general a nonseparable function of two electrons. But, its cylindrical symmetry is well suited for the description of diatomic molecules like H_2, it has cusps at both nuclei, and the electron

[71] W. Kolos and L. Wolniewicz, *J. Chem. Phys.*, **41**, 3663–3678 (1964); **43**, 2429–2441 (1965); **49**, 404–410 (1968).

[72] H. M. James and A. S. Coolidge, *J. Chem. Phys.*, **1**, 825 (1933).

repulsion integrals for simple cases such as H_2 are reasonably tractable. However, for general polyatomic systems the integrals are at least as complicated as when STO basis orbitals are used, which explains why elliptical basis orbitals have not been used as a basis in polyatomic calculations.

For the H_2 case, the basis is very well suited, and using this basis a CI wavefunction was formed, i.e.,

$$\Psi(1, 2) = \sum_i c_i \Phi_i(1, 2), \tag{12-118}$$

where the individual configurations are given by a symmetrized combination of basis functions, i.e.,

$$\Phi_i(1, 2) = \varphi_{r_i s_i \bar{r}_i \bar{s}_i \mu_i} + \varphi_{\bar{r}_i \bar{s}_i r_i s_i \mu_i}. \tag{12-119}$$

Depending on the choice of internuclear distance, different numbers of configurations were utilized. For example, in the vicinity of the equilibrium distance where high accuracy was needed, 100 configurations were used, while 35–55 configurations were used for other internuclear distances.

The net result, depicted in Fig. 12.4, is an extraordinarily accurate potential curve. For example, the nonrelativistic adiabatic dissociation energy predicted from these calculations is 36,118.1 cm^{-1}, which becomes 36,117.4 cm^{-1} when relativistic and radiative effects are added. This latter value differed only by 3.8 cm^{-1} from the experimental value known at the time, and served to stimulate more accurate experiments. More generally, it illustrated that with suitable computational techniques, essentially exact solutions of the Schrödinger equation could be obtained even when closed form solutions are not available.

Thus, while the basis functions and CI techniques used in this example are not generally applicable, they illustrate that computational quantum chemistry can provide a powerful tool for probing the chemistry of molecules. However, in order to extend these techniques to larger and more complex systems where the CI expansion typically converges rather slowly, both computational and conceptual modifications and improvements are needed. In the next several sections, we shall explore a variety of examples of techniques that have been developed in order to extend the scope of applicability of computational quantum chemistry to larger and more complex chemical systems, while seeking to maintain acceptable accuracy.

Before doing that, however, it is of interest to note an example that illustrates the point that, even with large basis sets and CI studies, unambiguous results are sometimes very difficult to attain. An example which illustrates this point nicely is the collection of studies attempting to describe the lowest $^1(\pi \rightarrow \pi^*)$ electron (Franck–Condon) transition in ethylene, i.e., from the ground state (called the N state) to the lowest singlet excited state (called the V state). This molecule is often viewed as a prototype for a π-system, and thus has received considerable attention, both experimentally and theoretically. However, as simple as this molecule may appear, obtaining a thorough understanding of even the lowest $^1(\pi \rightarrow \pi^*)$ transition has proved to be a very challenging problem.

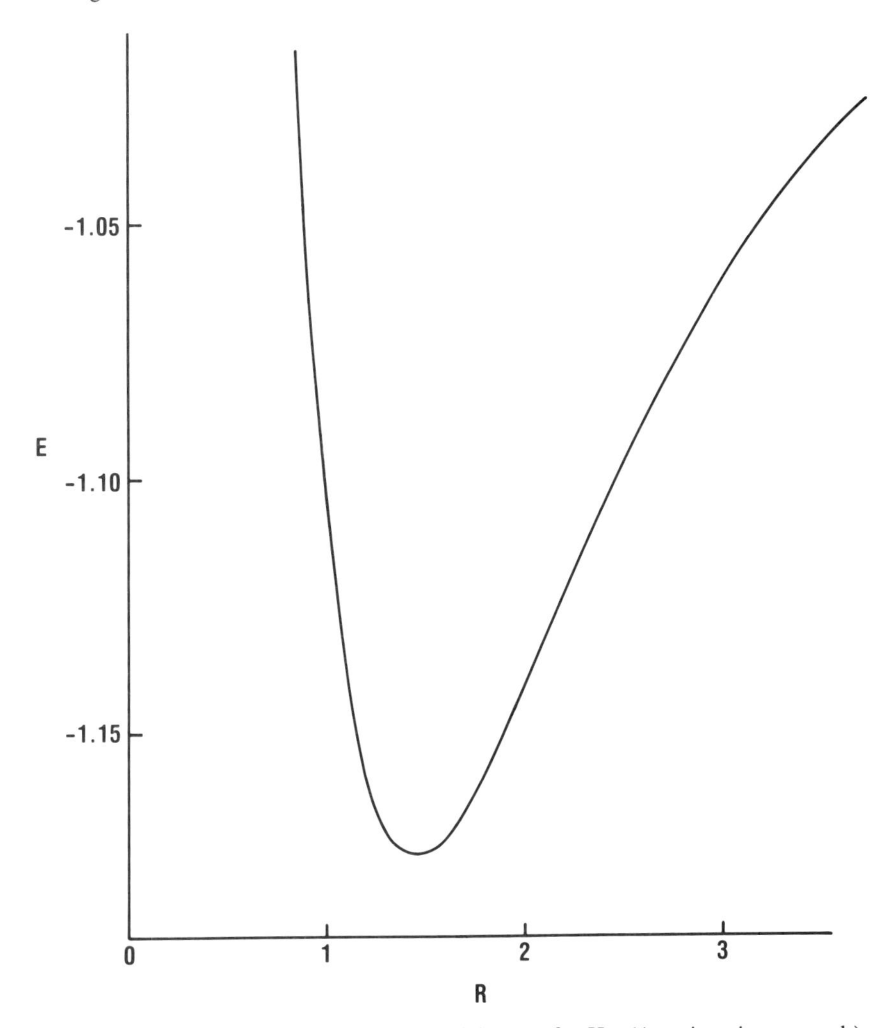

Figure 12.4. Calculated ground state potential curve for H_2. (Atomic units are used.)

From an experimental point of view,[73] the lowest singlet transition maximum intensity occurs at 7.65 eV. However, SCF and CI studies[74] found a diffuse V state (i.e., Rydberg-like) at the SCF level, plus transition energies above 9 eV. Lowering of the calculated transition energies (to ~8.3 eV) was achievable only

---

[73] A. J. Merer and R. S. Mulliken, *Chem. Rev.*, **69**, 639 (1969); D. F. Evans, *J. Chem. Soc.*, 1735 (1960). In our discussions only the Franck–Condon transition will be discussed. However, it should be noted that the lowest $^1(\pi \rightarrow \pi^*)$ transition is complicated further by twisting of the geometry in the V state causing a transition origin to appear at ~5.0 eV.

[74] M. B. Robin, H. Bosch, N. Kuebler, B. E. Kaplan, and J. Meinwold, *J. Chem. Phys.*, **48**, 5037 (1968); J. M. Schulman, J. W. Moskowitz, and C. Hollister, *J. Chem. Phys.*, **46**, 2759 (1967); U. Kaldor and I. Shavitt, *J. Chem. Phys.*, **48**, 191 (1968).

through addition of diffuse functions to the basis.[75] These latter results and the resulting diffuse charge distribution caused classification of the V state as a Rydberg state, but also led to a number of alternative studies and interpretations.

For example, a review of experimental results then available plus additional studies in solid and liquid rare gases prompted Miron and collaborators[76] to conclude that the V state was a "valence state," and not the diffuse state predicted theoretically.

The theoretical studies that followed are of particular interest, in part because of the insight they have revealed concerning the V state. However, they are of even greater interest in the current discussions because they reveal how quite different results can be obtained with different CI approaches, and the need for considerable care and "art" in CI studies in certain cases.

In considering the various theoretical studies, it is useful to note first that, while some differences in basis set choice are seen, each study usually has included a basis containing double zeta plus polarization plus diffuse functions in the basis. Since such basis sets are believed to be sufficiently large and flexible to describe states such as the V state in ethylene, the differences in results can be attributed usually to differences in the CI approach taken. In addition, all studies excluded the core (i.e., inner shell carbon) MOs from consideration in the CI studies. While such a constraint may affect the magnitude of the transition energy, it is not expected to have a significant effect on the results.

One of the first studies following the proposed Rydberg nature of the V state was that of Ryan and Whitten,[77] who suggested that the V state may have significant ionic character since CI studies based only on $\pi \rightarrow \pi^*$ excitations produced $N \rightarrow V$ transition energies that were too high (~ 8.94 eV). They then reoptimized the σ^* MOs and performed CI studies using variable occupancy of the five highest occupied MOs plus certain π^* and σ^* orbitals. The result (see Table 12.8) was a calculated $N \rightarrow V$ transition energy of 8.02 eV and a V state that appeared to be quite spatially contracted, which was thought to be due to the better treatment of ionic-type charge redistributions reflected through $\sigma \rightarrow \sigma^*$ interactions in the CI wavefunction. In any case, they showed that a relatively accurate transition energy could be obtained without requiring that the V state have significant Rydberg character.

However, in another study, Bender *et al.*[78] performed iterative natural orbital calculations in which (in the most extensive study) *all* filled MOs (except the

---

[75] T. H. Dunning, Jr., W. J. Hunt and W. A Goddard III, *Chem. Phys. Lett.*, **4**, 147 (1969); R. J. Buenker, S. D. Peyermihoff, and W. E. Kammer, *J. Chem. Phys.*, **55**, 814 (1971).

[76] E. Miron, B. Raz, and J. Jortner, *Chem. Phys. Lett.*, **6**, 563 (1970).

[77] J. A. Ryan and J. L. Whitten, *Chem. Phys. Lett.*, **15**, 119–123 (1972). The "spatially extended" orbitals included in the basis were less diffuse than in other studies reported here.

[78] C. F. Bender, T. H. Dunning, Jr., H. F. Schaefer III, W. A. Goddard, and W. J. Hunt, *Chem. Phys. Lett.*, **15**, 171–178 (1982). Details of the MC–SCF approach are given in Section 12-3A. The studies of Bender *et al.* did not include *d*-orbitals (polarization orbitals) in the basis, but the authors believed that the results would not be changed qualitatively by their inclusion.

Table 12.8. Configuration Interaction Studies of the V State in Ethylene

Nature of CI study	ΔE (eV)	$\langle \Psi \vert \Sigma x_i^2 \vert \Psi \rangle$ (a.u.)
σ^* optimization; single and double excitations from five filled MOs (Ryan and Whitten)	8.02	6.94[a]
All single and double excitations from six filled MOs; iterative NO approach (Bendu *et al.*)	8.1[b]	35.4
MC–SCF; four configurations (Basch)	—	41.35
Single and double excitations from multireference NO-based configurations (Buenker and Peyerimhoff)	8.13	30[c]
CI studies, special consideration to choice of configuration (McMurchie and Davidson)	7.93	17.4
Experimental	7.65	

[a] Calculated using the π^* MO only, and not the full wavefunction.

[b] Estimated.

[c] Reported in Brooks and Schaefer study.

inner shell MOs) were allowed variable occupancy and all single plus double excitations were included in the CI study. The result was an N → V transition energy estimate of ~8.1 eV, and a diffuse charge distribution (i.e., $\langle x^2 \rangle = 35.4$ a.u.). They concluded that the V state was a diffuse, "Rydberg-like" state, in direct contrast with the Ryan and Whitten results (even though both studies found essentially the same N → V transition energy).

Continuing the saga, an MC–SCF study[79] appeared soon after the work of Bender *et al.*, in which four configurations were included. In particular, the highest occupied σ and π MO plus lowest unoccupied σ^* and π^* MO were optimized within the CI wavefunction. The results (see Table 12-8) again indicate a diffuse, Rydberg-like character for the V state. However, concerns were expressed that basis set additions and larger CI expansions might be expected to reduce the extent of the V state charge distribution.

Next, an extensive all-valence electron CI study was reported several years later,[80] in which NO-based configurations were used to construct a large CI

[79] H. Basch, *Chem. Phys. Lett.*, **19**, 323–327 (1973).

[80] R. J. Buenker and S. D. Peyerimhoff, *Chem. Phys.*, **9**, 75–89 (1976).

wavefunction (15,933 configurations) consisting of all single and double excitations from multiple reference configurations.[81] An essentially identical transition energy was found as in other studies (8.13 eV), but the V state spatial description was decreased (~ 30 a.u., see Table 12.8) from the studies of Bender *et al.* and Basch. It was concluded that the V state was therefore an "essentially valence-shell species," but that long-range d–π functions in the basis plus "moderate" CI size is necessary to achieve satisfactory results.

The next study was a large CI study by McMurchie and Davidson,[82] designed to determine the limit to the spatial contraction of the V state. Separate SCF calculations were carried out on the ground and V states, followed by CI studies designed to account for essentially all of the π-electron correlation energy. They found (see Table 12.8) an excitation energy of 7.93 eV and a state whose charge distribution was quite contracted, from which they concluded that the V-state is "relatively valencelike."

In reaching that conclusion, they found that $\sigma \rightarrow \sigma^*$ excitations mixing with $\pi \rightarrow \pi^*$ excitations were essential in the CI wavefunction. In addition, methods of choosing *CI* configurations that are not "biased," i.e., that do not presuppose that the SCF result is the correct one, were deemed to be important in achieving correct results.

If nothing else, the ethylene spectral studies illustrate the "art" that is needed for CI studies, due in large part to the computational inability to routinely carry out sufficiently large and flexible SCF/CI studies with similarly large and flexible basis sets. As a result, alternative CI formulations and other approaches have been developed, and we shall explore several of these in the sections to follow.

12-3. Specialized CI Approaches

As we have seen in the previous section, large CI expansions are frequently employed in high accuracy studies, and a variety of choices exist for constructing correlation orbitals. Attempts to avoid the former and connect the latter to traditional chemical concepts continue to occupy the focus of substantial research efforts. In this section some of the alternative approaches will be outlined, at least conceptually, with particular focus on those approaches that incorporate "traditional" chemical concepts in their formulation. As we shall see, these approaches do reduce the length of CI expansions substantially, while at the same time increasing the ease of interpretation of the wavefunction in chemical terms. Of course, the price that typically is paid for these benefits is a loss in absolute accuracy, and the ultimate utility of an approach frequently depends on cancellation of errors, and can be assessed generally only by comparison with appropriate experiments.

---

[81] Multireference CI approaches are discussed in greater detail in Section 12-3C.

[82] L. E. McMurchie and E. R. Davidson, *J. Chem. Phys.*, **66**, 2959 (1959). The particular portion of their studies discussed here is referred to as "CI1" in the McMurchie/Davidson paper.

A. Multiconfiguration SCF Theory

The approach used in multiconfiguration SCF theory (MC–SCF) is based on the observation that, if conventional CI approaches are employed, the orbitals used for correlation purposes are not usually optimum. For example, the virtual orbitals resulting from Hartree–Fock calculations that are frequently used for correlation purposes in CI studies result essentially from orthogonalization requirements, and are not optimized for use in correlation studies. In order to avoid this difficulty, a number of efforts[83] have been made to optimize the orbitals used for correlation purposes while at the same time optimizing the linear coefficients of configurations.

Expressed mathematically, the total N-electron wavefunction in MC–SCF theory can be written as

$$\Psi(1, 2, \cdots, N) = \sum_a A_a \Phi_a(1, 2, \cdots, N), \qquad (12\text{-}120)$$

where the Φ_a are orthonormal configurations that are composed of single (or multiple) Slater determinants, and A_a is the CI coefficient of configuration Φ_a. In a conventional CI approach the energy would be minimized only with respect to the linear coefficients (A_a), and the orbitals used in the configurations would be assumed to be given (i.e., *fixed* with respect to the minimization). In the MC–SCF approach, however, the orbitals comprising each configuration are also allowed to vary in order to minimize the total energy. For example, if configuration Φ_a is a single Slater determinant of orbitals $\{\chi\}$, i.e.,

$$\Phi_a = N' \det\{\chi_1^a(1)\chi_2^a(2) \cdots \chi_N^a(N)\}, \qquad (12\text{-}121)$$

where N' is a normalization constant, and

$$\chi_j^a = \sum_k c_{jk}^a \phi_k^a, \qquad (12\text{-}122)$$

in which $\{\phi_a\}$ is the basis set used in the study, then the MC–SCF procedure determines optimum values for the coefficients (c_{jk}^a) of the basis functions as well as the linear coefficients (A_a). Furthermore, this process must be repeated for each configuration in the CI expansion of Eq. (12-120).

The result of simultaneous optimization of CI coefficients and orbital coefficients is *two* sets of equations that need to be satisfied simultaneously. The first of these, that is used to obtain total energies and CI coefficients (A_a), is the

[83] See, for example, A. C. Wahl and G. Das, in *Methods of Electronic Structure Theory*, H. F. Schaefer, III (ed.), Plenum Press, New York, 1977, pp. 51–78; G. Das, *J. Chem. Phys.*, **58**, 5104 (1973); G. Das, T. Janis, and A. C. Wahl, *J. Chem. Phys.*, **61**, 1274 (1974); J. Hinze, *J. Chem. Phys.*, **59**, 6424 (1973); B. Levy and G. Berthier, *Int. J. Quantum Chem.*, **2**, 307 (1968); **3**, 247 (1969); R. C. Raffenetti and K. Ruedenberg, *Int. J. Quantum Chem.*, **34**, 625 (1970); J. Olsen, D. L. Yeager, and P. Jorgensen, *Adv. Chem. Phys.*, **LIV**, 1-176 (1983); A. C. Hurley, *Electron Correlation in Small Molecules*, Academic Press, New York, 1976, pp. 22–30.

familiar secular equation:

$$\det|\mathbf{H}-E\mathbf{I}|=0. \tag{12-123}$$

The second set of equations related to finding the optimum orbitals is where the complications occur, since the optimum orbitals can, at least in principle, be different for each configuration. Specifically, minimization of the total energy of the wavefunction of Eq. (12-120) with respect to the c_{jk}^{a} of Eq. (12-122) to find optimum orbitals [for a given basis $\{\phi\}$] leads to equations that resemble the Fock equations in ordinary Hartree–Fock theory, but have additional terms that arise because of the dependence of the orbitals on the particular configuration being considered. In particular, the Fock equations to be solved are given by

$$\mathbf{F}_i\mathbf{c}_i=\sum_j \epsilon_{ij}\mathbf{S}\mathbf{c}_j, \tag{12-124}$$

where

$$\mathbf{F}_i=\sum_a n_{ia}A_a^2\left(\mathbf{h}+\sum_\nu\sum_j n_{ja}t_{ija}^{\nu}\mathcal{G}^{\nu}\mathbf{D}_j\right), \tag{12-125}$$

$$(\mathbf{D}_j)_{pq}=c_{jp}c_{jq}, \tag{12-126}$$

$$(\mathcal{G})_{pq,rs}^{\nu}=\langle\phi_p|\mathcal{G}_{rs}^{\nu}|\phi_q\rangle, \tag{12-127}$$

and

$$(\mathbf{h})_{pq}=\langle\phi_p|h|\phi_q\rangle. \tag{12-128}$$

S is the orbital overlap matrix, n_{ia} is the occupancy of orbital i in configuration a, h is the ordinary one-electron operator of Hartree–Fock theory, t_{ija} is a coupling coefficient, and $\mathcal{G}_{pqrs}$ is a usual electron repulsion integral. In Eq. (12-125) it is seen how different orbitals can arise for different configurations, through the sums over configurations (ν and a). Also, it is assumed that the configuration coefficients (A_a) are known in Eq. (12-125). Since the A_a cannot be known until the orbital coefficients are determined, this implies that a multiple iteration process is required, with self-consistency in the MO coefficients (c_{ij}) required in the solution of Eq. (12-124), and a continued "macroiteration" between Eqs. (12-123) and (12-124) until the A_a and c_{ij} are self-consistent from one iteration to the next.

Among the ways in which the MC–SCF procedure has been implemented in practice is a procedure that Das and Wahl called the "Optimized Valence Configuration" (OVC) approach.[84] In this approach, both the number of configurations included and the orbitals allowed to vary are limited. Specifically, the configurations included are intended to accomplish two main goals: (1) to assure that the wavefunction appropriately describes dissociation into single

[84] G. Das and A. C. Wahl, *J. Chem. Phys.*, **44**, 87 (1966); **47**, 2934 (1967); **56**, 1769 (1972); *Phys. Rev. Lett.*, **24**, 440 (1970). For other approaches to solving the coupled MC-SCF equations see J. Hinze, *J. Chem. Phys.*, **59**, 6424 (1973); T. L. Gilbert, *Phys. Rev.*, **A6**, 580 (1972).

Table 12.9. Calculated Spectroscopic Constants Using the OVC Approach Compared to Experimental Values

Molecule	ω_e (cm^{-1})[a]	R_e (Bohrs)[b]	D_e (eV)[c]
H_2			
Hartree–Fock	4561	1.39	3.64
OVC	4398	1.40	4.63
Experiment	4400	1.40	4.75
NaF			
Hartree–Fock	570	3.65	3.08
OVC	570	3.66	3.70
Experiment	536	3.64	4.94

[a] Lowest vibrational stretching frequency for H_2.

[b] Equilibrium bond distance.

[c] Dissociation energy.

configuration wavefunctions for products in their ground state (i.e., include those configurations that correct one of the important qualitative Hartree–Fock errors), and (2) to describe correlation among electrons outside a chosen "core" of electrons.

To illustrate the rapid improvement beyond Hartree–Fock in describing spectroscopic properties of interest using this approach, Table 12.9 provides data for H_2 and NaF. For H_2, which is a prototype of what Das and Wahl refer to as a "simple covalent" system, we see that the OVC results for both the equilibrium bond distance and lowest stretching frequency are in essentially complete agreement with experiment. Even the dissociation energy, which would be incorrectly calculated in Hartree–Fock theory, is found to be only 2.5% in error. Furthermore, only three types of configurations beyond the Hartree–Fock result were needed in order to obtain these results. These three types of configurations also correspond closely to chemical intuition. For example, one type of configuration describes "left–right" correlation (i.e., they describe the situation in which, when one electron is in the vicinity of one nucleus, the other will be in the vicinity of the second nucleus). This is the type of configuration that must be added to the Hartree–Fock wavefunction if proper dissociation is to be obtained. Another type of configuration describes "in–out" correlation, and the final type of configuration describes "angular" correlation. Thus, at least for the "simple covalent" case, only a few configurations beyond the Hartree–Fock wavefunction are needed, and can be identified relatively easily through chemical intuition.

The NaF molecule provides an example of an "ionic" molecule, as well as a system containing substantially more electrons. Furthermore, the dissociation products are not closed shell atoms, but are open shell (neutral) atoms. Even with these complications, we see from Table 12.9 that reasonable results are obtained for both the equilibrium bond distance and lowest stretching frequency.

However, we see that a 25% error is present in the calculated dissociation energy reflecting the more complex nature of the products.

This technique was also extended to the calculation of dispersion (van der Waals) forces[85] and to triatomic molecules,[86] but the difficulties associated with the choice of Slater-type orbitals as the basis set coupled with the difficulties associated with choice of configurations beyond the first few obvious ones and the convergence of the MC–SCF procedure slowed its initial use. However, substantial efforts have occurred during the last decade,[87] which have improved the applicability of the approach substantially.

B. Electron Pair Theories

As alternatives to large CI expansions or MC–SCF approaches, a number of limited CI formulations have been developed that utilize electron pairs as the fundamental structural entity. Such approaches clearly draw on the long importance and success that the electron pair concept has had in chemistry (e.g., in chemical bonds, lone pairs, etc.), but also recognize that theoretical approaches beyond one-electron (orbital) theories will be necessary if accuracy sufficient for application to a broad spectrum of chemical problems is to be achieved.

1. Separated Pairs

To distinguish wavefunctions based on electron-pair functions from those based on orbitals, Shull[88] has suggested use of the term *geminal* for electron-pair functions. Many approaches to the construction of such functions have been described,[89] and we shall develop the simplest electron-pair concept in this section. Modifications and improvements to this model will be given in the next section.

A *"separated pair"* electronic wavefunction for a $2N$ electron system is one

---

[85] See, for example, A. F. Wagner, G. Das. and A. C. Wahl, *J. Chem. Phys.*, **60**, 1885 (1974).

[86] See, for example, W. B. England, N. H. Sabelli, and A. C. Wahl, *J. Chem. Phys.*, **63**, 4596 (1975).

[87] See R. Shepard, *Ab Initio Methods in Quantum Chemistry*, Volume 2, K. P. Lawley (ed.), John Wiley, New York, 1987, pp. 64–196, for an extensive review of developments and for references additional studies using this approach.

[88] H. Shull, *J. Chem. Phys.*, **30**, 1405 (1959).

[89] For early work in this area see, for example, L. Pauling, *Proc. R. Soc. (London)*, **A196**, 343 (1949); A. C. Hurley, J. E. Lennard-Jones and J. A. Pople, *Proc. R. Soc. (London)*, **A220**, 446 (1953); L. A. Schmid, *Phys. Rev.*, **92**, 1373 (1953); J. M. Parks and R. G. Parr, *J. Chem. Phys.*, **28**, 335 (1958); **32**, 1657 (1960); E. Kapuy, *Acts Phys. Acad. Sci. Hung.*, **9** 237 (1958); **10**, 125 (1959); **11**, 409 (1960); **12**, 185 (1960); M. Karplus and D. M. Grant, *Proc. Natl. Acad. Sci. U.S.A.*, **45**, 1269 (1959); R. M. McWeeny and K. A. Ohno, *Proc. R. Soc. (London)*, **A255**, 367 (1960); T. Arai, *J. Chem. Phys.*, **33**, 95 (1960).

having the form[90]

$$\Psi(1, 2, \cdots, 2N) = \mathcal{A}'\{G_1(1, 2)G_2(3, 4) \cdots G_N(2N-1, 2N)\}, \tag{12-129}$$

where each geminal (G_μ) is antisymmetric, i.e.,

$$G_\mu(1, 2) = -G_\mu(2, 1), \tag{12-130}$$

and where $\mathcal{A}'$ is a partial antisymmetrizer that exchanges electrons only between geminals. In addition to overall wavefunction normalization, each geminal is also individually normalized, i.e.,

$$\int\int G_\mu^*(1, 2)G_\mu(1, 2)\, d\tau_1\, d\tau_2 = 1. \tag{12-131}$$

Since each geminal is a function of two electrons, correlation effects between electron pairs (i.e., intrapair correlation) can be accounted for in detail. In the simplest form of the theory, however, an additional constraint is made, primarily for computational purposes. In particular, it is assumed that each geminal can be described by *its own basis set* of spin orbitals, i.e., a total (complete) orthonormal basis set of spin orbitals $\{\chi_{ij}\}$ is partitioned into subsets for each geminal:

$$\begin{array}{llll} G_1\colon \chi_{11} & \chi_{12} & \cdots & \chi_{1,N1} \\ G_2\colon \chi_{2,N1+1} & \chi_{2,N1+2} & \cdots & \chi_{2,N2} \\ G_3\colon \chi_{3,N2+1} & \chi_{3,N2+2} & \cdots & \chi_{3,N3}, \text{ etc.} \end{array} \tag{12-132}$$

Thus, a spin orbital used in the basis subset for one geminal may not be used in the basis subset of another geminal. The consequence of this assumption[91] is that the resulting geminals are one-electron orthogonal, i.e.,

$$\int G_\mu^*(1, 2)G_\nu(1, 3)\, d\tau_1 = 0 \qquad (\mu \neq \nu). \tag{12-133}$$

The above assumption is known as the *strong orthogonality constraint*, and the overall wavefunction is sometimes referred to as an *antisymmetrized product of strongly orthogonal geminals* (APSG) wavefunction.

Also, since each geminal is a function of only two electrons, it is possible without loss of generality to separate each geminal into a product of space (Λ_μ) and spin (Θ) functions, i.e.,

$$G_\mu(1, 2) = \Lambda_\mu(1, 2)\Theta_\mu(1, 2). \tag{12-134}$$

For purposes of exposition of the theory, we shall consider the case in which each geminal is a singlet (and thus the overall wavefunction describes a singlet state), with normalized antisymmetric spin functions:

$$\Theta_\mu(1, 2) = \frac{1}{\sqrt{2}}\{\alpha(1)\beta(2) - \beta(1)\alpha(2)\}, \qquad \mu = 1, 2, \cdots, N, \tag{12-135}$$

---

[90] The case of an even number of electrons is chosen here for pedagogical convenience, but is not an inherent limitation of the approach.

[91] See Problem 12-17.

and normalized symmetric space functions:

$$\Lambda_\mu(1, 2) = \Lambda_\mu(2, 1), \qquad \mu = 1, 2, \cdots N. \tag{12-136}$$

The energy expression for an APSG wavefunction as described above is given by

$$E = \int \cdots \int \Psi^*(1, 2, \cdots, 2N)\mathcal{H}\Psi(1, 2, \cdots, 2N)\, d\tau_1 \cdots d\tau_{2N}, \tag{12-137}$$

where a nonrelativistic Hamiltonian operator ($\mathcal{H}$) within the Born–Oppenheimer approximation has been assumed. Using the relevant equations given above, the energy expression can be rewritten as

$$\begin{aligned} E = \sum_\mu^N \Bigg\{ & 2 \int\int \Lambda_\mu^*(1, 2)h(1)\Lambda_\mu(1, 2)\, d\mathbf{r}_1\, d\mathbf{r}_2 \\ & + \int\int \Lambda_\mu(1, 2)\frac{1}{r_{12}}\Lambda_\mu(1, 2)\, d\mathbf{r}_1\, d\mathbf{r}_2 \Bigg\} \\ & + \sum_{\mu<\nu}^N \Bigg\{ 4 \int\int\int\int \Lambda_\mu^*(1, 2)\Lambda_\nu^*(3, 4)\frac{1}{r_{13}} \\ & \cdot \Lambda_\mu(1, 2)\Lambda_\nu(3, 4)\, d\mathbf{r}_1\, d\mathbf{r}_2\, d\mathbf{r}_3\, d\mathbf{r}_4 \\ & - 2 \int\int\int\int \Lambda_\mu^*(1, 2)\Lambda_\nu^*(3, 4)\frac{1}{r_{13}} \\ & \cdot \Lambda_\mu(3, 2)\Lambda_\nu(1, 4)\, d\mathbf{r}_1\, d\mathbf{r}_2\, d\mathbf{r}_3\, d\mathbf{r}_4 \Bigg\}, \end{aligned} \tag{12-138}$$

where h is the ordinary one-electron operator and the spin integrations have been carried out.

If each geminal is now expanded in terms of its natural orbitals [see Eq. (12-132)], i.e.,

$$\Lambda_\mu(1, 2) = \sum_i^{N_\mu} c_{\mu i}\chi_{\mu i}^*(1)\chi_{\mu i}(2), \qquad \mu = 1, 2, \cdots N, \tag{12-139}$$

where

$$\int \chi_{\mu i}^*(1)\chi_{\nu j}(1)\, d\mathbf{r}_1 = \delta_{\mu\nu}\delta_{ij}, \tag{12-140}$$

and (from geminal normalization)

$$\sum_i^{N_\mu} c_{\mu i}^2 = 1, \tag{12-141}$$

then the energy expression can be rewritten as

$$E=\sum_{\mu}^{N}\left\{2\sum_{i}^{N_\mu} c_{\mu i}^2 \int \chi_{\mu i}^*(1)h(1)\chi_{\mu i}(1)\, d\mathbf{r}_1 + \sum_{i,j}^{N_\mu} c_{\mu i}c_{\mu j}\int\int \chi_{\mu i}^*(1)\chi_{\mu j}(1)\frac{1}{r_{12}}\chi_{\mu j}^*(2)\chi_{\mu i}(2)\, d\mathbf{r}_1\, d\mathbf{r}_2\right\}$$

$$+\sum_{\mu<\nu}^{N}\left\{\sum_{i}^{N_\mu}\sum_{j}^{N_\nu} c_{\mu i}^2 c_{\nu j}^2 \cdot\left[4\int\int \chi_{\mu i}^*(1)\chi_{\mu i}(1)\frac{1}{r_{12}}\chi_{\nu j}^*(2)\chi_{\nu j}(2)\, d\mathbf{r}_1\, d\mathbf{r}_2 - 2\int\int \chi_{\mu i}^*(1)\chi_{\nu j}(1)\frac{1}{r_{12}}\chi_{\nu j}^*(2)\chi_{\mu i}(2)\, d\mathbf{r}_1\, d\mathbf{r}_2\right]\right\} \quad (12\text{-}142)$$

To see the relationship to Hartree–Fock theory, we note that truncation of Eq. (12-139) to a single term gives

$$\Lambda_\mu(1, 2)=\chi_{\mu 1}(1)\chi_{\mu 1}(2), \quad (12\text{-}143)$$

with $c_{\mu 1} = 1$. For this case, it is easily seen that the total wavefunction [Eq. (12-129)] reduces to a single Slater determinant, and the optimum orbitals will be essentially Hartree–Fock orbitals.

In order to obtain an optimized geminal description, we see from Eq. (12-139) that it is necessary to optimize the energy with respect to both the occupation coefficients ($c_{\mu i}$) and the natural orbitals ($\chi_{\mu i}$). Using procedures similar to those used earlier to derive optimized CI coefficients,[92] optimization of the energy expression in Eq. (12-142) with respect to occupation coefficients gives rise to a set of coupled eigenvalue equations:[93]

$$\sum_{i}^{N_\mu} H_{ki}^{\mu} c_{\mu i}=\epsilon_\mu c_{\mu k} \qquad \begin{matrix} k=1, 2, \cdots, N_\mu \\ \mu=1, 2, \cdots, N \end{matrix}, \quad (12\text{-}144)$$

where

$$H_{ki}^{\mu}=\delta_{ki}\left\{2\int \chi_{\mu k}^*(1)h(1)\chi_{\mu k}(1)\, d\mathbf{r}_1 + \sum_{\nu(\neq\mu)}^{N} c_{\nu j}^2\left[4\int\int \chi_{\mu k}^*(1)\chi_{\mu k}(1)\frac{1}{r_{12}}\chi_{\nu j}^*(2)\chi_{\nu j}(2)\, d\mathbf{r}_1\, d\mathbf{r}_2 - 2\int\int \chi_{\mu k}^*(1)\chi_{\nu j}(1)\frac{1}{r_{12}}\chi_{\nu j}^*(2)\chi_{\mu k}(2)\, d\mathbf{r}_1\, d\mathbf{r}_2\right]\right\}$$

$$+\int\int \chi_{\mu i}^*(1)\chi_{\mu k}(1)\frac{1}{r_{12}}\chi_{\mu k}^*(2)\chi_{\mu i}(2)\, d\mathbf{r}_1\, d\mathbf{r}_2. \quad (12\text{-}145)$$

---

[92] See Chapter 10, Section 10-4.

[93] See Problem 12-18.

These equations [Eq. (12-144)] are coupled in the sense that H^{μ}_{kk} is dependent on the squares of the other occupation coefficients ($c^2_{\nu j}$), and Eq. (12-144) has to be solved iteratively. In practice, obtaining solutions of Eq. (12-144) using an iterative procedure typically does not lead to difficulty.

In order to optimize the energy with respect to natural orbitals, a basis set $\{\varphi_p\}$ of m orbitals is usually chosen, and each natural orbital is expanded in terms of the basis, i.e.,

$$\chi_{\mu i}=\sum_{p}^{m} a_{p,\mu i}\varphi_p=\boldsymbol{\varphi}\mathbf{a}_{\mu i}, \tag{12-146}$$

where $\boldsymbol{\varphi}$ is a row vector of basis orbitals and $\mathbf{a}_{\mu i}$ is a column vector. Thus, the problem of determination of optimum natural orbitals is now converted to a problem of determining optimum expansion coefficients ($a_{p,\mu i}$). If Eq. (12-146) is substituted into the energy expression [Eq. (12-142)], and the energy is minimized with respect to choice of the $a_{p,\mu i}$ (subject to maintaining orthonormality of the natural orbitals), the following equations result[94]:

$$\mathbf{F}_{\mu i}\mathbf{a}_{\mu i}=\mathbf{S}\sum_{\nu}^{N}\sum_{j}^{N_\nu}\mathbf{a}_{\nu j}\lambda_{\mu i,\nu j}, \tag{12-147}$$

where $\lambda_{\mu i,\nu j}$ is a Lagrangian multiplier and $\boldsymbol{\lambda}$ is Hermitian. Also,

$$(\mathbf{S})_{pq}=\int \varphi_p^*(1)\varphi_q(1)\,d\mathbf{r}_1, \tag{12-148}$$

and

$$\mathbf{F}_{\mu i}=c^2_{\mu i}\mathbf{H}+\sum_{j}^{N_\nu} c_{\mu i}c_{\mu j}\mathbf{K}_{\mu j}+\sum_{\nu(\neq\mu)}^{N}\sum_{j}^{N_\nu} c^2_{\mu i}c^2_{\nu j}(2\mathbf{J}_{\nu j}-\mathbf{K}_{\nu j}), \tag{12-149}$$

where

$$\mathbf{H}=\int \boldsymbol{\varphi}^\dagger(1)h(1)\boldsymbol{\varphi}(1)\,d\mathbf{r}_1, \tag{12-150}$$

$$\mathbf{J}_{\nu j}=\int\int \boldsymbol{\varphi}^\dagger(1)\boldsymbol{\varphi}(1)\frac{1}{r_{12}}\chi^*_{\mu j}(2)\chi_{\nu j}(2)\,d\mathbf{r}_1\,d\mathbf{r}_2, \tag{12-151}$$

and

$$\mathbf{K}_{\nu j}=\int\int \boldsymbol{\varphi}^\dagger(1)\chi_{\nu j}(1)\frac{1}{r_{12}}\chi^*_{\nu j}(2)\boldsymbol{\varphi}(2)\,d\mathbf{r}_1\,d\mathbf{r}_2. \tag{12-152}$$

It should be noted that $\mathbf{J}_{\mu i}$ and $\mathbf{K}_{\nu j}$ are each ($m \times m$) matrices, even though they are also labelled by the index "μi" or "νj."

Equation (12-147) represents a set of equations that appears to resemble those of Hartree–Fock theory, but there are several significant differences (and

[94] See, for example, W. Kutzelnigg, *J. Chem. Phys.*, **40**, 3640–3647 (1964).

complications). First, from a conceptual point of view, we see that determination of the optimal geminals via Eq. (12-147) can be interpreted as determining the *two-electron* geminal in the effective field of all the other electrons. This is a clear difference and obvious benefit of using this approach compared to the Hartree–Fock approach. However, application of a unitary transformation of the basis that diagonalizes λ cannot be applied as in Hartree-Fock theory, because each matrix $\mathbf{F}_{\mu i}$ may be different for different μi values. Also, the matrices $\mathbf{F}_{\mu i}$ are not invariant with respect to a unitary transformation of the basis set. Hence, the off-diagonal matrix elements $(\lambda_{\mu i, \nu j})$ cannot be easily eliminated, which complicates the solution of Eq. (12-147) considerably. However, a number of methods for the solution of Eq. (12-147) have been proposed,[95] and solutions can typically be obtained.

In order to illustrate the results that can be obtained using this approach, let us consider the LiH molecule. A number of studies have appeared for LiH using this approach,[96] and we shall focus on the study by Mehler *et al.*, since it is a relatively definitive study with respect to the advantages and limitations of the separated pair approach.

The basis orbitals chosen for their study were Slater-type orbitals (STO), and included 13 different types of STOs, i.e., $1s$, $2s$, $2p_\sigma$, $2p_\pi$, $3p_\sigma$, $3p_\pi$, and a set of $3d_\delta$ STOs for the inner shell on Li and $2s'$, $2p'_\sigma$, $2p'_\pi$, and $3s'$ orbitals for the outer shell (bonding orbital). The primes have been used to remind us that the orbitals for the outer shell must be different than those of the inner shell, which is a requirement of separated pair theory that we saw earlier in the discussion (and which is a significant limitation of the theory). For the H atom, a basis set of five different types of STOs, i.e., 1s, 2s, $2p_\sigma$, $2p_\pi$, and $3d_\delta$ STOs was chosen. This gives rise to 18 natural orbitals, and the wavefunction (at the equilibrium distance of $R = 3.015$ Bohr) that results can be written as

$$\Psi = \mathcal{A}[G_K(1, 2)\ G_B(3, 4)], \tag{12-153}$$

where G_K is a geminal describing the Li inner shell electrons, and G_B is a geminal describing the Li–H bond. The optimized description of the geminals in the above equation were found to be given as follows:

$$\begin{aligned} G_K = {} & 0.99883(K\sigma 1)^2 - 0.02463(K\sigma 2)^2 - 0.02158(K\sigma 3)^2 \\ & - 0.00383(K\sigma 4)^2 - 0.00142(K\sigma 5)^2 - 0.02411(K\pi 1)^2 \\ & - 0.00386(K\pi 2)^2 - 0.00383(K\pi 3)^2 - 0.00384(K\delta 1)^2, \end{aligned} \tag{12-154}$$

[95] See, for example, R. McWeeny and K. A. Ohno, *Proc. R. Soc. (London)*, **A255**, 367 (1960); E. Kapuy, *Acta Phys. Acad. Sci. Hung.*, **12**, 185, 351 (1960); **13**, 345, 461 (1961); D. D. Ebbing and R. C. Henderson, *J. Chem. Phys.*, **42**, 2225 (1965); K. J. Miller and K. Ruedenberg, *J. Chem. Phys.*, **48**, 3414, 3444, 3450 (1968); W. Kutzelnigg, *J. Chem. Phys.*, **40**, 3640 (1964); J. Loter and R. E. Christoffersen, *Int. J. Quantum Chem.*, **3**, 651–661 (1969).

[96] See, for example, D. D. Ebbing and R. C. Henderson, *J. Chem. Phys.*, **42**, 2225–2231 (1965); R. Alrichs and W. Kutzelnigg, *J. Chem. Phys.*, **48**, 1819–1832 (1968); E. L. Mehler, K. Ruedenberg, and D. M. Silver, *J. Chem. Phys.*, **52**, 1181–1205 (1970).

and

$$G_B = 0.98545(B\sigma 1)^2 - 0.12319(B\sigma 2)^2 - 0.05695(B\sigma 3)^2 - 0.01232(B\sigma 4)^2 - 0.00198(B\sigma 5)^2 - 0.00094(B\sigma 6)^2 - 0.07104(B\pi 1)^2 - 0.00947(B\pi 2)^2 - 0.00482(B\pi 3)^2. \quad (12\text{-}155)$$

When interpreting these results, it is of interest first to note that the inner shell geminal (G_K) strongly resembles the inner shell geminal description of the Li atom,[97] where the overlap between the two geminals is ~ 0.997. Thus, one of the hoped-for features of the theory is seen, i.e., geminals are transferable from one system to another (in this case, from an atom to a molecule).

In the case of the bonding geminal (G_B), the principal natural orbital ($B\sigma 1$) can be written approximately as

$$(B\sigma 1) \sim 0.17\ (\text{Li-}2s') + 0.21\ (\text{Li-}2p') + 0.17\ (\text{Li-}3s') + 0.66\ (\text{H-}1s) \quad (12\text{-}156)$$

Thus, the principal natural orbital exhibits a strong polarization toward the hydrogen, as indicated by the magnitude of the (H-1s) coefficient. A similar polarization is seen in the π-type natural orbitals. This polarization is usually interpreted to mean that the LiH molecule has a significant amount of Li^+H^- character.

From an energetic point of view, the separated pair wavefunction had a total energy of -8.05418 H at the equilibrium distance, compared to an experimental value of -8.0705 H. Also, the energy of the principal natural orbitals (which is essentially the Hartree–Fock description) is -7.98469 H. This means that the correlation energy recovery of the separated pair wavefunction is approximately 0.069 H, compared to the total correlation energy of 0.0858 H. Thus, approximately 80% of the total correlation energy has been recovered by the separated pair wavefunction.

Also, the binding energy (i.e., comparing LiH to Li + H) and various spectroscopic constants were calculated, and are summarized in Table 12.10. The accuracy of calculated spectroscopic constants is seen in general to be quite high except for the polarizability, suggesting further that the geminal idea (and perhaps even the "separated pair" formulation) may provide a satisfactory theoretical basis for general computational studies of chemical systems. However, this particular example is one in which the electron pairs are quite separable, and interpair electron interactions are small. Furthermore, the basis orbitals used to describe each pair are, in general, dissimilar, thus allowing the strong orthogonality constraint to be satisfied without placing undue strain on basis set flexibility.

On the other hand, the strong orthogonality constraint can be seen to be a significant drawback in cases that are not so ideal. For example, when Silver *et*

---

[97] A separate calculation was carried out to obtain an optimized Li atom description.

Table 12.10. Binding Energy and Spectroscopic Constants for LiH

Quantity	Calculated	Experimental[a]	Percentage deviation
Binding energy (H)	0.0847	0.0925	
	(2.30 eV)	(2.52 eV)	−8.4
B_e (cm^{-1})	7.381	7.513	−1.75
ω_e (cm^{-1})	1483	1405.6	5.5
$\omega_e\chi_e$ (cm^{-1})	24.45	23.20	5.4
α_e (cm^{-1})	0.2849	0.213	33.7
R_e (Å)	1.611	1.595	0.96

[a] Experimental data are taken from G. Herzberg, *Molecular Spectra and Molecular Structure*. I. *Spectra of Diatomic Molecules*, D. Van Nostrand, Princeton, NJ, 1950. Experimental quantities include the rotational constant (B_e), vibrational constants (ω_e and $\omega_e x_e$), and the polarizability (α_e).

al.,[98] extended the studies described above the the case of the imidogen molecule (N–H), they found substantially less favorable results. In this case, a total energy of −55.03352 H was calculated, compared to an experimental value of −55.252 H. Thus, the calculated total energy is in error by approximately 0.218 H (or 137 kcal/mol). Also, the SCF energy is found to be −54.9784 H, meaning that the separated pair wavefunction has a correlation energy recovery of 0.055 H, which is only 20% of the total correlation energy of 0.2736 H. Thus, the absolute accuracy of the calculated results is far less than needed for chemical accuracy of a few kcal/mol.

When examined in greater detail, it was found that the largest source of error was in the description of the lone pair geminal. The primary reason why the lone pair geminal is poorly described is because the $2p$ orbitals that would contribute strongly to the lone pair geminal have been used in the description of the bonding and triplet natural orbitals, and are therefore not available for use in the lone pair geminal (because of the strong orthogonality constraint). The other sources of error were believed to be somewhat evenly divided between the description of intergeminal correlation effects and intrageminal effects. In any case, it has become clear that, if accuracies in calculated total energies are to be of chemical utility (i.e., absolute accuracy to within a few kcal/mol), alternatives to the separated pair approximation to geminal theory are needed.

2. Other Electron Pair Approaches

A number of alternative approaches have been developed to go beyond the limitations of the strong orthogonality constraint. As we shall see, however, these approaches typically achieve size consistency but lose their variational nature. This latter point means that the calculated correlation energy may be

[98] D. M. Silver, K. Ruedenberg, and E. L. Mehler, *J. Chem. Phys.*, **52**, 1206–1227 (1970).

greater than 100% of the true value. In practice, between 80 and 120% of the exact correlation energy has typically been found.[99]

To see how these wavefunctions are constructed, we begin by deriving an explicit expression for the correlation energy that has also been noted by Szabo and Ostlund.[100] If Ψ is the (unnormalized) exact wavefunction for a system with Hamiltonian $\mathcal{H}$ and ground state energy $\mathcal{E}_0$, then

$$\mathcal{H}\Psi = \mathcal{E}_0 \Psi. \tag{12-157}$$

If Ψ is expanded in terms of configurations containing single, double, etc., excitations, we have:[101]

$$\Psi = \Phi_0 + \sum_{a,\alpha} c_a^\alpha \Phi_a^\alpha + \sum_{\substack{a<b \\ \alpha<\beta}} c_{ab}^{\alpha\beta} \Phi_{ab}^{\alpha\beta} + \sum_{\substack{a<b<c \\ \alpha<\beta<\gamma}} c_{abc}^{\alpha\beta\gamma} \Phi_{abc}^{\alpha\beta\gamma} + \cdots, \tag{12-158}$$

where the rotation is that introduced earlier [Eq. (12-15)], Φ_0 is the normalized Hartree–Fock wavefunction for the system, and the orbitals comprising the various configurations are assumed to be mutually orthonormal. If E_0 is the Hartree–Fock energy corresponding to Φ_0, we can write

$$\begin{aligned}(\mathcal{H} - E_0)\Psi &= (\mathcal{E}_0 - E_0)\Psi \\ &= E_{\text{correl}}\Psi,\end{aligned} \tag{12-159}$$

where the correlation energy (E_{correl}) is given by

$$E_{\text{correl}} = \mathcal{E}_0 - E_0. \tag{12-160}$$

If Eq. (12-160) is multiplied on the left by Φ_0 and integrated, we have

$$\langle \Phi_0 | \mathcal{H} - E_0 | \Psi \rangle = E_{\text{correl}}. \tag{12-161}$$

Using Eq. (12-158), we can use Brillouin's Theorem and other manipulations to rewrite the left-hand side of the above equation as

$$\begin{aligned}\langle \Phi_0 | \mathcal{H} - E_0 | \Psi \rangle &= \Big\langle \Phi_0 \Big| \mathcal{H} - E_0 \Big| \Phi_0 + \sum_{a,\alpha} c_a^\alpha \Phi_a^\alpha + \sum_{\substack{a<b \\ \alpha<\beta}} c_{ab}^{\alpha\beta} \Phi_{ab}^{\alpha\beta} \\ &\quad + \sum_{\substack{a<b<c \\ \alpha<\beta<\gamma}} c_{abc}^{\alpha\beta\gamma} \Phi_{abc}^{\alpha\beta\gamma} + \cdots \Big\rangle \\ &= \Big\langle \Phi_0 \Big| \mathcal{H} \Big| \sum_{\substack{a<b \\ \alpha<\beta}} c_{ab}^{\alpha\beta} \Phi_{ab}^{\alpha\beta} \Big\rangle .\end{aligned} \tag{12-162}$$

[99] For a discussion of pair correlation theories in greater detail see, for example, W. Kutzelnigg, *Methods of Electronic Structure Theory*, Volume 3, H. F. Schaefer III (ed.), **3**, Plenum Press, New York, 1977, pp. 129–188.

[100] See A. Szabo and N. S. Ostlund, *Modern Quantum Chemistry*, Macmillan, New York, 1982, Chapter 5, for additional discussion of pair and coupled-pair theories.

[101] The wavefunction Ψ in this form is said to be *intermediately normalized*, since it is not normalized to unity, but $\langle \Psi | \Phi_0 \rangle = 1$.

Combining Eq. (12-161) with Eq. (12-162) gives

$$E_{\text{correl}} = \sum_{\substack{a<b \\ \alpha<\beta}} c_{ab}^{\alpha\beta} \langle \Phi_0 | \mathcal{H} | \Phi_{ab}^{\alpha\beta} \rangle. \tag{12-163}$$

This result, while useful in a formal sense, is not as helpful computationally as it might appear. In particular, even though the result indicates that the entire correlation energy can be found from a knowledge of only double excitation configurations, the entire CI wavefunction must be known before the coefficients ($c_{ab}^{\alpha\beta}$) can be determined.

More important, however, is the idea suggested by Eq. (12-163) that pair-type correlations are likely to dominate the correlation energy. Not only is this consistent with chemical intuition and the geminal theory introduced earlier, it also suggests additional computational approaches can be developed that alleviate some of the constraints of separated pair theory discussed earlier.

The simplest approach based on Eq. (12-163) is to assume that the correlation energy can be expressed as a sum of pair contributions, i.e.,

$$E_{\text{correl}} = \sum_{a<b} e_{ab}, \tag{12-164}$$

where

$$e_{ab} = \sum_{\alpha<\beta} d_{ab}^{\alpha\beta} \langle \Phi_0 | \mathcal{H} | \Phi_{ab}^{\alpha\beta} \rangle. \tag{12-165}$$

Then, if the contributions to the correlation energy from all electrons except the pair under consideration are ignored, we obtain an approach to correlation energy estimation known as the *independent electron pair approximation*[102] (IEPA). Specifically, the trial wavefunction is taken to have the form

$$\Psi_{ab} = \Phi_0 + \sum_{\alpha<\beta} d_{ab}^{\alpha\beta} \Phi_{ab}^{\alpha\beta}, \tag{12-166}$$

where Φ_0 is the Hartree–Fock wavefunction for the system. As the form of the wavefunction implies, only double excitations from orbitals a and b are considered, and the result is a wavefunction that includes correlations within electron pairs only. Single excitations (e.g., from a or b individually) are ignored, and interactions between electron pairs are also ignored.

Proceeding by using analysis similar to that which led to Eq. (12-163), it can be shown[103] that the equations that determine the pair correlation energies (e_{ab}) are given by

$$\begin{pmatrix} \mathbf{0} & \mathbf{B}_{ab}^{\dagger} \\ \mathbf{B}_{ab} & \mathbf{D}_{ab} \end{pmatrix} \begin{pmatrix} 1 \\ \mathbf{d}_{ab} \end{pmatrix} = e_{ab} \begin{pmatrix} 1 \\ \mathbf{d}_{ab} \end{pmatrix}, \tag{12-167}$$

[102] O. Sinanoglu, *J. Chem. Phys.*, **36**, 706, 3198 (1962); *Adv. Chem. Phys.*, **6**, 315 (1964); **14**, 337 (1969); R. K. Nesbet, *Phys. Rev.*, **109**, 1632 (1958); *Adv. Chem. Phys.*, **9**, 321 (1965); **14**, 1 (1969).

[103] See Problem 12-19.

where

$$(\mathbf{d}_{ab})_{\alpha\beta} = d_{ab}^{\alpha\beta}, \tag{12-168}$$

$$(\mathbf{B}_{ab})_{\alpha\beta} = \langle \Phi_0 | \mathcal{H} | \Phi_{ab}^{\alpha\beta} \rangle, \tag{12-169}$$

and

$$(\mathbf{D}_{ab})_{\alpha\beta,\gamma\delta} = \langle \Phi_{ab}^{\alpha\beta} | \mathcal{H} - E_0 | \Phi_{ab}^{\gamma\delta} \rangle. \tag{12-170}$$

When these above equations are solved for each pair correlation energy (e_{ab}), the total correlation energy estimate is obtained from Eq. (12-164).

To illustrate the results that can be obtained using this approach, let us consider a series of molecules and an atom in which successively increasing numbers of electron pairs are present. In particular, let us consider BH_3, CH_4, NH_3 (pyramidal), H_2O, and Ne, using the study of Kutzelnigg[104] and colleagues as a basis for comparison. In these studies, basis sets consisting of five $1s$ functions, three $2p$ functions, two sets of d functions, and one f set were used for the heavy atom, plus three s functions and two p sets for each hydrogen. Inner shell correlation energies were neglected, since they have very little influence on molecular properties.

Results of these studies are given in Table 12.11. Considering individual pair correlation energies first, we see that the intrabond correlation energy (e_b) decreases from BH_3 to H_2O. This is usually explained in terms of decreasing availability of the $2p$ valence atomic orbital for intrabond correlation purposes as it is used elsewhere. Next, we note that the correlation energy of lone pairs (e_n) is smaller than intrabond correlation energies, presumably because lone pairs are less localized than intrabond paris. Interorbital correlation energies (${}^{1,3}e_{b,b'}$, ${}^{1,3}e_{b,n}$, ${}^{1,3}e_{n,n'}$) are all seen to be approximately the same magnitude, although an increase is seen in $e_{b,b'}$ from BH_3 to H_2O. This latter effect is attributed to the decreasing valence angle, and consequent increased differential overlap between bonds.

Considering total correlation energies, we see that the IEPA estimates are higher than the CI estimates in each case. However, the IEPA estimates are not necessarily larger than the exact values. Specifically, it was estimated that, in general, the IEPA total correlation estimates (without K shell contributions) represent approximately 75% of the "exact" values. Errors that remain are attributed primarily to truncation of the orbital basis ($\sim$0.033 a.u.) and to the lack of inclusion of contributions from "unlinked clusters"[105] ($\sim$0.02 – 0.05 a.u.). When corrections to the orbital basis are made, greater than 100% of the

[104] R. Alrichs, H. Lischka, V. Staemmler, and W. Kutzelnigg, *J. Chem. Phys.*, **62**, 1225–1234 (1975); R. Alrichs, F. Driessler, H. Lischka, V. Staemmler, and W. Kutzelnigg, *J. Chem. Phys.*, **62**, 1235–1247 (1975).

[105] These will be discussed in Section 12-4.

Table 12.11. Total Valence Correlation Energies ($E^{\text{CI}}_{\text{correl}}$ and $E^{\text{IEPA}}_{\text{correl}}$) and Pair Correlation Energies (e_b, e_n, ${}^{1,3}e_{bb'}$, ${}^{1,3}e_{bn}$, ${}^{1,3}e_{nn'}$) Calculated Using the IEPA Approach[a]

	$-E^{\text{CI}}_{\text{correl}}$	$-E^{\text{IEPA}}_{\text{correl}}$	$-e_b$	$-e_n$	$-{}^1e_{bb'}$	$-{}^3e_{bb'}$	$-{}^1e_{bn}$	$-{}^3e_{bn}$	$-{}^1e_{nn'}$	$-{}^3e_{nn'}$
BH_3	0.12346	0.13952	0.0340	—	0.0050	0.0075	—	—	—	—
CH_4	0.19779	0.24008	0.0330	—	0.0069	0.0112	—	—	—	—
NH_3 (pyramidal)	0.22414	0.28093	0.0326	0.0291	0.0088	0.0139	0.0116	0.0170	—	—
H_2O	0.24182	0.30690	0.0326	0.0275	0.0110	0.0169	0.0125	0.0185	0.0148	0.0198
Ne	0.25590	0.30379	—	0.0258	—	—	—	—	0.0139	0.0196

[a] All energies are given in atomic units.

correlation energy is found in the IEPA approach,[106] emphasizing again the nonvariational aspect of the approach.

In order to make more accurate approximations to the correlation energy, a number of improvements to the IEPA approach have been suggested. These approaches incorporate coupling between electron pairs, and have been called the "coupled cluster approximation" (CCA),[107] the "coupled electron pair approximation" (CEPA),[108] or the "independent-pair potential approximation" (IPPA).[109] In general, accuracies are improved, at the expense of computational complexity and increased difficulty of easy interpretation, and the need for further improvement remains.[110]

C. Large and Very Large Molecules

As we have seen in repeated ways, devising techniques and basis sets that allow calculation of molecular energies and properties whose absolute error is sufficiently small to assure applicability to chemical problems is not currently possible even for relatively small chemical systems. Furthermore, a great deal of current chemical and molecular level biology research efforts consider much larger molecules, thus providing further significant frustration to theoretical chemists. Essentially three options exist in this situation: (1) acknowledge that absolute accuracies cannot be obtained and work on other problems; (2) utilize semiempirical techniques that introduce experimental data and/or simplifying mathematical assumptions that allow applicable techniques to be created; or (3) utilize *ab initio* quantum mechanics to devise small basis sets and associated techniques that, while not providing satisfactory absolute accuracies, provide *balanced* results in which, e.g., relative energies, can be reliably calculated. It is the latter approach that will be described below, utilizing several examples to illustrate how such technologies can be developed and applied. One of the major advantages of such an approach is that it is substantially easier than with other

---

[106] R. K. Nesbet, T. L. Barr, and E. R. Davidson, *Chem. Phys. Lett.*, **4**, 203 (1969).

[107] J. Cizek and J. Paldus, *Phys. Scripta*, **21**, 251 (1980).

[108] W. Meyer, *Methods of Electronic Structure Theory*, Volume 3, H. F. Schaefer III, (ed.), Plenum Press, New York, 1977, pp. 413–446.

[109] E. L. Mehler, *Theoret. Chim. Acta*, **35**, 17–32 (1974).

[110] Other approaches have also been suggested, e.g., the *generalized valence bond* (GVB) approach of Goddard and collaborators. This is another electron pair-based approach that can be thought of as a generalization of valence bond theory as described in Chapter 10 (Section 10-1A). For further discussion see, for example, F. W. Bobrowicz and W. A. Goddard III, *Modern Theoretical Chemistry*, Volume 3, M. F. Schaefer III (ed.), Plenum Press, New York, 1977, pp. 79–127; W. A. Goddard, *Phys. Rev.*, **157**, 81–93 (1967); W. J. Hunt, P. J. Hay and W. A. Goddard III, *J. Chem. Phys.*, **57**, 738–748 (1972); W. A. Goddard III and L. B. Harding, *Annu. Rev. Phys. Chem.*, **29**, 363 (1978); W. A. Goddard III, T. H. Dunning Jr., W. J. Hunt, and P. J. Hay, *Acc. Chem. Res.*, **6**, 368 (1973); B. J. Moss and W. A. Goddard III, *J. Chem. Phys.*, **63**, 3523–3531 (1975); W. A. Goddard III and R. D. Ladner, *J. Am. Chem. Soc.*, **93**, 6750–6756 (1971).

methods to identify sources of error in an *ab initio* approach, thus allowing improvements to be made more easily.

Questions regarding one aspect of "balance," i.e., basis set balance in Hartree–Fock calculations, have already been addressed.[111] In those discussions it was noted that balance in the *relative* ordering of orbital energies was an important asset in the interpretation of properties usually addressed through Hartree–Fock theory, e.g., ionization potentials.

For properties that require consideration of correlation effects and the use of multiconfiguration wavefunctions, e.g., electronic spectra and reaction surfaces, it is necessary to achieve another level of balance. For example, a very accurate calculation of only the ground state energy without a relatively equivalent description of excited states is of little value if explanation of the electronic absorption spectra is the desired result.

As to what kinds of characteristics should be included in describing large and very large molecules such as polypeptides or DNA fragments, there is no single, agreed on set of criteria to be used. However, among the desirable criteria are that (1) the method should be size extensive, at least in principle; (2) the description of states should be balanced (e.g., the method/basis set should not describe only one or a few particular electronic transitions and describe others poorly); and (3) the basis sets and methodology should be applicable in a relatively convenient manner to systems containing hundreds of electrons and various first row nuclei.

Not surprisingly, there is not currently available a method that satisfies all of these criteria. However, a number of important steps have been taken in this direction. As an example of one of the earliest *ab initio* studies on a very large molecular system, the 1971 study by Clementi *et al.*[112] of the guanine–cytosine (GC) base pair is of interest. In that study, a basis set of 334 s- and p-type GTOs contracted to 105 functions was used to study the hydrogen bonding potential curves in the 136 electron GC pair. This set of basis functions corresponds to the use of contracted GTOs to provide a minimum basis set at the SCF level, i.e., $1s$, $2s$, $2p_x$, $2p_y$, and $2p_z$ functions were used on each nonhydrogen atom and one $1s$ function was used on each hydrogen atom. The calculations were restricted to the SCF level, using crystal data to fix the G–C distance and varying each of the three hydrogen bonding distances one at a time (see Fig. 12.5). Seven different HB1 distances were considered, while eight HB2 and eight HB3 distances were considered. At the time these calculations were undertaken, they represented major demands on computing resources. For example, the SCF studies on the 23 different geometric arrangements required 8 days of dedicated IBM 360/195 computer time! This included handling ~ 7 million integrals in the SCF process for each point.

The results obtained in this study did not indicate a double potential well for any of the three proton motions. This result was contrary to other computational

---

[111] See Chapter 11, Section 11-2C.

[112] E. Clementi, J. Mehl, and W. vonNiessen, *J. Chem. Phys.*, **54**, 508–520 (1971).

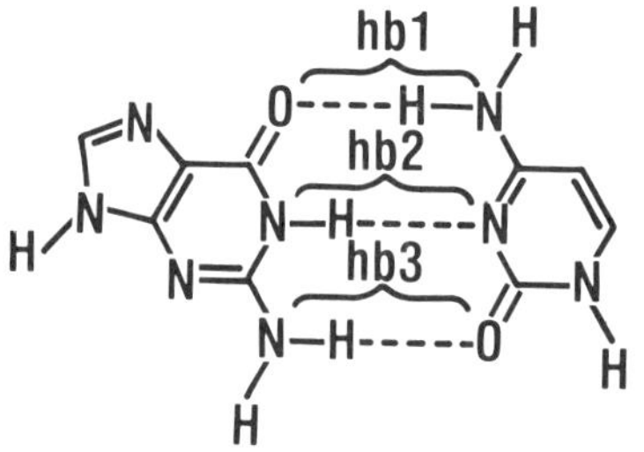

Figure 12.5. Hydrogen bonding distances varied in the GC base pair study.

studies and general theories of hydrogen bonding in DNA base pairs, but was probably due to one or more factors such as the assumption of rigid G–C geometry, the small basis sets used, the restrictions of Hartree–Fock theory, and the lack of consideration of concerted proton motion. Perhaps more important, however, was the fact that the calculations were possible at all at that point in time.

Another approach was also developed around the same time, that has been shown to be applicable both to CI as well as SCF studies of very large molecules. This approach, usually referred to as the *molecular fragment* approach,[113] constructs a "subminimal" basis set by utilizing floating spherical Gaussian orbitals (FSGO),[114] defined by

$$G_j(\mathbf{r}) = \left(\frac{2\pi}{\rho_j}\right)^{3/4} \exp\{-(\mathbf{r}-\mathbf{R}_j)^2/\rho_j^2\}, \tag{12-171}$$

where "ρ_j" is known as the "orbital radius," and $\mathbf{R}_j$ is the location of the FSGO relative to some arbitrary origin.

The location and size of these orbitals are determined by considering molecular fragments chosen to mimic anticipated bonding environments in larger systems. As an example, the simplest description of a carbon atom in an sp^3 environment would be obtained by considering the CH_4 molecule, with one FSGO located on the nucleus for the inner shell, and one FSGO in each of the CH bonding regions, as depicted in Fig. 12.6. Optimum values for location and size of each FSGO are determined by direct energy minimization.

To form large molecules at the SCF level, molecular fragments are suitably combined, and charge redistribution is allowed via determination of MO coefficients. For example, to form an ethane molecule, two CH_4 fragments are brought together, two hydrogen atoms are deleted,[115] and an SCF calculation is

[113] R. E. Christoffersen and G. M. Maggiora, *Chem. Phys. Lett.*, **3**, 419–423 (1969); R. E. Christoffersen, *Adv. Quantum Chem.*, **6**, 333–393 (1972); R. E. Christoffersen, D. Spangler, G. G. Hall, and G. M. Maggiora, *J. Am. Chem. Soc.*, **95**, 8526–8536 (1973); G. M. Maggiora and R. E. Christoffersen, *J. Am. Chem. Soc.*, **98**, 8325–8332 (1976).

[114] Use of these orbitals for small molecule descriptions was given by A. A. Frost, *J. Chem. Phys.*, **47**, 3707, 3714 (1967); *J. Am. Chem. Soc.*, **89**, 3064 (1967); **90**, 1965 (1967); *J. Phys. Chem.*, **72**, 1289 (1968).

[115] See Fig. 12.7 for a depiction of this process.

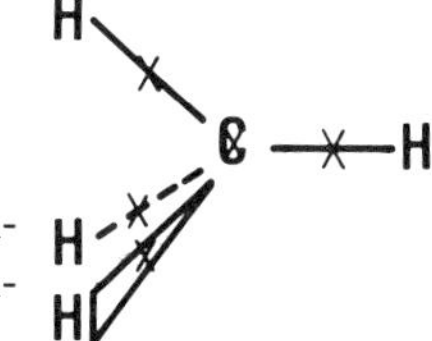

Figure 12.6. Depiction of a molecular fragment (CH_4) for determination of an FSGO basis describing an sp^3 environment. Approximate FSGO locations are indicated with an "X."

carried to determine MOs for ethane. In general, the formulation of MOs for large molecules from fragments can be written as

$$\varphi_i = \sum_A^P \sum_j^{N_A} c_{ji}^A G_j^A, \tag{12-172}$$

where the sums are over all fragments (P) and over all FSGO per fragment (N_A). The SCF wavefunction (in the closed shell case) is then written as

$$\Psi(1, 2, \cdots 2N) = [(2N)!]^{-1/2} \cdot \det\{\varphi_1(1)\alpha(1)\varphi_1(2)\beta(2) \cdots \varphi_{N/2}(N)\beta(N)\}, \tag{12-173}$$

and the charge redistribution that occurs in the formation of large molecules from fragments is reflected in the MO coefficients (c_{ji}^A). As discussed earlier,[116] basis sets constructed in this way exhibit remarkable balance in the description of relative MO energies and ordering, showing a linear relationship between large basis set studies and molecular fragment basis set descriptions.

However, of particular interest here is extension of this approach to the study of excited states and spectra, where correlation effects and multiconfiguration wavefunctions must also be included. The particular CI approach used was that devised by Whitten and Hackmeyer,[117] and represents a good example of a *multireference CI approach*.

In this approach, initial CI wavefunctions for M electronic states of interest are generated by performing single and double excitations from a set of "initial

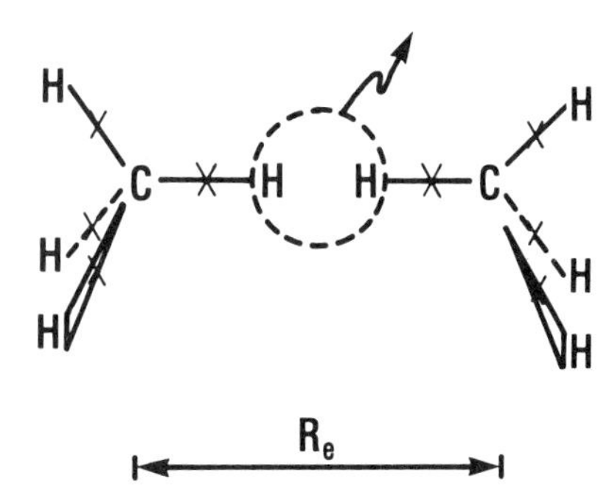

Figure 12.7. Formation of ethane from two CH_4 fragments.

[116] See Chapter 11, Section 11-2.C. and Section 11-2.D.

[117] J. L. Whitten and M. Hackmeyer, *J. Chem. Phys.*, **51**, 5584–5596 (1969).

parent configurations'' thought to be likely principal contributors to the states of interest. The initial configurations from which excitations are made generally include the SCF configuration plus singly and perhaps doubly excited configurations formed by replacing high-lying occupied MOs of the Hartree–Fock wavefunction by low-lying unoccupied MOs. The resulting initial CI expansions for each state (j) are written as

$$\Psi_j^{(1)} = \sum_k C_{kj}^{(1)} \Phi_k^{(1)}, \tag{12-174}$$

and include the initial *parent configurations* as well as generated configurations for which ΔE in Eq. (12-90) is greater than a prechosen threshold value, typically $\Delta E = 10^{-2}$–10^{-3} Hartrees. The expansion coefficients $C_{kj}^{(1)}$ are subsequently determined variationally.

Using these initial descriptions for each state, additional configurations are generated by forming single and double excitations from the parent configurations that are the principal contributions in the initial CI expansion $\Psi_j^{(1)}$ of each state of interest. Acceptance of a new configuration $(\Phi_k^{(2)})$ in the final CI description of the states of interest is governed by the following ''energy threshold criteria,'' i.e.,

$$\left| \frac{\langle \Phi_k^{(2)} | \mathcal{H} | \Psi_j^{(1)} \rangle}{\langle \Phi_k^{(2)} | \mathcal{H} | \Psi_k^{(2)} \rangle - \langle \Psi_j^{(1)} | \mathcal{H} | \Psi_j^{(1)} \rangle} \right| < \delta, \tag{12-175}$$

where δ is typically 10^{-4} to 10^{-5}. In this way multiple (e.g., triple, quadruple, ...) excitations from the ground state configuration are included in the final CI wavefunctions, and the method does not suffer from the ''size extensivity'' problems discussed in previous sections.

When applied to the description of excited states of prototype medium and large systems, a remarkable degree of balance is observed again, but this time in the relative order and energy of *excited* states (and not just ground state MO ordering). For example, the relationship between calculated and experimental transition energies for pyrazine[118] are displayed in Fig. 12.8. The balance in the CI description of the various excited states is apparent from the figure, where the observed linear relationship between calculated and observed transition energies applies equally well to $\pi \rightarrow \pi^*$, $n \rightarrow \pi^*$, singlet–singlet, and triplet–triplet transitions. Expressed more quantitatively, the following relationships are found:

$$\Delta E_i^{\text{exp}}\ (\text{eV}) = 0.6875 \cdot \Delta E_i^{\text{calc}}\ (\text{eV}) + 0.1504, \tag{12-176}$$

for pyrazine, with a standard deviation of 0.025 and a correlation coefficient of

[118] J. D. Petke, R. E. Christoffersen, G. M. Maggiora, and L. L. Shipman, *Int. J. Quantum Chem., Quantum Biol. Symp.*, **4**, 343 (1977).

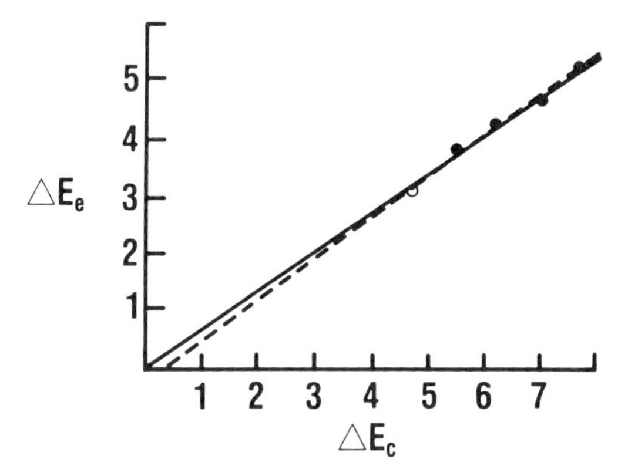

Figure 12.8. Calculated versus experimental low-lying singlet–singlet and singlet–triplet transitions (in eV) for pyrazine and carbazole. The solid line includes singlet transitions only, and the dashed line includes both singlet–singlet and singlet–triplet transitions. Solid circles refer to singlet states and the open circle refers to the $^3B_2$ state. [Reprinted with permission from L. E. Nitzche et al., *J. Am. Chem. Soc.,* **98**, 4798. Copyright 1978 American Chemical Society].

0.9980. For carbazole,[119] the relationship is:

$$\Delta E_i^{\exp}\,(\text{eV}) = 0.7177 \cdot \Delta E_i^{\text{calc}}\,(\text{eV}) - 0.252. \tag{12-177}$$

with a standard deviation of 0.0383 and a correlation coefficient of 0.9958. Thus, even though absolute values of transition energies cannot be obtained with these small basis sets, considerable insight into spectral features and excited state structure can be expected from such an approach.

As an illustration of how this can be applied to very large molecules of biological interest, the calculated spectra[120] for ethyl chlorophyllide **a** and ethyl pheophorbide **a** are given in Fig. 12.9, while the molecular structure of the molecules is given in Fig. 12.10. The calculated spectra have been scaled using a linear relationship derived from porphine studies,[121], i.e.,

$$\Delta E_i^{\exp} = 0.610 \cdot \Delta E_i^{\text{calc}} - 441.0, \tag{12-178}$$

where the energy differences are given in cm^{-1}, and the root mean square error is 640 cm^{-1}.

It is clear from even a cursory examination of Fig. 12.9 that the calculated spectrum for each molecule provides a good rationalization of the observed absorption spectrum. More quantitatively, is is seen that the region between approximately 23,000 and 30,000 cm^{-1} (called the Soret region) is rich with

[119] L. E. Nitzche, C. Chabalowski, and R. E. Christoffersen, *J. Am. Chem. Soc.*, **98**, 4797–4801 (1978).

[120] J. D. Petke, G. M. Maggiora, L. L. Shipman, and R. E. Christoffersen, *Photochem. Photobiol.*, **30**, 203–223 (1979).

[121] J. D. Petke, G. M. Maggiora, L. L. Shipman, and R. E. Christoffersen, *J. Mol. Spectrosc.*, **71**, 64–84 (1978).

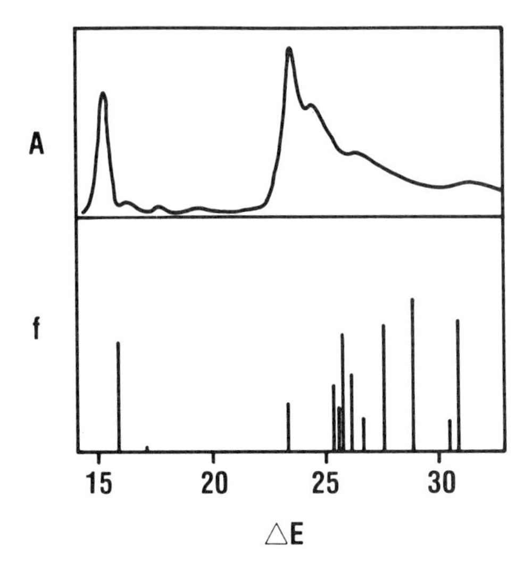

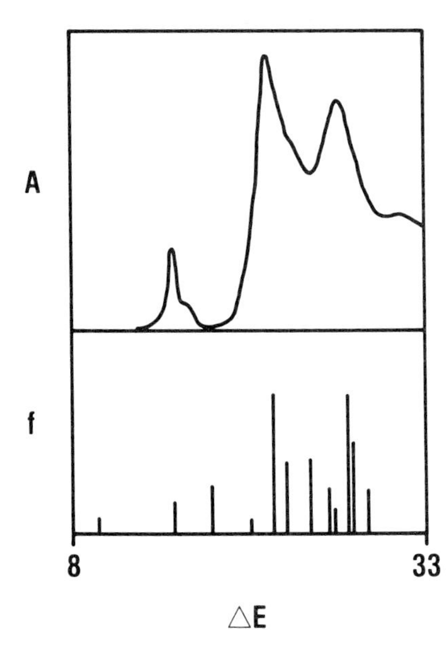

Figure 12.9. Top. Comparison of ethyl chlorophyllide **a** absorption spectra (in ethyl ether) with calculated transition energies (in cm^{-1}) and oscillator strengths (f). [Reprinted with permission from J. D. Petke, G. M. Maggiora, L. L. Shipman, and R. E. Christoffersen, *Photochem. Photobiol.,* **30**, 203–233. Copyright (1979) Pergamon Press plc.]. Bottom. Comparison of ethyl chlorophyllide **a** enol absorption spectra (in acetone) with calculated transition energies (in cm^{-1}) and oscillator strengths (f). [Reprinted with permission from J. D. Petke, G. M. Maggiora, L. L. Shipman, and R. E. Christoffersen, *J. Am. Chem. Soc.,* **103**, 4622–4623. Copyright (1981) American Chemical Society].

transitions, and that a variety of polarizations and oscillator strengths can be expected. The relevance of this observation is strengthened when it is recognized that previous rationalizations of the observed spectrum assumed that only a few states were present in this region (e.g., only two states were thought to be responsible for the Soret maximum in the chlorophyll case). In addition, analysis of the resulting wavefunctions allows understanding of the charge distribution and overall electronic structure of each of the excited states. Hence, while a

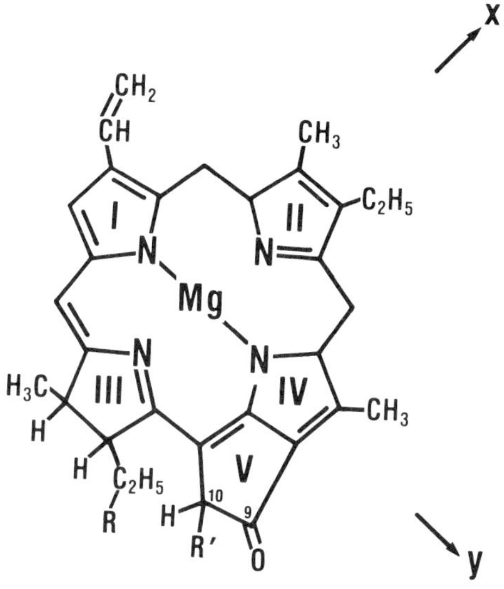

$R = CO_2C_2H_5$
$R = CO_2CH_3$

Figure 12.10. Molecular structure of ethyl chlorophyllide **a**. Ethyl pheophorbide **a** is obtained by replacing the Mg with hydrogens on the nitrogens in rings I and III.

great deal of improvements, extensions, modifications, etc., of techniques such as these is desirable, it is clear that SCF and CI techniques of *ab initio* quantum mechanics are applicable to large and very large molecular systems and provide important probes for study of electronic structural features of such systems.

In addition, there are many cases in which the chemical and/or biological questions to be answered are qualitative instead of quantitative, in which case the techniques described here are particularly useful. As an example of such a case, let us consider another problem related to photosynthetic systems. In particular, the electronic spectra of the enol form of ethyl chlorophyllide **a** looks similar to the spectra of the keto form of ethyl chlorophyllide **a** (see Fig. 12.9), but the spectra of the enol form does not exhibit fluorescence as expected and seen for the keto form.

The reason for such apparent anomolous behavior is easily seen from CI studies.[122] In particular, we see from Fig. 12.9 that the transition responsible for the lowest absorption is the $S_2 \leftarrow S_0$ transition, and *not* the $S_1 \leftarrow S_0$ transition as observed in ethyl chlorophyllide **a** (keto form). In the enol form, the $S_1 \leftarrow S_0$ transition is seen at lower energy ($\sim 9900\ \text{cm}^{-1}$), and with low oscillator

[122] J. D. Petke, L. L. Shipman, G. M. Maggiora, and R. E. Christoffersen, *J. Am. Chem. Soc.*, **103**, 4622 (1981).

strength ($f \sim 0.04$). Hence, while fluorescence from the S_1 state is expected and seen in the keto form, the enol form encounters absorption to S_2, essentially immediate transition from S_2 to S_1, but no fluorescence from S_1 due to the near-forbidden character of the $S_1 \leftarrow S_0$ transition. Thus, we see that, even when absolute energies cannot be quantitatively calculated, important insights not otherwise possible can be obtained even with small basis set SCF/CI studies on large molecular systems.

12-4. Many-Body Perturbation Theory and Coupled Cluster Theory

Thus far in this chapter we have focused on the use of CI techniques for understanding and estimating correlation effects and a variety of properties not well described using Hartree–Fock theory. This emphasis on CI techniques is due in large part to the wide usage of them and assurance of satisfying the Variation Principle. However, a number of alternative approaches are also possible, based on the use of perturbation theory. While such approaches are nonvariational in general, they possess other attributes such as size extensivity that offer advantages in some cases. Also, quantities such as relative energies on potential surfaces or excitation energies are calculated from the *difference* of energies, and hence have no variational bounds even if the individual energies are calculated variationally.

In this section we shall consider an example of perturbation theory approaches,[123] which are referred to usually as *Many Body Perturbation Theory*[124] (MBPT) or *Coupled Cluster Theory*[125] (CCT). This terminology arises primarily from historical reasons, since the theory was useful first to physicists interested in systems of infinite size, where size extensivity is essential.

The basis concepts underlying the approach are those of Rayleigh–Schrödinger perturbation theory, which were given in Chapter 9. Since we shall utilize a number of the results found earlier, it is useful to begin by summarizing

---

[123] For a review see, for example, R. J. Bartlett, *Annu. Rev. Phys. Chem.*, **32**, 359–401 (1981); see also K. F. Freed, *Annu. Rev. Phys. Chem.*, **22**, 771 (1967). Also, an excellent discussion of the use of the diagrammatic representational formulation of this approach can be found in A. Szabo and N. S. Ostlund, *Modern Quantum Chemistry*, MacMillan, New York, 1982, Chapter 6.

[124] For initial papers, see K. A. Brueckner, *Phys. Rev.*, **97**, 1353; **100**, 36 (1955); K. A. Brueckner, R. J. Eden, and N. C. Francis, *Phys. Rev.*, **93**, 1445 (1955); K. Brueckner and C. A. Levinson, *Phys. Rev.*, **97**, 1344 (1955); J. Goldstone, *Proc. R. Soc. (London)*, **A239**, 267 (1957); H. P. Kelly, *Adv. Chem. Phys.*, **14**, 129 (1969).

[125] F. Coester, *Nucl. Phys.*, **1**, 421 (1958); F. Coester and H. Kümmel, *Nucl. Phys.*, **17**, 477 (1960); H. Kümmel, *Nucl. Phys.*, **22**, 177 (1969); J. Cizek, *J. Chem. Phys.*, **45**, 4256 (1966); *Adv. Chem. Phys.*, **14**, 35 (1969); J. Paldus and J. Cizek, *Energy, Structure and Reactivity*, D. W. Smith and W. B. McRae (ed.), John Wiley, New York, 1973, p. 389.

them here in a convenient notation. For a nondegenerate system whose energy states (E_n) and associated wavefunctions (Ψ_n) we wish to determine via solution of

$$\mathcal{H}|\Psi_n\rangle = E_n|\Psi_n\rangle, \qquad n = 0, 1, 2, \cdots \tag{12-179}$$

we write

$$\mathcal{H} = \mathcal{H}_0 + \lambda V, \tag{12-180}$$

where $\mathcal{H}_0$ is an "unperturbed" Hamiltonian whose eigenvalues ($E_n^{(0)}$) and eigenfunctions ($|\psi_n^{(0)}\rangle$) are known, i.e.,

$$\mathcal{H}_0|\psi_n^{(0)}\rangle = E_n^{(0)}|\psi_n^{(0)}\rangle, \tag{12-181}$$

V is the perturbation potential, and λ is a parameter (to be set equal to unity). Expanding $|\Psi_n\rangle$ and E_n in terms of the unperturbed eigenfunctions and eigenvalues and corrections to them gives

$$|\Psi_n\rangle = |\psi_n^{(0)}\rangle + \lambda|\psi_n^{(1)}\rangle + \lambda^2|\psi_n^{(2)}\rangle + \cdots \tag{12-182}$$

$$E_n = E_n^{(0)} + \lambda E_n^{(1)} + \lambda^2 E_n^{(2)} + \cdots. \tag{12-183}$$

Expressions for the first several corrections to the energy are given by

$$E_n^{(1)} = \langle\psi_n^{(0)}|V|\psi_n^{(0)}\rangle, \tag{12-184}$$

$$E_n^{(2)} = \langle\psi_n^{(0)}|V|\psi_n^{(1)}\rangle = \sum_{\substack{m\\(\neq n)}} \frac{|\langle\psi_n^{(0)}|V|\psi_m^{(0)}\rangle|^2}{(E_n^{(0)} - E_m^{(0)})}, \tag{12-185}$$

and

$$E_n^{(3)} = \langle\psi_n^{(0)}|V|\psi_n^{(2)}\rangle. \tag{12-186}$$

Equation (12-186) can also be rewritten (see Problem 9-4) as

$$E_n^{(3)} = \sum_{\substack{m,p\\(\neq n)}} \frac{\langle\psi_n^{(0)}|V|\psi_m^{(0)}\rangle\langle\psi_m^{(0)}|V|\psi_p^{(0)}\rangle\langle\psi_p^{(0)}|V|\psi_n^{(0)}\rangle}{(E_n^{(0)} - E_m^{(0)})(E_n^{(0)} - E_p^{(0)})}$$
$$- E_n^{(1)} \sum_{\substack{m\\(\neq n)}} \frac{|\langle\psi_n^{(0)}|V|\psi_m^{(0)}\rangle|^2}{(E_n^{(0)} - E_m^{(0)})^2}. \tag{12-187}$$

The way in which this approach has been adapted for use in correlation energy estimation is by forming it in terms of Hartree–Fock theory and corrections to it[126]. In particular, we write the Hamiltonian as

$$\mathcal{H} = \mathcal{H}_0 + V \tag{12-188}$$

[126] This approach was described first by C. Moeller and M. S. Plesset, *Phys. Rev.*, **46**, 618 (1934), and as a result, chemical applications of this technique are sometimes known as *Moeller-Plesset Perturbation Theory* (MPPT).

but now we define $\mathcal{H}_0$ to be a "Hartree–Fock Hamiltonian," i.e.,

$$\mathcal{H}_0 \equiv \sum_i [h(i) + v^{\mathrm{HF}}(i)] = \sum_i f(i) \tag{12-189}$$

where $f(i)$ is the Fock operator,

$$h(i) = -\frac{1}{2}\nabla_i^2 - \sum_\alpha \frac{Z_\alpha}{r_{i\alpha}},$$

and

$$v^{\mathrm{HF}}(i) = \sum_j \left\{ \left[\int d\tau_2 \chi_j^*(2) \frac{1}{r_{12}} \chi_j(2) \right] \chi_i(1) - \left[\int d\tau_2 \chi_j^*(2) \frac{1}{r_{12}} \chi_i(2) \right] \chi_j(1) \right\} \tag{12-191}$$

where χ_j is a Hartree–Fock spin orbital. The perturbation term (V) then represents the correlation energy correction, and can be written as

$$V = \sum_{i<j} \frac{1}{r_{ij}} - V^{\mathrm{HF}}, \tag{12-192}$$

$$= \sum_{i<j} \frac{1}{r_{ij}} - \sum_i v^{\mathrm{HF}}(i), \tag{12-193}$$

where we have defined

$$V^{\mathrm{HF}} = \sum_i v^{\mathrm{HF}}(i). \tag{12-194}$$

It should be noted that, as defined above, $\mathcal{H}_0$ is a Hamiltonian for which the Hartree–Fock function ($\Psi^{(0)}$) is an eigenfunction, but with eigenvalue[127]

$$E_0^{(0)} = \sum_a \epsilon_a, \tag{12-195}$$

where the sum is over occupied orbitals, and ϵ_a is the Hartree–Fock orbital energy of orbital a. Thus, the zero order energy ($E_0^{(0)}$) is not equal to the total Hartree–Fock energy, which is given by

$$E_{\mathrm{HF}} = \sum_a \epsilon_a - \frac{1}{2} \sum_{a,b} [\langle ab|ab\rangle - \langle ab|ba\rangle] \tag{12-196}$$

---

[127] Throughout this discussion we shall assume (for simplicity) that we are dealing with the ground state of the system.

where the notation

$$\langle ab|cd\rangle = \int\int d\tau_1\, d\tau_2\, \chi_a^*(1)\chi_b(2)\frac{1}{r_{12}}\chi_c^*(1)\chi_d(2) \tag{12-197}$$

has been used.

However, it can be shown[128] that

$$E_0^{(1)} = \langle \Psi_0^{(0)}|V|\Psi_0^{(0)}\rangle \tag{12-198}$$

$$= -\frac{1}{2}\sum_{a,b}[\langle ab|ab\rangle - \langle ab|ba\rangle], \tag{12-199}$$

so that

$$E_{\text{HF}} = E_0^{(0)} + E_0^{(1)}. \tag{12-200}$$

In other words, using Hartree–Fock wavefunctions assures that zero and first order perturbation corrections are already included in the Hartree–Fock energy, and the first correction to the Hartree–Fock energy occurs is from second order perturbation theory (i.e., in $E_0^{(2)}$).

Turning to the second order correction we have, from Eq. (12-185)

$$E_0^{(2)} = \sum_{\substack{n\\(\neq 0)}} \frac{|\langle \psi_0^{(0)}|V|\psi_n^{(0)}\rangle|^2}{E_0^{(0)} - E_n^{(0)}}. \tag{12-201}$$

The sum in the above equation is taken over *all* states, at least in principle, and there is a question as to how such a sum can be evaluated. However, it can be shown[129] that

$$E_0^{(2)} = \sum_{\substack{a<b\\ \alpha<\beta}} \frac{\left|\left\langle \Psi_0 \left| \sum_{i<j}\frac{1}{r_{ij}} \right| \Psi_{ab}^{\alpha\beta}\right\rangle\right|^2}{\epsilon_a + \epsilon_b - \epsilon_\alpha - \epsilon_\beta} \tag{12-202}$$

$$= \sum_{\substack{a<b\\ \alpha<\beta}} \left[\frac{1}{\epsilon_a + \epsilon_b - \epsilon_\alpha - \epsilon_\beta}\right][\langle ab|\alpha\beta\rangle - \langle ab|\beta\alpha\rangle]^2, \tag{12-203}$$

where a and b refer to filled orbitals, and α and β refer to unfilled (virtual) orbitals. Higher order corrections can also be found, although the form of the expressions becomes somewhat cumbersome in this notation.[130]

As implied in the introduction to this section, the various order corrections to

[128] See Problem 12-20.

[129] See Problem 12-21.

[130] Simpler forms for third- and fourth-order corrections, for example, are obtainable through the "exponential ansatz" approach. See R. J. Bartlett, *Annu. Rev. Phys. Chem.*, **32**, 359–401 (1981).

Table 12.12. Percentage of Correlation Energy Recovered by Different Orders of Perturbation Theory

Molecule[a]	Second order	Third order	Fourth Order[b]	Higher order[c]
BH_3	80.0	16.5	3.0	0.50
H_2O	97.7	1.5	0.7	0.06
NH_3	94.3	5.0	0.6	0.12
CH_4	89.6	9.3	0.9	0.16
CO	100	−1.6	1.6	−0.09
CO_2	103.2	−4.1	0.9	0.00

[a] Data taken from R. J. Bartlett, *Annu. Rev. Phys. Chem.*, **32**, 359–401 (1981). Basis sets were double zeta plus polarization quality.

[b] Partial estimates.

the energy obtained via Rayleigh–Schrödinger perturbation theory have been shown to be size extensive, i.e., $E^{(n)}$ is proportional to the number of particles (n). This was proven by Goldstone,[131] and provides a quite useful attribute to the general approach.

The number and kinds of applications of this approach have been growing in recent years, and several high accuracy studies are also now available. The first molecular studies were those of Kelly[132] on H_2 and Miller and Kelly[133] on H_2O. Since then, however, applications to other small and medium-sized molecular systems have increased substantially.[134]

To illustrate the kinds of results that can be obtained, let us consider first the amount of correlation energy that can be recovered at various orders of perturbation theory. In Table 12.12, several molecules that illustrate this are listed, along with the ground state correlation energy recovered using second, third, fourth, and higher order perturbation theory. From these data we see that the second-order correction recovers the bulk of the correlation energy in all cases, although in the case of CO_2 the recovery is greater than 100%. Most of the remaining corrections, either positive or negative, are given in the third

[131] See the references preceding Eq. (12-156). The theorem Goldstone proved in this regard is generally known as the *Linked Cluster Theorem*.

[132] H. P. Kelly, *Phys. Rev. Lett.*, **23**, 455 (1969).

[133] J. H. Miller and H. P. Kelly, *Phys. Rev. Lett.*, **26**, 679 (1971); *Phys. Rev.*, **A4**, 480 (1971).

[134] See, for example, J. A. Pople, J. S. Binkley, and R. Seeger, *Int. J. Quantum Chem. Symp.*, **10**, 1 (1976); J. A. Pople. R. Seeger, and R. Krishnan, *Int. J. Quantum Chem. Symp.*, **11**, 149 (1977); J. A. Pople, R. Krishnan, H. B. Schlegel, and J. S. Binkley, *Int. J. Quantum Chem.*, **14**, 545 (1978); R. Krishnan and J. A. Pople, *Int. J. Quantum Chem. Symp.*, **14**, 91 (1978); R. Krishnan, M. J. Frisch, and J. A. Pople, *J. Chem. Phys.*, **72**, 4244 (1980); M. J. Frisch, R. Krishnan, and J. A. Pople, *Chem. Phys. Lett.*, **75**, 66 (1980). For examples of application to systems of biological interest, see H. Weinstein, R. Osman, W. D. Edwards, and J. P. Green, *Int. J. Quantum Chem.*, **QBS5**, 449 (1978); R. Osman, S. Topiol, H. Weinstein, and J. E. Eilers, *Chem. Phys. Lett.*, **73**, 399 (1978).

order contribution. Thus, for ground state correlation estimates, these methods can be quite useful. For ground state geometrics, similar results are obtained, e.g., Pople and colleagues[135] found that second-order corrections remove at least 50% of the error remaining in UHF geometry predictions.

In studies of dissociation energies, potential energy surfaces and, in general, properties other than the energy, encouraging results have been found in a number of cases. For example, the isomerization energy of methylisocyanide according to the reaction

$$CH_3NC \rightarrow CH_3CN$$

was estimated[136] using MBPT at 15 ± 2 kcal/mol, compared to an experimental estimate[137] of 10 ± 1 kcal/mol. Similar results were obtained on boron-containing molecules[138] such as B_2H_6, H_3BCO, and H_3BNH_3.

Thus, we see that MBPT studies that include as little as second order corrections typicaly recover ~90% of the correlation energy, and remove much of the SCF error in other ground state properties. Relatively efficient processes for obtaining higher order corrections are also available. On the other hand, treatment of excited states or reactions involving dissociation are examples where MBPT is typically less successful, usually because only a single reference state (i.e., the Hartree–Fock wavefunction) is used as a basis for excitations. However, whether CI, MBPT, or other techniques are used, it is clear that major progress in devising new conceptual approaches as well as computationally efficient algorithms has occurred, and can be expected to continue in the years ahead.

Problems

1. Using Hartree–Fock filled and virtual orbitals for an N-electron closed shell system and configuration "s" as the Hartree–Fock ground state configuration and configuration "t" to contain one double-excitation from the highest filled to lowest unfilled orbitals, determine the values of a_{ij}^{st} and b_{ij}^{st} in Eq. (12-7).
2. For a two-electron system described by a basis set containing four spin orbitals, show by direct construction that six Slater determinants are possible, consisting of the ground state (Hartree–Fock) configuration, four singly excited configurations, and one doubly excited configuration.

---

[135] J. A. Pople, J. S. Binkley, and R. Seeger, *Int. J. Quantum Chem. Symp.*, **10**, 1 (1976).

[136] J. A. Pople, R. Krishnan, H. B. Schlegel, and J. S. Binkley, *Int. J. Quantum Chem.*, **14**, 545 (1978); P. K. Pearson, H. F. Schaefer II, and U. Walgren, *J. Chem. Phys.*, **62**, 350 (1975).

[137] A. G. Maki and R. L. Sams, unpublished data referred to in R. J. Bartlett, *Annu. Rev. Phys. Chem.*, **32**, 359–401 (1981).

[138] L. T. Redmond, G. D. Purvis, and R. J. Bartlett, *J. Am. Chem. Soc.*, **101**, 2855 (1979).

3. Show that Eq. (12-15) can be rewritten equivalently as:

$$\Psi(1, 2, \cdots, N) = c_0\Phi_0 + \left(\frac{1}{1!}\right)^2 \sum_{a,\alpha} c_a^\alpha \Phi_a^\alpha + \left(\frac{1}{2!}\right)^2 \sum_{\substack{a,b\\ \alpha,\beta}} c_{a,b}^{\alpha,\beta} \Phi_{ab}^{\alpha\beta}$$

$$+ \left(\frac{1}{3!}\right)^2 \sum_{\substack{a,b,c\\ \alpha,\beta,\gamma}} c_{abc}^{\alpha\beta\gamma} \Phi_{abc}^{\alpha\beta\gamma} + \cdots.$$

4. a. Show that the number of n-tuply excited determinants that can be formed from a system having N electrons and M spin orbitals is

$$\binom{M}{n}\binom{M-N}{n}.$$

   b. A double-ζ basis for the LiH molecule contains 23 spatial orbitals (24 spin orbitals). Calculate the total number of possible configurations in such a basis, indicating the number of single, double, triple, and quadruple excitation configurations that are possible.
5. Prove Eq. (12-17).
6. Show that integration over the remaining electron coordinates of $\Gamma^{(p)}$ gives

$$\int \Gamma^{(p)}(\tau \mid \tau)\, d\tau = \binom{N}{p}.$$

   Also show that the first-order reduced density matrix is Hermitian.
7. Show that $\gamma_R(k \mid l)$ as defined in Eq. (12-33) is Hermitian, assuming that the orbitals in Eq. (12-32) are real and nonorthogonal.
8. Show that

$$0 \leq n_k \leq 1, \qquad k = 1, 2, \cdots, M \qquad (M \geq N),$$

   where n_k are the natural orbital occupation numbers for an N-electron wavefunction.
9. Show that the energy of a system of N electrons is completely determined by knowledge of the first- and second-order density matrices for the system.
10. Using the three orthogonal orbitals

$$\varphi_{ns} = L_{n+1}(2r)e^{-2r}, \qquad n = 1, 2, 3$$

   the following optimized wavefunction can be found for the ground state of the He atom:

$$\Psi = 0.963175\ \varphi_{1s}^2 - 0.250366\ \varphi_{1s}\varphi_{2s} - 0.032613\ \varphi_{2s}^2 + 0.092211\ \varphi_{1s}\varphi_{3s}$$
$$- 0.000310\ \varphi_{2s}\varphi_{3s} - 0.006692\ \varphi_{3s}^2.$$

   Show that, by forming and diagonalizing the first-order density matrix, the following normalized natural spin orbitals and occupation numbers can be obtained:

$$\chi_1 = 0.983545\ \varphi_{1s} - 0.168992\ \varphi_{2s} + 0.063880\ \varphi_{3s}, \qquad n_1 = 0.995660$$

$$\chi_2 = 0.178369\ \varphi_{1s} + 0.964488\ \varphi_{2s} - 0.194800\ \varphi_{3s}, \qquad n_2 = 0.004265$$

$$\chi_3 = -0.028690\ \varphi_{1s} + 0.202991\ \varphi_{2s} + 0.978760\ \varphi_{3s}, \qquad n_3 = 0.000075.$$

11. For a normalized two-electron wavefunction of the form

$$\Psi(\mathbf{r}_1, \mathbf{r}_2) = \sum_k \bar{C}_{kk}\psi_k(1)\psi_k(2) + \sum_{k<l} \bar{C}_{kl}[\psi_k(1)\psi_l(2) + \psi_l(1)\psi_k(2)]/\sqrt{2},$$

and where the orbitals $\{\psi_i\}$ are orthonormal, show that by defining

$$C_{kk} = \bar{C}_{kk}, \qquad C_{kl} = C_{lk} = 2^{-1/2}\bar{C}_{kl},$$

the first-order density matrix can be written as

$$\gamma(\mathbf{x}_1'|\mathbf{x}_1) = (\alpha_1'\alpha_1 + \beta_1'\beta_1)\sum_{k,l}\psi_k^*(r_1')\psi_l(r_1)\gamma_{lk},$$

where

$$\gamma_{lk} = \sum_m C_{lm}C_{km}^*, \qquad \text{or} \qquad \boldsymbol{\gamma} = \mathbf{C}\mathbf{C}^\dagger.$$

Also show that the same transformation that diagonalizes $\mathbf{C}$ also diagonalizes the first order density matrix, i.e., if

$$\mathbf{U}^\dagger\boldsymbol{\gamma}\mathbf{U} = \mathbf{n}$$

where $\mathbf{n}$ is diagonal, and

$$\mathbf{V}^\dagger\mathbf{C}\mathbf{V} = \mathbf{c}$$

where $\mathbf{c}$ is diagonal, then show that

$$\mathbf{U} \equiv \mathbf{V}, \qquad \text{and} \qquad c_k = \pm n_k^{1/2}.$$

12. Using the definitions of Ψ_A, Ψ_{I+}, and Ψ_{I-} in Eqs. (12-65)–(12-67) plus the choice of u and v in Eq. (12-70) and $\alpha = \beta$, show that Ψ_A, Ψ_{I+} and Ψ_{I-} are mutually orthogonal.
13. Using the choice of u and v in Eq. (12-70), show that

$$\chi_1 = \frac{(\cos\beta)u + (\sin\alpha)v}{\cos(\alpha-\beta)}$$

$$\chi_2 = \frac{(\sin\beta)u - (\cos\alpha)v}{\cos(\alpha-\beta)},$$

and verify Eqs. (12-74)–(12-77).
14. For the CH_4 molecule in an $S = S_z = 0$ spin state, consider a minimum

STO basis of $1s$, $2s$, $2p_x$, $2p_y$ and $2p_z$ orbitals on carbon, plus $1s$ orbitals on each hydrogen, and determine the following (assuming no spatial symmetry constraints):

a. The total number of possible determinants in a "full CI" study.
b. The number of each of the types of excitation determinants (i.e., single, double, triple, etc.) along with the total.
c. The number of each type of excitation determinants that remains after spin constraints have been applied to each determinant.

15. For a basis set $\{\phi\}$ of real orbitals, show that

$$\langle\phi_i|h|\phi_j\rangle=\langle\phi_j|h|\phi_i\rangle$$

$$\left\langle\phi_i\phi_j\left|\frac{1}{r_{12}}\right|\phi_k\phi_l\right\rangle=\left\langle\phi_j\phi_i\left|\frac{1}{r_{12}}\right|\phi_l\phi_k\right\rangle=\left\langle\phi_k\phi_j\left|\frac{1}{r_{12}}\right|\phi_i\phi_l\right\rangle$$
$$=\left\langle\phi_l\phi_k\left|\frac{1}{r_{12}}\right|\phi_j\phi_i\right\rangle=\left\langle\phi_i\phi_l\left|\frac{1}{r_{12}}\right|\phi_k\phi_j\right\rangle$$
$$=\left\langle\phi_k\phi_l\left|\frac{1}{r_{12}}\right|\phi_i\phi_j\right\rangle.$$

16. For a basis of three orbitals, determine the total number and explicit list of unique one- and two-electron integrals that are needed when the "lower triangle" form is used, assuming no molecular symmetry.
17. Prove Eq. (12-133) for a complete orthonormal basis set that is partitioned into separate subsets as illustrated in Eq. (12-132). Also show that, given Eqs. (12-132) and (12-133), the following two-electron orthogonality property is true:

$$\int\int G_\mu^*(1,2)G_\nu(1,2)\,d\tau_1\,d\tau_2=0\qquad(\mu\neq\nu).$$

18. Show that the expression for the geminal energy (ϵ_μ) in Eq. (12-144) is given by

$$\epsilon_\mu=\sum_{k,i}^{N_\mu}H_{ki}^\mu c_{\mu i}c_{\mu k}.$$

19. Show that, by starting with the energy eigenvalue equation for Ψ_{ab} and using Eqs. (12-164) and (12-165), multiplying successively on the left by Φ_0 and $\Phi_{ab}^{\alpha\beta}$ gives rise to Eqs. (12-167)–(12-170).
20. Prove Eq. (12-199).
21. Using Brillouin's Theorem plus the fact that V contains at most two particle interactions, prove Eq. (12-202).

Appendix 1

Selected Physical Constants, Units, and Conversion Factors[1]

Table A.1. Physical Constants and Units

Constant	Symbol	Magnitude (cgs units)
Speed of light (vacuum)	c	$2.99792458 \times 10^{10}$ cm/sec
Elementary charge	e	4.80324×10^{-10} esu
Electron rest mass	m_e	9.10953×10^{-28} g
Proton rest mass	m_p	1.672648×10^{-24} g
Plank's constant	h	6.62618×10^{-27} erg sec
Rydberg constant	R_∞	1.0973732×10^{5} cm^{-1}
Bohr radius	a_0	0.5291771×10^{-8} cm
Fine structure constant	α	7.297351×10^{3}

[1] For further discussion, see E. R. Cohen and B. N. Taylor, *J. Phys. Chem. Ref. Data*, **2**, 663 (1973).

Table A.2. Energy Conversion Factors

	1 atomic unit (Hartree)	ergs	Electron volts	cm^{-1}	kcal/mol
1 atomic unit (Hartree)[a] =	1	4.3598×10^{-11}	27.212	2.194746×10^{5}	627.51
1 erg =	2.2937×10^{10}	1	6.2415×10^{11}	5.0348×10^{15}	1.9393×10^{13}
1 electron volt =	3.6749×10^{-2}	1.6022×10^{-12}	1	8.0655×10^{3}	23.060
1 cm^{-1} =	4.55634×10^{-6}	1.9865×10^{-16}	1.23985×10^{-4}	1	2.8591×10^{-3}
1 kcal/mole =	1.5936×10^{-3}	6.9478×10^{-14}	4.3364×10^{-2}	3.4975×10^{2}	1

[a] See H. Shull and G. G. Hall, *Nature (London)*, **184**, 1559 (1959) for a discussion of this choice of units.

References

Abramowitz, M. and Stegun, I., *Handbook of Mathematical Functions*, Dover, New York, 1965, Chapter 22, p. 782.
Abscar, I., and vanWazer, J. R., *Chem. Phys. Letters*, **11**, 310 (1971).
Ahlrichs, R., *Theoret. Chim. Acta*, **35**, 59 (1974).
Ahlrichs, R., Dreissler, F., Lischka, H., Staemmler, V., and Kutzelnigg, W., *J. Chem. Phys.*, **62**, 1225–1234 (1975).
Ahlrichs, R., Dreissler, F., Lischka, H., Staemmler, V., and Kutzelnigg, W., *J. Chem. Phys.*, **62**, 1235–1247 (1975).
Ahlrichs, R., and Gruen, N., *Chem. Phys. Lett.*, **18**, 571 (1973).
Ahlrichs, R., Keil, F., Lischka, H., Zurawski, B., and Kutzelnigg, W. H., *J. Chem. Phys.*, **63**, 4685 (1975).
Ahlrichs, R., Kutzelnigg, W., and Bingel, W. A., *Theoret. Chim. Acta*, **5**, 289 (1966).
Ahlrichs, R., and Kutzelnigg, W., *J. Chem. Phys.*, **48**, 1819–1832 (1968).
Albat, R., and Gruen, N., *Chem. Phys. Lett.*, **18**, 571 (1973).
Alldredge, G. P., *Am. J. Phys.*, **38**, 1357 (1970).
Allen, T. L., and Shull, H., *J. Chem. Phys.*, **35**, 1644 (1961).
Amos, A. T., and Hall, G. G., *Proc. R. Soc. (London)*, **A263**, 483 (1961).
Anex, B. G., and Shull, H., in *Molecular Orbitals in Chemistry, Physics, and Biology*, P. O. Löwdin and B. Pullman (eds.), Academic Press, New York, 1964, p. 227.
Arai, T., *J. Chem. Phys.*, **33**, 95 (1960).
Arfken, G., *Mathematical Methods for Physicists*, Academic Press., New York, 1968, Sects. 2.2 and 17.6.
Bader, R. F. W., *Accts. Chem. Res.*, **8**, 34 (1975).
Bader, R. F. W., *J. Chem. Phys.*, **73**, 2871 (1980).
Bader, R. F. W., and Beddall, P. M., *J. Chem. Phys.*, **56**, 3320, (1972).
Bader, R. F. W., Beddall, P. M., and Peslak, J. J., *J. Chem. Phys.*, **58**, 557 (1973).
Bader, R. F. W., Duke, A. J., and Messer, R. R., *J. Am. Chem. Soc.*, **95**, 7715 (1973).

Bagus, P. S., Liu, B., McLean, A. D., and Yoshimime, M., in *Energy, Structure and Reactivity*, D. W. Smith and W. B. McRae (eds.), John Wiley, New York, 1973, p. 130.
Banyard, K. E., and Baker, C. C., *J. Chem. Phys.*, **51**, 2680 (1969).
Bardo, R. D., and Ruedenberg, K., *J. Chem. Phys.*, **60**, 918 (1974).
Barr, T. L., and Davidson, E. R., *Phys. Rev.*, **A1**, 644 (1970).
Bartlett, R. J., *Annu. Rev. Phys. Chem.*, **32**, 359 (1981).
Bartlett, R. J., Bellum, J. C., and Brändas, E., *Int. J. Quantum Chem.*, **57**, 449–462 (1973).
Bartlett, R. J., and Purvis, G. D., *Ann. N.Y. Acad. Sci.*, **367**, 62 (1981).
Basch, H., *Chem. Phys. Lett.*, **19**, 323–327 (1973).
Bastin, T. (ed.), *Quantum Theory and Beyond*, Cambridge University Press, Cambridge, 1971.
Bateman, H., in *Higher-Transcendental Functions*, Volume 1, Bateman Manuscript Project, A. Erdelyi (ed.), McGraw-Hill, New York, 1953, Section 3.8.
Bates, D. R., Ledsham, K., and Stewart, A. L., *Philos. Trans. R. Soc. (London)*, **A246**, 215 (1954).
Bauder, M., and Itzykson, C., *Rev. Mod. Phys.*, **838**, 330 (1966).
Bazley, N. W., *Proc. Natl. Acad. Sci. U.S.A.*, **45**, 850 (1959).
Bazley, N. W., *Phys. Rev.*, **120**, 44 (1960).
Bazley, N. W., and Fox D. W., *J. Res. Natl. Bus. Stand.*, **B65**, 105 (1961).
Bazley, N. W., and Fox, D. W., *Phys. Rev.*, **124**, 483 (1961).
Bazley, N. W., and Fox, D. W., *J. Math. Phys.*, **3**, 469 (1962).
Bender, C. F., *J. Comput. Phys.*, **9**, 547 (1972).
Bender, C. F., *J. Comput. Phys.*, **30**, 324 (1979).
Bender, C. F., and Davidson, E. R., *J. Phys. Chem.*, **70**, 2675 (1966).
Bender, C. F., and Davidson, E. R., *J. Chem. Phys.*, **47**, 4792 (1967).
Bender, C. F., and Davidson, E. R., *Phys. Rev.*, **183**, 23 (1969).
Bender, C. F., Dunning, T. H., Jr., Schaefer, H. F. III, Goddard, W. A., and Hunt, W. J., *Chem. Phys. Lett.*, **15**, 171–178 (1982).
Bernal, M. J. M., and Boys, S. F., *Philos. Trans. R. Soc. (London)*, **A245**, 139 (1952).
Bersohn, R., *J. Chem. Phys.*, **24**, 1066 (1956).
Berthier, G., *Compt. Rend. Acad. Sci.*, **238**, 91 (1954).
Berthier, G., *J. Chim. Phys.*, **51**, 363 (1954).
Berthier, G., in *Molecular Orbitals in Chemistry, Physics and Biology*, P. O. Löwdin and B. Pullman (eds.), Academic Press, New York, 1964, pp. 57–82.
Bethe, H. A., and Saltpeter, E. E., *Quantum Mechanics of One- and Two-Electron Atoms*, Springer-Verlag, Berlin, 1957, Section IIb, Section 45, and Chapter 2, Section 38.
Bingel, W. A., and Kutzelnigg, W., *Adv. Quantum Chem.*, **5**, 201 (1970).
Binkley, J. S., Pople, J. A., and Hehre, W. J., *J. Amer. Chem. Soc.*, **102**, 939 (1980).
Birss, F. W., and Fraga, S., *J. Chem. Phys.*, **38**, 2562 (1963).
Birss, F. W., and Fraga, S., *J. Chem. Phys.*, **40**, 3203, 3207, 3212 (1964).
Björna, N., *J. Phys.*, **36**, 1412 (1973).
Blinder, S. M., *Am. J. Phys.*, **33**, 431 (1965).
Bobrowicz, F. W., and Goddard, W. A., III, in *Modern Theoretical Chemistry*, Volume 3, H. F. Schaefer (ed.), Plenum Press, New York, 1977, pp. 79–127.
Bohm, D., *Quantum Theory*, Prentice-Hall, Englewood Cliffs, NJ, 1951, Chapter 23.

Bohm, D., *Quantum Theory*, Prentice-Hall, Englewood Cliffs, NJ, 1957, pp. 247–251.
Bohr, N., *Z. Phys.*, **6**, 1–9 (1921).
Bohr, N., *Proc. Cambridge Philos. Soc. (Supplement)*, Part I, Cambridge University Press, Cambridge, 1924, p. 22.
Bonaccorsi, R., Pullman, A., Scrocco, E., and Thomasi, J., *Theoret. Chim. Acta*, **24**, 52 (1972).
Born, M., *Nachr. Akad. Wiss. Göttigen*, 1 (1951).
Born, M., and Huang, K. *Dynamical Theory of Crystal Lattices*, Oxford University Press, London, 1954, Appendix 8.
Born, M., and Oppenheimer, J. R., *Ann. Phys.*, **84**, 457 (1927).
Boutique, J. P., Riga, J., Verbist, J. J., Fripiat, J. G., Haddon, R. C., and Kaplan, M. L., *J. Am. Chem. Soc.*, **106**, 312–318 (1984).
Boys, S. F., *Proc. R. Soc. (London)*, **A200**, 542–554, (1950).
Boys, S. F., *Proc. R. Soc. (London)*, **A217**, 136, 235 (1953).
Boys, S. F., in *Quantum Theory of Atoms, Molecules and the Solid State*, P. O. Löwdin (ed.), Academic Press, New York, 1966, p. 253.
Boys, S. F., and Bernardi, F., *Mol. Phys.*, **19**, 553 (1970).
Brandon, B. H., *Rev. Mod. Phys.*, **39**, 771 (1967).
Breit, G., *Phys. Rev.*, **34**, 553 (1929).
Breit, G., *Phys. Rev.*, **36**, 383 (1930).
Breit, G., *Phys. Rev.*, **39**, 616 (1932).
Brillouin, L., *Actualites sci. et ind.*, No. 71, Hermann et Cie, Paris, 1933.
Brillouin, L., *Actualites sci. et ind.*, No. 159, Hermann et Cie, Paris, 1934.
Brillouin, L., *J. Phys. Radium*, **3**, 373 (1935).
Brueckner, K. A., *Phys. Rev.*, **97**, 1353 (1955).
Brueckner, K. A., Eden, R. J., and Francis, N. C., *Phys. Rev.*, **93**, 1445 (1955).
Brueckner, K. A., and Levinson, C. A., *Phys. Rev.*, **97**, 1344 (1955).
Buenker, R. J., and Peyerimhoff, S. D., *J. Chem. Phys.*, **53**, 1368 (1970).
Buenker, R. J., and Peyerimhoff, S. D., *Chem. Reviews*, **74**, 127–188 (1974).
Buenker, R. J., and Peyerimhoff, S. D., *Chem. Phys.*, **9**, 75–89 (1976).
Buenker, R. J., and Peyerimhoff, S. D. and Kammer, W. E., *J. Chem. Phys.*, **55**, 814 (1971).
Burran, O., *Kgl. Danske Videnskab. Selskab. Mat. Fys. Medd.*, **7**, No. 14, 1 (1927).
Byron, F. W., and Fuller, R. W., *Mathematics of Classical and Quantum Physics*, Volume 1, Addison-Wesley, Reading, MA, 1969, pp. 214–215.
Caballol, R., Carbo, R., Gallifa, R., and Riera, J. M., *Int. J. Quantum Chem.*, **8**, 373 (1974).
Cade, P. E., and Huo, W., *J. Chem. Phys.*, **47**, 614 (1967).
Cade, P. E., Sales, K. D., and Wahl, A. C., *J. Chem. Phys.*, **44**, 1973 (1966).
Caldow, G. L., and Coulson, C. W., *Proc. Cambridge Phil. Soc.*, **57**, 341 (1961).
Carbo, R., Gallifa, R., and Riera, J. M., *Chem. Phys. Lett.*, **30**, 43 (1975).
Carbo, R., and Riera, J. M., *A General SCF Theory*, Lecture Notes in Chemistry, Volume 5, Springer-Verlag, Berlin, 1978.
Carlson, B. C., and Keller, J. M., *Phys. Rev.*, **121**, 659 (1961).
Carrington, A., and McLachlan, A. D., *Introduction to Magnetic Resonance*, Harper & Rowe, New York, 1967, Section 2.2 and Chapter 11.
Christianson, J. J., Rytting, J. T., and Izatt, R. M., *Biochemistry*, **9**, 4907 (1970).
Christoffersen, R. E., *J. Chem. Phys.*, **41**, 960 (1964).
Christoffersen, R. E., *Adv. Quantum Chem.*, **6**, 333–393 (1972).

Christoffersen, R. E., and Baker, K. A., *Chem. Phys. Lett.*, **8**, 4–9 (1971).
Christoffersen, R. E., Hagstrom, S., and Prosser, F. P., *J. Chem. Phys.*, **40**, 236 (1964).
Christoffersen, R. E., and Maggiora, G. M., *Chem. Phys. Lett.*, **3**, 419–423 (1969).
Christoffersen, R. E., and Ruedenberg, K., *J. Chem. Phys.*, **49**, 4825 (1968).
Christoffersen, R. E., and Shull, H., *J. Chem. Phys.*, **48**, 1790 (1968).
Christoffersen, R. E., Spangler, D., Hall, G. G., and Maggiora, G. M., *J. Am. Chem. Soc.*, **95**, 8526–8536 (1973).
Cizek, J., *J. Chem. Phys.*, **45**, 4256 (1966).
Cizek, J., *Adv. Chem. Phys.*, **14**, 35 (1969).
Cizek, J., and Paldus, J., *Phys. Scripta*, **21**, 251 (1980).
Claverie, P., Diner, S., and Malrieu, J. P., *Int. J. Quantum Chem.*, **1**, 751 (1967).
Clementi, E., *IBM J. Res. Devel.*, **9**, 2 (1965).
Clementi, E., and Corongin, G., *Chem. Phys. Lett.*, **90**, 359 (1982).
Clementi, E., and Davis, D. R., *J. Comput. Phys.*, **1**, 223 (1966).
Clementi, E., Mehl, J., and vonNiessen, W., *J. Chem. Phys.*, **54**, 508–520 (1971).
Clementi, E., and Roetti, C., *Atom. Data and Nuclear Data Tables*, **14**, 177 (1974).
Coester, F., *Nucl. Phys.*, **1**, 421 (1958).
Coester, F., and Kümmel, H., *Nucl. Phys.*, **17**, 477 (1960).
Cohen, E. R., and Taylor, B. N., *J. Phys. Chem. Ref. Data*, **2**, 663 (1973).
Cohen, I., *J. Chem. Ed.*, **40**, 256 (1963).
Cohen, M., and Dalgarno, A., *Proc. Phys. Soc. London*, **77**, 748–750 (1961).
Coleman, A. J., and Eidahl, R. M. (eds.), *Reduced Density Matrices with Applications to Physical and Chemical Systems, Queen's Papers on Pure and Applied Mathematics*, No. 11, Queens University, Kingston, Ontario, 1968.
Compton, A. H., *Phys. Rev.*, **18**, 96 (1921).
Compton, A. H., *Phys. Rev.*, **19**, 267 (1922).
Compton, A. H., *Phys. Rev.*, **21**, 483 (1923).
Condon, E. U., *Phys. Rev.*, **32**, 858 (1928).
Cooke, R. G., *Infinite Matrices and Sequence Spaces*, Macmillan, London, 1950.
Costain, C. C., *Phys. Rev.*, **108** (1951).
Cotton, F. A., *Chemical Applications of Group Theory*, John Wiley, New York, 1963, Appendix 2 and Chapter 3, Section 10.
Coulson, C. A., *Trans. Faraday Soc.*, **38**, 433 (1942).
Coulson, C. A., *Electricity*, Interscience/Oliver and Boyd, Edinburgh, 1953, Chapter 14.
Courant, R., and Hilbert, D., *Methods of Mathematical Physics*, Volume I, Interscience Publishers, New York, 1953, pp. 69–81, 293–294.
Czerny, M., *Z. Phys.*, **34**, 227 (1925).
Das, G., *J. Chem. Phys.*, **58**, 5104 (1973).
Das, G., Janis, T., and Wahl, A. C., *J. Chem. Phys.*, **61**, 1274 (1974).
Das, G., and Wahl, A. C., *J. Chem. Phys.*, **44**, 87 (1966).
Das, G., and Wahl, A. C., *J. Chem. Phys.*, **47**, 2934 (1967).
Das, G., and Wahl, A. C., *Phys. Rev. Lett.*, **24**, 440 (1970).
Das, G., and Wahl, A. C., *J. Chem. Phys.*, **56**, 1769 (1972).
Davidson, E. R., *J. Chem. Phys.*, **37**, 577 (1962).
Davidson, E. R., *J. Chem. Phys.*, **39**, 875 (1963).
Davidson, E. R., *Adv. Quantum Chem.*, **6**, 235–266 (1972).
Davidson, E. R., *J. Chem. Phys.*, **57**, 1999 (1972).

Davidson, E. R., *Rev. Mod. Phys.*, **44**, 451 (1972).
Davidson, E. R., *Chem. Phys. Lett.*, **21**, 565 (1973).
Davidson, E. R., *Int. J. Quantum Chem.*, **8**, 83 (1974).
Davidson, E. R., in *The World of Quantum Chemistry*, R. Daudel and B. Pullman (eds.), Dodrecht, Holland, 1974, p. 17.
Davidson, E. R., *J. Comp. Phys.*, **17**, 87–94 (1975).
Davidson, E. R., *Reduced Density Matrices in Quantum Chemistry*, Academic Press, New York, 1976.
Davidson, E. R., and Feller, D., *Chem. Rev.*, **86**, 681-696 (1986).
Davidson, E. R., and Jones, L. C., *J. Chem. Phys.*, **37**, 2966 (1962).
Davidson, E. R., and Silver, D. W., *Chem. Phys. Lett.*, **52**, 403 (1977).
Davisson, C., and Gerner, L. H., *Phys. Rev.*, **30**, 705–740 (1927).
DeBroglie, L., *Compt. Rend.*, **177**, 507-510 (1923).
DeBroglie, L., *Philos. Mag.*, **47**, 446–458 (1924).
DeFrees, D. J., Raghavachari, K., Schlegel, H. B., and Pople, J. A., *J. Am. Chem. Soc.*, **104**, 5576-5580 (1982).
Deutchman, P. A., *Am. J. Phys.*, **39**, 952 (1971).
Diamond, J. J., Segal, G. A., and Wetmore, R. W., *J. Phys. Chem.*, **88**, 3532-3538 (1984).
Diercksen, G. H. F., *Theoret. Chim. Acta*, **33**, 1 (1974).
Dirac, P. A. M., *Proc. R. Soc. (London)*, **A112**, 661-677 (1926).
Dirac, P. A. M., *Quantum Mechanics*, 4th ed., Oxford University Press, Oxford, England, 1958, pp. 58-62 and Sections 72 and 73.
Dunning, T. H., *J. Chem. Phys.*, **53**, 2383 (1970).
Dunning, T. H., *J. Chem. Phys.*, **53**, 2823 (1970).
Dunning, T. H., *J. Chem. Phys.*, **55**, 716–723 (1971).
Dunning, T. H., *J. Chem. Phys.*, **55**, 3958-3966 (1971).
Dunning, T. H., Jr., and Hay, P. J., *Methods of Electronic Structure Theory*, Volume 3, Plenum Press, New York, 1977, p. 1–27.
Dunning, T. H., Jr., and Hay, P. J., in *Modern Theoret. Chemistry*, Volume 3, H. F. Schaefer III (ed.), Plenum Press, New York, 1977, p. 1.
Dunning, T. H., Hunt, W. J. and Goddard, W. A. III, *Chem. Phys. Lett.*, **4**, 147 (1969).
Ebbing, D. D., and Henderson, R. C., *J. Chem. Phys.*, **42**, 2225 (1965).
Eckhart, C. E., *Phys. Rev.*, **36**, 878 (1930).
Edmiston, C., and Krauss, M., *J. Chem. Phys.*, **45**, 1833 (1966).
Edmiston, C., and Ruedenberg, K., *Rev. Mod. Phys.*, **35**, 457 (1963).
Edmiston, C. and Ruedenberg, K., *J. Chem. Phys.*, **43**, S95 (1965).
Edmiston, C. and Ruedenberg, K., *J. Chem. Phys.*, **43**, 597 (1965).
Edmiston, C., and Ruedenberg, K., in *Quantum Theory of Atoms, Molecules and the Solid State*, P. O. Löwdin (ed.), Academic Press, New York, 1966, p. 263.
Edmonds, A. R., *Angular Momentum in Quantum Mechanics*, Princeton University Press, Princeton, 1960.
Eisenhart, L. P., *Phys. Rev.*, **45**, 427 (1934).
Ellison, F. E., and Shull, H., *J. Chem. Phys.*, **21**, 1420 (1953).
Ellison, F. E., and Shull, H., *J. Chem. Phys.*, **23**, 2348 (1955).
England, W. B., Sabelli, N. H., and Wahl, A. C., *J. Chem. Phys.*, **63**, 4596 (1975).
Erdelyi, A., (ed.) *Higher Transcendental Functions*, Volumes I-IV, McGraw-Hill, New York, 1953.

Erdelyi, A. (ed.), *Tables of Integral Transforms*, Volumes I and II, McGraw-Hill, New York, 1954.
Evans, D. F., *J. Chem. Soc.*, 1735, (1960).
Eyring, H., Walter, J., and Kimball, G. E., *Quantum Chemistry*, John Wiley, New York, 1944, Chapter 4, pp. 103–106 and Chapter 6.
Falk, S., *Z. Angew. Math. Mech.*, **53**, 73 (1973).
Fano, G., *Mathematical Methods of Quantum Mechanics*, McGraw-Hill, New York, 1971, Chapter 4.
Fliazar, S., *Charge Distribution and Chemical Effects*, Springer-Verlag, Berlin, 1983.
Fock, V., *Z. Phys.*, **61**, 126 (1930).
Fock, V., *Z. Phys.*, **63**, 855 (1930).
Foster, J. M., and Boys, S. F., *Rev. Mod. Phys.*, **32**, 300 (1960).
Fowler, R., and Guggenheim, E. A., *Statistical Thermodynamics*, Cambridge University Press, Cambridge, 1952, p. 72.
Fox, D., Handy, N., Saxe, P., and Schaefer, H. F., III, 183rd American Chemical Society Meeting, Las Vegas, Nevada, 1982.
Franck, J., *Trans. Faraday Soc.*, **21**, 536 (1925).
Franck, J., and Hertz, G., *Phys. Z.*, **17**, 409 (1916).
Franck, J., and Hertz, G., *Verhand. Deutsch. Phys. Gesell.*, **16**, 457 (1914).
Franck, J., and Hertz, G., *Phys. Z.*, **20**, 132 (1920).
Freed, K. F., *Annu. Rev. Phys. Chem.*, **22**, 771 (1967).
Freed, K. F., *Chem. Phys. Lett.*, **2**, 255–256 (1968).
Frisch, M. J., Krishnan, R., and Pople, J. A., *Chem. Phys. Lett.*, **75**, 66 (1980).
Fröman, A., *J. Chem. Phys.*, **36**, 1490 (1962).
Frost, A. A., *J. Am. Chem. Soc.*, **89**, 3064 (1967).
Frost, A. A., *J. Am. Chem. Soc.*, **90**, 1965 (1967).
Frost, A. A., *J. Chem. Phys.*, **47**, 3707–3714 (1967).
Frost, A. A., *J. Chem. Phys.*, **47**, 3714–3716 (1967).
Frost, A. A., *J. Phys. Chem.*, **72**, 1289 (1968).
Frost, A. A., in *Methods of Electronic Structure Theory*, Volume 3, Plenum Press, New York, 1977, pp. 29–50.
Frost, A. A., in *Modern Theoretical Chemistry*, Volume 3, H. F. Schaefer III (ed.), Plenum Press, New York, 1977, Chapter 2, pp. 29–49.
Gallup, G. A., *Int. J. Quantum Chem.*, **8**, 267 (1974).
Gamba, A., *Nuovo Cimento*, **7**, 378 (1950).
Ghosh, P. K., *Introduction to Photoelectron Spectroscopy*, John Wiley, New York, 1983.
Gilbert, T. L., *Phys. Rev.*, **A6**, 580 (1972).
Goddard, W. A., *Phys. Rev.*, **157**, 81–93 (1967).
Goddard, W. A., Dunning, T. H., and Hunt, W. J., *Chem. Phys. Lett.*, **4**, 231 (1969).
Goddard, W. A., Dunning, T. H., Hunt, W. J., and Hay, P. J., *Acc. Chem. Res.*, **6**, 368 (1973).
Goddard, W. A., and Harding, L. B., *Annu. Rev. Phys. Chem.*, **29**, 363–396 (1978).
Goddard, W. A., and Ladner, R. D., *J. Am. Chem. Soc.*, **93**, 6750–6756 (1971).
Goldstein, H., *Classical Mechanics*, Addison Wesley, Reading, MA, 1950, pp. 255–258.
Goldstone, J., *Proc. R. Soc. (London)*, **A239**, 267 (1957).
Golebiewski, A., *Mol. Phys.*, **20**, 481 (1971).
Goodisman, J., and Klemperer, W., *J. Chem. Phys.*, **38**, 721–725 (1963).

Gordon, M. S., Binkley, J. S., Pople, J. A., Pietro, J., and Hehre, W. J., *J. Am. Chem. Soc.*, **104**, 2797 (1982).
Green, H. S., *Matrix Methods in Quantum Mechanics*, Barnes & Noble, New York, 1965.
Guest, M. F., and Saunders, V. R., *Mol. Phys.*, **28**, 819 (1974).
Gutowski, M., van Lenthe, J. H., Veerbeck, J., and Chalasinki, G., *Chem. Phys. Lett.*, **124**, 370 (1985).
Hackmeyer, M., and Whitten, J. L., *J. Chem. Phys.*, **54**, 3739–3750 (1971).
Hagstrom, S., and Shull, H., *J. Chem. Phys.*, **30**, 1314 (1959).
Hagstrom, S., and Shull, H., *Rev. Mod. Phys.*, **35**, 624 (1963).
Hall, G. G., *Proc. R. Soc. (London)*, **A205**, 541 (1951).
Hall, G. G., *Applied Group Theory*, Longmans, Green & Co., London, 1967.
Hall, G. G., *Symposium of the Faraday Society*, No. 2, The University Press, Aberdeen, 1968, p. 69.
Hall, G. G., *Chem. Phys. Lett.*, **20**, 501 (1973).
Hall, G. G., *Adv. Atom. Mol. Phys.*, **20**, 41 (1985).
Hall, G. G., Hyslop, J., and Rees, D., *Int. J. Quantum Chem.*, **3**, 195 (1969).
Hall, G. G., and Lennard-Jones, J. E., *Proc. R. Soc. (London)*, **A202**, 155 (1950).
Hall, G. G., and Lennard-Jones, J. E., *Proc. R. Soc. (London)*, **A205**, 357 (1951).
Hall, G. G., and Lennard-Jones, J. E., *Proc. Roy. Soc. (London)*, **A205**, 541 (1951).
Hall, L. H., *Group Theory and Symmetry in Chemistry*, McGraw-Hill, New York, 1969, Chapter 4.
Halmos, P. R., *Finite-Dimensional Vector Spaces*, VanNostrand, New York, 1958, pp. 3–4.
Hariharan, P. C., and Pople, J. A., *Theoret. Chim. Acta*, **28**, 213 (1973).
Harris, F. E., *Adv. Chem. Phys.*, **13**, 205 (1967).
Harris, F. E., in *Energy, Structure and Reactivity*, D. W. Smith and W. B. McRae (eds.), John Wiley, New York, 1973, p. 112.
Hartree, D. R., *Proc. Cambridge Philos. Soc.*, **24**, 89 (1928).
Hartree, D. R., *Proc. R. Soc. (London)*, **A154**, 588 (1936).
Hauser, W., *Introduction to The Principles of Mechanics*, Addison-Wesley, Reading, MA, 1965, Section 4-9, p. 99.
Hausman, R. F., Jr. and Bender, C. F., in *Methods of Electronic Structure Theory*, H. F. Schaefer III (ed.), Plenum Press, New York, 1977, Chapter 8, pp. 319–338.
Hausman, R. F., Jr., Bloom, S. D., and Bender, C. F., *Chem. Phys. Lett.*, **32**, 483 (1975).
Hay, P. J., *J. Chem. Phys.*, **59**, 2468 (1973).
Hegarty, D., and VanderVelde, G., *Int. J. Quantum Chem.*, **23**, 1135–1153 (1983).
Hehre, W. J., *Modern Theoretical Chemistry*, Volume 4, H. F. Schaefer III (ed.), Plenum Press, New York, 1977, Chapter 7.
Hehre, W. J., Ditchfield, R., Stewart, R. F., and Pople, J. A., *J. Chem. Phys.*, **52**, 2769 (1970).
Hehre, W. J., Radon, L., Scheyer, P., and Pople, J. A., *Ab Initio Molecular Orbital Theory*, Wiley-Interscience, New York, 1986.
Heisenberg, W., *Z. Phys.*, **38**, 411–426 (1926).
Heitler, W., and London, F., *Z. Phys.*, **44**, 455 (1927).
Hellwig, G., *Differential Operators of Mathematical Physics*, Addison-Wesley, Reading, MA, 1967, pp. 39–49.
Herzberg, G., *Molecular Spectra and Molecular Structure*. I. *Spectra of Diatomic Molecules*, D. Van Nostrand, Princeton, NJ, 1950, p. 58.

Herzberg, G., *Atomic Spectra and Atomic Structure*, Dover Publications, New York, 1944, pp. 11–12.
Hildebrand, F. B., *Introduction to Numerical Analysis*, McGraw-Hill Book, New York, 1956, p. 427.
Hinze, J., *J. Chem. Phys.*, **59**, 6424 (1973).
Hirao, K., *J. Chem. Phys.*, **60**, 3215 (1974).
Hirao, K., and Nakatsuji, H., *J. Chem. Phys.*, **59**, 1457 (1973).
Hirschfelder, J. O., *J. Chem. Phys.*, **33**, 1462 (1960).
Hirschfelder, J. O., and Kincaid, J. F., *Phys. Rev.*, **52**, 658 (1937).
Hirschfelder, J. O., and Linnett, J. W., *J. Chem. Phys.*, **18**, 130 (1950).
Hirschfelder, J. O., and Meath, W. J., *Adv. Chem. Phys.*, **12**, 3–106 (1967).
Hirschfelder, J. O., and Nazaroff, G. V., *J. Chem. Phys.*, **34**, 1666 (1961).
Hoffman, K., and Kunze, R., *Linear Algebra*, Prentice-Hall, Englewood Cliffs, NJ, 1961.
Houlden, S. A., and Csizmadia, I. G., *Theoret. Chim. Acta*, **30**, 209 (1973).
Houlden, S. A., and Csizmadia, I. G., *Theoret. Chim. Acta*, **36**, 275 (1973).
Houston, W. V., *Principles of Mathematical Physics*, McGraw-Hill, New York, 1948, Chapter 16.
Hsu, H. L., Davidson, E. R., and Pitzer, R. M., *J. Chem. Phys.*, **65**, 609 (1976).
Hund, E., *Z. Elektrochem.*, **34**, 437 (1928).
Hund, E., *Z. Phys.*, **51**, 759 (1928).
Hund, E., *Z. Phys.*, **73**, 1 (1931).
Hund, E., *Z. Phys.*, **73**, 565 (1932).
Hunt, W. J., and Goddard, W. A., III, *Chem. Phys. Lett.*, **3**, 414 (1969).
Hunt, W. J., Hay, P. J., and Goddard, W. A. III, *J. Chem. Phys.*, **57**, 738–748 (1972).
Hurley, A. C., *Electron Correlation in Small Molecules*, Academic Press, New York, 1976, pp. 22–30.
Hurley, A. C., Lennard-Jones, J. E., and Pople, J. A., *Proc. R. Soc. (London)*, **A220**, 446 (1953).
Huzinaga, S., *Phys. Rev.*, **120**, 866 (1960).
Huzinaga, S., *Phys. Rev.*, **122**, 131 (1961).
Huzinaga, S., *J. Chem. Phys.*, **42**, 1293–1302 (1965).
Huzinaga, S., *J. Chem. Phys.*, **51**, 3971 (1969).
Huzinaga, S., *Comp. Phys. Rep.*, **2**, 279 (1985).
Huzinaga, S., Andzelm, J., Klobukowski, M., Radzio-Andzelm, E., Sakai, Y. and Tatewaki, H., *Gaussian Basis Sets for Molecular Calculations*, Elsevier, Amsterdam, 1984.
Huzinaga, S., and Arnan, C., *Phys. Rev.*, **A1**, 1285 (1970).
Huzinaga, S., and Arnan, C., *J. Chem. Phys.*, **54**, 1948 (1971).
Huzinaga, S., Klobukowski, M., and Tatewaki, H., *Can. J. Chem.*, **63**, 1812 (1985).
Hylleraas, E., *Z. Phys.*, **54**, 347 (1929).
Hylleraas, E., *Z. Phys.*, **65**, 759 (1930).
Hylleraas, E., *Z. Phys.*, **71**, 739 (1931).
Hylleraas, E. A., *Mathematical and Theoretical Physics*, Volume 1, Wiley-Interscience, New York, 1970, Section 38.
Hylleraas, E. A., and Undheim, B., *Z. Phys.*, **65**, 759 (1930).
Indritz, J., *Methods in Analysis*, Macmillan, New York, 1963, p. 22.
Jackson, J. D., *Classical Electrodynamics*, John Wiley, New York, 1975.
Jaffe, G., *Z. Phys.*, **87**, 535 (1934).
James, H. M., and Coolidge, A. S., *J. Chem. Phys.*, **1**, 825 (1933).

Jammer, M., *The Conceptual Development of Quantum Mechanics*, McGraw Hill, New York, 1966, Chapter 7 and pp. 196–280.
Jepson, D. W., and Hirschfelder, J. O., *J. Chem. Phys.*, **32**, 1323 (1960).
Kahn, L. R., Hay, P. J., and Shavitt, I., *J. Chem. Phys.*, **61**, 3530 (1974).
Kaldor, U., and Shavitt, I., *J. Chem. Phys.*, **48**, 191 (1968).
Kaplan, W., *Advanced Calculus*, Addison-Wesley, Cambridge, MA, 1952, p. 434.
Kaplan, W., *Ordinary Differential Equations*, Addison-Wesley, Reading, MA, 1961, Chapter 4.
Kapuy, E., *Acta Phys. Acad. Sci. Hung.*, **9**, 237 (1958).
Kapuy, E., *Acta Phys. Acad. Sci. Hung.*, **10**, 125 (1959).
Kapuy, E., *Acta Phys. Acad. Sci. Hung.*, **11**, 409 (1960).
Kapuy, E., *Acta Phys. Acad. Sci. Hung.*, **12**, 185, 351 (1960).
Kapuy, E., *Acta Phys. Acad. Sci. Hung.*, **13**, 345, 461 (1961).
Karplus, M., and Grant, D. M., *Proc. Natl. Acad. Sci. U.S.A.*, **45**, 1269 (1959).
Kastler, D., *J. Chim. Phys.*, **50**, 556 (1953).
Keaveny, I. T., and Christoffersen, R. E., *J. Chem. Phys.*, **50**, 80–85 (1969).
Keeton, M., and Santry, D. P., *Chem. Phys. Lett.*, **1**, 105 (1970).
Keil, F., and Alrichs, R., *J. Chem. Soc.*, **98**, 4787 (1976).
Kelly, H. P., *Phys. Rev.*, **136**, B896 (1964).
Kelly, H. P., *Adv. Chem. Phys.*, **14**, 129 (1969).
Kelly, H. P., *Phys. Rev. Lett.*, **23**, 455 (1969).
Killingbeck, *J. Math. Phys.*, **11**, 2268 (1970).
Klessinger, M., *Theoretic. Chim. Acta*, **15**, 353 (1969).
Kolos, W., and Wolniewicz, L., *J. Chem. Phys.*, **41**, 3663–3678 (1964).
Kolos, W., and Wolniewicz, L., *J. Chem. Phys.*, **43**, 2429–2441 (1965).
Kolos, W., and Wolniewicz, L., *J. Chem. Phys.*, **49**, 404–410 (1968).
Koopmans, T. A., *Physica*, **1**, 104 (1933).
Kotani, M., Amemiya, A., Ishiguro, E., and Kimura, T., *Tables of Molecular Integrals*, 2nd ed., Maruyen, Tokyo, 1963.
Krauss, M., *Compendium of Ab Initio Calculations of Molecular Energies and Properties*, NBS Tech. Note 438, U.S. Dept. of Commerce, Washington, D.C., 1967.
Krishnan, R., Frisch, M. J., and Pople, J. A., *J. Chem. Phys.*, **72**, 4244 (1980).
Krishnan, R., and Pople, J. A., *Int. J. Quantum Chem.*, **14**, 91 (1978).
Kümmel, H., *Nucl. Phys.*, **22**, 177 (1969).
Kutzelnigg, W., *J. Chem. Phys.*, **40**, 3640–3647 (1964).
Kutzelnigg, W., *Methods of Electronic Structure Theory*, H. F. Schaefer III (ed.), Plenum Press, New York, 1977, pp. 129–188.
Kutzelnigg, W., Ahlrichs, R., Labib-Iskander, I., and Bingel, W. A., *Chem. Phys. Lett.*, **1**, 447 (1967).
Kutzelnigg, W., Meunier, A., Levy, B., and Berthier, G., *Int. J. Quantum Chem.*, **12**, 77 (1977).
Laidlaw, W., and Birss, F. W., *Theoret. Chim. Acta*, **2**, 181 (1964).
Lancaster, P., *Theory of Matrices*, Academic Press, New York, 1969, pp. 45–49.
Langhoff, S. R., and Davidson, E. R., *Int. J. Quantum Chem.*, **8**, 61 (1974).
Lavery, R., and Pullman, A., in *New Horizons of Quantum Chemistry*, P. O. Löwdin and B. Pullman (eds.), D. Reidel, Dordrecht, Holland, 1983, pp. 439–451.
Lennard-Jones, J. E., *Trans. Faraday Soc.*, **25**, 688 (1928).
Lennard-Jones, J. E., *Proc. R. Soc. (London)*, **A198**, 1, 14 (1949).
Lennard-Jones, J. E., and Pople, J. A., *Proc. R. Soc. (London)*, **A202**, 166 (1950).

Lennard-Jones, J. E., and Pople, J. A., *Proc. R. Soc. (London)*, **A210**, 190 (1951).
Levy, B., and Berthier, G., *Int. J. Quantum Chem.*, **2**, 307 (1968).
Levy, B., and Berthier, G., *Int. J. Quantum Chem.*, **3**, 247 (1969).
Lin, J., Yu, C., Peng, S., Akiyama, I., Li, K., Lee, L. K., and LeBreton, P. R., *J. Phys. Chem.*, **84**, 1006-1012 (1980).
Longstaff, J. V. L., and Singer, K., *Proc. R. Soc. (London)*, **A258**, 421 (1960).
Longuet-Higgins, H. C., *J. Chim. Phys.*, **46**, 175 (1949).
Longuet-Higgins, H. C., *J. R. Inst. Chem.*, **77**, 197 (1953).
Loter, J., and Christoffersen, R. E., *Int. J. Quantum Chem.*, **3**, 651-661 (1969).
Loudin, R., *Am. J. Phys.*, **27**, 649 (1959).
Löwdin, P. O., *J. Chem. Phys.*, **18**, 365 (1950).
Löwdin, P. O., *J. Chem. Phys.*, **19**, 1396 (1951).
Löwdin, P. O., *Phys. Rev.*, **97**, 1474-1490, (1955).
Löwdin, P. O., *Phys. Rev.*, **97**, 1509 (1955).
Löwdin, P. O. *Calcul des Fonctions d'Onde Moléculaire*, CNRS, Paris, 1958, p. 23.
Löwdin, P. O., *Phys. Rev.*, **110**, 1466 (1958).
Löwdin, P. O., *Adv. Chem. Phys.*, **2**, 207-322 (1959).
Löwdin, P. O., *J. Math. Phys.*, **3**, 969 (1962).
Löwdin, P. O., *Rev. Modern Phys.*, **34**, 520 (1962).
Löwdin, P. O., *Adv. Quantum Chem.*, **3**, 324 (1967).
Löwdin, P. O., *Rev. Mod. Phys.*, **39**, 259 (1967).
Löwdin, P. O., and Pullman, B. Eds., *Molecular Orbitals in Chemistry, Physics, and Biology*, Academic Press, New York, 1964.
Löwdin, P. O., and Shull, H., *Phys. Rev.*, **101**, 1730 (1956).
MacDonald, J. K. L., *Phys. Rev.*, **43**, 830 (1933).
MacDonald, J. K. L., *Phys. Rev.*, **46**, 828 (1934).
Maggiora, G. M., and Christoffersen, R. E., *J. Am. Chem. Soc.*, **98**, 8325-8332 (1976).
Maki, A. G., and Sarns, R. L., unpublished data referred to in R. J. Bartlett, *Annu. Rev. Phys. Chem.*, **32**, 359-401 (1981).
Margenau, H., and Murphy, G. M., *The Mathematics of Physics and Chemistry*, Volume 1, D. Van Nostrand, New York, 1961, Section 3.7.
Martin, D., and Hall, G. G., *Theoret. Chim. Acta*, **59**, 281 (1981).
Matsen, F. A., *Adv. Quantum Chem.*, **1**, 59-114 (1964).
McConnell, H. M., *J. Chem. Phys.*, **24**, 764 (1956).
McMurchie, L. E., and Davidson, E. R., *J. Chem. Phys.*, **66**, 2959 (1959).
McWeeny, R., *Symmetry—An Introduction to Group Theory and Its Applications*, Pergamon, London, 1963.
McWeeny, R. M., and Ohno, K. A., *Proc. R. Soc. (London)*, **A255**, 367 (1960).
Meckler, A., *J. Chem. Phys.*, **21**, 1750 (1953).
Mehler, E. L., *Theoret. Chim. Acta*, **35**, 17-32 (1974).
Mehler, E. L., Ruedenberg, K., and Silver, D. M., *J. Chem. Phys.*, **52**, 1181-1205 (1970).
Merzbacher, E., *Quantum Mechanics*, John Wiley, 1961, pp. 80-82.
Merzbacher, E., *Quantum Mechanics*, 2nd ed., John Wiley, New York 1970, pp. 82-85.
Melvin, M. A., *Rev. Mod. Phys.*, **28**, 18 (1956).
Mely, B., and Pullman, A., *Theoret. Chim. Acta*, **13**, 278 (1969).
Merer, A. J., and Mulliken, R. S., *Chem. Rev.*, **69**, 639 (1969).
Messiah, A., *Quantum Mechanics*, Volume 1, John Wiley, New York, 1958, p. 8.

Messiah, A., *Quantum Mechanics*, Volume 1, Appendix A, North Holland Publishing Co., Amsterdam, 1965, pp. 169, 188.
Meyer, W., *J. Chem. Phys.*, **58**, 1017 (1973).
Meyer, W., *Methods of Electronic Structure Theory*, Volume 3, H. F. Schaefer III (ed.), Plenum Press, New York, 1977, pp. 413–446.
Miller, J. H., and Kelly, H. P., *Phys. Rev. Lett.*, **26**, 679 (1971).
Miller, J. H., and Kelly, H. P., *Phys. Rev. Lett.*, **A4**, 480 (1971).
Miller, K. J., and Ruedenberg, K., *J. Chem. Phys.*, **48**, 3414, 3444, 3450 (1968).
Miron, E., Raz, B. and Jortner, J., *Chem. Phys. Lett.*, **6**, 563 (1970).
Moccio, R., *Theoret. Chim. Acta*, **7**, 85 (1967).
Møeller, C. and Plesset, M. S., *Phys. Rev.*, **46**, 618 (1934).
Moore, W. J., *Physical Chemistry*, Prentice-Hall, New York, 1962, p. 655.
Morokuma, K., Kato, S., Kitaura, K., Obara, S., Ohta, K., and Hanamura, M., in *New Horizons of Quantum Chemistry*, P. O. Löwdin and B. Pullman (eds.) D. Reidel, Dordrecht, Holland, 1983, pp. 221-241.
Moss, B. J., and Goddard, W. A. III, *J. Chem. Phys.*, **63**, 3523-3531 (1975).
Mueller, C. R. and Eyring, H., *J. Chem. Phys.*, **19**, 1495 (1951).
Mulliken, R. S., *Phys. Rev.*, **32**, 186, 388, 761 (1928).
Mulliken, R. S., *J. Chem. Phys.*, **3**, 573 (1935).
Mulliken, R. S., *Rev. Mod. Phys.*, **14**, 204 (1942).
Mulliken, R. S., *J. Chim. Phys.*, **46**, 497, 675 (1949).
Mulliken, R. S., *J. Chem. Phys.*, **23**, 1833, 1841, 2338, 2343 (1955).
Mulliken, R. S., *J. Chem. Phys.*, **36**, 3428 (1962).
Mulliken, R. S., *Molecular Orbitals in Chemistry, Physics and Biology*, P. O. Löwdin and B. Pullman (eds.), Academic Press, New York, 1964.
Musher, J. I., *Am. J. Phys.*, **34**, 267 (1966).
Musher, J. I., *J. Phys. (Paris)*, **31**, Suppl. C4, 51 (1970).
Nesbet, R. K., *Proc. R. Soc. (London)*, **A230**, 312 (1955).
Nesbet, R. K., *Phys. Rev.*, **109**, 1632 (1958).
Nesbet, R. K., *J. Math Phys.*, **2**, 701 (1961).
Nesbet, R. K., *Adv. Chem. Phys.*, **9**, 321 (1965).
Nesbet, R. K., *J. Chem. Phys.*, **43**, 311 (1965).
Nesbet, R. K., *Adv. Chem. Phys.*, **14**, 1 (1969).
Nesbet, R. K., *J. Comput. Phys.*, **17**, 87 (1975).
Nesbet, R. K., Barr, T. L., and Davidson, E. R., *Chem. Phys. Lett.*, **4**, 203 (1969).
Neumann, D. B., and Moskowitz, J. W., *J. Chem. Phys.*, **50**, 2216 (1969).
Neumann, J. V., *Mathematical Foundations of Quantum Mechanics*, Princeton University Press, Princeton, 1955.
Newell, G. F., *Phys. Rev.*, **78**, 711 (1950).
Nitzche, L. E., Chabalowski, C., and Christoffersen, R. E., *J. Am. Chem. Soc.*, **98**, 4797–4801 (1978).
Ogryzalo, E. A., *J. Chem. Ed.*, **42**, 150 (1965).
Ohno, K., and Morokuma, K., *Quantum Chemistry Literature Data Base*, Elsevier, New York, 1982.
O'Leary, B. Duke, B. J., and Eilers, J. E., *Adv. Quantum Chem.*, **9**, 1-68 (1975).
Olsen, J., Yeager, D. L., and Jorgensen, P., *Adv. Chem. Phys.*, **LIV**, 1-176 (1983).
O-Ohata, K. O., Taketa, H., and Hizinaga, S., *J. Phys. Soc. Jpn*, **21**, 2306 (1966).
Oppenheimer, J. R., *Phys. Rev.*, **35**, 461 (1930).
Osman, R., Topiol, S., Weinstein, H., and Eilers, J. E., *Chem. Phys. Lett.*, **73**, 399 (1978).

Pal, B. C., *Biochemistry*, **1**, 558 (1962).
Paldus, J., in *Theoretical Chemistry: Advances and Perspectives*, Volume 2, H. Eyring and D. G. Henderson (eds.), Academic Press, New York, 1976, p. 131.
Paldus, J., and Cizek, J., *Energy, Structure and Reactivity*, D. W. Smith and W. B. McRae (eds.), John Wiley, New York, 1973, p. 389.
Panofsky, W. K. H., and Phillips, M., *Electricity and Magnetism*, Addison-Welsey, Reading, MA, 1955, Chapter 13.
Parks, J. M., and Parr, R. G., *J. Chem. Phys.*, **28**, 335 (1958).
Parks, J. M. and Parr, R. G., *J. Chem. Phys.*, **32**, 1657 (1960).
Parlett, B. N., *The Symmetric Eigenvalue Problem*, Prentice-Hall, Englewood Cliffs, NJ, 1980.
Parr, R. G., and Brown, J. E., *J. Chem. Phys.*, **49**, 4849 (1968).
Pauli, W., *Z. Phys.*, **31**, 765–785 (1925).
Pauli, W., *Z. Phys.*, **43**, 601 (1927).
Pauling, L., *J. Chem. Phys.*, **4**, 673 (1936).
Pauling, L., *Nature of the Chemical Bond*, Cornell University Press, Ithaca, NY, 1939.
Pauling, L., *Proc. R. Soc. (London)*, **A196**, 343 (1949).
Pauncz, R., *Alternant Molecular Orbital Method*, Saunders, Philadelphia, 1967.
Pauncz, R., *Spin Eigenfunctions*, Plenum Press, New York, 1979.
Pearson, P. K., Schaefer, H. F. III, and Walgren, U., *J. Chem. Phys.*, **62**, 350 (1975).
Peck, E. R., *Electricity and Magnetism*, McGraw-Hill, New York, 1953, p. 201.
Pendergast, P., and Fink, W. H., *J. Comput. Phys.*, **14**, 286 (1974).
Petke, J. D., *Chem. Phys. Lett.*, **126**, 26 (1986).
Petke, J. D., Christoffersen, R. E., Maggiora, G. M., and Shipman, L. L., *Int. J. Quantum Chem., Quantum Biol. Symp.*, **4**, 343 (1977).
Petke, J. D., Maggiora, G. M., Shipman, L. L., and Christoffersen, R. E., *J. Mol. Spectrosc.*, **71**, 64–84 (1978).
Petke, J. D., Shipman, L. L., and Christoffersen, R. E., *Photobiochem. and Photobiol.*, **30**, 203–223 (1979).
Petke, J. D., Shipman, L. L., Maggiora, G. M., and Christoffersen, R. E., *J. Am. Chem. Soc.*, **103**, 4622 (1981).
Pietro, W. J., and Hehre, W. J., *Comput. Chem.*, **4**, 241 (1981).
Pietro, W. J., Levi, B. A., Hehre, W. J., and Stewart, R. F., *Inorg. Chem.*, **19**, 2225 (1980).
Pilar, F. L., *Elementary Quantum Chemistry*, McGraw-Hill, New York, 1968, p. 463.
Platt, J. R., Ruedenberg, K., Scherr, C. W., Ham, N. S., Labhart, H., and Lichten, W., *Free Electron Theory of Conjugated Molecules*, John Wiley, New York, 1964.
Poirer, R., Kari, R., and Csizmadia, I. G., *Handbook of Gaussian Basis Sets: A Compendium for Ab Initio Molecular Orbital Calculations*, Elsevier, New York, 1985.
Pople, J. A., in *Modern Theoretical Chemistry*, Volume 4, H. F. Schaefer III (ed.), Plenum Press, New York, 1977, pp. 1–28.
Pople, J. A., Binkley, J. S., and Seeger, R., *Int. J. Quantum Chem. Symp.*, **10**, 1 (1976).
Pople, J. A., Krishnan, R., Schlegel, H. B., and Binkley, J. W., *Int. J. Quantum Chem.*, **14**, 545 (1978).
Pople, J. A., and Nesbet, R. K., *J. Chem. Phys.*, **22**, 571 (1954).
Pople, J. A., Schneider, W. G., and Bernstein, H. I., *High-Resolution Nuclear Magnetic Resonance*, McGraw Hill, New York, 1959.

Pople, J. A., Seeger, R., and Krishnan, R., *Int. J. Quantum Chem. Symp.*, **11**, 149 (1977).
Powell, J. L. and Crasemann, B., *Quantum Theory*, Addison-Wesley, Reading, MA, 1961, Section 6-13.
Powell, R. E., *J. Chem. Ed.*, **45**, 45 (1968).
Preuss, H., *Z. Naturforsch*, **11**, 283 (1956).
Primas, H., in *Modern Quantum Chemistry*, Volume 2, O. Sinanaglu, (ed.), Academic Press, New York, 1965, p. 45.
Pulay, P., Forarasi, G., Pang, F., and Boggs, J. E., *J. Am. Chem. Soc.,* **101**, 2550 (1979).
Radom, L., *Modern Theoretical Chem*, Volume 4, Plenum Press, New York, 1977, pp. 333–356 and Chapter 8.
Raffenetti, R. C., *J. Comp. Phys.*, **32**, 403–419 (1979).
Raffenetti, R. C., and Ruedenberg, K., *Int. J. Quantum Chem.*, **34**, 625 (1970).
Ralston, A., and Wilf, H. S., *Mathematic Methods for Digital Computers*, Volume 1, John Wiley, New York, 1960.
Ransil, B. J., and Sinai, J. J., *J. Chem. Phys.*, **46**, 4050–4074 (1967).
Rauk, A., Allen, L. C., and Clementi, E., *J. Chem. Phys.*, **52**, 4133 (1970).
Redmond, L. T., Purvis, G. D., and Bartlett, R. J., *J. Am. Chem. Soc.*, **101**, 2855 (1979).
Reeves, C. M., and Fletcher, R., *J. Chem. Phys.*, **42**, 4073 (1965).
Richards, W. G., Walker, T. E. H., and Hinkley, R. K., *A Bibliography of Ab Initio Molecular Wavefunctions*, Oxford University Press, NY, 1971.
Richards, W. G., Walker, T. E. H., and Hinkley, R. K., *A Bibliography of Ab Initio Molecular Wavefunctions*, Supplement for 1970–1973 and 1974–1977, Oxford University Press, NY.
Roach, G. F., *Green's Functions*, van Nostrand Reinhold, New York, 1970.
Roberts, E. M., Foster, M. R., and Selig, F. F., *J. Chem. Phys.*, **37**, 485 (1962).
Robin, M. B., Bosch, H., Kuebler, N., Kaplan, B. E., and Meinwold, J., *J. Chem. Phys.*, **48**, 5037 (1968).
Roos, B. O., *Chem. Phys. Lett.*, **15**, 153 (1972).
Roos, B., and Siegbahn, P., *Theoret. Chim. Acta*, **17**, 208 (1970).
Roos, B. O., and Siegbahn, P. E. M., in *Methods of Electronic Structure Theory*, Volume 3, H. F. Schaefer III (ed.), Plenum Press, New York, 1977, p. 277.
Roothaan, C. C. J., *Rev. Mod. Phys.*, **23**, 69 (1951).
Roothaan, C. C. J., *Rev. Mod. Phys.*, **32**, 179 (1960).
Rosen, N., *Phys. Rev.*, **38**, 2099 (1931).
Rosenberg, B. J. and Shavitt, I., *J. Chem. Phys.*, **63**, 2162 (1975).
Rothenberg, S., and Davidson, E. R., *J. Chem. Phys.*, **45**, 2560 (1966).
Ruedenberg, K., and Poshusta, R. D., *Adv. Quantum Chem.*, **6**, 267 (1972).
Ruedenberg, K., Raffenetti, R. C., and Bardo, R. D., in *Energy, Structure and Reactivity*, Proc. of 1982 Boulder Conf., D. W. Smith (ed.), John Wiley, New York, 1973, p. 164.
Ryan, J. A., and Whitten, J. L., *Chem. Phys. Lett.*, **15**, 119–123 (1972).
Salmon, W. I., *Adv. Quantum Chem.*, **8**, 37 (1974).
Sansone, G., *Orthogonal Functions*, Interscience Publ., New York, 1959, Chapters 3 and 4.
Saunders, V. R., in *Methods in Computational Molecular Physics*, G. H. F. Diercksen and S. Wilson (eds.), D. Reidel, Boston, MA, 1983, pp. 1–36.
Saunders, V. R., and Hillier, I. H., *Int. J. Quantum Chem.*, **7**, 699–705 (1973).

Saxe, P., Schaefer, H. F. III, and Handy, N. C., *Chem. Phys. Lett.*, **79**, 202 (1981).
Schaad, J. L., and Hicks, W. V., *J. Chem. Phys.*, **53**, 851 (1970).
Schaefer, H. F., III, *J. Chem. Phys.*, **54**, 2207 (1971).
Schiff, L. I., *Quantum Mechanics*, McGraw-Hill, New York, 1968, Chapter 6.
Schlichter, D. P., *Principles of Magnetic Resonance*, Harper & Row, New York, 1963, Chapter 8 and pp. 238–243.
Schmeidler, W., *Linear Operators in Hilbert Space*, Academic Press, New York, 1965.
Schmid, L. A., *Phys. Rev.*, **92**, 1373 (1953).
Schrödinger, E., *Ann. Phys.*, **79**, 361, 489 (1926).
Schrödinger, E., *Ann. Phys.*, **80**, 437 (1926).
Schrödinger, E., *Ann. Phys.*, **81**, 109 (1926).
Schulman, J. M., and Kaufman, D. N., *J. Chem. Phys.*, **53**, 477 (1970).
Schulman, J. M., Moskowitz, J. W., and Hollister, C., *J. Chem. Phys.*, **46**, 2759 (1967).
Schwartz, M. E., in *Modern Theoretical Chemistry*, Volume 4, H. E. Schaefer III (ed.), Plenum Press, New York, 1977, pp. 333–356, 357–380.
Schwarz, H. R., Rutishauser, H., and Steifel, E., *Numerical Analysis of Symmetric Matrices*, Prentice-Hall, Englewood Cliffs, NJ, 1973.
Schwenke, D. W., and Truhlar, G. D., *J. Chem. Phys.*, **82**, 2418 (1985).
Schwenke, D. W., and Truhlar, G. D., *J. Chem. Phys.*, **84**, 4113 (1985).
Scrocco, E. and Tomasi, J., *Topics Curr. Chem.*, **42**, 97–120, (1973).
Segal, G. A., *Modern Theoretical Chemistry*, Volumes 7 and 8, Plenum Press, New York, 1977.
Segal, G. A., and Wetmore, R. W., *Chem. Phys. Lett.*, **32**, 556–560 (1975).
Shankar, R., *Principles of Quantum Mechanics*, Plenum Press, New York, 1985, Chapter 20.
Shavitt, I., *Meth. Comput. Phys.*, **2**, 1 (1963).
Shavitt, I., *Meth. Comput. Phys.*, **3**, 1–45 (1963).
Shavitt, I., in *Methods of Electronic Structure Theory*, Volume 3, H. F. Schaefer (ed.), Plenum Press, New York, 1977, pp. 189–275.
Shavitt, I., in *New Horizons of Quantum Chemistry*, P. W. Löwdin and B. Pullman (eds.), D. Reidel, Holland, 1983, pp. 279–293.
Shavitt, I., Bender, C. F., Pipano, A., and Hosteny, R. P., *J. Comp. Phys.*, **11**, 91–108 (1973).
Shavitt, I., Rosenberg, B. J., and Palalikit, S., *Int. J. Quantum Chem.*, **S10**, 33 (1976).
Shepard, R., *Ab Initio Methods in Quantum Chemistry*, Volume 2, K. P. Lawley, (ed.), Wiley, New York, 1987, pp. 64–196.
Shih, S., Buenker, R. J., and Peyerimhoff, S. D., *Chem. Phys. Lett.*, **16**, 244 (1972).
Shilov, G. E., *An Introduction to the Theory of Linear Spaces*, Prentice-Hall, Englewood Cliffs, NJ, 1961.
Shipman, L. L., *Chem. Phys. Lett.*, **31**, 361 (1975).
Shipman, L. L., and Christoffersen, R. E., *Chem. Phys. Lett.*, **15**, 469–474 (1972).
Shull, H., *J. Chem. Phys.*, **30**, 1405 (1959).
Shull, H., *J. Am. Chem. Soc.*, **82**, 1287–1295 (1960).
Shull, H., *J. Phys. Chem.*, **66**, 2320 (1962).
Shull, H., *J. Am. Chem. Soc.*, **86**, 1469 (1964).
Shull, H., and Ebbing, D. D., *J. Chem. Phys.*, **28**, 866 (1958).
Shull, H., and Hall, G. G., *Nature (London)*, **184**, 1559 (1959).

Shull, H., and Löwdin, P. O., *J. Chem. Phys.*, **23**, 1362 (1955).
Shull, H., and Löwdin, P. O., *J. Chem. Phys.*, **23**, 1565 (1955).
Shull, H., and Löwdin, P. O., *Phys. Rev.*, **110**, 1466 (1958).
Shull, H., and Löwdin, P. O., *J. Chem. Phys.*, **30**, 617 (1959).
Shull, H., and Prosser, F., *J. Chem. Phys.*, **40**, 233 (1964).
Sigiura, Y., *Z. Phys.*, **45**, 484 (1927).
Silver, D. M., and Nieuwport, W. C., *Chem. Phys. Lett.*, **57**, 421 (1978).
Silver, D. M., Ruedenberg, K., and Mehler, E. L., *J. Chem. Phys.*, **52**, 1206–1227 (1970).
Silver, D. M., Wilson, S., and Nieuwport, W. C., *Int. J. Quantum Chem.*, **14**, 635 (1978).
Silverstone, H. S., *Phys. Rev. (A)*, **5**, 1092 (1972).
Silverstone, H. S., *J. Chem. Phys.*, **67**, 4172 (1977).
Sinanouglu, O., *Proc. Natl. Acad. Sci. U.S.A.*, **47**, 1217 (1961).
Sinanouglu, O., *J. Chem. Phys.*, **36**, 706, 3198 (1962).
Sinanouglu, O., *Adv. Chem. Phys.*, **6**, 315 (1964).
Sinanouglu, O., *Adv. Chem. Phys.*, **14**, 337 (1969).
Siu, A. K., and Hayes, E. F., *J. Chem. Phys.*, **61**, 37 (1974).
Slater, J. C., *Phys. Rev.*, **34**, 1293–1322 (1929).
Slater, J. C., *Phys. Rev.*, **35**, 210 (1930).
Slater, J. C., *J. Chem. Phys.*, **1**, 687 (1933).
Slater, J. C., *Phys. Rev.*, **51**, 846 (1937).
Slater, J. C., *Quantum Theory of Atomic Structure*, Volume 1, McGraw-Hill, New York, 1960, Chapter 12.
Slater, J. C., *Quantum Theory of Atomic Structure*, Volume 2, McGraw-Hill, New York, 1960, Chapters 21 and 24.
Slater, J. C., *Quantum Theory of Molecules and Solids*, Volume 3, McGraw-Hill, New York, 1967, p. 9.
Slichter, D. P., *Principles of Magnetic Resonance*, Harper & Row, New York, 1963, Chapter 8 and pp. 238–243.
Smirnov, V. I., *A Course of Higher Mathematics*, Volume V, Addison-Wesley, Reading, MA, 1964.
Smith, C. M., and Hall, G. G., *Int. J. Quantum Chem.*, **31**, 685 (1987).
Snyder, L. C., and Amos, A. T., *J. Chem. Phys.*, **42**, 3670–3683 (1965).
Spangler, D., and Christoffersen, R. E., *Int. J. Quantum Chem. Quantum Biol. Symp.*, **5**, 127 (1978).
Spangler, D., and Christoffersen, R. E., *Int. J. Quantum Chem.*, **17**, 1075 (1980).
Spangler, D., Maggiora, G. M., Shipman, L. L., and Christoffersen, R. E., *J. Am. Chem. Soc.*, **99**, 7478–7489 (1977).
Stanton, R. E., *J. Chem. Phys.*, **36**, 1298 (1962).
Steiner, E., *J. Chem. Phys.*, **54**, 1114 (1971).
Stern, O., and Gerlach, W., *Z. Phys.*, **8**, 110 (1922).
Stern, O., and Gerlach, W., *Z. Phys.*, **9**, 349 (1922).
Stevenson, A. F., and Crawford, M. F., *Phys. Rev.*, **54**, 375 (1938).
Stewart, G. W., *Introduction to Matrix Computations*, Academic Press, New York (1973).
Stewart, G. W. in *Proceedings IFIP Congress 74*, North-Holland, Amsterdam, 1974.
Stratton, J. A., Morse, P. M., Chu, L. J., and Hunter, R. A., *Elliptic, Cylinder, and Spheroidal Wavefunctions*, John Wiley, New York, 1941.

Switkes, E., Stevens, R. M., Lipscomb, W. N., and Newton, M. D., *J. Chem. Phys.*, **51**, 2085 (1969).
Szabo, A. and Ostlund, N., *Modern Quantum Chemistry*, Macmillan, New York, 1982, Chapter 6.
Szego, G., *Orthogonal Polynomials, Am. Math. Soc. Colloq. Publications*, Volume 23, American Math. Soc., Providence, RI, 1939.
Taketa, H., Huzinaga, S., and O-Ohata, K., *J. Phys. Soc. Jpn.*, **21**, 2313 (1966).
Taylor, A. E., *Advanced Calculus*, Ginn and Co., New York, 1955, Chapter 16, and p. 184.
Taylor, G. R., and Parr, R. G., *Proc. Natl. Acad. Sci. U.S.A.*, **38**, 154 (1952).
Teller, E., *J. Chem. Phys.*, **41**, 109 (1937).
Temple, G., *Proc. R. Soc. (London)*, **A119**, 276 (1928).
Thole, B. T., and van Duijnen, P. T., *Theoret. Chim. Acta*, **63**, 209–222 (1983).
Tinkham, M., *Group Theory and Quantum Mechanics*, McGraw-Hill, New York, 1964, p. 20 and Chapter 3.
Tolman, R. C., *Relativity, Thermodynamics and Cosmology*, Oxford University Press, New York, 1934.
Tong, K. C., and Edmiston, C., *J. Chem. Phys.*, **52**, 997 (1970).
Uhlenbeck, G. E., and Gaudsmit, S., *Naturwissenschaften*, **13**, 953 (1925).
Von Neumann, J., *Mathematical Foundations of Quantum Mechanics*, Princeton University Press, Princeton, 1955.
Von Neumann, J., and Wigner, E., *Z. Phys.*, **30**, 467 (1929).
Wagner, A. F., Das, G., and Wahl, A. C., *J. Chem. Phys.*, **60**, 1885 (1974).
Wahl, A. C., *J. Chem. Phys.*, **41**, 2600 (1966).
Wahl, A. C., *Science*, **151**, 961 (1966).
Wahl, A. C., and Das, G., in *Methods of Electronic Structure Theory*, H. F. Schaefer, III (ed.), Plenum Press, New York, 1977, p. 51–78.
Walsh, A. D., *J. Chem. Soc.*, 2260 (1953).
Wang, S. C., *Phys. Rev.*, **31**, 579 (1928).
Weinbaum, S., *J. Chem. Phys.*, **1**, 593 (1933).
Weinhold, F., *Adv. Quantum Chem.*, **6**, 299–331 (1972).
Weinstein, D. H., *Phys. Rev.*, **40**, 797 (1932).
Weinstein, D. H., *Phys. Rev.*, **41**, 839 (1932).
Weinstein, D. H., *Proc. Nat. Acad. Sci. U.S.A.*, **20**, 529 (1934).
Weinstein, H., Osman, R., Edwards, W. D., and Green, J. P., *Int. J. Quantum Chem.*, **QBS5**, 449 (1978).
Weinstein, H., Osman, R., Green, J. P., and Topiol, S., in *Chemical Applications of Atomic and Molecular Electrostatic Potentials*, P. Politzer and D. G. Truhler (eds.), Plenum Press, New York, 1981, pp. 309–334.
Weinstein, H., Osman, R., Topiol, S., and Green, J. P., *Ann. NY Acad. Sci.*, **367**, 434–451 (1981).
Weiss, M. J., *Higher Algebra*, John Wiley, New York, 1949, p. 80.
Weissman, S. J., *J. Chem. Phys.*, **25**, 890 (1956).
Whitten, J. L., *J. Chem. Phys.*, **39**, 349 (1963).
Whitten, J. L., *J. Chem. Phys.*, **44**, 359 (1966).
Whitten, J. L., and Hackmeyer, M., *J. Chem. Phys.*, **51**, 5584–5596 (1969).
Wigner, E., *Math. Naturw. Anz. ungar. Akad. Wiss*, **53**, 477 (1935).
Wigner, E., *Group Theory*, Academic Press, New York, 1959, Chapter 9.

Wilkinson, J. H., *The Algebraic Eigenvalue Problem*, Oxford University Press, London, 1965.
Wilkinson, J. H. and Reinsch, C., *Linear Algebra*, Springer-Verlag, New York, 1971.
Wilson, S., *Methods of Computational Physics*, G. H. Dierksen and S. Wilson (eds.), Reidel, Dordrecht, The Netherlands, 1983.
Wilson, S., *Adv. Chem. Phys.*, **67**, 429 (1987).
Wind, H., *J. Chem. Phys.*, **42**, 2371–2373 (1965).
Wittel, K., and McGlynn, S. P., *Chem. Rev.*, **77**, 745–771 (1977).
Yarkony, D. R., Schaefer, H. F., III, and Rothenberg, S., *J. Am. Chem. Soc.*, **96**, 656–659 (1974).
Ziman, J. M., *Elements of Advanced Quantum Theory*, Cambridge University Press, Cambridge, 1969, Chapter 2.

Index

Ab initio, 640, 647
Abelian group, 405
Addition theorem, 287–288
Adenine molecule, 544–545
Adiabatic approximation, 347
Ammonia molecule, 404, 416, 531–533, 615, 639, 652
  tunneling, 248–250
Angular momentum
  classical, 4, 9
  coupling, 460–471
  general angular momentum operator, 273
  general considerations, 273–282
  general eigenfunctions, 279–281
  intrinsic, 17
  orbital angular momentum eigenfunctions, 282
  orbital angular momentum quantization, 17, 278
  quantization, 277
  spin quantization, 278–279, 288–291
  square of general angular momentum operator 273
  step-down operator, 273
  step-up operator, 273
  total, 461
Anharmonic oscillator, 378–381
Associated Laguerre polynomial, 300, 514
Associated Legendre functions, 284–288
Asymmetry parameter, 533–536
Atomic units, 298
Atomization energy, 534
Atoms in molecules, 544
Aufbau principle, 324
  for molecules, 492–498
Average local potential, 485
Average natural orbitals, 600

Balmer series, 16
Basis sets, 503, 512–530
  balance, 640
  bond function, 519
  cartesian Gaussian orbitals, 516
  complete sets, 514
  contractions, 523–530
  cusp behavior, 516
  double zeta basis, 515, 529
  Dunning contraction, 523
  elliptical basis orbitals, 619
  even-tempered orbitals, 517
  floating spherical Gaussian orbitals (FSGO), 519–523
  Gaussian-type orbitals, 516
  hydrogen-like functions, 514

lobe-functions, 519
minimal basis set, 515
*M-N*1G contraction, 529–530
orbital radius, 519
polarization functions, 519
primitive functions, 513–523
single zeta basis, 515
Slater-type functions, 515
split-valence basis, 515, 529
STO-nG contraction, 526–530
superposition error, 601
Bessel's inequality, 52
BH_3 molecule, 639, 652
Binding energy, 400
Bohr radius, 298
Bond dissociation, 555
Bond functions, 519
Bond order, 541
Born–Oppenheimer approximation, 343–347
Bose–Einstein statistics, 431
Bosons, 431
Bound state, 16
Bra vector, 169
Breit equation, 341–343
Brillouin's theorem, 557, 581–582, 604, 636

Carbazole molecule, 645
Carbon dioxide molecule, 652
Carbon monoxide molecule, 531–533, 652
Center of mass, 7, 294, 315
Changes of basis, 138–147
active, 136
passive, 142
Charge and bond order matrix, 506, 561
Circular frequency, 25
Classical mechanics, 2–11
Closed shell configuration, 499
Closure relation, 180
Complete sets of functions, 71–78
Compton scattering experiments, 22
Configuration interaction (CI) theory, 577–624
basis set superposition error, 601
computational considerations, 607–615
conceptual considerations, 597–607
configuration (configuration state function) 577
convergence, 586–589
direct configuration interaction approach, 611–615
doubly-excited configuration, 580
external space, 611
full configuration interaction, 581, 618
general formulation, 577–581
Hartree–Fock configuration, 579
internal space, 611
multireference configuration interaction, 643
orbitals for CI studies, 598–601
reference state, 612
singly-excited configuration, 580
size consistency, 605–606
size extensivity, 606–607
total number of configurations, 601–602
Conjugate variables, 8
Constant of motion, 3, 198–204
Continuum functions, 79–82
Contour maps, 540–541
Convergence in the mean, 68
Coordinates
confocal elliptical, 218, 318
curvilinear, 213, 214
generalized, 8, 176
spherical polar, 215, 282
Correlation
angular, 627
in–out, 592
interpair, 531
Correlation diagrams, 492–498
Correlation energy, 491, 531, 552, 636–637
Correspondence principle, 327–328
Coulomb integral, 483, 563
Counterpoise method, 601

Coupled cluster approximation, 640
Coupled-cluster theory, 648-653
Cramer's Rule, 114

Davisson–Germer experiments, 24
DeBroglie relation, 24
Degeneracy, degree of, 149, 381
Density matrix, 507, 583–590
  first-order reduced density matrix, 583, 585
  *p*th order reduced density matrix, 583
  spinless *p*th order reduced density matrix, 583
Determinants, 109–116
  cofactor, 110
  Laplace expansion, 110
  minor, 110
  unrestricted, 468
Different orbitals for different spins, 468
Dipole moment, 533–536, 548
Dirac delta function, 79, 547
Dirac equation, 335–340
Dirac notation, 169–170
Dissociation energy, 531
Double well potential, 241–250
Double zeta basis, 515
Doubly-excited configuration, 580, 602

Ehrenfest's Theorem, 218
Eigenfunction (eigenvector), 97
  antisymmetric, 234, 245
  parity, 233–235
  simultaneous, 193–195
  symmetric, 234, 243
Eigenvalue, 97, 189
  degenerate, 149
  degree of degeneracy, 149
  matrix eigenvalue problems, 147–167
  nondegenerate, 190
Electron affinities, 491–492, 525
Electron density contours, 312–313
Electron impact experiments, 13–16
Electron pair theories, 628–640
Electron repulsion integrals, 483, 505, 511, 515
Electron volt, 14, 658
Electronic configuration, 495
Electrostatic potential maps, 542–545
Energy
  conservative, 9
  continuum, 16
  conversion factors, 658
  emission, 15
  error, related to wavefunction error, 357
  expression for a single Slater determinant, 447
  levels, quantum mechanical quantization, 11, 12
  total classical, 7
Equation of motion
  Hamilton, 9, 206
  Heisenberg, 206
  interaction, 210
  Lagrangian, 331
  Schrödinger, 182
Ethyl chlorophyllide *a* molecule, 645–647
Ethyl pheophorbide *a* molecule, 645–647
Ethylene molecule, 620–624
Euler's theorem, 449–450
Even-tempered orbitals, 517
Exact wavefunction, 437
Exchange integral, 483, 563
Excited state, 16
Exclusive orbitals, 550
Expectation value (average value), 187, 188, 232

Fermi–Dirac statistics, 431
Fermi hole, 553
Fermions, 431
Field gradient, 533–536
Floating spherical Gaussian orbitals (FSGO), 509, 519–523, 546
Fluorescence, 15
F_2 Molecule, 531–533

Force
  conservative, 6
  constant, in harmonic oscillator, 251
  single classical particle, 3
  total external classical, 7
Formaldehyde molecule, 533–537
Fourier integral formula, 77, 80
Fourier series
  generalized, 68, 69
  trigonometric, 71
Fourier transforms, 77, 78
Franck–Condon principle, 617
Franck–Hertz experiment, 13
Free electron model, 268
Frozen-orbital approximation, 491
Full configuration interaction, 581
Function space, 66–69
Functions, homogeneous, 450

Gauge invariance, 348
Gaussian basis orbitals, 369, 457, 509
  cusp behavior, 371
  floating spherical Gaussian orbitals, 509
Gaussian-type orbitals (GTO), 516
Geminal, 628
Generalized valence bond approach, 640
Generating function, 269, 285, 292
Gradient operator, 6
Gram–Schmidt orthogonalization procedure, 55–61, 153
Gramian (Gram determinant), 115
Gross atomic population, 541
Gross charge, 541–542
Ground state, 16
Group theory, 402–429
  axis of symmetry, 403
  character of a matrix, 410
  character table, 414
  class, 406
  conjugate matrices, 411
  dimension of representation, 410
  direct product, 418
  direct product representation, 419
  great orthogonality theorem, 412
  identity operation, 403
  improper rotation, 404
  inversion operation, 403
  irreducible representation, 410–412
  law of combination, 404
  little orthogonality theorem, 413
  matrix representation of symmetry operators, 409
  multiplication table, 405
  number of elements in a class, 411
  point groups, 403
  reducible representation, 410–412
  reflection operation, 403
  symmetry element, 403
  symmetry operations, 403
Guanine-cytosine base pair, 641–642

Hall–Roothaan LCAO-MO-SCF theory, 498–513
Hamiltonian operator, 211–216, 327–347
  external fields (electric and magnetic), 328–333, 384
  relativistic effects, 333–334
  vector and scalar potentials, 330
Harmonic oscillator, 28, 250–265
  average momentum, 264
  average position, 264
  eigenfunctions, 261–263
  energy levels, 257
Hartree atomic units, 298
Hartree–Fock equations (closed shell), 486
Hartree–Fock orbitals (closed shell), 486, 503
Hartree–Fock theory (closed shell), 481–556
  approximate Hartree–Fock orbitals, 508
  canonical orbitals, 489
  conceptual approach, 481
  convergence, 508
  deficiencies of the theory, 552–556

energy and molecular properties, 530–538
Fock matrix, 511
Fock operator, 486, 503
Hartree–Fock limit, 531
interpretation of wavefunctions, 538–552
matrix elements, 506
molecular orbitals, 503
population analysis, 541–543
restricted closed shell, 500
SCF equations, 503–509
Hartree–Fock Theory (open shell), 556–573
restricted Hartree–Fock theory (RHF), 562–570
unrestricted Hartree–Fock theory (UHF), 556–562
Helium atom, 359–363, 588
Hellmann–Feynman theorem, generalized, 452
Hermite polynomials, 261, 379
Hilbert space, 61–66, 176
H_2^+ molecule, 317–325
H_3^+ molecule, 406, 421–426
HF molecule, 531–533
Hydrogen atom
energy levels, 299–300
ionization potential, 240
radial distribution function, 304
real representation of wavefunctions, 311
relative electron motion equation, 297–303
scaling, 456–457
separation of center of mass motion, 293–297
spectrum, 16, 17
Variation Theorem examples, 357–359, 369–372
wavefunction cusp, 307, 371
wavefunction shape, 303–313
wavefunction symmetry, 321
wavefunctions, 302
Hydrogen molecule
bond length, 532
dissociation energy, 627
Hartree–Fock energy, 531
Hartree–Fock example, 509–512
ionization potential, 533
molecular orbital method, 401–402, 555
Natural orbital analysis, 592–593
potential surface, 619–621
valence bond method (Heitler–London), 398–401
Wang function, 400
Weinbaum function, 400
Hydrogen-like functions, 514
Hypervirial theorem, 447–448

Imidogen molecule, 635
Independent electron pair approximation (IEPA), 637
Independent particle model, 433
Inelastic collisions, 15
Inner product spaces, 42–49
Intermediate normalization, 636
Ionization, 16
Ionization potential, 490, 525, 533
Isomorphism
between Hilbert space and Function space, 69–70
between matrices and operators, 128, 130–134
between vector spaces, 42
Iterative natural orbital method, 597, 600

Ket vector, 169
Kinetic energy
classical, 7, 10
Klein–Gordon equation, 335
Koopmans theorem, 489
Kronecker delta, 50

Lagrange method of undetermined multipliers, 82–84, 564
Lagrangian, classical, 8, 331
Lagrangian multipliers, 84, 487, 568, 632
Larmor's theorem, 20

LCAO, 504
LCAO-MO-SCF theory, 498–513
Legendre polynomials, 285–286
Linear independence
  vectors, 36, 51, 115
Linear variation method (Raleigh–Ritz Method), 363–368
Linked cluster theorem, 652
Lithium hydride molecule, 633–635
Local potential, 485
Localized orbitals, 549–552, 599
Lorentz invariance, 333, 335, 338
Lower bounds, 372–374
  Temple formula, 373
  Weinstein formula, 373

MacDonald's theorem, 367–368
Magnetic field, 20
Magnetogyric ratio, 20
Many-body perturbation theory, 648–653
Matrices
  addition, 103, 104
  adjoint, 127
  banded, 128
  complex conjugate, 127
  conformation, 104
  equality, 103
  Hermitian, 128
  idempotent, 128
  infinite, 167–168
  inversion, 121–127
  involutory, 128
  linear matrix transformation, 131
  lower triangular, 128
  matrix definition, 102
  matrix multiplication, 104
  metric, 164
  nilpotent, 128
  nonsingular, 117
  norm, 109
  normal, 128
  orthogonal, 128
  overlap, 164, 364
  partitioning, 107–109, 390
  positive-definite Hermitian quadratic form, 164
  positive-definitive, 165
  power of a square matrix, 109
  pure imaginary, 128
  rank, 119
  real, 128
  rectangular, 102
  scalar matrix, 106
  scalar multiplication, 104
  skew-Hermitian, 128
  skew-symmetric, 128
  square, 102
  symmetric, 128
  trace, 120, 584
  transpose, 127
  tridiagonal, 128
  unit, 103
  unitary, 128
  upper triangular, 128
Matrix elements, 164
  between Slater determinants, 437–447
  for one-electron operators, 442–443
  for two-electron operators, 443–446
  for zero-electron operators, 440–441
Matrix representation of linear operators, 129–138
Maxwell's equations, 329
Mean square deviation, 188
Methane molecule, 404, 531–533, 639, 652
Methoxide radical, 570–571
Methylisocyanide molecule, 653
Moeller–Plesset perturbation theory, 649
Molecular fragment method, 522, 537–538, 642–643
Molecular orbitals, 401, 503
Moment of inertia, 18
Momentum
  internal, 7

linear, 3, 7
Morse potential, 269, 270
Mulliken population analysis, 541–543
Multiconfiguration SCF theory (MC-SCF), 625–628
Multiplicity, 468

Natural expansion of a wavefunction, 586
principal natural orbital, 634
Natural orbital, 586–590
Nature of the chemical bond, 590–597
atomic wavefunction, 594
fractional atomic character, 595
fractional ionic character, 595
ionic wavefunction, 594
Neon atom, 639
Net atomic population, 541
Newton's equations of motion, 3
Nitrogen molecule, 531–532
Nonadiabatic approximation, 347
Noncrossing rule, 323–324, 493
Nondegeneracy theorem, 221–223
Nonlocal operator, 485
Normalization constant, 226, 231, 261, 300

Observable, 178
compatible, 192, 195
Occupation number, 586, 632
Occupied orbitals, 489, 504
Open shell wavefunction, 469
Operand, 85
Operators
addition of linear operators, 88
adjoint, 91
annihilation, 254, 573
antisymmetrization, 433, 437
bounded, 99
commutator, 89, 274
commuting, 195
Coulomb, 485
creation, 254
exchange, 485
Fock, 486
general, 84
Hamiltonian, 211–216
Hermitian (self-adjoint), 93
idempotent, 96
identity, 86, 405
inverse, 86, 405
kinetic energy, 212
ladder, 273
Laplacian, 214, 295–296
linear, 84–97
lowering, 256
nonlocal, 485
normal, 97
Null, 86
permutation, 433
position, 179
potential energy, 211
power series, 89
product of linear operators, 88
projection, 95, 420, 433, 573
raising, 258
time evolution (time-development), 184
width, 188, 196
Optimized valence orbital approach, 626
Orbital, 323
antibonding, 323, 494
bonding, 323, 494
doubly occupied, 432
polarization, 519
singly occupied, 432
transformations, 471–476
valence, 517
Orbital angular momentum, 4, 271–272
total, 460
Orbital energy, 488, 512, 534
Orbital exponent, 358, 400, 515, 518
Orbital radius, 519
Orthonormality, 49–61
Gram–Schmidt orthogonalization procedure, 55–58

orthogonality of vectors, 50
orthonormality of vectors, 50
symmetric orthogonalization procedure, 365
Oscillator orbitals, 550
Oscillator strength, 646
Overlap matrix, 505
Oxygen atom, 525

Pair correlation energies, 639
Parent configurations, 644
Parity of a permutation, 112
Parseval's Equation, 54
Particle on a ring, 223–228
energy levels, 225
Particle trapped in a box, 228–233
energy levels, 230
Pauli exclusion principle, 429–431
Pauli spin matrices, 290, 338
Pentadienyl radical, 572
Perturbation theory
Brillouin–Wigner perturbation theory, 386–392
degenerate states, 381–384
effective perturbation potential, 391
first order energy correction, 376
first order equation, 376
first order wavefunction correction, 377
nondegenerate states, 374–378
Raleigh–Schrodinger perturbation theory, 374–386, 604, 648–649
second and third order energy correction, 376
unperturbed wavefunction, 376
Phase of a wave, 25
Phase velocity, 25
Phosphoresence, 15
Photons, 12
momentum, 22
wavelength, 15
Physical constants, 657
Planck's constant, 1, 657
Plane waves, 27, 297
Poisson bracket, 206
Population analysis, 507
Porphyrin, 549
Positrons, 336
Potential curve, 553–555
Potential Energy
classical, 6
conservative, 211
Principal axes, 162
Probability
associated with a measurement, 185–190
density, 178, 303, 540
most probable value of a measurement, 233
Propagation number, 25
Propagation vector, 26
Pseudonatural orbitals, 600
Pyrazine molecule, 644–645

Quadratic form, 161
Quadrupole moment, 533–536
Quantum mechanics, 1
Heisenberg representation, 204
postulates, 175–191
Schrödinger representation, 204
Quantum number
Azimuthal, 301
''good'' quantum numbers, 320
magnetic, 19, 301
orbital angular momentum, 17
principal, 16, 301
spin, 17

Radial nodes, 516
Recurrence relation, 260, 261
Reduced mass, 18, 294
Resolution of the identity, 96
Rest mass, 333
Rigid rotor, 17, 314–317
energy levels, 316
Rodrigues generating formula, 285
Rotational constant (rigid rotor), 316
Rydberg constant, 16
Rydberg states, 522, 616, 622

Scale factor, 454
Schoenflies notation, 406
Schrodinger equation
  time-dependent, 182
  time-indepedent, 182
Schwartz' inequality, 45
Secular determinant, 148
Secular equations, 148
Secular polynomial, 148
Selection rule, 18, 317
Self-consistent-field (SCF) approach, 508
Separated atom limit, 320, 493
Separated pair theory, 628–635
  antisymmetrized product of strongly orthogonal geminals (APSG) wavefunction, 629
Series representation of functions, 68
Similarity transformation, 145, 205
Simultaneous equations
  homogeneous, 114
  inhomogeneous, 114
Singlet state, 468
Singly-excited configuration, 580, 602
Slater determinant, 432, 482
Slater-type functions, 515
Sodium chloride, 627
Spectra
  continuous, 179
  discrete, 179
Spherical harmonics, 288, 514
Spherical wave, 27
Spin angular momentum, 288–291
  total, 461
Spin density matrix, 562
Spin operators, 338, 341
Spin orbitals, 432, 483
Spin state projection operator, 573
Spinor components, 338
Spin-orbit coupling, 340, 342–343
Spin-spin coupling, 342–343
Square well potential, 235–241
Stark effect, 384–386
States
  eigenstate, 186
  immediately after measurement, 190
  stationary, 183
  time evolution, 181–185
Stern–Gerlach experiment, 19
Strong orthogonality constraint, 629
Superposition
  waves, 26

Torque, 4
Transition probability, 200–202
Triangle inequality, 45
Triplet state, 469
Tunneling, 246–250
Turnover rule, 91

Unbound states, 16
Uncertainty principle, 28, 195–198, 265–267
Undetermined multipliers, method of, 82–84, 483
Unitary transformation, 147, 365, 487
United atom limit, 320, 493
Unrestricted Hartree–Fock theory (UHF), 556–562

Valence bond theory, 398–401
  covalent terms, 400
  ionic terms, 400
Valence electrons, 497
Variation principle, 352–355
Variation-Perturbation theory, 387
Variation theorem, 354
Vector
  active transformations, 136
  addition, 31, 34
  associativity, 33
  closure under addition, 33, 47, 62
  column, 102, 103
  commutivity under addition, 33
  components, 32
  coordinates, 32
  length, 33, 43

length, 33, 43
linear combination, 36
norm, 33
position, 3
row, 102, 103
scalar multiplication, 32, 34, 48, 63
scalar product (inner product), 33, 43
unit, 39
vector product, 33
zero vector, 33, 35
Vector space, 31, 33
basis, 39
complex, 43
comprised of functions, 47–49
dimension, 41
dual vectors, 170
Euclidean, 43
inner product, 44
spanning, 38
unitary, 44
Vertical transition, 617
Virial theorem, 448–449
Virtual orbitals, 504, 579, 598

Walsh's rules, 495–498
Water molecule, 403, 531–533, 590, 616–619, 639, 652
Wave amplitude, classical, 24
Wavefunction (state function, eigenfunction), 176–178
antisymmetry requirement, 431
boundary conditions, one-point, 221, 316
boundary conditions, two-point, 222
closed shell, 467, 472–473
continuity, 220
criteria, 220–221
determinantal, 432
discontinuity, 231
expansion theorem, 435–437
factorization, 434
open shell, high spin, 473–475
open shell, singlet, 475–476
piecewise continuous, 220
quadratic integrability, 220
scaling, 453–460
single configuration, 467
single Slater determinant, 467
singlet, 468
single-valued, 220
triplet, 469
unrestricted determinant, 468, 471–472
Wavelength, classical, 24
Wave-particle duality, 23, 24
Work, 5
Wronskian, 222

Zero point energy, 265–267